Atlas of Science

Visualizing What We Know

Katy Börner

The MIT Press
Cambridge, Massachusetts
London, England

For information about special quantity discounts, please e-mail special_sales@mitpress.mit.edu

This book was set in Adobe Caslon Pro by Elisha F. Hardy (graphic design and layout), Tracey Theriault (design advice), and Katy Börner (concept), Cyberinfrastructure for Network Science Center, School of Library and Information Science, Indiana University. Printed and bound in Malaysia.

Library of Congress Cataloging-in-Publication Data

Börner, Katy.
Atlas of science : visualizing what we know / Katy Börner.
p. cm.
Includes bibliographical references and indexes.
ISBN 978-0-262-01445-8 (hardcover : alk. paper) 1. Classification of sciences—Atlases. 2. Science—Atlases. 3. Communication in science—Data processing. 4. Digital mapping. I. Title.
Q177.B67 2010
501.2—dc22

2009048451

10 9 8 7 6 5 4 3

I dedicate this atlas to my parents, Monika Börner and Heinz Börner.

Contents

Part 1: Introduction

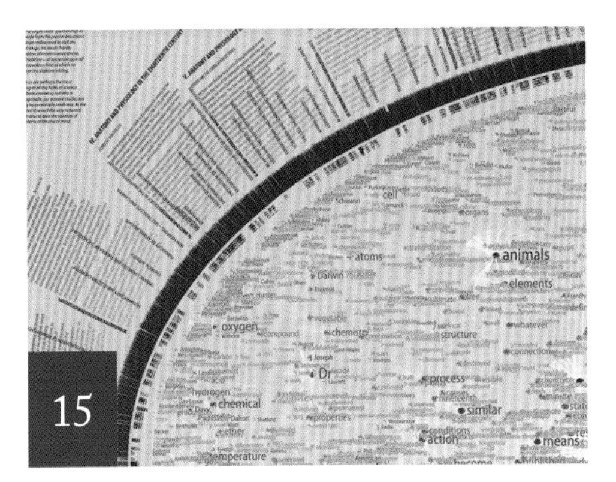

Part 2: The History of Science Maps

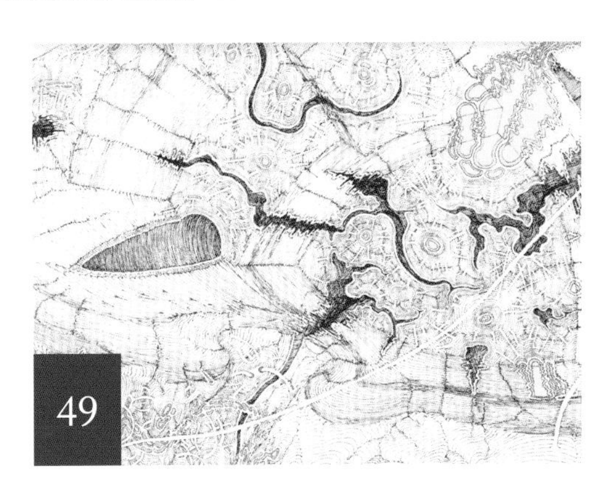

Part 3: Toward a Science of Science

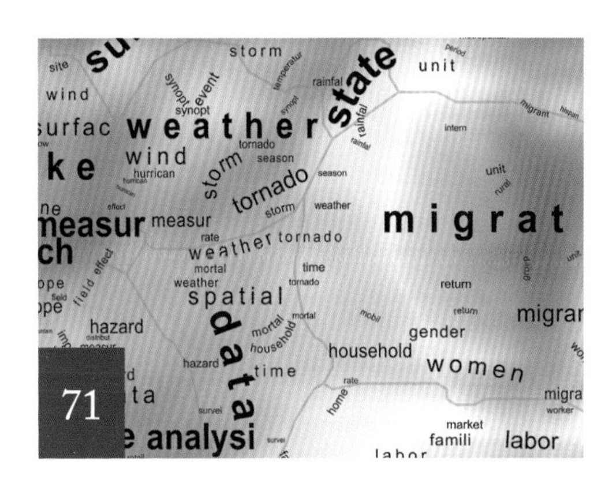

Part 4: Science Maps in Action

Part 5: The Future of Science Maps

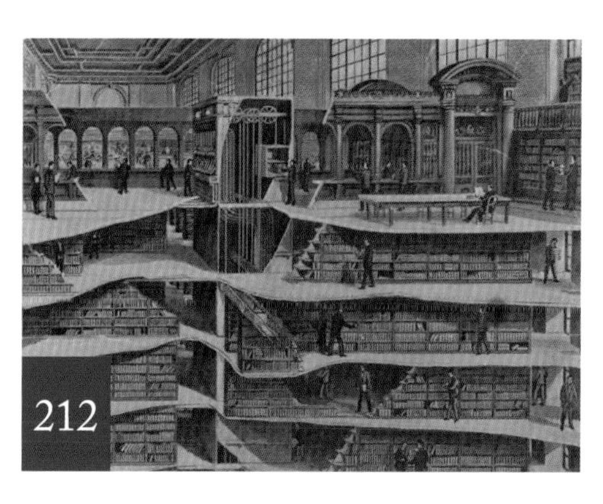

Foreword

Those of us who make maps for a living like to think we face difficult challenges in our ongoing efforts to represent the world. To depict a complex, constantly changing, three-dimensional world on a two-dimensional page or computer screen, we simplify and symbolize by following well-established cartographic traditions, many of which date back centuries. We strive to accurately reflect the spatial alignments and interrelationships that for the most part reflect undeniable facts in the "real" world.

The explorers whose work is represented in the pages of this rich and fascinating volume face challenges far more daunting. First, the world they strive to represent is an abstract and intellectual one, not a physical reality that can be imaged from space, surveyed on the ground, and depicted in miniature on a map. The interrelationships among the landmarks of this abstract world are real, but they are not easily represented in the simple, straightforward ways that one can convey the distances between, say, three cities.

Second, there is no equivalent in the cartography of science to the standards and conventions upon which we mappers of the physical world comfortably depend. There's no agreed-upon notion of north-as-up, of systems of latitude and longitude, of symbols, scale, and projection. Mapping the world of science requires the invention of a brand-new geography. Not only that, but the new geography then needs to be represented visually using colors, lines, and symbols for which no conventions exist. Science mapmaking is a rather young human endeavor. Although many visual renderings of science have been created over the last hundred years, it was only in 1999 that a map of all sciences was rendered by automatic means. Since then, a variety of efforts have used very different data sets, resulting in a rich variety of maps. Clearly, however, no consensus yet exists on how to best represent the world of science. We can only hope and assume that the process will evolve more quickly than did the ancient conventions on which we traditional cartographers rely.

Third, the world that is being mapped in this book is changing at a dizzying rate. It's a fact of twenty-first-century science that whole realms of inquiry bloom into existence almost overnight, creating new places and spaces in ways that are alien to "normal" cartography. It is as if entire continents and archipelagoes were to constantly erupt on the roiling surface of a map even as that map was being drawn for the first time.

So it is with wonder and admiration that I peruse this important and groundbreaking (pun intended) book. This detailed review of past work, evolving conventions, and possible futures will likely facilitate the diffusion, adoption, and development of data analysis and mapping techniques that will inform the design of useful and insightful science maps. These maps of science are doing the hard work that all good maps do: They're documenting a landscape in ways that will serve as a useful historical record, which is especially important when the landscape is changing rapidly. They're showing us interesting patterns and interrelationships, revealing a surprising, even revelatory, new geography of science. They're helping us navigate a realm that for many of us has been terra incognita. They're allowing us to explore, to imagine, to understand. Which, come to think of it, is precisely why we cartographers love our work.

Allen Caroll
Chief Cartographer
National Geographic Society

Preface

This atlas attempts to introduce maps of science to a global audience. Through maps of science, we can begin to see all that we know as landscape—viewed as if from above or from a great distance. Science maps provide guidance for navigating, understanding, and communicating the dynamic and changing structure of science and technology. The career trajectories of individual researchers and their professional networks, the intellectual footprint of any given institution or country, and emerging research frontiers or bursts of activity can be projected onto science maps and animated over time. Science maps complement local fact retrieval via search engines by providing information used to determine context and relevance. They serve as visual interfaces to immense amounts of data—depicting perhaps millions of data records that we can perceive rapidly to effectively discern apparent outliers, clusters, and trends.

The *Atlas of Science* is designed to accompany the *Places & Spaces: Mapping Science* exhibit, which introduces large-scale maps of science to the public. All 30 maps from the first three (of what will be ten) exhibit iterations are included. It also features a history of the exhibit—with its curators and advisory board—as well as biographies of the mapmakers (current as of the date of map creation). Many more maps continue to be created, and milestone works are featured in the exhibit as well as online at **http://scimaps.org**.

The atlas is organized into five parts. The 200+ pages contain more than 35 full-page maps of science, 50 data charts, and 500 full-color images, including portraits of renowned mapmakers. Part 1 discusses the growth and dynamics of our collective scholarly knowledge and the utility of science maps in navigating and managing this flood of information. Part 2 situates the development of science studies and science maps in the complex network of visionary thinkers and their inventions. It also includes a 22-page comprehensive timeline of milestone algorithm, visualization, tool, and book contributions that helped to advance the state of mapping science. Part 3 reviews major techniques used to map science on an individual, local, or global scale in a temporal, geographic, semantic, or network fashion. Part 4—the heart of this atlas—presents a visual feast of science maps, the stories behind them, and the biographies of their makers. In Part 5, we conclude with a forecast of the future of science mapping. Every part consists of spreads, each with a synergistic combination of text and imagery to communicate a specific topic or theme. Neither the text nor the imagery can stand alone—they complement each other.

Meant to serve as a visual index to the rich scholarly work that exists on the mapping of science and its practical applications, the atlas contains acknowledgments of contributions for every spread—more than 80 in some instances. With a total of more than 1,650 citation references, such information could not be furnished alongside each page layout. Instead, the **References and Credits (page 212)** section lists references by section for each spread, followed by a subject **Index (page 247)**. Besides citation references, the **References and Credits** section also provides image credits, data credits, software credits, and acknowledgments of the contributions of scholars with extensive expertise in bibliometrics, scientometrics, webometrics, informetrics, information science, library science, history of science, communication sciences, social sciences, geography, cartography, Internet research, economics, physics, and related areas.

The Web site **http://scimaps.org/atlas** serves as a complement to this atlas, enabling access to sources, EndNote and BibTeX files for all references, and extended search functionality. The site also provides links to high-resolution maps, particularly those included in the *Places & Spaces: Mapping Science* exhibit.

Just as an atlas of the world needs to be continually updated, an atlas of science must incorporate new developments as they emerge. It is a "living document," amended to reflect comments, corrections, and recommendations, as well as new science. As such, the supporting Web site invites comments and offers frequent updates of all materials in preparation for future editions.

More than four years in the making, the *Atlas of Science* constitutes a unique collaboration of the Cyberinfrastructure for Network Science Center team at the School of Library and Information Science, Indiana University; many of our colleagues around the globe; ESRI Press initially; and the MIT Press subsequently. It is our hope that this atlas provides actionable information and insight for many levels of application in study, work, and life.

Katy Börner
Cyberinfrastructure for Network Science Center
School of Library and Information Science
Indiana University

April 16, 2010

Acknowledgments

The atlas and the exhibit from which it was born would not have been possible without the financial support of National Science Foundation awards IIS-0238261, IIS-0513650, IIS-0534909, IIS-0715303, IIS-0724282, IIS-0750993, CBET-0831636, CHE-0524661, CHE-0723989, SCI-0533892, and SBE-0738111; National Institutes of Health awards R21DA024259 and U24RR029822; two James S. McDonnell Foundation grants; the Cyberinfrastructure for Network Science Center, the School of Library and Information Science, and University Information Technology Services all three at Indiana University. Additional financial support was provided by the Indiana 21st Century Research and Technology Fund.

Generous support for the *Places & Spaces: Mapping Science* exhibit was also provided by Thomson Reuters (formerly Thomson Scientific or The Thomson Corporation); the Science, Industry and Business Library at the New York Public Library; infoUSA; Discovery Logic; Gale; Blair; and Elsevier. Thomson Reuters also provided much of the data used to create the science maps.

Scientific meetings, workshops, and conferences were instrumental to recent advances in mapping science research and the design of this atlas. These included two workshops on Visual Interfaces to Digital Libraries and six symposia on Knowledge Domain Visualizations, both organized by Chaomei Chen and Katy Börner, and a Sackler Symposium on Mapping Knowledge Domains, organized by Richard Shiffrin and Katy Börner.

Five focused workshops on the topics of scholarly databases and data integration, science mapping, and forecasting science in 2005 and 2006 were of special importance. It is the expertise and courage of the attendees (pictured below) who freely shared their knowledge across disciplinary and cultural boundaries that helped to bring both the exhibit and this atlas into being.

The mapping of science draws from a rich body of cartographic mapmaking. Rebecca C. Cape served as a living index to the amazing holdings of the Lilly Library at Indiana University. Jim Flatness and Mike Klein at the Library of Congress provided expert advice and support. Anne J. Haynes and Collette Mak were instrumental in gaining access to WorldCat.

For the design of the atlas, I drew inspiration from children's books, such as *Millions to Measure*, the *Big Book of Time*, and *Zoom City*; books by Scott McCloud on understanding, reinventing, and making comics; science publications from Usborne and DK; and scholarly works such as Heinrich Berghaus's *Physikalischer Atlas*, works by Edward R. Tufte, and maps by National Geographic.

Beyond the contributions listed in the **References and Credits (pages 212–246)**, I would like to thank the many students, friends, colleagues, and bloggers who provided informal feedback over the last six years. Their honest feedback and expert advice helped to improve the readability and utility of the science maps as well as the communication of the techniques involved in their creation.

The atlas "dream team"—at the Cyberinfrastructure for Network Science Center within the School of Library and Information Science at Indiana University—included Elisha F. Hardy, who designed many of the images and nearly 1,000 layout mockups needed to ensure the synergistic interplay of imagery and text; Russell J. Duhon, who did much of the custom data analysis and visualization for the atlas; Qizheng (Stanley) Bao, Angela M. Zoss, Jennifer Coffey, and Mark A. Price, who formatted the more than 1,000 references and cred-

December 1 and 2, 2005: Mapping Science Workshop
Thomson Scientific, Philadelphia, Pennsylvania

April 4, 2006: Mapping Science Workshop
New York Academy of Science, New York City, New York

May 21, 2006: Modeling Science Workshop
Indiana University, Bloomington, Indiana

its and prepared them for the Web site; and Kristin Reed, Roxana Cazan, Richard S. Pinapati, Benjamin Ray Gonzalez Jr., Marla Fry, Bryan Hook, and Mark A. Price, who managed to secure the more than 1,100 copyright permissions. Our team benefited enormously from the supportive research environment at Indiana University, particularly at the School of Library and Information Science, directed by Blaise Cronin.

Gordana Jelisijevic was instrumental in shaping and fine-tuning the language of the atlas. Her writing mastery and editing expertise brought eloquence as well as clarity to this material. Teresa Elsey performed the final proofread of the atlas on behalf of the MIT Press.

Elisha F. Hardy designed many of the images in the atlas as well as the layout of all text and imagery in close collaboration with Katy Börner. This atlas attests to her diligence, stamina, and design skills. Tracey Theriault contributed her professional expertise to the final design of the atlas.

Four colleagues provided detailed feedback on penultimate draft of the atlas: André Skupin, Peter A. Hook, Deborah MacPherson, and Alex Soojung-Kim Pang. Their comments combined with feedback from anonymous reviewers provided by the MIT Press were instrumental in finalizing the atlas.

The *Atlas of Science* was originally designed for publication as an 11-inch x 13-inch (28 centimeter x 33 centimeter) full-color book in landscape format by ESRI Press. ESRI applied common standards of geography: for each map, piece of data and software, and image, credits need to be provided; all text on a map needs to be legible; and the accompanying text must fully support the interpretation of the map. Ideally, all science maps published hereafter would adhere to these standards for improving replicability, readability, and rigor. Shortly after I submitted the complete manuscript in August 2008, I learned that ESRI needed data and software copyrights for each of the more than 400 science maps—in addition to copyrights for the final maps. Due to the lack of formal standards for the data sets and tools used to generate science maps, it proved impossible to acquire all of these copyright levels. ESRI was thus unable to publish the atlas and reverted all rights back to me.

I am thankful to the MIT Press for taking on this demanding project. Transferring the material generated by our team to the MIT Press constituted an impressive challenge on both sides. The ambition to make maps of science understandable to a general audience; the complexity of the atlas in terms of synergistic text and image placement; and the sheer number of images, references, and credits all posed major challenges. Marguerite Avery, Erin K. Shoudy, Abby Roake, and Mel Goldsipe at the MIT Press played an important role in this process. They were instrumental in translating my vision for the atlas to the MIT Press, as well as their production needs to my team, while omitting irrelevant complexities.

The MIT Press also supported a larger than originally agreed upon page size for the atlas that considerably improved the legibility of many featured maps. However, all exhibit maps are sized 30 inches x 24 inches (about 76 centimeters x 61 centimeters) and several have a very high information density that cannot be reproduced in a book. I invite interested readers to visit the Web site for the atlas (**http://scimaps.org/atlas**) to explore full resolution versions of all maps.

The atlas would not have been possible without the support of my colleagues, who accepted my apologies for not reviewing papers and proposals and for not serving on committees; my students, who read and commented on many parts of the atlas while needing to schedule meetings instead of simply visiting impromptu; the authors of the more than 10,000 e-mails that were never read nor processed as I worked full-time on the atlas; and last but not least, my friends, who saw much of my husband and kids but very little of me.

Extraordinary women—like Monika Börner, Bonnie DeVarco, Monika Herzig, Janice M. Hicks, Deborah MacPherson, Weixia (Bonnie) Huang, Anne Prieto, Andrea Scharnhorst, Maria Zemankova, and many others—empowered my thinking and provided the physical and moral support needed for this undertaking.

Last but not least, I would like to thank Melanie B. Goldstone, Eleanor B. Goldstone, and Robert L. Goldstone for their love and support.

August 10 and 11, 2006: Scholarly Data and Data Integration Workshop
Indiana University, Bloomington, Indiana

May 26, 2007: Forecasting Science Workshop
New York Hall of Science, Queens, New York

AVONIUS · ZEPHIRUS ·
mare glaciale
scitlanda
5 10 15 20
occeanus
ibernia
albion insula
germanicū
aquitā
cantabricū
Gallia
hispania
Europa
Germania magna
Sarmatia
pannonia
DATIA
Italia
herculeū
ibericū
gallicū
balearicū
affricū
Athlanticum
numidicū
adriaticū
macedonia
Pontus

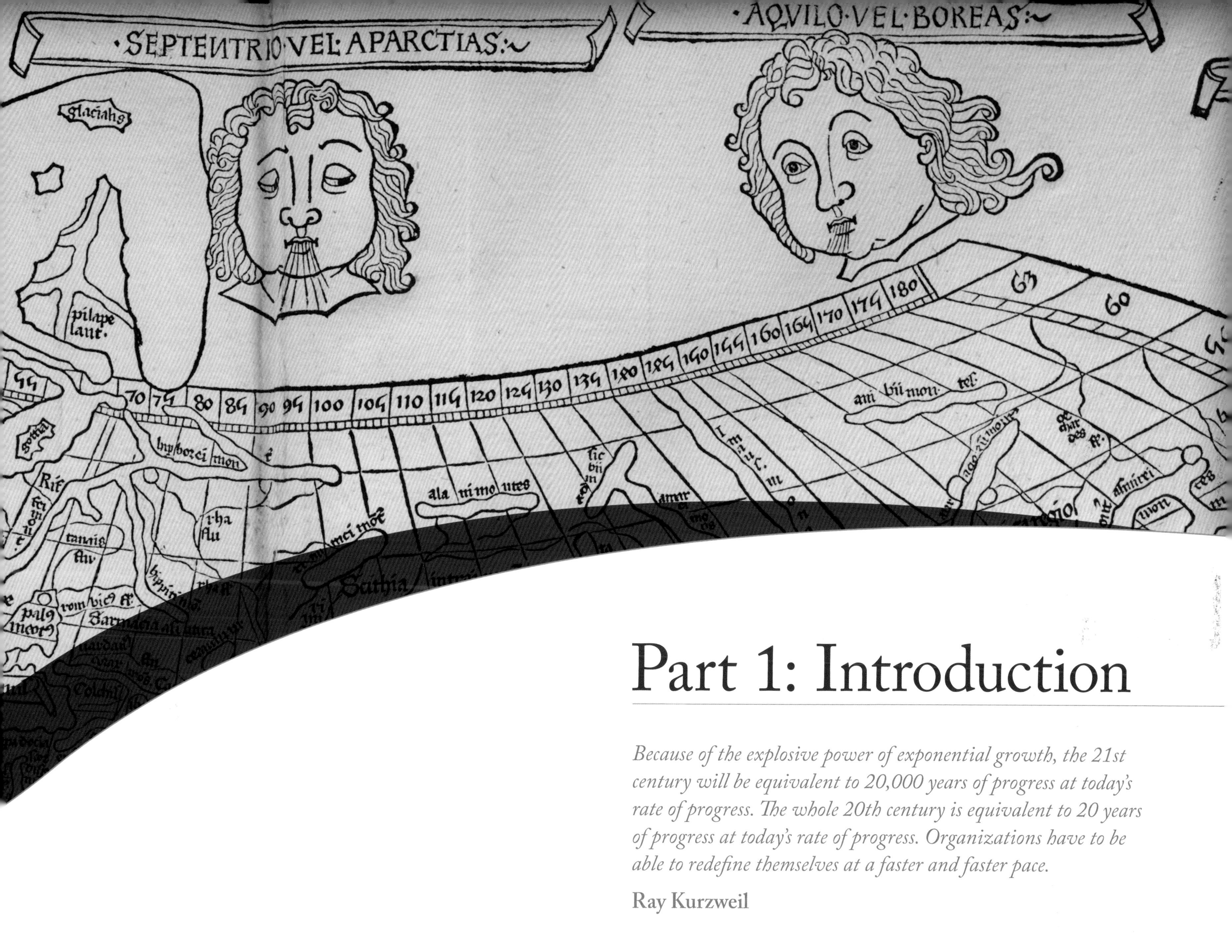

Part 1: Introduction

Because of the explosive power of exponential growth, the 21st century will be equivalent to 20,000 years of progress at today's rate of progress. The whole 20th century is equivalent to 20 years of progress at today's rate of progress. Organizations have to be able to redefine themselves at a faster and faster pace.

Ray Kurzweil

Knowledge Equals Power

Access to high-quality data, knowledge, and expertise tends to lend one authority and power. Yet there is so much to learn in so little time—tough deadlines abound. This spread discusses the dramatic increase in human population together with a history of the type and quantity of information that we have managed to produce. It also illustrates the uneven distribution of population density, urbanization, scientific productivity, and technological development on our planet. The next two pages focus on the accelerated growth of science and technology and the impact of tool development. The remainder of this introduction reviews the knowledge needs of diverse stakeholders and the evolution of geographic and science maps as guides to the navigation and exploration of physical places and abstract spaces.

Population Growth

The timeline below plots the dramatic increase in human population—estimated, recorded, and predicted—from 1,000,000 BC to AD 2200. According to United Nations estimates, the world population reached the 6 billion mark in 1999 and the 7 billion mark in 2011. It is expected to cross the 8 billion mark in 2028 and the 9 billion mark in 2054; it will nearly stabilize, at just above 10 billion, after 2200.

Significant events that have positively influenced our species, such as tool development and technological revolutions–as well as events that have negatively influenced it, such as epidemics and wars (in bold)–are listed below the population timeline. Major ages are indicated.

Knowledge and Technology Overload

The number of currently active researchers exceeds the number of researchers who ever lived previously. Researchers publish or perish. Some areas of science produce more than 40,000 papers a month. We are expected to know more works than one can possibly read in a lifetime. We receive many more e-mails per day than can be processed in 24 hours. We are supposed to be intimately familiar with data sets, tools, and techniques that are continually changing and increasing in number and complexity—all this while being reachable 24 hours per day, 7 days per week.

Libraries and storage facilities are being filled more quickly than they are being built. Scientific data sets, algorithms, and tools need to be mastered to advance science. The figure on the right depicts just how much information exists. No single man or machine can process and make sense of such an enormous stream of data, information, knowledge, and expertise.

All this leads to a quickly increasing specialization of researchers, practitioners, and other knowledge workers; a disconcerting fragmentation of science; a world of missed opportunities for collaboration; and a nightmarish feeling that we are doomed to keep reinventing the wheel for eternity.

Us from Above

The four maps on the right depict the **2005 World Population** (top left), city lights in **Earth at Night** as a proxy for urbanization (top right), **2003 Scientific Productivity** (bottom left), and **2007 IP Address Ownership** as a proxy for technological development (bottom right). All four use the same equidistant cylindrical projection. World population correlates strongly with urbanization (see Earth at Night) scientific productivity, and IP address ownership in economically developed countries. In other areas of the world, there exist major contrasts: many densely populated areas are black in the Earth at Night map, and scientific productivity and Internet access are scarce and often limited to urban areas.

The Web of Knowledge

Over the last few centuries, our collective knowledge has been preserved and communicated via scholarly works such as papers and books. Works might herald a novel algorithm or approach, report experimental results, or review one area of research among others. Some report several small results; others proclaim a single large result, like the decoding of the human genome. Some confirm results; others disprove them. Different areas of science vary greatly in their publishing formats and quality standards. The description of a single set of results, such as the development of a novel algorithm, could be published in a computer science journal, a biology journal, or a physics journal—and would look different in each publication.

Citation references came into existence in 1850. Citation networks show who consumes, elucidates, or cites whose work. Networks of coauthors can be extracted by counting how often two authors have collaborated. The quality of a paper and the reputation of its author are commonly estimated via citation and download counts. Acknowledgments provide links to experts, data sets, equipment, and funding information. Experts may be challenged when asked to identify all the claims a scholarly

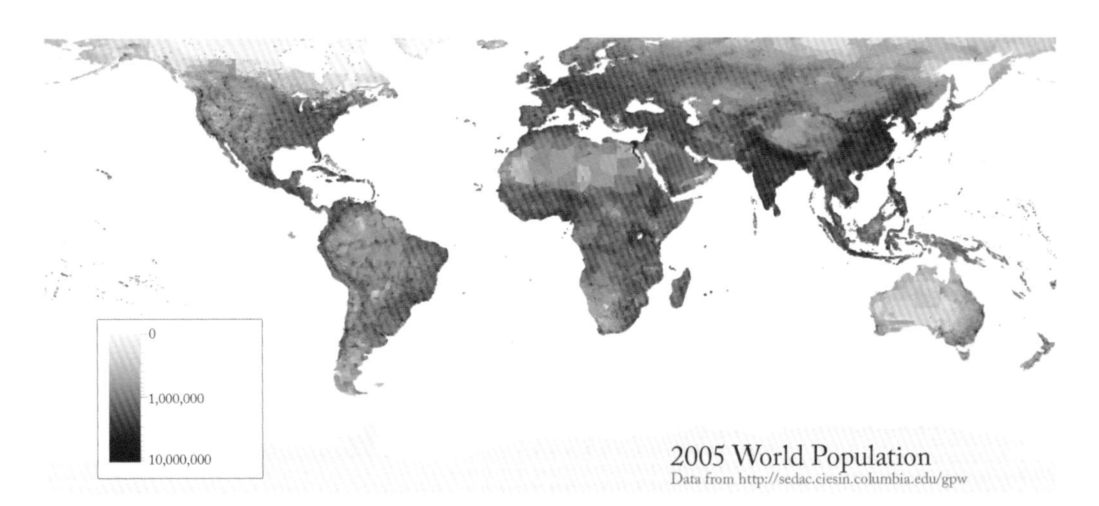

The population map uses a quarter-degree box resolution. White boxes represent a count of zero people. Darker shades of red indicate higher population counts per box, using a logarithmic interpolation. The highest density boxes appear in Mumbai (Bombay), with 11,687,850 people in the quarter-degree block; Kolkata (Calcutta), 10,816,010; and Shanghai, 8,628,088. China and India are the only two countries to have more than 1 billion inhabitants.

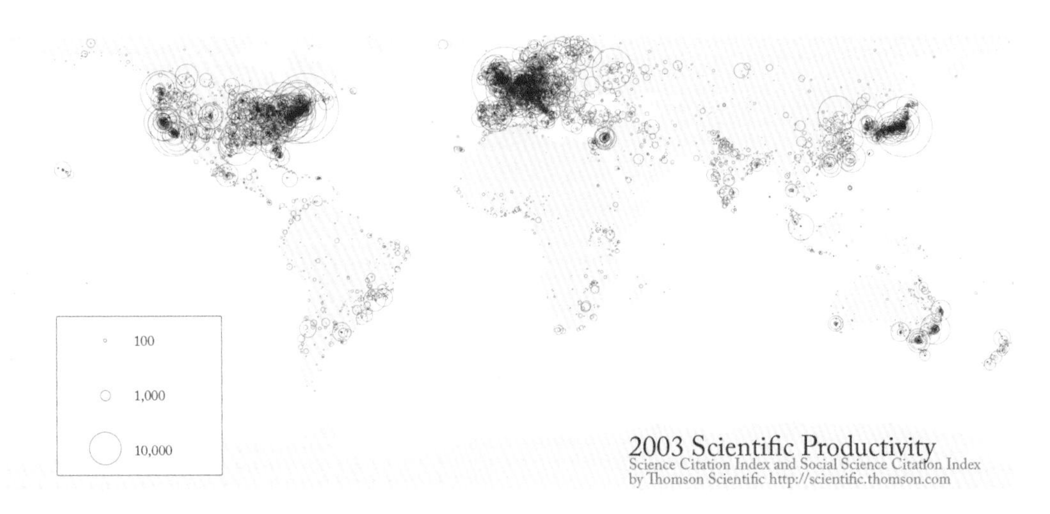

This figure shows where science research is performed. Each circle indicates a geographic location where scholarly papers are published—the larger the circle, the more papers produced. The top three paper-production areas are Boston, Massachusetts; London, United Kingdom; and New York, New York. Note how this map compares to the *Earth at Night* and *2007 IP Address Ownership* maps, while contrasting with the *2005 World Population* map.

Classical Period — *The Middle Ages*

1,000,000 — 100,000 — 10,000 — 1,000 — 100 — 10 — BC | AD — 500

500,000 *Fire discovered*
150,000 *Man evolves to Homo Sapiens*

30,000 *Language appears*
19,000 *Altamira cave paintings*
11,000 *Humans appear in the Americas*

9,000 *Animal domestication*
9,000 *First cities*
3,000 *Math beginnings and first written code of law*
2,000 *Maps drawn*
1,050 *Phoenician alphabet*

776 *Olympic Games founded*
753 *Rome founded*
447 *Parthenon completed*
200 *Papermaking in China*

79 *Mt. Vesuvius destroys Pompeii and Herculaneum*

255 *Eratosthenes measures the circumference of the Earth*

476 *Fall of Rome and the Western Empire*

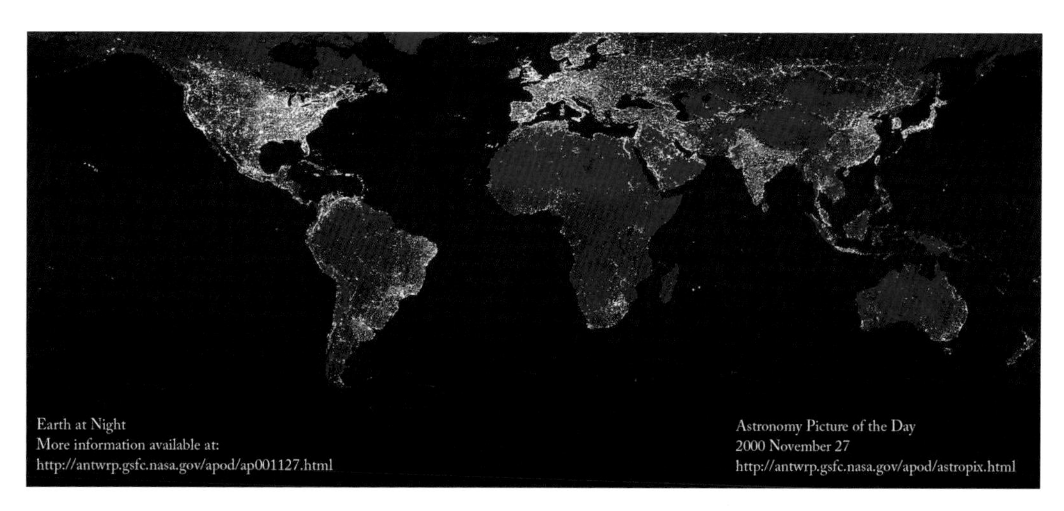

This image shows city lights at night. It was composed from hundreds of pictures made by orbiting satellites. The coasts of Europe, the eastern United States, and Japan are particularly well lit. Many cities exist near rivers or oceans so that goods can be exchanged at less expense by boat. The central parts of South America, Africa, and Asia are rather dark despite their high population densities (see map to left).

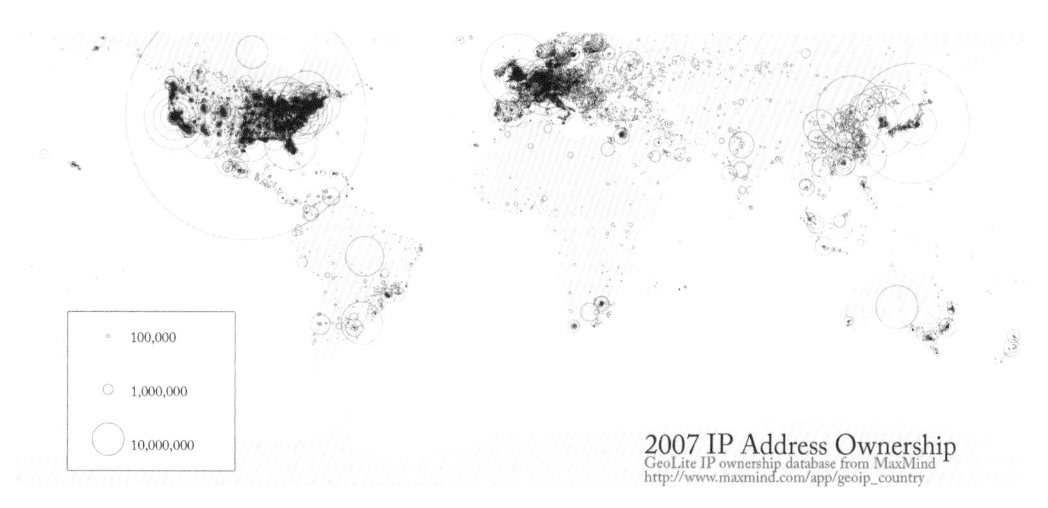

This map shows IP address ownership by location. Each owner is represented by a circle, and each circle's size corresponds to the number of IP addresses owned. The largest circle denotes MIT's holdings of an entire class-A subnet, which equates to 16,581,375 IP addresses. The countries that own the most IP addresses are the United States (560 million), Japan (130 million), and Great Britain (47 million).

work makes. In some cases it is simply not known how important today's discovery will be for tomorrow's society. It is an even harder task to determine how discoveries are interlinked with the complex network of prior (supporting and contradicting) results manifested in papers, books, patents, data sets, software, and tools.

Consequently, it is difficult for researchers, educators, and practitioners to keep up with the increasing quantity and the accelerating speed of our knowledge production. This is a major concern, as scientific results are needed to enable all human beings to live healthy, productive, and fulfilling lives. We need better tools to access, track, manage, and utilize our collective scholarly knowledge and expertise. Maps of science that guide our scholarly endeavors may well make a difference.

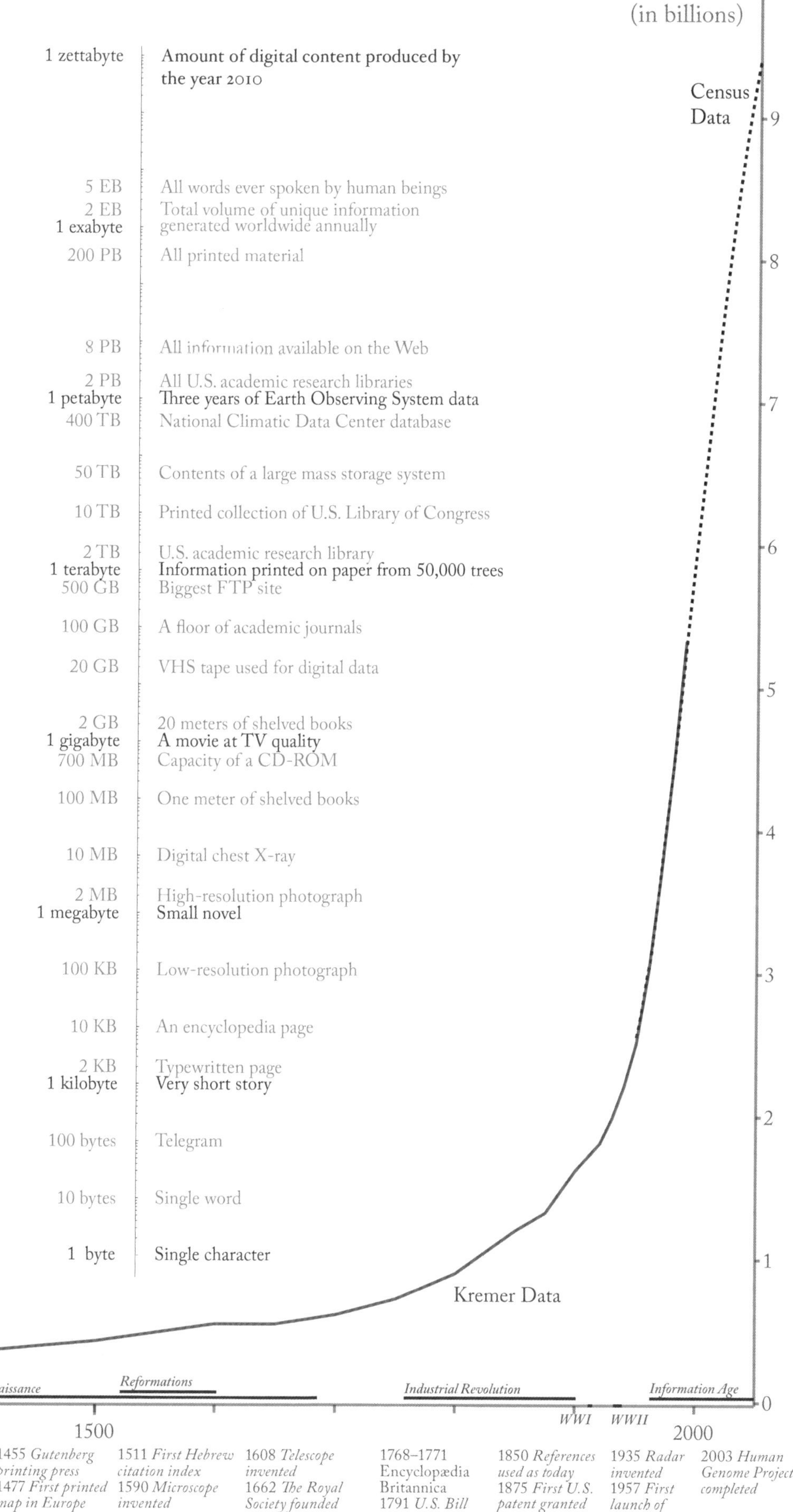

The Rise of Science and Technology

The data and graphics presented here document the tremendous increase observed in the number of books, journals, papers, and patents produced, as well as in the numbers of researchers and engineers. They also reveal the impact of funding, policy decisions, and historical events on the developments of science. We conclude with a discussion of science and society in equilibrium.

The Rise of the Creative Class

In the agricultural age, value was created from land and physical labor. In the industrial age, it was the product of raw materials and again of labor. In the information age, value is the result of these but even more so of creativity, imagination, and intelligence. Knowledge-based professionals, or members of the "creative class," include architects, artists, designers, engineers, musicians, and scientists, among others.

In the United States around 1900, there were 38,000 professional engineers, 9,000 chemists, and 12,000 other scientists, constituting 0.26 percent of the labor force. In 1970, there were a total of 2 million scientists and engineers, making up 2.5 percent of the labor force. That is a tenfold increase in only 70 years. The steady increase in engineers, physical scientists, mathematicians and information technologists, social scientists, and life scientists in relation to the total U.S. population is shown in the **People** graph below. Note the 500-fold increase in mathematicians and information technologists over the 50-year time span.

In 2008, knowledge-based professionals made up 20 to 30 percent of the labor force in developed countries—yet their salaries accounted for 50 percent of all recorded earnings. In the United States, on average, a high school diploma will lead to lifetime earnings of $1,100,000; a bachelor's degree, $2,100,000; a master's degree, $2,500,000; and a doctorate, $4,400,000.

Enormous amounts of effort and money are spent on gaining access to the brightest minds of our population. For example, a ticket to the TED (Technology, Entertainment, Design) conference—featuring "inspired talks by the world's greatest thinkers and doers"—costs about $6,000. The tickets for TED2008 sold out more than a year in advance, and more than 3,000 people were on the TED2009 waiting list when tickets were sold out.

Growth of Science

Since the early 1800s, there have been numerous studies on the progress of scientific research and development, including the quantity of results produced. Here, we consider the history of scientific publications, including books, journals, and scholarly papers, as well as Wikipedia entries.

Books

Records indicate that 8 million books were published prior to the 19th century, while 1 million books were published in 2008 alone. Since the invention of movable type, an estimated 100 million books have been printed. The **Books** graph below shows the number of books indexed per publication year by providers such as WorldCat, Bookman/Publishers Weekly/Bowker, and Chemical Abstracts Service (CAS).

Various efforts to scan all existing books and make them freely available online as e-text books (e-books) are underway. Two projects are discussed here and included in the **Books** graph. The Million Book Project, launched by several professors from Carnegie Mellon University, offers scans of 10,850 books. Project Gutenberg, the first producer of free e-books (see **page 18, Knowledge Collection**) featured more than 22,000 e-books in November 2007 and plans to offer an additional 1 million e-books by 2015. The chart shows the number of books digitized (not published) per year. Both projects use the Internet Archive (see **page 18, Knowledge Collection**) as a backup distribution site. At the rate of less than one thousand books per year, however, it will take 100,000 years to digitize all existing books. To that end, it would be necessary to capture born-digital publications at the time of creation and to scan books at a rate of 1 million books per year, or 3,000 each day.

Papers and Journals

The growth of science as a whole is often estimated by counting the number of scientific journals in existence. If we begin in 1750 with a count of 10 journals and end in 1950 with a count of 100,000, we find a steady doubling time of 15 years. The graph on the right, by Derek John de Solla Price, shows the total number of scientific journals and abstract journals as a function of date. It was originally published in de Solla Price's *Science Since Babylon* (1962).

In *Science: Growth and Change* (1971), Henry W. Menard reported the growth rates for different types of scientific records. He identified a tenfold increase in the number of papers every 50 years, abstracts every 30 years, and computer indexes every 10 years.

The **Papers and Wikipedia Entries** graph below shows the number of papers—or abstracts in the case of the National Federation of Advanced Information Services (NFAIS) abstracting service—served by different databases per publication year. The highest numbers belong to NFAIS. Data provided by Thomson Reuters includes the Science Citation Index Expanded (1900–present; 6,126 journals), Social Sciences Citation Index (1956–present; 1,802 journals), and Arts and Humanities Citation Index (1975–present; 1,136 journals), among others. Thomson Reuters's Web of Science (WoS; established 1981), includes all three data sets. The third-highest number of holdings belongs to Elsevier's Scopus, which indexes 15,000 peer-reviewed journals from more than 4,000 publishers. Google Scholar is missing from this graph, as detailed data was not available.

Papers from the Royal Society (established 1665) and Physical Review (established 1887) journals record the scholarly history of their respective research communities in a very comprehensive and longitudinal fashion. JSTOR, with its strong archi-

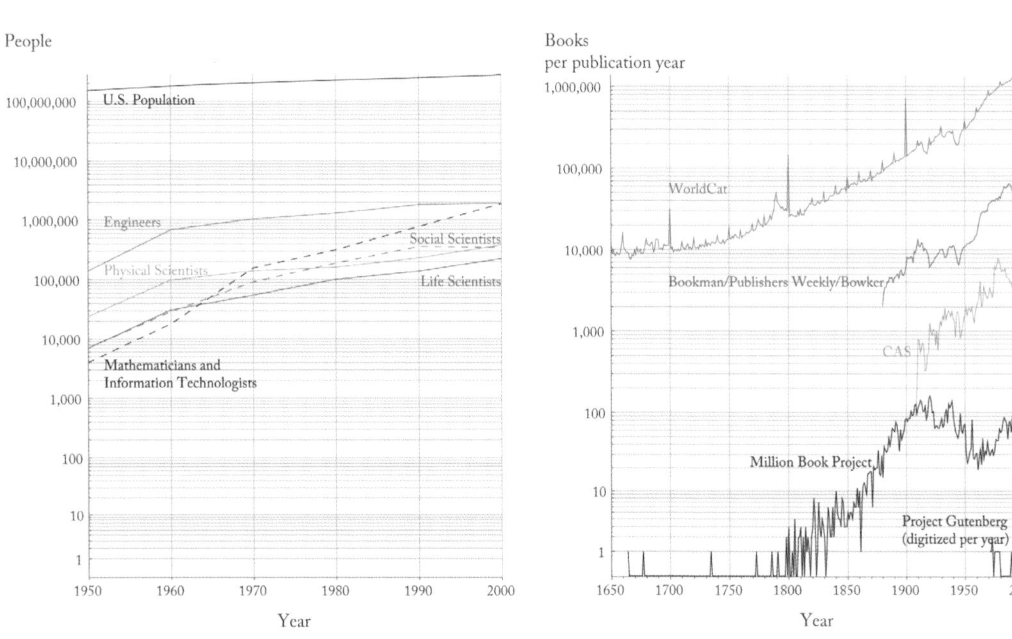

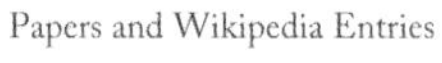

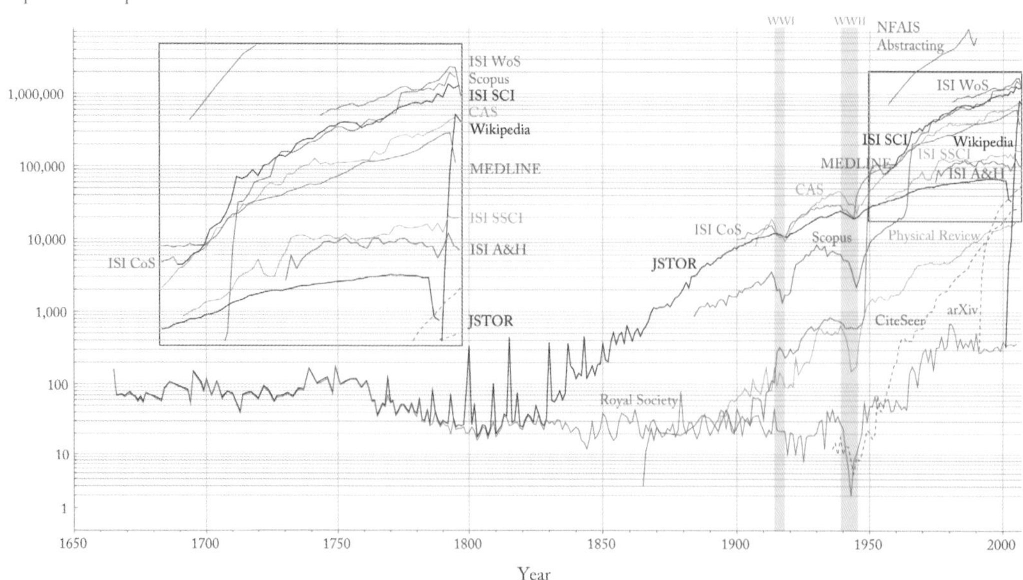

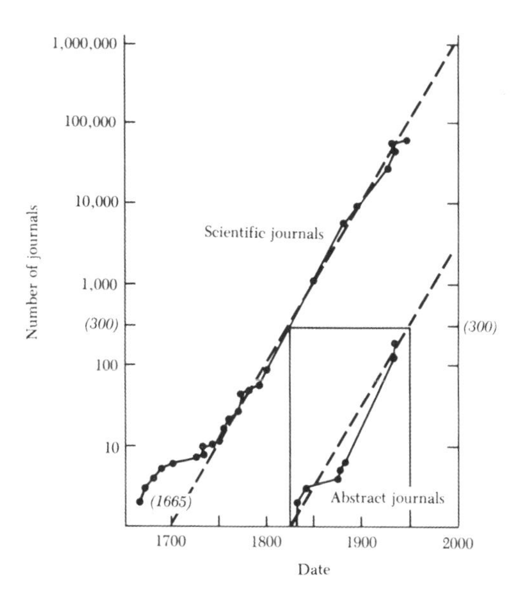

val mission, includes the Royal Society papers and offers broad coverage of the arts and humanities. Note the effect of external events on scholarly publication activity, for example, the decrease in papers printed during World War I and World War II.

MEDLINE, CiteSeer, and arXiv provide free access to full-text articles. MEDLINE (established 1971) is sponsored by the National Library of Medicine. It currently offers more than 19 million records—mostly biomedical—and in 2008 offered more records than Scopus for the years 1949–1965. ArXiv and CiteSeer are grassroots efforts of the physics and computer science communities, respectively. Paul Ginsparg created arXiv in 1991. Steve Lawrence, C. Lee Giles, and Kurt Bollacker at NEC Research made CiteSeer available internally in 1997 and started serving it to the world in 1998.

Data from Jimmy Wales's Wikipedia (see **page 19, Jimmy Wales**) was included to hint at the possibility of a different grain size for recording scholarly results (see **page 20, Paul Otlet**) and the power of the "million minds" approach (see **page 24, Global Brain**). An encyclopedia entry cannot be directly compared to a research paper, and the editing process Wikipedia uses is clearly different from the peer-review process that scholarly papers undergo. Nevertheless, the number and quality of entries written by volunteers around the globe for the English version of Wikipedia is impressive (see also **page 166, Science-Related Wikipedian Activity**).

Patents

Intellectual property rights have been claimed and protected since 500 BC. Lawful patent protection came into existence in the Republic of Venice as early as 1474. The total number of patents granted by main patent offices between 1883 and 2005 is shown in the **Patents** graph. Between 1883 and 1959, patenting activity was concentrated in the United States, Germany, the United Kingdom, and France. The average annual growth rate was about 2 percent. Starting in 1960, new states and nations, such as Japan and the Soviet Union, increased the number of filings. The European Patent Convention came into existence in 1977 and led to a decline in filings at national offices in Germany, France, and the United Kingdom. Since 1980, the patent offices of the United States, Europe, and China have all experienced significant growth in filings, at a rate of around 3.35 percent per year. The Patents graph shows the vast and sharp increase in the number of applications filed, particularly by the United States and Japan. In 2005, the European and Chinese patent offices filed nearly the same number of patents.

Highly cited patents are assumed to have higher technological impact. Patent citations indicate connections between companies and technological areas, as well as between industry and academia. Patent filings are used to compute indicators of activity, association, and impact, then applied to competitor assessment, merger/acquisition targeting, and investment strategy decisions.

Investing in the Future

Nations and regions fund research and education to improve their scientific wealth and economic power. In 2002, the public and private sector investment in research and development (R&D) as a percentage of the gross domestic product (GDP) was 1.86 percent in the United Kingdom, 2.20 percent in France, 2.51 percent in Germany, and 2.67 percent in the United States.

United States R&D expenditures by funding source since 1953, in constant 2000 dollars, are shown to the left in the **U.S. R&D Expenditures** graph. Note the crossing of federal government and industry funding in 1980 and the decline of industry funding in 2002 after the technology stock bubble burst.

In 2006, the Wellcome Trust spent $800 million on funding, with an average project cost of about $375,000. A typical project produces three major papers, so the price of one paper can be estimated at $125,000.

The amount of money required to perform cutting-edge R&D continues to increase. In 1970, the cost of one scientist—including salary, office, and laboratory support—was calculated to be $41,000 per year. Since then, science has become ever more interdisciplinary and dependent on technology. In 2010, the Large Hadron Collider and the Hubble Space Telescope, and similar such socio-technical infrastructures are vital sources of information and inspiration in our quest to expand our knowledge of the world.

Science and Society in Equilibrium

In *Little Science, Big Science* (1963), de Solla Price predicted that the growth of science would eventually decelerate, as the number of scientists could not possibly exceed the population count and the dollars devoted to R&D could not exceed the gross national product (GNP). He calculated that the number of scientists doubles every 15 years, while the U.S. population doubles every 40 to 50 years.

The pair of graphs below are derived from Joseph P. Martino's *Science and Society in Equilibrium* (1969). The left graph compares the number of U.S. scientists to the U.S. population between 1940 and 1969. During those nearly 30 years, the proportion of scientists in society increased from less than 0.5 percent to 1.0 percent. In 2008, of the 300 million U.S. residents, scientists numbered 5 million or 1.7 percent of the population (see **page 4, The Rise of the Creative Class**). The right graph shows the increase in U.S. GNP since 1946 and the dollar resources expended on R&D since 1953. As shown below, the U.S. GNP devoted to R&D had doubled from slightly less than 1.5 percent to 3 percent. Of the $13 trillion GNP in recent years, 2.3 percent, or $0.3 trillion, has been devoted to R&D (see previous section, **Investing in the Future**). While the number of scientists steadily increases in society, the increase in funding parallels the increase in GNP.

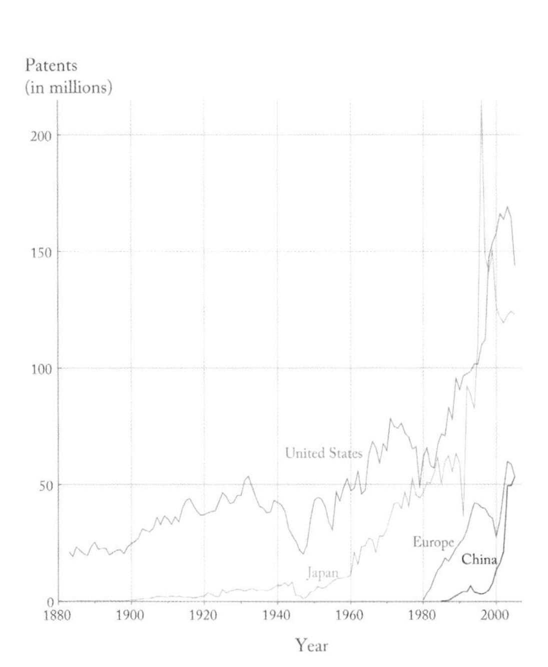

U.S. R&D Expenditures
Constant 2000 dollars (millions)
Industry
Federal
Other
Year

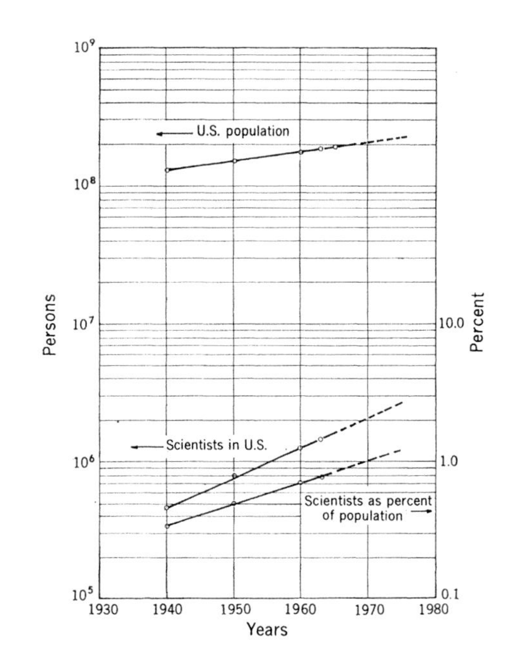

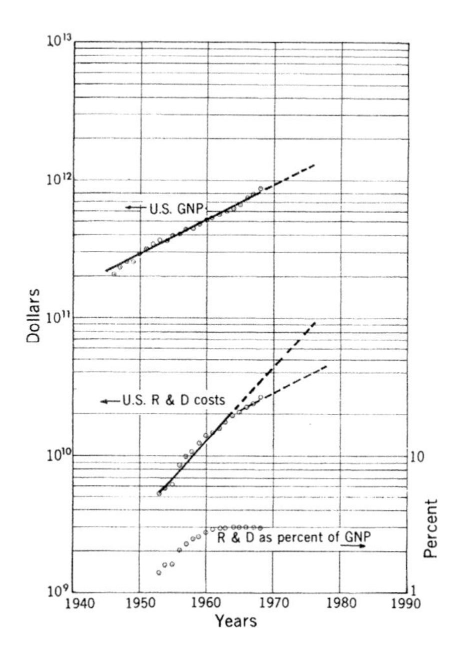

Addictive Intelligence Amplifiers

In a knowledge- and innovation-driven world, survival of the fittest is determined by one's inventiveness, expertise, and powers of influence. Creativity and innovation tend to be stimulated by the sociotechnical environment enjoyed by an individual or community. Tools that amplify our intelligence and allow us to engage in symbiotic relationships with other experts and technology play an important role here. Given the limitations of human perception and cognition, these tools improve our fitness.

Shrinking Planet

Increases in travel and communication speeds reduce the relative size of our world. This is made tangible in the chart by Buckminster Fuller (bottom right; see also **page 25, Buckminster Fuller**) which shows transportation times from 500,000 BC to AD 1965. The shaded area represents population growth; the solid line, transportation speed; and the dashed line, communication speed. The entire time span is divided into 9 epochs. For each epoch, the travel times, means of transportation, distance traveled per day, state size, and means of communication are given. Four Dymaxion maps—a projection of the World map onto the surface of a polyhedron then unfolded in two dimensions—show the relative size of the world as transportation and communication technology advance.

An extended population graph, covering 1,000,000 BC to AD 2005, can be found on **page 2, Knowledge Equals Power**. Extensions of Fuller's chart are shown in the **U.S. Transportation** graph (opposite page), which shows passenger miles for national railroad, intercity automobile, and domestic airline travel in the United States. Note the rise and fall of railroad and car traffic around 1945 and the impact of the oil crises in 1973 and 1979.

The **U.S. Communication** graph (opposite page) shows tremendous increases in communication, first by telegraph, then by radio and television, and most recently by cell phones and the Internet. Since the World Wide Web came into existence in 1989 (see **page 20, Tim Berners-Lee**) it has been growing at a phenomenal rate. Due to the Web's decentralized nature, the existing number of static Web pages is difficult to compute. Plus, many Web pages are generated on the fly using databases; the size of these dynamically generated Web pages, the so-called deep Web, is much more extensive. Even larger than the stock of Web pages is the number of blog posts and e-mails sent. In 2008 alone, 210 billion e-mails were sent each day by 1.3 billion users worldwide and 900,000 news blogs were posted.

From Little Boxes to Big Boxes

According to Barry Wellman, communities evolved from "little boxes" (clusters of tightly knit "door-to-door" neighbors) to "glocalized" networks (with dense local and weak global links) to sparse groups with "networked individualism" (dynamically changing links, regardless of locality).

The developed world is in the midst of a paradigm shift both in the ways in which people and institutions are actually connected. It is a shift from being bound up in homogenous "little boxes" to surfing life through diffuse, variegated social networks. Although the transformation began in the pre-Internet 1960s, the proliferation of the Internet both reflects and facilitates the shift.
Barry Wellman

This evolution is driven by the complexity and dynamics of our activities and the capabilities of our tools. Teams form, storm, norm, and perform at a high pace. While we might be happier and more productive offline, many of us spend a considerable amount of time online, weaving our increasingly larger and more dynamic networks.

Accelerating the Rate of Change

The tools we build work to accelerate the rate of change. The more knowledge we generate, the more we have to manage. The more efficiently we can link to others, the larger the networks we need to operate, navigate, and maintain.

The more tools, the faster the change—reaching a point where it is questionable whether the environment can sustain it. Major tools are atomic bomb, telephone, production-line system of manufacture, aircraft, plastics, guided rocket, television. Each has enormous potential for man's benefit or his destruction.
James Burke

On-the-Fly Assembly

We are entering an era where units of interest are becoming reduced and less defined. Books are being replaced by e-mails, memos, and endlessly evolving wiki items. Data warehouses and cathedral-like

SHRINKING OF OUR PLANET BY MAN'S INCREASED TRAVEL AND COMMUNICATION SPEEDS AROUND THE GLOBE

YEAR	500,000 BC	20,000 BC	300 BC	500 BC	1,500 AD	1900 AD	1925	1950	1965
Required time to travel around the globe	A few hundred thousand years	A few thousand years	A few hundred years	A few tens of years	A few years	A few months	A few weeks	A few days	A few hours
Means of transportation	Human on foot (over, ice bridges)	On foot and by canoe	Canoe with small sail or paddles or relays of runners	Large sail boats with oars, pack animals, and horse chariots	Big sailing ships (with compass), horse teams, and coaches	Steam boats and railroads (Suez and Panama Canals)	Steamships, transcontinental railways, autos, and airplanes	Steamships, railways, auto jet and rocket aircraft	Atomic steamship, high speed railway auto, and rocket-jet aircraft
Distance per day(land)	15 miles	15-20 miles	20 miles	15-25 miles	20-25 miles	Rail 300-900 miles	400-900 miles	Rail 500-1,500	Rail 1000-2000
Distance per day (sea or air)		20 by sea	40 miles by sea	135 miles by sea	175 miles by sea	250 miles by sea	3,000-6000 air	6000-9500 air	408,000 air
Potential state size	None	A small valley in the vicinity of a small lake	Small part of a continent	Large area of a continent with coastal colonies	Great parts of a continent with trans-oceanic colonies	Large parts of a continent with transoceanic colonies	Full continents & Transocean Commonwealths	The Globe	The globe and more

Communications							
Word of mouth, drums, smoke, relay runners, and hand printed manuscripts prior to 1441 A.D.	① The Gutenberg 1441 printing press	② The rapid print Web 1863 newspaper press	③ The Bell 1876 telephone	④ The Marconi 1895 telegraph	⑤ First commercial 1920 radio broadcast	⑥ National 1950 Television	⑦ Transcontinental T. V. 1965 with the introduction of Early Bird satellite

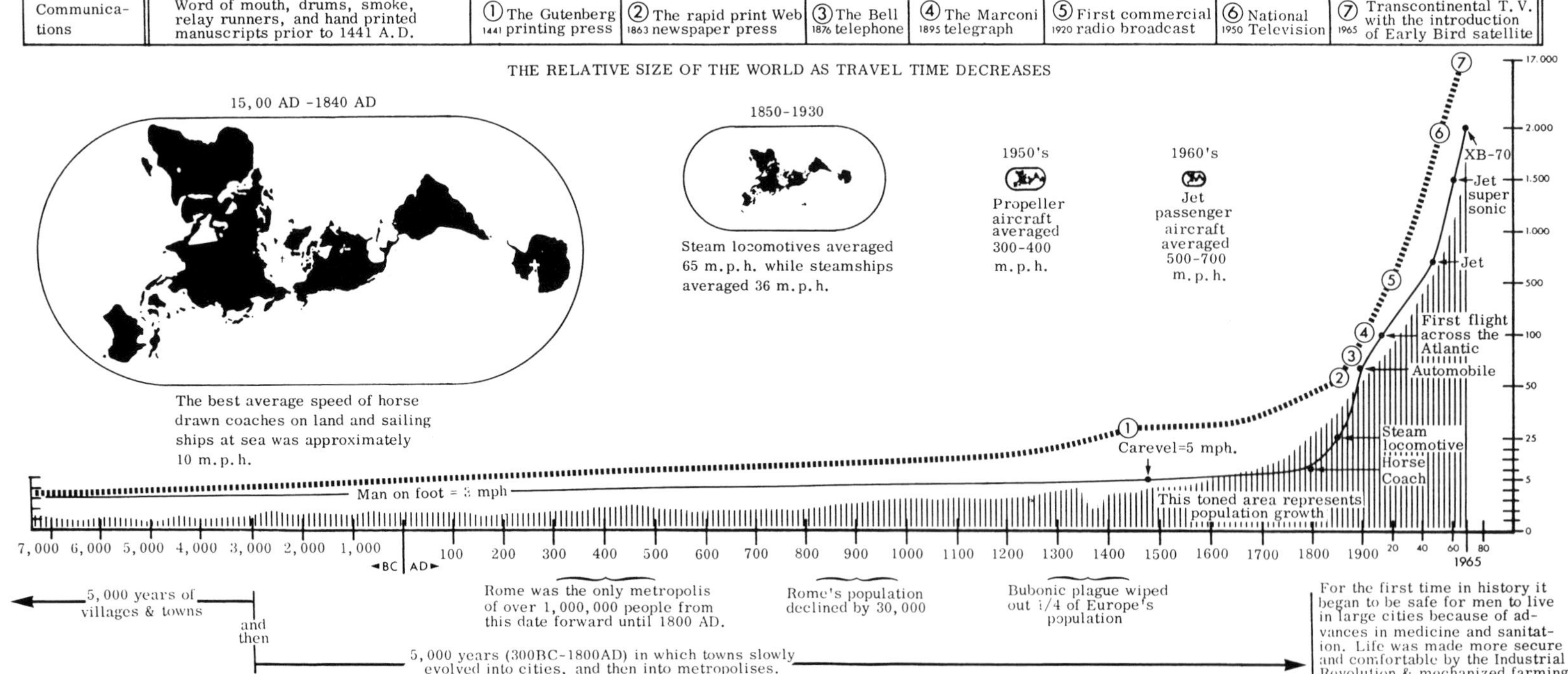

supercomputing centers are superseded by highly modular plug-and-play architectures and marketplaces of scholarly data sets, algorithms, and other resources. Project teams, and even whole companies, are assembled on the fly for a few days or a few years. There is no longer any lifelong job guarantee. Social and emotional intelligence, lifelong learning skills, and the ability to communicate across disciplinary and cultural boundaries become ever more important. Disciplined, synthesizing, creative, respectful, and ethical minds are promised the brightest futures.

Urban Species

Spatially, we are grouped into clusters called cities. About 500 years ago, there were only 5 cities with more than 100,000 people. In 2008, there were about 300 cities with populations of more than 1 million. In 1800, 50 million people lived in cities; in 2000, 3 billion did—more than half of the global population. For the first time in human history we are an urban species. In 2009, cities occupy near 3 percent of the world's land but they support more than 50 percent of the global population and use 75 percent of the world's resources.

This urban migration and population growth was discussed by George K. Zipf in *Human Behavior and the Principle of Least Effort* (1949). He ranked U.S. cities in decreasing order of population size for the years 1790 to 1930, and plotted population size versus rank in a log-to-log plot

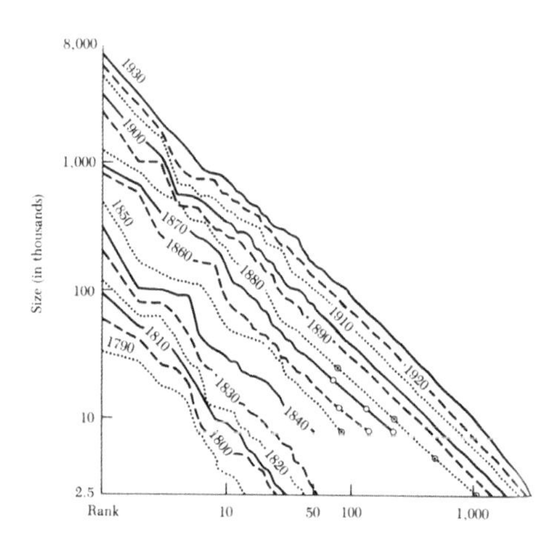

(above). The resulting graphs follow a power law (see also **page 57, Power Laws**) as curving lines for different years move parallel to one another.

Natural-Born Cyborgs

While science and technology increase at a rapidly accelerating pace, human perception and cognition appear to stay nearly constant. Lifelong learning and training increase our knowledge and expertise, but not by several orders of magnitude. This is where tools come in—they amplify our intelligence by letting us offload tasks, such as keeping time, doing routine jobs (for example, calculations), and connecting to people (for example, via phone or e-mail). They augment our environments by interlinking the many everyday physical objects and bits and pieces of information that populate our homes and offices. The Internet and the World Wide Web are the beginnings of a global "world brain," as envisioned by many far-seeing thinkers (see **page 24, Global Brain**) which intimately links the unique capabilities of people and machines. To fully exploit the power of tools, we will need to

... appreciate what we already are: creatures whose minds are special precisely because they are tailor-made to mix and match neural, bodily, and technological ploys. ... Cognitive technologies are best understood as deep and integral parts of the problem-solving systems that constitute human intelligence. They are best seen as proper parts of the computational apparatus that constitutes our minds.
Andy Clark

Moths to the Flame

We are drawn to technology like moths to a flame, as evidenced by our dependence on watches, cell phones, cars, and the Internet. If one of our technologies breaks down—a laptop on a business trip, for instance—our capacity and efficacy too are impaired. Handling calls and processing e-mails keeps many of us in the flow, and lets us effectively operate in a complex web of social, professional, and other networks.

One day, perhaps soon, we'll create mobile, semi-intelligent beings to do our dull, dirty, dangerous work. Soon after that, they'll become so useful and so competent that we'll keep them as pets and as companions for our children. ... How we treat them, how we employ them, even whether they live or die, all will be up to us. Yet for that very reason, how we use them—these creations of our genius, these children of our minds—will determine how the future judges us.
Gregory J. E. Rawlins

Man vs. Machine

In the 1980s, microcomputers reached the consumer market. Since then, the amount of information that computers are capable of processing and the rate at which they do so has doubled about every two years, a phenomenon known as Moore's law. At this rate, some predict that computing speed will reach a limit in about 80 years. Other researchers cite physical laws and predict that this doubling rate could proceed for another 600 years.

Computers are universal machines, their potential extends uniformly over a boundless expanse of tasks. Human potentials, on the other hand, are strong in areas long important for survival, but weak in things far removed.
Hans Moravec

Machines, and particularly computers, will remain instrumental in augmenting our collective intelligence. A true symbiosis of human perception and cognition and mechanistic and computational implants seems on the near horizon (see **page 24, J.C.R. Licklider**).

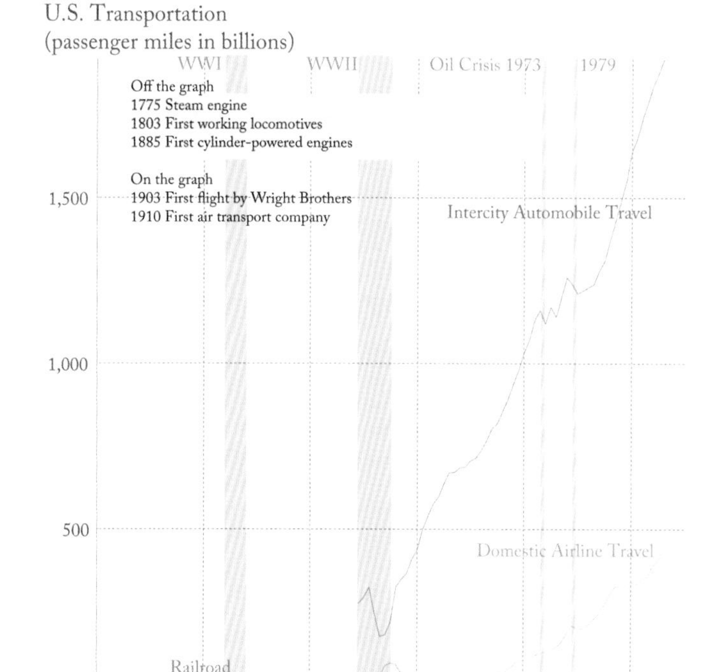

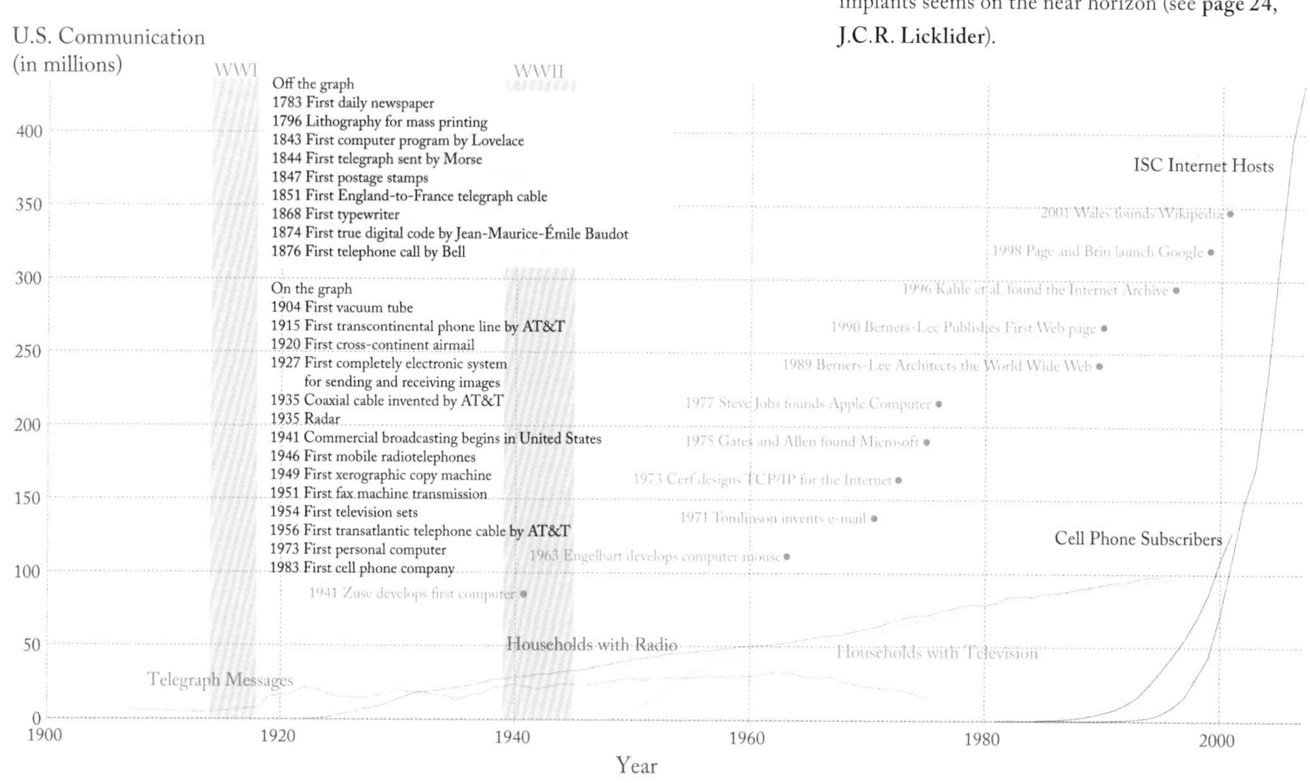

Knowledge Needs and Desires

While people have highly diverse needs and priorities, many find themselves drowning in data and information. They simply cannot read all the e-mails, papers, or books they are supposed to know; explore the many new datasets and tools that become available each day; collaborate closely with relevant experts and colleagues; and devote sufficient time and attention to the many important decisions they make each day. The concrete needs and desires of six prototypical groups are reviewed here with a special emphasis on those that might be addressed by maps of science. Sample science map displays are given in the lower part of this spread, along with descriptions of how to interpret or use them. **Part 5: The Future of Science Maps (page 197)** presents approaches and tools that are specifically tailored to the user groups discussed here.

What information consumes is rather obvious. It consumes the attention of its recipients. Hence a wealth of information creates a poverty of attention.
Herbert A. Simon

Data Providers and Librarians

The mission of data providers and digital libraries is to support access to our collective knowledge. Excepting the fact that commercial providers charge for information access and public libraries are free, they share many goals. Generally, both aim for highly usable and accessible collections. They respect copyrights and help create and promote best practices, standards, and open systems to ensure the longevity of and ongoing access to their holdings. In many cases, they add value to materials by supplying contextual information or metadata. They develop tools and services to promote enhanced interpretation, context, and understanding. Both have to deal with the fact that the need for quality has by far outshadowed the need for quantity.

Developing a high-quality yet affordable library on a specific topic for a scholarly community is not a trivial matter. Since the early 19th century, research librarians have systematically applied citation analysis in their collection development; the quality of a journal was determined by the number of citations it received. With the implementation of the Thomson Reuters citation indexes, it has become possible to automatically extract citation data at the journal and individual paper level to guide the search, evaluation, and use of knowledge.

The number and variety of available databases is overwhelming. Databases vary considerably in their temporal, geographic, and topical coverage. Database quality ranges from "downloaded from the Web" to "manually curated by experts." Visual interfaces to digital libraries provide an overview of the holdings—as well as indexes to the records—of a library or publisher. They apply powerful data analysis and layout techniques to generate visualizations or maps of large document sets. Visual interfaces can be understood as a value adding service intended to help humans mentally organize, electronically access, and manage large, complex information spaces.

Industry

A deep understanding of technology, governmental decision-making, and societal dynamics is required to make informed economic choices that can ensure survival in highly competitive markets. Discontinuities caused by disruptive technologies have to be determined and relevant innovations detected, grasped, and used. Companies need to look beyond technical feasibility to identify the value of new technologies, predict diffusion and adoption patterns, and discover new market opportunities as well as threats.

The absorptive capacity of a company—in other words, its ability to attract the sharpest minds and "play with the best"—has a major impact on its survival. The importance of social networking tools and network visualizations increases with the demand for an understanding of the "big picture" in a rapidly changing global environment. (See **Claiming Intellectual Property Rights via Patents** below).

Competitive technological intelligence analyses, technology foresight studies, and technology road mappings are used to master these tasks. Easy access to the most current and cutting-edge results, data, tools, and expertise is key to success. Last, but not least, companies need to communicate their vision and goals to a diverse set of stakeholders in order to hire and cultivate experts, attract venture capital, and continue to promote their products.

Collaborative Visual Interfaces to Digital Libraries
A three-dimensional virtual world is used to organize papers spatially. Each paper is represented as a node in a semantic network of related papers. Citation bars represent the number of citations a paper received. Users can click on a paper to retrieve details, which are displayed as a pop-up description (shown in the window at right). Multiple users can explore this space together, communicate via a chat window, and combine results.

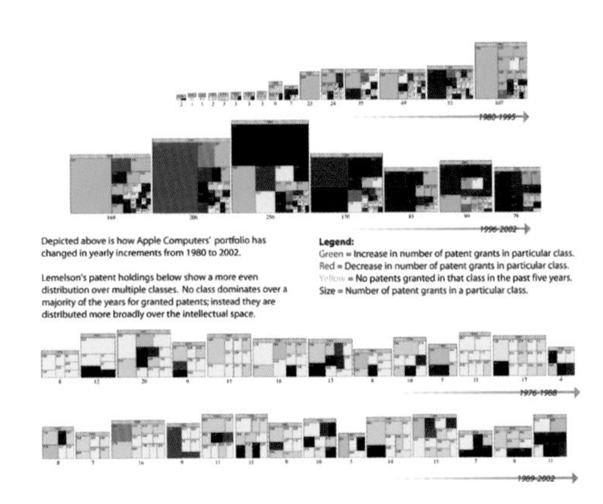

Claiming Intellectual Property Rights via Patents
The evolving patent portfolios of Apple (1980–2002) and Jerome Lemelson (1976–2002) are shown here. The number of patents granted per year matches the size of the square. Each square is further subdivided into color-coded patent classes: green signifies an increase in the number of patents, red a decrease in the number of patents, and yellow that no patent was granted in that class over the last five years. While Apple claims more and more space in the same patent classes, Lemelson's patent holdings are distributed more broadly over the intellectual space.

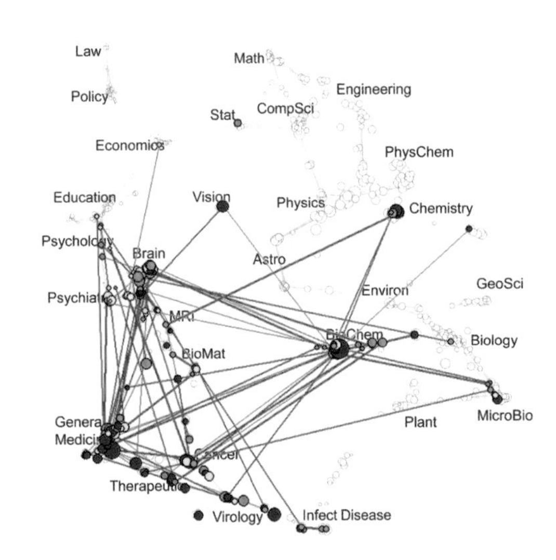

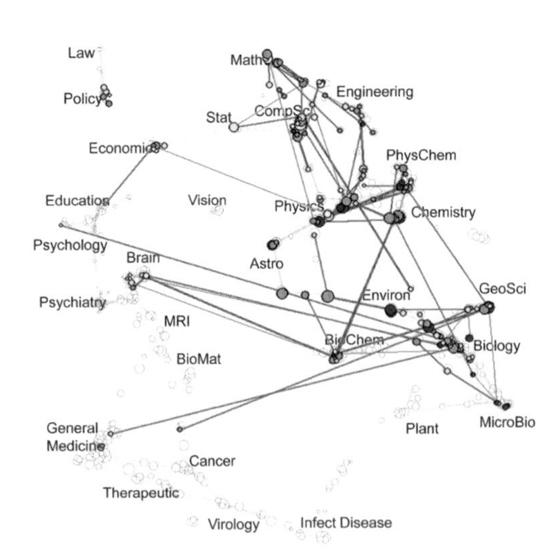

Funding Profiles of NIH (left) and NSF (right)
Using a base map of science (see **page 12, Toward a Reference System for Science**), the core competency of institutes, agencies, or countries can be mapped and visually compared. Shown here are funding profiles of the National Institutes of Health (NIH) and the National Science Foundation (NSF). As the base map represents papers, funding was linked by matching the first author of a paper and the principal investigator using last name and institution information. A time lag of three years between funding of the grant and publication of the paper was assumed. While the NIH mostly funds biomedical research, the NSF focuses on math, physics, engineering, computer science, environmental and geo sciences, and education. Overlaps exist in chemistry, neuroscience, and brain research (see **page106, The Structure of Science** for details).

Science Policy Makers

Increasing demands for accountability require decision makers to assess outcomes and impacts of science and technology policy. There is an urgent need to evaluate the impact of funding and research on scientific progress: to monitor long-term money flow and research developments, to evaluate funding strategies for different programs, and to determine project durations and funding patterns. Should science be supported by an approach that is "Newtonian" (curiosity-driven), "Baconian" (application-driven), or a "Jeffersonian" compromise? Should small grants be distributed to many scholars, or should large-scale funding be concentrated in a small number of centers?

In addition, professional science managers are keen to identify areas for future development, to stimulate new research areas, and to increase the flow of ideas into products. Hence, they need to detect emerging research areas; understand how various science fields are interlinked; examine multidisciplinary areas; measure collaborations and knowledge flows at the personal, institutional, national, and global levels; identify and compare core competencies of economically competing institutions and countries; and identify and fund central, rather than peripheral, research centers.

Researchers

Most researchers wear multiple hats: they are authors, editors, reviewers, teachers, mentors, and science administrators to varying degrees.

As researchers and authors, they need to strategically tap their expertise and resources to enhance their reputations. Expertise refers both to the knowledge they already have or can quickly obtain, as well as the expertise that can be acquired via collaborations. Resources refer to data sets, software, and tools, as well as to people supervised or paid.

Researchers and authors also need to keep up with novel research results; examine potential collaborators, competitors, and related projects; weave a strong network of collaborations; ensure access to high-quality resources; and monitor funding programs and their success rates. Last but not least, they need to review and incorporate findings and produce and disseminate superior research results in pursuit of citation counts, download activity, and press coverage.

As editors and reviewers, researchers act as gatekeepers of science. They need detailed expertise in their own domains and in related domains of research to ensure that only the most valuable and unique works are added to the growing mountain of scholarship.

As teachers and mentors, they provide students with a deep understanding of the structure and evolution of their fields as well as the peculiarities of a domain of research and practice. They might give an initial overview of the material to be covered, then highlight prominent scientists, important papers, and key events. They provide pathways to help students discover, understand, and interrelate details.

As science administrators, they are responsible for decisions regarding hiring and retention, promotion and tenure, internal funding allocations, budget allocation, and outreach. Toward this end, a global overview of the major entities and processes in their areas, as well as their temporal, spatial, and topical dynamics is vital.

Children

In school, children learn about many different sciences. However, they never get to see them as landscape or topography "from above." Hence, they have a limited understanding of the breadth and the complex interrelations of these many sciences.

How will they be able to answer these fundamental questions: What intellectual travels did prominent inventors undertake? Why is mathematics necessary to succeed in almost all sciences? How do the different sciences build upon one another? And how do I find my own place in science?

Imagine a map of our collective scholarly knowledge hangs beside the map of the world in each classroom. Imagine students can not only travel our planet online, but also explore the web of knowledge. How might this change the way we learn, understand, and create? **Hands-On Science Maps for Kids** (**page 186**) introduces physical renderings of science maps and world maps that invite children, and adults alike, to learn where major inventors and scientists pursued their work and where inventions and discoveries came into existence.

Society

Science is public rather than private knowledge—yet scientific publications are rarely accessed or understood by the general public. Placing all existing knowledge into a format that is easy to navigate has the potential to dramatically improve access to scientific knowledge and expertise. Ubiquitous, free, and simple access to information would thus advance the circulation and application of knowledge. Imagine a daily broadcast—like a weather forecast—that communicates breakthrough results and discoveries in science and technology to a wider audience.

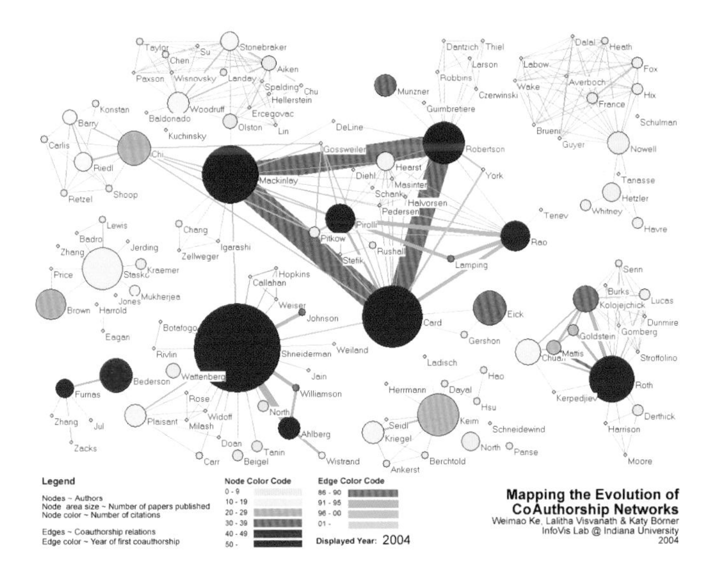

Mapping the Evolution of Coauthorship Networks
This is the last frame of an animation sequence that shows the evolution of authors (nodes) and their coauthorship relations (links) in the domain of information visualization. Coding of node area size reflects the number of papers an author has published; color denotes the number of citations these papers received. Link width equals the number of coauthorships, and link color denotes the year of the first collaboration. Large dark nodes are preferable. Large light nodes indicate papers that have not (yet) been cited. Ben Shneiderman, working in a student-dominated academic setting, experienced a different collaborative environment than did Stuart K. Card, Jock Mackinlay, and George G. Robertson, who worked at Xerox Parc for most of the time captured here.

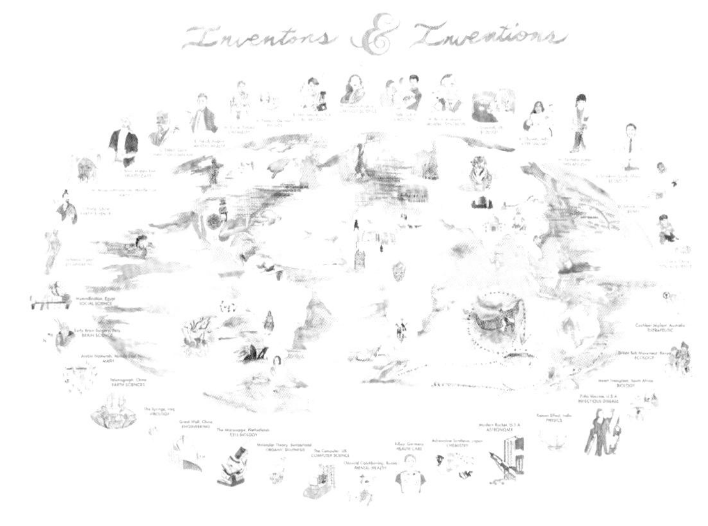

Hands-On Science Maps for Kids
The **Hands-On Science Maps for Kids** discussed and shown on **page 186** invite children to see, explore, and understand science "from above." This map shows our world and the places where science is practiced or researched. A complementary map shows the major areas of science and their complex interrelationships. Children and adults alike are invited to help solve the puzzle by placing the images of prominent scientists, inventors, and inventions in their proper places. Children are encouraged to look for the many hints hidden in the drawings to find the perfect place for each puzzle piece. What other inventors and inventions exist? Where would a favorite science teacher and science experiment go? What area of science to explore next?

Places & Spaces: Mapping Science Exhibit
This exhibit is a 10-year effort to introduce science maps to the general public. It aims to demonstrate the power of maps to help us navigate and make sense of physical places and abstract topic spaces. In 2007, the exhibit featured the first three of 10 mapping iterations—The Power of Maps, The Power of Reference Systems, and The Power of Forecasts—as well as an Illuminated Diagram display, WorldProcessor globes, and Hands-On Science Maps for Kids. The Web site **http://scimaps.org** features many more maps, together with references, the schedule of showings, and information on how to host the exhibit.

The Power of Maps

Cartographic maps have guided our explorations for centuries. They have enabled the discovery of new worlds even while marking territories inhabited by legendary sea creatures. Convenient to use (flat and small for easy handling), they are credible tools (made by established mapmakers and institutions) that simplify our world (showing key points) with clear, direct, and lasting visual impact. Only after millennia of observation and charting did we begin to arrive at a scientifically accurate map of the world. Much later, the first thematic maps, statistical graphics, and data visualizations came into existence. Fortunately, the mapmakers of science can learn and benefit from prior work to arrive at visual depictions of scientific networks and semantic spaces that help us navigate and make sense of the world of science.

I sense that humans have an urge to map—and that this mapping instinct, like our opposable thumbs, is what makes us human.
Katharine Harmon

The Power of Stories

For thousands of years, stories were the primary means of imparting wisdom from one generation to another. With the invention of writing, these stories began to be recorded on smooth and enduring surfaces, from stone and vellum to papyrus and paper—all of which could be preserved, enabling continued communication of their contents. Many stories were accompanied by imagery. One example is the 20-inch x 230-foot (approximately 50 centimeters x 70 meters) Bayeux Tapestry (portion shown below)—an embroidered cloth providing, in part, a visual narrative of events surrounding the 1066 Norman Conquest of England.

Early Mapmaking

Maps of real and imagined worlds helped our ancestors to externalize, navigate, and comprehend the world and their place in it. The first maps were more descriptive than scientific, with neither reference systems nor standard symbols and formats.

Prehistoric maps have been found on cave walls. More than 4,000 years ago in Babylon, people drew maps on clay. Ancient Egyptians made maps from papyrus and developed ways of surveying the land.

Ancient maps were primarily based on travelers' descriptions of the lands they had seen. Maps like the one above, by Hecataeus of Miletus (circa 560–550 BC), were extremely schematic. Nevertheless, they were a great advance, as the relative positions of the continents could be communicated.

An example of the quintessential travel map is the Peutinger map (above right), a medieval copy of an ancient Roman map (circa 64–12 BC). It shows roads as jagged lines, where every node is a stopping place. Symbols depict different types of settlements, such as cities and military camps; the distance between them is also shown, as well as important features in the landscape. Most fascinating is the way this itinerary is conceived: the entire Roman Empire is compressed into a 20-foot x 1-foot (about

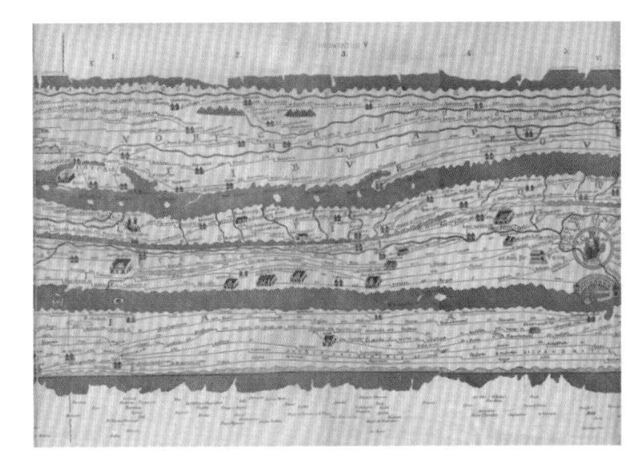

60 meters x 30 centimeters) scroll—ideal for the traveler to roll up and take with him on his voyages.

During the Middle Ages (400 to the late 1400s), mapmaking progressed in China and the Arab world. The Chinese printed the first map in 1155, more than 300 years before maps were printed in Europe.

The first detailed coastline maps, known as portolan charts, were developed around 1300 and have been found to be amazingly accurate. Marshall Islands stick charts, developed around 1500, represent ocean currents by stick patterns and island locations with shells and were used to aid navigation and fishing between islands.

Historical maps are great reflections of what was known at the time of their creation and also of what was valued. Physical maps, for instance, show natural features such as rivers and rocks. Human maps show farms, transportation, and homes.

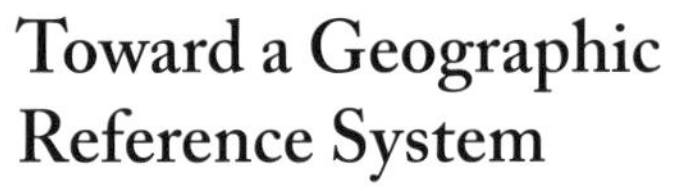

Toward a Geographic Reference System

Hundreds of years were required for the emergence of a common geographic reference system that was useful for navigation and mapping of commercial, social, political, and other entities.

The Greeks were the first to realize that the Earth is a sphere. They calculated the planetary dimensions and defined the locations of the equator and the poles. Hipparchus was the first to plot lines of latitude and longitude. Ptolemy compiled long tables of place names with their geographic features and coordinates, largely based on the descriptions of travelers. These tables, together with the Ptolemaic projection, facilitated the design of more accurate world maps in AD 160 (see example on **page 78, Cosmographia World Map**).

Nevertheless, because of errors and gaps in the data, most maps were out of proportion, inconsistent in their determination of orientation, and included mainly well-known coastal towns. As late as 1700, for instance, mapmakers did not yet have complete information about the coastlines of continents such as Australia (see **page 82, A New Map of the Whole World with the Trade Winds According to Ye Latest and Most Exact Observations**).

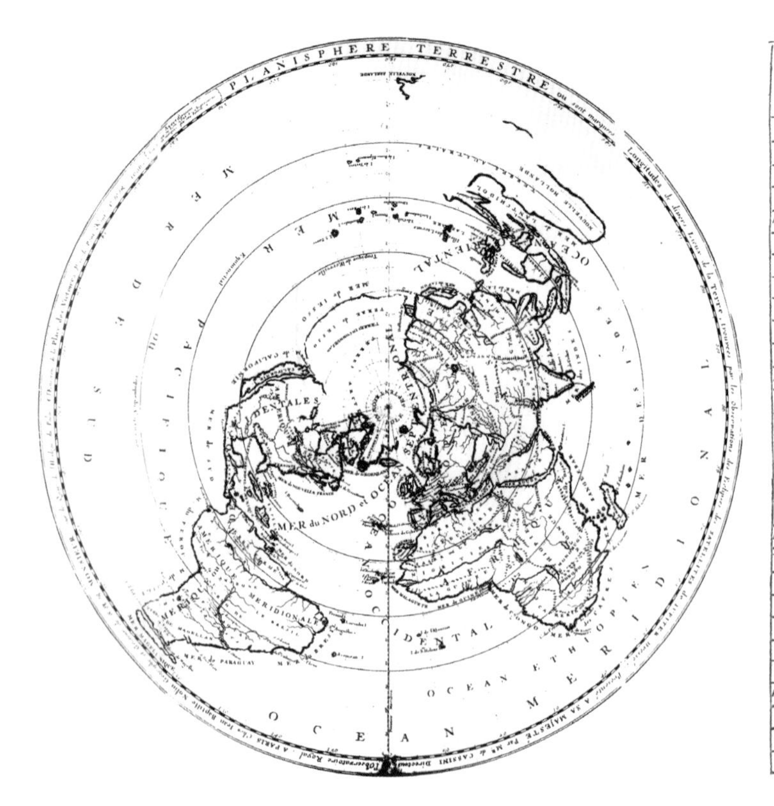

Februarius. 1668.
Configurationes Mediceorum.
Hora 7. P.M.

Scientific Mapmaking

Advances in timekeeping and surface measurements (triangulation) enabled the accurate mapping of space. Between 1600 and 1700, land surveying techniques were developed.

In 1696, the first accurate map of the Earth was drawn by Giovanni Domenico Cassini, based on 40 points of accurate latitude and longitude (below left; points shown in red). The north-south position (latitude) of any point on Earth could be determined via star paths. To measure east-west position (longitude), exact time measurement was essential: one minute of uncertainty implied a 10-mile margin of error in location. Inspired by Galileo Galilei's work, mapmakers used the planet Jupiter as a "clock in the sky." They carefully recorded the motions of Jupiter's moons (see Cassini's 1668 table of the eclipses of Jupiter's moons, opposite page, lower right). Based on these tables, events such as eclipses could be predicted and correctly timed to within a minute. By comparing the time at a certain location to local time in Paris, the site's longitude could be determined.

In 1744, Cassini's son César-François and his team started to map France in a rigorous fashion using triangulation. In the late 1700s, the world's first national land survey of France was completed. In 1870, Sir George Everest set out to map India by triangulation. For generations, a vast network of repeating sightline triangles was meticulously measured and recorded (see **Mapping the Highest Mountain,** below). What resembles a pattern of eyelashes on the northern border represents the sightlines to stations built above treetops. While analyzing the triangles in the calculating offices of Calcutta, the mapmakers discovered the highest peak in the world: Mount Everest.

In 2008, there were more than 20 global positioning system (GPS) satellites in orbit, each on its own track. While revolving around the planet, each transmits a radio signal to the ground. GPS receivers pick up these signals from multiple satellites and derive the exact location of the receiver. The World Geodetic System provides a standard coordinate frame for the Earth for use in cartography, geodesy, and navigation, despite continental plate movement, volcanic activity, earthquakes, tsunamis, and human activity upon the Earth's surface.

MAPPING THE HIGHEST MOUNTAIN

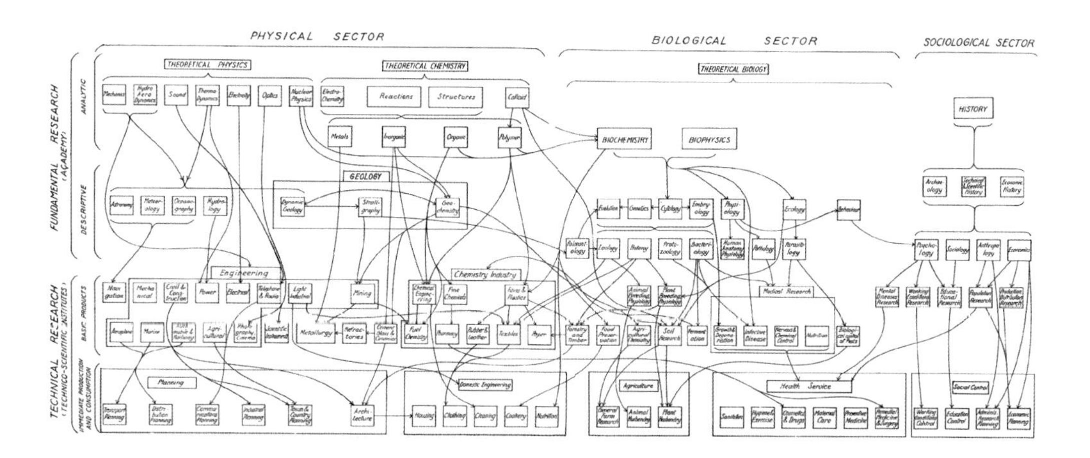

Thematic Mapmaking, Statistical Graphics, and Data Visualization

In the 18th century, mapmakers began to show more than just geographic locations on a map. New graphic forms such as isolines and contours were invented, allowing geologic, economic, historical, political, and medical data to be overlaid on maps. Edmond Halley's 1700 isogonic map is presumed to be the first thematic map in the modern sense. Many of the first statistical graphs had a strong geographic component. For example, John Snow's cholera map of 1854 used dashes next to each patient's home address to indicate the number and location of cholera cases in one London neighborhood—showing a high density of cholera cases near a well on Broad Street. Once the pump handle of that well was removed, the epidemic ended.

Charles Joseph Minard's 1861 map of **Napoleon's March to Moscow** (see **page 84**) combines a cartographic map and a timeline. It shows the terrible fate of Napoleon's army in Russia via six variables: line width represents the size of the army; line path shows the army's latitude and longitude; line direction is color coded for advance and retreat; and temperature on various dates during the retreat from Moscow.

Harry Beck's 1931 map of the London Underground sacrifices accuracy of both location and scale in favor of readability. In Beck's original map, and in nearly all recent subway maps, Tube lines are drawn in exclusively horizontal, vertical, or diagonal fashion and subway stations are shown to be equidistant, regardless of their actual locations.

Data Charts and Network Visualizations

Another line of mapping work involved the invention of data graphs, charts, and network visualizations without a geographic component.

In 1786, engineer and political economist William Playfair published *The Commercial and Political Atlas* in London. It contained 43 time-series plots and the first known bar chart. Playfair also invented the line graph, circle graph, and pie chart and he is also known as the inventor of statistical graphics.

In 1858, nursing pioneer Florence Nightingale designed charts that plotted the number of deaths due to preventable diseases versus the number of deaths due to wounds and other causes. These statistical comparison charts led to greater social awareness of preventable diseases and ultimately to reform in hospital sanitation methods.

In 1934, Jacob L. Moreno, known as the father of sociometry, published his first network diagram in the *New York Times.* Soon thereafter, many other social scientists and scholars began to map social and scholarly networks (see **page 26, Sociometry map**).

In 1939, John D. Bernal, a physicist, historian, and sociologist of science, designed one of the first "maps of science." This map divided science into physical, biological, and sociological sectors; it also distinguished technical and fundamental research (see above map).

Many more examples of the history and power of drawn maps have been beautifully compiled in the works of Edward R. Tufte, Howard Wainer, and Jacques Bertin and references to major books are given on **page 216, in References and Credits**.

Recent advances in computer technology and software development have made possible the algorithmic creation of maps from large-scale data sets, allowing the mapping of, for example, Web sites or cyberspace. Information visualizations support the interactive exploration of large amounts of abstract data. Knowledge visualizations focus on the creation or transfer of knowledge among people. Knowledge domain visualizations, also called science maps, are discussed next.

Science Maps and Their Makers

In 1963, Derek John de Solla Price suggested that science be studied using scientific methods. His visionary work led to the creation of the field of scientometrics. Early scientometric studies were completed by hand, as citation index databases did not yet exist and computers were not yet widely available. Today, computational scientometrics refers to the processing of terabytes of scholarly data by means of interconnected computers running advanced software. The communication of the structure and evolution of science at an individual, local, and global scale, however, is nontrivial. Tables and timelines are easy to read and understand, yet they fail to convey the complex interdependencies of scholarly entities and the feedback loops in which they are involved. The design of reference systems and visual vocabulary to depict science at different scales for different stakeholders is a major research topic.

Maps, even more than the printed word, impress people as authentic. We tend to accept the information on maps without question.
Jon A. Kimerling

Mapping Science

Science maps aim to visually encode the structure and evolution of scholarly knowledge (see **page 26-47, Milestones in Mapping Science** for more than 100 maps of science and technology). Science maps are also known as scientographs, literature maps, domain maps, and knowledge domain visualizations. The mapmakers of science are scientometricians, bibliometricians, visualization researchers, and graphic designers, among other experts. Recently, a number of cartographers and geographers have devoted their extensive talents and techniques to the mapping of nonphysical knowledge spaces.

Historically, science maps were created to navigate, understand, and communicate the structure of scientific knowledge. Recent work aims to use science maps to understand and communicate the dynamics of science. Science maps complement local fact retrieval via search engines by providing global views of large amounts of knowledge. They can be used to objectively identify premier research areas, experts, institutions, collections, grants, papers, journals, and ideas in a domain of interest.

It is important to note that science maps promote improvements in data quality and coverage. While we do not expect a search engine to retrieve every existing publication on a given subject, we do require a map of science to be accurate and complete. Fortunately, scholars want to be on the map and are therefore amenable to supplying the information necessary. Easy to use, Wiki-like interfaces empower scholars to add, organize, and interlink data items improving data quality and coverage (see also **page 68, Scholarly Marketplaces**).

The Utility of a Science Reference System

Science spaces differ from physical places in that they are abstract and cannot be touched or visited. As a result, most people find science to be alien and inaccessible. Science maps aim to make the structure and evolution of science tangible, appealing, and navigable.

Just as centuries were needed to arrive at a geographic reference system (see **page 10, The Power of Maps**) we must allow for the time and patience necessary to eventually agree upon a formally grounded and practically validated reference system for science. Certainly, having a science reference system to organize our collective scholarly knowledge is well worth the necessary disputes and battles, as it will ultimately help us settle on common terminology and standards. The envisioned reference system resembles a library classification system in that it can be used to organize all knowledge in space. However, instead of being used exclusively by librarians and classification-savvy library patrons, it would be used by the general public, both children and adults, to navigate and utilize the world of knowledge.

The design of a science reference system must take our current knowledge of science into account. It requires a conceptualization of the structure and growth of science (see **page 52-59, Conceptualizing Science**) and needs to be derived from the best and most comprehensive set of scholarly data in existence (**page 60, Data Acquisition and Preprocessing**). It must apply the most scalable and advanced algorithms (**page 62, Data Analysis, Modeling, and Layout**) and must be validated against existing classification systems and data sets in a formal and practical sense.

Different user groups and insight needs (see **page 8, Knowledge Needs and Desires** and **page 197, Part 5: The Future of Science Maps**) will demand very different maps. However, given a common reference system, the science map used in the classroom will be comparable to the maps used in financial decision-making or in library organization.

Toward a Reference System for Science

As the amount of information available increases (see **page 4, The Rise of Science and Technology**) our ability to bring order to the existing and continuous stream of new scholarly data needs to improve. Automated approaches that can analyze and map millions of scholarly records must be developed and applied. Results need to be confirmed for local and global accuracy and completeness. Manually derived classification systems have a leading role to play in the development and validation of a shared reference system.

Recent work by Kevin W. Boyack and Richard Klavans (see maps to the right) aspires to the creation of a global map of and spatial reference system for all sciences. The maps are based on a large subset of papers from the most comprehensive databases in existence: the Science Citation Index (SCI), Social Sciences Citation Index (SSCI), and Arts and Humanities Citation Index (AHCI), all by Thomson Reuters, and Scopus, provided by Elsevier (see **page 60, Data Acquisition and Preprocessing**). The four maps were generated using the following steps:

- A set of scholarly journals or papers is taken as input.
- The similarity between pairs of journals or papers is calculated based on either direct citation, cocitation (the number of times they are jointly cited by another paper), or bibliographic coupling (the number of references they share).
- The resulting similarity matrix is normalized. The network of paper/journal nodes and their linkages is analyzed to retain only the strongest links for each node. The spatial layout aims to place similar nodes close to each other and to minimize the crossing of linkages.
- Journals/papers are assigned to clusters.
- The result is interpreted and labeled manually.

This process is explained in detail in the following: **In Terms of Geography, Map of Scientific Paradigms**, and **Maps of Science: Forecasting Large Trends in Science** on **pages 106, 136**, and **170** respectively. The distinguishing factor of this work is that the clustering of papers or journals is not based on the original correlation matrix but on the spatial layout, or the position of nodes in a two-dimensional space.

Alternative approaches to science mapping exist and are featured in the timeline of milestone events in **Milestones in Mapping Science, pages 26-47**.

The **Backbone of Science** map (top left) has been used in extensive studies that aim to validate and optimize the processing pipeline applied to generate this map and maps discussed later. Specifically, the regional accuracy, scalability, accuracy of different similarity algorithms, and the readability of the layouts were examined. As the map is based on journal-level data, it is understood to be a disciplinary map.

The **2002 Base Map** (top right) expands on the backbone map, in that bibliographic coupling counts have been reaggregated at the journal cluster level to calculate the (x, y) position for each journal cluster and by association the (x, y) positions for each journal. Journal names can now be used to "science locate" individuals, institutions, countries, and scientific fields based on their publication records. The map was included in the first iteration of the *Places & Spaces: Mapping Science* exhibit (see **page 106, The Structure of Science**). In 2006, it was the most comprehensive map of science ever generated. The map was also used for diverse overlays of funding amounts per science area (see **page 202, Science of Science Policy Maps for Government Agencies**).

The **Paradigm Map** (bottom left) was generated by recursively clustering the 820,000 highly cited papers referenced in 2003. The resulting map has 776 paradigms or active research areas, each of which is indicated by a circle. The map appears in the second iteration of the Mapping Science exhibit (**page 136, Map of Scientific Paradigms**). It also served as a base map for the Illuminated Diagram display (**page 180, Illuminated Diagrams**) and the Hands-On Science Maps for Kids (**page 186, Hands-On Science Maps for Kids**).

The **UCSD Map of Science** (bottom right) shows a three-dimensional layout of disciplines (groups of journals). It places those disciplines on the surface of a sphere; the spheric layout is then flattened using a Mercator projection to create a two-dimensional version of the map. Each node is labeled and has an extensive list of key phrases as metadata, which can be used to "science locate" nonjournal data, such as patents or grants. That is, key phrases from each patent or grant (titles and abstracts) are extracted; fractional assignment to map nodes proceeds by matching the associated metadata. Thus, each grant or patent is fractionally assigned to multiple nodes. Adding the fractions allows for the number of grants, dollars by agency, or patents associated with each node to be computed. Drilldowns for each of the more than 550 nodes are underway. The UCSD map is part of the third iteration of the exhibit and is shown in large form on **page 170, Maps of Science: Forecasting Large Trends in Science**.

Although different in final form when published or exhibited, the maps are represented here using the same visual encoding. In all four maps, nodes represent clusters of journals. The nodes are size-coded by the number of papers they contain. Nodes in the UCSD map are color-coded according to the major fields of science they represent. Labels are used to indicate the general positions of these fields. While different data sets—paper- or journal-level data—and similarity measures are used, the placement of major science areas and the ways they interrelate is very similar across the four maps.

The 2002 Base Map is used on **page 106, The Structure of Science** to indicate which areas of science are captured by each exhibit map.

Mapmakers of Science

The old mapmakers are typically depicted as elderly, often bearded men, formally dressed in tweeds and gabardine, sitting at high desks with pens in hand, surrounded by stacks of books and maps, cloistered in towers of wisdom.

The new generation of computational scientometricians includes men and women with backgrounds in cartography, history (of science), psychology (perception and cognition), education, visualization, data mining, (digital) library science, scientometrics, bibliometrics, informetrics, or webometrics. They dress as they please and enjoy access to large-scale data sets, major cyberinfrastructures, and other experts around the globe. There are about 300 of them in the world, and several of them are driven to bring knowledge not only to other academic and government institutions but to every person on this planet.

Backbone of Science

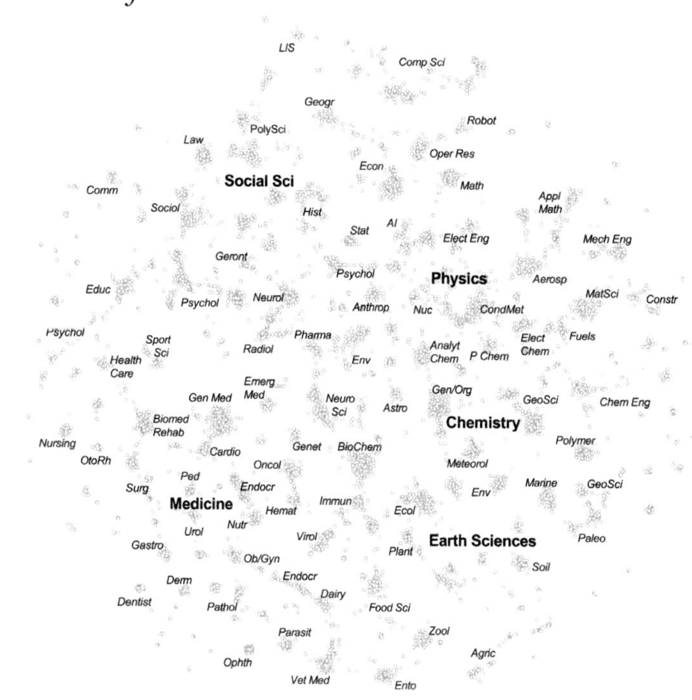

Data: Combined SCI/SSCI from 2000, 7,121 journals
Similarity Metric: Cocitation
Number of Disciplines: 212

2002 Base Map

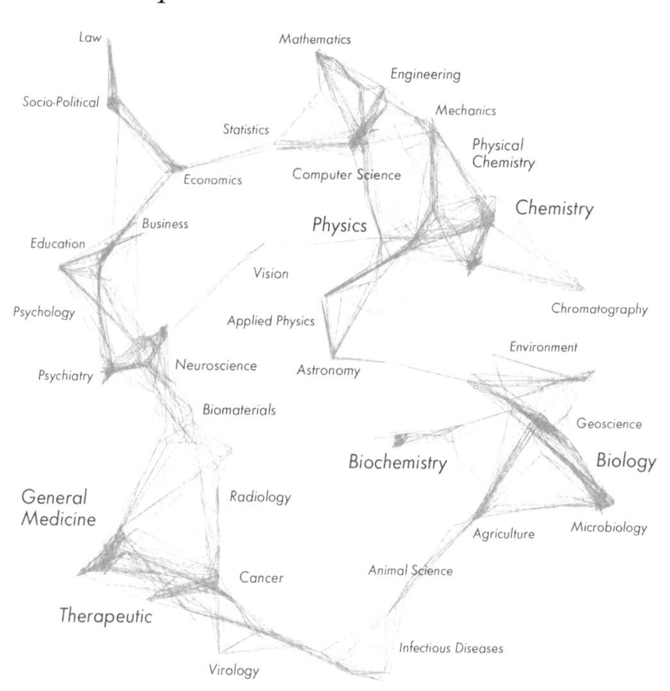

Data: Combined SCI/SSCI from 2002, about 1.07 million papers, 24.5 million references, 7,300 journals
Similarity Metric: Bibliographic Coupling
Number of Disciplines: 671

Paradigm Map

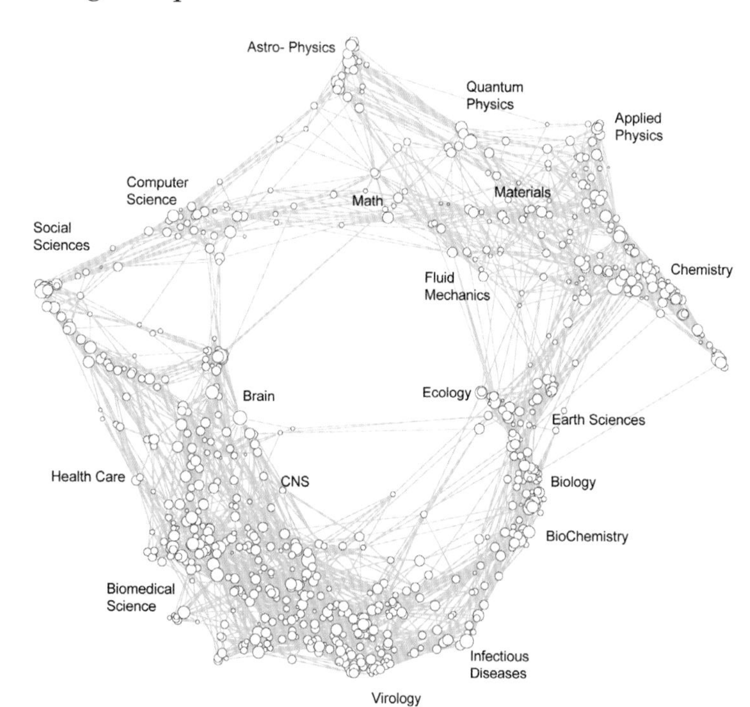

Data: Combined SCI/SSCI from 2003, about 820,000 highly cited reference papers
Similarity Metric: Cocitation
Number of Paradigms: 776

UCSD Map of Science

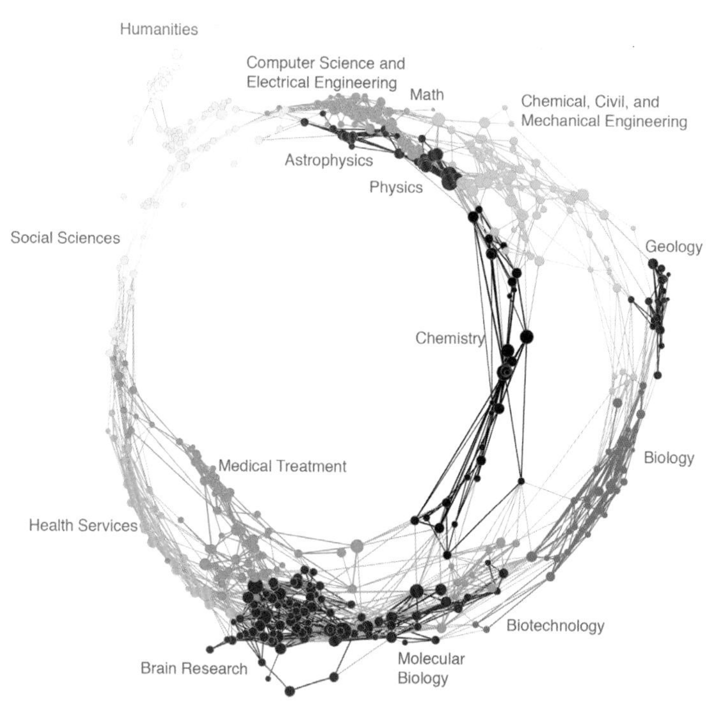

Data: WoS and Scopus for 2001–2005, 7.2 million papers, more than 16,000 separate journals, proceedings, and series
Similarity Metric: Combination of bibliographic coupling and keyword vectors
Number of Disciplines: 554

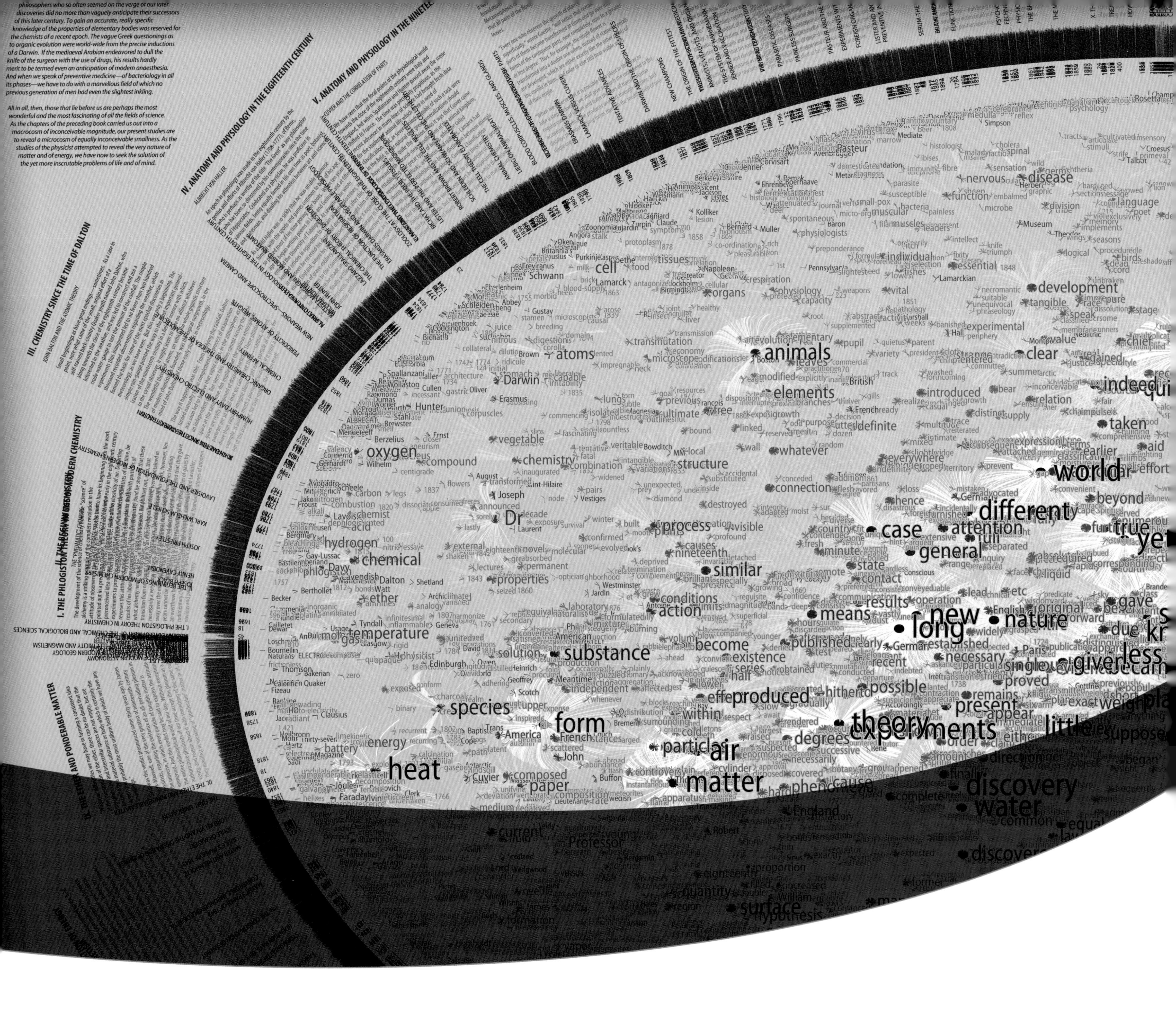

philosophers who so often seemed on the verge of our later
discoveries did no more than vaguely anticipate their successors
of this later century. To gain an accurate, really specific
knowledge of the properties of elementary bodies was reserved for
the chemists of a recent epoch. The vague Greek questionings as
to organic evolution were world-wide from the precise inductions
of a Darwin. If the mediaeval Arabian endeavored to dull the
knife of the surgeon with the use of drugs, his results hardly
merit to be termed even an anticipation of modern anaesthesia.
And when we speak of preventive medicine—of bacteriology in all
its phases—we have to do with a marvellous field of which no
previous generation of men had even the slightest inkling.
All in all, then, those that lie before us are perhaps the most
wonderful and the most fascinating of all the fields of science.
As the chapters of the preceding book carried us out into a
macrocosm of inconceivable magnitude, our present studies are
to reveal a microcosm of equally inconceivable smallness. As the
studies of the physicist attempted to reveal the very nature of
matter and of energy, we have now to seek the solution of
the yet more inscrutable problems of life and of mind.
IV. ANATOMY AND PHYSIOLOGY IN THE EIGHTEENTH CENTURY
V. ANATOMY AND PHYSIOLOGY IN THE NINETEE
III. CHEMISTRY SINCE THE TIME OF DALTON
I. THE PHLOGISTON THEORY IN CHEMISTRY
animals
elements
world
different
case
general
long
new
nature
theory
experiments
heat
form
species
matter
air
substance
chemical
oxygen
hydrogen
temperature
discovery
water
surface

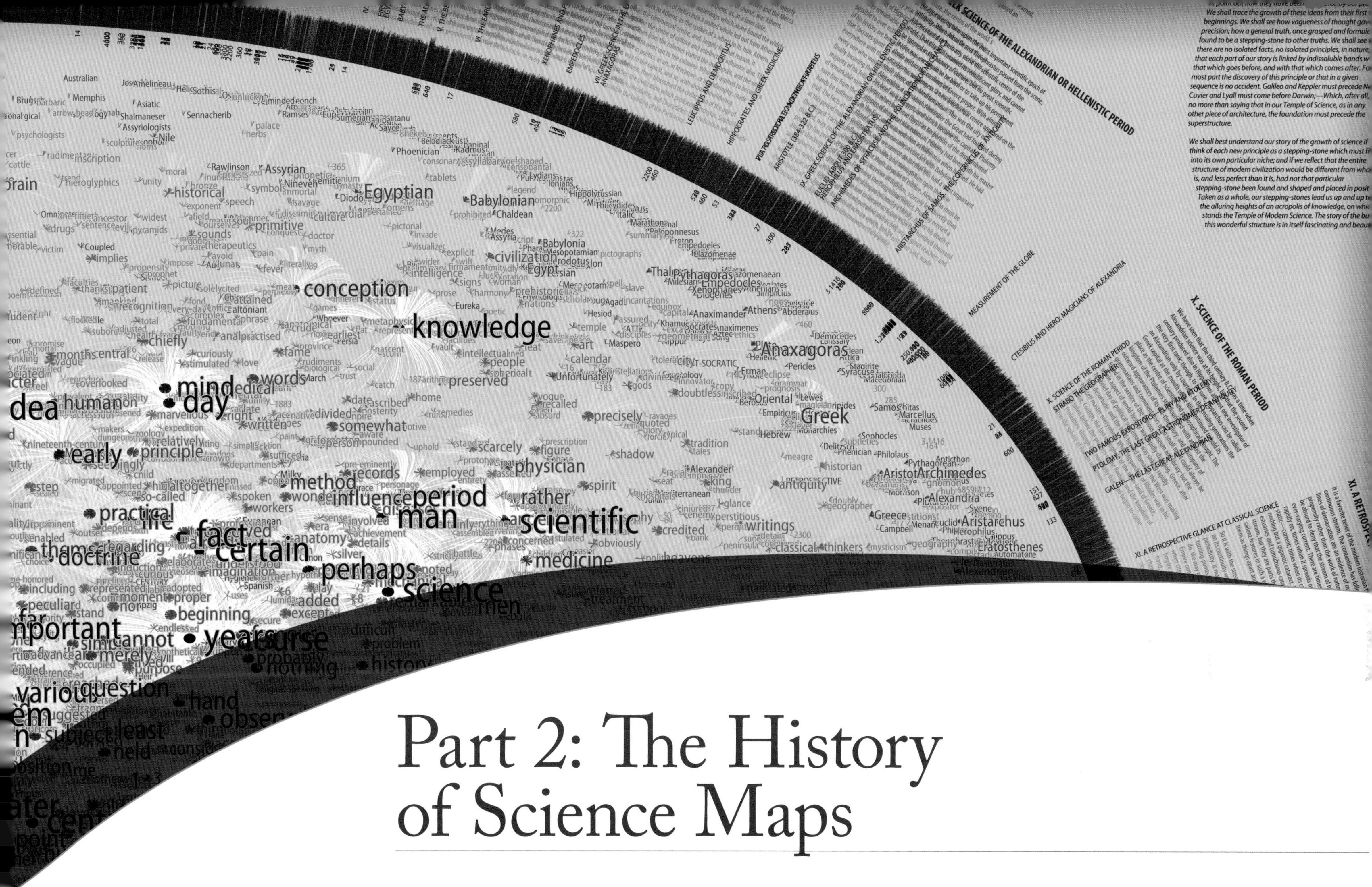

Part 2: The History of Science Maps

Noise becomes data when it has a cognitive pattern. Data becomes information when assembled into a coherent whole, which can be related to other information. Information becomes knowledge when integrated with other information in a form useful for making decisions and determining actions. Knowledge becomes understanding when related to other knowledge in a manner useful in anticipating, judging and acting. Understanding becomes wisdom when informed by purpose, ethics, principles, memory and projection.

George Santayana

Visionary Approaches

The question of how to store, classify, visualize, and disseminate scholarly knowledge is as old as the production of knowledge itself. This section provides an introduction to visionaries who have had a profound impact on the history or development of **(1) Knowledge Collection**, including the compilation of **(2) Encyclopedias**; **(3) Knowledge Dissemination**; **(4) Knowledge Classification**; **(5) Knowledge Interlinkage**; **(6) Knowledge Visualization**; and **(7) Man-Machine Symbiosis**, which might ultimately result in a decentralized **(8) Global Brain** that interlinks minds and machines on this planet.

The timeline shown here lists the eight major categories at right. For each category, we selected an exemplary group of visionaries who were successful in igniting the imaginations of the masses and in harnessing the resources needed to manifest their dreams. The name and image of each visionary is paired with a red life line. Johannes Gutenberg's life line is off the chart and shown in a separate box. The life lines of early and contemporary visionaries fade in and out, respectively. Gray dots indicate the dates of major contributions—accomplished either early or late in life. Some life lines overlap; others do not. For example, Eugene Garfield was born more than 15 years after Frank Shepard's death—they had no opportunity to meet in person. Cross linkages are employed to communicate the complex network of inspirations (light brown), support (dark brown), and facilitation (dark green). Many of the visionaries extensively promoted education, which is indicated by a red frame.

The portraits of the individuals and their achievements are intended to inspire further exploration and study of these and other visionaries and their complex interrelations. Through this biographical mapping, one sees how the design of science maps is not an isolated activity but one that is deeply embedded in the theoretical, practical, and technological advances discussed here.

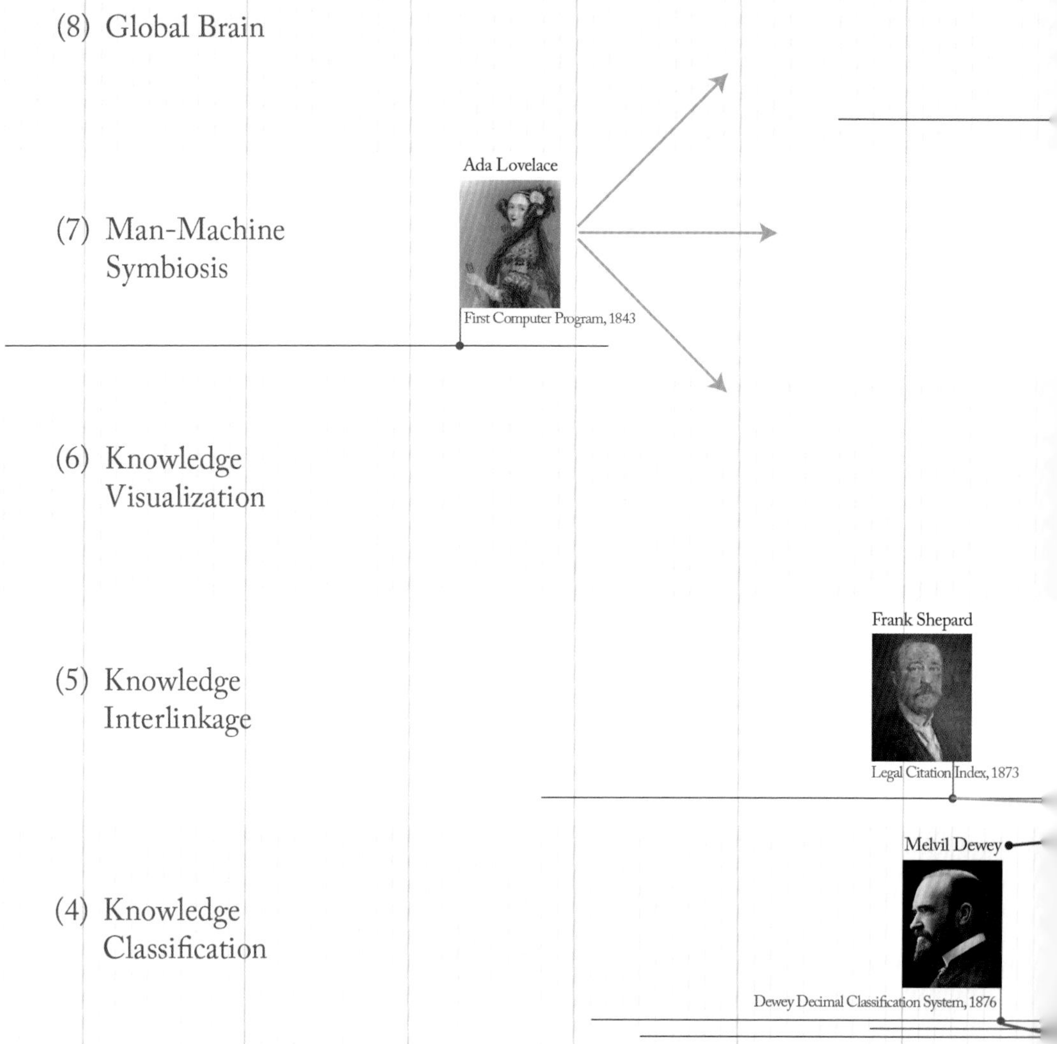

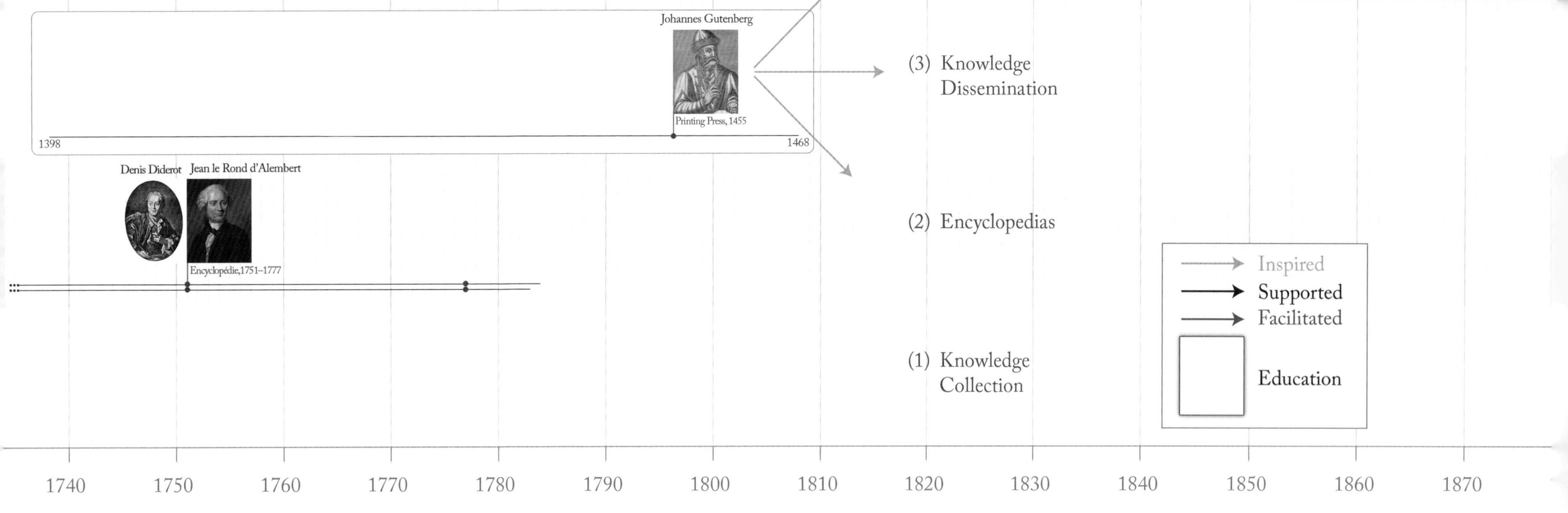

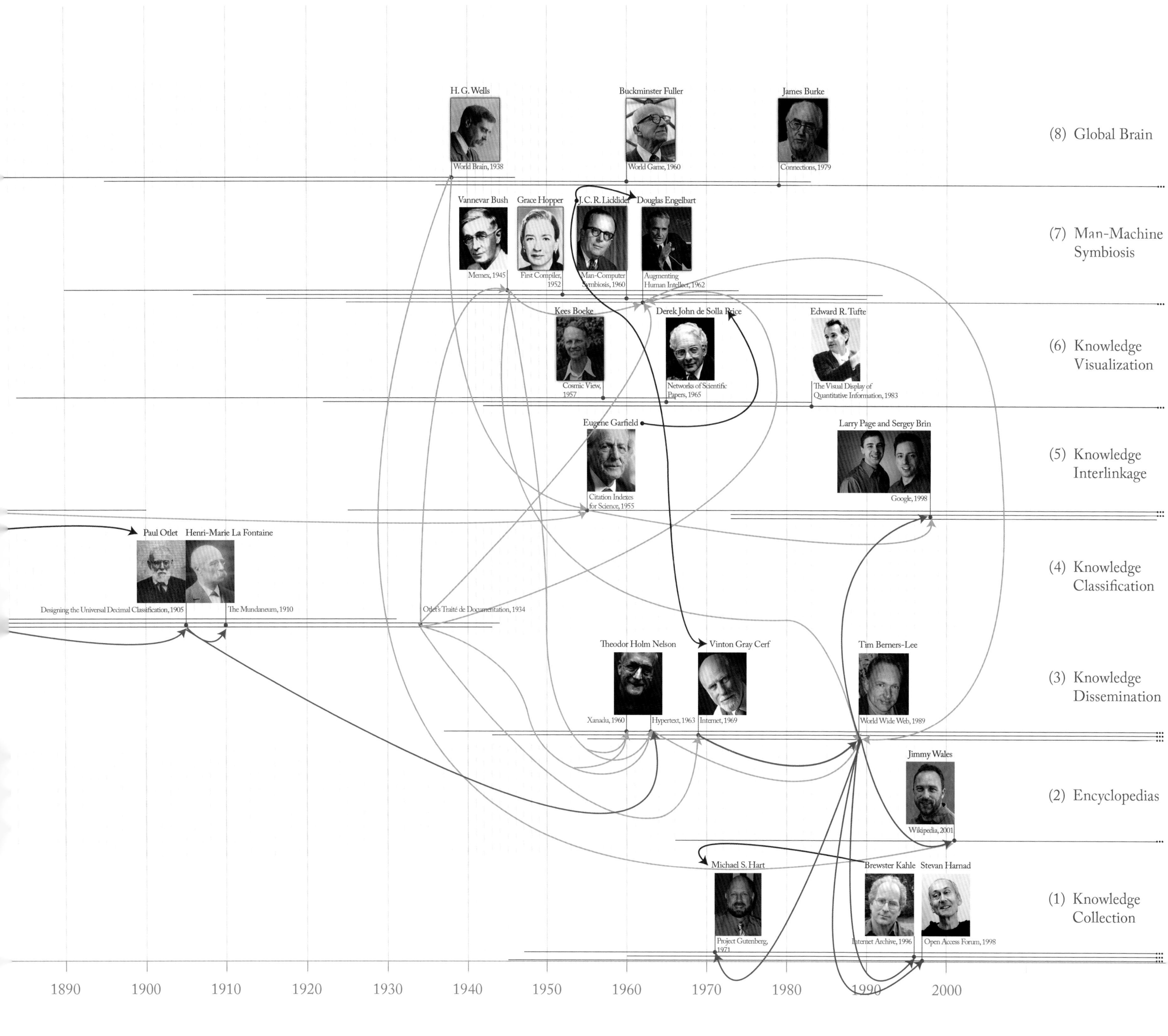

H. G. Wells
World Brain, 1938
Buckminster Fuller
World Game, 1960
James Burke
Connections, 1979
(8) Global Brain
Vannevar Bush
Memex, 1945
Grace Hopper
First Compiler, 1952
J. C. R. Licklider
Man-Computer Symbiosis, 1960
Douglas Engelbart
Augmenting Human Intellect, 1962
(7) Man-Machine Symbiosis
Kees Boeke
Cosmic View, 1957
Derek John de Solla Price
Networks of Scientific Papers, 1965
Edward R. Tufte
The Visual Display of Quantitative Information, 1983
(6) Knowledge Visualization
Eugene Garfield
Citation Indexes for Science, 1955
Larry Page and Sergey Brin
Google, 1998
(5) Knowledge Interlinkage
Paul Otlet
Henri-Marie La Fontaine
Designing the Universal Decimal Classification, 1905
The Mundaneum, 1910
Otlet's Traité de Documentation, 1934
(4) Knowledge Classification
Theodor Holm Nelson
Xanadu, 1960
Hypertext, 1963
Vinton Gray Cerf
Internet, 1969
Tim Berners-Lee
World Wide Web, 1989
(3) Knowledge Dissemination
Jimmy Wales
Wikipedia, 2001
(2) Encyclopedias
Michael S. Hart
Project Gutenberg, 1971
Brewster Kahle
Internet Archive, 1996
Stevan Harnad
Open Access Forum, 1998
(1) Knowledge Collection
1890
1900
1910
1920
1930
1940
1950
1960
1970
1980
1990
2000

Knowledge Collection

The design of a comprehensive map of science requires access to a complete and continually updated digital copy of all scholarly knowledge—in all languages, in all publication formats, and for all publication years.

There have been several prior attempts to amass all of humanity's collective knowledge. The Ancient Library of Alexandria in Egypt was one of the first libraries to aspire toward this universality: it aimed to compile in one physical place a copy of every work ever written. Its collection was estimated at 400,000 volumes (scrolls). Begun in the third century BC in the kingdom of the Ptolemies in Egypt, it did not survive past the fifth century AD. It has only recently been rebuilt as Bibliotheca Alexandrina, inaugurated in 2003 near the original library site. As of 2008, the U.S. Library of Congress holds 26 million of the estimated 100 million books ever printed. (See also Paul Otlet's and Henri-Marie La Fontaine's efforts to index all existing knowledge, **page 20, The Mundaneum**).

The visionaries discussed here aim to create free and universally accessible libraries of books, videos, audio materials, Web pages, and scholarly works (see also efforts such as **Ginsparg's arXiv, Giles's CiteSeer**, and others discussed on **page 4, The Rise of Science and Technology**). The collected material can then be consumed via computer screens, handheld reading devices such as Amazon's Kindle, or on-demand printing (see also **page 23, Man-Machine Symbiosis**).

Michael S. Hart (b. 1947): *Project Gutenberg, 1971*

On Independence Day, 1971, Michael S. Hart, then a student at the University of Illinois, keyed the U. S. Declaration of Independence, signed on July 4, 1776, into his computer. He typed the text in uppercase because lowercase was not yet available. Sending the resulting 5 kilobyte file to the 100 users of the embryonic Internet would have crashed the network, so he simply told others where the e-text was stored (hypertext links would be developed 20 years later). It was downloaded by six users. Project Gutenberg (**http://www.gutenberg.org**) was born. Its mission is "to encourage the creation and distribution of e-books" by everybody and by every possible means.

The idea of Project Gutenberg is to bring the source of all information, and civilization, to the masses in the same way the Gutenberg press did in the middle of the second millennium, only in a modern manner.

Jeffrey Thomas

In August 1989, the project provided access to 10 e-books; in August 1997 there were 1,000 e-books; and in 2009 there were 100,000 e-books in over 50 languages (see **pages 26-47, Books in Milestones in Mapping Science**). With the staff of Project Gutenberg developing and implementing new ideas, methods, and software, and with volunteers contributing e-books, the plan is to have one million books by 2015.

Brewster Kahle (b. 1960): *Internet Archive, 1996*

Brewster Kahle is an inventor and entrepreneur specializing in technologies that advance universal access to all knowledge. After graduating from the Massachusetts Institute of Technology in 1982, he helped start the supercomputer company Thinking Machines, which built systems for searching large text collections. In 1989 he invented the Wide Area Information Server (WAIS), the Internet's first publishing and distributed-search system, which created an online presence for many of the world's largest publishers. In 1996, Kahle cofounded Alexa Internet, which provides the search and discovery services included in more than 90 percent of Web browsers. In the same year, he founded the nonprofit Internet Archive (**http://www.archive.org**), a digital library that offers permanent access to digital historical collections for researchers, historians, and scholars. Kahle is now the digital librarian and director of the Internet Archive. As of 2008, the archive serves snapshots of the Web taken every two months since 1996 (circa 85 billion Web pages consuming about 2 petabytes of storage space); recordings from 20 television channels (about 1 petabyte); circa 300,000 texts; 180,000 books; and 200,000 audio, software, and educational items, constituting one of the largest digital archives in the world.

We can pull off Universal Access to All Knowledge. This is where Wiki is going toward, one of the great things that humanity will be remembered for, up there with a Man on the Moon in the mythology of humanity.

Brewster Kahle

According to Kahle, it costs about $10 to scan a book (3 cents per page in China, 10 cents in the United States). It would cost $800 million to scan all 26 million books in the U.S. Library of Congress. Each book is about 1 megabyte of text. Print on demand costs about $1 per book. A comparison to current library expenses—$3 to rent/return a book, $30 per book per library building—proves digital archives to be cost-effective.

Stevan Harnad (b. 1945): *American Scientist Open Access Forum, 1998*

Stevan Harnad, a Hungarian-born psychologist, is Canada Research Chair in Cognitive Sciences at Université du Québec à Montréal and professor of electronics and computer science at the University of Southampton. A specialist in cognition, categorization, communication, and consciousness, Harnad helped lay the groundwork for the open access (OA) initiative, which aims to make scholarly material available as a free, immediate, permanent, full-text, online accessible, Web-wide resource for anybody, maximizing its use and impact.

In 1978, Harnad founded the Open Peer Commentary journal *Behavioral and Brain Sciences*; and in 1989 he launched *Psycoloquy*, one of the first peer-reviewed OA journals. In 1990, he advocated "Scholarly Skywriting" (online publication and commentary, before and after peer review), and in 1994 he posted his "Subversive Proposal" to authors to heighten research usage and impact by self-archiving all their scientific and scholarly research to be free and publicly available online.

In the online age, journals will need only to provide the peer-review service. Authors will self-archive their papers, both before and after peer review, in their institutional E-print Archives, which will all be interoperable with one another, providing open access to all peer-reviewed research output as if it were all in one global archive.

Stevan Harnad

In 1997, Harnad commissioned Cogprints, an OA e-print archive in the cognitive sciences hosted at the University of Southampton. In 1998, he founded the American Scientist Open Access Forum. In 1999, Cogprints was made compliant with the newly launched Open Archives Initiative (OAI) protocol for metadata harvesting, and in 2000 its software was made freely available as generic GNU e-prints so that all universities could create their own OA Institutional Repositories (IRs).

In 2001, Harnad's Southampton doctoral student Tim Brody created Citebase Search, the first scientometric engine for navigating and ranking OA research space, which links research access, impact, and assessment.

Encyclopedias

Encyclopedias aim to collect the world's knowledge into a single work or compendium. Those we use today were developed according to the dictionary model, focusing on words and definitions, and then expanded to include deeper explorations of their subjects and their context or relation to a broader field of knowledge. Encyclopedias also include maps and illustrations as well as bibliographies and statistics.

The work generally considered to be the first encyclopedia is the *Historia Naturalis* by Pliny the Elder, compiled during the first century AD. Its 37 volumes contain some 20,000 facts about the natural world gathered from 2,000 books. They were widely copied in Western Europe through much of the Middle Ages. The *Encyclopédie* of Denis Diderot and Jean le Rond d'Alembert was published over the course of nearly three decades, from 1751 to 1777. The *Encyclopædia Britannica* came into existence in 1768, and its entire contents—32 print volumes containing more than 100,000 articles—are available online today. As of 2009, Jimmy Wales's Wikipedia comprises more than 29 million articles in more than 250 languages. Historically, encyclopedias and dictionaries have been created by well-educated and well-informed content experts. In contrast, millions of volunteers contribute to Wikipedia.

Encyclopedia entries, with their extensive interlinkage, facilitate easy access to and use of knowledge. Recent analyses and maps (see **page 166, Science-Related Wikipedian Activity**) show the quality and content coverage of encyclopedias generated without editorial control. The mapping of our collective scholarly knowledge benefits from high-quality, high-coverage data with many interlinkages in support of local and global traversal and discovery. Maps with Wikipedia-like annotation and editing functionality have the potential to improve the quality of science maps and their underlying data (see **page 69, The Wiki Way**).

Denis Diderot (1713–1784) *and* Jean le Rond d'Alembert (1717–1783): *Encyclopédie, 1751–1777*

Denis Diderot (French philosopher and writer) and Jean Le Rond d'Alembert (French mathematician, mechanician, physicist, and philosopher) edited the *Encyclopédie*. Many consider it the most famous early encyclopedia because of its scope, quality, and political and cultural impact in the years leading up to the French Revolution. The 32 volumes of the *Encyclopédie* were first published over the course of nearly three decades (1751–1777). The 21 volumes of text contain more than 70,000 articles on subjects ranging from asparagus to the zodiac. In the remaining 11 volumes, beautiful engravings accompany many articles. Diderot and d'Alembert, who initially joined the project as subordinates, soon advanced to become editors and expanded the project's scope in an attempt to encompass all existing knowledge. Ultimately, more than 140 individuals contributed articles or illustrations to the *Encyclopédie*.

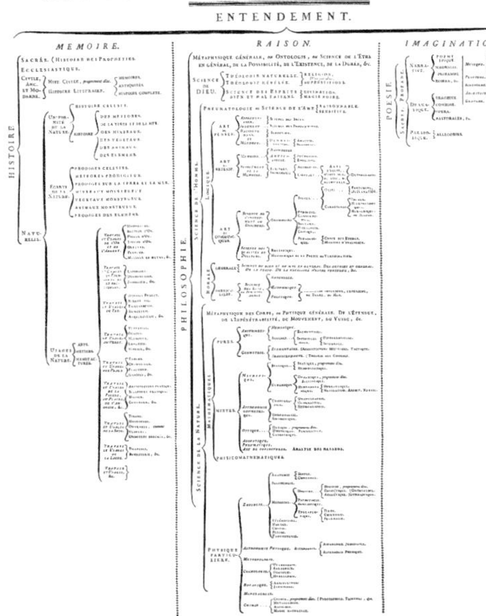

This is a work that cannot be completed except by a society of men of letters and skilled workmen, each working separately on his own part, but all bound together solely by their zeal for the best interest in the human race and a feeling of mutual good will.
Denis Diderot

The "Système Figurè" chart (previous column, bottom) shows the structure into which the *Encyclopédie* organized all human knowledge. The three main branches are memory, reason, and imagination.

Jimmy Wales (b. 1966): *Wikipedia, 2001*

Jimmy Donal "Jimbo" Wales is the founder of Wikipedia. He received a master's degree from the University of Alabama and then took PhD-level courses in the finance programs at the University of Alabama and Indiana University. He went on to become a futures and options trader in Chicago, which made him independently wealthy. In 1996, Wales moved to San Diego to form the Internet search engine startup Bomis. In 2000, he and Larry Sanger, an Ohio State philosophy student, created Nupedia, a first attempt at an open online encyclopedia. Progress on Nupedia was sluggish because of preoccupations with scholarly vetting. In 2001, Wales started Wikipedia, a multilingual free-content encyclopedia on the Internet. By the end of 2001, about 20,000 articles and 18 language editions existed. In September 2007, the English Wikipedia passed the 2 million-article mark making it the largest encyclopedia ever assembled (see graph on **page 4, The Rise of Science and Technology**). Wales is now president of the Wikimedia Foundation, the nonprofit corporation that operates Wikipedia and several other wiki projects.

Imagine a world in which every single person on the planet is given free access to the sum of all human knowledge. That's what we're doing.
Jimmy Wales

Knowledge Dissemination

Early knowledge dissemination took place through stories passed from one generation to the next. Later, cave paintings and clay tablets, then papyrus rolls and animal-skin parchment, were used to capture, communicate, and preserve knowledge (see significant events in **population graph, pages 2-3**). The introduction of inexpensive paper as a replacement for costly animal skin considerably improved the dissemination of knowledge. Inventions such as the printing press, hypertext, Internet, and the World Wide Web further improved knowledge access and diffusion.

Johannes Gutenberg (1398–1468): *Printing Press, 1455*

Johannes Gutenberg's invention of the printing press started with the insight that words, writings, and languages are expressed using a small number of different letters—one alphabet in various combinations. Given a large number of letters properly set together, a whole page of text could be printed at once. The letters could then be reset, the next text printed, and large books swiftly multiplied. After 1436, Gutenberg devoted his energy and fortune to inventing (1) type metal that was harder than lead, softer than iron, and perfect for the production of the moveable letter, (2) a press to quickly and uniformly press metal letters onto paper, modeled after the wine press, and (3) an ink mixture made from lampblack and boiled linseed oil that would not collect in drops and blot the paper but that was exceptionally black and would not fade.

It would take more than 20 years from Gutenberg's seminal insight to the printing of the first book—a Latin-language Bible—in Mainz, Germany, around 1455. Early documentation states that a total of 200 Bibles were scheduled to be printed on rag cotton linen paper and 30 copies on vellum (animal skin). Only 22 of these Gutenberg Bibles are now known to exist, of which 7 are on vellum. Upon completion of the printing, Gutenberg's moneylender, Fust, seized the molds, the type, the presses, and all the printed Bibles. Gutenberg died burdened by debts and with few friends by his side. His invention, however, led to the mass availability of printed books, ultimately fueling the 16th-century Protestant Reformation in Germany.

Theodor Holm Nelson (b. 1937): *Xanadu, 1960; Hypertext, 1963*

Theodor Holm Nelson is an American sociologist, philosopher, and pioneer of information technology. He has been visiting professor in the department of electronics and computer science at the University of Southampton and a visiting fellow at Oxford Internet Institute, working in the fields of information, computers, and human-machine interfaces.

In 1960, he founded Project Xanadu with the mission of designing a system that would support two-way unbreakable links, deep version management, incremental publishing, document-to-document intercomparison, copyright simplification and softening, and origin connection. By retrieving quoted contents from the virtual original of the author or rightsholder, for example, exact royalty payment for each download can be calculated.

In 1963, Nelson coined the words "hypertext" and "hypermedia" to describe his vision of worldwide hypertext, i.e., a universe of interactive literary and artistic works and personal writings deeply intertwined via hyperlinks (see also **page 23, Vannevar Bush**).

In 1960, I had a vision of a world-wide system of electronic publishing, anarchic and populist, where anyone could publish anything and anyone could read it. … With some wonderful companions, I have worked relentlessly for that vision for nearly fifty years, and do not intend to quit until the world has decent, serious, deep electronic literature.
Theodor Holm Nelson

Vinton Gray Cerf (b. 1943): *Internet, 1969*

Vinton Gray "Vint" Cerf is an American computer scientist who is known as one of the founding fathers of the Internet. He holds a B.S. in mathematics from Stanford University, and a master's degree (1970) and Ph.D. (1972) in computer science from UCLA.

The Internet, then known as ARPANET, was brought online in 1969 under a contract let by the renamed Advanced Research Projects Agency. The first ARPANET node was created at UCLA and the second at the Stanford Research Institute and the first host-to-host message was sent.

In 1973, while working with Robert E. Kahn, Cerf developed the Internet Protocol Suite (commonly known as TCP/IP), which set the transmission standard for data communications on the Internet. Though the design of the Internet was published in a 1974 paper, years of intensive labor were needed to make the Internet globally available to many.

While it took the telephone 75 years and the television 13 years to acquire 50 million users, it has taken the Internet only 5 years. In 2007, there were more than 1.2 billion Internet users.

The Internet is based on a layered, end-to-end model that allows people at each level of the network to innovate free of any central control. By placing intelligence at the edges rather than control in the middle of the network, the Internet has created a platform for innovation.
Vinton Gray Cerf

Tim Berners-Lee (b. 1955): *World Wide Web, 1989*

Sir Timothy John (Tim) Berners-Lee, a British computer scientist, is 3Com (Computer Communication Compatibility) Founders Chair at the Computer Science and Artificial Intelligence Laboratory at MIT as well as professor of electronics and computer science at the University of Southampton.

He invented the World Wide Web in 1989 while working for the European Organization for Nuclear Research (CERN). Drawing on the work of Douglas Engelbart, Vannevar Bush, and Theodor Holm Nelson, he wrote the Hypertext Transfer Protocol (HTTP)—the language that computers would use to exchange Web resources via the Internet. He also designed the Universal Resource Identifier (URI), today called the Uniform Resource Locator (URL)—a scheme for identifying each Web document with a unique address. In 1990, he wrote the first browser, or client program, for retrieving and viewing documents, and he christened this first-browser-editor the *WorldWideWeb* (later renamed Nexus to avoid confusion with the abstract information space known as World Wide Web).

Next he implemented the first Web server, that is, the software that handles the storing and transmitting of Web pages. Ultimately he created HTML, the Hypertext Markup Language that describes the organization of a Web page. Even with its fragile one-way links and no systematic management of version or contents, the Web—comprised of billions of online pages authored and interlinked by millions of people worldwide—is the largest "knowledge web" in existence today.

In 1993, Berners-Lee formed the World Wide Web Consortium (W3C), an international organization that develops protocols and guidelines to ensure the long-term growth of the Web and its availability to all. His recent work is devoted to the promotion of the Semantic Web and scholarly research about the Web. He is committed to making the Web universally accessible, without patents or royalties.

The Semantic Web is not a separate Web but an extension of the current one, in which information is given well-defined meaning, better enabling computers and people to work in cooperation.
Tim Berners-Lee

Knowledge Classification

Generations of philosophers such as Francis Bacon attempted to classify human knowledge. Carl Linnaeus's classification of living things, John Wilkins's universal classification scheme, and Ephraim Chambers's comprehensive *Cyclopaedia, or, An universal dictionary of arts and sciences* are among the early attempts to classify and hence bring order to the knowable universe. Librarians like Anthony Panizzi, Melvil Dewey, and Shiyali Ramamrita Ranganathan designed rules for cataloging and elaborate classification schemes. As examples, we here discuss the invention of the Dewey Decimal Classification and the Universal Decimal Classification, as well as associated visions for globalization and universalism. Classification systems bring structure to massive amounts of knowledge that might be stored in books, papers, and other formats. These systems, however, are difficult for nonlibrarians to use. Maps of science should continue to incorporate and support existing organizational schemas such as classifications, ontologies, and taxonomies, yet be as practical and simple to use as cartographic maps.

Melvil Dewey (1851–1931): *Dewey Decimal Classification System (DDC), 1876*

Melville Louis Kossuth (Melvil) Dewey was an American librarian. In 1876, while working as an assistant librarian, he anonymously published the first detailed general classification scheme for libraries, which now bears his name: the Dewey Decimal Classification (DDC) system. In developing the DDC, Dewey had two goals: first, to provide a method for arranging books on library shelves in an order helpful to users; and second, to provide a method of arranging surrogates of the books in a catalogue to facilitate information access. Today, the second need is ever more urgent, partially because books are now supplemented and replaced by smaller information units, such as periodicals, technical reports, or science blogs. The DDC uses decimals for its categories: there are ten main classes that are each subdivided into ten divisions, and each division has ten sections; this subdivision continues and is infinitely hierarchical. In the DDC, each book has a unique number that corresponds to its subject. Books on the same subject have the same classification number but differ in the second line of their call numbers, which contain part of the author's last name or the work's title. Over the last 130 years, the DCC has been greatly modified and expanded through more than 20 major revisions. It is used worldwide for cataloging library books.

The time was when a library was very much like a museum, and a librarian was a mouser in musty books ... The time is when a library is a school, and the librarian is in the highest sense a teacher.
Melvil Dewey

Also in 1876, Dewey helped establish the American Library Association (ALA), and he served as its secretary and president. He cofounded and edited *Library Journal* and was a strong promoter of library standards and reformed spelling. To many, Dewey is the father of modern librarianship.

Paul Otlet (1868–1944) *and* Henri-Marie La Fontaine (1854–1943): *The Mundaneum, 1910; Otlet's Traité de Documentation, 1934*

Paul Marie Ghislain Otlet was a Belgian lawyer, bibliographer, and internationalist. He pioneered the field of what is today known as information science (which he then called documentation). He is often considered the father of information management. Françoise Levie's 2002 documentary named him *The Man Who Wanted to Classify the World*. Otlet's obsession was to classify, encode, and unify books and documents published all over the world. His classification system can be seen as the predecessor of hypertext (see **page 19, Theodor Holm Nelson**). Otlet conceived of a library with no physical books whose contents could be viewed on a screen (see **page 24, Douglas Engelbart**). Several decades before the development of the Internet (see **page 19, Vinton Gray Cerf**), Otlet used terms like "web of knowledge," "link," and "knowledge network" to describe his vision for a central repository of all human knowledge.

Henri-Marie La Fontaine was one of Belgium's leading jurists and a renowned bibliographer. He received a doctorate in law from the Free University of Brussels and was a professor of international law as well as senator in the Belgian legislature for 36 years. He served as president of the International Peace Bureau from 1907 to 1943. La Fontaine was known for his fervent and total internationalism. He received the Nobel Peace Prize in 1913.

Designing the Universal Decimal Classification (UDC), 1905

In 1895, Otlet and La Fontaine established the Repertoire Bibliographique Universel (RBU) to create a master bibliography of the world's accumulated knowledge. Otlet knew from the start that their success would depend largely on the practical usefulness of the classification system. Upon examining prior systems, he found they could lead readers only to book titles, not to subjects, sources, or conclusions. Otlet asked Dewey for permission to extend his scheme and developed the Universal Decimal Classification (UDC), a complete implementation of a faceted classification supporting the assignment of multiple classifications to an object. In addition to tables of subject headings, the UDC supports a series of auxiliary tables that add such facets as notations for place, language, and physical characteristics of objects to be classified. As of 2008, the UDC comprised more than 62,000 individual classifications, translated into more than 30 languages.

The Mundaneum, 1910

In 1910, Otlet and La Fontaine founded the Palais Mondial, later called the Mundaneum, contained in an immense physical structure that was part of the Palais du Cinquantenaire in Brussels, Belgium. It was an effort to gather and classify all the world's knowledge on standard 3×5 index cards filed away in a sprawling array of cabinets. The Mundaneum was meant to be the heart of a "city of the intellect." It eventually included more than 12 million individual index cards and documents, cataloged using the

UDC. In 1934, the Belgian Government discontinued funding the Mundaneum and it was closed.

Otlet's Traité de Documentation, 1934

In 1934, 40 years before the invention of the Internet, Otlet published a magisterial work of synthesis, the *Traité de Documentation* (reprinted in 1989). "Documentation"—a term he coined in 1904—was not limited in his mind to the management of written documents or books. Otlet believed that the book was an inconvenient and inefficient carrier of information that had to be decomposed and dissected into its contents, images, schemas, charts, and tables for efficient information usage. He argued that no document could be properly understood by itself, but rather that a document's meaning becomes apparent only through its influence on other documents, and vice versa. "[A]ll bibliological creation," he wrote, "no matter how original and how powerful, implies redistribution, combination and new amalgamations."

The figure by Otlet (left, top) shows his vision of how classification can bring order: books can be reproduced as dossiers, atlases, or microfilm and be found via index cards; books are stored in libraries; parts of books can be used by the different sciences; and pieces of knowledge are assembled in new combinations by scientists to help us understand the world surrounding us.

In the process of developing his grandiose ideas of globalization and universalism, Otlet envisions an Internet-like network that links an individual to the world's rich information sources and the "civitas mundaneum"—the gobal community.

The other figure by Otlet (left, bottom) shows an expert sitting on a desk, creating a book that represents the exterior world. Below that image is the reader of that book, who will become an expert on the representation of the world.

Otlet also visualized a teleconferencing session involving a gramophone, film, radio, and television, anticipating what would later be called hypermedia. Everything a scholar might need would be available via devices on his work desk.

Cinema, phonograph, radio, television—these instruments considered to be substitutes for the book have become in fact the new book, the most powerful of means for the diffusion of human thought. By radio not only will one be everywhere able to hear, one will everywhere be able to speak. By means of television not only will one be able to see what is happening everywhere, but everyone will be able to view what he would like to see from his own vantage point. From his armchair, everyone will hear, see, participate, will even be able to applaud, give ovations, sing in the chorus, add his cries of participation to those of all the others.
Paul Otlet

Otlet continued developing new visions of a future high-technology information society until his death in 1944. His unfinished *Atlas Mundaneum (Encyclopedia Universalis Mundaneum)* brought visual expression to his ideas on knowledge organization, visualization, and dissemination.

Knowledge Interlinkage

The interlinkage of scholarly records to improve access and sensemaking has a long history. The earliest Hebrew manuscript citation index, ascribed to Maimonides, dates from the 12th century; the first Hebrew citation index to a printed book is dated 1511. The Talmud embeds a similar index. Here we discuss how Shepard's legal citation system inspired Garfield's Science Citation Index and later Google's link-based retrieval algorithm.

Linkages are an important vehicle for analyzing and mapping scholarly records originating from different areas of science that have various writing styles, formats, and word usages. While semantic analyses work well when applied to a specific domain, the mapping of all sciences requires the study of networks of paper citation or coauthorship relations.

Frank Shepard (1848–1902): *Legal Citation Index, 1873*

In 1873, Frank Shepard began what would become known as Shepard's Citations (1951), the only citation index of legal cases "to serve the Bench and Bar." Once a starting case or statute was located, a legal researcher could use Shepard's system to find linkages to other authorities that had cited the case or statute and provided editorial analysis. Each year Shepard printed a list of all appellate decisions that had cited any previous case. Initially, links had to be updated manually by affixing references on gummed paper next to the appropriate case in casebook volumes (see below). Today, the legal index is available in electronic format, complete with advanced navigation features.

Shepard's Citations served as a model for what is now Thomson Reuters's Science Citation Index and many other systems in existence today. In 1955, Eugene Garfield—inspired by browsing through Shepard's Citations in a public library—wrote a paper titled "Shepardizing the Literature of Science."

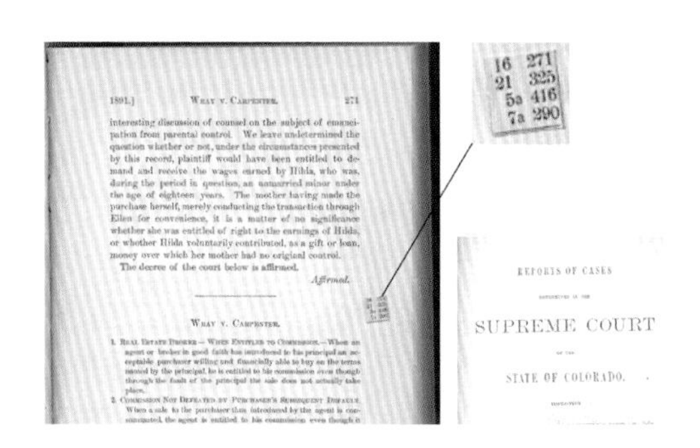

Eugene Garfield (b. 1925): *Citation Indexes for Science, 1955*

The American scientist and entrepreneur Eugene Garfield is a pioneer in bibliometrics, scientometrics, and information retrieval systems. He is the inventor and creator of Index Chemicus, Science Citation Index (SCI), Social Sciences Citation Index (SSCI), Arts and Humanities Citation Index (AHCI), and services such as Current Contents. Garfield is the founding publisher and editor of *The Scientist* and past president of the American Society for Information Science and Technology.

In 1955, Garfield outlined the idea of a unified citation index to the literature of science that was deeply inspired by Shepard's Citations. He envisioned a citation index for science as an "association-of-ideas index." In 1958, Garfield proposed "A Unified Index to Science" to the National Science Foundation (NSF), with the aim of creating a unified and standardized approach to scientific literature searches, regardless of subject. He was planning to utilize machines for the compilation of indexes and to broaden the number of potential users to reduce production costs. For Garfield, the unified index meant a step toward realizing H. G. Wells's "World Brain" (see **page 25, H. G. Wells**).

While legal cases and statutes had standard reference formats, scientific references had yet to be standardized and the amount of scientific literature was daunting. In 1961, Garfield's index covered only 5 percent of the 800 journals included in the *Physics Abstracts* and hence was essentially useless for physicists.

In 1964, the Science Citation Index was introduced as a five-volume set covering 613 journals and 1.4 million citations. It was followed by the Social Sciences Citation Index in 1972, the Arts and Humanities Citation Index in 1978, and the Web of Science in 1997 (see **page 4, The Rise of Science and Technology**).

We may have ivory towers, but they are all connected by cables under the ground or by invisible channels in the air. Scientific progress—the accumulated effect of your individual contributions—depends upon a free flow of information, of thousands of minute facts, of millions of seemingly unrelated observations made and reported by scientists in diverse specialties.
Eugene Garfield

Among Garfield's many contributions was the development of tools to study science. In 1964, he proposed to write the history of science by algorithmic means. His 1964 map and his recent software HistCite are featured on **page 120, HistCite Visualization of DNA Development**.

In 1973, Henry G. Small and Irina V. Marshakova-Shaikevich independently developed and published the idea of cocitation analysis. Cocitation assumes that the subject similarity of two papers increases with the number of times they are cited by other papers. Small later used cocitation analysis to cluster and map documents as research front specialties, see **page 55, Research Fronts**. In 1981, Garfield published the *ISI Atlas of Science: Biochemistry and Molecular Biology, 1978/80 (Including Minireviews of 102 Research Front Specialties)* and in 1984, his *ISI Atlas of Science: Biotechnology and Molecular Genetics, 1981/82 (Including 127 Research Front Specialties)* and *Bibliographic Update for 1983/84* appeared. Garfield's Current Contents papers—titled "ISI's Atlas of Science May Help Students in Choice of Career in Science" (1974), "Mapping the World of Nutrition: Citation Analysis helps Digest the Menu of Current Research" (1987), "Mapping the World of Epidemiology. Part 2. The Techniques of Tracking Down Disease" (1988)—communicated the value of science maps for diverse stakeholders.

Larry Page (b. 1973) *and* Sergey Brin (b. 1973): *Google, 1998*

American entrepreneurs Lawrence Edward (Larry) Page and Sergey Brin launched the Google search engine in September 1998, while at Stanford University—and soon thereafter founded Google, Inc. Since then, Google has grown to more than 10,000 employees worldwide. The Google search engine is based on Google's patented PageRank technology, which relies on the structure of links between Web sites to determine the ranking of an individual site in search results. PageRank is a link analysis algorithm that assigns a numerical weight to each node of a network; that is, the hyperlinks among Web pages are used to "measure" each page's rank within a set. The algorithm and associated databases have been carefully tuned for access and advertisement needs and are closely held company secrets.

Google now offers more than just Web search. Google Scholar (**http://scholar.google.com**) uses autonomous citation indexing introduced in 1998 by C. Lee Giles, Steve Lawrence, and Kurt Bollacker to interlink and help users navigate the network of scholarly literature. Google Books (**http://books.google.com**) makes books full-text searchable. End of 2009, Google had digitized some ten million books. While library catalogues index only whole books, Google Books interlinks book chapters and sections via their references. Otlet's vision of information items "smaller than a book" that can be recompiled, reused, and understood in their context, comes within reach through Google Books.

Google Maps (**http://maps.google.com**) and Google Earth (**http://earth.google.com**) have opened the eyes of many to the beauty of our planet, and they demonstrate the utility of georeferenced search for finding restaurants, hotels, and other places of interest. As our knowledge development depends strongly on access to existing resources, Google holds the key to both our current and future knowledge.

Knowledge Visualization

The design of legible and usable science maps draws on art, photography, architecture, geography, design, and many other fields of study (see also **page 50, Data Communication—Visualization Layers**). Here, we feature works by educator and pacifist Kees Boeke; physicist and science historian Derek John de Solla Price; and political scientist, statistician, and information designer Edward R. Tufte.

Kees Boeke (1884–1966): *Cosmic View: The Universe in 40 Jumps, 1957*

Cornelis (Kees) Boeke was a Dutch educator and pacifist who believed that a better society could be built through educating children. He founded the school De Werkplaats (The Working Place), which used Maria Montessori's methods and was a great success. In one of his last works, *Cosmic View: The Universe in 40 Jumps* (1957), he sought to provide both children and adults with an understanding of the relationships between our world and the cosmos in terms of scale and interlinkages, helping us also to realize the enormousness of the powers that humanity has begun to harness—as well as misuse and exploit.

When we thus think in cosmic terms, we realize that man, if he is to become really human, must combine in his being the greatest humility with the most careful and considerate use of the cosmic powers that are at his disposal.
Kees Boeke

Cosmic View begins with a simple photograph of a girl sitting outside her school, holding a cat. We then seem to ascend with ever-expanding aerial views of the girl's environment. First we see her "yeard," then her "neghborhood," Bilthoven a suburb of Utrecht in the central part of the Netherlands, and Western Europe. The eighth picture shows Earth. Further zooming out, our solar system and then the Milky Way come into view. The journey stops at picture 26 depicting the inconceivably large number of galaxies that surround the Milky Way, illuminating how the Earth is located in an unfathomably infinite universe. Reversing the direction of travel, the essay gradually returns us to the original photo, the graphics showing ever-smaller areas in increasing detail. We see the girl's hand, then a mosquito sitting on her finger, different types of bacteria on her skin, the structure of a salt crystal, sodium and chlorine atoms of the salt crystal, the nucleus of the sodium atom and twelve electrons whirling around it. The magnification continues until the nucleus of the sodium atom is reached. Boeke writes commentary alongside each graphic, as well as introductory and concluding remarks.

In 1977, Charles and Ray Eames, two of the most important American designers, produced *Powers of Ten*, a short documentary film with music by Leonard Bernstein, which depicts the relative scale of the universe in factors of ten in a manner even more elaborate and detailed than Boeke's book. In 1983, Philip Morrison and Phylis Morrison, with the office of Charles and Ray Eames, published *Powers of Ten: A Book About the Relative Size of Things in the Universe and the Effect of Adding Another Zero.*

Derek John de Solla Price (1922–1983): *"Networks of Scientific Papers," 1965*

Derek John de Solla Price was an English writer and educator with doctorates in both physics and the history of science. After holding positions with Cambridge, the Smithsonian Institution, and Princeton University, among others, he became the first Avalon Professor of the History of Science at Yale University and the first chair of the university's department of science and medicine.

In 1947, he began a three-year teaching position at Raffles College, now the University of Singapore. During his stay at Raffles he made his now-famous observation of the exponential growth of science.

The college had acquired a complete set of the Philosophical Transactions of the Royal Society of London, which had its inception in 1665. De Solla Price stored the bound volumes in his home while the college library was under construction. Taking the opportunity to read them cover-to-cover—and thereby gaining his initial education in the history of science—he noticed that the chronologically stacked volumes formed an exponential curve against the wall.
Eugene Garfield

De Solla Price subsequently surveyed all the other sets of journals he could find and concluded that exponential growth was an apparently universal phenomenon in scientific literature. He presented his observations to the Sixth International Congress for the History of Science in Amsterdam in 1950, which marked his transition from research in physics and mathematics to the history of science.

In 1965, de Solla Price published "Networks of Scientific Papers," which gave an overview of the total world network of scientific papers. Building on works and data by Eugene Garfield, Michael M. Kessler, John W. Tukey, Charles E. Osgood, and others, he discussed the uneven distributions of references and citations.

De Solla Price used a matrix visualization (see **Research Fronts figure, page 28**) to show the evolution of a particular field—the phenomenon of

N-rays, circa 1904. During the time N-rays were studied, the field was both small and closed. The 200 papers, constituting the complete record of research results, are represented chronologically from left to right and top to bottom. Dots represent references between papers. Full rows denote highly cited ("classic") papers. Full columns indicate review articles, which seem to appear every 30 or 40 papers. The review papers cite earlier papers that have been lost from sight behind the research front. There is also a higher dot density in a strip near the diagonal and extending over the 30 or 40 papers, immediately preceding each paper in turn. That is, about half of the references are to a research front of recent papers, and the other half are to papers scattered uniformly through the literature.

Since only a small part of the earlier literature is knitted together by the new year's crop of papers, we may look upon this small part as a sort of growing tip or epidermal layer, an active research front.
Derek John de Solla Price

With his work, de Solla Price wished to create a "calculus of science," and he suggested that it be modeled on econometrics and thermodynamics. He had a passion for scientific apparatus, and his work shows the importance of technology and methodology in the advancement of science.

Edward R. Tufte (b. 1942): *The Visual Display of Quantitative Information, 1983*

Edward Rolf Tufte is professor emeritus at Yale University, where he has taught courses in statistical evidence, information design, and interface design. He is a fellow of the American Statistical Association, the Guggenheim Foundation, the Center for Advanced Study in the Behavioral Sciences, and the American Academy of Arts and Sciences. Tufte has written, designed, and self-published books on analytical design—including *The Visual Display of Quantitative Information, Envisioning Information, Visual Explanations,* and *Beautiful Evidence*—for which he has received more than 40 awards for content and design.

The Visual Display of Quantitative Information, his first book, was completed in 1982. No publisher, however, would print it to his exacting standards—precisely those design principles he had articulated in the book.

Publishers seemed appalled at the prospect that an author might govern design.
Edward R. Tufte

Tufte wanted lavish, abundant, high-resolution images; beautiful typesetting; and footnotes placed adjacent to the corresponding text—all printed on thick, creamy paper at a rather low sales price. He decided to publish the book himself, took out a second mortgage at nearly 18 percent interest, and founded Graphics Press in 1983. His book was instantly hailed as a classic, and he repaid the mortgage loan within six months.

Since then, 1.4 million Tufte books have been printed, 200,000 people have attended Tufte's workshops, and he has inspired millions to communicate data and knowledge more effectively. Tufte has consulted on information design and statistical matters for the CBS and NBC television networks, the *New York Times, Newsweek,* Hewlett-Packard, the Bureau of the Census, and the Bureau of Justice Statistics. He has also prepared evidence for several jury trials. Tufte has been called the "Leonardo da Vinci of data" and "the world's leading analyst of graphic information."

Though Tufte has not been involved in the mapping of science, his writings and teachings represent a treasure trove of theories and practices on how to communicate data, information, and knowledge in ways both effective and beautiful (see **page 84, Napoleon's March to Moscow**).

Man-Machine Symbiosis

Do human brains have the capacity to process, manage, and employ the ever-increasing flood of information? New tools are required that can help augment not only our physical strength but also our intellectual and visual capacities. While writing things down can help us hone and sharpen our memory, computing machines are required to help us process more data. The first interfaces between analog "wetware" (that is, human brains) and digital software left much to be desired.

Ada Lovelace (1815–1852): *First Computer Program, 1843*

Augusta Ada King, Countess of Lovelace—the only legitimate child of Lord Byron—called herself "an Analyst (and Metaphysician)." In 1833, at age 17, she met Charles Babbage, inventor of the Difference Engine. They began a voluminous correspondence on mathematics, logic, and ultimately all subjects. In 1834, Babbage made plans for a new kind of calculating machine, an Analytical Engine. While his Parliamentary sponsors refused to support a second machine, he found sympathy for his new project abroad. When, in 1842, Italian mathematician Louis Menebrea published a memoir in French on the subject of the Analytical Engine, Babbage enlisted Lovelace as translator. During a nine-month period from 1842 to 1843, Lovelace worked feverishly on the article. When she showed Babbage her translation, he suggested that she add her own notes, which turned out to be three times the length of the original article.

Letters between Babbage and Lovelace flew back and forth, filled with fact and fancy. Lovelace understood the plans for the device as well as Babbage, but she was better at articulating its promise. She rightly saw the device as what we would call a general-purpose computer. In her 1843 article, Lovelace predicts that such a machine might be used to compose complex music and produce graphics and that it could be used for both practical and scientific purposes. She was correct.

It was suited for "developping [sic] and tabulating any function whatever ... the engine [is] the material expression of any indefinite function of any degree of generality and complexity."
Ada Lovelace (quoted in *Ada: Countess of Lovelace* by Doris Langley Moore)

Lovelace suggested to Babbage that she would write a plan for how the engine might calculate Bernoulli numbers. This plan is now regarded as the first "computer program," and Lovelace is credited with being the first computer programmer.

Vannevar Bush (1890–1974): *Memex, 1945*

Vannevar Bush was an American engineer, prominent policymaker, and public intellectual. He was a leading figure in the development of the military-industrial complex in the United States. His unwavering support for federal funding of science and engineering helped to improve higher education; it also maintained American supremacy in military and civilian technology. Bush's management of atomic weapons research was a model for later "big science" projects.

In 1945, he published "As We May Think," an article in which he describes a technology that (1) reduces the written record to manageable size by means of microphotography, (2) replaces writing with talking to a "vocoder" that takes dictation and functions as a "mechanical supersecretary", (3) uses a "**Cyclops Camera**" worn on the forehead (pictured below) that records anything you see and want to archive, (4) reduces simple repetitive thought by using a "thinking machine" and the laws of logic to convert premises into conclusions, and (5) stores trails of thought in the "Memex" for eternity.

The "**Memex**," a portmanteau of "memory extender," supports the automatic storage of and efficient access to books, records, and individual communications (see **page 20, Paul Otlet**). The idea was based on Bush's prior work, which involved

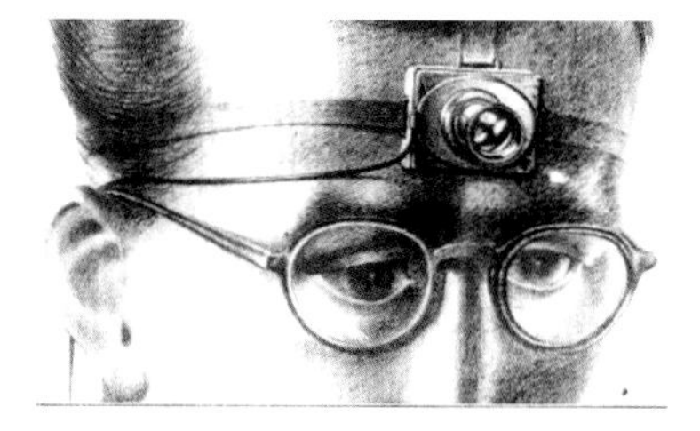

Cyclops Camera

the development of an improved **Photoelectric Microfilm Selector**, an electronic retrieval technology pioneered by Emanuel Goldberg of Zeiss Ikon (Dresden) in the 1920s. The Memex, as a theoretical proto-hypertext computer system, later influenced the development of hypertext (see **page 19, Theodor Holm Nelson**) and the World Wide Web (see **page 20, Timothy Berners-Lee**).

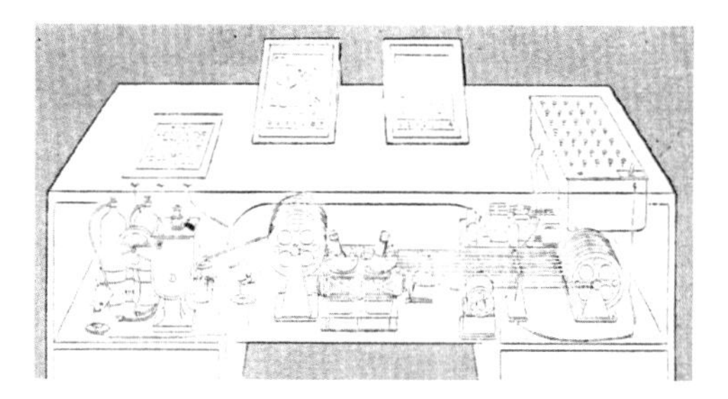

Memex

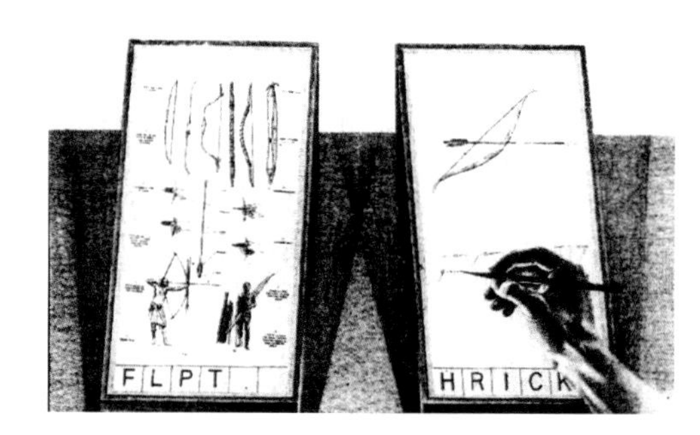

Photoelectric Microfilm Selector

Importantly, "As We May Think" listed a set of problems that knowledge-extending technologies must solve while enabling the general public to benefit from those technologies.

Science offers a largely unexplored hinterland for the pioneer who has the tools for his task. The rewards of such exploration both for the nation and the individual are great.
Vannevar Bush

Grace Hopper (1906–1992): *First Compiler, 1952*

Grace Murray Hopper was an American computer pioneer, as well as a rear admiral, who spent a half-century helping to keep the United States on the leading edge of high technology. Hopper received a BA in mathematics and physics from Vassar College and an MA and PhD in mathematics from Yale University. Although she wanted to join the military when the United States entered World War II, she was not accepted; instead, she was successful in persuading the Navy Reserve to accept her in 1943.

In 1949, Hopper joined the Eckert-Mauchly Computer Corporation as senior mathematician and worked with John Presper Eckert and John Mauchly on the UNIVAC computer. At that time, programmers had to write lengthy instructions in binary code, that is, long sequences of 0s and 1s. This made programming a time-consuming and error-prone task. Hopper believed that computers could be made to understand programs written in English and started to work on a "compiler" that would translate English-language instructions into the language of the target computer. Hoping that "the programmer may return to being a mathematician," she and her staff developed the B-O compiler, later known as FLOW-MATIC, for the UNIVAC. Using FLOW-MATIC, they were able to make the UNIVAC I and II "understand" 20 statements in English. Based on their success, Hopper recommended that an entire programming language be developed using English words, but she was told that this was impossible because "computers didn't understand English."

I had a running compiler and nobody would touch it. They told me computers could only do arithmetic.
Grace Hopper

After three years, Hopper's idea was finally accepted, and she published her first compiler paper in 1952. Hopper and her team subsequently developed many important concepts for mathematical and business compilers, including subroutines, formula translation, relative addressing, the linking loader, code optimization, and the kind of symbolic manipulation embodied in mathematics that is now at the core of digital computing.

In the 1960s, Hopper led the effort to develop COBOL (Common Business-Oriented Language), a programming language still in use today. In all her software development efforts, she promoted the international standardization of computer languages and validation procedures. Throughout her life, Hopper strongly advocated the education and training of men and women for careers in computer science and data processing. In 1991, she became the first individual woman to receive America's highest technology award: the National Medal of Technology (now called the National Medal of Technology and Innovation).

J.C.R. Licklider (1915–1990): *"Man-Computer Symbiosis," 1960*

Joseph Carl Robnett Licklider, affectionately known as J.C.R. or "Lick," was a visionary American computer scientist. He held a BA in physics, mathematics, and psychology and an MA in psychology from Washington University in St. Louis; he received a PhD from the University of Rochester for his early work in psychoacoustics. In 1950, he joined MIT and began working on his growing interest in information technology.

In 1960, Licklider wrote "Man-Computer Symbiosis," in which he argued that a linkage between computers and the human brain would surpass the capabilities of either alone. Applying psychological principles to the design of human-computer interfaces, he envisioned networked computers, software that migrated on the computer network to wherever it was needed, and user interfaces that offered graphical computing and point-and-click input. At a time when punched cards and paper tape were the common interface to computers, he foresaw highly interactive networked computer applications in digital libraries, e-commerce, and online banking.

In the anticipated symbiotic partnership, men will set the goals, formulate the hypotheses, determine the criteria, and perform the evaluations. Computing machines will do the routinizable work that must be done to prepare the way for insights and decisions in technical and scientific thinking.
J.C.R. Licklider

Later, while working at the Defense Advanced Research Projects Agency (DARPA), Licklider was instrumental in funding and managing the research that led to modern personal computers and the Internet. He funded early efforts in time-sharing and application development, most notably the work of Douglas Engelbart (see **page 24, Douglas Engelbart**) and his famous "oN Line System," and early networking research, such as that for the ARPANET (see **page 19, Vinton Gray Cerf**). In 1968, Licklider published "The Computer as a Communication Device" (coauthored with Robert Taylor), predicting the use of computer networks to support distributed information resources and communities of common interest independent of their locations.

Licklider was also instrumental in creating the first PhD programs in computer science, sponsoring research at four universities, allowing them to offer graduate computer science degrees. These departments in turn provided role models for other PhD programs in the United States.

Douglas Engelbart (b. 1925): *"Augmenting Human Intellect: A Conceptual Framework," 1962*

Douglas C. Engelbart is an American inventor of Swedish and Norwegian descent. He was Licklider's contemporary at the Defense Advanced Research Projects Agency (DARPA) and also his protégé, influenced by Vannevar Bush's ideas (see **page 23, Vannevar Bush**). His work was strongly influenced by Ivan Sutherland's work on the "Ultimate Display" (1965). In 1962, Engelbart published "Augmenting Human Intellect: A Conceptual Framework," in which he envisioned computer-based technologies that supported direct data manipulation and computation while improving individual and group processes for knowledge work.

Man's population and gross product are increasing at a considerable rate, but the complexity of his problems grows still faster, and the urgency with which solutions must be found becomes steadily greater in response to the increased rate of activity and the increasingly global nature of that activity. Augmenting man's intellect, in the sense defined above, would warrant full pursuit by an enlightened society if there could be shown a reasonable approach and some plausible benefits.
Douglas Engelbart

With his team, Engelbart developed a suite of sophisticated software called the "oN Line System" (NLS), which explored the limits of human–machine interaction, including such elements as hypertext, object addressing, dynamic file linking, shared-screen collaboration, multiple windows, on-screen video teleconferencing, and the mouse as an input device. NLS was the focus of the "mother of all demos" in San Francisco on December 8, 1968. Many of the technologies and computer-usage paradigms Engelbart envisioned then are still in use today.

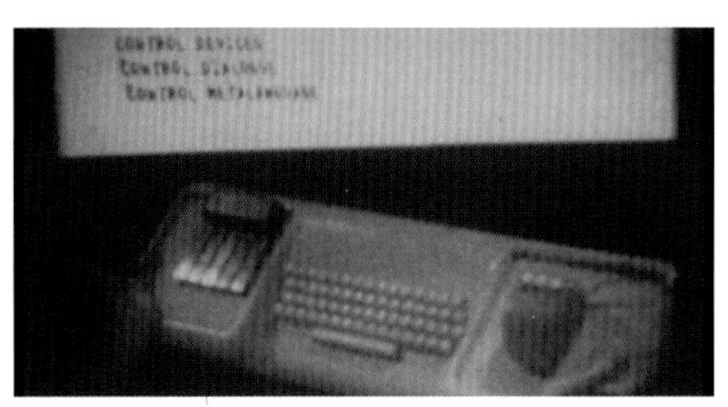

NLS Demo

Global Brain

How should citizens be educated and prepared to collectively manage, utilize, and contribute to our collective knowledge? What bridges might enable the free flow of knowledge from sky-high ivory towers to the taxpayers who finance them? How can we grow an operational "global brain" that efficiently uses sensors, computers, and the Internet to ensure the survival of our species?

The three visionaries discussed here have used futuristic fiction, games, and television as effective means to providing a global understanding of existing and predicted complex systems.

H. G. Wells (1866–1946): *"World Brain," 1938*

Herbert George Wells was a giant of English literature and one of the foremost thinkers of his time. No field of writing was foreign to him, as he made forays into history, social science, commentary, and futurism. His first novel, *The Time Machine* (1895), is viewed as a radical, insightful, and scholarly discourse on the science of time-space relations. Entertaining and provocative, it is believed by many to be the best short novel ever written.

Wells's other futuristic fictions include *The Island of Dr. Moreau* (1896), which depicts the folly of taking science too far; *The Invisible Man* (1897), which shows that science has its limits; *War of the Worlds* (1898), an epic tale of technically advanced Martians invading Victorian England and virtually conquering civilization; and *First Men in the Moon* (1901), a cautionary tale about mankind making first contact with an alien race.

> *Human history becomes more and more a race between education and catastrophe.*
> H. G. Wells

In 1938, Wells outlined his vision of a future information center in his book *World Brain*. At the beginning of the book he describes a "World Encyclopedia" that draws existing ideas and knowledge together into "a comprehensive conception of the world" that "holds the world together mentally." This World Encyclopedia would describe

> *. . . in clear understandable language, and kept up to date, the ruling concepts of our social order, the outlines and main particulars of all fields of knowledge, and exact and reasonably detailed picture of our universe, a general history of the world, and . . . a trustworthy and complete system of references to primary sources of knowledge. . . . It would be alive and growing and changing continually under revision, extension and replacement from original thinkers in the world everywhere.*
> H. G. Wells

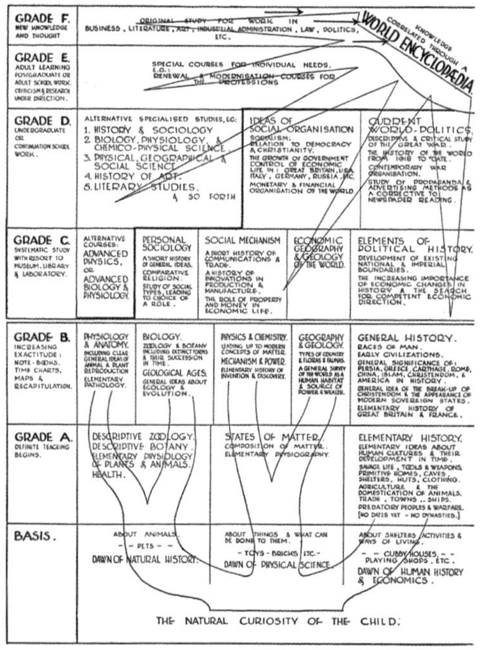

Wikipedia (see **page 19, Jimmy Wales**) can be seen as a first implementation of Wells's 1938 vision.

In *World Brain*, Wells also estimated that it would take about 2,400 hours (6 hours per week, 40 weeks per year, over 10 years of schooling) to teach what a common citizen should know. He proposed to teach the topics shown in the World Encyclopedia figure (below left), and this classification of knowledge subsequently inspired numerous efforts to create a global repository of interlinked knowledge.

Buckminster Fuller (1895–1983): *World Game, 1960s*

Richard Buckminster "Bucky" Fuller was an American designer, architect, engineer, poet, inventor, and visionary. He considered himself an average individual and had neither special monetary means nor an academic degree.

When he was 32, his young daughter died from polio and spinal meningitis. Bankrupt and jobless, and feeling responsible for her death, Fuller found himself at the verge of suicide. At the last moment he decided instead to embark on what he described as "an experiment, to find what a single individual can contribute to changing the world and benefiting all humanity."

The experiment proved successful: Fuller wrote more than 30 books; was awarded 28 U.S. patents and numerous honorary doctorates; coined and popularized terms such as "spaceship earth," "ephemeralization," and "synergetics"; and contributed numerous inventions to design and architecture, such as the geodesic dome.

In all his works, Fuller aimed to answer one main question: "Does humanity have a chance to survive lastingly and successfully on planet Earth, and if so, how?" Defining wealth in terms of knowledge, he argued that the amount of existing knowledge together with the quantities of recyclable resources already extracted from the Earth made competition for necessities obsolete. He saw cooperation as the optimum survival strategy and wrote:

> *It no longer has to be you or me. Selfishness is unnecessary and hence-forth unrationalizable as mandated by survival. War is obsolete."*
> Buckminster Fuller

In the 1960s, Fuller proposed and implemented the World Game—or World Peace Game—a comprehensive, anticipatory, design-science approach to the problems of the world. The playing of World Game was intended to "make the world work for 100 percent of humanity in the shortest possible time through spontaneous cooperation without ecological damage or disadvantage to anyone."

Fuller envisioned that the game would be accessible to anyone. Individuals or teams could join and compete or cooperate to find solutions to mankind's challenges. The best solutions would be disseminated to the masses through a free press. Ultimately, the political process would be forced to move in the direction that the values, imagination, and problem-solving skills of those playing the democratically open game dictated. In order to work, the players would need access to the world's vital statistics: availability and production of minerals, manufactured goods and services, and human labor and expertise, as well as its unmet needs and capabilities. The current state of the world had to be monitored to bring vital news into the "game room" live. All this was done prior to the existence of personal computers and the Internet.

James Burke (b. 1936): *Connections, 1979*

James Burke is a British science historian, educator, author, and media producer. Educated at Oxford University, he received an MA in English literature and lectured in English at universities in Bologna and Urbino, Italy. In 1966, he joined the BBC's Science and Features Department in London. He was fascinated by the use of television to educate and entertain with science and technology.

Burke became interested in the dependence of the so-called great minds of the past on the work of their predecessors and the social and commercial requirements of their times. Innovation of all kinds in science and the humanities was almost always multidisciplinary in origin—yet modern education now taught subjects in ring-fenced silos to which access was limited, above all, by vocabulary.

> *People tend to become experts in highly specialized fields, learning more and more about less and less. . . . Unfortunately, so much specialization falsely creates the illusion that knowledge and discovery exist in a vacuum, in context only with their own disciplines, when in reality they are born from interdisciplinary connections. Without an ability to see these connections, history and science won't be learnable in a truly meaningful way and innovation will be stifled.*
> James Burke

Convinced that our ability to interconnect existing knowledge would be the key to solving the problems that plagued mankind, Burke developed the widely popular documentary television series *Connections* (1979), followed by the sequels *Connections*² (1994) and *Connections*³ (1997). These series on the interconnected history of science and technology were first aired on the BBC, subsequently on PBS channels in the United States, and then in 50 more countries. Burke's subsequent book, *Connections*, was a bestseller on both sides of the Atlantic.

In 1996, Burke started the Knowledge Web project (**http://www.k-web.org**), a digital incarnation of his TV series and publications. His book *The Knowledge Web* was published in 2000. Today, the Knowledge Web is an interactive learning tool that supports the exploratory traversal of a dense network of thousands of history's key figures. A journey across this web reveals the interdisciplinary connectedness of all knowledge. It encourages users to create their own knowledge webs, mapping the connections inherent in their own classrooms, communities, and corporations. Eventually the Knowledge Web will become an immersive, inhabited virtual reality of historical people and places "akin to a learning, growing brain—information will be added to the site constantly as visitors explore it."

Milestones in Mapping Science

This timeline shows major advances in the development of **Algorithms, Visualizations, Tools,** and **Books** that aim to increase our understanding of the inner workings of science and technology. The **References and Credits (pages 212-246)** provide more than 350 references to the original papers, reviews, and books featured in this timeline.

Algorithms

Listed here are the names and authors of analysis and visualization algorithms commonly used to generate science maps such as those shown in the visualization design, mapmaking, and tool development section of the timeline and in **Part 4: Science Maps in Action**.

Algorithmic rather than conceptual contributions are included. For example, graph theory was successfully applied by Euler to solve the Königsberg bridges problem in 1735. However, it was not until 1963 that an algorithm was developed that could lay out a set of nodes and edges automatically.

Algorithms that were immediately applied to generate a specific type of science map (for example, Historiographs, HistoryFlow, or TextArc) and maps that were packaged as proprietary tools (for example, VxOrd or IN-SPIRE) are listed in the visualizations and tools sections exclusively.

Visualizations

When selecting visualizations of science and technology, we concentrated on those that are considered major technical contributions, led to major insights, or have been widely used. Early work on social networks was included, as science is driven by social actors and their personal relationships. Only peer-reviewed works are shown. When publishers requested very high fees for the reproduction of figures, author-supplied figures were used instead.

The scope and type of each map is indicated by the symbols at the upper left corner of the map (see legend). Science maps that were generated entirely by hand are underlined. The exhibit science maps featured in **Part 4** are indicated by a green bounding box. Portraits of mapmakers are featured above the maps when available and approved by the authors.

Some early maps are outside the time period captured here but are discussed in the **Visionary Approaches (pages 16-25)** section.

Tools

Many tools exist that can potentially be applied to analyze scholarly data. Here we list those tools that were specifically designed to study science and technology and tools developed for other communities (for example, social network research) but widely adopted for science and technology studies. Tools might be developed in academic settings, by government institutions, or by companies. Given the speed of science and technology progress and competition, few tools last for longer than 10 years.

Books

This listing shows the rich diversity of existing scholarly work. Titles are given in their original languages. Translations are listed in the **References and Credits** when available. A more extensive and continuously updated list of suggested readings can be found at **http://scimaps.org/atlas**.

Algorithms

Visualizations

Tools

Books

1930-1960

Factor Analysis (FA)
Thurstone

Cluster Analysis
Tryon

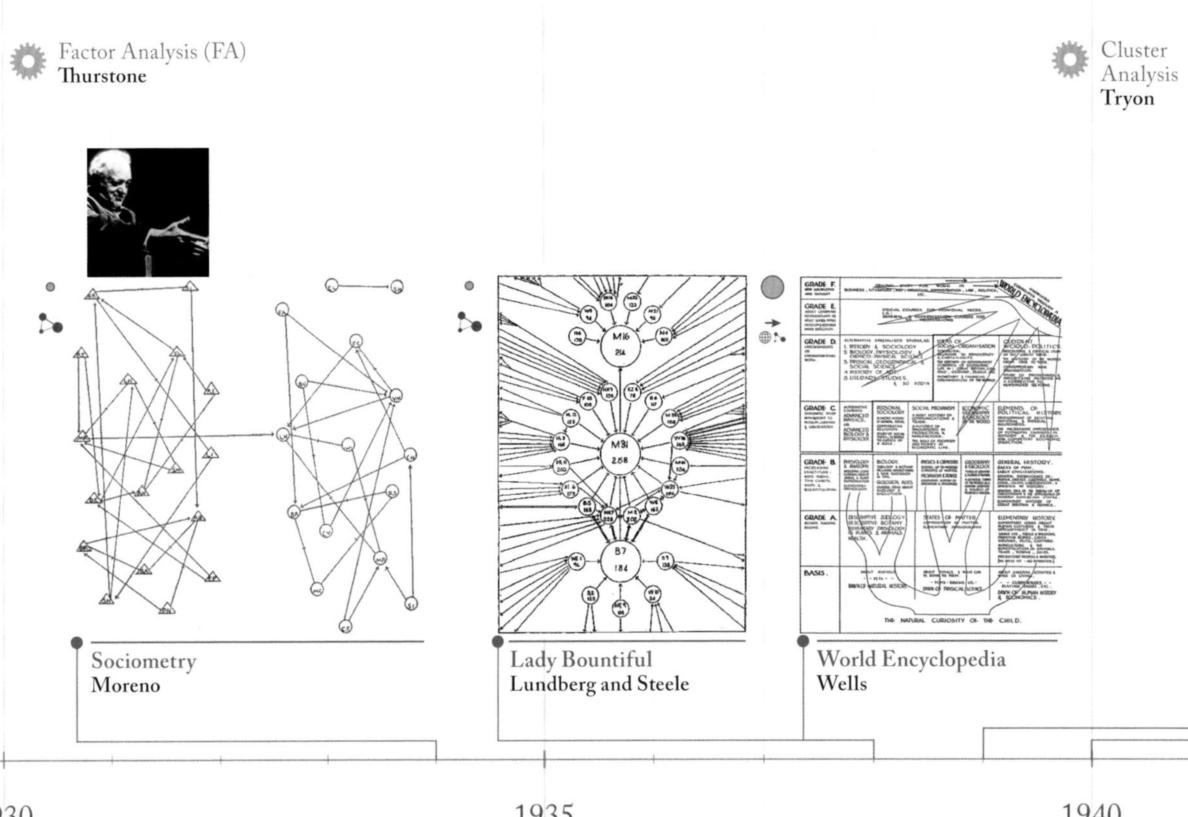

Sociometry
Moreno

Lady Bountiful
Lundberg and Steele

World Encyclopedia
Wells

1930 1935 1940

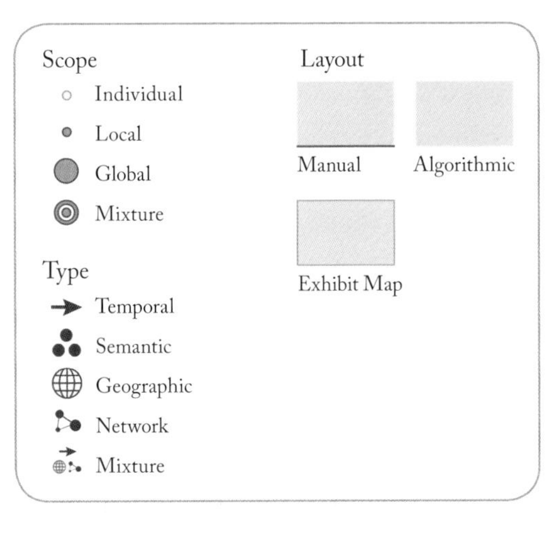

Traité de Documentation: Le Livre sur le Livre—Théorie et Pratique
Otlet

Who Shall Survive?
Moreno

World Brain
Wells

The Social Function of Science
Bernal

Information Theory
Shannon

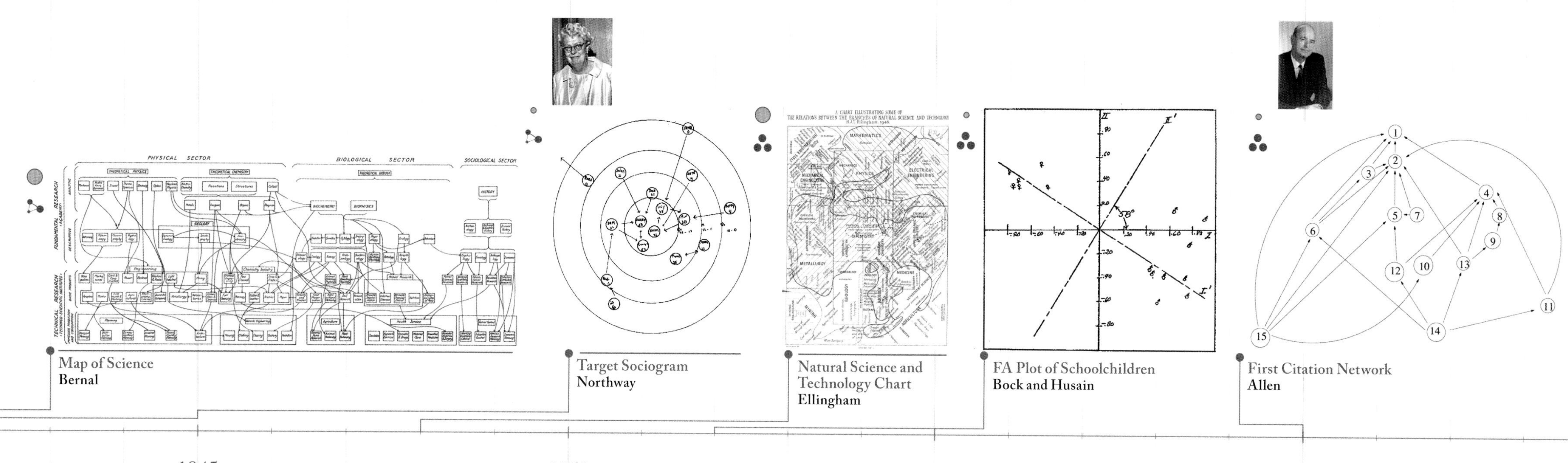

Documentation
Bradford

Science and Information Theory
Brillouin

Science Since Babylon
Price

The Mechanisation of the World Picture
Dijksterhuis

1960-1982

Algorithms

Visualizations

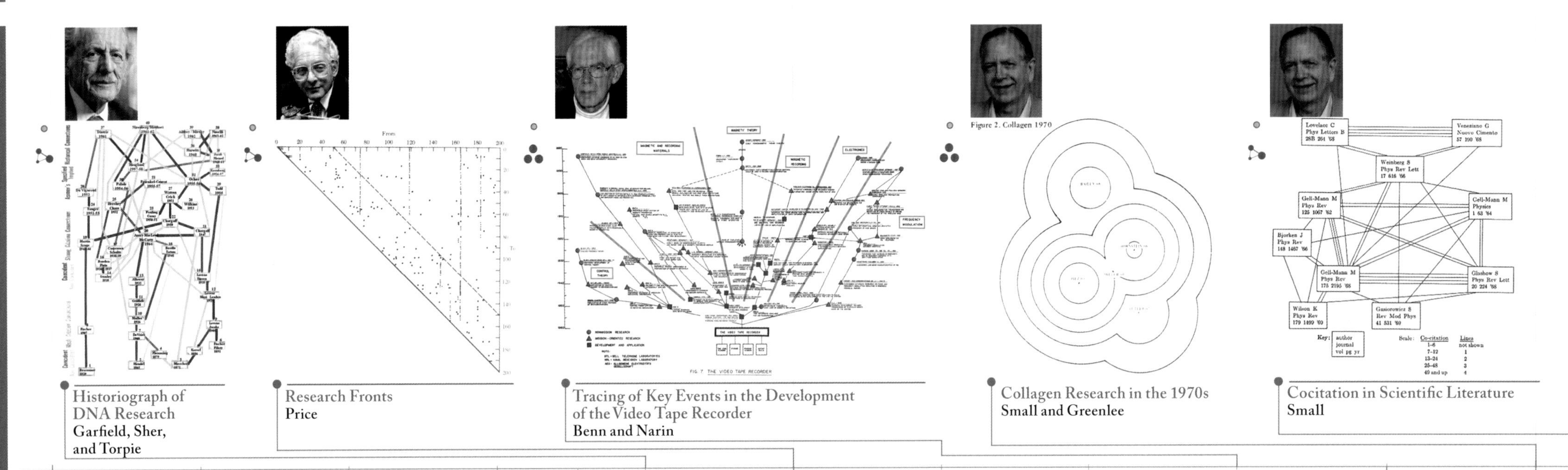

Historiograph of DNA Research
Garfield, Sher, and Torpie

Research Fronts
Price

Tracing of Key Events in the Development of the Video Tape Recorder
Benn and Narin

Collagen Research in the 1970s
Small and Greenlee

Cocitation in Scientific Literature
Small

Tools

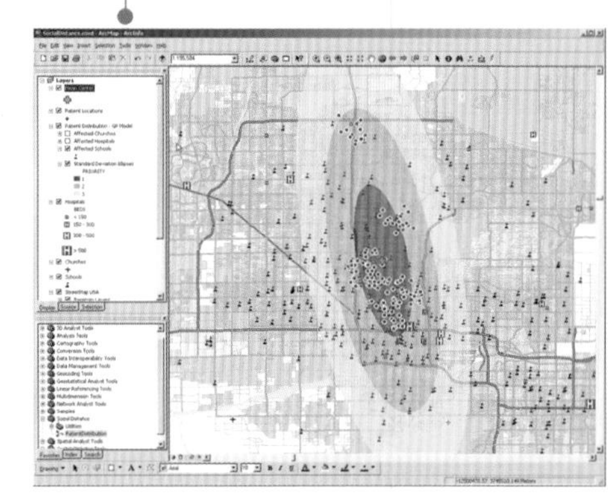

Geographic Information Systems (recent screenshot)
ESRI

Books

The Structure of Scientific Revolutions
Kuhn

Diffusion of Innovations
Rogers

Little Science, Big Science
Price

Libraries of the Future
Licklider

Annual Review of Information Science and Technology (ARIST)
Published 1966-2011

Innovation Diffusion as a Spatial Process
Hägerstrand

Naukometriya
Nalimov and Mulchenko

Capacity of Science
Dobrov et al.

Aktuelle Probleme der Wissenschaftswissenschaft
Dobrov et al.

Betweenness Centrality (BC)
Anthonisse

Automatic Citation Indexing
Salton

Cocitation Analysis (CC)
Independently proposed by Small and Marshakova-Shaikevich

Vector Space Model (VSM)
Salton, Yang, and Wong

Triangulation Method
Lee, Slagle, and Blum

Author Cocitation (ACC)
White and Griffith

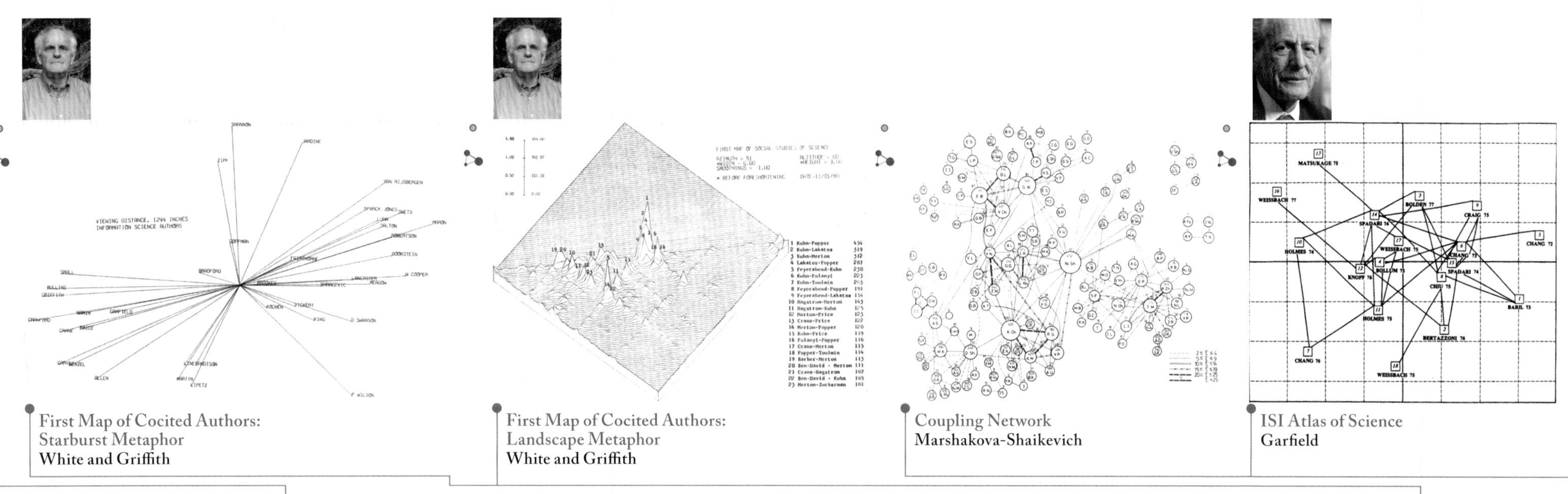

First Map of Cocited Authors: Starburst Metaphor
White and Griffith

First Map of Cocited Authors: Landscape Metaphor
White and Griffith

Coupling Network
Marshakova-Shaikevich

ISI Atlas of Science
Garfield

1975

1980

Science: Growth and Change
Menard

Invisible Colleges: Diffusion of Knowledge in Scientific Communities
Crane

First Science Indicators Report
NSF

The Sociology of Science
Merton

Wissenschaft als allgemeine Arbeit
Laitko

Citation Indexing: Its Theory and Application in Science, Technology, and Humanities
Garfield

Truth, Rationality, and the Growth of Scientific Knowledge
Popper

ISI Atlas of Science –Biochemistry and Molecular Biology 1978/80 including Minireviews of 102 Research Front Specialties
Starchild et al. (eds.)

Critical Path
Fuller

1982-1998

Algorithms

Quantitative Validation
McCain

Coword Analysis
Callon et al.

Cluster Tracking and Mapping
Garfield

Spring Graph Layout
Eades

Self-Organizing Map (SOM)
Kohonen

Kamada-Kawai Graph Layout
Kamada and Kawai

Identifying Scientific Frontiers
Garfield and Small

Visualizations

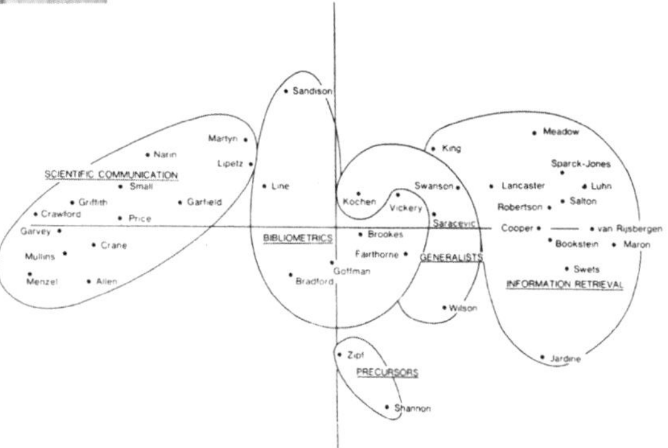

Map of Information Science
White and Griffith

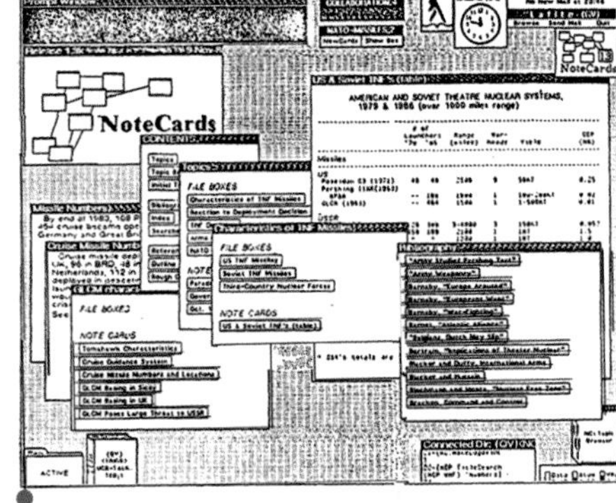

NoteCards
Halasz, Moran, and Trigg at Xerox PARC

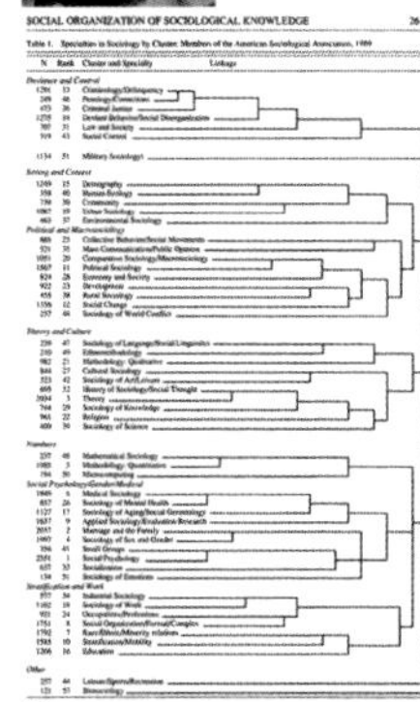

Specialties in Sociology
Ennis

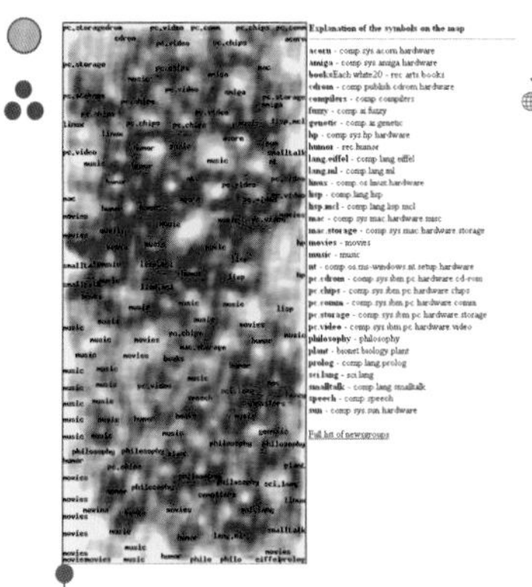

SOM of Newsgroup Postings
Kohonen

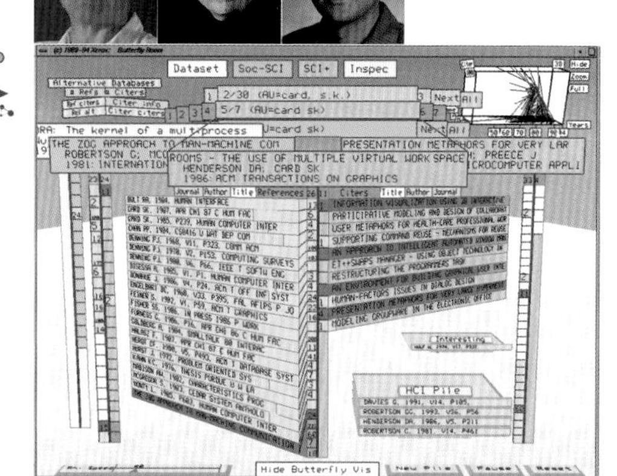

Butterfly Citation Browser
Mackinlay, Card, and Rao at Xerox Research

1985

Tools

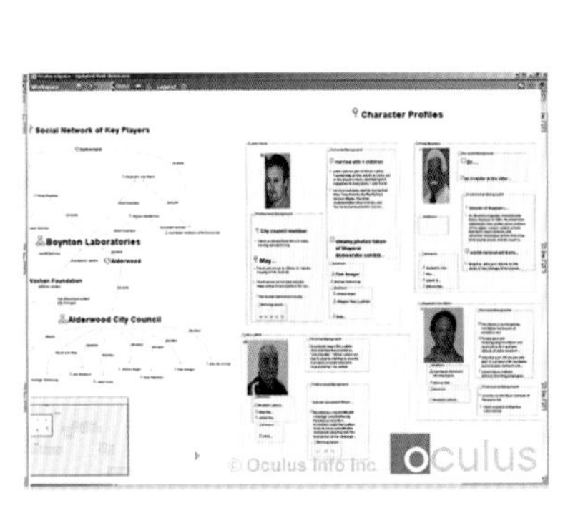

NSpace Sandbox
Oculus Info, Inc.

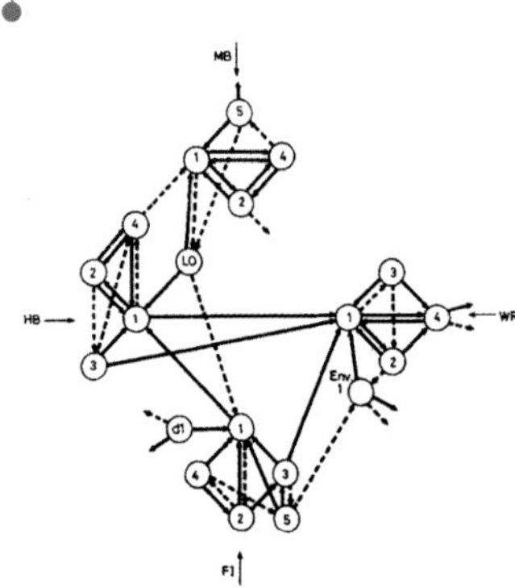

Science and Technology Dynamics Toolbox
Leydesdorff

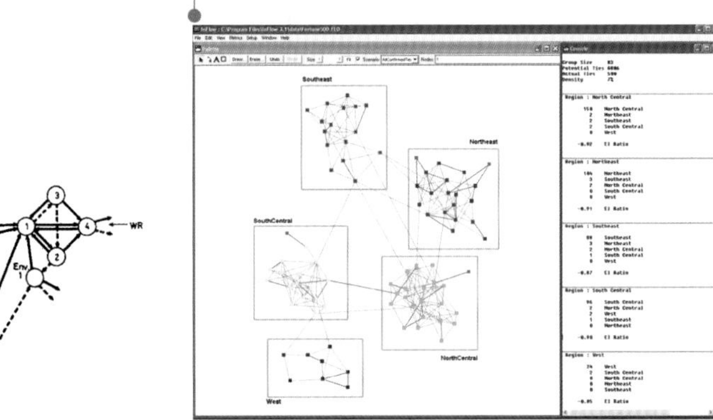

In Flow
Krebs

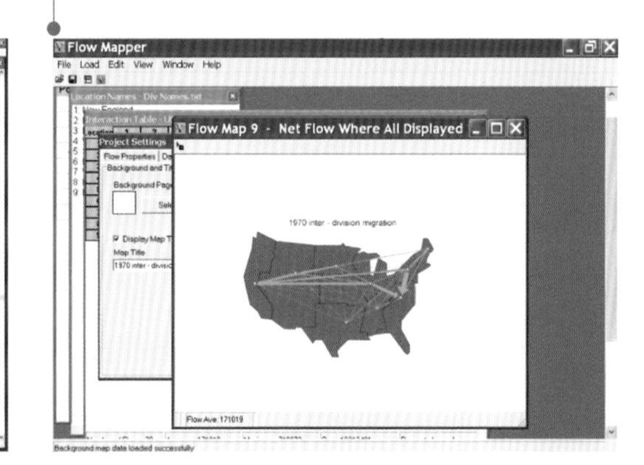

Flow Mapper
Tobler

Books

The Discoverers
Boorstin

Foresight in Science: Picking the Winners
Irvine and Martin

The Citation Process: The Role and Significance of Citations in Scientific Communication
Cronin

The Intellectual Organization of the Sciences
Whitley

ISI Atlas of Science: Biotechnology and Molecular Genetics 1981/82 and Bibliographic Update for 1983/84
Garfield et al. (eds.)

Homo Academicus (French)
Bourdieu

Little Science, Big Science and Beyond
Price

Laboratory Life: The Construction of Scientific Facts
Latour and Woolgar

Mapping the Dynamics of Science and Technology: Sociology of Science in the Real World
Callon, Law, and Rip

Essays of an Information Scientist: Toward Scientography
Garfield

Matematicheskie Modeli v Issledovanii Nauki (Mathematical Models in the Exploration of Science)
Jablonski

Science in Action: How to Follow Scientists and Engineers Through Society
Latour

Seti nauchnykh kommunikatsii (Networks of Scientific Communication)
Dumenton

Handbook of Quantitative Studies of Science and Technology
van Raan (ed.)

Sistema Tsitirovaniya (System of Citing)
Marshakova-Shaikevich

Problemy intensifikatsii nauki: tekhnologiya nauchnykh issledovanii (Problems of Intensification of Science: Scientific Research Technology)
Kara-Mursa

Maps Are Territories: Science Is an Atlas
Turnbull

Research Foresight: Priority-Setting in Science
Martin and Irvine

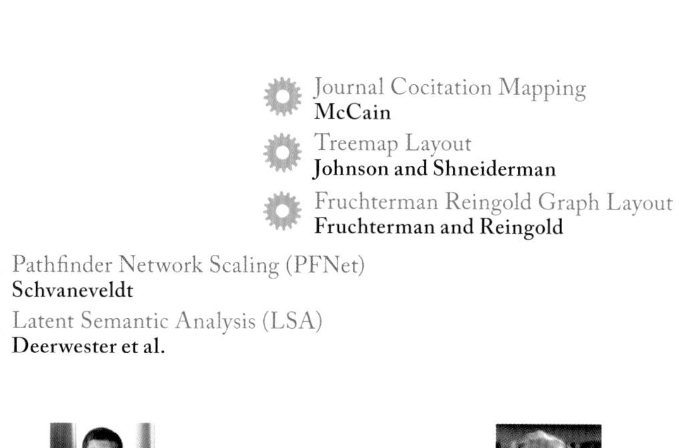

Journal Cocitation Mapping
McCain

Treemap Layout
Johnson and Shneiderman

Fruchterman Reingold Graph Layout
Fruchterman and Reingold

Pathfinder Network Scaling (PFNet)
Schvaneveldt

Latent Semantic Analysis (LSA)
Deerwester et al.

Longitudinal Coupling
Small

Finding Complementary Literatures
Swanson and Smalheiser

Combined Linkage
Small

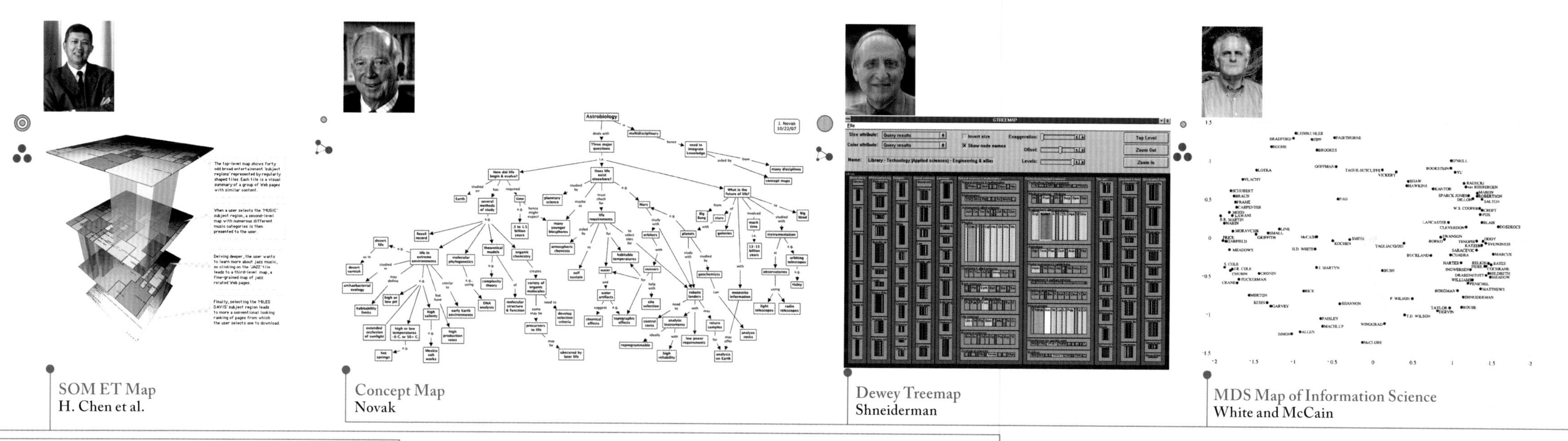

SOM ET Map
H. Chen et al.

Concept Map
Novak

Dewey Treemap
Shneiderman

MDS Map of Information Science
White and McCain

1990

1995

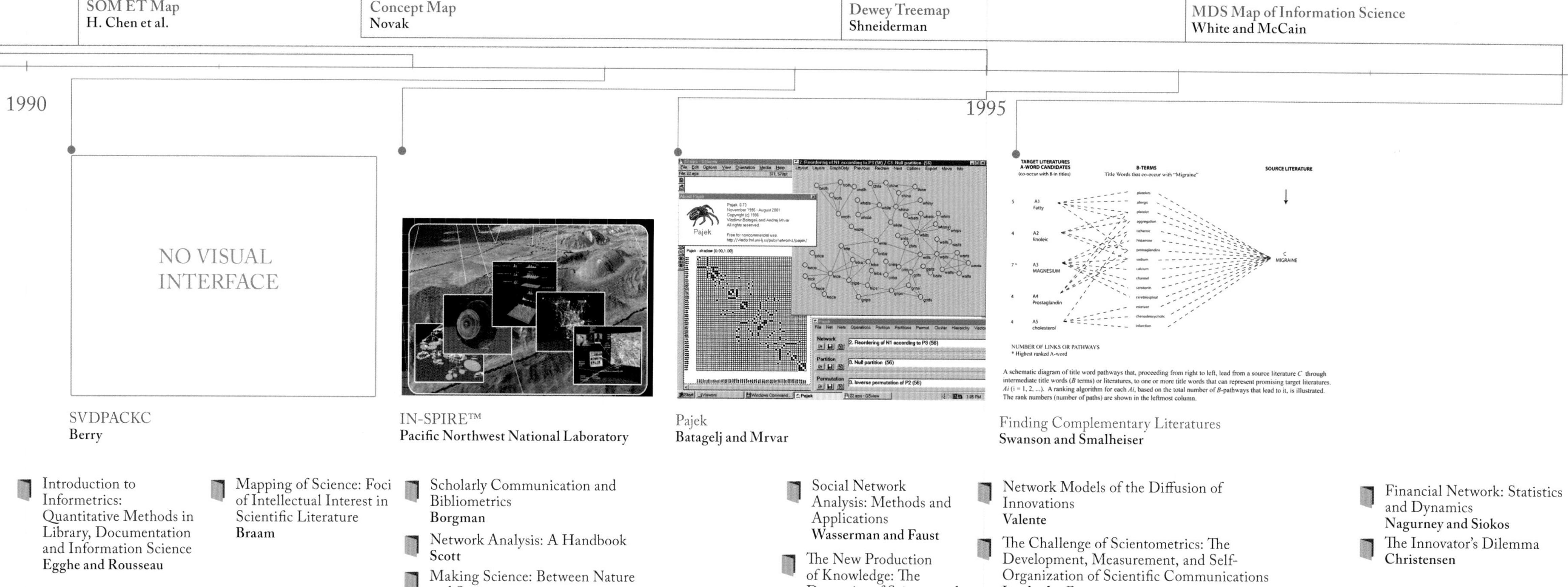

SVDPACKC
Berry

IN-SPIRE™
Pacific Northwest National Laboratory

Pajek
Batagelj and Mrvar

Finding Complementary Literatures
Swanson and Smalheiser

- Introduction to Informetrics: Quantitative Methods in Library, Documentation and Information Science
  Egghe and Rousseau
- Mapping of Science: Foci of Intellectual Interest in Scientific Literature
  Braam
- Scholarly Communication and Bibliometrics
  Borgman
- Network Analysis: A Handbook
  Scott
- Making Science: Between Nature and Society
  Cole
- Social Network Analysis: Methods and Applications
  Wasserman and Faust
- The New Production of Knowledge: The Dynamics of Science and Research in Contemporary Societies
  Gibbons et al.
- Network Models of the Diffusion of Innovations
  Valente
- The Challenge of Scientometrics: The Development, Measurement, and Self-Organization of Scientific Communications
  Leydesdorff
- Financial Network: Statistics and Dynamics
  Nagurney and Siokos
- The Innovator's Dilemma
  Christensen

1998

1999

Algorithms

Automatic Citation Indexing
Giles, Bollacker, and Lawrence

PageRank
Brin and Page

Hubs and Authorities
Kleinberg

Visualizations

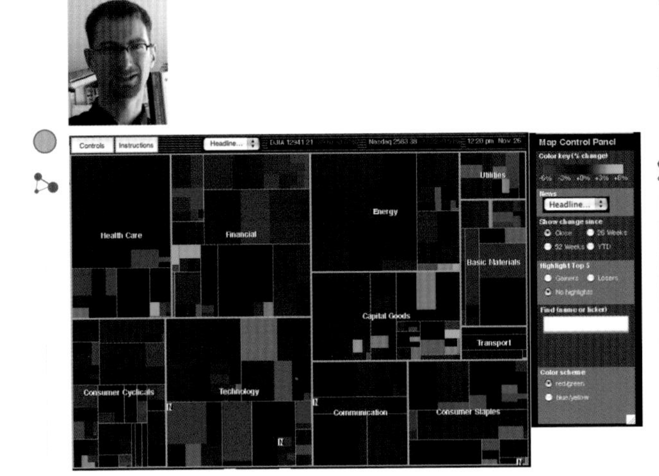

Map of the Market
Wattenberg

© 1999 IEEE

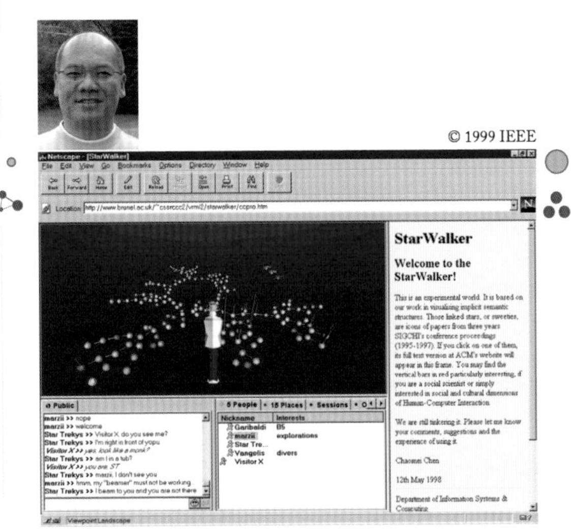

Collaborative StarWalker
C. Chen et al.

PLACES & SPACES 1st Iteration **(see page 76)**

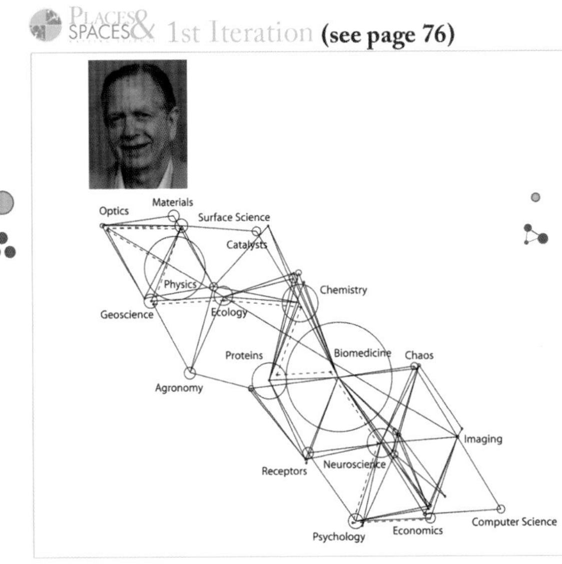

1996 Map of Science
Small

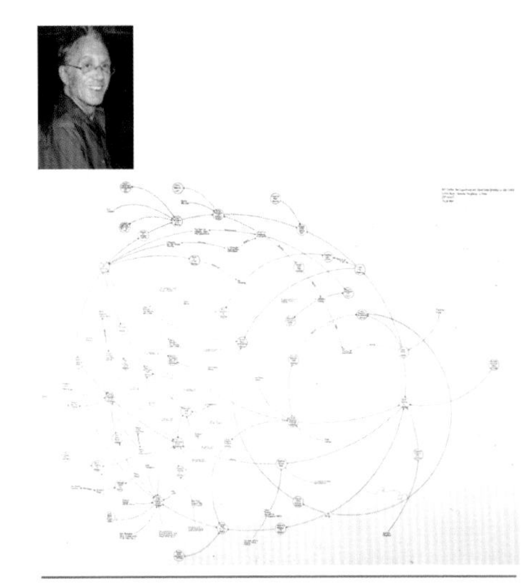

Bill Clinton, Lippo Group and China Ocean Shipping Co. aka COSCO
Lombardi

Tools

1998

1999

DrL: Distributed Recursive (Graph) Layout
Sandia National Laboratories

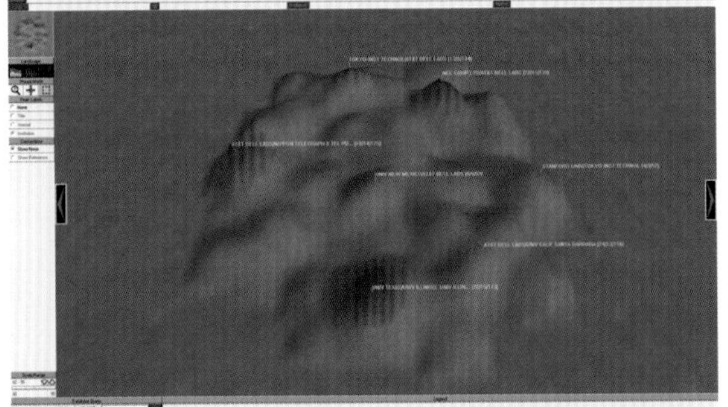

VxInsight Tool
Sandia National Laboratories

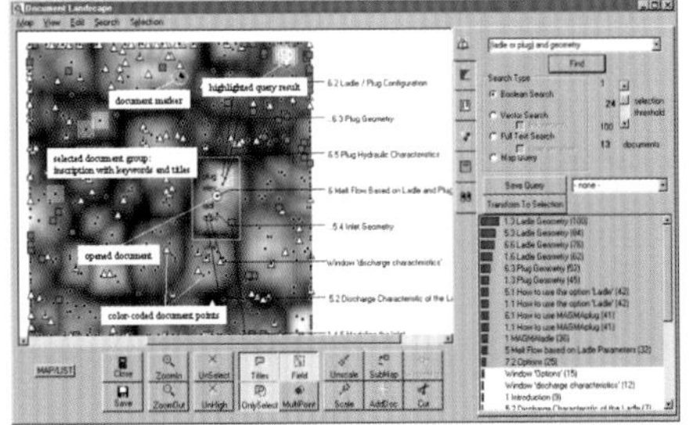

DocMiner Tool
Aachen University

Colt Library
Hoschek et al.

Books

Consilience: The Unity of Knowledge
Wilson

The Competitive Advantage of Nations
Porter

The Citation Culture
Wouters

Cultural Boundaries of Science: Credibility on the Line
Gieryn

Small Worlds
Watts

2000

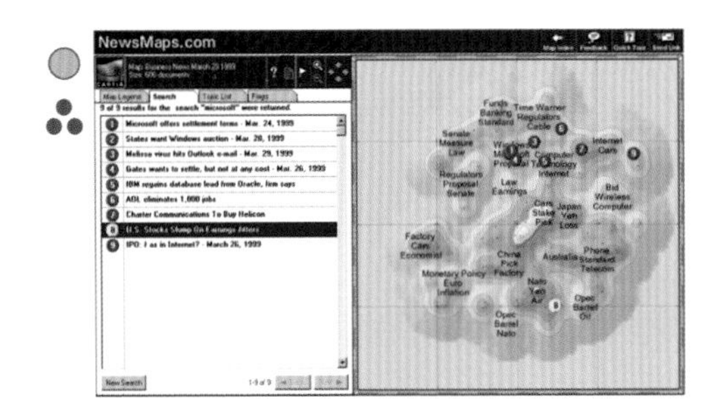

Aureka
Thomson

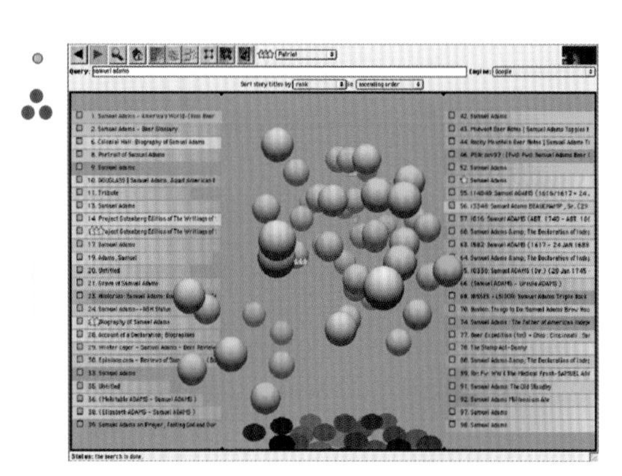

Lighthouse Clustering Visualization
Leuski and Allan

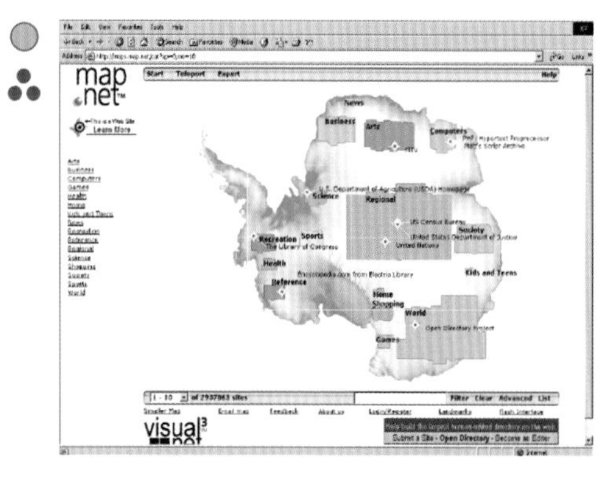

Visual Net™
Bray at Antarcti.ca System Inc.

2000

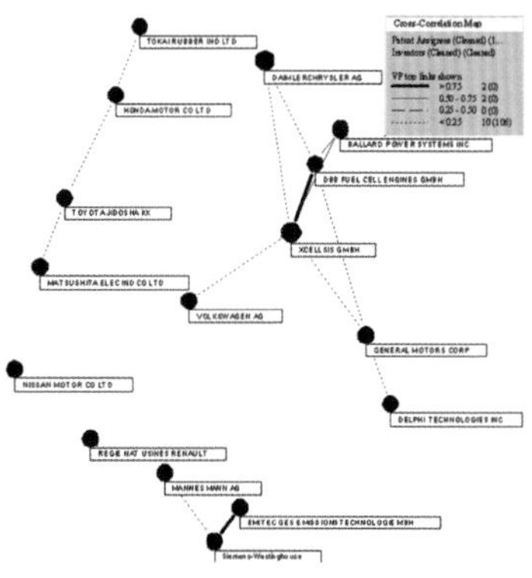

VantagePoint Tool
Search Technology, Inc.

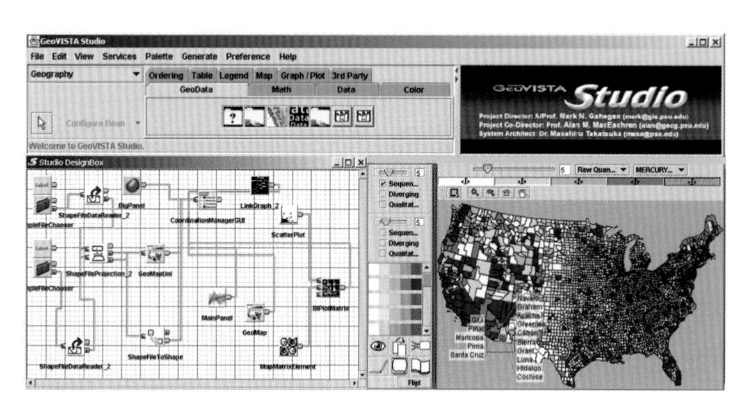

GeoVISTA Studio
Gahegan, Takatsuma, and Hardisty

- Managing Global Innovation: Uncovering the Secrets of Future Competitiveness
  Boutellier, Gassman, and Zedtwitz
- The Web of Knowledge: A Festschrift in Honor of Eugene Garfield
  Cronin and Atkins (eds.)
- Global Brain: The Evolution of Mass Mind from the Big Bang to the 21st Century
  Bloom
- When Information Came of Age: Technologies of Knowledge in the Age of Reason and Revolution, 1700-1850
  Headrick
- The Metrics of Science and Technology
  Geisler

Algorithms

Visualizations

Tools

Books

2000

2002

Faster Algorithm for BC
Brandes

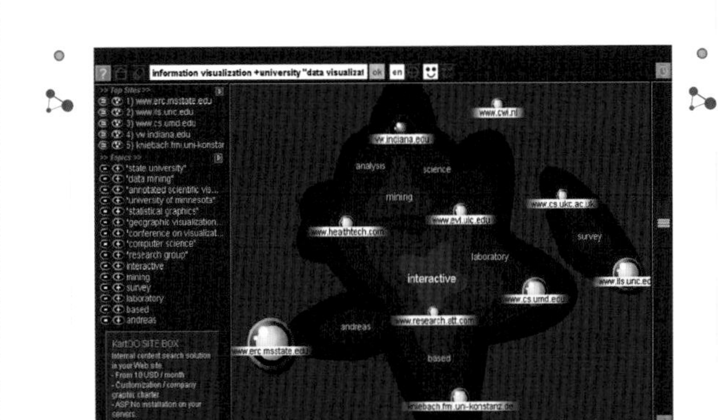

Kartoo Visual Interface to Search Results
Kartoo.net

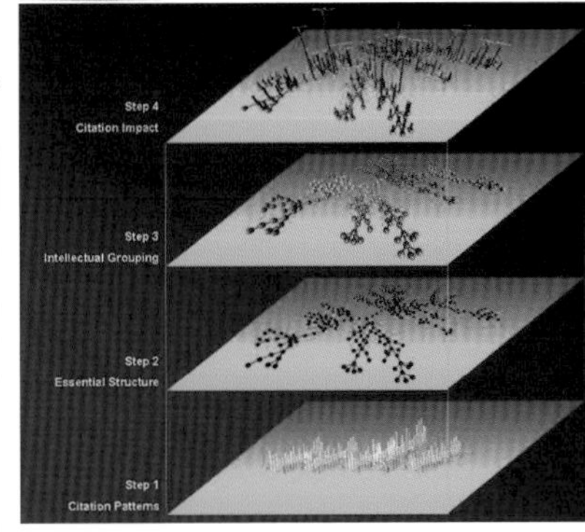

Multilayer Science Map
Chen and Paul

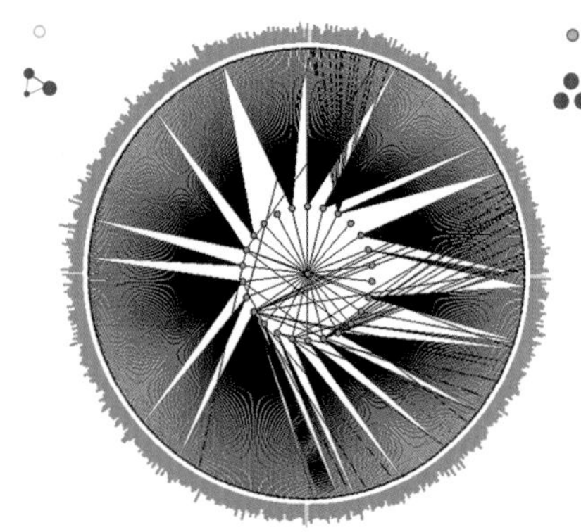

Coauthor Environment
Newman

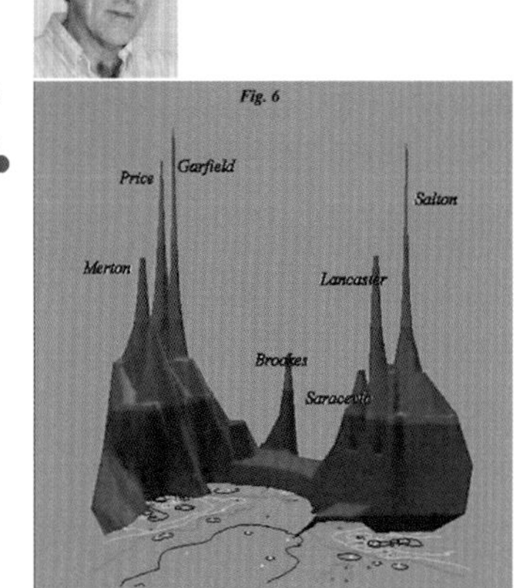

GIS Map of White and McCain
Old

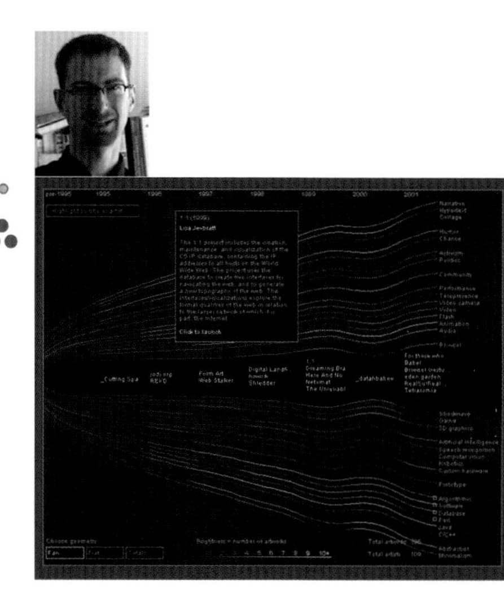

Idealine
Wattenberg

2000

2001

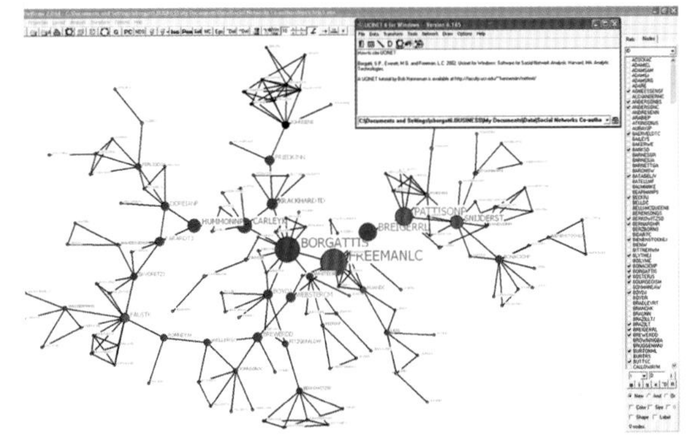

UCINet
Borgatti et al.

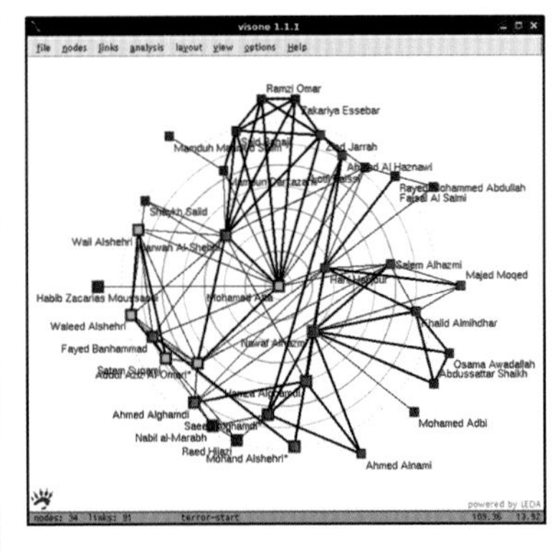

Visone Tool
Brandes and Wagner

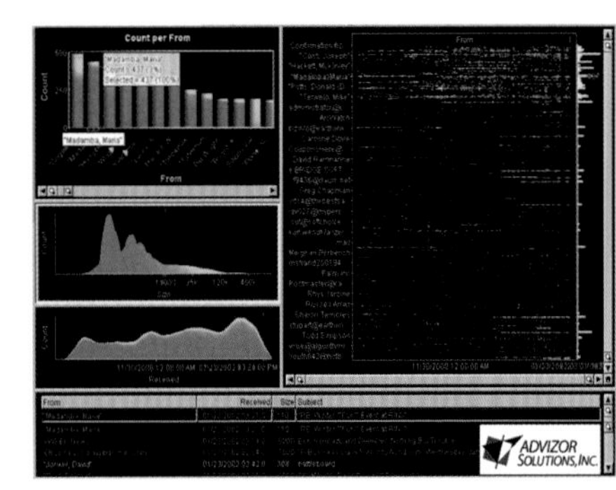

E-Mail Analysis
ADVISOR

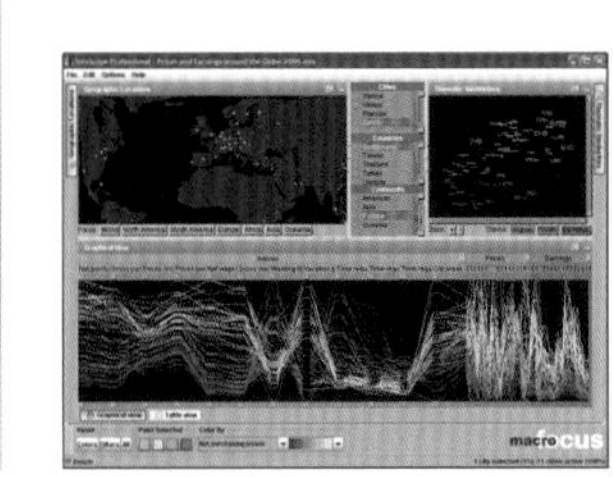

Infoscope
Macrofocus GmbH

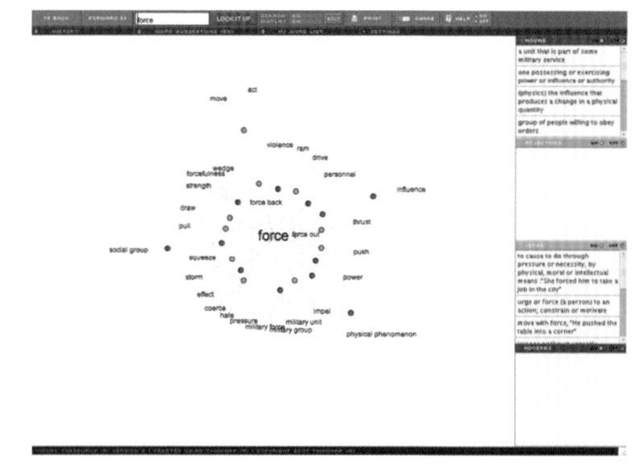

Visual Thesaurus
Thinkmap, Inc.

A Sociological Theory of Communications: The Self-Organization of the Knowledge-Based Society
Leydesdorff

2002

Burst Detection
Kleinberg

Topics Model
Griffith and Steyvers

Combine Link and Content Information
Richardson and Domingos

Community Detection
Girvan and Newman

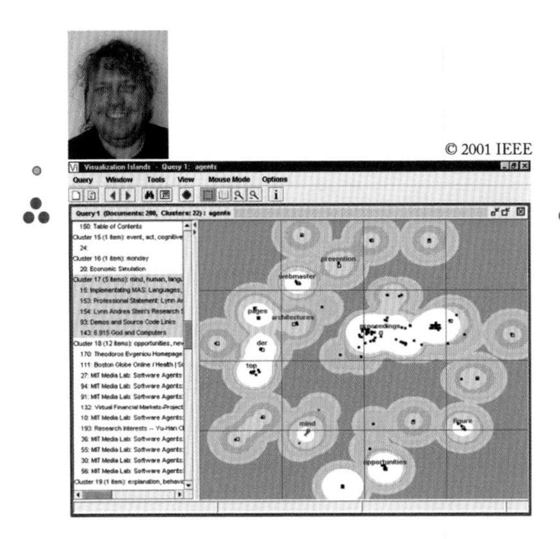

VisIslands Search
Andrews et al.

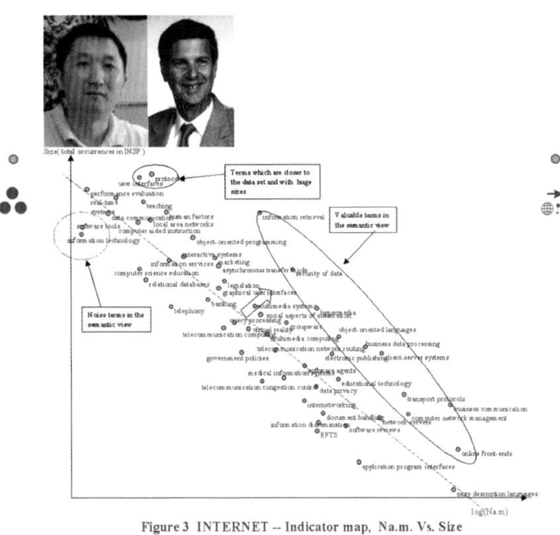

Figure 3 INTERNET -- Indicator map, Na.m. Vs. Size

Science and Technology Forecasting
Zhu and Porter

Illuminated Diagram Display
Paley

2002

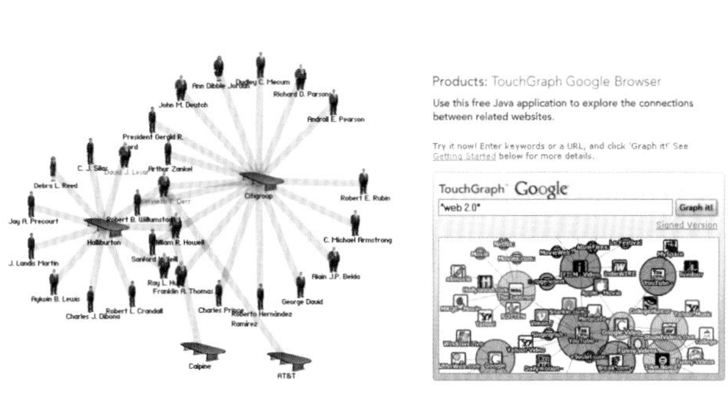

They Rule
Josh On

TouchGraph
Shapiro

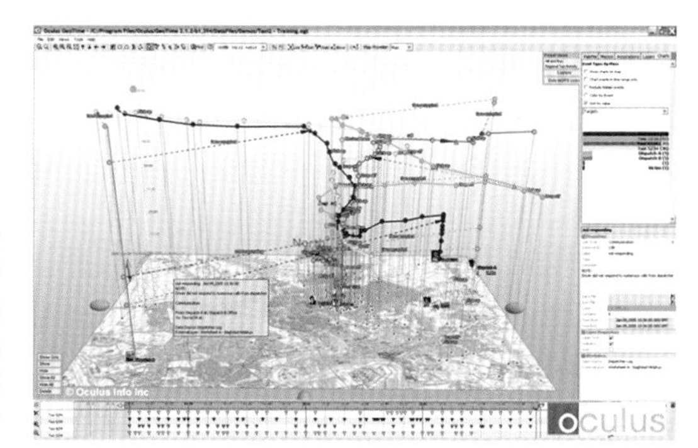

GeoTime
Kapler and Wright @ Oculus

Issue Crawler
Govcom.org Foundation

- Networks of Innovation: Change and Meaning in the Age of the Internet
  Tuomi
- Patents, Citations and Innovations: A Window on the Knowledge Economy
  Jaffe and Trajtenberg
- Visual Interfaces to Digital Libraries (Vol. LNCS 2539)
  Börner and C. Chen
- Mapping Scientific Frontiers
  C. Chen
- Linked: How Everything Is Connected to Everything Else and What It Means
  Barabási

2002

2003

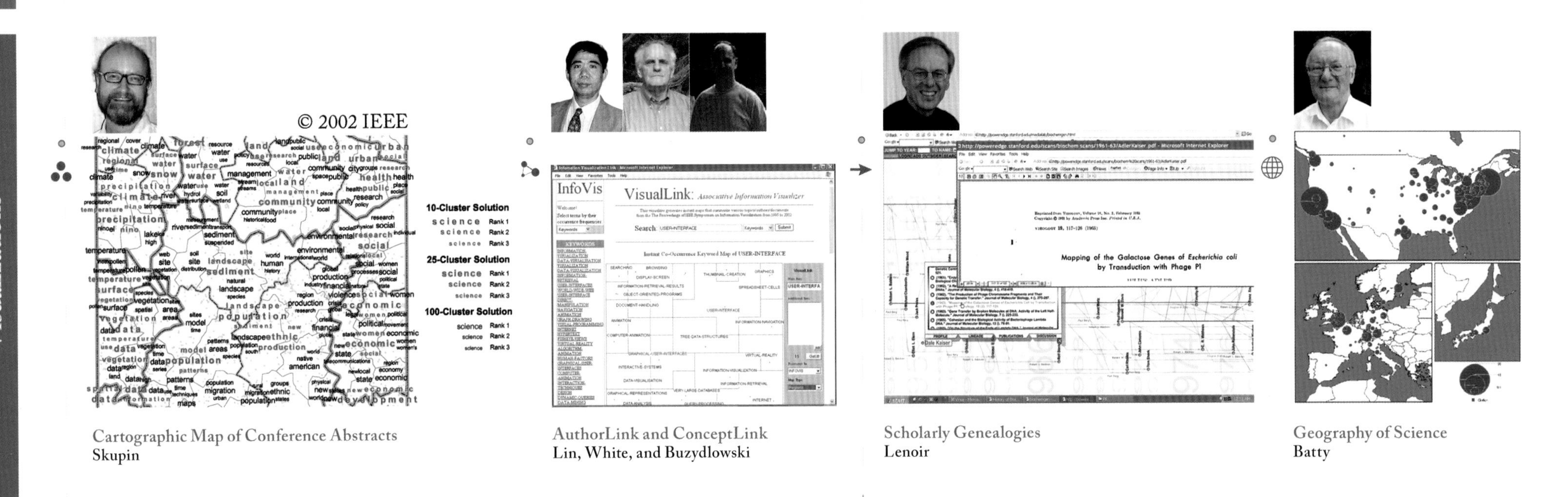

Cartographic Map of Conference Abstracts
Skupin

AuthorLink and ConceptLink
Lin, White, and Buzydlowski

Scholarly Genealogies
Lenoir

Geography of Science
Batty

2002

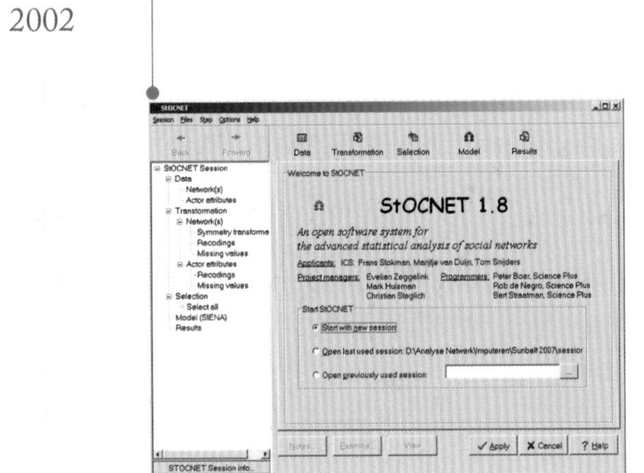

StOCNET
Huisman and Duijn

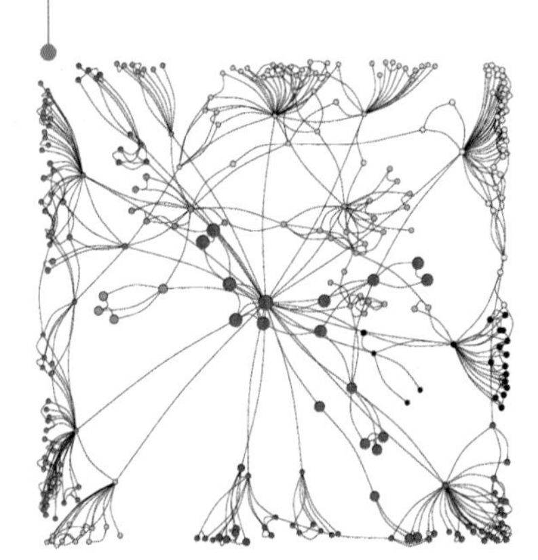

JUNG
O'Madadhain, Fisher, and White

2003

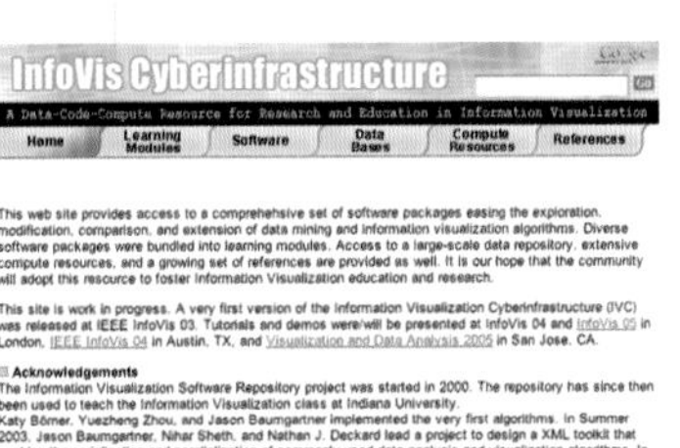

Information Visualization Cyberinfrastructure
Baumgartner et al.

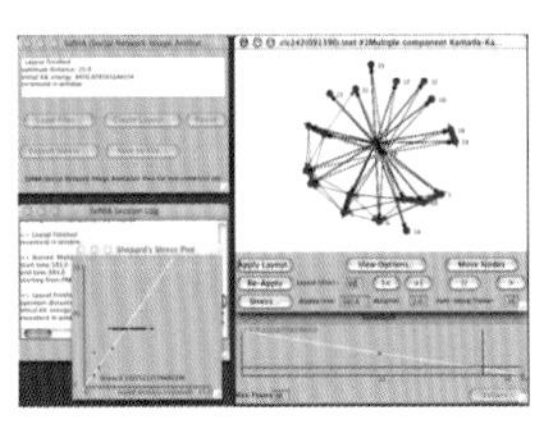
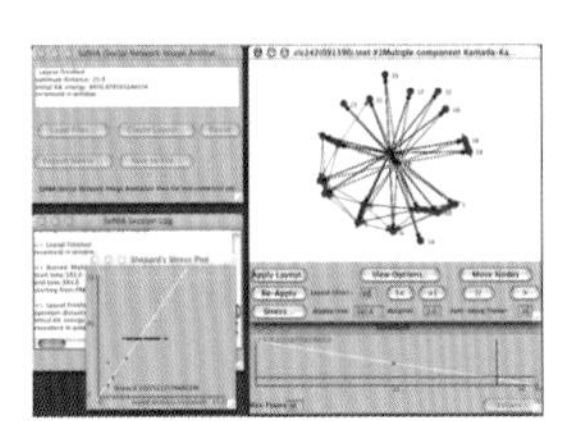

Sonia Social Network Image Animator
Bender-deMoll and McFarland

RefViz Overviews of Literature Search Results
Thomson ISI ResearchSoft

- Theories of Communication Networks
  Monge and Contractor
- Six Degrees
  Watts
- Putting Science in Its Place: Geographies of Scientific Knowledge
  Livingstone

2004

Weighted PageRank Algorithm
Wenpu and Ghorbani

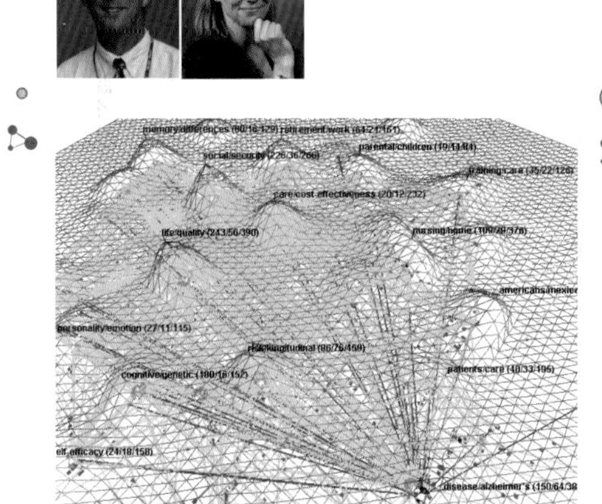

Linking Papers and Funding
Boyack and Börner

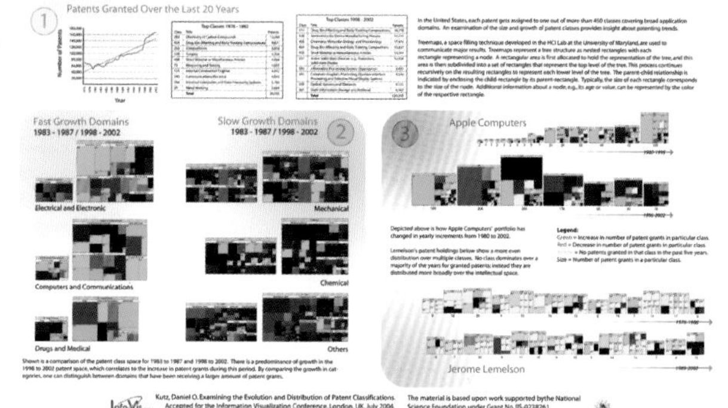

Evolution and Distribution of Patent Classifications
Kutz

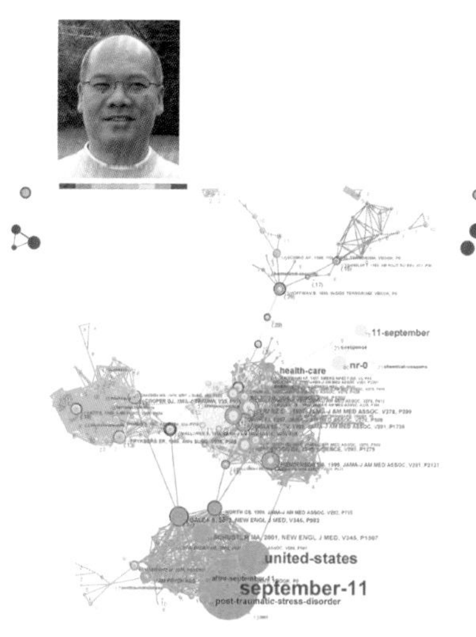

Mapping Scientific Frontiers
C. Chen

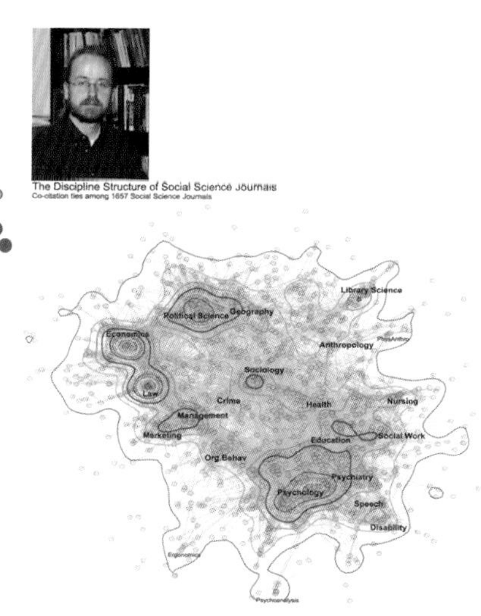

Contour Sociogram of Social Science Journals
Moody

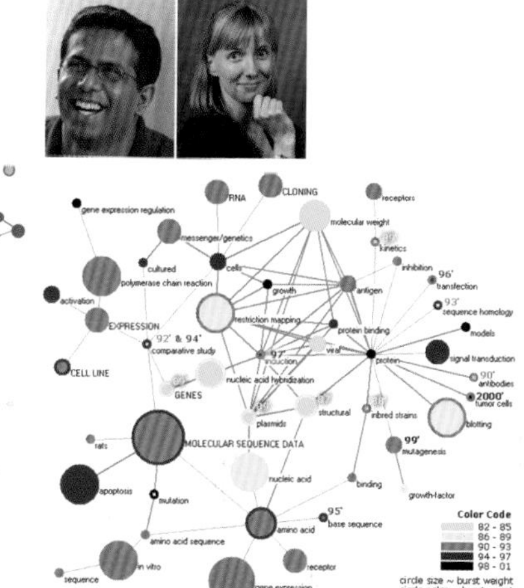

Mapping Topic Bursts
Mane and Börner

2004

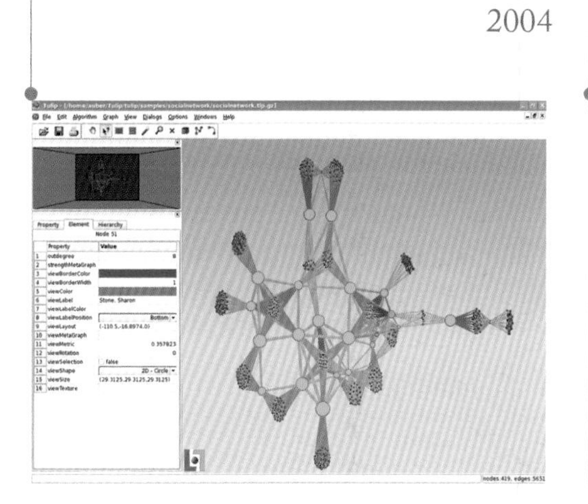

Tulip Software
Auber

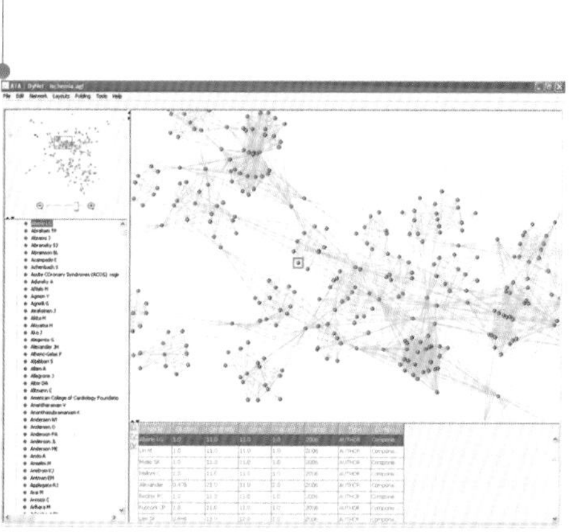

DyNet Dynamic Network Software Package
ATA

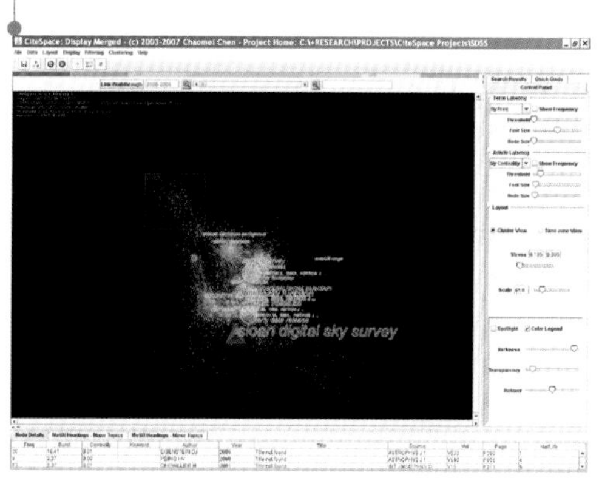

CiteSpace Trend Analysis
C. Chen

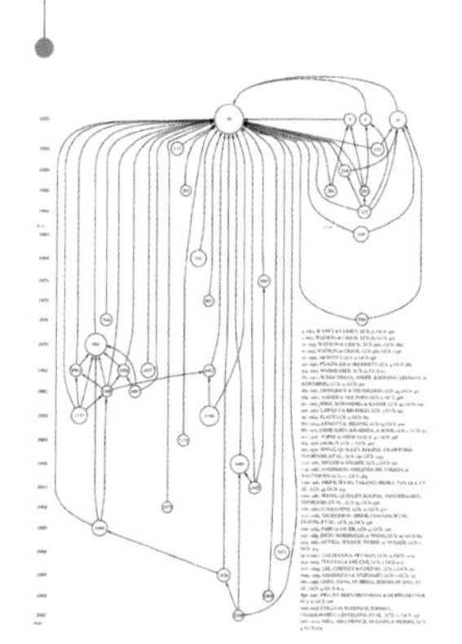

HistCite Historiographs
Garfield et al.

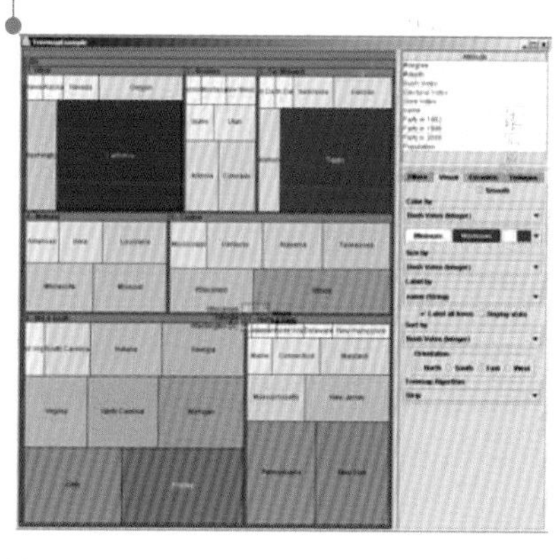

The Infovis Toolkit
Fekete

- Mapping Knowledge Domains
  Shiffrin and Börner (eds.)
- Handbook of Quantitative Science and Technology Research: The Use of Publication and Patent Statistics in Studies of S&T Systems
  Moed, Glanzel, and Schmoch
- The Development of Social Network Analysis: A Study in the Sociology of Science
  Freeman
- Innovation and Incentives
  Scotchmer

Algorithms

Visualizations

Tools

Books

2004

2005

Non-Euclidean Spring Embedders
Kobourov and Wampler

Flow Map Layout
Phan et al.

Acknowledgment Indexing
Councill et al.

Places & Spaces 1st Iteration **(see page 76)**

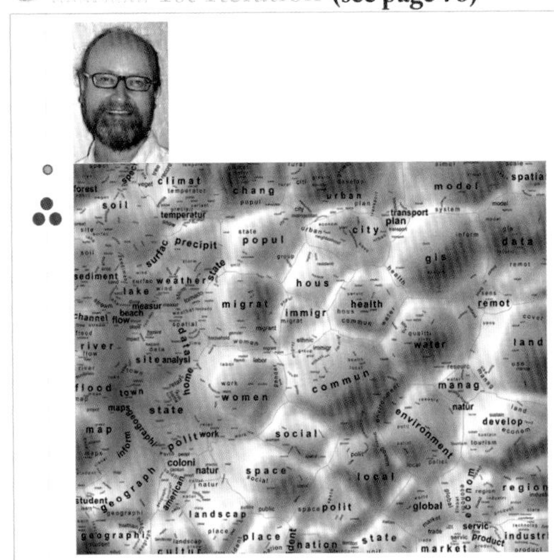

GIS Map of Geography
Skupin

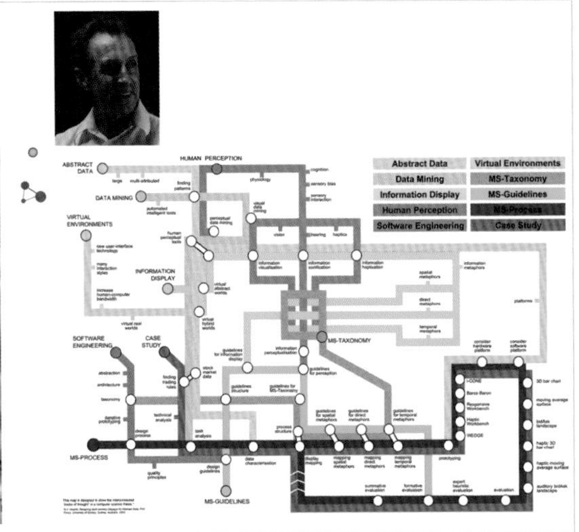

Subway Domain Map
Nesbitt

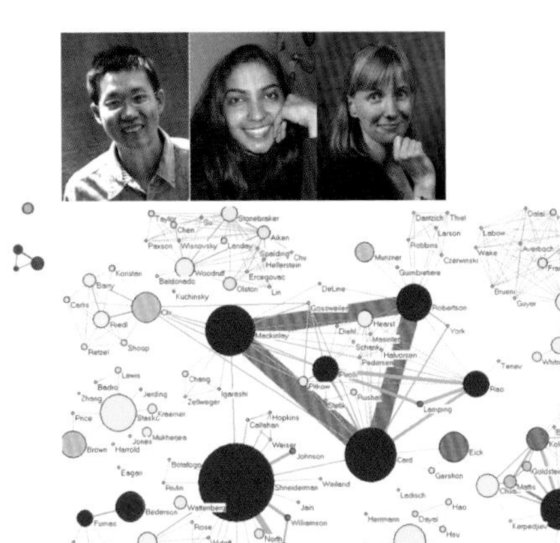

Evolving Coauthor Networks
Ke, Viswanath, and Börner

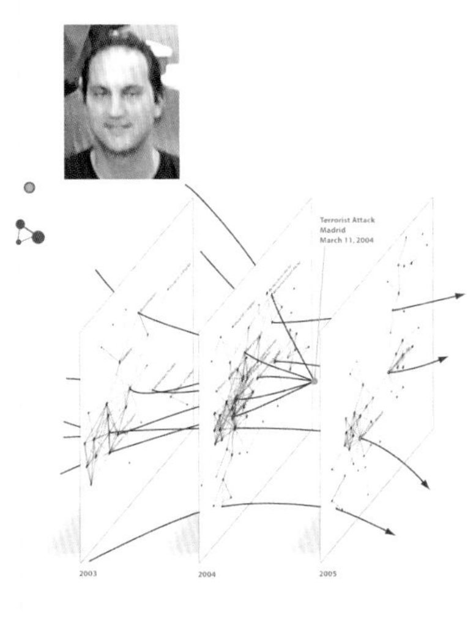

Critical Paths and Trajectories of Individuals
Bender-deMoll, Nottoli, and Rodriguez

2004

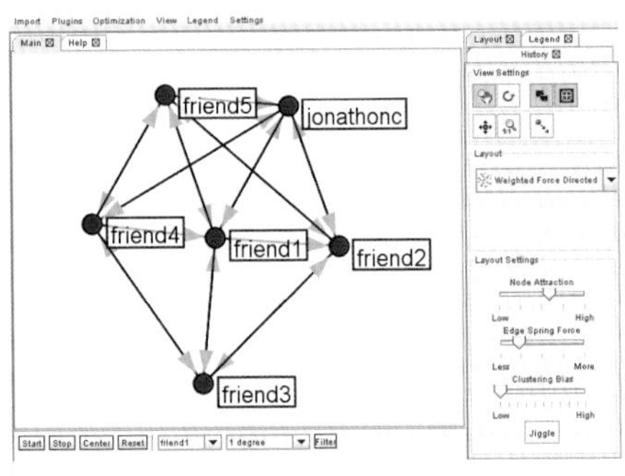

NetVis Dynamic Visualization of Social Networks
Cummings

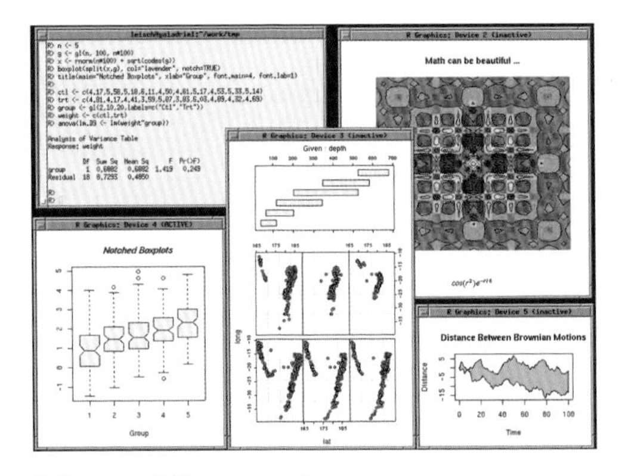

R Statistical Computing Language
R Foundation

2005

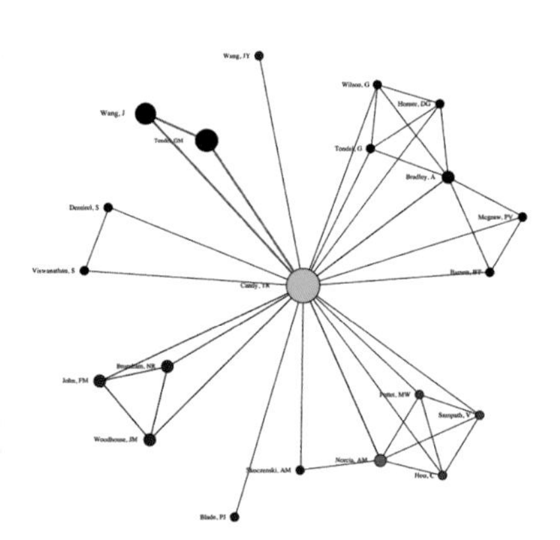

Prefuse Visualization API
Heer, Card, and Landay

Models and Methods in Social Network Analysis
Carrington and Wasserman (eds.)

The Hand of Science: Academic Writing and Its Rewards
Cronin

Measurement and Statistics on Science and Technology: 1920 to the Present (Routledge Studies in the History of Science, Technology, and Medicine).
Godin

1st Iteration (see page 76)

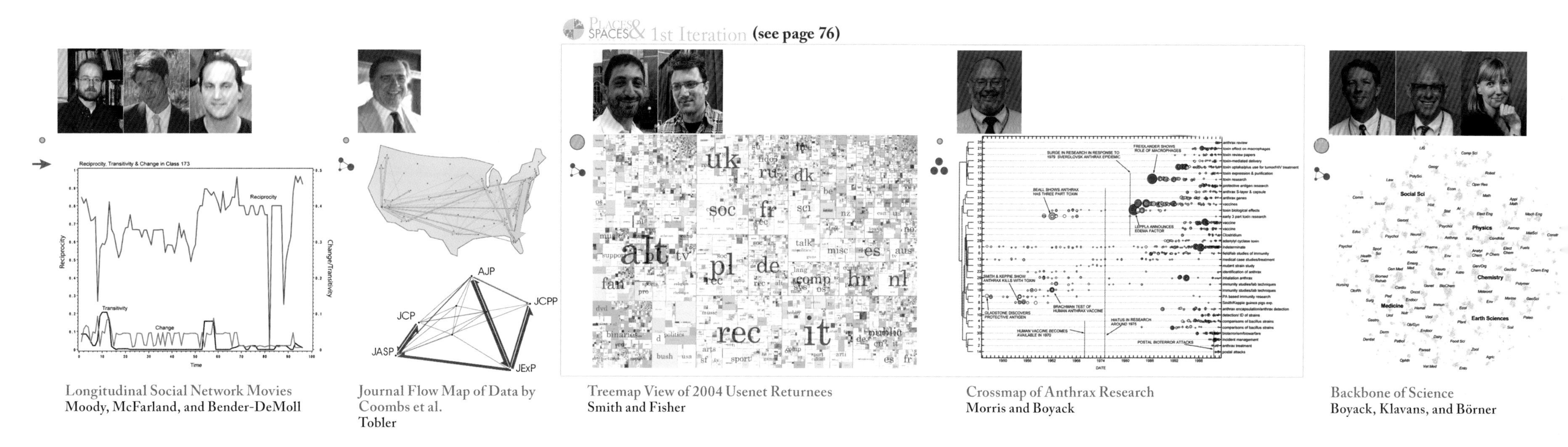

Longitudinal Social Network Movies
Moody, McFarland, and Bender-DeMoll

Journal Flow Map of Data by Coombs et al.
Tobler

Treemap View of 2004 Usenet Returnees
Smith and Fisher

Crossmap of Anthrax Research
Morris and Boyack

Backbone of Science
Boyack, Klavans, and Börner

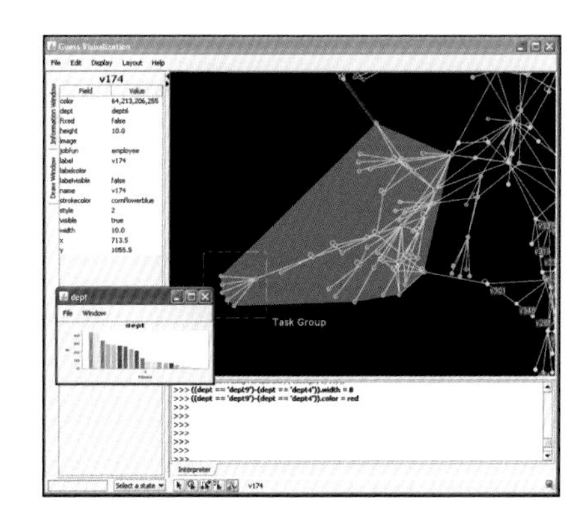

GUESS Graph Exploration System
Adar

Author-Name-Disambiguation
Author-ity
Torvik et al.

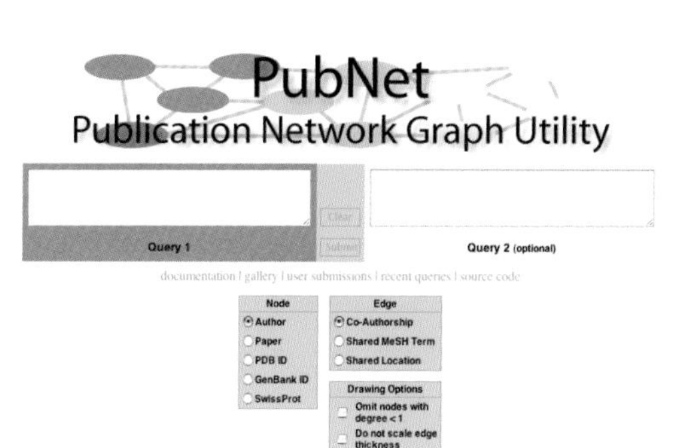

PubNet
Douglas et al.

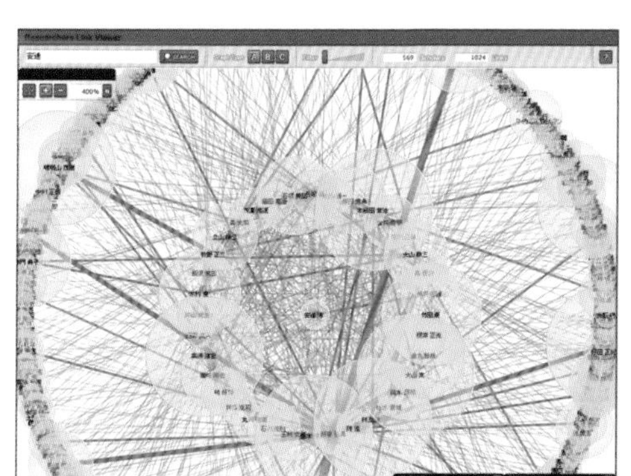

CiNii Researchers Link Viewer
Ichise et al.

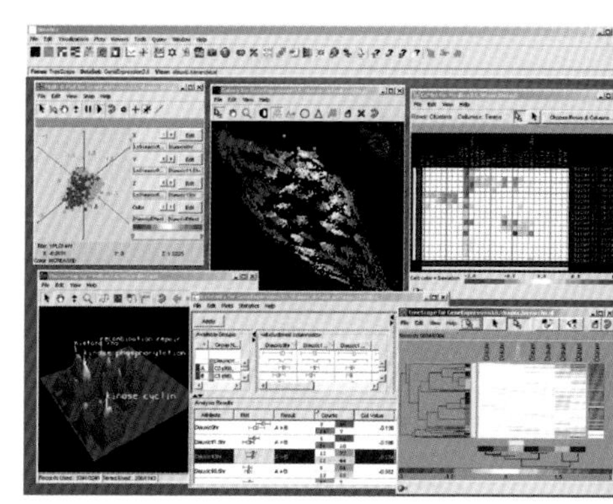

OmniViz
BioWisdom

Algorithms

Visualizations

Tools

Books

2006

Places & Spaces 1st Iteration **(see page 76)**

Places & Spaces 2nd Iteration **(see page 110)**

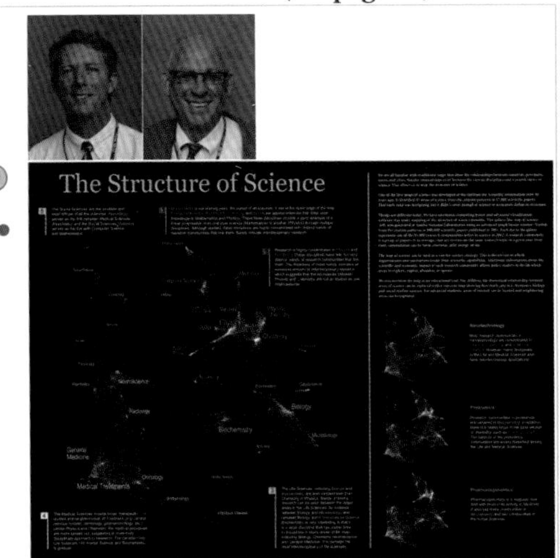

2002 Structure Map
Boyack and Klavans

HistCite Visualization of DNA Development
Garfield et al.

History Flow
Viegas and Wattenberg

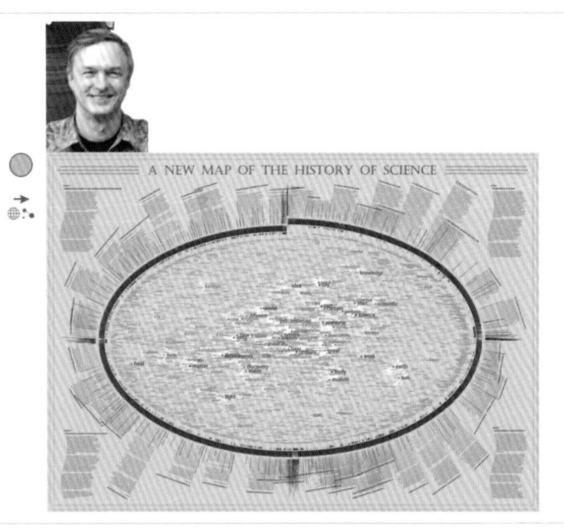

TextArc Visualization of "The History of Science"
Paley

2006

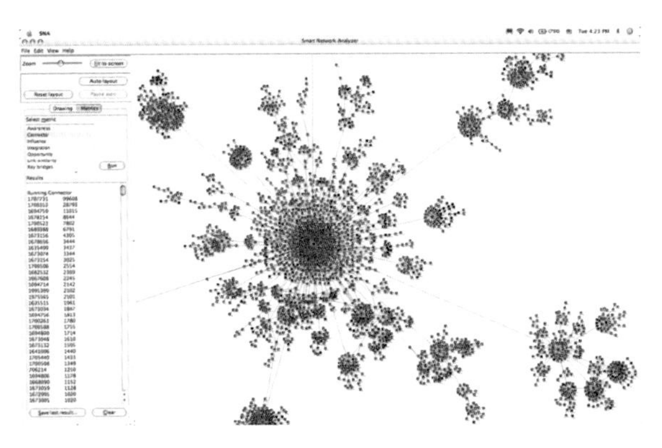

Smart Network Analyzer
Krebs

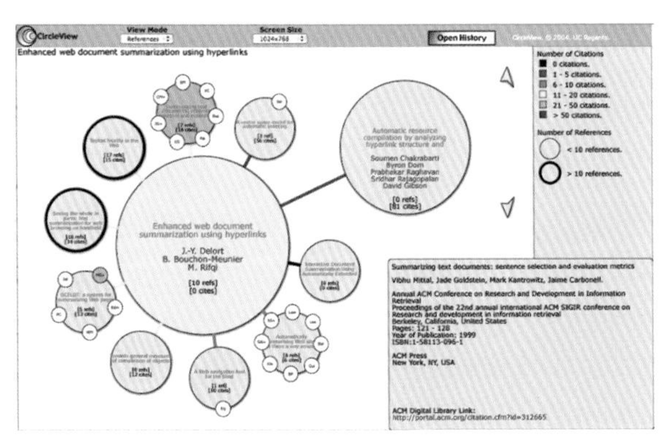

CircleView Document Visualization
Bergström and Whitehead

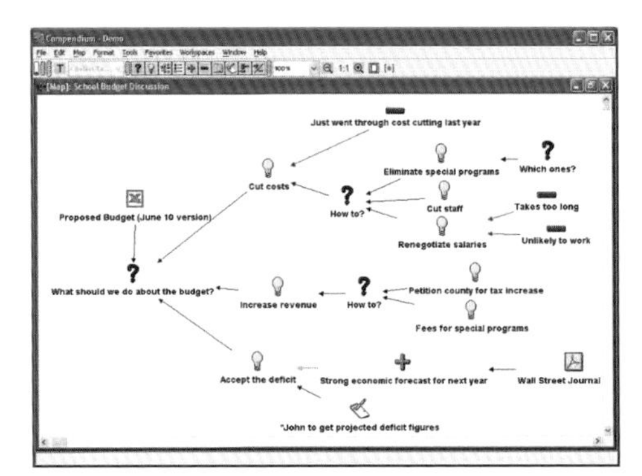

Compendium Dialog Mapping
Conklin

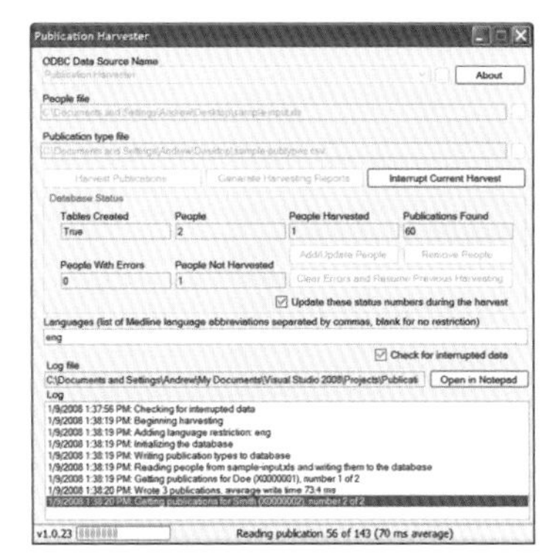

PublicationHarvester
Azoulay, Stellman, and Zivin

The Knowledge-Based Economy: Modeled, Measured, Simulated
Leydesdorff

Evaluations of Individual Scientists and Research Institutions. Part I and II. A Selection of Papers Reprinted from the Journal *Scientometrics*
Braun (ed.)

Places & Spaces 3rd Iteration (see page 144)

Taxonomy Visualization
Börner et al.

Map of Scientific Paradigms
Boyack and Klavans

Zones of Invention—Patterns of Patents
Günther

Science and Technology Outlook: 2005-2055
Soojung-Kim Pang et al.

2006

Places & Spaces Additional Elements (see page 178)

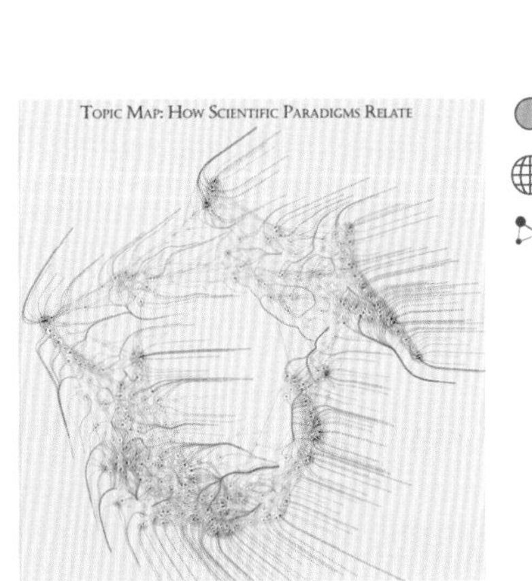

Illuminated Diagram Display of Science
Boyack at el.

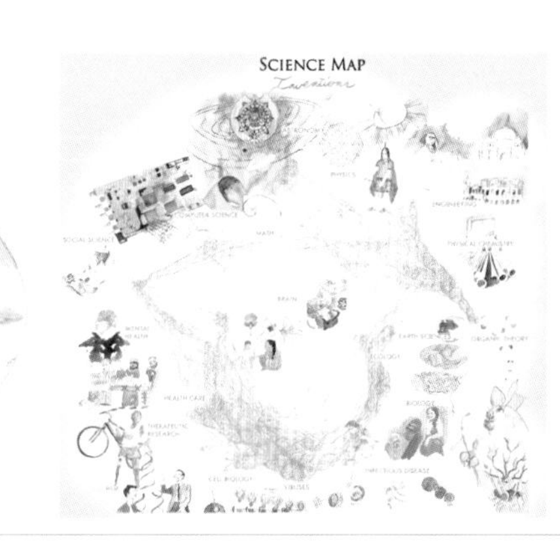

Hands-On Science Maps for Kids
Palmer et al.

2006

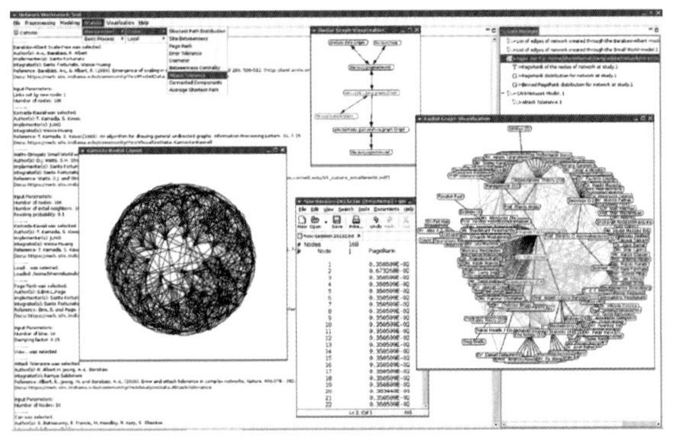

Network Workbench
Huang et al.

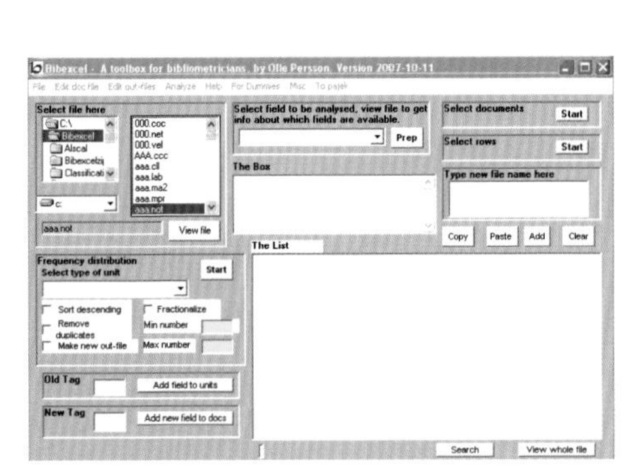

BibExcel Bibliographic Data Analyzer
Persson

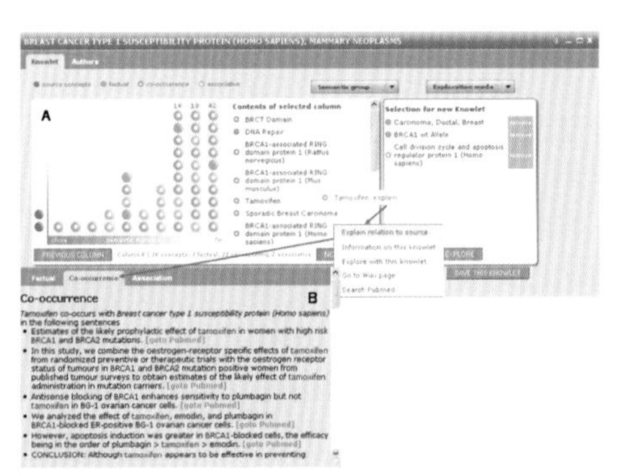

Knowlet Spaces for Collective Annotation
Mons et al.

WorldProcessor Globe of Patents
Günther

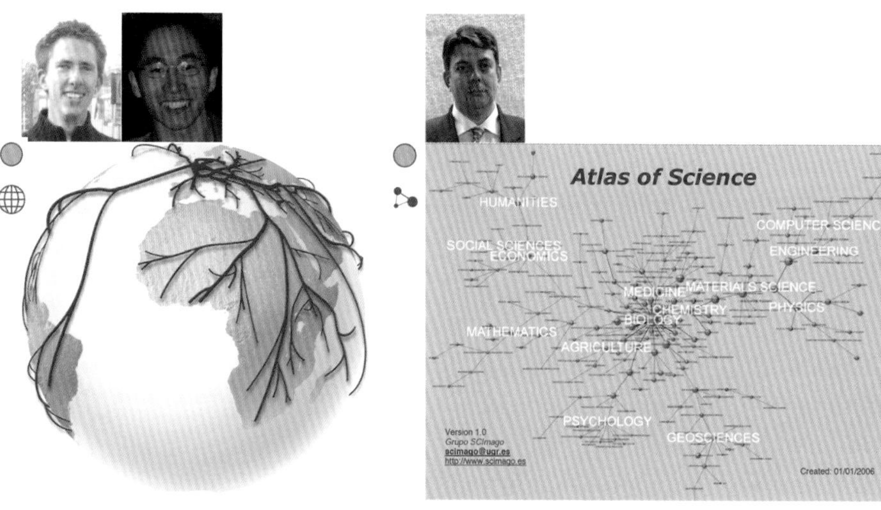

UK's Global Ecological Footprint
Moran and Phan

Atlas of Science
Moya-Anegón et al.

2006-2007

Places & Spaces 3rd Iteration (see page 144)

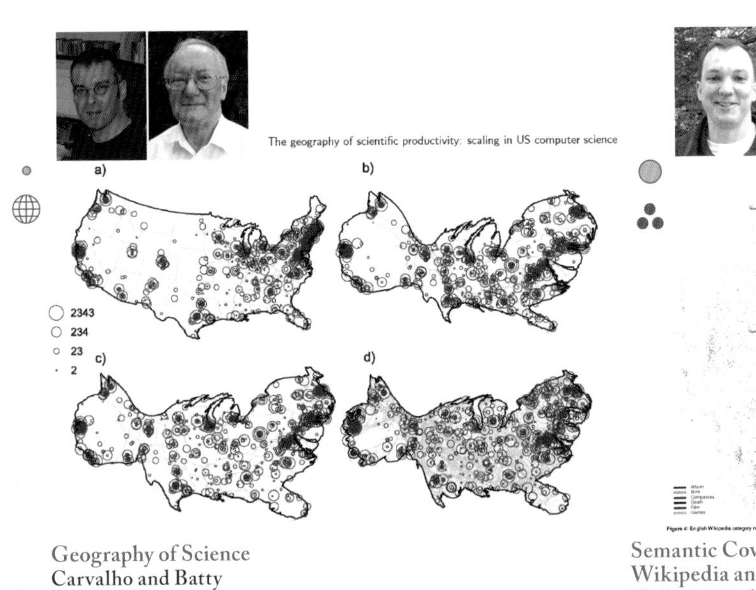

Geography of Science
Carvalho and Batty

Semantic Coverage of Wikipedia and its Authors
Holloway et al.

113 Years of Physical Review
Herr et al.

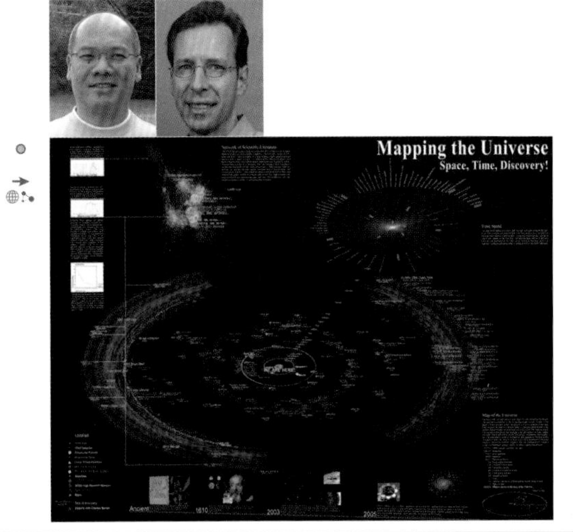

Mapping the Universe: Space, Time, and Discoveries!
Chen et al.

2006

2007

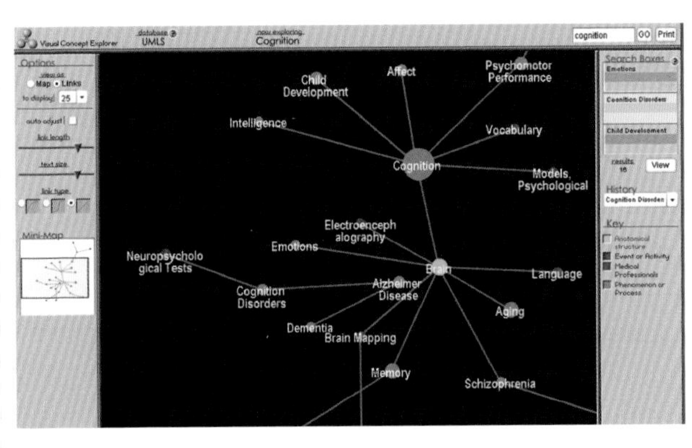

Visual Concept Explorer
Lin

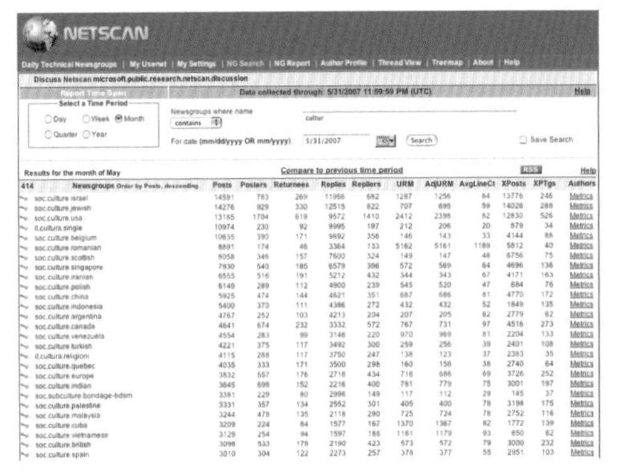

Netscan Usenet Visualization
Smith and Fisher @ Microsoft

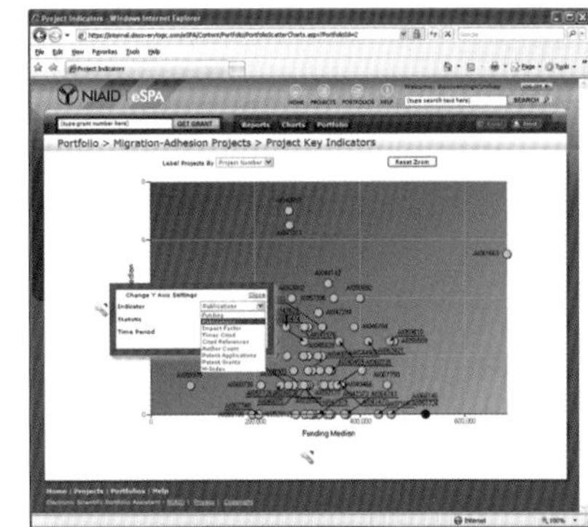

Electronic Scientific Portfolio Assistant: Project Key Indicators
DiscoveryLogic

Visualizing the Structure of Science
Quesada et al.

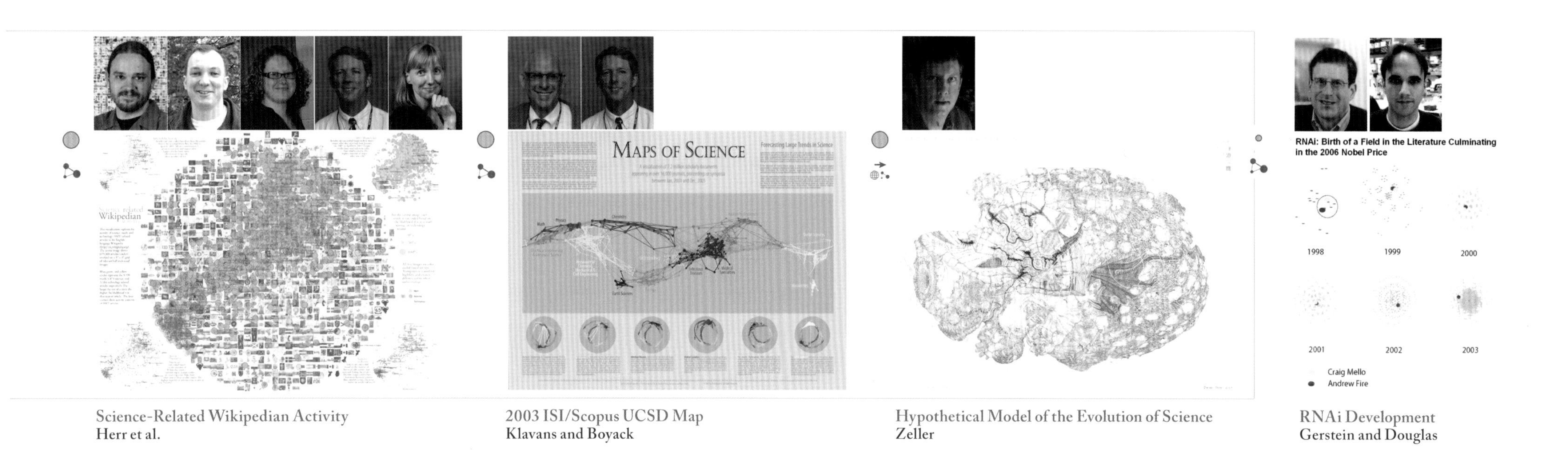

Science-Related Wikipedian Activity
Herr et al.

2003 ISI/Scopus UCSD Map
Klavans and Boyack

Hypothetical Model of the Evolution of Science
Zeller

RNAi Development
Gerstein and Douglas

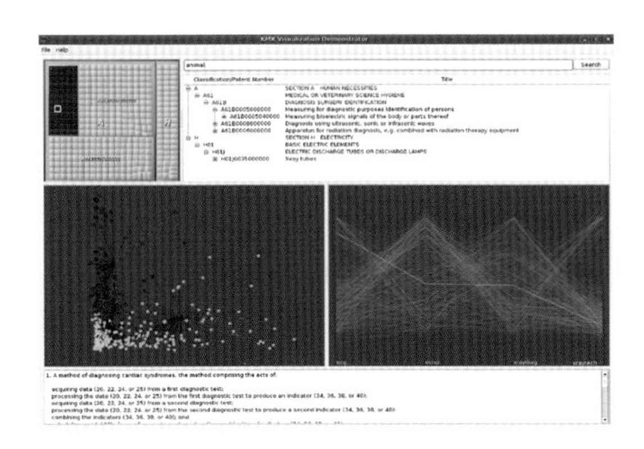

Treparel Patent Visualization
Heijs

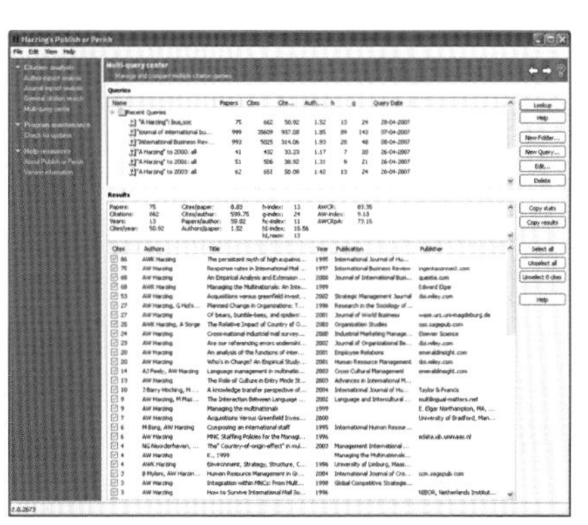

Publish or Perish Academic Citations Analyzer
Harzing.com

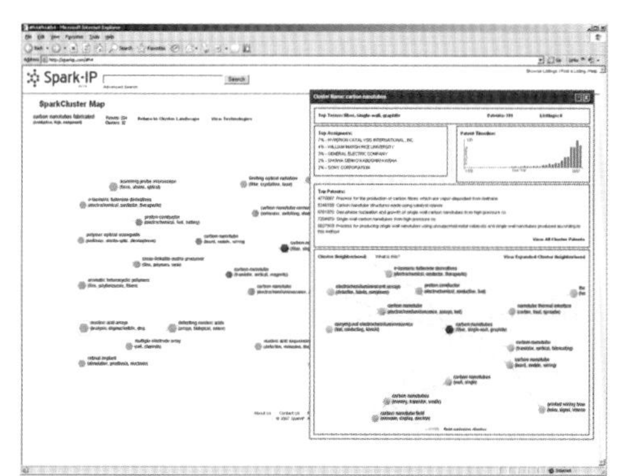

SparkCluster Map
SparkIP

SciTrends
Cokol

2007

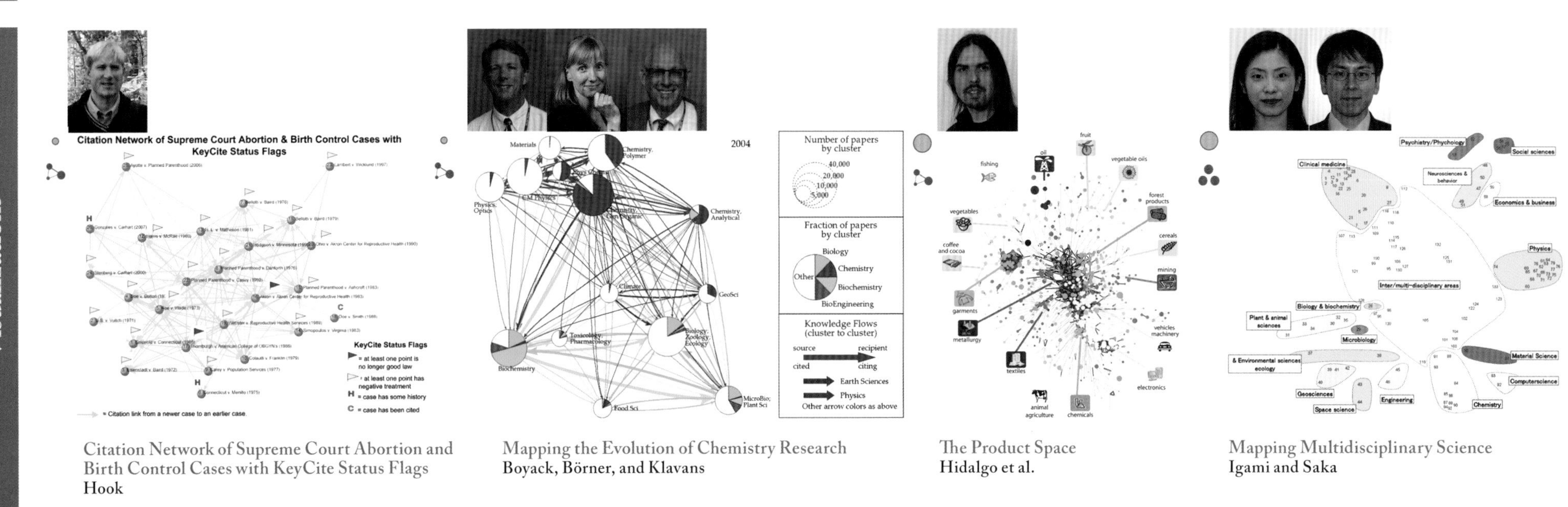

Citation Network of Supreme Court Abortion and Birth Control Cases with KeyCite Status Flags
Hook

Mapping the Evolution of Chemistry Research
Boyack, Börner, and Klavans

The Product Space
Hidalgo et al.

Mapping Multidisciplinary Science
Igami and Saka

2007

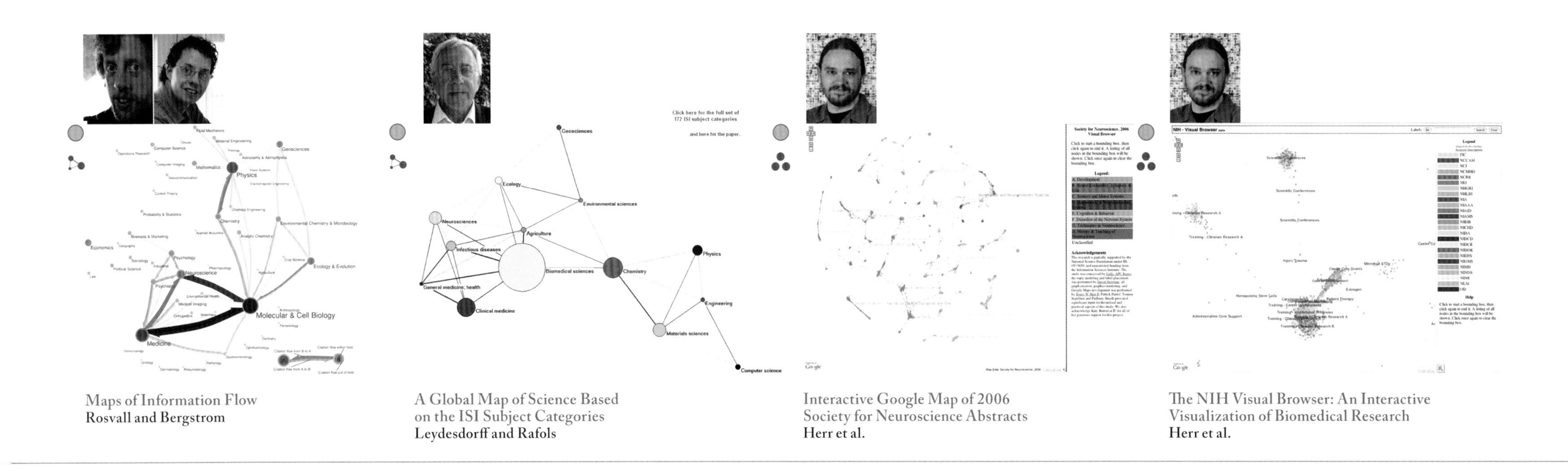

Maps of Information Flow
Rosvall and Bergstrom

A Global Map of Science Based on the ISI Subject Categories
Leydesdorff and Rafols

Interactive Google Map of 2006 Society for Neuroscience Abstracts
Herr et al.

The NIH Visual Browser: An Interactive Visualization of Biomedical Research
Herr et al.

Discussion

Early maps were generated by hand. With the exception of works by Harold Johann Thomas Ellingham, Eugene Garfield, and Francis Narin, maps were not color-coded. The first computer-generated maps appeared in the 1970s, the first interactive maps in the 1980s. Maps at that time remained two-dimensional.

In 1994, the Pacific Northwest National Laboratory developed IN-SPIRE, one of the first systems to map data in a three-dimensional space. Sandia National Laboratories followed suit with VxInsight in 1998. In 1987, Xerox PARC developed NoteCards, one of the first interactive data spaces, and in 1995 they pioneered the first three-dimensional user interface. Also in 1995, Hsinchun Chen generated one of the first multilevel data maps. In 1999, Chaomei Chen studied the utility of multiuser, collaborative, three-dimensional data spaces for information access.

In 1995, Teuvo Kohonen developed an algorithmic means to analyze and map more than a million records. In 1998, Martin Wattenberg's **Map of the Market** was one of the first real-time visualizations of a data stream. By 2000, many tools and approaches were being developed for the interactive exploration of search results.

In 1999, Henry G. Small published the first data-driven, algorithmically generated map of all of science; it remained the only such map until 2005. Between 2005 and 2007, eight different maps of all of science were generated by five different international teams in Japan, the Netherlands, Spain, and two U.S. locations.

New work is built on existing work. Each of the examples below cites a series of works featured in the timeline that developed in a progressive fashion:

- Eugene Garfield's original **Historiograph of DNA Research** (1962); his long-term development of **HistCite** (first published in 2004); and his exhibit map (2006), which incorporates a rerendering of the 1962 historiography and the application of HistCite.
- Howard White et al.'s **Maps of Cocited Authors** (1982), **Map of Information Science** (1998), and the interactive **AuthorLink** (2002).
- Waldo Tobler's early works on the visualization of flow, his **Flow Mapper** tool (1987), and the tool's application in geographic and network journal data (2005).
- Ben Shneiderman's **Treemap Layout** (1992), its utilization in Hsinchun Chen's **SOM ET Map** (1995) and **Dewey Treemap** (1997), and later Wattenberg's **Map of the Market** (1998) and Marc A. Smith et al.'s **Treemap View of 2004 Usenet Returnees** (2005).
- White and Kate W. McCain's **Map of Information Science** (1998) and John Old's GIS rendering of same (2001).
- Chaomei Chen's **Collaborative StarWalker** information spaces (1999), **Multilayer Science Maps** (2001), **Mapping Scientific Frontiers** (2004), and **Mapping the Universe** (2007) and his continuous development of the **CiteSpace Trend Analysis** tool (2004).
- Michael Batty et al.'s work on the **Geography of Science** (2003 and 2006).
- James Moody et al.'s **Contour Sociogram of Social Science Journals** (2004) and **Longitudinal Social Network Movies** (2005).
- Kevin Boyack and Richard Klavans's work toward a base map of science, followed by the creation of a series of maps (2005–2007).

Over time, former tools are subsumed by new tools, software APIs, and libraries. Examples include the **Information Visualization Cyberinfrastructure** (2003), Jean-Daniel Fekete's **The Infovis Toolkit** (2004), and the **Network Workbench** (2006). Mashups also emerge, such as Bruce W. Herr II et al.'s **Interactive Google Map of 2006 Society for Neuroscience Abstracts** and **The NIH Visual Browser: An Interactive Visualization of Biomedical Research.**

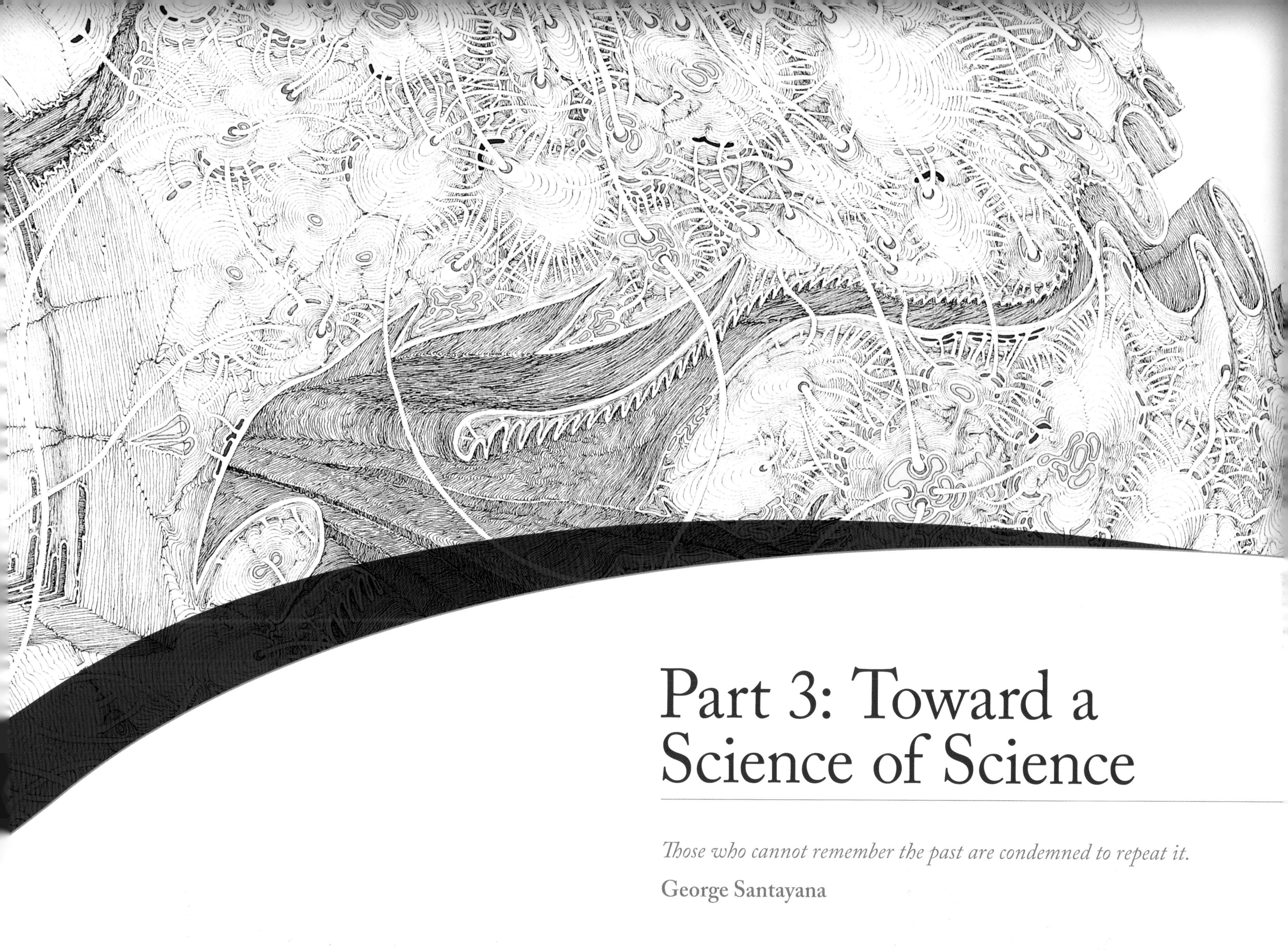

Part 3: Toward a Science of Science

Those who cannot remember the past are condemned to repeat it.

George Santayana

Building Blocks

A science of science has to be based on existing theories and practical knowledge of the basic and aggregate units of science and their interrelations and dynamics. It requires a deep care for the needs of diverse stakeholders, detailed knowledge about the quality and coverage of existing data, and expertise on the capabilities and limitations of different approaches, algorithms, tools, and computing infrastructures. Results have to be communicated, taking the capabilities and limitations of human visual perception and cognitive processing into account, and they need to be validated and interpreted in close collaboration with domain experts. Outcomes should be used to inform theory building, to fine-tune system design, and to communicate science to a broad audience.

[Maps,] like speeches and paintings, are authored collections of information and also are subject to distortions arising from ignorance, greed, ideological blindness, or malice.
Mark Monmonier

Overview

The major building blocks of a science of science, including the ways they connect and interact, are shown on the opposite page. A typical science study starts with a Needs Analysis of a selected stakeholder group. The selection of a specific approach toward Conceptualizing Science then shapes the subsequent workflow design, involving Data Acquisition and Preprocessing; Data Analysis, Modeling, and Layout; and Data Communication—Visualization Layers. All data sets, algorithms, and parameter values used in a study have to be documented in detail, so that the study can be replicated and interpreted. After that, Validation and Interpretation should proceed in collaboration with domain experts and stakeholders. Insights gained might generate additional needs, inspire changes to the workflow, or influence the conceptualization of science itself. This process is highly incremental, often demanding many cycles of revision and refinement to ensure the best data sets are used, optimal algorithm parameters are applied, and the clearest insight is achieved.

Needs Analysis

The goal of a needs analysis is to identify the needs and desires of stakeholders (see **page 8, Knowledge Needs and Desires**). A needs analysis results in an understanding of the stakeholder group (age, gender, profession, and technical and subject expertise); its current use of metaphors and tools; its information needs, priorities, and preferences (in context and scenarios of use); the type, format, and sequence of information needed; and the physical and social work context. Numerous theoretical frameworks (as well as introductory and advanced publications, including textbooks) exist to guide a formal needs analysis (**see References and Credits, pages 212-246**. Note that the stakeholder could also be a researcher with a deep interest in testing a scientific hypothesis on the structure and dynamics of science.

Conceptualizing Science

There exists a rich body of literature relevant to science of science studies and applications (see **pages 26-47, Milestones in Mapping Science**). Yet, the theories and results are scattered across thousands of papers published in different disciplines, including (but not limited to) the sociology of science and psychology of the scientist; the economics of science; scientometrics, informetrics, webometrics, and history of science; operational research on science; the diffusion of scientific information; physics models of science; and the planning of science.

A true science of science must capture the current theory of the inner workings of science and interlink it with practical needs and tool developments. The following four spreads are devoted to the conceptualization of science, including a discussion of the basic anatomy of science (**page 52, Conceptualizing Science: Basic Anatomy of Science**); an identification and characterization of the basic units, aggregate units, and linkages (**page 54, Conceptualizing Science: Basic Units, Aggregate Units, and Linkages**); a review of basic properties, indexes, and laws (**page 56, Conceptualizing Science: Basic Properties, Indexes, and Laws**); and last, but not least, an overview of models of science that aim to increase our understanding of science dynamics (**page 58, Conceptualizing Science: Science Dynamics**).

Data Acquisition and Preprocessing

About 80 percent of a typical project's total effort is spent on data acquisition and preprocessing; well-prepared data is mandatory for high-quality results. Data sets might be acquired via questionnaires, crawled from the Web, downloaded from a database, or accessed as a continuous data stream. Data sets differ by their coverage and resolution of time (days, months, years), geography (languages or countries considered), and topics (disciplines and selected journal sets). Their size ranges from several bytes to petabytes of data. They might contain high-quality materials curated by domain experts or random content retrieved from the Web. Based on a detailed needs analysis and deep knowledge about existing databases, the best suited yet affordable data sets have to be selected, filtered, integrated, and augmented. It may also be necessary to extract networks or dynamic features (see **page 60, Data Acquisition and Preprocessing**).

Data Analysis, Modeling, and Layout

Data analysis ranges from simple rankings based on original attribute values (such as number of times cited) to the application of sophisticated and computationally demanding algorithms. There are algorithms that analyze temporal patterns (age, growth, latency to peak, decay rate, and burst), geographic features (areas, surfaces, density, and boundaries), semantic properties (dimensionality reduction and clustering), and network features (density, degree centrality, and reachability). Data models are grouped into two major types: descriptive models and process models. *Descriptive models* aim to illustrate the major features of a (typically static) data set, such as statistical patterns of article citation counts, networks of citations, individual differences in citation practice, the composition of knowledge domains, or the identification of research fronts as indicated by new yet highly cited papers. *Process models,* or predictive models, aim to simulate, statistically describe, or formally reproduce statistical and dynamic characteristics of interest.

Data layout refers to the spatial layout of a data set, for example, the overlay of a data set on a geographic map or the layout of a network. The selection of algorithms has to be driven by the insight needs and their alignment with a conceptualization of science as well as the properties of the data, programming resources, and computing resources available (see **page 62, Data Analysis, Modeling, and Layout** for details). Frequently, many different algorithms have to be applied to make sense of a data set. The selection and combination of algorithms with specific input and output data formats and parameter values is also called workflow design.

Data Communication—Visualization Layers

The results of an analysis are communicated via visuals such as a graph or maps. Any visual depiction of data can be decomposed into multiple Visualization Layers (see figure). Each layer has a distinct function, yet changes in one layer tend to affect other layers. The **Reference System** organizes the space. **Projections/Distortions** of the reference system help emphasize certain areas or provide focus and context. Placing the **Raw Data** in a reference system reveals spatial patterns. **Graphic Design** refers to the visual encoding of data attributes using qualities such as size, color, and shape coding of nodes, linkages, or surface areas. Frequently, **Aggregation/Clustering** techniques are applied to identify data entities with common attribute values or dense connectivity patterns. Sometimes it is beneficial to show multiple simultaneous views of the data, here referred to as **Combination**. And in many cases, it is desirable to have **Interaction** with the data, that is, to zoom, pan, filter, search, and request details on demand. Selecting a data entity in one view might highlight this entity in other views. The **Legend Design** delivers guidance on the purpose, generation, and visual encoding of the data. Mapmakers should proudly sign their visualizations, adding credibility as well as contact information. **Deployment** of results is enabled through paper printouts, online animations, or interactive, three-dimensional, audiovisual environments. The different layers communicate with one other via a **Shared Data Model**. Details and examples are given in **Data Communication—Visualization Layers (page 64)** and **Exemplification (page 66)**.

Validation and Interpretation

Study results have to be validated by means such as testing the impact of different parameter values; comparing different (combinations of) analysis, modeling, and layout algorithms; determining the robustness of results when introducing noise to the data; and judging results against additional data. Result interpretation requires input from domain experts and stakeholders. It often includes the identification of patterns, trends, and outliers and their relationships to external events, such as changes in funding. Validation and interpretation not only provide critical input to the revision and fine-tuning of the workflow design, but they also inspire follow-up studies.

Scholarly Marketplaces

A large percentage of science studies use proprietary tools to analyze proprietary data sets. This makes it difficult, if not impossible, to replicate and confirm results, compare alternative approaches, or run the same algorithms on a new data set. Ideally, a Scholarly Marketplace, or **Data-Algorithm-Compute-Cloud**, would exist that supports the easy upload, download, annotation, and rating of data sets and algorithms but also the trading of storage and computing resources.

Data sets might be small, static, and stored in text files. Others might be served from multiple Web sites; they might reside in a database or be streamed in real time from one or multiple sources.

Algorithms include parsers, file loaders, converters, and filters, as well as algorithms used to analyze, model, and visualize data sets. They may take the form of algorithm plugins, Web services, or standalone tools. They may be proprietary, or they may be free; they may be open source or not.

Storage and computing resources might come as second-hand data disks, computers, and servers but also as cloud services.

The online interface to the envisioned marketplace might resemble Craigslist or other classifieds sites in that it provides an easy means to advertise or acquire resources and services. Instead of buying or selling clothes or cars, looking for housing, or finding local services, however, scholars would use the marketplace to advertise and exchange scientific data, algorithms, services, and resources. Some transactions might involve exchange of money, access to datasets and algorithms might require citations to the paper in which the dataset was first detailed or the algorithm was first published. Other offers might result in close collaborations of data providers and algorithms developers and those that are interested to use them in their research.

The most useful resources in terms of reliability, documentation, compatibility, and insight are likely to attract high demand as well as download and citation counts. Hence, the marketplace becomes a means to share and access the best data sets and algorithms in existence together with storage and computational resources needed for comprehensive study. Components of the envisioned marketplace are discussed on **page 68, Scholarly Marketplaces**.

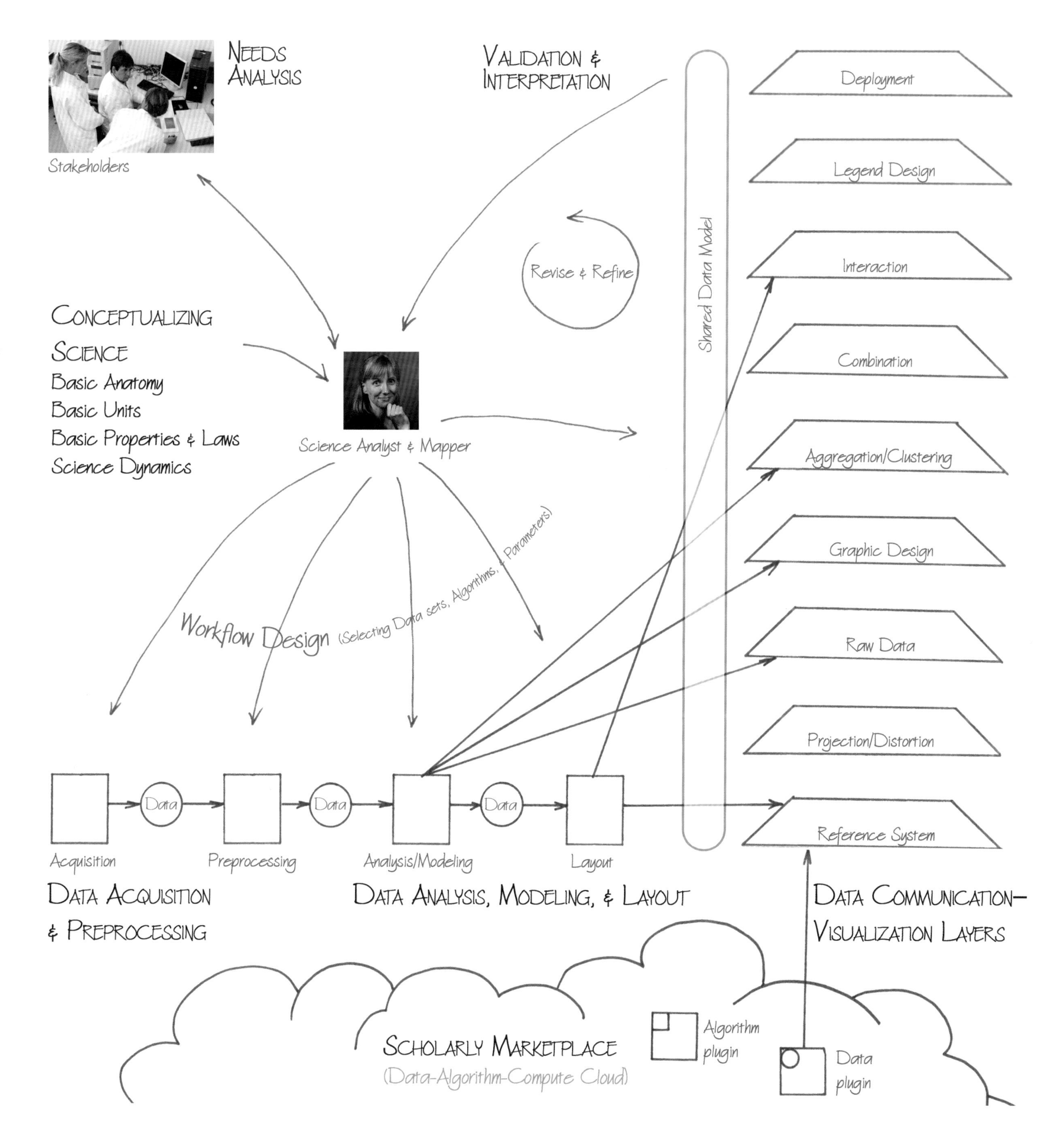

Conceptualizing Science: Basic Anatomy of Science

The study of science and the design of valid science maps both demand a scientific approach. The basic units of analysis, including papers and journals, with their properties and interrelations, need to be defined. The aggregation of these units into "invisible colleges" or scientific research fronts has to be operationalized. Existing indexes and laws need to be considered. Last but not least, the dynamics of science have to be understood and communicated both visually and intellectually. The following four spreads introduce a basic science conceptualization, with discussion of key components and prototypical examples. Relevant books and papers are listed in the timeline in **Milestones in Mapping Science** (**pages 26-47**) and the **References and Credits** (**pages 212-246**).

Grandiose schemes always meet with excessive resistance, not because they are impossible to achieve, but because there are only a few with sufficient persistence to materialize their dreams and even fewer to carry them out. Ultimately, most large endeavors must fall by the wayside, to be replaced by others. However, their value at a particular stage of history cannot be disputed.

Eugene Garfield, proposing a Unified Index to Science

Toward a Science of Science

The application of the scientific method to science itself was first proposed in 1923 and named *science of science* in 1928. It refers to the design of research-based theories, the testing of hypotheses, and the employment of qualitative and quantitative techniques and models to analyze empirical data in order to increase our knowledge of the structure and evolution of science.

Science of science studies draw from diverse fields such as scientometrics, informetrics, webometrics, history of science, sociology of science, psychology of the scientist, operational research on science, the economics of science, the analysis of the flow of scientific information, and the planning of science.

Scientometrics refers to the quantitative study of science and technology using mathematical, statistical, and data-analytical methods and techniques for gathering, handling, interpreting, and predicting a variety of features of the science and technology enterprise, including scholarly communication, performance, development, and dynamics. The practice of scientometrics often requires the use of bibliometrics, the measurement of texts and information. Modern scientometrics is based primarily on the work of Derek John de Solla Price and Eugene Garfield (see **page 16, Visionary Approaches**). Given the interplay of science and technology and the impact of governmental decisions, it is important that science of science studies take into account scholarly, economic, and government data.

The history of science aims to describe the structure and evolution of science. Relevant books are listed in the "Books" section in **Milestones in Mapping Science Timeline** (**pages 26-47**).

The *sociology of science* is the subfield of sociology that studies science as a social activity, focusing on the social conditions and effects of science, as well as the social structures and processes of scientific activity. While many of these contributing scientific fields are centuries old, it is only today that sufficiently large and comprehensive repositories of scholarly data exist in digital format, methods and algorithms scale to such large data sets, and computational infrastructures are available to process the data.

Features and Norms of Science

Many scholars have attempted to characterize the features and norms of science and to capture and promote what it means to be a scientist.

Robert K. Merton's norms of science include communism (that is,common ownership of discoveries), universalism of scientific truths, disinterestedness or selflessness, and organized skepticism (that is, all ideas must be tested rigorously). He later added originality or novelty in research contributions.

According to Edward O. Wilson, the distinguishing features of science, as opposed to pseudoscience, are repeatability, economy, mensuration, heuristics, and consilience.

Repeatability refers to the fact that any scientific analysis, model, or map can successfully be rerun or regenerated by a scholar other than the author. This requires that data sets be accessible and documented in a way that they can be recompiled. Software must also be made available or documented in sufficient detail so that it can be reimplemented and run with identical parameter settings.

Economy demands that results be presented in the simplest and most informative manner, for both the expert and the general audience.

Mensuration refers to proper measurement, using universally accepted scales in an unambiguous fashion.

Heuristics refers to the principle that the best science stimulates further discovery.

Consilience suggests that consistent explanations and results are most likely to survive.

The Power of Visual Conceptualizations

James Watson and Francis Crick's three-dimensional model of the DNA structure served as a central visual metaphor for the transfer and combination of theories and data from chemistry and genetics. The model represents a real chemical entity, which was based on Maurice Wilkins and Rosalind Franklin's diffraction data, and it was successfully applied to test and explain many of the fundamental processes of genetics.

A theoretically grounded, data-driven, and shared conceptualization of science would prove similarly useful. It could provide the intellectual framework needed to interlink and piece together the puzzle of existing conceptualizations and models of science. Leading scientists have given various forms to these models, including philosophical concepts (John D. Bernal, Thomas S. Kuhn, Karl Popper), utopian stories (Herman G. Wells,

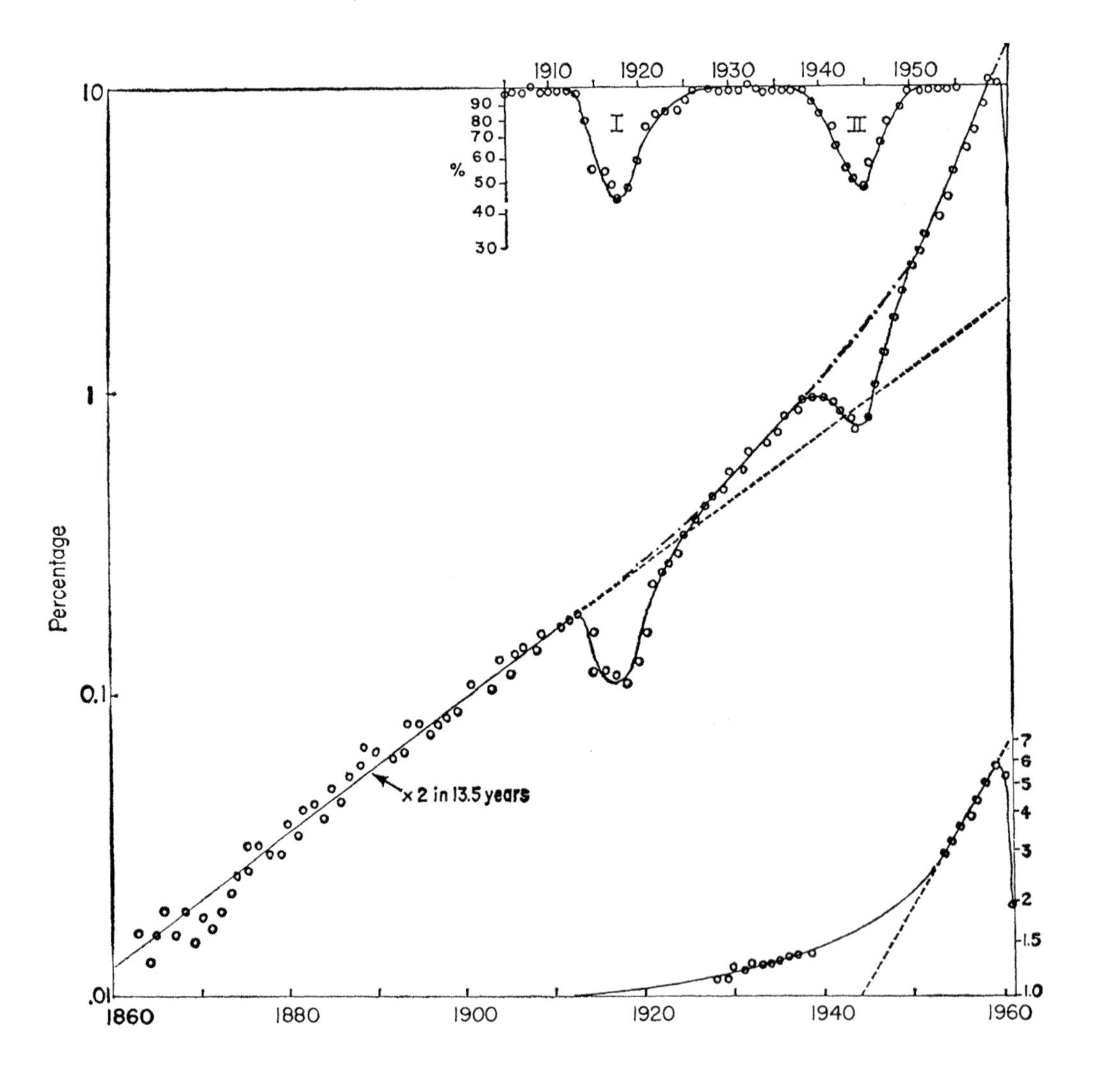

Stanisław Lem), visual drawings (Paul Otlet), empirical measurements (de Solla Price, Garfield), and mathematical theories (William Goffman, Anatolij Ivanovich Yablonskij) among others. The design of any conceptualization requires the identification of the boundaries of the system or object in question, the definition of basic building blocks of science (units of analysis or key actors), interactions of building blocks (via coupled networks), basic mechanisms of growth and change, and existing laws (static and dynamic). Ideally, the conceptualization can be presented in a visual form so that disciplinary and cultural boundaries can be bridged more easily. An initial attempt at a partial visual conceptualization is shown below and extended in the subsequent spread.

Visual Conceptualization of Science

Science can be conceptualized as a self-organizing process that consumes inputs like money and labor force and generates outputs such as scientific results and expertise. The goal is the systematic generation of knowledge that makes a difference.

An initial visual conceptualization of the global structure and evolution of science, together with patchwork inserts of major concepts, is explained and depicted here. The conceptualization draws heavily upon the work of de Solla Price and Menard.

(I) First Scientific Concept

The beginning of science is defined by the first scientific concept. That first concept is the nucleus from which future scientific knowledge radiates and expands in all directions. All active scientific concepts evolve continuously, frequently changing both name and structure in the process. Few concepts stay active for long; most are subsumed by new concepts or become obsolete. All concepts are situated in time, geographic place, and scientific context.

(II) Cumulative Structure

The depth of a layer reflects the amount of knowledge added. Just as tree rings indicate growth years, cumulative layers suggest good years for science—rich in funding, bright minds, and scientific breakthroughs. War years and economic collapses also take their toll (see the impact of the world wars on the number of publications produced on **page 4, The Rise of Science and Technology**, and graph on opposite page).

De Solla Price referred to the citation of well-regarded scientists' work as "scholarly bricklaying." Newton wrote, "If I have seen further it is by standing on ye shoulders of Giants." New layers of scientific products are endlessly being added to this three-dimensional manifestation of scholarly knowledge.

(III) Research Specialty Tubes

Fields of research correspond to outward-growing tubes. They may begin as the result of a seminal idea or a split of older fields, or they may be interdisciplinary, emerging from current fields. Typically, they expand over time and split when they reach a critical size. Some research fields do decline or even die, perhaps due to scientific revolutions.

(IV) Research Frontier Epidermis

The outer layer of science comprises all recent scholarly works and closely resembles the global map of science shown on **page 170, Maps of Science: Forecasting Large Trends in Science**. The research specialties vary considerably in vitality and average reference age (see **page 136, Map of Scientific Paradigms**)—the darker the area, the faster the advance of the research front. As de Solla Price wrote, "Science grows very regularly, in a very structured way, and from its epidermis rather than from its body." Scholars act as "containers of the research front tradition and state of the art in science and technology."

(V) Scholars

In this conceptualization, scholars tirelessly operate on the crust of science: consuming resources, producing scholarly results, and weaving their scientific networks. They receive funding and produce papers, data, software, and hardware. As they coauthor, cite, and acknowledge contributions by others, their contributions to the growing "termite hill" of scholarly results become immortal and the basis for future scholarly works. Some scholars spend their entire professional lives—perhaps 30 to 50 annual layers—at the center of one scientific field (tube). Others change fields frequently as they contribute scientific results, leaving trails in multiple layers and tubes. Although authors tend to frequently change affiliations and locations, they typically remain in the same or a related area of research, and they are judged by their peers from this area. Hence, their main allegiance is to peers rather than to institutions.

(VI) Linkages

Within and between the many layers of science are linkages that interconnect basic and aggregate units of science. Citation links connect papers. Coauthorship relations interconnect authors. Papers might acknowledge authors, funding, hardware, and software. Most linkages are to entities that are related in terms of time, geographic locations, and scientific areas. However, there are some key "weak links" that play an important role in bridging rather distant scientific communities and fields. Note how citation and coauthor behaviors vary between fields: Mathematicians often work alone and tend not to cite many publications. Biomedical researchers frequently have hundreds of coauthors and references. Historians also have extensive references but use different selection criteria.

(VII) Clustering in Time, Geography, and Topic Space

Scientific units naturally cluster in time, as only living authors can coauthor and new papers tend to reference recent papers. They cluster in geographic locations because of the impact of social networks, shared institutions, and the effects of different languages, cultures, and science policy systems (see **page 66, Exemplification**). They also group in topic space, as interdisciplinary research requires considerably more effort than collaborations/citations within one area (tube) of research. In de Solla Price's words, "Scientists tend to congregate in fields, in institutions, in countries, in the use of certain journals. They do not spread out uniformly." See Garfield's historiographs of science in **HistCite Visualization of DNA Development (page 120)** and in the image below.

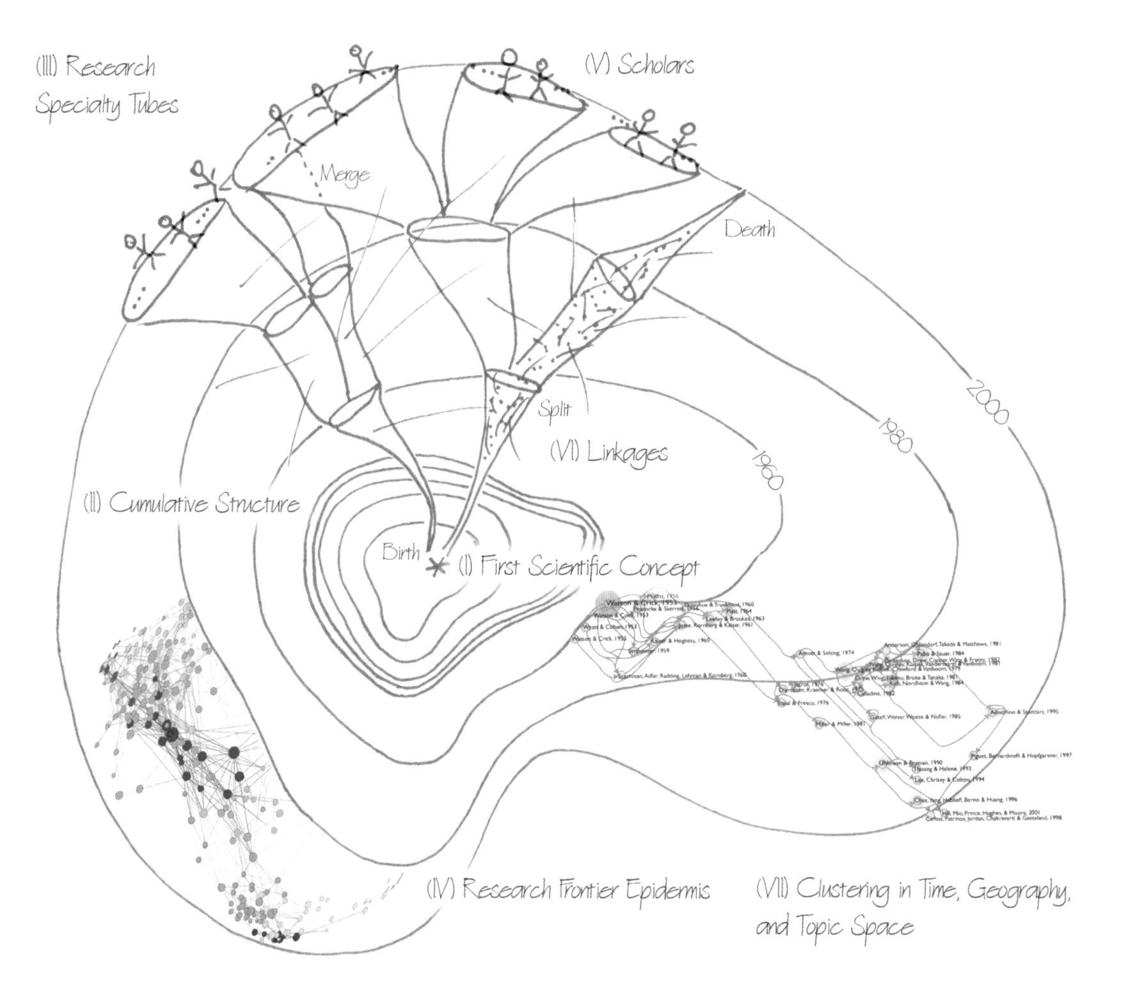

Science Dynamics

Some fields experience extensive and rapid growth, attracting a large number of new scholars in a short span of time. They effectively build on existing knowledge communicated via electronic journals and are cited by others within days of their publication. Other fields report results in books, and several years may pass before the book is published and the knowledge spread. It takes even longer for subsequent books that might cite the original book to be published. According to Menard, subfields grow at highly variable rates, with 5 to 45 years of doubling time. See the discussion in **Conceptualizing Science: Science Dynamics (page 58)**.

Conceptualizing Science: Basic Units, Aggregate Units, and Linkages

How can the conceptualization introduced above help us to review and interlink our existing knowledge? What are the atomic elements or basic units of science? How do basic units of science group into aggregate units such as countries and scientific fields? How do they relate to one another and what do we know about them? How do basic and aggregate units grow and evolve? This spread discusses basic units, aggregate units, and their interlinkages. Major properties, indexes, and existing laws are considered on the next two pages. Dynamic aspects are reviewed thereafter.

... depicting the connections between papers, the cumulating structure of science has a texture full of short-range connections like knitting, whereas the texture of a humanistic field of scholarship is much more of a random network with any point being just as likely to be connected with any other.

Derek John de Solla Price

Basic Units

Science of science studies commonly focus on authors and their institutions and countries of origin or operation; scientific products such as books, papers, journals, patents, data sets, hardware, and software; and resources such as funding.

Authors

A person who has published a scientific paper is called an author. Each author has a biological age and a professional age (determined by the number of active research years), a level of productivity, and a reputation. Reputation can be measured by number of published papers, received citations, download counts, success of students, founding or leading of a school, founding or editing of a journal, officiating in a professional society, presiding over a science government agency, receiving the Nobel Prize or other prizes, and elections to prestigious organizations (such as the National Academies or the Royal Society).

Scientific Papers

Historically, the scientific paper has been at the core of scientometric studies. It can be seen as a peer-reviewed intellectual property claim that attributes a scientific contribution to a set of authors, their institutions, and countries. Each paper includes an author's name and address; a title, abstract, perhaps keywords, and full text (used in textual topic analyses); and references and acknowledgments (used in citation and acknowledgement analyses). Last but not least, each paper is published in a journal, proceedings, or elsewhere.

Author Name(s) and Address(es)

Early Royal Society papers were published anonymously with approval of the gentleman members. Today, nearly every paper has a set of authors. The number of authors strongly depends on the field of research. Mathematicians typically publish as single authors. Papers in biology and physics frequently have more than 100 authors. The institution, country, and geolocation of an author can be derived from the address field. Some data sets provide only the address of the first author. Others include all addresses but no exact matching of authors to addresses (see **page 60, Data Acquisition and Preprocessing**).

Title, Abstract, Keywords, and Full Text

The scholarly contributions of a paper are encoded in the words occurring in the title, abstract, keywords, and full text. Words can be semantically analyzed to identify the topic of a paper or to identify topic bursts and trends in a set of papers (see **page 62, Data Analysis, Modeling, and Layout**).

References and Acknowledgments

Scientific papers are neither produced in isolation nor do they exist in isolation. Rather, they build on, extend, and approve or disprove the works referenced, while benefiting from the support listed in the acknowledgments.

The concept of references derives from the use of ancient footnotes called *scholia*, which enabled readers to trace backward in time to the antecedents of the paper being perused. References to a paper, also called citations, enable the reader to look forward in time. Papers that are highly downloaded or cited, having been deemed useful by many authors, are presumed to be more valuable. In essence, citations serve as the currency of science (see **page 58, Conceptualizing Science: Science Dynamics**). Papers that become highly cited after a long period of low citation counts are also called *sleeping beauties* (see figure on opposite page).

Acknowledgments recognize the value of a colleague's social, intellectual, or material (for example, gene samples or instruments) contribution or funding. They appear in the introduction of a paper, as footnotes, or as endnotes, constituting a valuable bibliometric indicator.

Journals

Upon publication, a scientific paper inherits the topical coverage and reputation of the journal in which it appears. Journals such as *Science*, *Nature*, and *Proceedings of the National Academy of Sciences* (*PNAS*) are highly interdisciplinary. Others cover a specific topic area and represent a continuous line of research with many citations pointing to papers previously published in that journal. The number of papers published in a journal and issues of a journal published per year vary.

Patents and Technical Reports

Patents are written to protect intellectual property claims. Technical reports document the state of the art of any technology such as a machine, drug, product, or process. Whereas scientists are keen to write, technologists prefer to read, safeguarding their own work and intellectual property to ensure they are the first to bring a new product to the market. As de Solla Price proclaimed, science is "papyrocentric" while technology is "papyrophobic."

Hardware, Software, and Data Sets

Some sciences require expensive facilities, such as telescopes or particle accelerators. Computational sciences require substantial computing power, storage capacity, and technical expertise. Consequently, infrastructures, software, and data sets are becoming significant products and enablers of science.

Funding

The pursuit of science requires salaries, buildings, and equipment (see **page 4, The Rise of Science and Technology**). "Big science" conducted in biology, astronomy, or physics often requires state-of-the-art facilities and a generous budget to pay for advanced tools and cyberinfrastructure, and for the immense number of researchers involved in one experiment. Other sciences simply require a modest salary and access to books.

Aggregated Units

Basic scientific units are frequently aggregated over time, geographical space, topic fields, or organizations for means of data reduction and anonymization. For example, scholars might be organizationally grouped into teams, institutes, universities, and nations. Geographic locations might be grouped into states, nations, and continents. Papers are aggregated into journals or discipline-specific journal sets. In addition, there are some well-studied aggregate units of analysis, such as "invisible colleges," research specialties, and research fronts, discussed here.

Invisible Colleges

The term *invisible college* was first coined by and for the scholars who founded the Royal Society in 1660 "to gain an appreciative audience for their work, to secure priority, and to keep informed of work being done elsewhere by others."

In the 1960s, de Solla Price revived the term to describe informal communities of about 200 scholars who perform work in the same research specialty, share theories and ideas, read and publish in the same journals, attend the same conferences, are frequently associated with the same research laboratories or grant-giving institutions, collaborate in joint research efforts, cite one another's work frequently, and also have social relationships.

Research Specialties

Research specialties have been studied from a sociological, bibliographical, communicative, and cognitive approach to identify their social structure, the scientific paradigm they share, the "base knowledge" they are drawing from, the research topics they are tackling, the relations among entities, and changes occurring over time. A typical research specialty is assumed to have fewer than 100 core authors/scientists and fewer than 5,000 papers in its corresponding research literature. Authors within a research specialty tend to belong to the same invisible colleges.

According to Kuhn, mature science is characterized by successive transitions from one scientific paradigm to another via scientific revolutions. Long phases of "normal science" are suddenly disrupted by "technological revolutions" due to technical advances, such as the discovery of RNA interference by Andrew Fire and Craig C. Mello, or by "conceptual revolutions" due to theory development, such as the discovery of the structure of DNA by James D. Watson and Francis Crick. In both cases, an old theory X is replaced by a new theory Y, which can explain all that X can plus something more or which can explain the same set of things in simpler ways. A new research specialty is born.

The new specialty and paradigm might draw from one or many existing specialties and it will grow as it attracts new scholars. It will split into more specialized areas when it becomes too large. Ultimately, it might be rendered obsolete by a new scientific revolution or be at least partially subsumed by new research specialties.

Research Fronts

According to de Solla Price, invisible colleges focus on and cite a relatively small number of articles (about 40 or 50), effectively creating a "research front" of the research specialty. As scientists mainly tend to cite new work, the papers in a research front are continually changing (see **page 22, Derek John de Solla Price;** also see description of the outermost "epidermis" layer of science on **page 53, (IV) Research Frontier Epidermis**).

Research fronts in evolving citation networks can be identified via (1) clusters of cocited papers (see **page 86, 1996 Map of Science,** and **page 162, Mapping the Universe: Space, Time, and Discovery!**), (2) a cluster of cocited papers and all articles that cite the cluster, or (3) a cluster of papers that cite a common group of articles (bibliographic coupling). The papers that are cited and cocited by a research front are also called its intellectual base.

Linkages and Derived Networks

In many respects, basic and aggregate units of science are defined and shaped by their connections. We recognize them by *who they speak to* as much as by *what they speak about*. Different types of connections and their empirical manifestations are explained and depicted below. In general, three types of linkages are distinguished: *direct linkages* such as paper citation linkages; *co-occurrences* of words or references; and *citations* of authors or papers. Linkages might be among units of the same type, such as coauthorship linkages, or between units of different types, such as authors and the papers they produce. Units of the same type can be interlinked via different link types; for instance, papers can be linked according to coword, cocitation, or bibliographic coupling analysis. Linkages might be directed and/or weighted. Nodes and their linkages can be represented as an adjacency matrix or edge list and visually as structure plot or graph. Each nonsymmetrical occurrence matrix has two associated (symmetrical) co-occurrence matrices; for instance, for each paper citation matrix exists a bibliographic coupling and a cocitation matrix.

Direct Linkages

Paper-Paper (Citation) Linkages

Papers cite other papers via references, forming a nonweighted, directed paper citation graph (see **page 120, HistCite Visualization of DNA Development**, and HistCite graph insert in figure on previous spread). It is beneficial to indicate the direction of information flow from older to younger papers via arrows (see figure below). References enable readers to search the citation graph backward in time. Citations to a paper support the forward traversal of the graph. Citing and being cited can be seen as two vital roles of a paper.

Author-Paper (Consumed/Produced) Linkages

There are active and passive units of science. Active units might be authors who produce and consume passive units, such as papers, patents, data sets, and software. The resulting networks have multiple types of nodes, including authors and papers. Directed edges indicate the flow of resources from sources to sinks, for example, from an author to his or her written/produced paper A to the author who reads/consumes paper A.

Co-Occurrence Linkages

Author-Author (Coauthor) Linkages

Having the names of two authors (or their institutions and countries) listed on one paper, patent, or grant is an empirical manifestation of scholarly collaboration. The more often two authors collaborate, the greater the weight of their joint coauthor link. Weighted, undirected coauthorship networks appear to have a high correlation with social networks that are themselves shaped by geographic proximity.

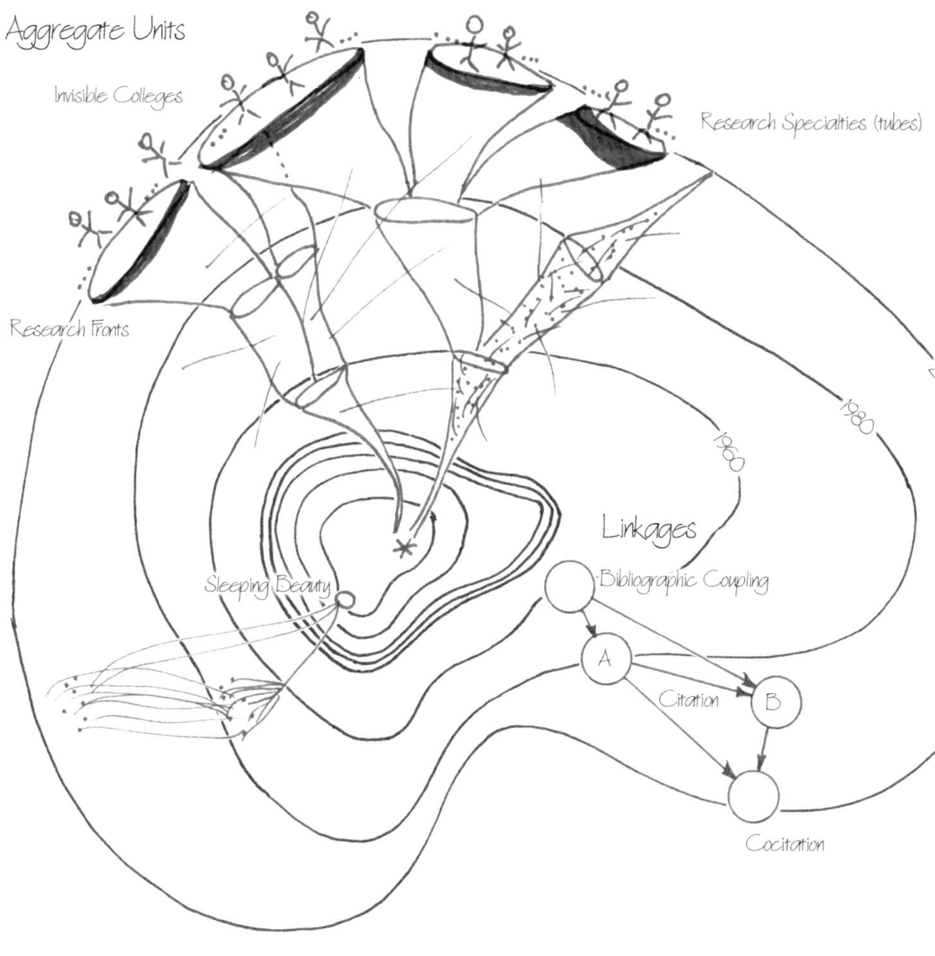

Word Co-Occurrence Linkages

The topic similarity of basic and aggregate units of science can be calculated via an analysis of the co-occurrence of words in associated texts (see **page 62, Data Analysis, Modeling, and Layout**). Units that share more words are assumed to have higher topic overlap and are connected via linkages and/or placed in closer proximity on a topic map. Co-occurrence networks are weighted and undirected.

Reference Co-Occurrence (Bibliographic Coupling) Linkages

Papers (or patents) that share common references are said to be coupled bibliographically. The bibliographic coupling (BC) strength of two basic or aggregate units of science can be calculated by counting the number of times that papers associated with them jointly reference the same third work in their bibliographies. The coupling strength is assumed to reflect topic similarity. Weighted, undirected networks can be derived from BC matrices.

Cocitation Linkages

Two basic or aggregate units of science are said to be cocited if papers associated with them jointly appear in the list of references of a third paper. The more often two units are cocited, the more they are expected to have something in common.

Document Cocitation Analysis

Document cocitation analysis (DCA) was simultaneously and independently introduced by Henry G. Small and Irina V. Marshakova-Shaikevich independently in 1973. It is the logical opposite of bibliographic coupling. The cocitation frequency equals the number of times two papers are cited together (in other words, when they appear together in one reference list of a third paper).

Author Cocitation Analysis

Authors of works that are repeatedly juxtaposed in references or citation lists are assumed to be related. Clusters in author cocitation analysis (ACA) networks often reveal shared schools of thought or methodological approach, common subjects of study, collaborative and student-mentor relationships, ties of nationality, or other relationships. ACA gives analysts the power to depict regions of scholarship as densely crowded and interactive or as sparsely populated and isolated (see sample maps on **pages 26-47, Milestones in Mapping Science**).

Conceptualizing Science: Basic Properties, Indexes, and Laws

Many properties, indexes, and laws aim to describe the structure and inner workings of science. Indexes can be applied to quickly derive basic quantitative measures for a unit of science. Laws describe general properties of specific units of science that seem to hold true universally. They are discussed in the sequence of their discovery.

An investigation of the distribution of quality in men and in papers shows immediately that there exists greater inequality than in any country's distribution of economic wealth and a bigger rat race than in any other competitive activity of mankind.
Derek John de Solla Price

Basic Properties

Early studies of science aimed to conceptually understand and quantify the interrelations and the dynamics of major units of science. Examples are de Solla Price's plot of the exponential increase in the number of journals (**page 4, Papers and Journals**) and his depiction of the evolution of a closed research field and its research front (**page 22, Derek John de Solla Price**). Additional results by de Solla Price and Menard are given here.

Balance of Papers and Citations

The figure to the right shows the balance of papers and citations for a given "almost closed" field in a single year, based on quantitative data analysis. The field in question had grown exponentially and has **100 old papers** and **7 new papers**. The **7 new papers** have on average 13 references, and the figure shows how those **91 references** link to existing papers. Here, **10** references are made to papers outside of the field. From the 100 old papers in the field, **40** are not cited, **50** are cited once, and **10** are cited more than once.

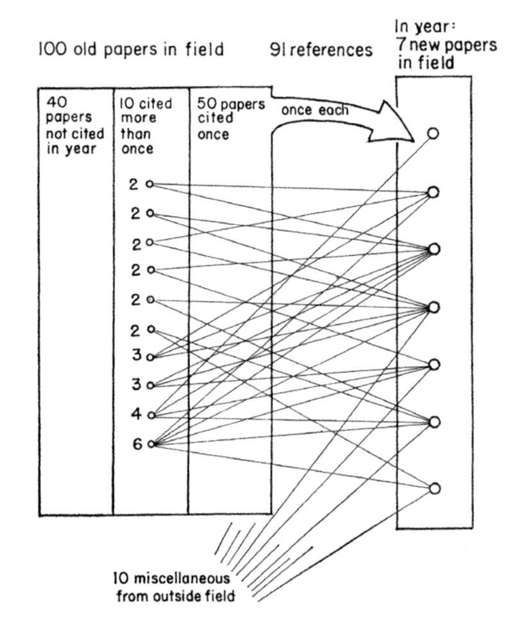

Author Trajectories

The figure to the right shows de Solla Price's rendering of the flow of authors through annual indexes representing different scientific disciplines. Four major types of authors are distinguished based on their publication patterns: **(1) Newcomers** at the beginning of their careers, **(2) Transients** who published only once, **(3) Continuants** who publish large numbers of papers, and **(4) Terminants** at the end of their careers. Percentages for each type were derived from an empirical data analysis.

Coauthorships between newcomers, terminants, and transients are often mediated by continuants. Continuants are likely to coauthor and rarely appear as single authors. Continuants and terminants typically play the roles of hubs or gatekeepers.

Scientific Growth

Menard's work showed that it is advancement and growth, not the size of a field, that determines the number of citations per paper. In his thought experiment, he envisioned two fields: one with few existing papers and rapid growth, the other with many papers and slow growth. While more papers are generated in the larger field, a new paper has to share new citations with many existing papers. In the smaller field, a new paper has a quickly increasing number of opportunities to be cited, and it shares those citations with fewer papers.

Menard also examined the impact of scientific growth on the career trajectory of a young researcher, considering three scenarios: (1) An established, slow-growing field with 200 people and 27,000 papers: Young researchers entering this field will go through a long apprenticeship. It will take them 20 to 40 years and about 57 papers to achieve the power of middle-aged researchers. (2) The average subfield, with about 100 people and 6,000 papers: Young authors may achieve influence in their thirties after publishing 36 papers. (3) A young but fast-growing subfield with about 10 people and 190 papers: Here, young authors (in their twenties and thirties) are put in charge. They spend 38 days reading the existing works and then start contributing their own. Six papers suffice to become a major expert. There are abundant job offers, rapid promotion, youthful honors, and the possibility of highly distinguished careers. Note that the growth rate of a field affects the age distribution of authors in that field. The mean author ages for the slow, average, and fast fields are 42, 35, and 25, respectively.

Indexes

Indexes help to calculate one number per basic unit of science for means of ranking and thresholding. Three commonly used indexes are discussed here: an index for an author's productivity, an index for the vitality or innovativeness of a research field, and an economic index that measures the market share held by particular suppliers in a market. While these three indexes are far from exhaustive, these examples aim to communicate the value of indexes for the study of science.

Herfindahl-Hirschman Index, 1945/1950

The Herfindahl-Hirschman Index (HHI) measures the relative size, distribution, and resulting competition of firms in a market. It can also be applied to characterize research competition in a discipline—for instance, in terms of research output or funding intake. The HHI is defined as the sum of the squares of the market shares of each individual firm. For example, in a market consisting of two firms with shares of 20 percent and 80 percent, the HHI is $20^2 + 80^2 = 6800$. The HHI increases both as the number of firms in the market decreases and as the disparity in size between those firms increases. Markets with an HHI between 1000 and 1800 are considered to be moderately concentrated, and those in which the HHI is in excess of 1800 points are considered to be concentrated. Firm mergers that increase the HHI of a concentrated market by more than 100 points presumptively raise antitrust concerns under the Horizontal Merger Guidelines issued by the U.S. Department of Justice and the Federal Trade Commission.

Price Index, 1970

The Price index measures the recency of the literature cited by a paper, journal, or research field and can be used to characterize the vitality or innovativeness of a research field. It is calculated by dividing the number of references published in the last 5 years by the number of all references in a paper and multiplying the result by 100. A field that is all research front and no general archive might have an index of 75 to 80 percent.

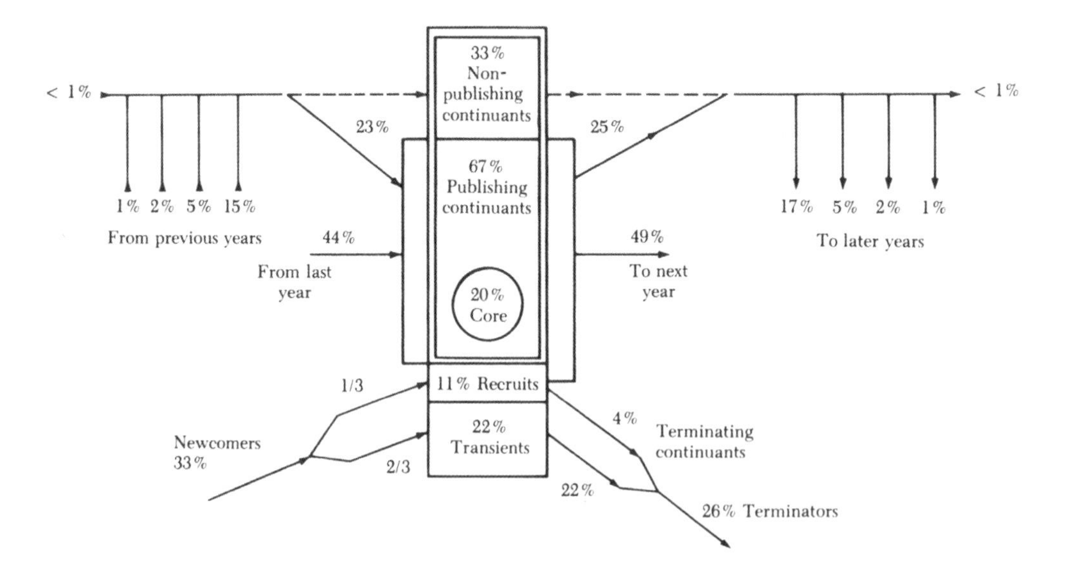

De Solla Price also used the index to distinguish hard science (with an index of 60 to 70 percent), soft science (around 42 percent), and humanities (10 to 30 percent), which might be an oversimplification due to the impact of publishing mechanisms (e-prints versus books).

Hirsch Index, 2005

The Jorge E. Hirsch index (*h*-index) aims to quantify an individual's scientific research output. It equals the highest number *n* of papers by one author, which have each amassed at least *n* citations. For example, if an author has published five papers with 0, 2, 4, 5, 7 citations respectively, the *h*-index is 3. It was adopted by Thomson Reuters and Scopus to supplement other measurements of author impact. The Royal Society of Chemistry recently released the *h*-index for the top 2000 living chemists. Given the very different dynamics of scientific disciplines, *h*-indexes cannot be compared across disciplinary boundaries.

The *h*-index favors those authors with older publications who have had time to acquire citation counts. In fact, it is more reasonable to divide the *h*-index by the number of years an author was active in a field and to assume he or she was productive if the resulting value is above 1. References to papers that suggest other improvements are provided in the **References and Credits** (**page 228**).

General Laws

The exemplary laws of science discussed here describe general properties that exist in scholarly data. They can be seen as facts that create a foundation on which future studies can be built.

Power Laws

Scholarly networks such as coauthor and paper-citation networks show the "rich get richer" phenomenon or effect, also known as the Matthew edict, the Matthew index/effect, cumulative advantage, or "preferential attachment" (see models on **page 58, Conceptualizing Science: Science Dynamics**). For example, papers or patents with many citations are likely to attract even more citations. Authors with many coauthors will attract even more coauthors. Authors with many papers will produce even more papers (see graphs below). The name of the effect is derived from Matthew 25:29: "For to everyone who has, more shall be given."

Mathematically, the concept of preferential attachment was introduced by the mathematician George Pólya in 1923, and taken up and popularized by George U. Yule (in 1925) and Herbert Simon (around 1950) as an explanation for the power law distributions observed empirically in the population sizes of cities and the distribution of wealth among members of a society. Robert K. Merton later wrote on "The Matthew Effect in Science," and de Solla Price derived "A General Theory of Bibliometric and Other Cumulative Advantage Processes." In 1999, Albert-László Barabási and Réka Albert proposed a model that simulates preferential attachment.

De Solla Price's studies revealed that 10 percent of all papers have no references, and 10 percent are never cited. About 1 percent of papers have neither references nor citations by other papers; are unconnected in a pure citation network and could only be found by topical searches.

Recent work shows that connectivity distributions of coauthor and paper citation networks can best be fitted by a power law form with an exponential cutoff.

Pareto Principle, 1897

The well-known Pareto principle—also known as the 80/20 rule or the law of the vital few—is a simplistic representation of the power law effect. It states that, for many events, 80 percent of the effects come from 20 percent of the causes. It was originally discovered in the late 1800s by economist and avid gardener Vilfredo Federico Damaso Pareto, who noticed that 80 percent of the land in Italy was owned by 20 percent of the population. Later, while gardening, he observed that 20 percent of the peapods in his garden yielded 80 percent of the peas that were harvested. The principle also applies to income distribution (a large percentage of wealth is concentrated in a small proportion of the entire population) and citation patterns (20 percent of all papers receive 80 percent of all citations).

Lotka's Law, 1926

Alfred J. Lotka's inverse square law describes the productivity of authors in terms of scientific papers. Specifically, the number of authors producing *n* papers is proportional to $1/n^2$. The graph below left shows a log-log plot of the number of papers over the number of authors who publish that many papers for two journals: *Philosophical Transactions* of the Royal Society (o) and *Chemical Abstracts* (x). The full line shows Lotka's inverse square law modified to take into account the overestimation of highly prolific authors.

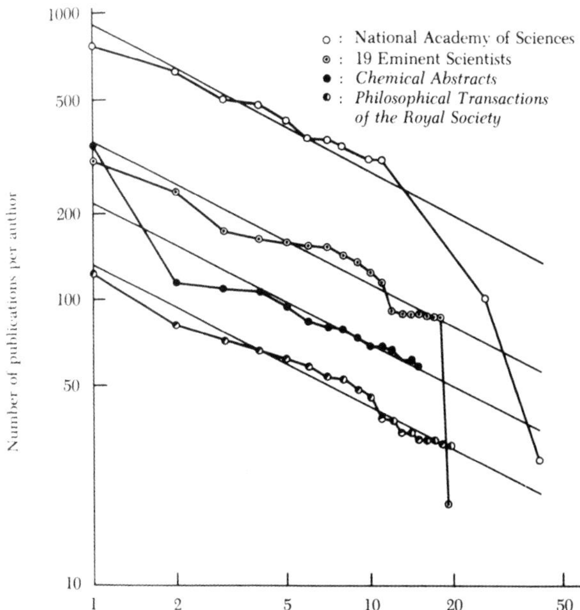

Bradford's Law, 1934

Samuel C. Bradford's law characterizes the scattering of papers over different journals or fields. Garfield compared this scattering to a comet, which has a concentrated core and widely fanned tail. Depending on the field and task, there are exponentially diminishing returns of extending a search for papers or references in journals. Bradford's law might help to explain why a science citation index is feasible. It is frequently applied by librarians to select the most central information sources in a scientific field.

Zipf's Law, 1949

The linguist George K. Zipf noticed that the frequency of occurrences of any word in a natural language text is inversely proportional to its rank in the frequency table; few words occur very often while most words occur rarely. The law has also been applied to characterize the number of inhabitants in cities (see figure of communities of 2,500 or more inhabitants ranked in decreasing order of population size on **page 6, Addictive Intelligence Amplifiers**). The graph below right shows different author groups, each rank ordered by the number of their publications in a log-log plot. Again, the curves for the different groups are parallel to one another and seem to exhibit power law behavior.

Moore's Law, 1965

Gordon Moore predicted as early as 1965 that the number of transistors on a chip would double about every two years. Processing speed, memory capacity, and the number and size of pixels in digital cameras are strongly linked to Moore's law and are improving at (roughly) exponential rates as well, leading to a dramatic increase in the utility of digital devices (see **page 6, Addictive Intelligence Amplifiers**).

Metcalfe's Law, circa 1980

Robert Metcalfe developed his law to describe the value of the Ethernet. In 1993, George Gilder extended it to communication networks. The law states that both types of networks grow with the square of the number of devices or people they connect. As the number of people with access to the Internet grows, for example, the number of unique connections that a new person can make grows exponentially.

Conceptualizing Science: Science Dynamics

As scientific networks coevolve, their structures and dynamics interact and change over time. Statistical mechanics, agent-based approaches, population and compartmental models, and other approaches have been successfully applied to uncover dynamic properties of basic units, evolving networks, diffusion patterns, and feedback cycles. Subsequently, we discuss a set of models selected to show the richness and diversity of existing approaches.

Scholars try to maximize the yield from their foraging activities by incorporating (ingesting) fresh ideas and insights culled from their natural habitats. Academic reputations are based, in part, on the accumulation of certifiable novelty.

Blaise Cronin and Carol A. Hert

Model Design and Validation

Process models, also called predictive models, use a mathematical formulism with certain parameters or an algorithm together with an input script to simulate a certain process or data set under study. Models are validated by comparing simulated data sets with the empirically measured data set. Formulas and parameters, algorithms and their inputs are iteratively revised until simulated data resemble measured data—within a certain confidence interval.

Properties of Basic Units

There are diverse models that aim to capture the structure and evolution of basic units of science, such as authors or ideas manifested in papers. Two are reviewed here as examples.

Scholarly Information Foraging

Pamela E. Sandstrom applied optimal foraging theory (OFT)—a deductive, microeconomic approach—to predict the composition of a forager's diet given personal and environmental constraints. In her model, a scholar's "diet" includes the range of authors or colleagues, with their ideas and works, consumed for the purpose of sustaining research and a publishing career. OFT prey-choice predicts that a particular resource will be included in the optimal diet if its energy return per unit of handling time remains greater than the return rate (including search time) averaged for all higher-ranking resource types. Adding more resource types increases handling costs but lowers searching costs, as more types are judged acceptable. An improvement in foragers' skills or technology also lowers searching costs. Other OFT models relate resource distribution patterns to the costs and benefits of solitary foraging compared to foraging in social groups. Many scholars have narrow diets—that is, they are specialists. This strategy considerably lowers handling costs as a familiar set of resources can be used repeatedly. However, when a favored resource type becomes depleted (former colleagues or funding sources are unavailable), scholars tend to broaden their diet.

Hypes, Fads, and Fashions

The Internet has a pronounced influence on the dynamics of information access. Higher interconnectivity and bandwidth increase the number of access and download hits one site can acquire per time unit while decreasing the amount of time attention is paid to one site, news item, or idea.

In 2007, the Spice Girls reunion concert in London sold out within 38 seconds. According to the show's promoters, more than one million people in the United Kingdom and more than five million worldwide signed up for the ticket ballot on the band's official Web site. Zoltán Dezsö and colleagues studied visitation patterns for news documents and concluded that hits tend to decay following a power law. In general, hits to most news items significantly decay 36 hours after posting. Lada Adamic clustered blog popularity profiles and identified, for example, different decay rates for Slashdot than for BoingBoing.

Decay rates also exist for scholarly results. For example, the probability that a paper written n years ago is cited today can be approximated by a Weibull function with three parameters that control the height, width, and rightward extension of the curve. As the rightward extension increases, so too does the probability that a paper is cited for a longer period of time. See citation probability graphs (a) and (b) below and discussion in **Diffusion of Innovations** on opposite page.

Evolving Networks

Basic units of science exist in a delicate ecology of scholarly networks. Diverse models aim to capture the dynamics of evolving networks and network activity patterns. Three sample models are discussed here.

The "Rich Get Richer" Effect

Scale-free networks (see **page 57, Power Laws**) can be simulated by processes of incremental growth, rewiring, and preferential attachment. Among the best-known statistical network models is the Barabási-Albert preferential attachment model, which assumes that networks grow continuously by the addition of new nodes and that new nodes attach preferentially to nodes that are already well connected. The model starts with a small initial set of randomly connected nodes. In each modeling step, a new node is created and preferentially connected to the most highly connected node. This model has been successfully applied to coauthor networks and paper-citation networks as well as to social, biological, and technological networks.

Coevolving Author-Paper Networks

Scholarly networks coevolve. There is no coauthor network without papers being created, and there are no paper-citation networks if there are no authors. Each basic/aggregate unit of science exists in a delicate ecology of coevolving scholarly networks. There are positive and negative feedback cycles that might be the true reason for "rich get richer" effects, as well as for collapses. Diverse agent-based models were developed to simulate coevolving networks.

Nigel Gilbert's simulation of the structure of academic science assumes that papers generate future papers. Authors play a rather incidental role. The model was successfully validated based on the number and distribution of citation counts.

The TARL model (topics, aging, and recursive linking) by Katy Börner and colleagues simulates the coevolution of paper-author networks. It attempts to capture the roles of authors and papers in the production, storage, and dissemination of knowledge. The model also assumes the following: coauthor and paper-citation networks coevolve; though authors are mortal, papers are immortal; only living authors are able to coauthor; all existing (but no future) papers can be cited; and information diffusion occurs directly via coauthorships and indirectly via the consumption of other authors' papers. For details see **page 66, Exemplification (Study III)**.

Dynamics of Innovation Networks

Gilbert proposed an agent-based model that simulates the dynamics of innovation network formation. Agents represent firms, policy actors, and research labs, among other entities. Each agent has a knowledge base from which it generates "artifacts" that might become innovations. The innovativeness of the artifacts is judged by an oracle. Agents can improve their artifacts incrementally or radically. They can also seek partners who may contribute additional knowledge. The model was empirically shown to reproduce qualitatively the characteristics of innovation networks in two sectors: personal and mobile communications and biotechnology.

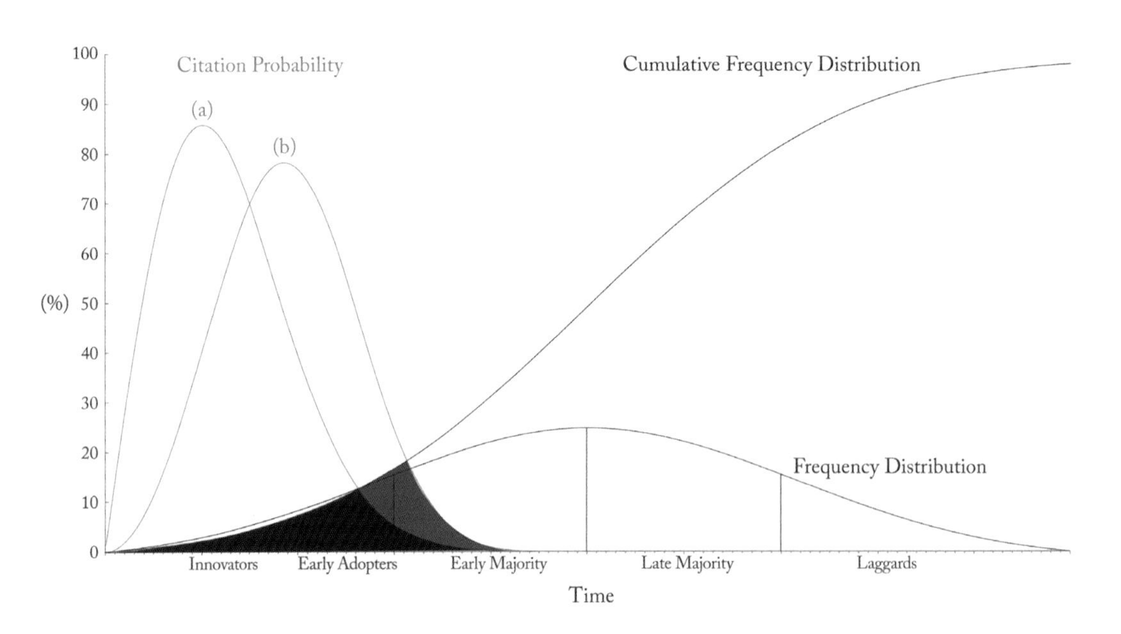

Diffusion Patterns

Diffusion of knowledge is commonly measured via the diffusion of tangible objects (such as people, books, materials, or devices) or intangible objects (such as knowledge spread via coauthor/inventor/investigator linkages or manifested in citations and acknowledgments). The majority of scholarly knowledge is diffused via scholarly networks rather than geometric constraints. Channel capacity (not territory) and resource limits influence the amount and spread of diffusion. Highly connected nodes and nodes/edges that interconnect subnetworks play the role of hubs and gatekeepers, respectively. Their removal (or vaccination) has a major impact. Relevant models aim to (1) identify and track the diffusing entity, often called the concept symbol or meme, (2) analyze the impact of the diffusion substrate, including network structure or geographic area, (3) study diffusion dynamics, and (4) examine the possibility of changing the rate at which diffusion processes happen. Diffusion research has historically been performed in biology, physics, economics, and social science.

Diffusion of Innovations

Everett M. Rogers identified five types of adopters for any new innovation or idea: **Innovators** (2.5 percent of the population), **Early Adopters** (13.5 percent), **Early Majority** (34 percent), **Late Majority** (34 percent), and **Laggards** (16 percent). The respective bell-shaped **Frequency Distribution** and S-shaped **Cumulative Frequency Distribution** curve are shown at left.

The speed of technology diffusion/adoption is determined by the rate at which adoption takes off and the rate at which later growth occurs. An inexpensive, easy-to-understand technology might take off faster than a more complex technology. Computers, however, started off relatively slowly, yet they took off more quickly as value-added synergies with other technologies began to emerge.

It is interesting to note the interplay of innovation diffusion and innovation decay. When overlaying popularity profiles or citation patterns (see **page 58, Hypes, Fads, and Fashions**), the window of opportunity for an innovation or idea becomes visible. Only those technologies and ideas that get adopted before they decay have a chance for adoption—thus, a paper with **Citation Probability** distribution **(a)** will experience less change than one with distribution **(b)**. Those that are not adopted in time will have to be rediscovered or reinvented.

Idea Spreading as Epidemic Process

William Goffman was the first to model the spread of ideas within a population of scientists as a time-dependent, epidemic process. An epidemic process can be characterized by a set of actors and a set of states—such as susceptible, infectious, and removed. At any time, each actor is in one of the given states. Exposure to infectious material, like an idea, changes the state from susceptible to infectious. Death or shift in interest results in removal. The process itself might be in a stable or unstable state (see also **page 150, Impact of Air Travel on Global Spread of Infectious Diseases**).

Luís M. A. Bettencourt and colleagues applied population models to study the impact of author contact rates on the speed of research. They found that a doubling of the contact rate between nanotube researchers led to a time savings of about four years in innovation. In the case of research on bird flu (H5N1), a doubling of the contact rate showed discovery time savings of approximately one decade.

Feedback Cycles

Many growth and diffusion patterns and laws observed in the science system are caused by feedback cycles such as those below. Understanding, modeling, and optimizing these cycles—the key to the inner workings of science—are expected to be at the core of future modeling work.

"Rich Get Richer" Effect

The "rich get richer" phenomenon previously discussed can be seen in the impact high quality papers have on opportunities to receive funding; that affects the hiring of postdoctoral scholars and other staff, as well as buyout of teaching time, which in turn affects research quality and quantity, closing the cycle. Clearly, a scholar's cycle does not exist in isolation. The polarity of one cycle—whether positive, stable, or negative—strongly depends on the speed and effectiveness of the cycles of one's peers and colleagues (see **Cycle of Credibility,** below).

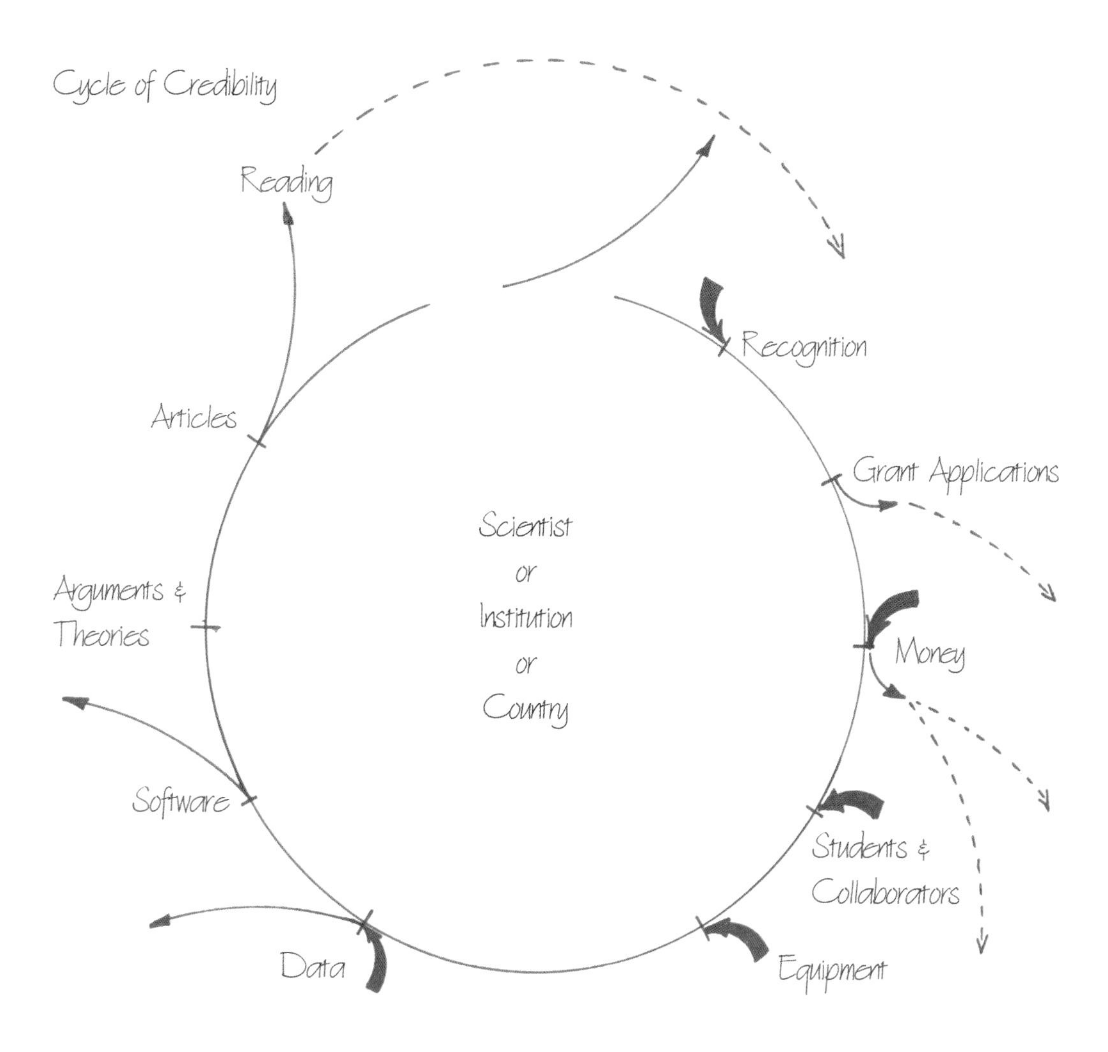

Educational Supply and Demand Cycles

Universities supply the scientists that industry demands. Sudden increases in demand during boom times lead to high salaries for those that have the required expertise. Sudden decreases in demand result in recessions with high unemployment, idle plants, and a continuing surplus of supply over demand.

Symbolic Capital and the Cycle of Credibility

Pierre Bourdieu identified three fundamental categories of symbolic capital in science: social capital based on the networks of social relationships and influence, economic capital because of ownership of money and stocks, and cultural capital. Cultural capital appears as three types: embodied in a scholar's expertise; objectified if manifest as art, books, or other physical objects; and institutionalized if recognized and legitimized by an institution in the shape of a degree, award, or prestigious appointment. He argued that a scholar's value is defined by the quantity of each kind of capital he or she possesses.

Extending this work, Bruno Latour and Steve Woolgar propose the **Cycle of Credibility,** shown here in adapted form. It illustrates the conversion of different types of capital—such as **Recognition, Grant Applications, Money, Students and Collaborators, Equipment, Data, Software, Arguments and Theories,** and **Articles**—into one another. The forces of supply and demand create the value of the commodity. As with monetary capital, the speed and amount of capital converted determines the profit made.

In this conceptualization, scientists are investors of credibility. They are keen to collaborate with leading experts; to publish in the most influential journals; to work with the best data sets, software, and instruments; and to acquire funding. All this is likely to improve their own production of credible information. Reward and credibility are based on peer evaluation, requiring scientists to conform to the scientific norms in their respective research specialties.

Data Acquisition and Preprocessing

The study of science requires access to data sets of the highest quality and with the best coverage available. Ideally, it is based on freely available data, algorithms, and tools so that anyone can replicate or improve upon results. Often, data sets of different types—for instance, funding and publication output—and from different providers must be combined. This poses serious data federation and integration challenges. Given the enormous number and diversity of scholarly records, no single institution can keep track of all existing data. Instead, data acquisition, federation, curation, and preservation must happen in a distributed and decentralized yet concerted fashion.

The ideal information resource would feature high quality of content (i.e., be accurate and complete) as well as high accessibility (i.e., excel in availability, ease of use, and interactivity).
Lawrence M. Sanger

Data Types, Sizes, and Formats

Data sets come in raw, preprocessed, summarized, and aggregated formats. Records in one data set might have a similar structure (for example, papers or patents) or come in diverse formats and sizes (for example, software programs written for different purposes in diverse programming languages). The data set size may be small (less than 1,000 records); medium (between 1,000 and 100,000 records); or large (more than 100,000 records). Data sets are stored as delimited text files, as XML files, in databases, or as streaming sources (such as RSS feeds).

Data Quality

The quality of a data set determines both the quality of results and the kinds of analysis that can be performed. Some data sets are expertly curated to ensure the highest quality; others are merely retrieved from the Web using software programs called crawlers. Some data sets provide links to other types of data sets; a patent, for example, might reference a paper. Some data sets provide full text that can be used to improve text search, analysis, and the examination of citation contexts. There are data sets with standard identifiers that ease the identification and interlinkage of unique records (see **page 61, Data Integration and Federation**). Very few data sets provide unique author identifiers. Misspellings of author names cause up to 45 percent of underestimation of citation counts—a major problem if the counts are used for promotion decisions or quality judgments. As an example, in the Thomson Reuters database in 1999, the name of Derek John de Solla Price was recorded as DeSollaD, DeSollaPD, DeSollaPDJ, Price D, Price DD, Price DDS, Price DJ, Price DJD, and Price DS. If these misspellings are not corrected, then Price's works will be attributed to nine instances of his name, and nine nodes will be used to represent him in a coauthor network affecting the size and structure of the network. Addresses and geolocations are often only available for the first author. Data sets with citation linkages support more types of analysis, such as citation analyses.

The term "data provenance"—or lineage—refers to the process of tracing and recording the origins of data and its movement between databases. Ideally, detailed data provenance records would be available for all data used in a science study.

Data Coverage

Data sets differ in their coverage of publication categories, including journals, books, conferences, patents, publication years, geographic aspects, language, and topical coverage. Data sets might come with usage data, such as citation or download counts. The graphs on **page 4, The Rise of Science and Technology**, show the temporal coverage of publication years for major databases. The maps on **page 106, The Structure of Science**, show the topical coverage of funding by the U.S. Department of Energy, the National Science Foundation, and the National Institutes of Health. To arrive at valid conclusions, the coverage of a data set must fit the scope of a science study.

Shown below is a science map with an overlay of the topic coverage of Web of Science (WoS) for nanoscience researchers at Indiana University in Bloomington and Indianapolis. Circle size represents the number of researchers per topic area. The temporal coverage of this data set (in terms of the number of papers published per year) is shown below the topic map. A geographic map of this data set shows two dots—one for each city.

Most science studies use the databases of either Thomson Reuters or Scopus, as each constitutes a multidisciplinary, objective, and internally consistent database. A number of recent studies have examined and compared the coverage of WoS, Scopus, Ulrich's Periodicals Directory, and Google Scholar (GS)—finding a rather small overlap in records. The overlap between WoS and Scopus was only 58.2 percent. The overlap between GS and the union of WoS and Scopus was a mere 30.8 percent. While Scopus covers nearly twice as many journals and conferences as WoS, it covers fewer journals in the arts and humanities. A comprehensive analysis requires access to more than one database.

Data Acquisition

Depending on the scope of a study a data set might be compiled by hand (for individual-level studies) or retrieved via keyword search or cited reference search (for local-level studies); alternatively, an entire database might be used (for global-level studies). Keyword-based search exploits content similarities, while cited reference search exploits linkages. Both approaches can be used in combination or iteratively. That is, a set of records can be retrieved based on keywords; new keywords can then be extracted from the result set and used in a second search to retrieve more records. Similarly, all papers that cite a set of seminal papers can be retrieved and in turn, all papers that reference the citing papers can be retrieved as well. PageRank, used by Google, weights linkages according to their rank in the network.

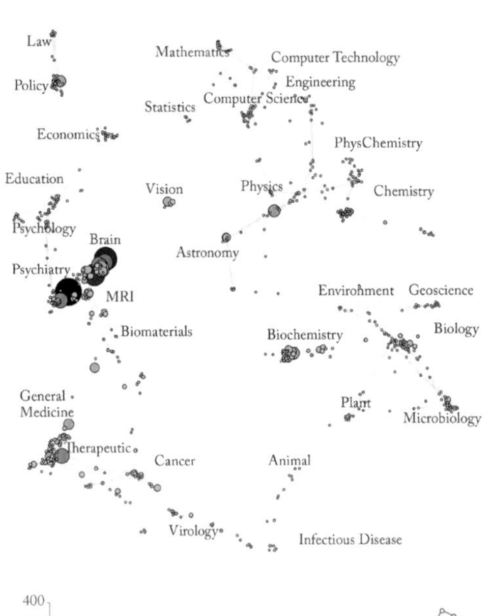

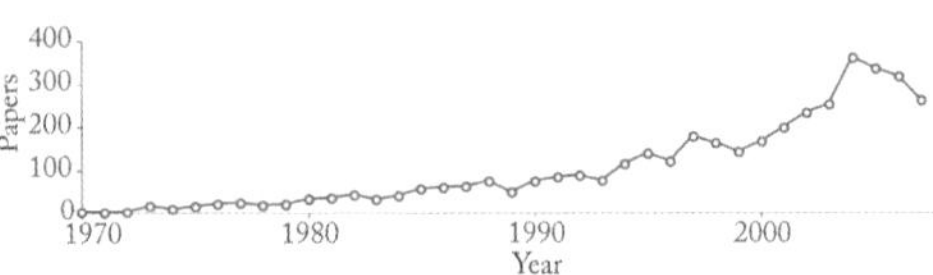

Data Preprocessing

Data may be continuous (time) or discrete (topics). It may be ordinal/numeric (counted and ordered but not measured) or nominal/categorical (counted but not ordered or measured). A first step is often the identification of unique records, naming the authors, investigators, or inventors with their institutions, countries, or geolocations. The high frequency of self-citations can be exploited to identify multiple spellings (for example, of author names). If text needs to be analyzed, words are often stemmed; for example, "scientific," "science," and "scientifically" are reduced to "scien." This considerably reduces the number of unique terms and often leads to higher accuracy in semantic analysis (see **page 63, Topical Analysis**). Data normalization is of vital importance when merging data sets or comparing results across databases.

Often, raw data has to be processed to derive attributes and linkage information required in subsequent analysis. Most basic units of science have a publication date, publication type (journal paper, book, patent, or grant), and topics (keywords or classifications assigned by authors or publishers). Authors have an address with information on their affiliations and geolocations. Because authors and papers are associated, the authors' geolocations and affiliations can be attributed to their papers. Similarly, publication dates, publication types, and topics can be associated with the authors. Diverse networks can be derived and formatted for subsequent analysis (see **page 54, Conceptualizing Science: Basic Units, Aggregate Units, and Linkages**).

Data Augmentation

Additional data can enhance the utility of a data set considerably. To locate authors on a geographic map, the geolocations (longitude and latitude for each author affiliation) have to be derived. To "science locate" scholarly units such as papers or authors, the associated journal names or topics have to be identified and located on a science map. The citation impact of a paper (or its authors or its authors' institutions) can be estimated via the impact factor (IF) of its journal.

Baseline Statistics

Whenever a new data set is acquired, baseline statistics should be run. Simple plots, such as graphs of the number of records or scholarly entities over time, may quickly reveal missing data. It is always a good idea to compare total counts with the numbers given by the data provider. Examining frequency distributions of unique scholarly entities or alphabetically sorted lists of terms can help to define the quality of data.

Data Integration and Federation

Often data sets from different sources have to be merged. For example, import-export studies require comparisons of funding intake to publication/patent output; thus funding, publication, and patent data must be interlinked. Impact studies often consider citation and download data, which may be collected by MEtrics from Scholarly Usage of Resources (MESUR) from six significant publishers, four large institutional consortia, and four significant aggregators.

Unfortunately, many scholarly data sets (such as papers, patents, and grants) are stored in data silos, which are "vertically locked"—they are neither interoperable nor interlinked. A scholarly "interoperability substrate" that supports persistent and unique identifiers for authors, institutions, countries, papers, patents, grants, data sets and services, and their complex interlinkages is needed. This interoperability substrate would make possible the horizontal traversal and interlinkage of existing data silos. Currently, there are several different approaches to combining and federating data—with different formats, languages, and licenses—from various sources into one single virtual data source or data service.

Centralized Approaches

All data are stored in a central data warehouse. Examples are Google Scholar or the Scholarly Database at Indiana University.

Centralized (Deep Web) Search Approaches

All data are kept at their original location but provide interfaces to multiple data sets for cross-referencing. Examples include *Science.gov* and *WorldWideScience.org*. Science.gov is a gateway to more than 50 million pages of authoritative science information, including research and development results, provided by U.S. government agencies. WorldWideScience.org is a global science gateway meant to accelerate scientific discovery and progress. Its multilateral partnerships enable federated searching of national and international scientific databases. Both these sites use deep Web technology for their federated searches.

Peer-to-Peer Approaches

Decentralized databases are used with decentralized services, which function like popular music file-sharing applications.

Evolving Interoperability Standards and Tools

Existing standards include the Digital Object Identifiers (DOI) system for persistent object identification; the Resource Description Framework (RDF) as a means of describing and interchanging metadata; OpenURL for linking information resources to library services; the Open Archives Initiative Protocol for Metadata Harvesting (OAI-PMH) for the exchange of XML-formatted data; and the Open Archives Initiative Object Reuse and Exchange (OAI-ORE), which defines standards for the description and exchange of aggregations of Web resources. In addition, there are registry efforts such as the NSF Digital Library Service Registries, OpenURL Registry, and Info URI Registry.

CrossRef Web Services

These services offer structured cross-publisher metadata for indexing and linking services. They are used by authorized partners to collect metadata in order to streamline crawling, indexing, and linking services. In May 2008, more than 27 million metadata records were registered via the OAI-PMH interface. Many publishers check references against CrossRef and get DOIs to reduce their error rate for citations.

The W3C SWEO Linking Open Data Community Project

This project aims to extend the Web with a data commons, by publishing various open data sets in Resource Description Framework (RDF) format on the Web and by setting RDF links between data items from different data sources. Data records from diverse open data sets such as Wikipedia, Wikibooks, Geonames, MusicBrainz, WordNet, the DBLP bibliography, and others published under Creative Commons or Talis licenses can be navigated across data set boundaries via RDF links using a Semantic Web browser. In October 2007, more than three billion RDF triples were interlinked by 680,000 RDF links (see figure below). The below image shows the Linked Open Data cloud datasets as of March 2009. Each dataset is colored according the application area, e.g., publishers (green), government (yellow), music (blue), biomedical (red). In November 2009, 13.1 billion RDF triples were interlinked by about 142 million RDF links.

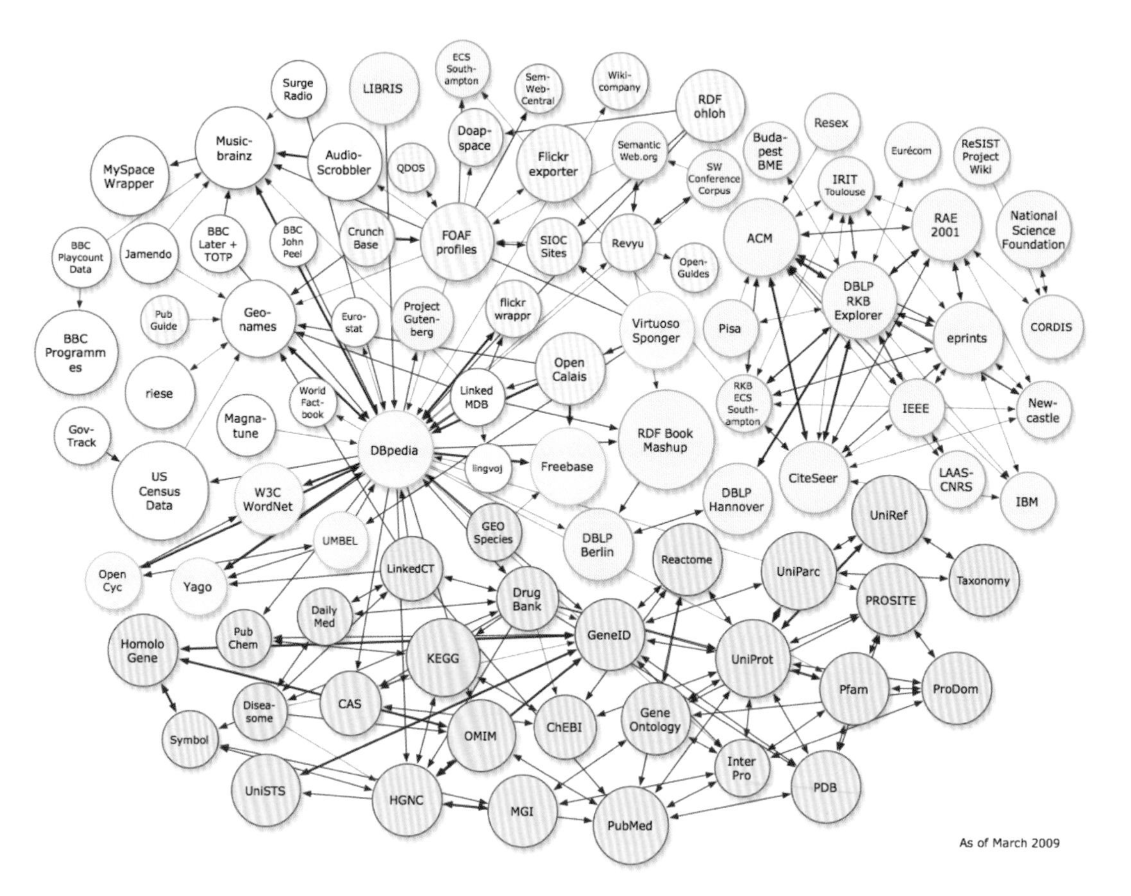

Practical Ontology of Scholarly Data

The MESUR project at the Research Library of the Los Alamos National Laboratory, now at Indiana University, is an ontology of basic scholarly units that covers bibliographic, citation, and usage data. The ontology can be represented as a semantic network of heterogeneous artifacts connected to one another by qualified or typed relationships. The ontology is compliant with the current standards in metadata archiving and harvesting, usage recording, and representational frameworks for semantic networks, and it might become a valuable means of federating scholarly data sets.

Digitometric Services and Tools

Value is added to open-access e-print archives by using the OAI protocol for metadata harvesting. Celestial is an OAI cache and gateway tool. *Citebase Search* enhances OAI harvested metadata with linked references extracted from the full text to provide a Web service for citation navigation and research impact analysis. Digitometrics uses data harvested via OAI to provide advanced visualization and hypertext navigation for the research community. Together, these services provide a modular, distributed architecture for building a semantic Web for research literature.

Data Preservation

Clay tablets and books can be preserved by benign neglect—they easily survive hundreds of years upon a dry shelf. Digital data is more numerous, mutable, diverse, and bound to ever-changing technical infrastructure. Its preservation requires reliable backups, database updates, and data reformatting to meet new standards. In the distant past, religious bodies were the owners and gatekeepers of knowledge; in the last century, reviewers and editors played a dominant role in quality control and knowledge dissemination. Today, anyone can produce and publish data online. Scholarly marketplaces (as discussed on **page 68, Scholarly Marketplaces**) seem to be a viable solution to capture, interlink, and preserve the most valuable assets.

Data Analysis, Modeling, and Layout

The study of science is rich and diverse. The timeline in **Milestones in Mapping Science (pages 26-47)** lists major algorithms, approaches, and tools that have been applied to increase our understanding of science. A review of work in statistics, geography, cartography, data mining, network science, social network analysis, physics, and biomedicine—related to the analysis, modeling, and communication of science—can be found in two reviews titled "Knowledge Domain Visualizations" and "Network Science" (and the references therein). Here we present a general discussion of the functionality, scopes, and implementations of algorithms and their assembly into workflows. Temporal, topical, geographic, and network analysis and visualization approaches are discussed on the opposite page. For other models of science, see **page 58, Conceptualizing Science: Science Dynamics**, and **page 66, Exemplification.**

The purpose of computing is insight—not numbers.
R. W. Hamming

Algorithm Functionality

Algorithms differ in their functionality: Some help to prepare data by filtering, sorting, or reformatting. Others analyze data by extracting features, clusters, trends, or outliers. Algorithms can be grouped by the type of data on which they operate: temporal (*when*), geographic (*where*), topical (*what*), or network (*with whom*) data. Temporal algorithms work on time-series data, such as e-mail or publication records, and the relationships between these events (*when*). Geographic analysis algorithms help make sense of spatial data such as spatial distributions or trajectories (*where*). Topical or semantic approaches are applied to analyze texts (*what*). Network science approaches help in the study of networks (*with whom*). See opposite page for details. Most studies involve multiple types of analysis. Algorithms can also be grouped according to the scale on which they are typically used—individual, local, or global.

Data models aim to simulate, describe, or formally reproduce the statistical and dynamic characteristics of interest (*why and how*). They are validated by comparing simulated and empirical data sets. Of particular interest are models that "conform to the measured data not only on the level where the discovery was originally made but also at the level where the more elementary mechanisms are observable and verifiable." Layout algorithms help to communicate analysis and modeling results (see next spread).

Algorithm Scopes

Most algorithms do not scale well and can only be applied to data sets that are small (less than 1,000 records) or medium-size (1,000 to 100,000 records). Some are optimized for larger data sets, as is necessary for studies of global scope. Data sets and algorithms for different scopes can be mixed to provide focus and context. For example, a global map of science with overlays of nanotechnology research might be shown together with a high-resolution insert of key nanotechnology experts.

Algorithm Implementations

Algorithms are implemented in many different programming languages. Depending on the users' programming skills, there can be huge differences in terms of the memory footprints and processing time of algorithms, as well as in their accuracy and ease of use. It is highly desirable to have standalone, modular algorithms that can be "plugged and played" in a flexible manner. Frequently, the output of a certain algorithm is incompatible with the input format required by the next algorithm in line. It is common to have almost as many data converter algorithms as analysis, modeling, and layout algorithms. Standards and techniques that support the easy and flexible plug and play of algorithms are discussed on **page 68, Scholarly Marketplaces**.

Workflow Design

A workflow—or processing pipeline—is a repeatable pattern of actions, enabled by a systematic organization of resources and information flows, that can be documented and reused. A typical science study workflow involves 5 to 20 approaches/algorithms/tools and the same range of different (intermediate) data formats. The selection of valid workflow elements—from the hundreds of scholarly databases and thousands of different approaches/algorithms/tools developed in the different sciences, together with their composition and parameterization—is a complex and demanding task.

Data types, sizes, formats, quality, scope, and coverage differ widely (see **page 60, Data Acquisition and Preprocessing**). Algorithm functionalities, scopes, and implementations are even more diverse.

Exemplary workflows (compiled by experts) and general workflow schemes are used to guide the design of new workflows and studies. A general workflow for mapping science appears in the table below (adopted from the Börner et al. 2003 "Knowledge Domain Visualizations" review). The major steps are: (1) data extraction, (2) definition of the unit of analysis, (3) selection of measures, (4) calculation of similarity between units, (5) ordination, or the assignment of coordinates to each unit, and (6) display for analysis and interpretation. Steps 4 and 5 of this process are often distilled into one operation. For example, self-organizing maps (see **page 102, In Terms of Geography**) conduct a semantic analysis and produce a spatial map.

For details on step 1, see **page 60, Data Acquisition and Preprocessing.** Step 6 is discussed on **page 64, Data Communication – Visualization Layers**. Steps 2 through 5 are explained below.

Units of Analysis

We discussed major units of science, such as authors, papers, and journals, on **page 56, Conceptualizing Science: Basic Units, Aggregate Units, and Linkages**. To summarize, these entities generally have a publication date and publication type (journal paper, book, patent, or grant) and cover a range of topics (keywords or classifications assigned by authors or publishers). Authors have addresses providing information on their affiliations and geolocations. Because authors and records are associated, author geolocations and affiliations can be attributed to the authors' papers. Similarly, the publication dates, publication types, and topics can be associated with the authors.

Selection of Measures

Statistics such as the number of papers, grants, coauthorships, or citations per author (over time); bursts of activity (number of citations, patents, collaborators, or funding sources); or changes in topics and geolocations for authors and their institutions over time can be computed. Derived networks are examined to count the number of papers or coauthors per author or the number of citations per paper or journal, as well as to determine the strength or success of coauthor/inventor/investigator relationships. The geographic and topic distribution of funding input and research output; the structure and evolution of research topics; evolving research areas (based on young yet highly cited papers); and the diffusion of information, people, and money over geographic and topic space can be studied. For details, consult the review articles mentioned above and cited in the **References and Credits (page 234)**.

Layout

The selection of an appropriate layout or reference system strongly depends on the insight need and type of analysis to be performed (see opposite page).

(1) Data Extraction	(2) Unit of Analysis	(3) Measures	Layout (often one code does both similarity and ordination steps)		(6) Display
			(4) Similarity	(5) Ordination	
Searches *WoS* *Scopus* *Google Scholar* *MEDLINE* *Patents* *Funding* **Broadening** *By citation* *By terms*	**Common Choices** *Journal* *Document* *Author* *Term*	**Counts/ Frequencies** *Attributes (e.g., terms)* *Author citations* *Cocitations* *By year* **Thresholds** *By counts*	**Scalar (unit by unit matrix)** *Direct citation* *Cocitation* *Combined linkage* *Coword/coterm* *Coclassification* **Vector (unit by attribute matrix)** *Vector Space Model (words/terms)* *Latent Semantic Analysis (LSA)* *Singular Value Decomposition (SVD)* **Correlation (if desired)** *Pearson's R on any of above*	**Dimensionality Reduction** *Eigenvector/Eigenvalue Solutions* *Factor Analysis (FA)* *Principal Components Analysis (PCA)* *Multidimensional Scaling (MDS)* *Pathfinder Networks (PFNet)* *Self-Organizing Maps (SOM)* *Topics Model* **Cluster Analysis** *Partition* *Hierarchical* **Spatial Placement** *Triangulation* *Force-Directed Placement (FDP)*	**Interaction** *Browse* *Pan* *Zoom* *Filter* *Query* *Detail on demand* **Analysis and Interpretation**

Temporal Analysis

Science evolves over time. Attribute values of scholarly entities and their diverse aggregations increase and decrease at different rates and respond with different latency rates to internal and external events. Temporal analysis aims to identify the nature of phenomena represented by a sequence of observations such as patterns, trends, seasonality, outliers, and bursts of activity.

Data

A time series is a sequence of events or observations that are ordered in time. Time-series data can be continuous (there is an observation at every instant of time; see figure to the right) or discrete (observations exist for regularly or irregularly spaced intervals). Temporal aggregations—over journal volumes, years, or decades—are common.

Algorithms

Frequently, some form of filtering is applied to reduce noise and make patterns more salient. Smoothing (averaging using a smoothing window of a certain width) and curve approximation might be applied. The number of scholarly records is often plotted to get a first idea of the temporal distribution of a data set. It might be shown in total values or as a percentage of those. One may find out how long a scholarly entity was active; how old it was at a certain point; what growth, latency to peak, or decay rate it has; what correlations with other time series exist; or what trends are observable. Data models such as the least squares model (available in most statistical software packages) are applied to best fit a selected function to a data set and to determine if the trend is significant. Kleinberg's burst detection algorithm is commonly applied to identify words that have experienced a sudden change in frequency of occurrence.

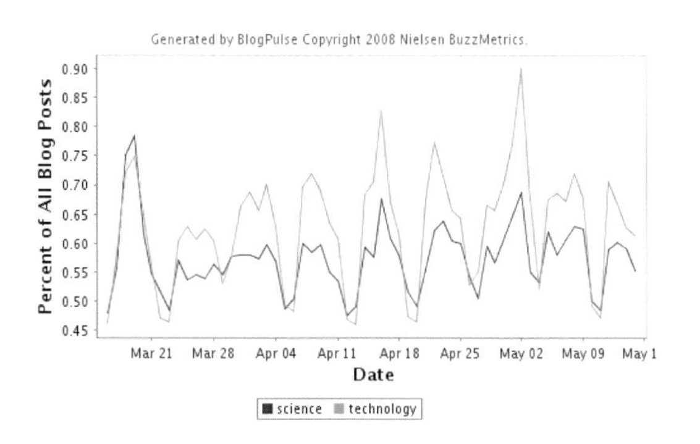

Geographic Analysis

Geographic analysis aims to answer the question of where something happens and what impact it has on neighboring areas.

Data

Geographic analysis requires spatial attribute values or geolocations for authors and their papers, extracted from affiliation data or spatial positions of nodes, generated from layout algorithms. Geographic data can be continuous (each record has a specific position) or discrete (a position or area exists for sets of records, like the number of papers per country). Spatial aggregations (for example, merging data via postal codes, counties, states, countries, and continents) are common (see **page 66, Exemplification**).

Algorithms

Cartographic generalization refers to the process of abstraction. This includes (1) graphic generalization: the simplification, enlargement, displacement, merging, or selection of entities without enhancement or effect to their symbology and (2) conceptual symbolization: the merging, selection, and symbolization of entities, including enhancement (such as representing high-density areas with a city symbol).

Geometric generalization aims to solve the conflict between the number of visualized features, the size of symbols, and the size of the display surface. Cartographers dealt with this conflict intuitively in part until researchers like Friedrich Töpfer attempted to solve them with quantifiable expressions.

Flow maps use line thickness and direction to show the number of tangible or intangible entities that diffuse over a geographic location or science space (see Chinese Academy of Sciences coauthor network, below, and **page 158, 113 Years of Physical Review**).

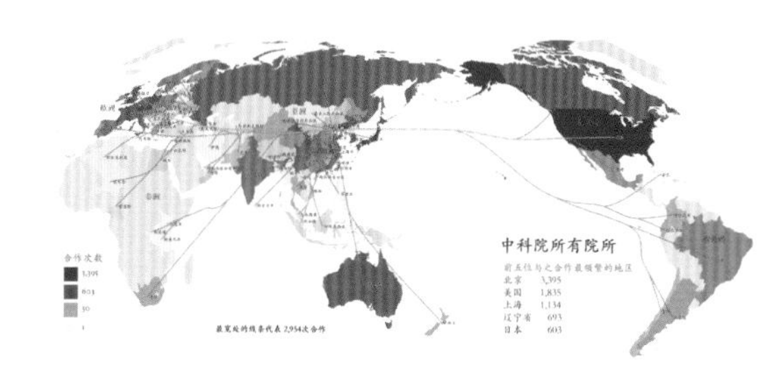

Topical Analysis

The topic coverage and topical similarity of basic and aggregate units of science (authors or institutions) can be derived from the units associated with them (papers, patents, or grants).

Data

The topic or semantic coverage of a unit of science can be derived from the text associated with it. Topical aggregations (for example, over journal volumes, scientific disciplines, or institutions) are common.

Algorithms

Topic analysis extracts the set of unique words or word profiles and their frequency from a text corpus. Stop words, such as "the" and "of," are removed. Stemming can be applied. Coword analysis identifies the number of times two words are used in the title, keyword set, abstract, or full text of a paper. The space of co-occurring words can be mapped, providing a unique view of the topic coverage of a data set (see **page 66, Exemplification**). Similarly, units of science can be grouped according to the number of words they have in common.

Salton's term frequency inverse document frequency (TFIDF) is a statistical measure used to evaluate the importance of a word in a corpus. The importance increases proportionally to the number of times a word appears in the paper but is offset by the frequency of the word in the corpus.

Dimensionality reduction techniques (see table on opposite page) are commonly used to project high-dimensional information spaces (for example, the matrix of all unique papers multiplied by their unique terms) into a low, typically two-dimensional space.

The self-organizing map below shows the topic landscape of geography abstracts; see **page 102, In Terms of Geography**.

© 2002 IEEE

Network Analysis

The study of networks aims to increase our understanding of natural and manmade networks. It builds on social network analysis, physics, information science, bibliometrics, scientometrics, informetrics, webometrics, communication theory, sociology of science, and several other disciplines.

Data

Authors, institutions, and countries, as well as words, papers, journals, patents, and funding, are represented as nodes and their complex interrelations as edges (see **page 52, Conceptualizing Science: Basic Anatomy of Science**). Nodes and edges can have time-stamped attributes.

Algorithms

Diverse algorithms exist to calculate specific node, edge, and network properties (see "Network Science" review). Node properties include degree centrality, betweenness centrality, or hub and authority scores. Edge properties include durability, reciprocity, intensity (weak or strong), density (how many potential edges in a network actually exist), reachability (how many steps it takes to go from one "end" of a network to another), centrality (whether a network has a "center" point), quality (reliability or certainty), and strength. Network properties refer to the number of nodes and edges, network density, average path length, clustering coefficient, and distributions from which general properties such as "small-world," "scale-free," or "hierarchical" can be derived. Identifying major communities via community detection algorithms and calculating the "backbone" of a network via pathfinder network scaling or maximum flow algorithms helps to communicate and make sense of large-scale networks. See the coauthor network of information visualization researchers below.

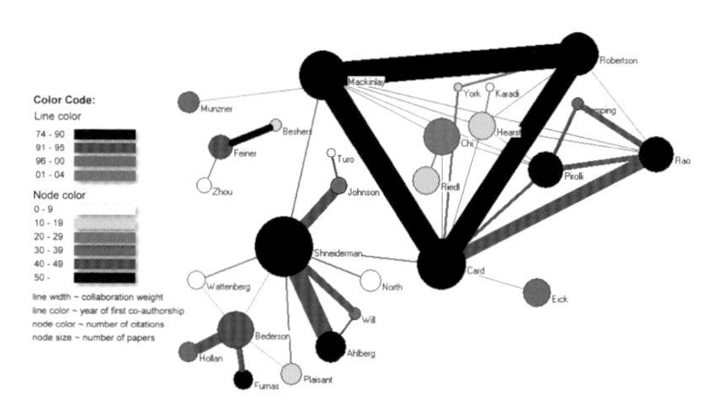

Data Communication—Visualization Layers

We are trained to read data graphs, charts, maps, and diagrams. But how to best represent something as abstract as science? Typically, many different data sets need to be merged, and diverse preprocessing, analysis, and layout algorithms have to be employed in order to arrive at insightful results. The effects of data and algorithms choices have to be communicated to help with the interpretation of the results. Here we discuss visual perception and cognitive processing principles, the importance of flow experiences, sociable interfaces, and visual scalability issues as guides to workflow and visualization design. The visualization layers inspired by geographic information systems and introduced on **page 50, Building Blocks**, are detailed and exemplified on the opposite page.

The First Law of Geography: Everything is related to everything else, but closer things are more closely related.
Waldo Tobler

Visual Perception and Cognitive Processing

The human body and mind—having evolved over millions of years—are optimized for mating, hunting, and gathering, but not necessarily for making global sense of enormous scientific data streams. Our genetic makeup equips us with a brain that is a powerful pattern-matching engine. About 60 to 75 percent of our brain's cortex is devoted to visual information processing. Human vision is a highly dynamic, continuous process that constructs short-lived models of the external world. Structures, groups, and trends can be discovered among hundreds of objects (data). Patterns of moving points can be perceived easily and rapidly. Gestalt laws define how we group objects and perceive forms. Bottom–up perception combined with top-down, concept-driven processes enable us to quickly make sense of our surroundings in the context of existing mental models.

Relevant work comes from cognitive science, psychology, psychophysics, education, and graphic design. Major works by Rudolf Arnheim, Christopher G. Healey, John Maeda, Stephen E. Palmer, Colin Ware, and others are listed in the **References and Credits (page 235)**. A deep understanding of visual perception and cognitive processing is mandatory for the design of effective visualizations.

Usable Interfaces and Flow Experiences

Some activities—like extreme sports or proving a math theorem—place us into a mental state of operation that psychologist Mihaly Csikszentmihalyi called "flow." A person in flow is fully immersed in an activity and feels energized focus, full involvement, and success. Flow experiences are some of the most intensely rewarding and enjoyable moments in life, and many of us devote considerable amounts of energy and resources to achieve them. Ideally, data exploration and communication keeps users in flow by balancing **challenges** and **skills** to avoid **boredom** and **anxiety** (see figure below).

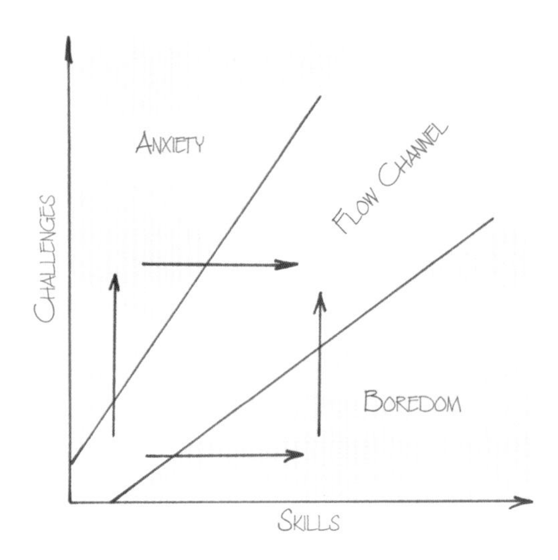

Sociable Interfaces

Most visualizations are designed for individual consumption. Increasing task complexity, however, poses the need for collaborative exploration and interpretation of data analysis and modeling results. Experts from different domains of research and different geolocations need to collaborate in order to arrive at urgently needed solutions. Old-fashioned paper printouts allow geographically colocated experts to examine, annotate, and discuss data. Shared handheld/desktop interfaces and collaborative virtual reality environments support the interactive exploration of data. See **page 186, Hands-On Science Maps for Kids** for a discussion of multiuser environments in education.

Computational and Visual Scalability Issues

Scalability issues apply to bandwidth size and speed, efficient data structures and algorithm implementations, and visual presentation. Visual scalability is affected by the hardware and software used as well as by our own visual perception and cognitive processing capabilities. Improvements are needed to develop data aggregation strategies, visual metaphors, interactivity, perspectives, and visual design patterns.

The development of science mapping tools is likely to resemble the evolution of geographic information systems (GIS), which made possible the efficient design and production of affordable geographic maps from huge databases. Cartographers had to adapt their design processes and production methods to match the functionality of these tools. Even today, automatically generated maps do not yet match the information density, insightfulness, and beauty of the hand-drawn masterpieces of traditional mapmakers. How can we design tools that enable the design of insightful visualizations while remaining flexible enough to support creativity in power users?

Modular Visualization Design

It appears to be desirable to have a visual (design) pattern language that provides modularity at the interface design level and guides the iterative design of "visual workflows." This language would generate informative yet aesthetically pleasing visualizations. "Data layers" are used in GIS systems to support the visual layering and coordination of different data sets, such as water pipes, streets, and electricity lines. Design layers are supported by graphic design software systems such as Adobe Photoshop and Illustrator, which enable the separate design and modular composition of design elements.

The Data Communication—Visualization Layers introduced on **page 50, Building Blocks**, and detailed on the opposite page define distinct parts with highly specific functionality that collectively define a visualization. The layers can be instantiated in any order; a typical sequence from **Reference System** to **Deployment** is shown at right. The instantiation of any one layer tends to strongly affect the remaining layers. For example, changing the size of nodes in a network may result in overlapping nodes; therefore, the reference system may need to be adjusted. The decision to have multiple windows requires that graphic design and layout still allow for legibility at a smaller window size. Algorithms and other instantiation choices for layers are provided by different sources; for instance, graph theorists provide effective graph layout algorithms for the **Reference System** layer; geographers and others develop geographic **Projections** and **Distortions**; and designers develop type fonts and color palettes used in **Graphic Design**.

Communication Versus Exploration

Visualizations aim to communicate or transfer information, to prompt visual thinking, and to support exploration, but they vary considerably in which aspect is emphasized. Early mapmaking, visualization design, and science map making focused on the use of static maps that effectively communicate specific pieces of information to a public audience. Recent works are more likely to be highly interactive maps that support individuals and small groups interested in hypothesis generation, data exploration, and decision support.

Interpretation

Science studies and resulting maps of science are not objective, neutral artifacts. They are created. They include or omit information. They communicate particular messages. Commonly, the messages are those of the powerful who pay for the visualizations. Therefore, it is of utmost importance that the complete workflow used to create these studies and maps be documented in a fashion that enables anybody with access to the same data sets and algorithms to arrive at the very same results. Repeatability also eases the comparison of approaches and improves the verification and credibility of results.

Well-designed studies and the resulting maps represent true data in an informative way, engaging both visual perception and cognitive processing. They are easy to interpret as well as aesthetically pleasing.

Visualization Layers

1. Reference System

Different needs, data types, and analysis results demand different reference systems. Some of the most commonly used reference systems are two-dimensional, such as two axes (for example, number of publications per year or number of citations over an alphabetically or rank-ordered list of institutions). Hierarchies are frequently plotted as indented lists or as a treemap. Parallel coordinate systems plot data sets over multiple axes arranged in parallel. Circular reference systems can be used to examine the network structure (see Mark E. J. Newman's coauthor network in **Milestones in Mapping Science, page 34**) or the semantic structure of a data set (see W. Bradford Paley's **TextArc Visualization of "The History of Science", page 128**). Geographic reference systems plot spatial data at different scales. Science reference systems support the "science location" of data (see **page 12, Towards a Reference System of Science** and **Part 4: Science Maps in Action**). Note that some references have no well-defined interpretation. Examples are network layouts or self-organizing maps that attempt to place similar nodes closer in space, but are invariant to rotation and mirroring.

2. Projection/Distortion

Many cartographic projections exist. The three main kinds are planar/azimuthal (circular), conic (fan-shaped), and cylindrical (rectangular). Projections are chosen such that distortions are minimized in accordance with map purpose.

Distortion techniques such as equal-area cartograms, also known as density-equalizing maps, are widely used for distorting the surface areas of countries according to given variables (for example, number of papers published). Given our familiarity with the world or U.S. map, these maps can be easily interpreted despite their distortion. Polar coordinates and hyperbolic spaces are sometimes used to provide focus and context.

3. Raw Data

A reference system allows for the placement of data. Density patterns and outliers may become visible, but data records having identical coordinates will appear as one data point.

4. Graphic Design

Most data records have multiple attributes, which can be represented by size-, color-, and shape-coding. Size-coding is made with the same coordinates; however, different attribute values make multiple records visible. Textual labels for major graphical elements help interpret a map. Landmarks ease navigation and exploration.

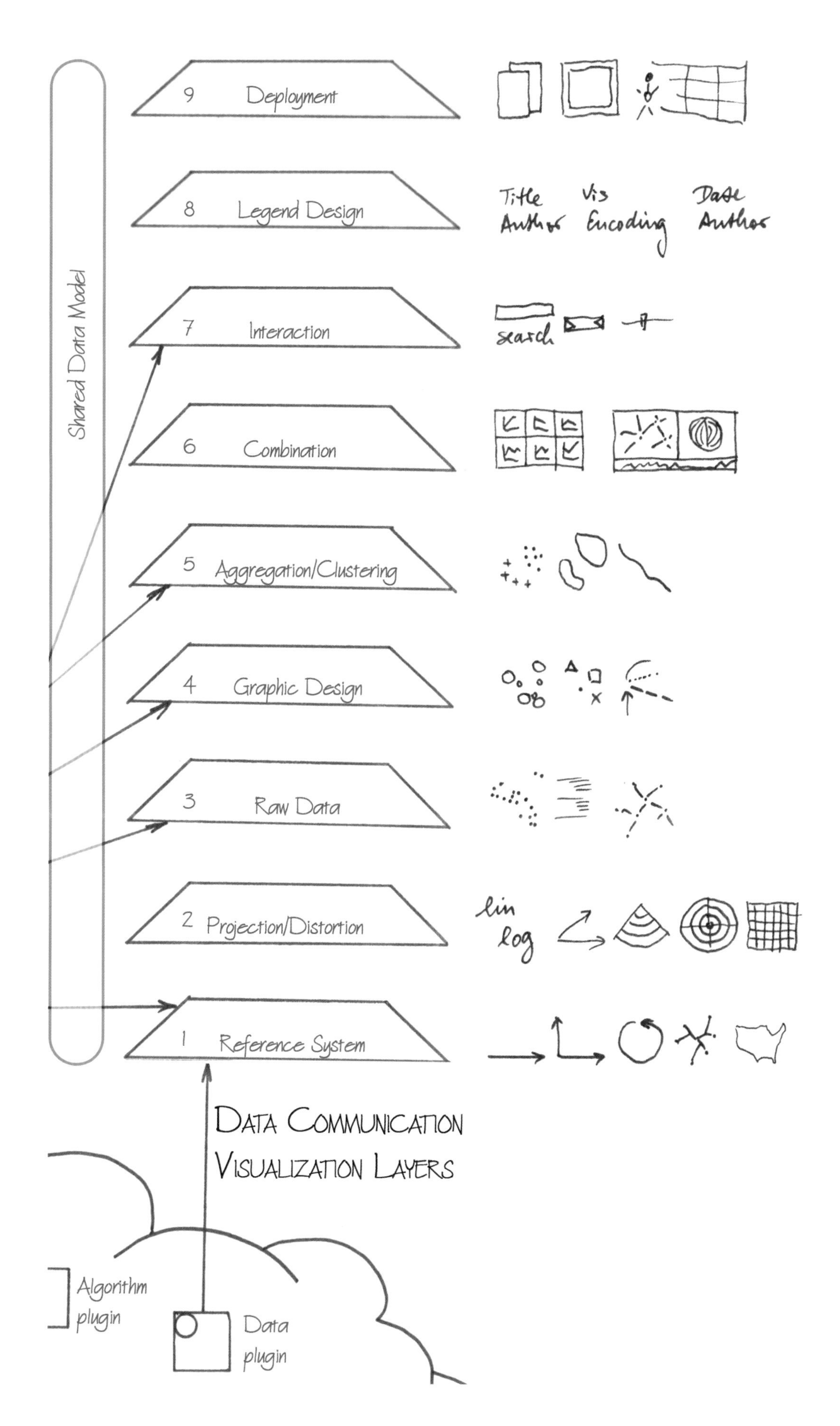

5. Aggregation/Clustering

High-density areas and limited screen/paper space often mandate the grouping of records into higher-level structures (for example, authors can be grouped by their geographic location or institution). Semantic spaces are often split into topic areas or network communities. Cluster boundaries can help to visually separate them. Network layouts often benefit from the identification of communities using betweenness centrality clustering, and the highlighting of backbone structures is calculated using pathfinder network scaling.

6. Combination

It is often beneficial to examine a data set from different perspectives—using multiple, coupled windows. For example, to look at the growth of a nation it might be beneficial to examine a geographic map of exported goods and a science map of federal funding with resulting patents. Small multiples are graphical depictions of different attributes of a data set using the identical reference system—for example, a scatterplot. They can be examined within a user's eye span to support comparisons.

7. Interaction

Often, data is too vast to be understood at once. Interaction via zooming and panning, exploration via brushing and linking, and access to details via search and selection becomes important. Ben Shneiderman's visual information seeking mantra—"Overview first, zoom and filter, then details-on-demand"—summarizes the major visual design guidelines.

8. Legend Design

No visualization is complete without information on what data is shown and how it was processed, by whom, and when. As more advanced data preprocessing and analysis algorithms are developed, it becomes necessary to educate viewers on the effect of parameters and visualization layer instantiation decisions, which add credibility and support interpretation.

9. Deployment

Visualizations can be deployed as printouts on paper, online animations, or interactive, three-dimensional, audiovisual environments. Printouts are inexpensive and high-resolution; they can be very large and are easy to examine and annotate. Interactive displays are more expensive yet lower-resolution; they support the direct manipulation of algorithm parameters and visual elements.

Exemplification

Data acquisition, preparation, analysis, and modeling, as well as the communication and interpretation of results, are exemplified here using a series of three science studies. Each of the studies uses the same 20-year data set from the Proceedings of the National Academy of Sciences (*PNAS*) but responds to a different question. Study I aims to identify the most highly cited institutions and their roles as knowledge producers and knowledge consumers; study II extracts the global topical structure of the data set as well as bursting topics; and study III uses a computational model to increase our understanding of the dynamics of coevolving author-paper networks. General workflows and a subset of the study results are presented for means of illustration and comparison.

Tell me, I forget, show me, I remember, involve me, I understand.
Benjamin Franklin

Data Acquisition

All three studies use a 20-year data set comprising 45,120 regular *PNAS* papers published by 105,915 unique authors in the years 1982–2001 (see gray area X in the chart below). Citation links are only available for papers that are cited by other papers in this data set. Citations given to or received from papers outside this data set were not considered. The *PNAS* data set covers high-quality interdisciplinary research with a strong focus on biomedical research. It is unlikely that it captures all works by any one author.

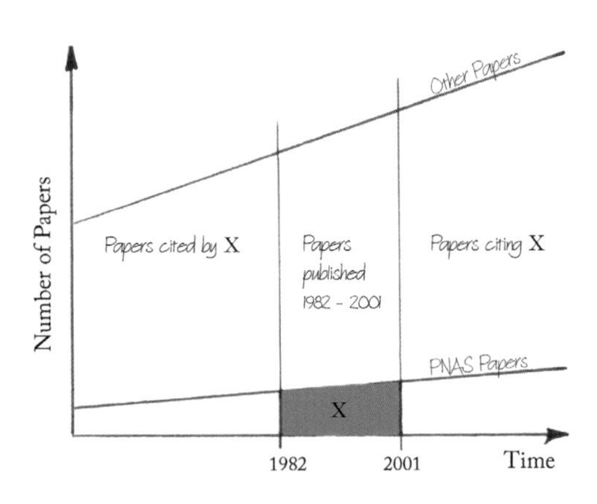

Data Preparation

Extensive data preparation consumed 80 percent of the project time. Major data preparation steps specific to each study are explained subsequently.

Study I: The unique set of U.S. institutions (such as universities, research labs, and corporate entities) with their inter-citation counts was identified. Compromising geographic identity and statistical significance, multiple geographically-near instances of any one institution were merged while distant campuses of the same institution were kept separate. The resulting sets of unique institutions were labeled by a concatenation of the original name with the smallest postal code within that county. Longitude and latitude coordinates for each postal code were derived using geographic lockup services.

Study II: The top 10 percent of the most highly cited documents for each of the 20 years, excluding papers without titles, were selected. From these 4,699 papers, a total of 34,299 unique topic words from paper titles, Institute for Scientific Information (ISI) keywords, and MEDLINE's controlled vocabulary (MeSH terms) were extracted.

Study III: Statistical properties and temporal distributions of the 45,120 regular *PNAS* papers were calculated.

Data Analysis, Modeling, and Layout

Study I: Inter-citation counts for all unique institutions were calculated and the top 500 institutions identified. Lists of top consumers/producers with their consumers/producers were calculated. Interestingly, the top five produced 30,572 papers (65 percent of the total) and received 195,889 (52 percent) of a total of 377,935 citations.

Study II: Jon M. Kleinberg's burst detection algorithm was applied to identify those words in the 34,299 topic words that experienced a sudden increase in usage in the selected 4,699 papers. Next, the intersection of the highest frequency and most highly bursting word sets was computed, the top 50 words were selected, and their co-occurrence matrix was calculated. Study III: A general process model that simultaneously grows coauthor and paper-citation networks was implemented. The model takes a set of input values and parameters and generates networks in which authors are interlinked via undirected coauthorship relations. Papers are interconnected via directed "provides input to" citation links. Authors and papers are interlinked via directed "consumed" links, denoting the flow of information from papers to authors, as well as directed "produced" links, representing the act of paper generation by authors. The statistical and dynamic properties of these networks were compared against the *PNAS* data set.

Data Communication—Visualization Layers

Study I: Figure I(a) shows characteristics of the top 5 consumers. A log-log plot of the number of institutions citing one another over geographic distance for four consecutive 5-year time periods can be seen in I(b). Citation activity for the top 500 and labels for the top 5 most highly cited institutions were overlaid on a geographic U.S. map in I(c). The height and color of the citation bars indicate the number of citations received (see legend).

Study II: The number of times the top 10 most frequently used words occur in the 4,699 papers is shown in II(a). The network of the top 50 highest frequency and most highly bursting words was plotted using the algorithm by Fruchterman-Reingold. Note that the network layout is invariant to rotation and mirroring. Photoshop was used to generate the final layout, II(c). Here, nodes are area-size-coded by burst weight, color-coded by burst onset, and circle-ring-color-coded by year of maximum word counts. The width of the links reflects the number of times two words co-occurred (see legend).

Study III: The statistical properties of the 45,120 regular *PNAS* papers are tabulated in III(a). III(b) and III(c) show the total number of actual and simulated papers, authors, and citations over time. In Figures III(a–c), values for the simulated data set (SIM) are highlighted in green while results for the *PNAS* data set are given with black background.

Consumers (citing institutions)	# citations made	Top ten producers (institutions cited by consumer), ordered by number of citations
Harvard U	13,552	MIT, Massachusetts Gen Hosp, Brigham & Women's Hosp, Johns Hopkins U, Stanford U, U Calif San Francisco, Yale U, Rockefeller U, U Washington, Washington U
U Calif SF	4,682	Harvard U, MIT, Stanford U, Johns Hopkins U, U Washington, Washington U, U Calif Berkeley, U Texas, U Calif SD, U Calif LA
MIT	4,655	Harvard U, Whitehead Inst Biomed Res, Johns Hopkins U, Stanford U, U Calif SF, Yale U, Rockefeller U, U Calif LA, Massachusetts Gen Hosp, U Calif Berkeley
NCI (zip: 20814)	4,519	Harvard U, NCI (zip: 20205), NCI (zip: 21701), MIT, Duke U, Johns Hopkins U, NIAID, NICHHD, Stanford U, U Calif SF
Yale U	4,464	Harvard U, MIT, Stanford U, Rockefeller U, Johns Hopkins U, Washington U, U Calif SF, U Washington, NCI, Massachusetts Gen Hosp

I(a) Top 5 consumers, their total number of citations made (excluding self-citations), and their top 10 producers

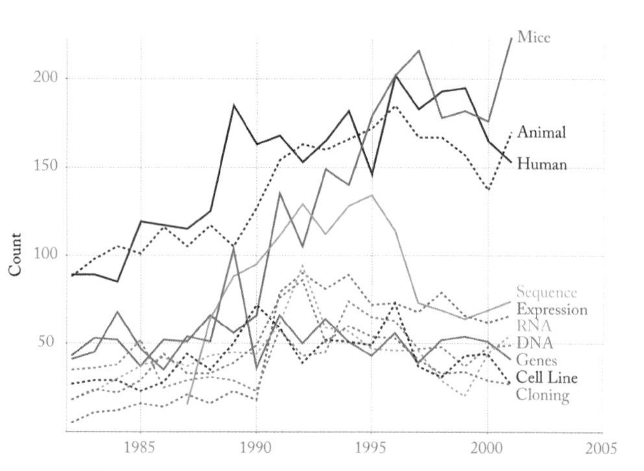

II(a) Frequency count for the most frequently used words

Number of nodes n, average node degree <k>, path length l, cluster coefficient C, and power law exponent γ.

Network	n	<k>	l	C	γ
Coauthorship networks					
LANL	52,909	9.7	5.9	0.43	--
MEDLINE	1,520,251	18.1	4.6	0.066	--
SPIRES	56,627	1.73	4.0	0.726	1.2
NCSTRL	11,994	3.59	9.7	0.496	--
Math.	70,975	3.9	9.5	0.59	2.5
Neurosci.	209,293	11.5	6	0.76	2.1
PNAS	105,915	8.97	5.89	0.399	2.54
Paper citation networks					
ISI	783,339	8.57	--	--	3
PhysRev	24,296	14.5	--	--	3
PNAS	45,120	3.53	--	0.081	2.29
SIM	37,114	2.13	--	0.074	2.05

III(a) Statistical properties

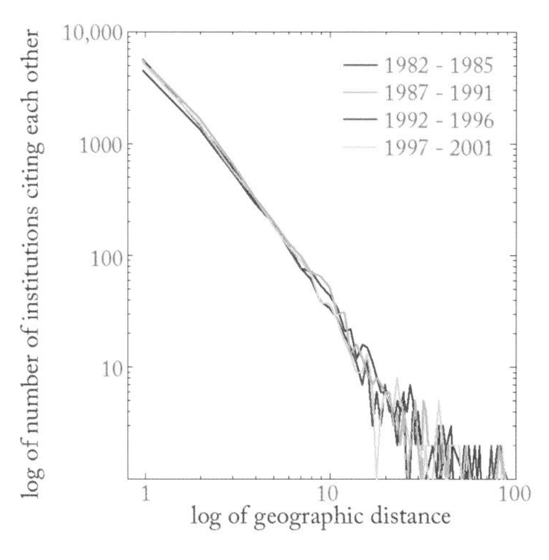

I(b) Log-log plot showing the variation of the number of institutions that cite one another over geographic distance between them for each of the four time slices

I(c) Geographic locations of the top 500 institutions with number of received citations indicated by height and color of bars

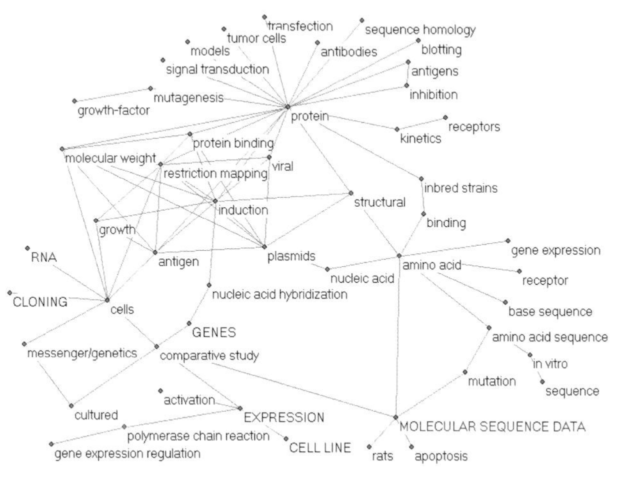

II(b) Fruchterman–Reingold layout of word co-occurrence matrix

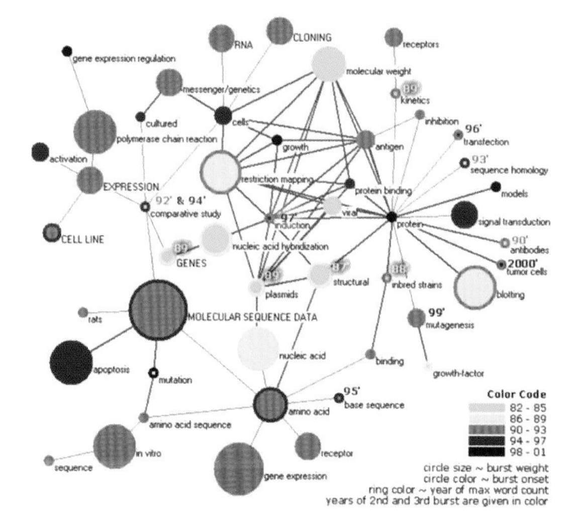

II(c) Final layout with size- and color-coding, labels, and legend

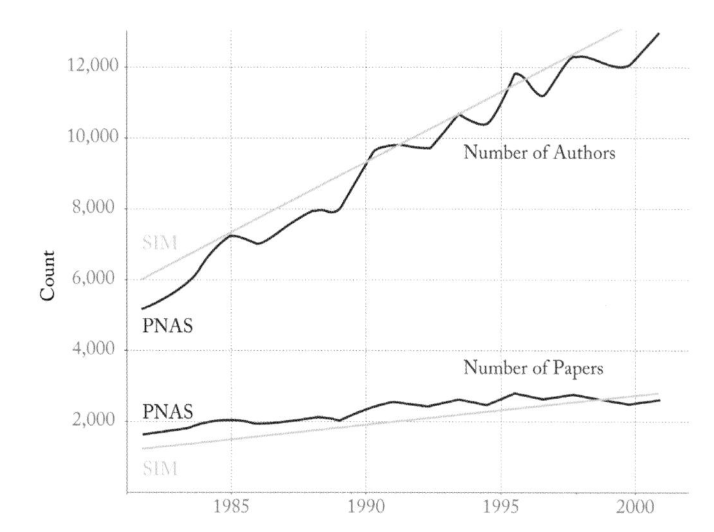

III(b) Number of authors and papers over time

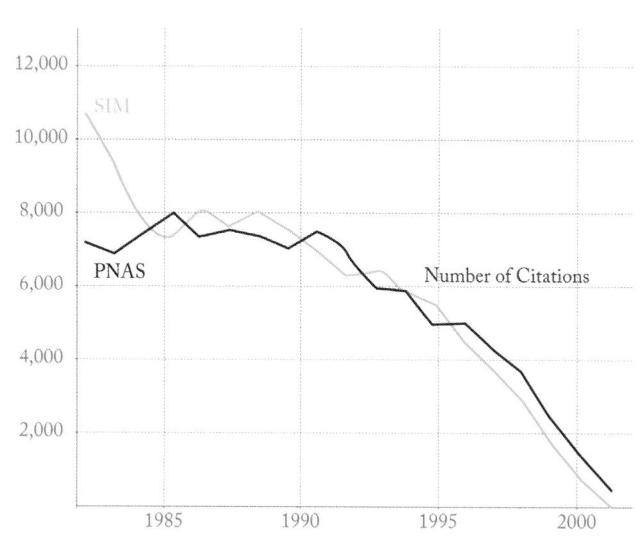

III(c) Number of received citations over time

Interpretation

Study I: Mapping Knowledge Diffusion and the Importance of Space

This study aimed to determine whether the Internet leads to more global citation patterns (that is, more citation links between papers produced at geographically distant research institutions). A novel approach to analyzing the dual role of institutions as information producers and consumers and to studying and visualizing the diffusion of information among them was developed. Surprisingly, the widespread adoption of the Internet does not seem to have affected the distance over which information diffuses as manifested by citation links. The citation linkages between institutions fall off with the distance between them, and there is a strong linear relationship between the log of the citation counts and the log of the distance that does not change over time. Reasons for local collaborations might include "winner takes all" funding schemes; the demands of complex, large-scale instrumentation; and the need to gain experience, train researchers, and sponsor protégés. The social component of citation seems to become more important as researchers are flooded with information, and spatial proximity eases the creation and continuation of close personal relationships.

Study II: Identifying Research Topics and Trends

Scientific research is highly dynamic. New areas of science are continually evolving; some may shift in importance, others merge or split. Because of the steady increase in the number of scientific publications, it is challenging to keep abreast of the structure and dynamic development of one's own field, let alone all scientific domains. However, knowledge of "hot" topics, emergent research frontiers, and change of focus in certain areas is a critical component of resource allocation decisions in research laboratories, government institutions, and corporations. This study aimed to increase understanding of the topic coverage and activity bursts of words in highly cited *PNAS* papers. Interestingly, the burst of words seems to precede their wide spread usage. "Protein" and "model" were among the highly "bursty" terms between 1998 and 2001, and they have become important research topics since then.

Study III: Modeling the Coevolution of Author-Paper Networks

Models of scientific structure and evolution can help us understand the inner workings of science (see **page 58, Conceptualizing Science: Science Dynamics**). The TARL model (topics, aging, and recursive linking) describes the coevolution of coauthor and paper-citation networks. Using an agent-based approach, TARL simulates nodes (authors or papers), their edges (undirected coauthor, directed consumed, and directed paper-citation), and their attributes (time and topics). Topics cluster papers and authors *topically*. Aging is an antagonistic force to preferential attachment. Even highly connected nodes receive a decreasing number of links over time. Aging clusters papers and authors *temporally*. Recursive linking refers to the tendency of authors to cite papers referenced in material they are currently reading, which provides a grounded mechanism for the "rich get richer" phenomenon as an emergent property of the elementary activity of authors. According to this model, the number of topics is linearly related to the clustering coefficient of the simulated paper citation network.

Scholarly Marketplaces

As discussed in **Part 1**, it is beyond the financial and intellectual resources of any one person or institution to create a repository of all scholarly knowledge, data, and software and to develop the tools needed to exploit such a repository. Instead, we should aim to design sociotechnical infrastructures that empower everyone to contribute to and exploit the common good of science. The envisioned infrastructures have to provide access to powerful data-algorithm-computational resources. More importantly, they have to fulfill basic human needs and exploit existing scholarly reputation mechanisms to get widely adopted. The required social engineering and supporting system architectures are discussed here.

The grandest projects of humanity took on the order of 100,000 people: the Panama Canal, the pyramids of Egypt. Now, for the first time in history, we can easily get more people than that working together. Imagine what we could do with 500 million people.
Luis von Ahn

Designing Vibrant Marketplaces

The value of a marketplace increases with the number and diversity of sellers, buyers, and exchanged goods. In the case of a scholarly marketplace, goods are basic scholarly units such as experts, publications, patents, data sets, algorithms, tools, specimens, and equipment. Sellers and buyers are scientists, educators, and practitioners, as well as the general public. Scholarly products are offered in exchange for resources or reputation (see **page 59, Symbolic Capital and the Cycle of Credibility**). For example, a researcher might publish a preprint or data set online to receive more citations and download counts that increase his reputation.

Using a Million Minds

Millions of people connecting via the Internet make possible the creation of libraries, software, and infrastructures that are greater then the sum of their parts; see also **Metcalfe's law (page 57)**. Rather than maintaining one grand Library of Alexandria or Mundaneum at one physical location (see **page 18, Knowledge Collection**) distributed resources are now being built to harvest our collective intelligence and make it publicly available.

Successful examples are social networking destinations like Facebook, online encyclopedias like Wikipedia, and photo- and video-sharing sites like Flickr and YouTube. Without the millions of user contributions, such sites would be "empty shells"—of little interest or value. Political groups like MoveOn.org have galvanized online grassroots organizing. News aggregators like Digg have given editing power to readers. There are peer pioneers like Spikesource and Zopa; "Ideagoras" like eBay, IdeaConnection, and crowdsourced software; prosumers like Second Life, wiki, and the New Alexandrians who support the sharing of science (see **page 198, Visual Interfaces to Digital Libraries**); open platforms, including Amazon API, Amazon Cloud, and Pikspot; and the global platform of Boeing: The peer producers. The ESP Game, licensed to Google, lets randomly paired players compete for the best labels that describe an image. It is estimated that 5,000 people continuously playing the game could assign a label to all images indexed by Google in 31 days. Combined, these Web sites have changed the landscape of countless industries, and some have become worth billions. Sites like Many Eyes and Swivel support the sharing of data and data graphs. Playful, social exploration of data leads to serious analysis. In this Web 2.0 world, people work while having fun (see **page 69, Application Design**, and opposite page).

Software Bazaars and Crowdsourcing

The "million minds" approach is also useful when designing software. As Eric S. Raymond puts it, in early times, important software (such as operating systems) was built like a cathedral, "carefully crafted by individual wizards or small bands of magi working in splendid isolation, with no beta to be released before its time." In recent times, however, some of the most successful software projects, such as Linux, have come to resemble a great babbling bazaar, with early and frequent releases, delegation of all things, and openness to the point of promiscuity. Anyone can contribute anything, and a coherent and stable system seems to emerge only by a succession of miracles.

Going one step further, "crowdsourcing" refers to the phenomenon of corporations creating goods, services, and experiences in close cooperation with consumers. Crowdsourcing turns product development into an open-source platform and allows companies to tap into the intellectual capital of amateurs. In exchange, end users have a direct say in what actually gets produced, manufactured, developed, designed, serviced, or processed.

Social Engineering

Human needs and reputation mechanisms have to be exploited to design vibrant scholarly marketplaces.

Understanding Human Needs

Abraham Maslow's hierarchy distinguishes three types of needs: basic, growth, and being needs. **Basic Needs** comprise the **physiological** (hunger, thirst, and bodily comforts); **safety** and **security** (being out of danger); **belongingness** and **love** (affiliation with and acceptance within community); and **esteem** (the self-regard that allows one to achieve goals and recognition). **Growth Needs** refer to the **cognitive** (knowing, understanding, and exploring) and the **aesthetic** (symmetry, order, and beauty). **Being Needs** include **self-actualization** (self-fulfillment; the realization of one's potential) and **self-transcendence** (moving beyond oneself to help others actualize their potential).

Growth forces, innate to all of us, are what gradually guide us up in this hierarchy of needs. We feel alive if we are in flow and connected. We want to belong to a family, team, or community. We like to be special and valued by others through our work and way of living, our kindness and generosity. We want to make a difference and strive to spend our lives on meaningful and rewarding projects.

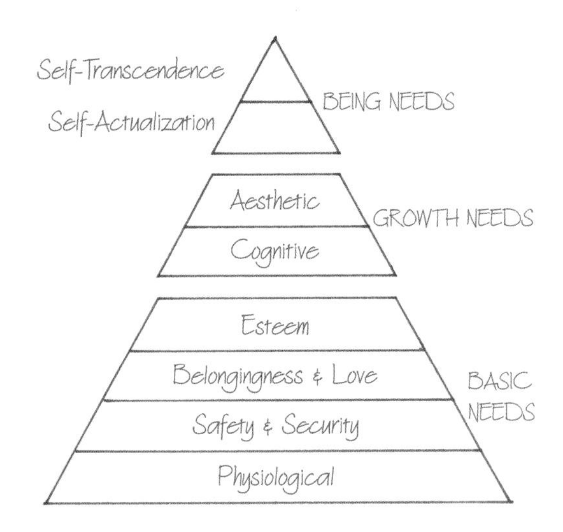

In today's developed countries, it is relatively easy to satisfy one's basic needs; there, most people are striving for growth needs and being needs. Illness or unemployment tend to return our focus to basic needs; when those are satisfied, we quickly refocus our attention on higher aspirations.

Serving Scholarly Needs

A successful scholarly marketplace has to serve scholarly needs (see **page 8, Knowledge Needs and Desires**) and exploit existing scholarly reputation mechanisms (see **page 59, Feedback Cycles**). It has to support easy access to and sharing of resources; rating, commenting on, and documenting these resources in support of their management and usage; effective compilation of resources and services into customized tools for use in projects that might last several days or many years; design of advanced visualizations of empirical or simulation data for research and publication purposes; collaboration; and rapid diffusion of results to scholars, education, government, and industry.

The marketplace needs to exploit existing reputation mechanisms: scholars who share resources or contribute to the annotation, rating, or interlinking of resources need to benefit in terms of visibility, citation and download counts, and ultimately via promotion, scientific awards, and funding. Ideally, the marketplace would show not only who achieves but also who shares and cares. It must be adaptable to new technological developments, discoveries, and grand challenges and customizable to different scientific areas, languages, and cultures. It has to scale (in terms of disk space, computing power, and diversity of scholarly entities) and be maintainable for many decades (so that effort spent on annotation today is not lost tomorrow).

Science today is global. Scientific conferences bring together the expertise and diversity of researchers and practitioners from different scientific disciplines and cultures. Scholars consider not only jobs in their home countries but also on other continents. A scholarly marketplace that addresses concrete needs in an easy-to-understand fashion will be used by many—further increasing its value.

Sociotechnical Cloud Design

Many bits and pieces of the envisioned scholarly marketplace already exist. They are depicted in the figure on the opposite page and discussed here.

Internet

The lowest level shown is the physical network of cables and computers, disks, printers, sensors, and routers that make up the physical Internet. Each component in the network has an IP address used to interconnect components via the TCP/IP in a decentralized fashion (see **page 19, Vinton Gray Cerf**).

Storage Clouds

Sets of hardware components can be aggregated into storage clouds using standard REST and SOAP Web service interfaces. They support the storage and retrieval of any amount of data from anywhere on the Web. An example is the Amazon Simple Storage Service (Amazon S3).

In 1992, 1 gigabyte of disk space cost $1,000. In 2001, the same disk space cost $9. In 2005, 1 terabyte cost $2,200; this price is expected to drop to $20 in 2010. Very soon it will be more cost-effective to pay per use for disk space than to own or manage private infrastructure.

Compute Clouds

Multiple computing resources can be aggregated so that users can create, launch, and terminate server instances on demand. Examples include the Amazon Elastic Compute Cloud (Amazon EC2), which makes possible the scalable deployment of applications, and XtremWeb, which supports the setup of desktop grids using compute cycles from idle home computers via a peer-to-peer infrastructure. Computing energy becomes a service supplied on demand—similar to electricity and water. Grid-computing technology interlinks some of the most powerful computers and databases to make their capabilities and services available to authenticated users. Examples include Grid.org projects researching cancer, smallpox, or anthrax and IBM's Deep Thunder applied to local, high-resolution, short-term weather forecasting.

Peer-to-peer architectures interconnect multiple personal computers in a decentralized fashion so that data and compute cycles can be shared. The resulting computing power is easily comparable to that of a supercomputer, and many applications (such as SETI@Home, Einstein@Home, Folding@Home, and Astropulse), have applied this technique for scientific computations.

Cyberinfrastructures

A storage-algorithm-compute cloud is also known as cyberinfrastructure (CI). CIs address the ever-growing need to connect researchers and practitioners to the data, algorithms, and massive disk space and computing power that many computational sciences require.

Service-Oriented Architectures

Service-oriented architectures (SOA), such as the Open Services Gateway initiative (OSGi) or mashups that combine data or functionality to create a new service, are an easy way to bundle and pipeline algorithms into "algorithm clouds." OSGi is a standardized, component-oriented computing environment for networked services. It has been successfully used in the industry, in applications from high-end servers to embedded mobile devices, since 1999, and is widely adopted in the open-source realm. Alliance members include IBM, Sun, Intel, Oracle, Motorola, and NEC, among others. CIShell is an open-source software specification for the integration and utilization of data sets, algorithms, tools, and computing resources. It extends OSGi by providing interfaces for the easy bundling of data sets (in diverse formats) and algorithms (written in different programming languages by diverse application holders) into workflows and tools.

Data or services from several external sources can be connected or "mashed up" via application programming interfaces (APIs). An API is a source-code interface provided by an operating system or library to support requests for services by other programs. For example, a program might read out eBay data and display sales locations via a Google map. ProgrammableWeb serves as an index to the steadily increasing number of open APIs and mashups across the Web. In April 2008, the APIs most often used in mashups were Google Maps (used in 49 percent of mashups), Flickr (11 percent), Amazon (8 percent), and YouTube (8 percent).

Web Services

Today, many applications are Web services, converting the Web browser into a universal canvas for information and service delivery. Services communicate using XML messages that follow the Simple Object Access Protocol (SOAP) or Representational State Transfer (REST). SOAP is the foundation layer of the Web services stack, providing a basic messaging framework upon which abstract layers can be built. Using SOAP over HTTP supports communication behind proxies and firewalls. Metadata about Web services is commonly encoded using the Universal Description Discovery and Integration (UDDI) protocol. Applications use UDDI to find services at design time or runtime.

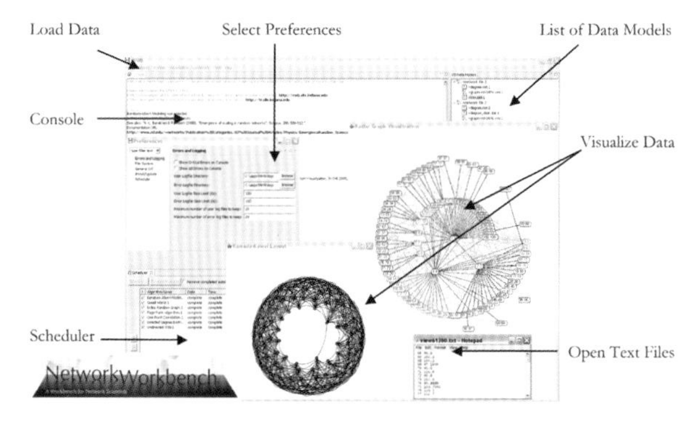

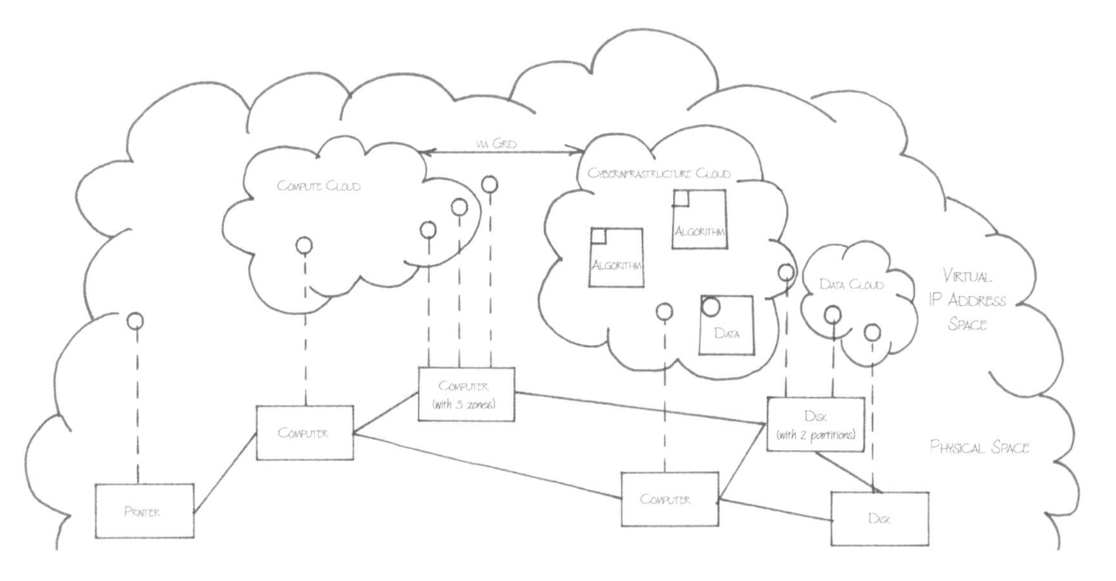

Application Design

The Wiki Way

WikiSpecies, WikiProfessionals, and WikiProteins combine wiki and Semantic Web technology in support of real-time community annotation of scientific data sets. The goal is to meld some of the most important biomedical databases into a single information resource.

WikiMapia was inspired by Google Maps and Wikipedia. It is a Web 2.0 project that describes the whole world—any place on Earth can be annotated on a Google map. Photos can be added and annotations translated.

OpenStreetMap utilizes Web 2.0, georeferencing, geotags, and broadband communication to empower citizens to create a free, editable map of the world, largely using global positioning system (GPS) traces.

Gapminder, Swivel, and Many Eyes

Gapminder provides a great way to interactively and intuitively explore data. Swivel and Many Eyes go one step further and allow all visitors to upload, visualize, and interact with their own data in novel ways. Both sites are highly interactive and socially oriented. Users can form communities, meet like-minded people, and browse the space of their interests. The sites are driven, populated, and defined by those who congregate there. Swivel empowers users to upload and combine data sets and to visualize them as scatter graphs, bar charts, and geomaps. Since its launch in December 2005, more than 1 million data sets have been uploaded. Many Eyes, from IBM, lets users upload and visualize their data in 18 different ways, including network diagrams and tag clouds. In January 2010, it had more than 90,000 data sets, of which there were almost 50,000 visualizations.

Plug and Play Data and Software

The Network Workbench (NWB) Tool uses OSGi and CIShell to support the sharing of network science data sets and algorithms. It can be deployed as a data-algorithms repository, a peer-to-peer architecture, a server-client architecture, or as a standalone tool for offline usage. In January 2010, more than 150 algorithm plugins exist and several tools adopted the OSGi/CIShell core adding more algorithm plugins. The NWB Community Wiki is a place for users to obtain, upload, and request new algorithms and data sets. The site is a sounding board visitors use to work collaboratively in creating a tool that benefits the entire scientific community.

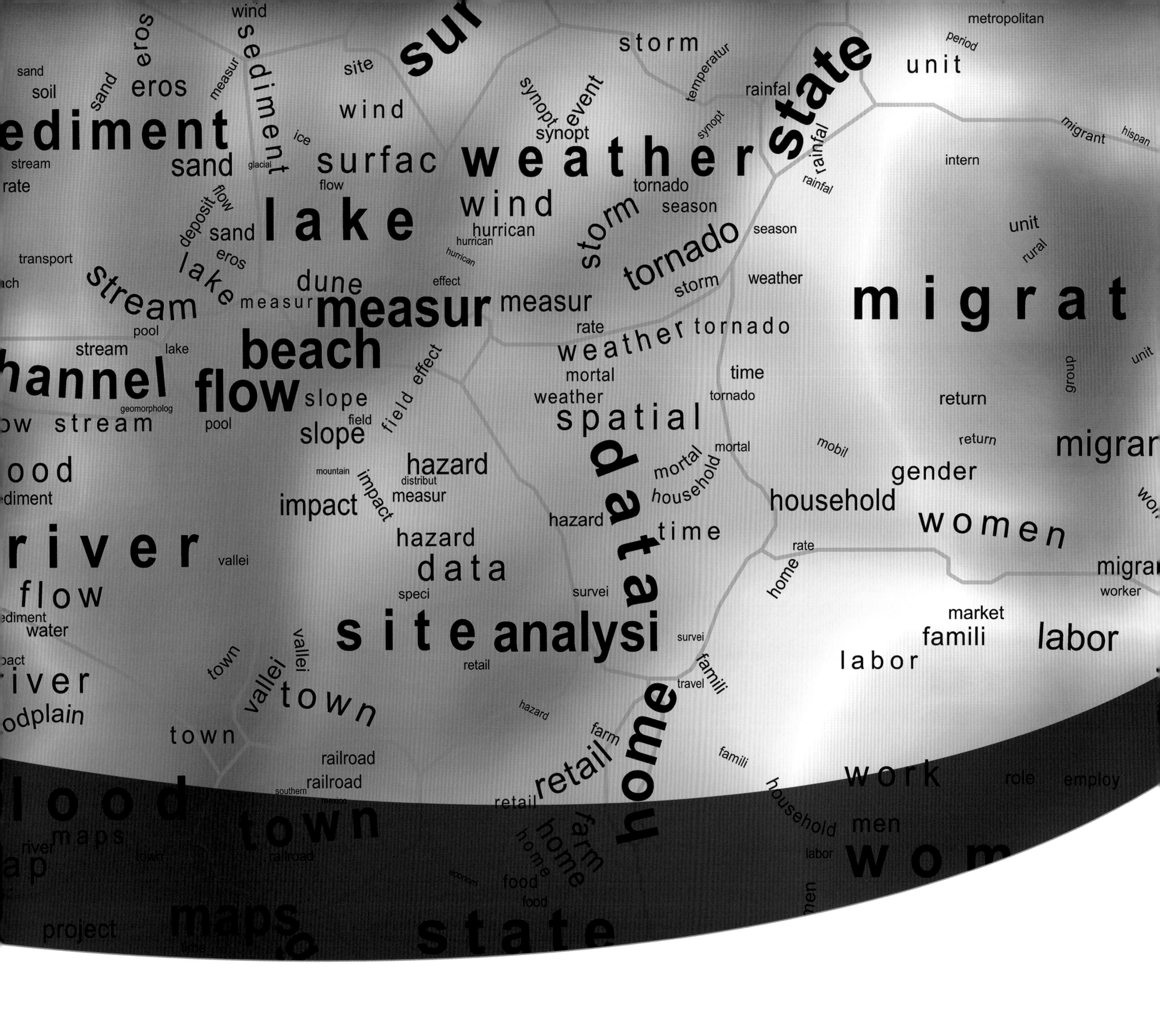

wind
sediment
eros
sand
soil
site
wind
storm
temperatur
unit
metropolitan
period
synopt
event
synopt
rainfal
state
migrant
hispan
stream
rate
sand
glacial
ice
surfac
weather
rainfal
rainfal
intern
flow
lake
wind
tornado
season
deposit
sand
hurrican
hurrican
storm
tornado
season
unit
rural
transport
eros
lake
stream
dune
effect
weather
storm
migrat
measur
measur
measur
pool
rate
tornado
weather
stream
lake
beach
channel
flow
slope
field
effect
mortal
time
tornado
unit
group
geomorpholog
pool
stream
slope
weather
spatial
return
return
mobil
mortal
mountain
impact
hazard
distribut
measur
data
mortal
household
gender
impact
household
hazard
time
women
river
vallei
hazard
data
rate
home
flow
speci
survei
market
worker
water
site analysi
survei
famili
labor
labor
vallei
town
vallei
town
retail
travel
famili
hazard
farm
town
railroad
railroad
southern
retail
famili
retail
home
household
work
role
employ
men
flood
town
maps
home
farm
food
food
labor
state
project
maps

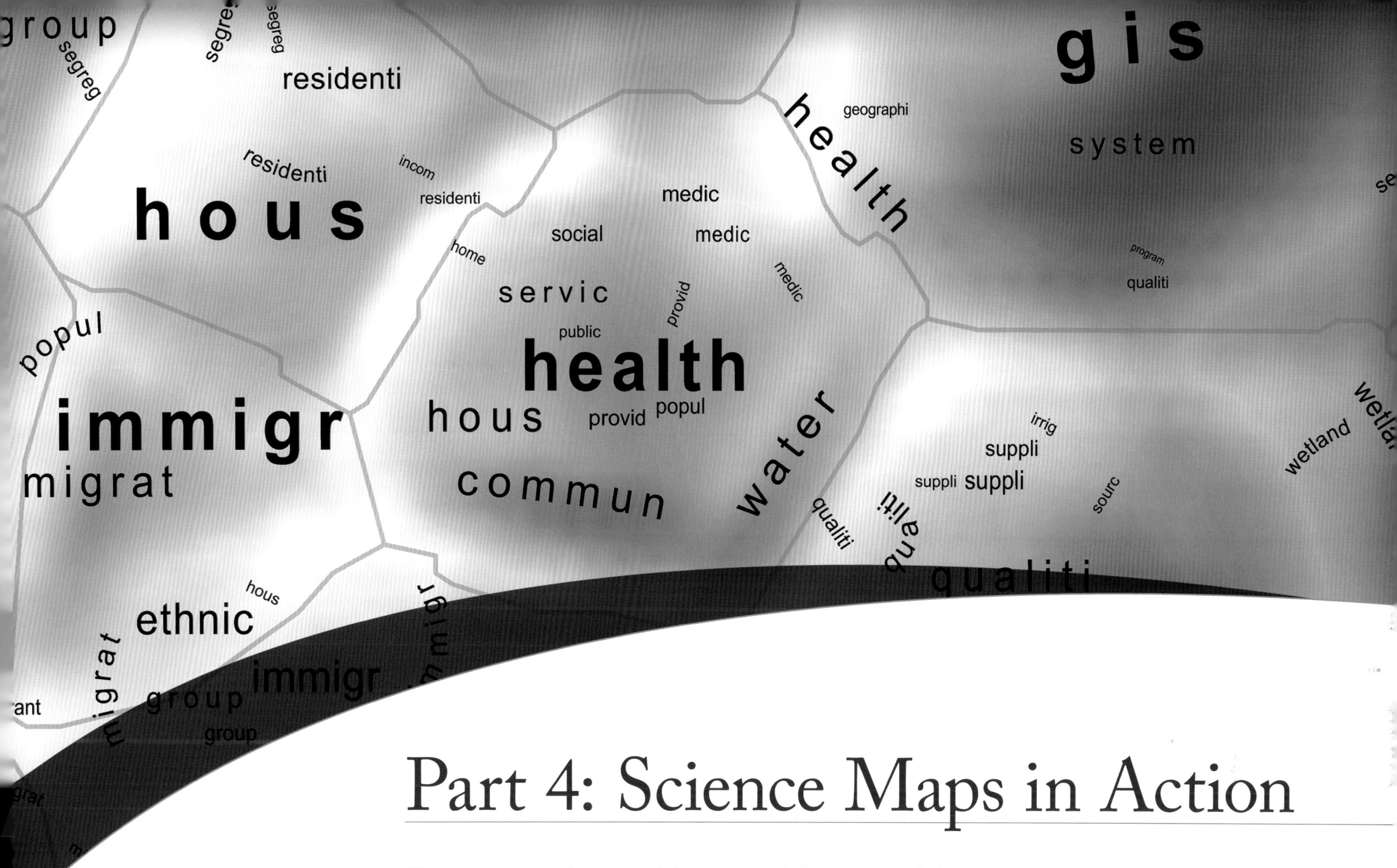

Part 4: Science Maps in Action

If we ever get to the point of charting a whole city or a whole nation, we would have … a picture of a vast solar system of intangible structures, powerfully influencing conduct, as gravitation does in space. Such an invisible structure underlies society and has its influence in determining the conduct of society as a whole.

Jacob L. Moreno

Places & Spaces: Mapping Science

Motivation and Goal

The *Places & Spaces: Mapping Science* exhibit aims to introduce science mapping techniques to the general public and to experts across diverse disciplines for educational, scientific, and practical purposes.

While science mapping techniques are extensively used by those who can pay for them, most people do not know they exist. This exhibit brings a unique selection of existing and newly designed maps of science to conferences, libraries, and education centers. It is meant to inspire cross-disciplinary discussion on how to best track and communicate scholarly activity and scientific progress on a global scale.

Visitors to the exhibit are often surprised to see their own field of research from a height of 10,000 feet—and keen to learn about the data and techniques involved. In turn, their comments on how they experience and interpret the maps are invaluable for the future design of more useful and effective science maps.

Ten Iterations in Ten Years

The exhibit has been envisioned as 10 iterations over a 10-year period. Ten new maps are added each year, so there will be a total of 100 maps by 2014. Every iteration of the exhibit attempts to learn from the best examples of visualization design. We therefore compare and contrast four existing maps in other disciplines with six maps of science. Existing maps may be hand-drawn or computer-generated, old or new; they may come from any area of scholarly activity. Maps of science are typically computer-generated and depict the structure or dynamics of science; many were specifically redesigned or created for the exhibit.

The first iteration of the exhibit (2005), *The Power of Maps,* helps us understand, navigate, and manage both physical places and abstract knowledge spaces. We contrast four early cartographic maps with six early maps of science. The second iteration (2006), *The Power of Reference Systems,* seeks to inspire discussion about a common reference system for all of our scientific knowledge. The third iteration (2007), *The Power of Forecasts,* shows how complex dynamic processes can be simulated to predict their structural evolution and dynamics. We contrast four existing maps of dynamic processes with six potential science forecasts.

Six exhibit iterations (2008-2013) will feature maps that are customized for diverse user groups, such as industry managers, science policy makers, researchers, children, and the general public. The final iteration (2014), *Telling Lies with Science Maps,* will help raise awareness of possible science map abuse.

The exhibit was envisioned by Katy Börner (left, top picture), Kevin W. Boyack (left, bottom), Sarah I. Fabrikant, Deborah MacPherson, and André Skupin in January 2005. Katy Börner and Deborah MacPherson (left, middle) curated the first two iterations of the exhibit. In September 2006, Julie M. Davis took over Deborah's position.

Advisory board members include Deborah MacPherson, Accuracy & Aesthetics (knowledge management and architecture); Kevin W. Boyack, SciTech Strategies (chemistry and business intelligence); Sara Irina Fabrikant, University of Zürich, Switzerland (geography and psychology); Peter A. Hook, Indiana University (law and library science); André Skupin, San Diego State University (geography and information science); Bonnie DeVarco, MediaTertia—Emerging Technologies in Education (writer and educator); Chaomei Chen, Drexel University (information science); and Dawn Wright (aka "Deepsea Dawn"), Oregon State University (geology and physical geography). Their biographies are available at **http://scimaps.org**.

Elements of the Exhibit

The exhibit has a physical presence, as well as a virtual online one. As of 2009, the physical exhibit consisted of an introductory panel, 50 museum-quality maps sized 30-inch x 24-inch (about 76 centimeters x 61 centimeters), the Illuminated Diagrams, two Hands-On Science Maps for Kids, three WorldProcessor globes, and an introductory video with interviews of science mapmakers (see figures). These elements come in eight shipping crates that have a total weight of over 2000 pounds (about 907 kilograms) and require a forklift to move.

A 30-foot x 3-foot (about 9 meter x 1 meter) poster version of the 50 maps is much easier to transport and install, and hence travels more widely (see pictures of previous venues on the next page).

The accompanying Web site (**http://scimaps.org**) provides details on the maps in the exhibit, as well as numerous additional maps, including portraits of the mapmakers and explanations of how the maps work. There is a page devoted to science maps for kids and a page referencing key papers, books, and online resources on science mapmaking. Finally, the site lists an exhibit tour schedule and advisory board information.

Introductory Panel

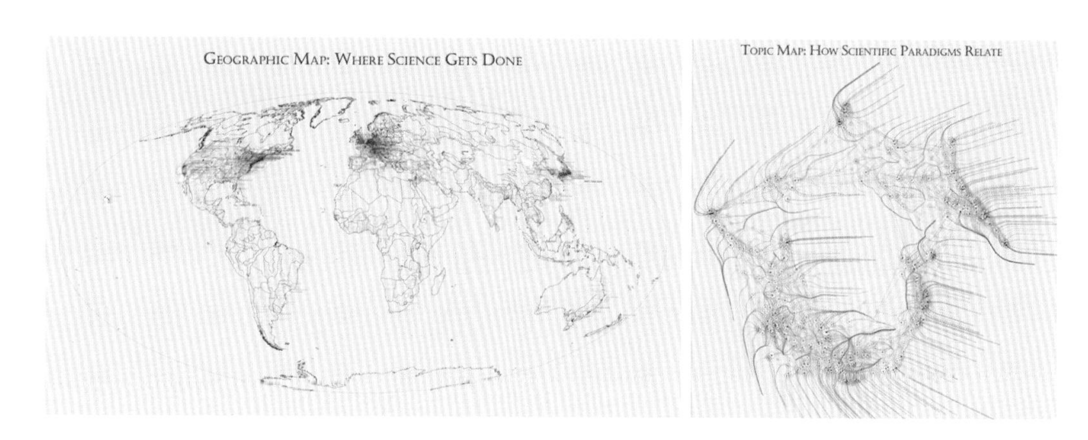

Illuminated Diagrams

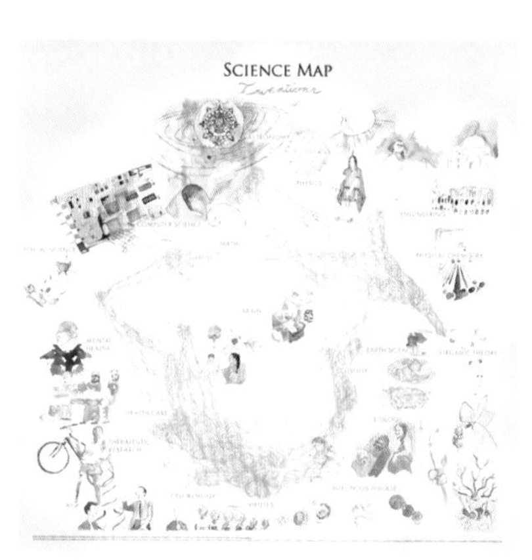

Hands-On Science Maps for Kids

WorldProcessor Globes

DVD

Web Site

Venues

As of 2009, the exhibit has been displayed at more than 175 venues in over 15 countries and 40 scientific disciplines. Venues for the the first three iterations (2005 to 2007) of the framed version of the exhibit were overlaid on a **World Map** (this page) and a **Science Map** (opposite page, see legend below).

The **first iteration** (light brown trail) debuted at the Annual Meeting of the Association of American Geographers in April 2005. It went on to six more venues and is now on permanent display at the National Science Foundation building in Arlington, Virginia.

A completely new set of 20 maps, with a larger introductory panel, the Illuminated Diagrams, and three WorldProcessor globes were created for the opening of the **second iteration** at the Science, Industry and Business Library of the New York Public Library (NYPL) in April 2006 (green trail). The exhibit then went on to the New York Hall of Science and the Hands-On Science Maps for Kids were added in February 2007.

The **third iteration,** with 30 reprinted maps and new crates, opened in September 2007 at the American Museum of Science and Energy in Oak Ridge, Tennessee (dark brown trail).

Locations of poster holders are indicated in red.

ESRI DC Office, Vienna, Virginia

American Chemical Society National Meeting and Exposition, Chicago, IL

World Map with Exhibit Venues

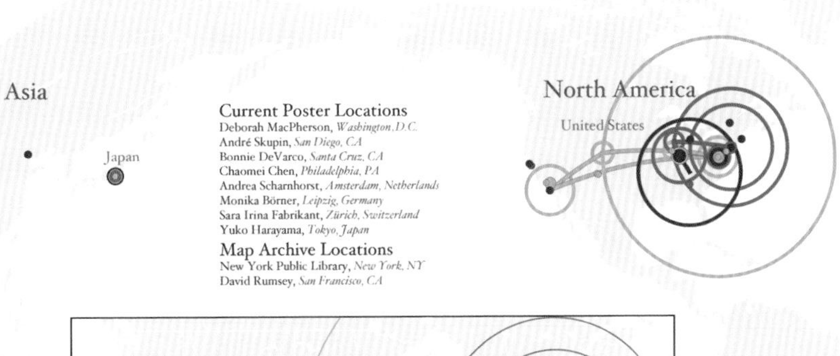

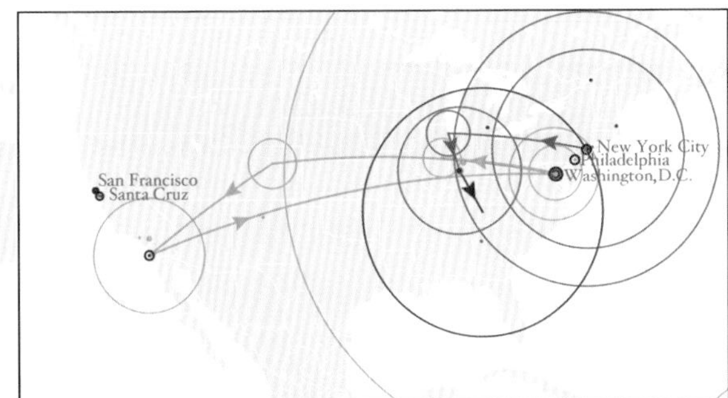

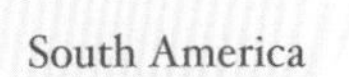

Netherlands
Germany
Switzerland
Europe
Africa

Expedition Workshop, NSF board room, Arlington, VA

ISSI Conference, Madrid, Spain

New Network Theory Conference, Amsterdam, The Netherlands

ESRI User Conference, San Diego, CA

SIGGRAPH 2007, San Diego, CA

New York Hall of Science, Queens, NY

Hall of Flags, NIST, Gaithersburg, MD

NSF building, Arlington, VA

Monroe County Public Library, Bloomington, IN

SIBL branch of NYPL, New York, NY

AAG Conference, Denver, CO

Science Map with Exhibit Venues

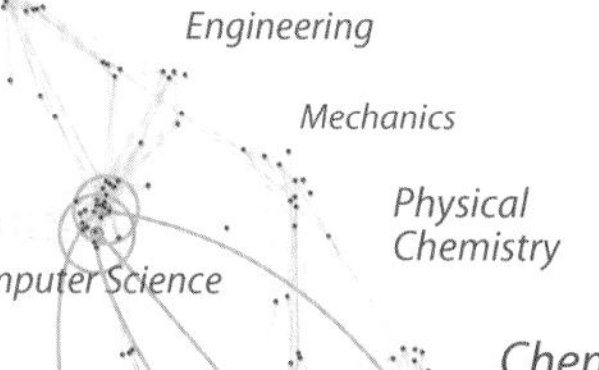
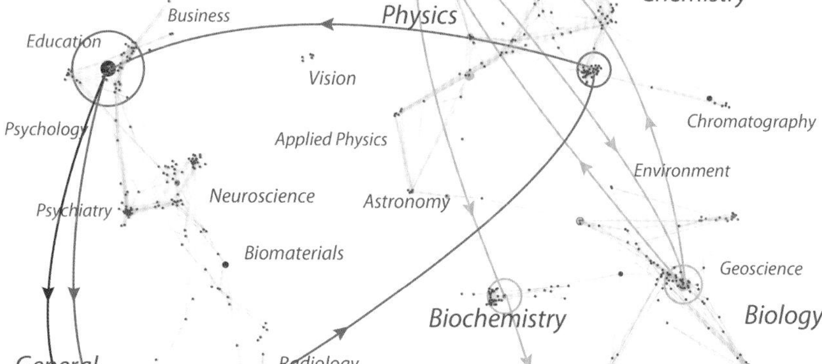
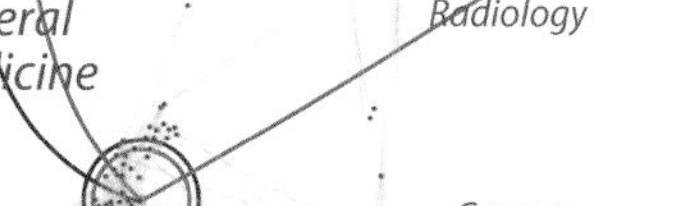
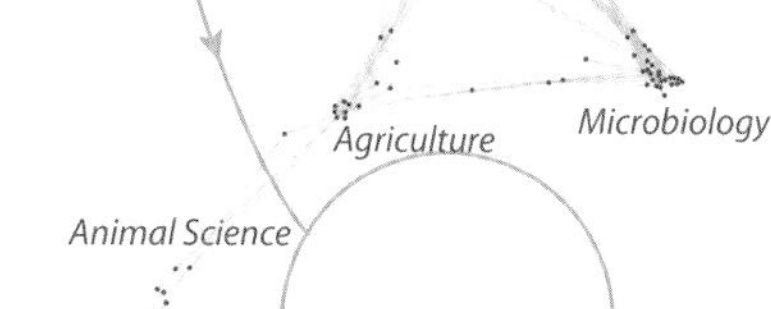

American Museum of Science and Energy, Oak Ridge, TN

Natcher Conference Center, NIH, Bethesda, MD

Legend for Science & Geography Maps

Node Fill	Node Size
Poster Exhibit	Duration
Physical Exhibit	Smallest: One Day
	Largest: 620 Days
Node Color	
1st Iteration	Travel Direction
2nd Iteration	
3rd Iteration	Poster Holders

Organization of Part Four

This part of the atlas provides a visual feast of the 30 maps featured in the first three iterations of *Places & Spaces: Mapping Science*. The maps are presented in the order in which they were exhibited. Each existing map is presented on two pages. Science maps are discussed in more detail, with four pages devoted to each map. The first (left) page gives the map title, the name and biography of the mapmaker(s) as of 2008, the aims of the mapmaker(s), and a brief interpretation of major insights gained from the map. Areas of the world and of science that each map covers are also shown, together with the time captured. The second (right) page opens to a full-page print. As the exhibit maps are substantially larger than this book, enlargements are provided whenever feasible. The third and fourth page for each science map describes the data, analysis technique, reference system, data overlays, and unique features. Additional elements of the exhibit are detailed as well.

First Iteration of Exhibit (2005): The Power of Maps

Four Early Maps of Our World
versus
Six Early Maps of Science

The first exhibit iteration on *The Power of Maps* demonstrates how maps help us to understand, navigate, and manage both physical places and abstract knowledge spaces.

Early maps of our planet were certainly neither complete nor perfect, yet they proved invaluable for explorers. As keys to navigation, exploration, and communication, maps helped explorers find promising new lands while avoiding sea monsters.

Maps of science today are based on limited knowledge and are therefore imperfect. In order to generate comprehensive maps that are entirely accurate and reliable, we must first have proper coverage and interlinkage of multilingual, multidisciplinary, and multimedia scholarly knowledge.

The first pictures of Earth from space were experienced by many as transformative of their perceptions of life and the cosmos. It is our hope that maps of science will increase our appreciation and application of knowledge while serving as useful navigational tools.

The Power of Maps features four cartographic maps, including one of the earliest global maps of our world by Ptolemy, an early map of the new world by Johannes Janssonius, an early map of the whole world by Herman Moll, and an early statistical graph by Charles Joseph Minard. Each of the six maps of science employs a different metaphor: a node-link diagram; a treemap; a self-organizing map rendered using geographic information systems (GIS); a subway map; a crossmap; and a galaxy view. Which metaphor is most effective in designing a visual index of our collective science and technology knowledge and expertise?

Note that the makers of the early cartographic maps all had their own printing presses, while the makers of the first maps of science all hold PhDs.

Four Early Maps of Our World

Claudius Ptolemy, Cosmographia World Map, 1482

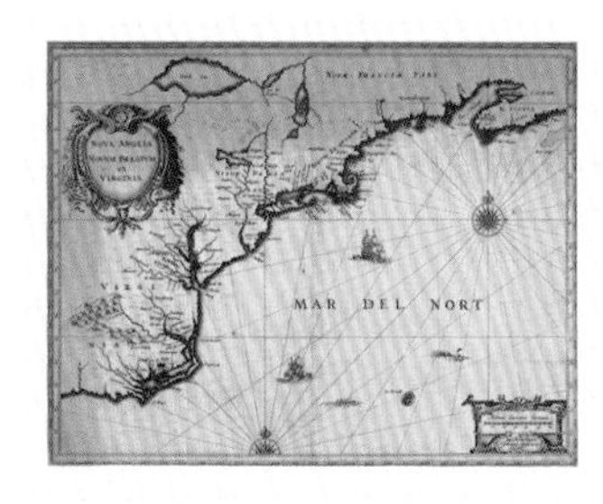

Johannes Janssonius, Nova Anglia, Novvm Belgivm et Virginia, 1642

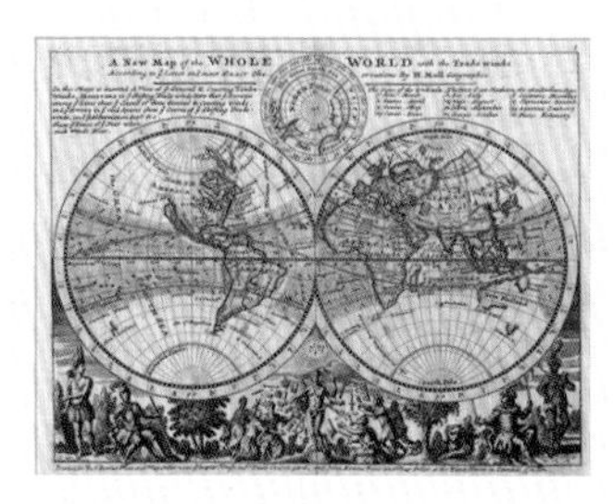

Herman Moll, A New Map of the Whole World with the Trade Winds According to Ye Latest and Most Exact Observations, 1736

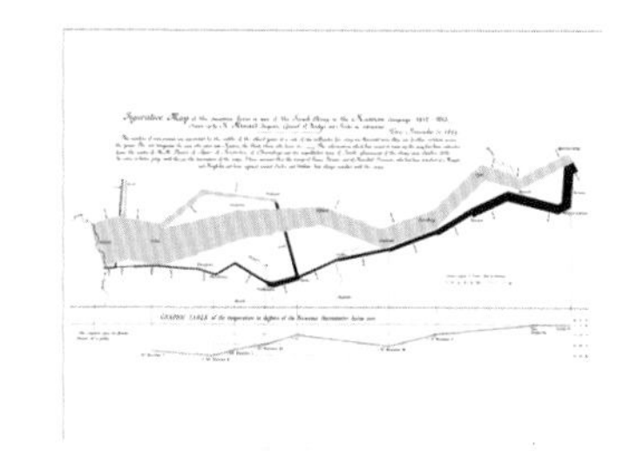

Charles Joseph Minard, Napoleon's March to Moscow, 1869

Six Early Maps of Science

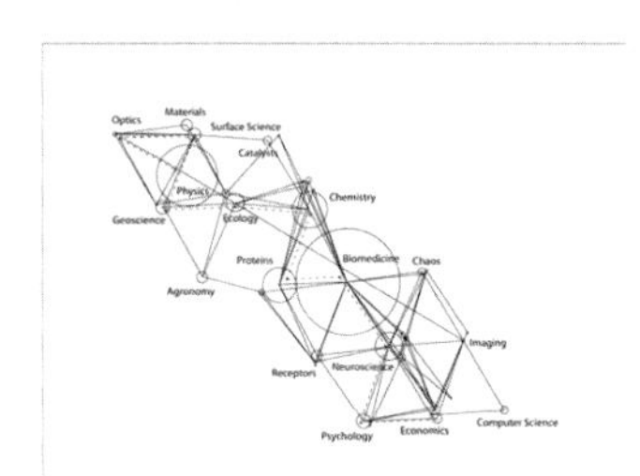

Henry G. Small, 1996 Map of Science: A Network Representation of the 43 Fourth-Level Clusters Based on Data from the 1996 Science Citation Index, 1999

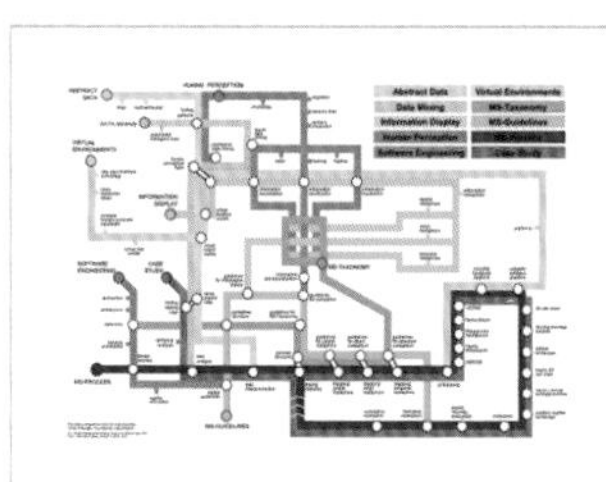

Keith V. Nesbitt, PhD Thesis Map, 2004

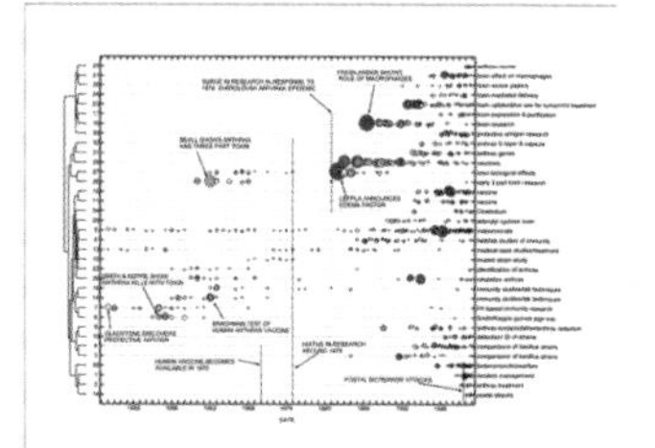

Steven A. Morris, Timeline of 60 Years of Anthrax Research Literature, 2005

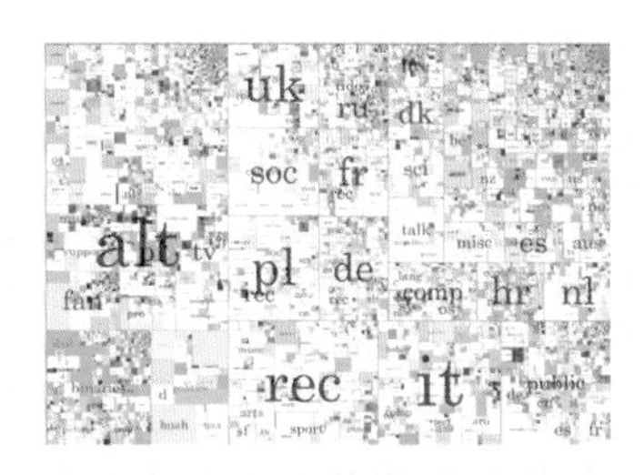

Marc A. Smith, Danyel Fisher, Treemap View of 2004 Usenet Returnees, 2005

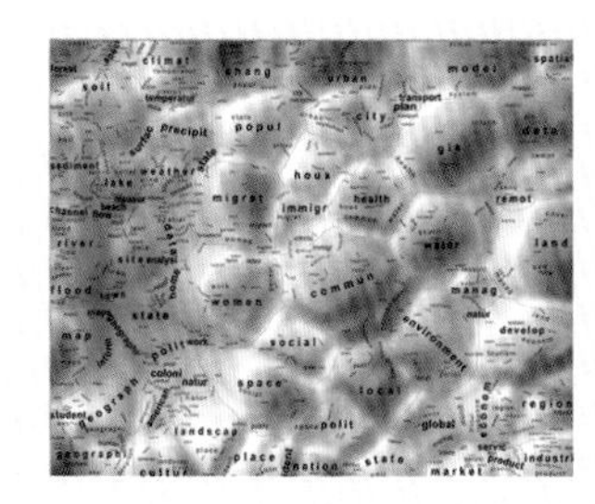

André Skupin, In Terms of Geography, 2005

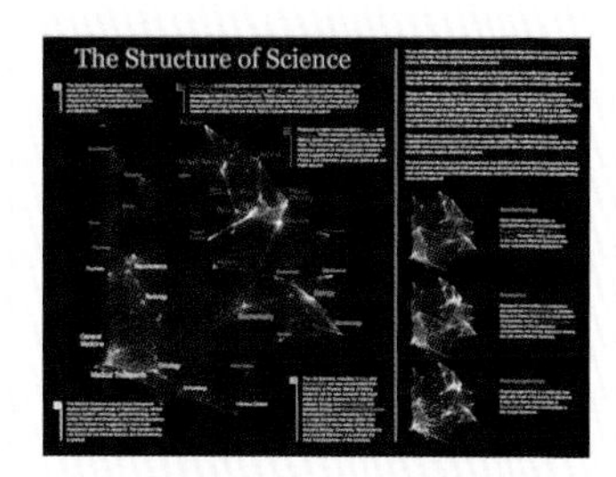

Kevin W. Boyack and Richard Klavans, The Structure of Science, 2005

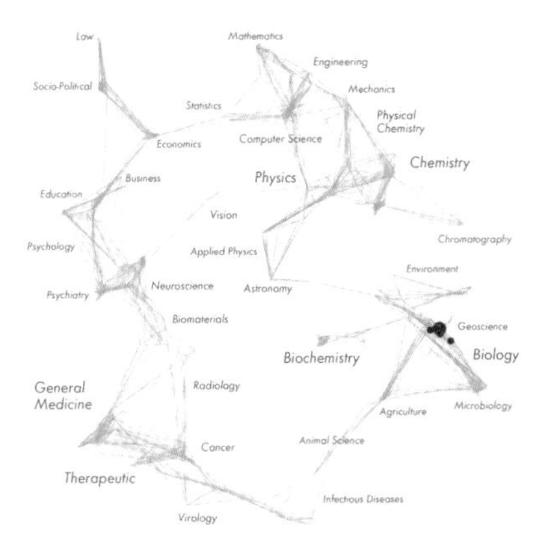

1482

Cosmographia World Map

By Claudius Ptolemy

ULM, GERMANY, 1482

Courtesy of the James Ford Bell Library, University of Minnesota, Minneapolis

This world map is reconstituted from Ptolemy's *Cosmographia* (or *Geographia*). While there are no surviving maps that can actually be attributed to Ptolemy, the geographic data and cartographic approach are indisputably his. This world map was compiled primarily from an extensive database that forms the major part of "Books" 2–7 of the *Cosmographia*. The books provide many pages of tables with the latitudes and longitudes of places all over the ancient world.

Ptolemy's works were lost to Western Europe after the fall of the Roman Empire during the Middle Ages. They were rediscovered more than a thousand years later, in the late 12th century, during the Renaissance. His *Cosmographia* was translated into Latin around 1405, and numerous manuscript copies were made throughout the 15th century.

The map shown here is taken from an edition of *Cosmographia* published in Ulm, Germany in 1482. This edition is noticeably different from previous Italian editions (see sample below), as it was printed from carved wood blocks rather than copperplate engravings.

While being a great achievement, Ptolemy's map of the world is in many ways imperfect. The differences from today's depictions of the world are numerous. For instance, the map shows **AFFRICA** as an extended southern land and **ASIA** as exaggerated in size, extending too far eastward. The latter error may have been a factor in the decision of Christopher Columbus (1451–1506) to sail west for the Indies. In fact, Columbus never accepted that he had discovered a new continent in 1492, instead insisting that he had reached the Indies and China. The Indian Ocean, **MARE INDICVM**, is here drawn as an enclosed body of water—which was disproved in 1497 by Bartholomeu Dias of Portugal when he successfully sailed around the Cape of Good Hope.

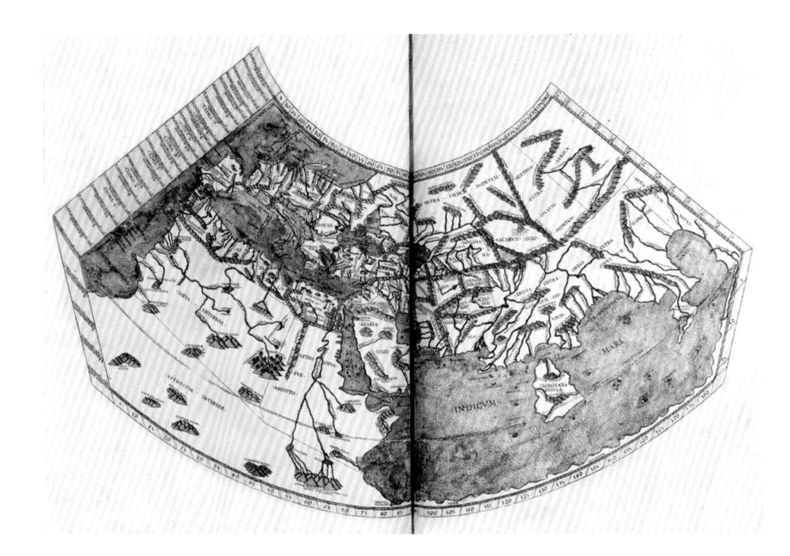

Claudius Ptolemy

(circa AD 90–168) was a profoundly influential astronomer, astrologer, geographer, and mathematician. Also known as Claudius Ptolemaeus, Ptolomaeus, Klaudios Ptolemaios, and Ptolemeus, he lived and worked in Alexandria, Egypt. His curiosity about the dynamic relationships between celestial bodies and the causes and effects of climate led him to invent a longitude and latitude grid system to construct maps of the world. His mathematical proofs stating that the Earth is a sphere are still accepted today, in spite of the fact that he placed this sphere as a fixed point in the center of a universe that revolved around it daily. Ptolemy synthesized and extended Hipparchus's system of epicycles and circles to develop a geocentric theory of the solar system. His theory required more than 80 epicycles to explain the motions of the sun, the moon, and the five planets known in his time. It is described in *Mathematical Syntaxis*—widely called the *Almagest*—a 13-book mathematical treatment of the phenomena of astronomy, providing a means of calculating positions. The *Almagest* included a star catalog of 48 constellations, listing the names we still use today. Ptolemy also made major contributions to the history of geography and cartography. Ptolemy's is the first known projection of the sphere onto a plane. His geography remained the principal work on the subject until the work of Columbus.

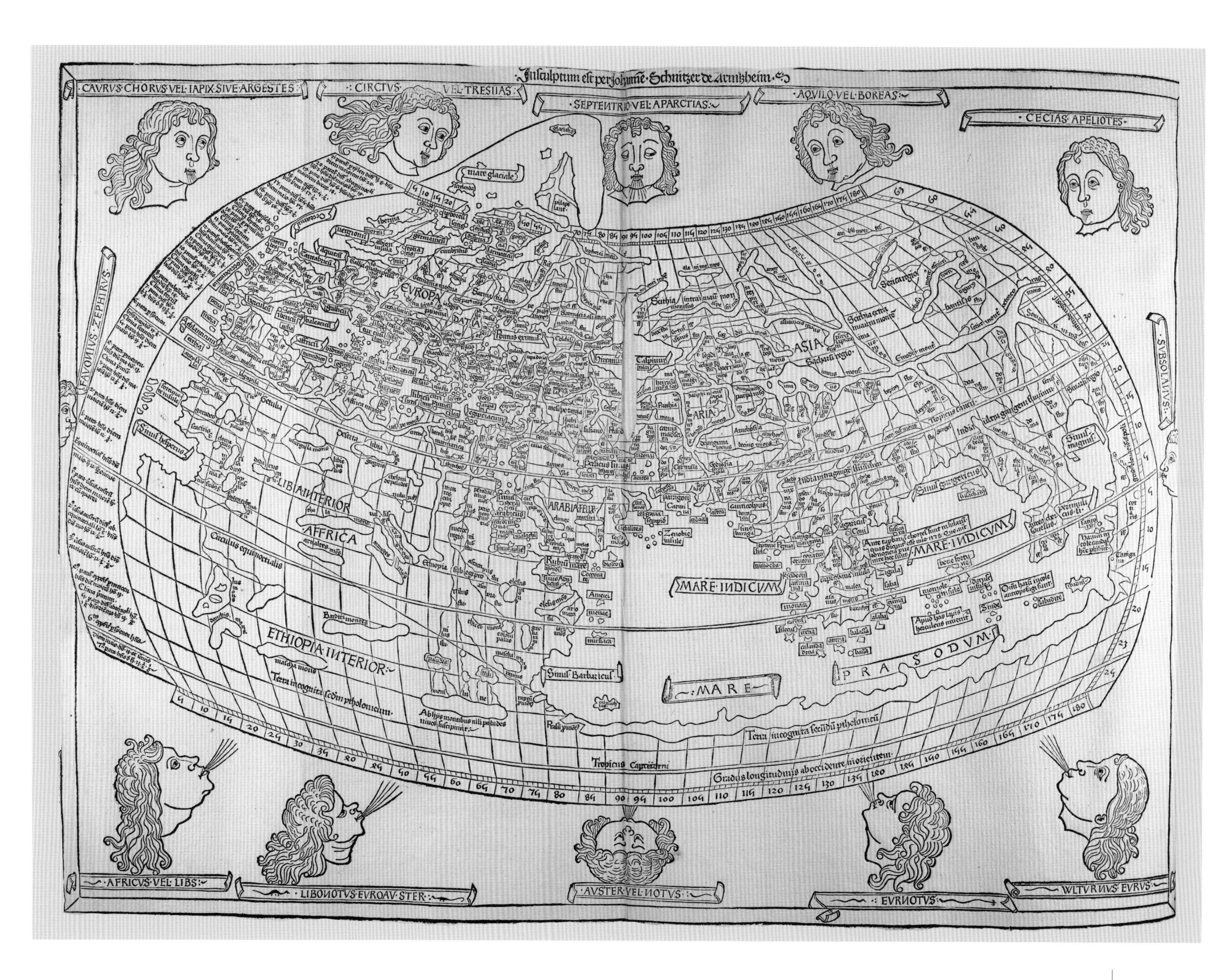

CAVRVS·CHORVS·VEL·IAPIX·SIVE·ARGESTES
CIRCIVS VEL·TRESIIAS
SEPTENTRIO·VEL·APARCTIAS
AQVILO·VEL·BOREAS
CECIAS·APELIOTES
SVBSOLANVS
FAVONIVS·ZEPHIRVS
mare glaciale
EVROPA
ASIA
AFFRICA
LIBIA INTERIOR
ARABIA FELIX
MARE INDICVM
MARE INDICVM
Sinus gangeticus
Circulus equinoctialis
ETHIOPIA·INTERIOR
Sinus Barbaricus
MARE
PRASODVM
Terra incognita secundum pitholomeum
Tropicus Capricorni
Gradus longitudinis ab occidente in orientem
AFRICVS·VEL·LIBS
LIBONOTVS·EVROAVSTER
AVSTER·VEL·NOTVS
EVRNOTVS
VVLTVRNVS·EVRVS

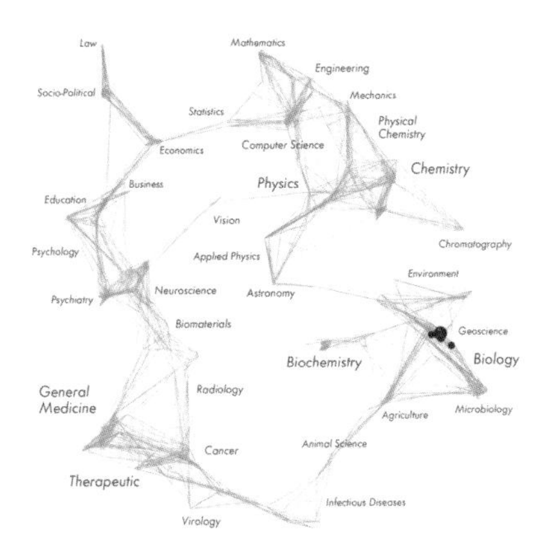

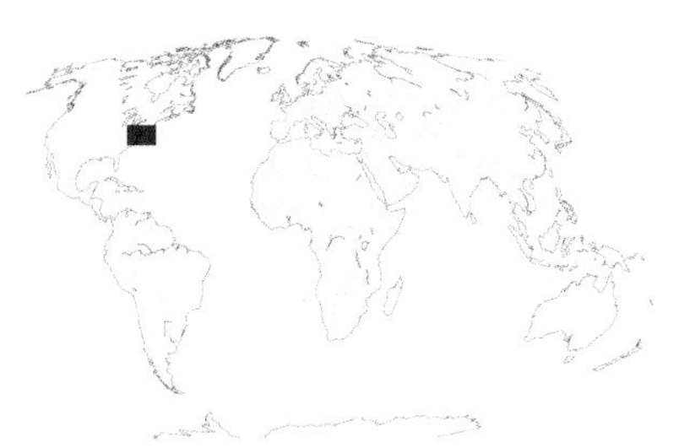

1642

Nova Anglia, Novvm Belgivm et Virginia

By Johannes Janssonius

AMSTERDAM, HOLLAND, 1642

Courtesy of the Library of Congress, Geography and Map Division, Washington, DC

The first maps of the new world were made by Native Americans on skins, bark, rock, and the earth itself. Jansson's maps were some of the first to be created in Europe. Maps of the New World from this time, created in Holland, England, and France, typically show the east coast of what is today North America.

Nova Anglia, Novvm Belgivm et Virginia is one of the finest depictions of the Americas published in the 17th century. It was the first map to use the names Manbattes (Manhattan) and N. Amsterdam (New York). Extending from the Cape of Fear in the south to Nova Scotia in the north, this map of the New World was influential because it showed all of the current Dutch holdings from New England to Virginia. It is based on Johannes De Laet's rare map of 1630 and is actually the third iteration of Jansson's 1636 *Nova Anglia, Novvm Belgivm et Virginia.* Unlike De Laet's map, Jansson's map was very successful and widely used with regular updates over 75 years. Europeans' increased interest in the colonization of North America after 1600 is concisely shown here and increased partially because of this map.

The map of Virginia below, also from that time, was extensively copied among mapmakers. It was drafted by John Smith around 1607 and engraved in 1612 in England. Many of the place names cited by Smith are still in use today.

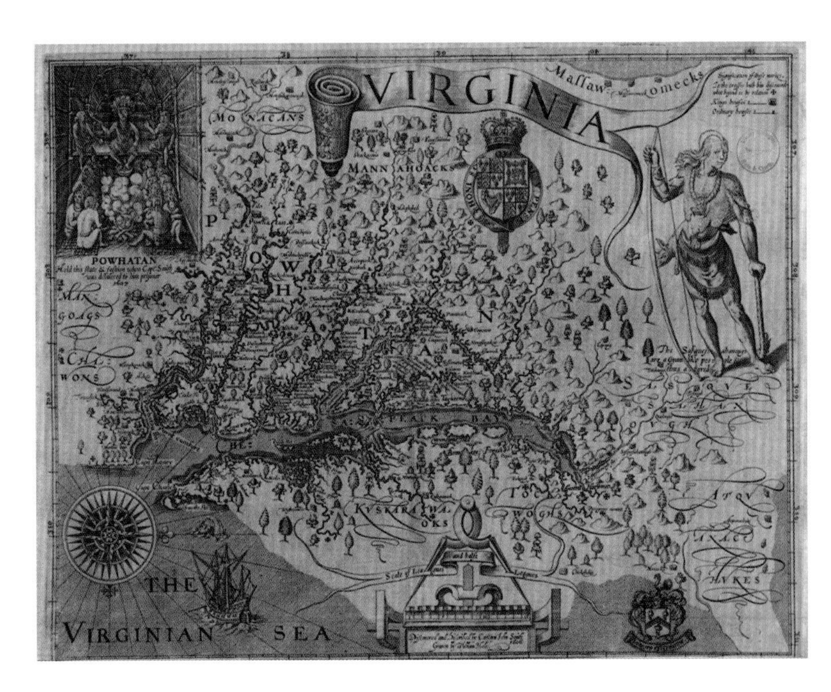

Johannes Janssonius (1588–1664), more commonly known as Jan Jansson, is recognized today for having compiled and published the 17th-century *Atlas Major.* His father, Jan Janszoon the Elder, was a bookseller and publisher. In 1612, Jansson married the daughter of the cartographer Jodocus Hondius and set up business in Amsterdam as a book publisher. In 1616, he published his first maps of France and Italy, and he remained prolific thereafter. Jansson's maps closely resemble those of Willem Janszoon Blaeu and, in fact, were often copied from them. However, Jansson's maps, to many eyes, tend to be more flamboyant and decorative. From about 1630 to 1638, Jansson worked in partnership with his brother-in-law Henricus Hondius, issuing further editions of the Mercator/Hondius atlases to which his name was added. On the death of Henricus, he took over the business, expanding the work still further, until eventually he published the 11-volume *Atlas Major.*

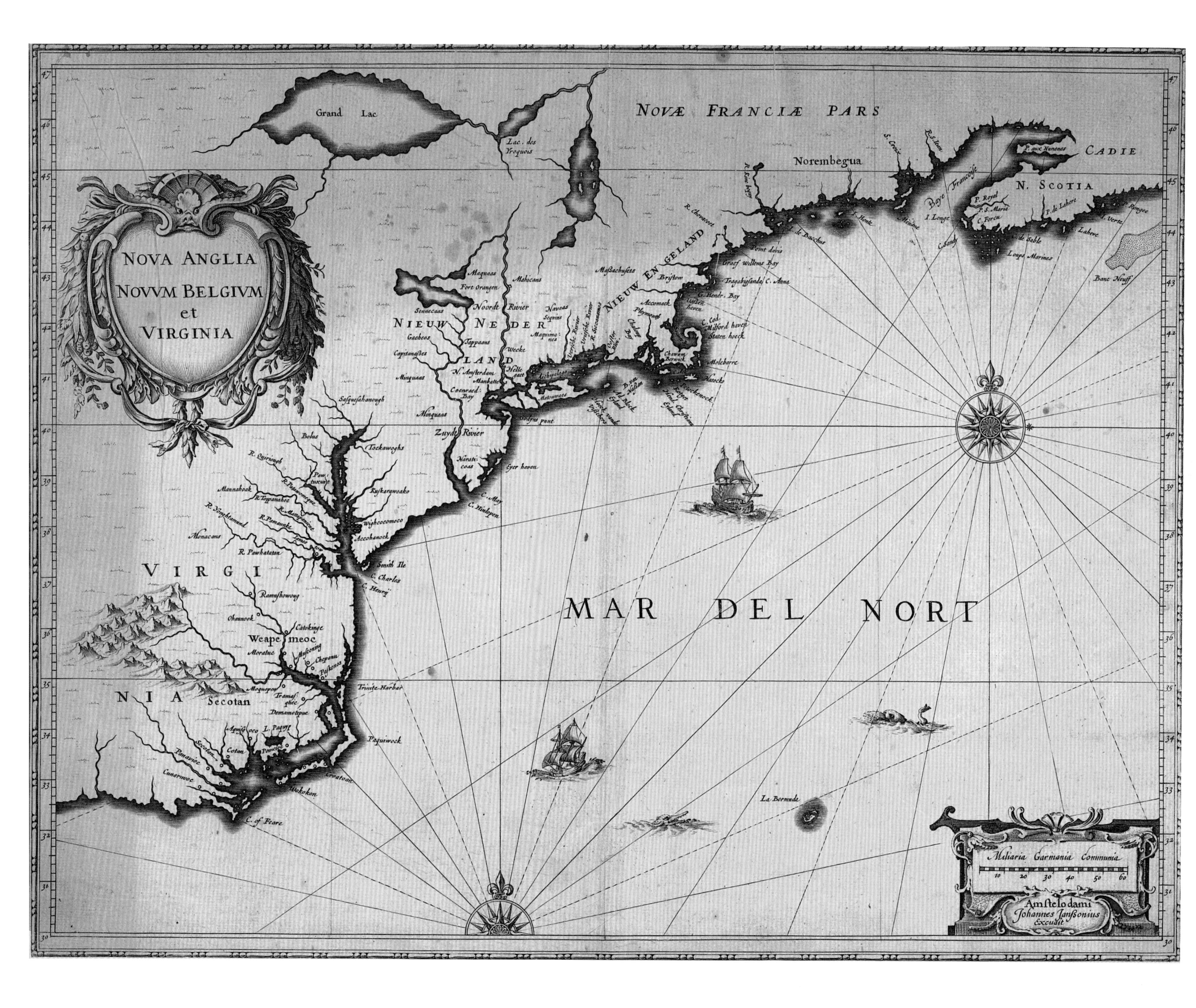
NOVA ANGLIA
NOVVM BELGIVM
et
VIRGINIA
NOVÆ FRANCIÆ PARS
Grand Lac
Norembegua
CADIE
N. SCOTIA
NIEUW EN GELAND
NIEUW NEDER LAND
VIRGINIA
MAR DEL NORT
La Bermuda
Miliaria Germanica Communia
10 20 30 40 50 60
Amstelodami
Johannes Janssonius
Excudit

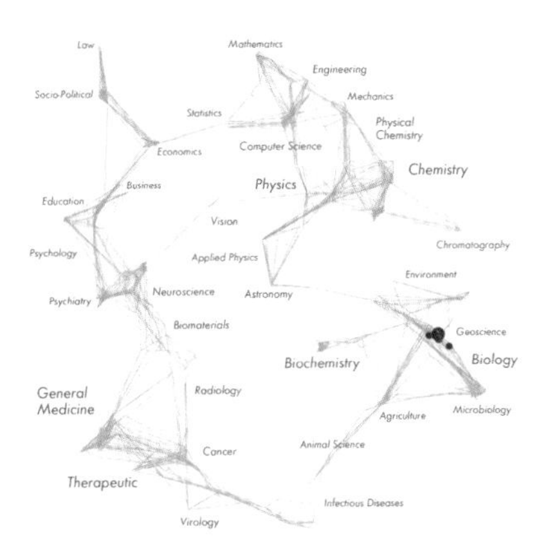

1736

A New Map of the Whole World with the Trade Winds According to Ye Latest and Most Exact Observations

By Herman Moll

LONDON, 1736

Courtesy of the David Rumsey Map Collection, Cartography Associates, San Francisco, California

This is a hand-colored engraved double-hemisphere map of the world. There is an inset of the Arctic and a table of signs of the "Zodiack." The long note at top left discusses the trade winds indicated by arrows throughout the map. The continents are represented by 12 allegorical figures, who are surrounded by plants native to these lands, with a lion observing them all.

This map features California as an island (see detailed map below), a popular misconception at the time. Early maps showed California as a peninsula, but a later exploration by Sebastian Vizcaino in 1602 claimed that California was separated by sea from the American continent. Antonio de Herrera was one of the first to illustrate this in his map of 1622; later maps by Abraham Goos in 1624 and Henry Brigg in 1625 continued to replicate this error. The misconception firmly took hold when California was drawn as an island in maps published by the Dutch, including those by such illustrious mapmakers such as Jan Jansson in 1636, followed by Nicolas Sanson, Guillaume Blaey, Peirre Duvall, and Moll. An end came to this tradition in 1747, when King Ferdinand of Spain issued a royal decree declaring California part of the mainland.

Note that many coastlines, such as Australia, "New Holland," "New Zieeland," and the northwestern part of North America are incomplete. The old mapmakers captured the known, leaving undiscovered regions as open spaces and labeling them with "Unknown L.," as in the lower right of the western hemisphere.

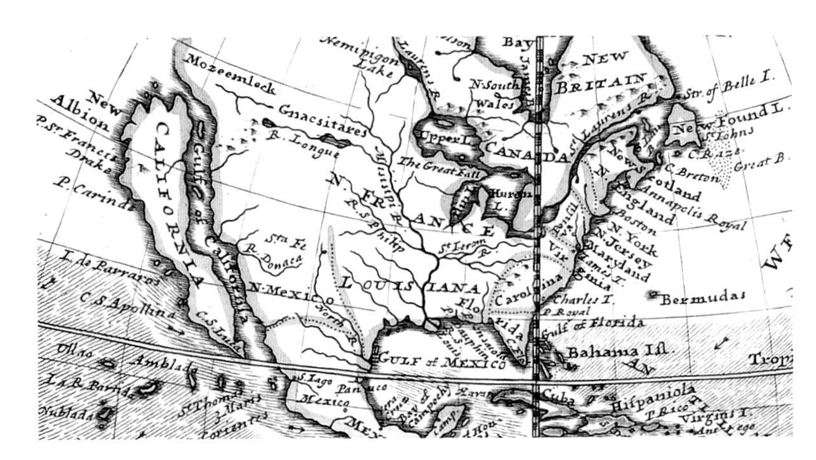

Herman Moll (1654–1732) was known to be a Dutch geographer and cartographer, who may have lived in Germany and then settled in London, United Kingdom, in 1698. At the turn of the 17th century, Moll was the most famous map publisher in England. He was also the first cartographer to create an elegant map of England that correctly portrayed its shape. His style combined elaborate embellishment with clear, bold lettering—the latter used for important details. Moll lived in a time of great discoveries, amid the Anglo-Dutch wars to control the North Sea and the rise of the New Netherlands and North America. It was a time of great explorers and stories like those of Robert Hooke, Daniel Defoe, Jonathan Swift, Christian Huygens, and Antonie van Leeuwenhoek. Moll prided himself on his work and publicly rebuked mapmakers who republished preexisting maps under new titles without having investigated their accuracy or completion—as this could have proved fatal in cases where known depths of water and sands were omitted.

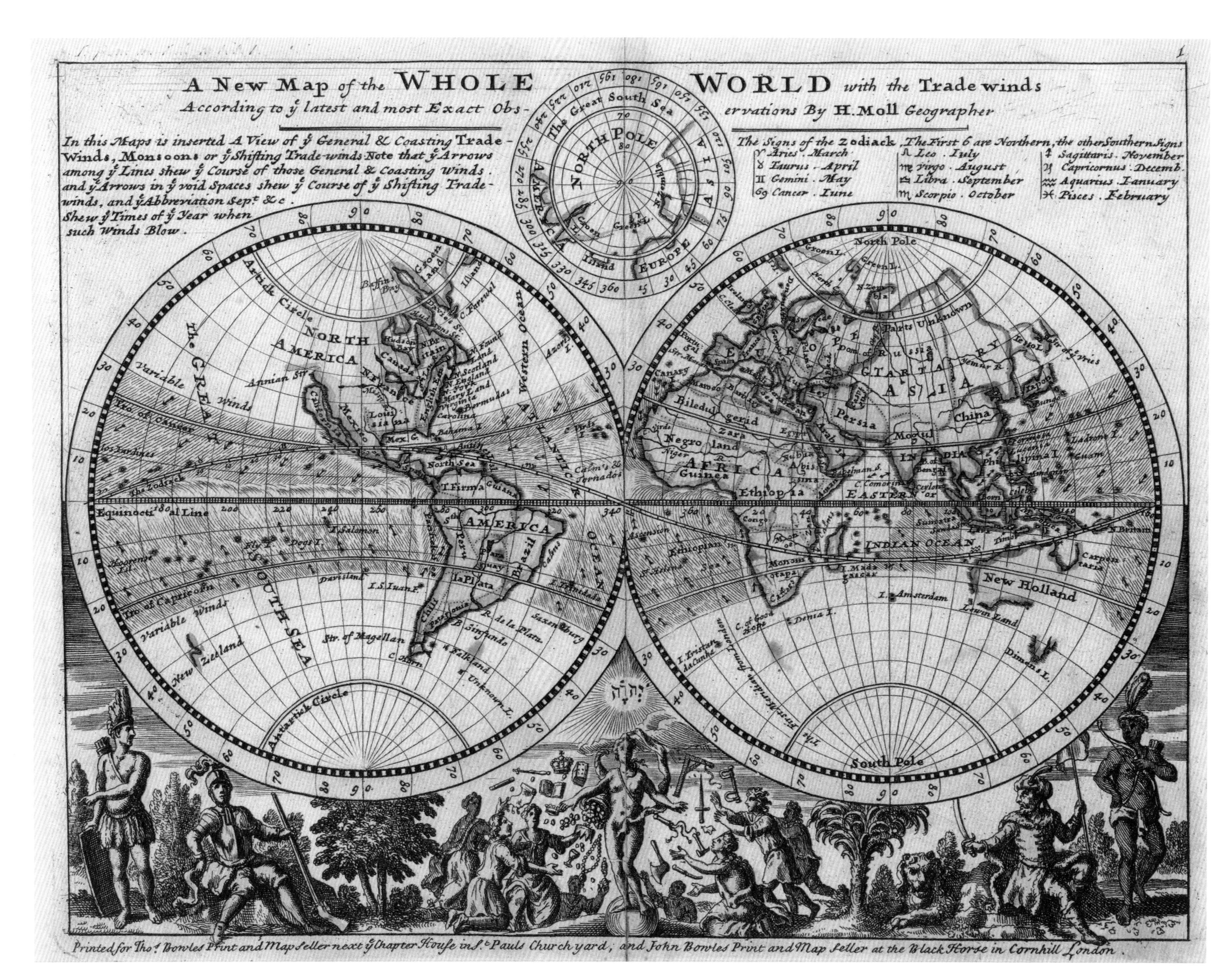
1
A New Map of the WHOLE WORLD with the Trade winds
According to ye latest and most Exact Observations By H.Moll Geographer
In this Maps is inserted A View of ye General & Coasting Trade-Winds, Monsoons or ye Shifting Trade-winds Note that ye Arrows among ye Lines shew ye Course of those General & Coasting Winds and ye Arrows in ye void Spaces shew ye Course of ye Shifting Trade-winds, and ye Abbreviation Sept. &c Shew ye Times of ye Year when such Winds Blow.
The Signs of the Zodiack, The First 6 are Northern, the other Southern Signs
♈ Aries . March
♉ Taurus . April
♊ Gemini . May
♋ Cancer . Iune
♌ Leo . Iuly
♍ Virgo . August
♎ Libra . September
♏ Scorpio . October
♐ Sagittaris . November
♑ Capricornus . Decemb.
♒ Aquarius . Ianuary
♓ Pisces . February
The Great South Sea
NORTH POLE
AMERICA
ASIA
EUROPE
NORTH AMERICA
Western Ocean
Artick Circle
Tro. of Cancer
Equinoctial Line
Tro. of Capricorn
Variable Winds
The Zodiack
SOUTH SEA
AMERICA
Brazil
Str. of Magellan
New Zeeland
Antartick Circle
North Pole
EUROPE
Russia
TARTARY
ASIA
China
Persia
INDIA
AFRICA
Negro land
Guinea
Ethiopia
INDIAN OCEAN
New Holland
South Pole
Printed for Thos. Bowles Print and Map seller next ye Chapter House in St. Pauls Churchyard; and John Bowles Print and Map seller at the Black Horse in Cornhill London.

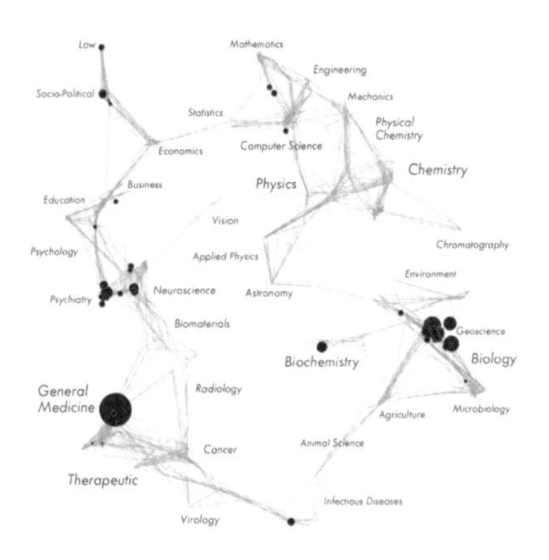

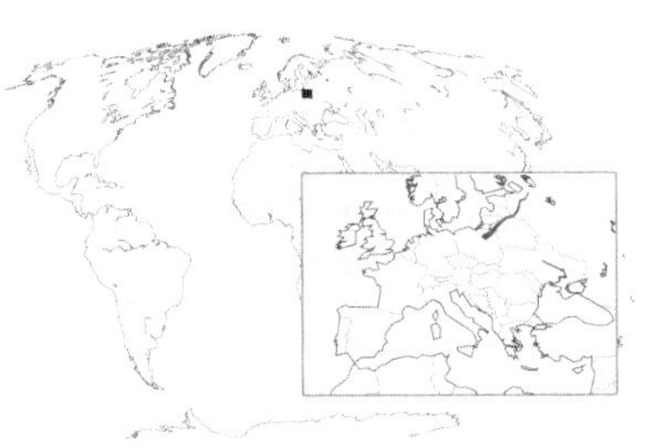

1869

Napoleon's March to Moscow

By Charles Joseph Minard
PARIS, 1869
Courtesy of Edward Tufte, Graphics Press, Cheshire, Connecticut

Minard's thematic map of Napoleon's march is perhaps the single best-known statistical graphic of the 19th century, hailed today by statisticians, geographers, and historians alike. It shows the futility of Napoleon's attempt to invade Russia and the utter destruction of his Grande Armée in the last months of 1812.

The story begins in June 1812, at the Polish-Russian border (left) with 422,000 men. The diminishing size of the army is shown by the width of a brown band, overlaid on the map of Russia. A fraction of the troops splits off from the main group and pauses at Polotzk. The crossing of the Berezina River alone takes the lives of more than 80,000. In September 1812, an army of 100,000 successfully reaches Moscow. However, Czar Alexander I and the residents of Moscow had fled and burned the city, leaving little to conquer. A frustrated Napoleon had little choice but to return to the part of Europe he controlled for food, shelter, and supplies. Minard then traces the remnants of the Grande Armée as it makes its way back toward the Neiman River. In doing so, the parallel tracks of the advancing and retreating army are set next to one another, making the continuing deterioration of the army all the more visible and heartwrenching. As the army slowly made its way across barren earth (the Russians had burned food along this path while blocking other escape paths), one of the worst winters of the time set in. Minard tracks the plummeting temperature against this trek on a horizontal axis at the bottom of the page, even more profoundly capturing the dire straits in which the retreating army found itself. Not surprisingly, the pitiful band of troops that returned from Russia marked the onset of the collapse of Napoleon's Continental Empire. (Minard's original graphic does not mention Napoleon's name once.)

Tufte, in his praise of Minard's map, identifies six separate variables captured within it. First, the line width continuously marks the size of the army. Second and third, the line itself shows the latitude and longitude of the army as it moved. Fourth, the lines show the direction in which the army was traveling, both in advance and retreat. Fifth, the location of the army with respect to certain dates is marked. Finally, the temperature along the path of retreat is displayed. Few, if any, maps before or since have been able to weave so many variables into one captivating whole in such a coherent, compelling fashion.

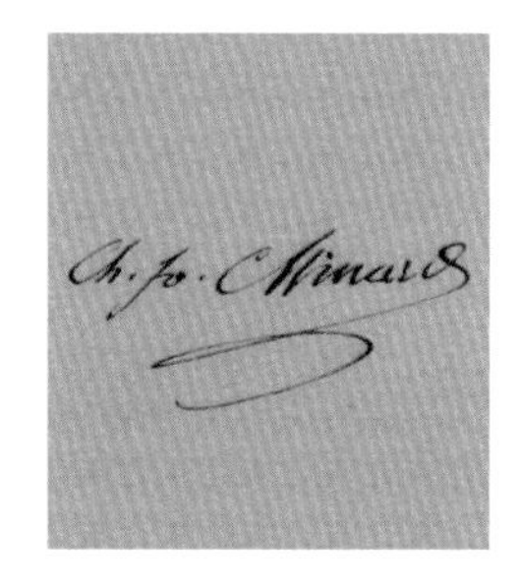

Charles Joseph Minard (1781–1870) was a French civil engineer who liked to study streams and physics. In the course of his long career as an engineer, he took part in many prominent public works and the restoration of European canals, bridges, and roads after the Napoleonic Wars. Before his retirement, he published treatises on construction, including unique memoirs on partial routes, distances, traveling, and expenses. When he was nearly 65 years old, he began to publish *cartes figuratives* (figurative maps). Most of his early maps focused on the flow of goods and passengers along railroad, river, and sea routes of commerce. He used innovative techniques like pie charts, flow maps, and choropleth maps to substitute mathematically proportioned images for dry and complicated columns of statistical data; as a result, a panoptic view became possible at first glance, and unprecedented comparisons could be made. By the early 1860s, at nearly 80 years of age, his interests shifted from economic phenomena to historical subjects such as the movements of famous armies. In the last year of his life, he created the *Carte figurative des pertes successives en hommes de l'Armee Français dans la campagne de Russe 1812–1813*, also known as *Napoleon's March to Moscow*.

Edward Tufte reproduced Minard's *Napoleon's March to Moscow* in *The Visual Display of Quantitative Information*, with the conviction that it might well be the best statistical graphic ever drawn.

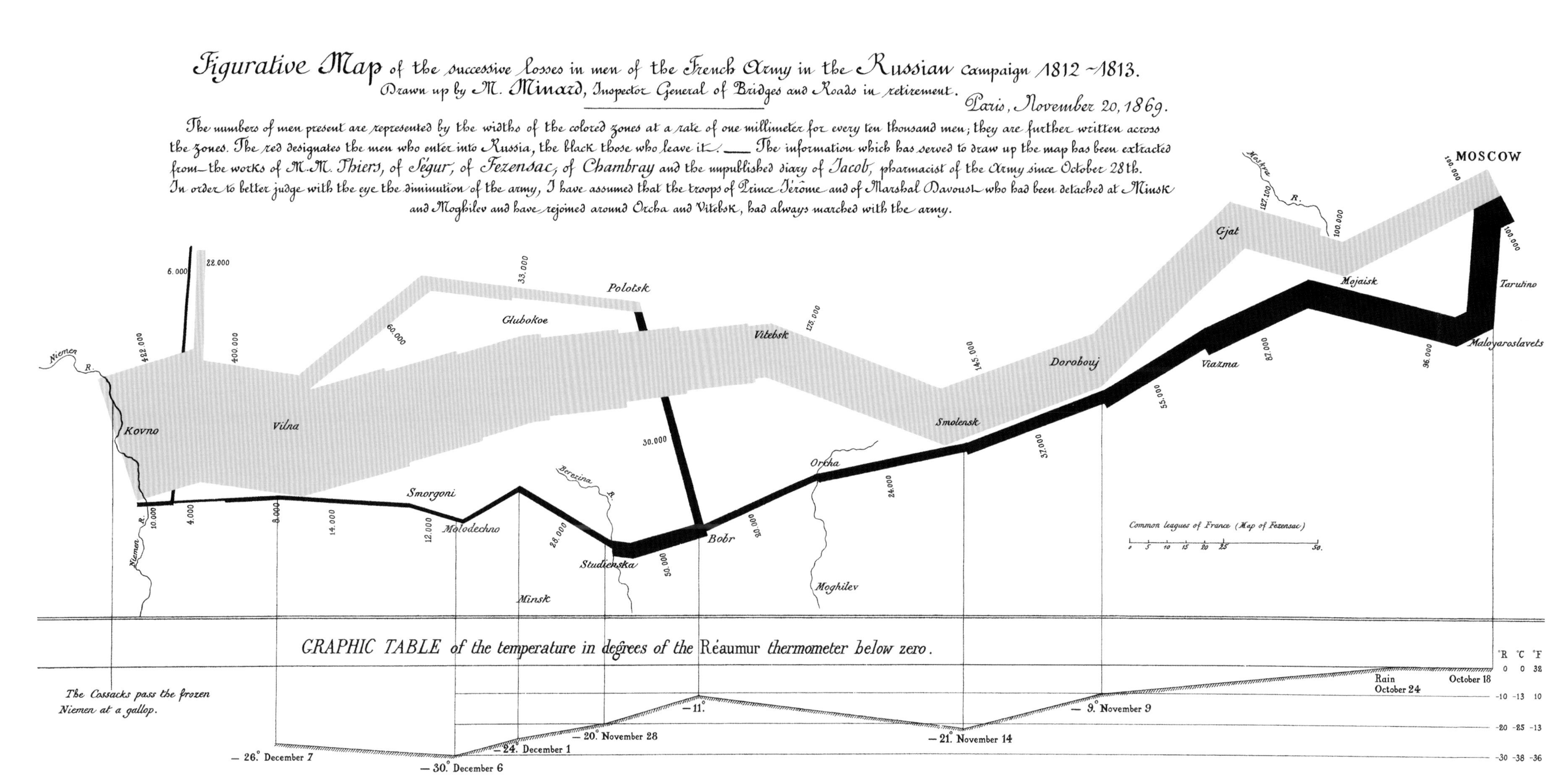

Figurative Map of the successive losses in men of the French Army in the Russian campaign 1812 – 1813.
Drawn up by M. Minard, Inspector General of Bridges and Roads in retirement.
Paris, November 20, 1869.
The numbers of men present are represented by the widths of the colored zones at a rate of one millimeter for every ten thousand men; they are further written across the zones. The red designates the men who enter into Russia, the black those who leave it. —— The information which has served to draw up the map has been extracted from the works of M.M. Thiers, of Ségur, of Fezensac, of Chambray and the unpublished diary of Jacob, pharmacist of the Army since October 28th.
In order to better judge with the eye the diminution of the army, I have assumed that the troops of Prince Jérôme and of Marshal Davoust who had been detached at Minsk and Moghilev and have rejoined around Orcha and Vitebsk, had always marched with the army.
MOSCOW
Niemen R.
Kovno
Vilna
Glubokoe
Polotsk
Vitebsk
Smolensk
Dorobouj
Gjat
Viazma
Mojaisk
Tarutino
Maloyaroslavets
Moskva R.
Orcha
Bobr
Studienska
Berezina R.
Smorgoni
Molodechno
Minsk
Moghilev
422.000
400.000
22.000
6.000
60.000
33.000
175.000
145.000
127.100
100.000
100.000
100.000
96.000
87.000
55.000
37.000
24.000
20.000
50.000
30.000
28.000
12.000
14.000
8.000
4.000
10.000
Common leagues of France (Map of Fezensac)
GRAPHIC TABLE of the temperature in degrees of the Réaumur thermometer below zero.
°R °C °F
0 0 32
-10 -13 10
-20 -25 -13
-30 -38 -36
The Cossacks pass the frozen Niemen at a gallop.
— 26.° December 7
— 30.° December 6
— 24.° December 1
— 20.° November 28
— 11.°
— 21.° November 14
— 9.° November 9
Rain October 24
October 18

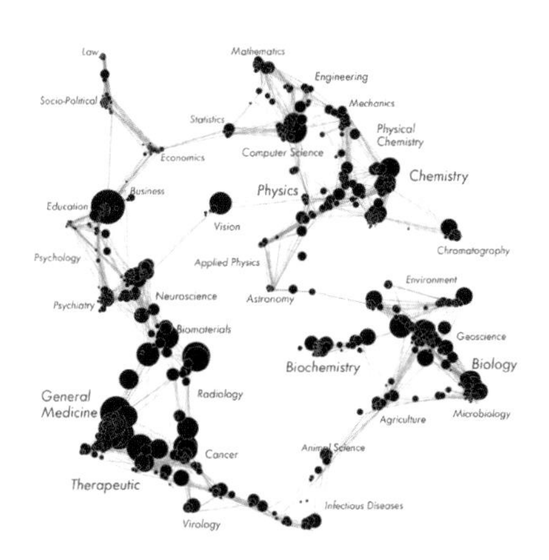

1999

1996 Map of Science: A Network Representation of the 43 Fourth-Level Clusters Based on Data from the 1996 *Science Citation Index*

By Henry G. Small

PHILADELPHIA, PENNSYLVANIA, 1999

Courtesy of Henry G. Small, Thomson Reuters

Aim

While it is easy to count and chart the number of publications authored or coauthored by different experts, institutions, or nations, it is far more difficult to describe the type of research they conduct. Prior classification systems are of little help, as research topics change so rapidly. Small aims to map science on a large scale and at multiple levels, based on fractional-citation counting and bottom-up cocitation clustering. He is interested in charting pathways that link scientific disciplines, facilitating the import and export of scientific results across disciplinary boundaries.

Interpretation

At right is the fourth and most general level of Small's *1996 Map of Science*, which includes 43 major areas of research. Each area is represented by a circle with a size proportional to the number of papers published. Labels for major areas are based on a frequency analysis of paper titles and journal category names. Distance between the circle's center is determined by the number of cocitations between the research areas. Lines represent strong cocitation links between areas. The interactive version of the map supports zooming into less aggregated, more detailed areas of science (see next spread).

General areas range from **Optics** on the upper left to **Computer Science** on the lower right. **Biomedicine** is the largest and most central area of research. **Biomedicine** and **Proteins** are close to each other, as they are jointly cited many times. The circle for **Computer Science** appears small, as computer science research is commonly published in conference and workshop proceedings that are not covered in this data set. The dotted line indicates the trail of papers and respective research areas that would be encountered when traveling through the map of science from **Economics** (start) to *Wave Function of the Universe* in **Physics** (finish).

Henry G. Small received a joint PhD in chemistry and the history of science from the University of Wisconsin. After working as a historian of science at the AIP's Center for History of Physics, in 1972 he joined the staff of the Institute for Scientific Information (now part of Thomson Reuters), where he is currently chief scientist. Small has written more than 100 papers on topics in citation analysis and the mapping of science. His 1973 paper in the *Journal of the American Society for Information Science and Technology* (*JASIST*) on cocitation in scientific literature is one of the most cited papers in *JASIST* and was later designated a Citation Classic. He received the Derek J. de Solla Price Memorial Medal from the journal *Scientometrics* in 1987, the Best *JASIST* Paper Award in 1987, and the Award of Merit from the American Society for Information Science and Technology in 1998. He is a fellow of the American Association for the Advancement of Science (AAAS) and of the National Federation of Abstracting and Information Services (NFAIS). He is a past president of the International Society for Scientometrics and Informetrics.

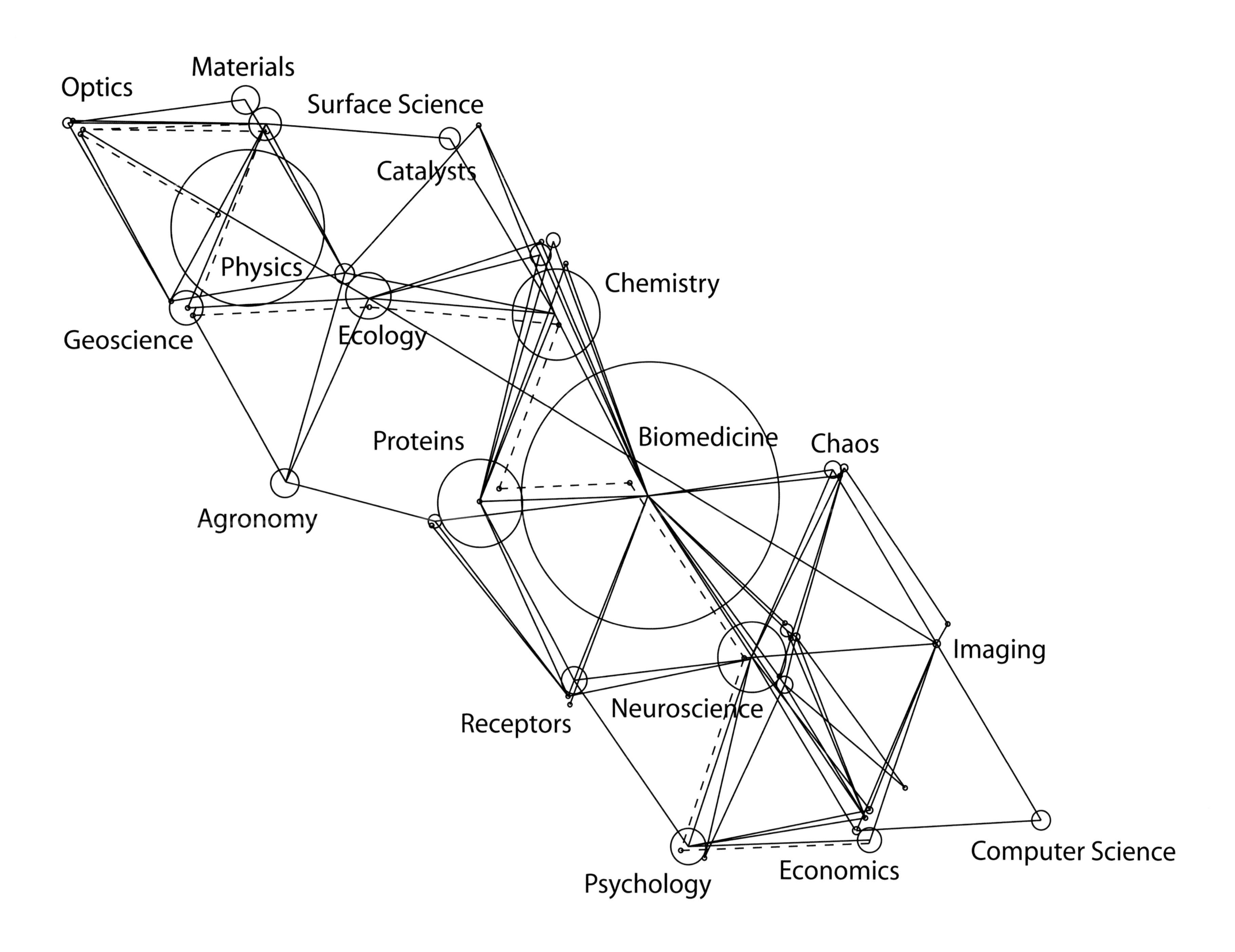
Materials
Optics
Surface Science
Catalysts
Physics
Chemistry
Geoscience
Ecology
Proteins
Biomedicine
Chaos
Agronomy
Imaging
Neuroscience
Receptors
Computer Science
Economics
Psychology

Data

Small examined 36,720 highly cited multidisciplinary papers from Thomson Reuters' Science Citation Index (SCI) for the 15 years from 1981 to 1995.

Technique

Map generation proceeds bottom-up via a multistep data analysis and layout process. Highly cited papers are subjected to a combination of fractional-citation counting and bottom-up cocitation clustering via an optimized single-linkage approach. Here, the cocitation similarity of two papers equals the number of times they appear together in the reference lists of all papers in the 1996 SCI. The result is a nested hierarchy that groups papers at four levels of aggregation, see 1996 Map of Science on opposite page. Clusters at each level of aggregation are laid out in a two-dimensional space using an order-dependent geometric triangulation process that produces a unified ordination of the hierarchical grouping of papers.

The leaf nodes of the hierarchy are highly cited papers. At the lowest level, each research front consists of a group of cocited core papers. Some research fronts have only two papers. No front has more than an arbitrary maximum of 50 papers. The first-level clustering partition groups those papers into rather specific research areas. The second-level clustering partition groups specific research areas and their papers into more general areas. Aggregation of specific into more general areas of research continues for the third- and fourth-level partitions. At each level, the distance between the circle's center corresponds to the number of cocitations between the clusters. The result is a nested hierarchy of research areas (see also opposite page), in which each area of research contains a map of similar construction at a lower level of aggregation. The original map facilitated interactive exploration.

Reference System

Triangulation is a layout technique that maps points from an n-dimensional space into a typically two-dimensional one. It starts by placing a randomly selected first object (here, a paper) at the origin of the coordinate system. Next, the most similar object (paper) is determined and placed at a specified distance from the first object. The location of the third object is defined by its distance from the preceding two objects—hence the name triangulation. Subsequently, the notion of repulsion from the origin is used to select the quadratic solution furthest from the origin—the spatial layout grows outwards. The resulting map exactly represents the local distances between single data objects but lacks global optimization. Just like classical ordination methods, triangulation results in layouts that are invariant to rotation and mirroring.

Data Overlays

The size of each circle is proportional to the number of papers published in the research area. Area labels are based on a frequency analysis of paper titles and journal category names. Links between circles represent aggregate paper cocitations. Pathways from one area of research to another are indicated by dotted lines. Two arbitrary areas of science can be chosen and an algorithmic procedure returns the path with the strongest cocitation links along the route. The strongest links at the top level are determined first. The process is repeated at each lower level of the hierarchical maps—down to the fourth ground level of published core papers. The dashed pathway connecting economics to physics has exactly 331 core papers and can be written as:
Economics (start) > **Psychology** > **Psychiatry** > **Neuroscience** > **Neurophysiology** > **Immunology and HIV** > **Crystallography** > **Chemistry** > **Plant Physiology** > **Ecology** > **Climatology** > **Glaciology** > **Geology** > **Seismology** > **Physics of Earth** > **Diamonds** > **Fractals** > **Quantum Physics** > **Wave Function of the Universe** (finish).

Details

In recent work, Small has explored the notion of a pathway through science in greater detail. Pathways linking the literature of different disciplines are likely to also transfer methods between them. This is known as cross-disciplinary fertilization. As Small has noted, this reaching out or stretching can import or export methods, ideas, models, or empirical results from one field to another. This requires scientists to have not only a broad awareness of the literature, but also the creative imagination to foresee how outside information fits with the problem at hand. Small developed algorithms to blaze a magnificent trail of several hundred papers across the literatures of different scientific disciplines.

Unique Features

The Institute for Scientific Information (ISI) of Thomson Reuters serves interactive research front maps as part of the Special Topics Web site (**http://sciencewatch.com**), designed to complement Essential Science Indicators. Each month, a new scientific research area that has experienced notable recent advances or is of special current interest is selected. Citation analyses, interviews, and commentaries are compiled for this topic. Results are presented as citation rankings for scientists, institutions, nations, and journals, and also via interviews and essays by prominent scientists.

The image below shows the research front map for the special topic of "Organic Thin-Film Transistors," published in July 2007. Core, highly cited papers composing this research front are shown as blue circles. A paper's bibliographic information is displayed when the user clicks on the circle. The solid lines between circles represent the strongest cocitation links for each paper, indicating that the papers are frequently cited together. Weaker links are shown by dashed lines. Papers close to each other on the map are generally more highly cocited. The most recent papers are indicated in red.

Clicking on "John A. Rogers" in the Interviews menu opens a Web page (below) that provides further details about Rogers's work on "elastomeric transistor stamps" and their significance for the field of organic thin-film transistors. Only the top of the Web page is shown.

Organic Thin-Film Transistors

RESEARCH FRONT MAPS | Map methodology

Additional Research Front Maps for the special topic of Organic Thin-Film Transistors

Map	Interviews
N-TYPE Thin-Film TRANSISTORS	Interview with John E. Anthony regarding his paper, Anthony JE, *et al.*, "Functionalized pentacene: improved electronic properties from control of solid-state order," *J. Am. Chem. Soc.* 123(38): 9482-3, 26 September 2001.
REGIOREGULAR POLYTHIOPHENE	Michael D. McGehee talks about his paper, Kline RJ, *et al.*, "Controlling the field-effect mobility of regioregular polythiophene by changing the molecular weight," *Advan. Mater.* 15(18): 1519-+, 16 September 2003.
SINGLE-CRYSTAL THIN FILM TRANSISTORS	John A. Rogers **discusses his featured Research Front Map paper (below), Sundar VC, *et al.*, "Elastomeric transistor stamps: reversible probing of charge transport in organic crystals," *Science* 303(5664): 1644-6, 12 March 2004.**
	Vitaly Podzorov **talks about his paper (below), "Field-effect transistors on rubrene single crystals with parylene gate insulator," *Applied Physics Letters* 82[11]: 1739-41, 17 March 2003.**

SINGLE-CRYSTAL Thin-Film TRANSISTORS

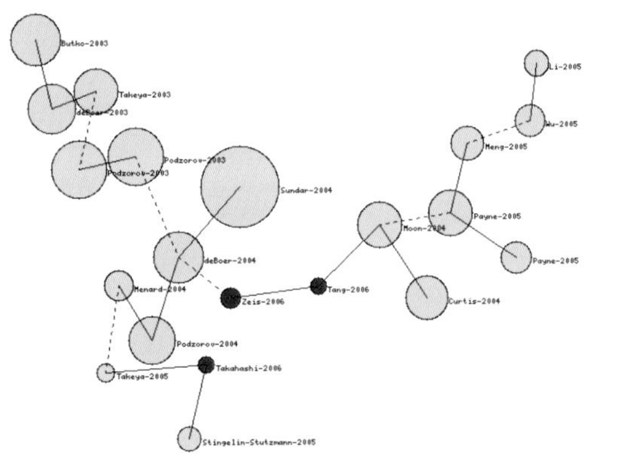

Organic Thin-Film Transistors

ESI Special Topics: J
Citing URL: http://esi-topics.com/otft/interviews/rfm3_JohnARog

A Research Front Map INTERVIEW with Professor John A. Rogers

<• Return to Research Front Map

In the interview below, we talk with Professor John Rogers about his paper, "Elastomeric transistor stamps: reversible probing of charge transport in organic crystals" (Sundar VC, et al., Science 303[5664]: 1644-6, 12 March 2004). This paper is a core paper in the Research Front on Single-Crystal Thin-Film Transistors in our Special Topic on Organic Thin-Film Transistors, and currently has 156 citations to its credit. In Essential Science Indicators[SM], Professor Rogers's citation record includes 188 papers cited a total of 4,484 times to date. These papers can be found in the fields of Physics, Chemistry, and Materials Science. Professor Rogers is a Founder Professor of Engineering, Professor of Materials Science and Engineering, and Professor of Chemistry at the University of Illinois at Urbana-Champaign.

ST: Would you please describe the significance of your paper and why it is highly cited?

The work is highly cited, we believe, for two reasons: (1) the techniques that we described are useful to others for study of these kinds of systems and (2) the large, directionally anisotropic mobilities that we reported in rubrene are of interest. Widespread adoption of these and related methods provide evidence for the first reason. Other groups reproducing these results and extending them in interesting ways support the significance of the latter reason.

Professor John A. Rogers's most-cited paper with 564 cites to date:
Xia YN, *et al.*, "Unconventional methods for fabricating and patterning nanostructures," *Chem. Rev.* 99(7): 1823-48, July 1999.

Professor John A. Rogers's paper(s) represented in the Research Front map with 156 cites to date:
Sundar VC, *et al.*, "Elastomeric transistor stamps: reversible probing of charge transport in organic crystals," *Science* 303(5664): 1644-6, 12 March 2004.

Source: *Essential Science Indicators.*

ST: How did you become involved in this research and were there any particular successes or obstacles that stand out?

We have been interested in organic semiconductor materials for some time, mainly for their use in flexible electronics. The study of single-crystal samples of rubrene reported in this paper involved a collaboration with a materials physics group at Rutgers, aimed at exploring upper limits in transport properties for more technologically relevant thin-film materials. The mechanical and chemical fragility of the crystals demanded new approaches to examine their electrical properties, in this case through the use of field-effect transistor structures. The soft contact, "transistor stamp" approach that we reported enables non-destructive and reversible measurement of the intrinsic properties of transport on the pristine surfaces of such samples.

ST: Where do you see your research and the broader field leading in the future?

The organic electronics community continues to make impressive progress, not only on the basic science of organic semiconductors but also on their transition to commercial technologies. In fact, a large fraction of the research and development in this area now occurs at companies. Many university groups, including ours, are shifting their attention to newer classes of advanced materials that have the potential to offer improved performance.

A detailed description of the methodology applied to identify research fronts is available via this page and shown below.

Overview

The idea of thin-film transistors based on organic materials is now over 20 years old, during which time the technology has evolved from laboratory curiosity to real-world applications. The promise of ultra-low-cost, lightweight, flexible electronic devices means that the organic thin-film transistor may someday be as ubiquitous as its silicon-based competitor.

Our Special Topics ranking of the most influential papers of the past decade is led by a seminal 1999 review article in *Nature*—"Electroluminescence in conjugated polymers" (Friend RH, *et al.*, *Nature* 397[6715]: 121-8, 14 January 1999)—that was coauthored by 11 of the field's leading innovators and has since racked up almost 2,000 citations. Other hot papers of the decade include a half-dozen review articles charting the evolution of the field, another five on the molecular, electronic, and chemical characteristics of pentacene-based thin-film transistors, and articles on such applications as paper-like electronic displays and high-performance all-polymer integrated circuits.

The compilation of the hottest papers over the last two years is much more democratic and has yet to reveal a definitive pattern. Several review papers discuss advances in organic field-effect transistors, while others characterize the performance and characteristics of thin-film transistors based on semiconductors of a dizzying variety of organic materials.

Methodology

To construct this database, papers were extracted based on topic-supplied keywords for Organic Thin-Film Transistors. The keywords used were as follows:

```
organic Thin-Film transistor* OR organic field effect transistor*
```

The baseline time span for this database is 1997-February 28, 2007. The resulting database contained 2,199 (10 years) and 1,141 (2 years) papers; 4,860 authors; 51 countries; 250 journals; and 980 institutions.

Rankings

Once the database was in place, it was used to generate the lists of top 20 papers (two- and ten-year periods), authors, journals, institutions, and nations, covering a time span of 1997-February 28, 2007 (first bimonthly, a 10-year plus 2-month period).

The top 20 papers are ranked according to total cites. Rankings for author, journal, institution, and country are listed in three ways: according to total cites, total papers, and total cites/paper. The paper thresholds and corresponding percentages used to determine scientist, institution, country, and journal rankings according to total cites/paper, and total papers respectively are as follows:

Entity:	Scientists	Institutions	Countries	Journals
Thresholds:	16	9	8	15
Percentage:	1%	1%	50%	10%

Exploring the *1996 Map of Science*

In his 1998 AAAS presentation on "Mapping the World of Science," Eugene Garfield used a series of maps prepared by Henry G. Small to demonstrate the interactive exploration of science at multiple levels. Starting from the global map, he rendered exploratory close-ups of BIOMEDICINE and ECONOMICS.

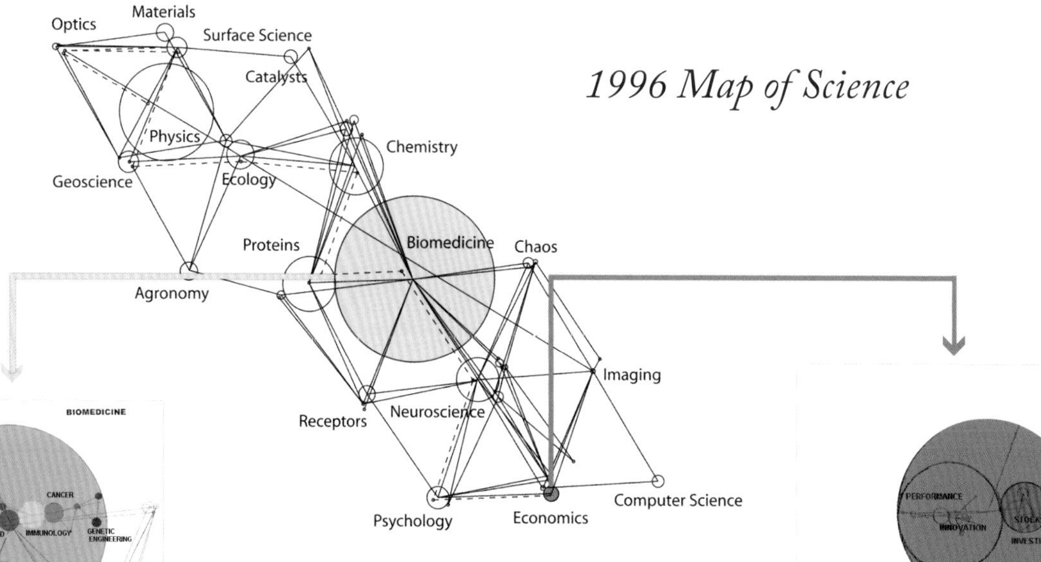

1996 Map of Science

The fourth-level map of BIOMEDICINE shows a large number of specialties, including CARDIOLOGY, IMMUNOLOGY, and CANCER. We zoom into IMMUNOLOGY.

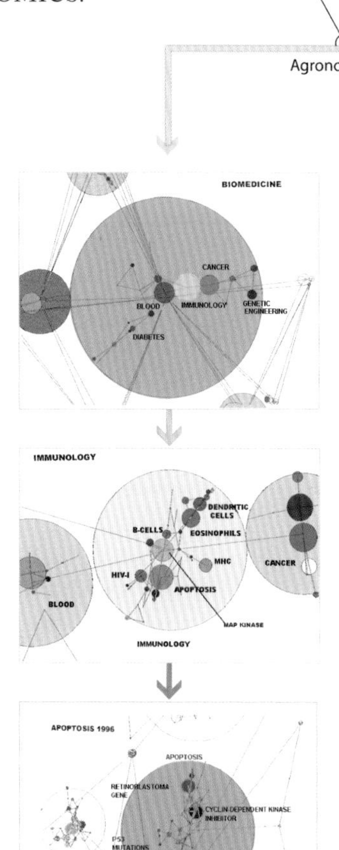

< Fourth Level of Aggregation >

Next, we zoom into ECONOMICS, revealing the fourth-level map with the BUSINESS and INVESTING clusters, topics familiar to Wall Street analysts. The clusters are close to each other because of their many cocitations.

IMMUNOLOGY consists of subspecialties like DENDRITIC CELLS, EOSINOPHILS, and APOPTOSIS; all three represent a large number of publications.

< Third Level of Aggregation >

Zooming into INVESTING reveals clusters STOCK MARKET, GROWTH, and CYCLES.

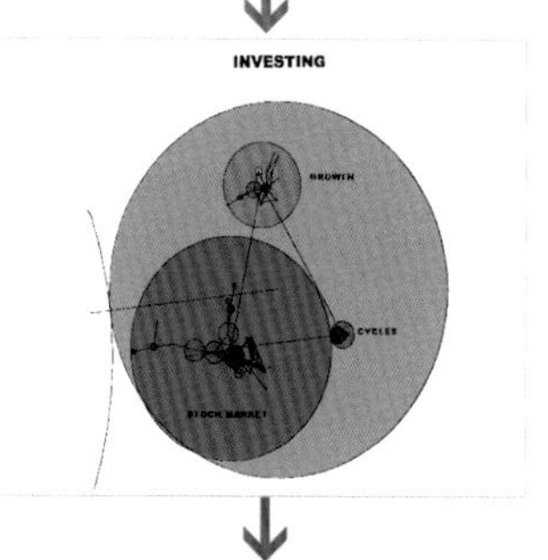

APOPTOSIS itself consists of dozens of research areas, including one which is described simply as APOPTOSIS and P53.

< Second Level of Aggregation >

The STOCK MARKET cluster includes literature on RISK, RETURNS, VOLATILITY, and COINTEGRATION.

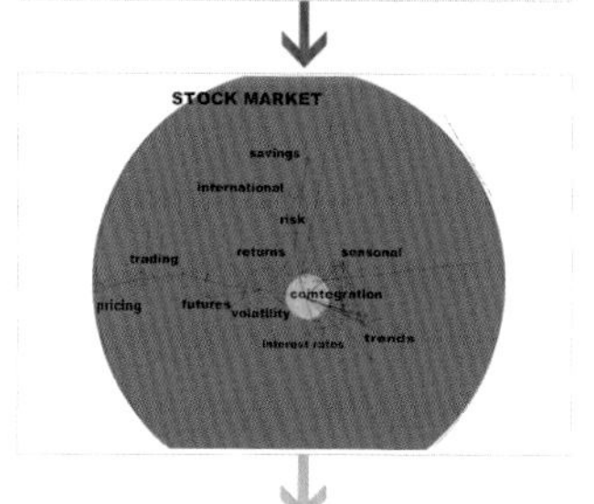

Zooming into this area, we get to see the dozens of highly cited and cocited core papers in this subspecialty.

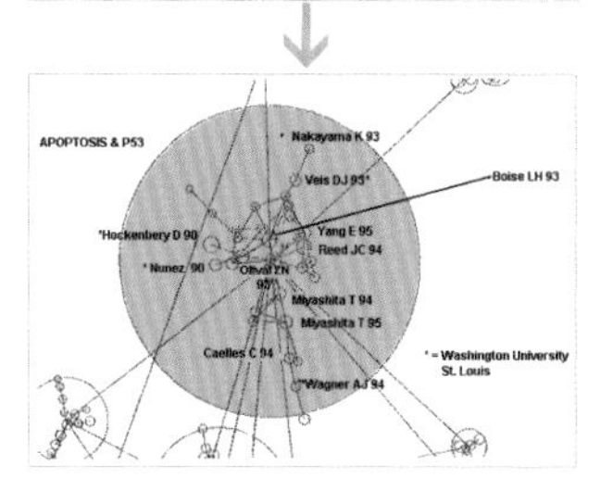

Research Fronts
< First Level of Aggregation >

The amount of economic literature on COINTEGRATION, a mathematical and statistical tool frequently used to study association between two time series, is vast.

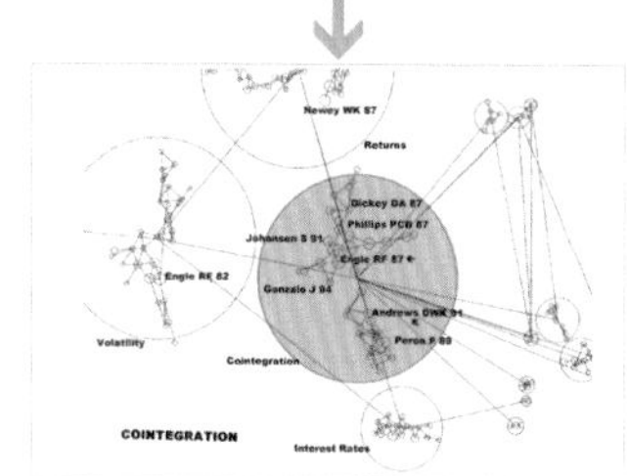

Highly Cited Papers
from First Level of Aggregation

Among those papers are Hockenberry's 1990 *Nature* paper and Boise's 1993 paper in *Cell*, which in 1996 had 312 and 272 citations, respectively.

Apoptosis and P53 Core Papers

Cite	Name	Year	Journal
272	**Boise LH**	**93**	**Cell**
102	Caelles C	94	Nature
88	Chittenden T	95	Nature
76	Farrow SN	95	Nature
47	Gonzalezgarcia M	94	Development
30	Hanada M	95	J Biol Chem
90	Hermeking H	94	Science
312	**Hockenbery D**	**90**	**Nature**
86	Kiefer MC	95	Nature

Cointegration Core Papers

Cite	Name	Year	Journal
12-45	Andrews DWK	91-94	Econometric
18	Banerjee A	92	J Bus Econ
15	Cheung YW	93	Ox Bus Econ
12	Christiano LJ	92	J Bus Econ
15	Cooley TF	85	J Monet Ec
14	Davies RB	87	Biometrika
15	Dickey DA	87	J Bus Econ
10	Dufour JM	82	J Economet
225	**Engle RF**	**87**	**Econometric**

The work of R. F. Engle at the University of California, San Diego, published in 1987 as "Cointegration and error correction," is at the center of this front. This appeared in the journal *Econometrica*. While it was cited 225 times in 1996, it has been explicitly cited more than 1,700 times since publication.

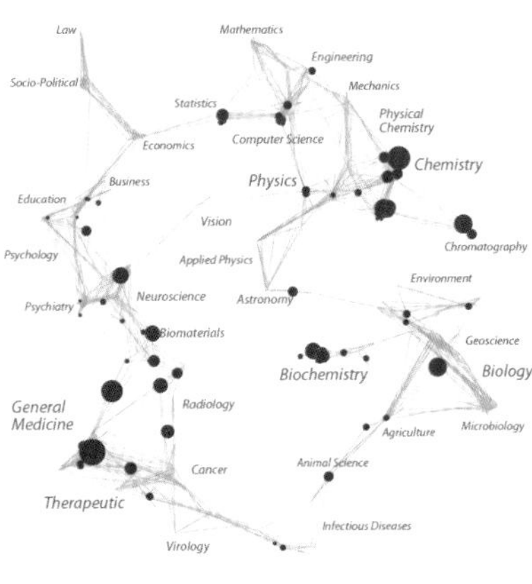

2004

PhD Thesis Map

By Keith V. Nesbitt

NEWCASTLE, AUSTRALIA, 2004

Courtesy of IEEE and Keith V. Nesbitt, Charles Sturt University, Australia

Aim

While at Sydney University, Nesbitt worked under the supervision of Professor Peter Eades, a scholar famous for developing automatic drawing algorithms for graphs. After several years of research, Nesbitt had completed a first draft of his PhD thesis and arranged a meeting with Eades—at which it became clear that communicating the broad scope, complex character, and many ideas involved in the work was proving too unwieldy. Eades left him with the suggestion that a clear path or roadmap might help a reader navigate the many interlinked threads while helping Nesbitt salvage his thesis. With this in mind, Nesbitt traveled home by train—when he spotted the Sydney metro map on the carriage wall. He realized that, instead of a roadmap, he could make a metro map of his thesis. Immediately, using colored pencils, he sketched the map he would later rerender with drawing software—to the satisfaction and surprise of Eades, a great fan of metro maps.

Interpretation

This manually drawn map shows the main interconnecting themes or tracks that run through Nesbitt's PhD thesis. It uses the familiarity of a metro map to help readers navigate and explore. Each separate "track of abstract thought" in the thesis is shown in a different color. As if using the metro system, readers can follow a single track and see the main stations (ideas) upon it. Some stations are more important than others, as they join the different themes. Connecting stations are the perfect place to change one's track of reading and thought. As the space in which the tracks are laid is free of orientation, it is possible to read the map in any direction. Nevertheless, one might begin at the top left stations to travel across or down the map in any progressive fashion.

Keith V. Nesbitt is a senior lecturer in the School of Design, Communication, and Information Technology at the University of Newcastle, Australia. Previously, he worked at Charles Sturt University in Bathurst; he also spent 10 years at the applied research facility of Broken Hill Proprietary Company Ltd., where he applied advanced computing technologies to a cross-section of business domains, including exploration, mining, manufacture, logistics, and finance. He studied medicine before developing an interest in computing and subsequently receiving a bachelor's degree in mathematics from the University of Newcastle and a master's in computing and a PhD in computer science from Sydney University. His doctoral work investigated new conceptual models for designing multisensory user interfaces and included the design and evaluation of visual, auditory, and haptic displays for finding patterns in stock market data. Nesbitt summarizes his work as tool building that helps people find useful patterns in data. Related to this, Nesbitt also studies how the mind works and how people create and use ideas. His current scientific interests include designing multisensory displays, perception and cognition, complex systems, software engineering, user-interface design, and conceptual modeling.

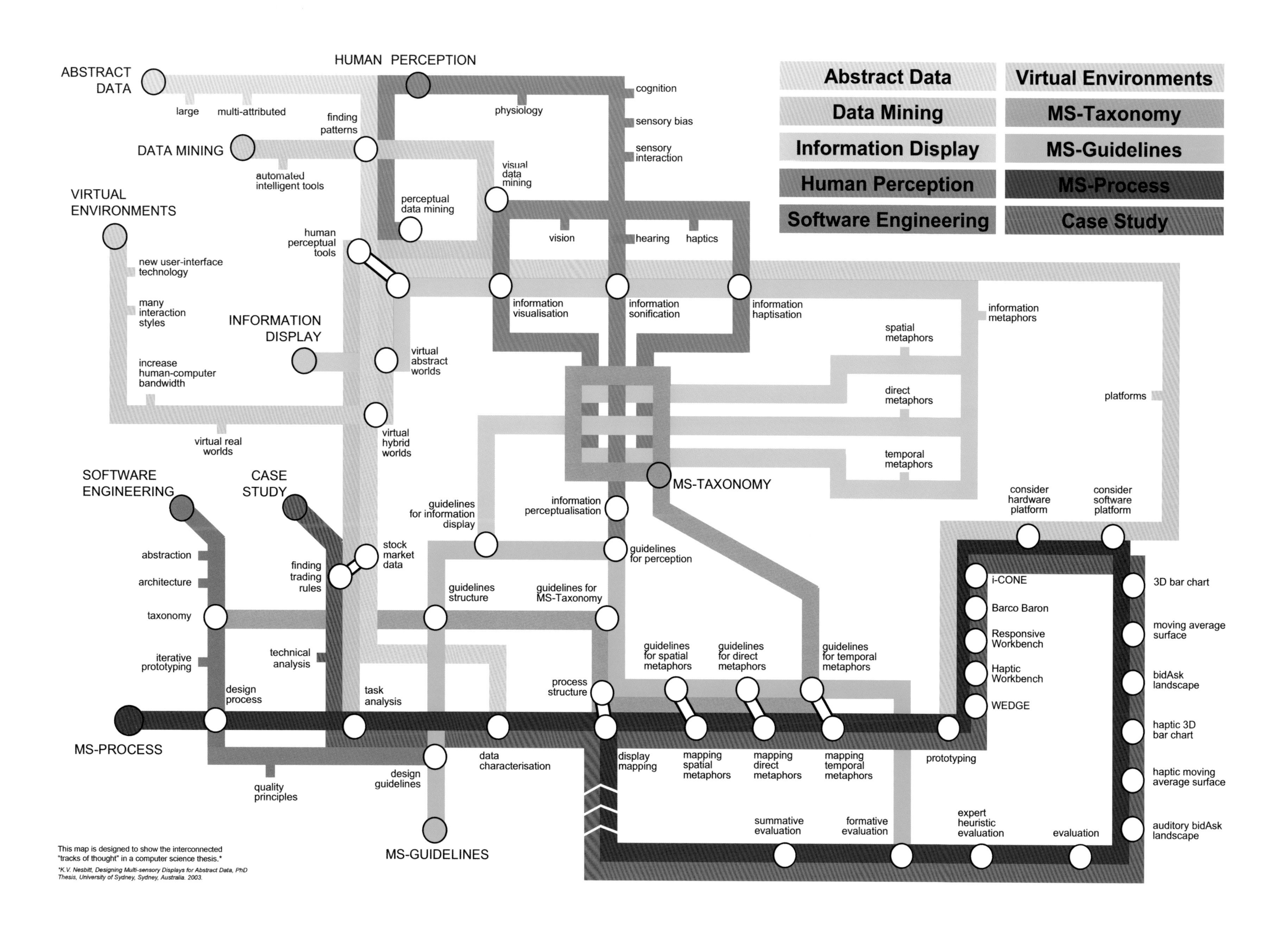

ABSTRACT DATA
large
multi-attributed
HUMAN PERCEPTION
cognition
physiology
sensory bias
sensory interaction
finding patterns
DATA MINING
automated intelligent tools
visual data mining
perceptual data mining
VIRTUAL ENVIRONMENTS
human perceptual tools
vision
hearing
haptics
new user-interface technology
many interaction styles
information visualisation
information sonification
information haptisation
information metaphors
INFORMATION DISPLAY
virtual abstract worlds
spatial metaphors
increase human-computer bandwidth
direct metaphors
platforms
virtual real worlds
virtual hybrid worlds
temporal metaphors
SOFTWARE ENGINEERING
CASE STUDY
MS-TAXONOMY
guidelines for information display
information perceptualisation
consider hardware platform
consider software platform
abstraction
stock market data
guidelines for perception
architecture
finding trading rules
i-CONE
3D bar chart
guidelines structure
guidelines for MS-Taxonomy
Barco Baron
taxonomy
moving average surface
Responsive Workbench
technical analysis
guidelines for spatial metaphors
guidelines for direct metaphors
guidelines for temporal metaphors
iterative prototyping
bidAsk landscape
Haptic Workbench
design process
task analysis
process structure
WEDGE
haptic 3D bar chart
MS-PROCESS
display mapping
mapping spatial metaphors
mapping direct metaphors
mapping temporal metaphors
prototyping
data characterisation
design guidelines
quality principles
haptic moving average surface
summative evaluation
formative evaluation
expert heuristic evaluation
evaluation
auditory bidAsk landscape
MS-GUIDELINES
Abstract Data
Data Mining
Information Display
Human Perception
Software Engineering
Virtual Environments
MS-Taxonomy
MS-Guidelines
MS-Process
Case Study
This map is designed to show the interconnected "tracks of thought" in a computer science thesis.*
*K.V. Nesbitt, Designing Multi-sensory Displays for Abstract Data, PhD Thesis, University of Sydney, Sydney, Australia. 2003.

Data

The data depicted in the map are the interconnecting themes that run through Nesbitt's PhD thesis, as detailed below.

Pink and gray tracks

The thesis develops a number of new tools to help people design multisensory displays of data. The aim is to take lots of abstract data and then build a computer interface that uses pictures, sound, and also touch to show the most important parts of the data. These interfaces take advantage of our perceptual abilities to find patterns. For example, we may want to find useful patterns in stock market prices. Another way to describe this is *data mining*, because we can use the display to mine the data for useful information. The two tracks called **Abstract Data** (pink) and **Data Mining** (gray) thus provide the motivation for the work in the thesis.

Light blue, dark green, brown, and orange tracks

Building a multisensory display can be quite complex because it is not always clear how to present abstract data. For instance, what would stock market data look, feel, and sound like? There are many fields of scientific study that are important to understand when building such displays. Those scientific fields that proved highly influential on this thesis are shown in the tracks called **Information Display** (light blue), **Human Perception** (dark green), **Software Engineering** (brown), and **Virtual Environments** (orange). The stations along these tracks are relevant topics or concepts that occur within these fields.

Light purple, green, and blue tracks

The thesis develops three main theoretical tools: a new way to categorize the parts that make up a multisensory display (**MS-Taxonomy**); a set of guidelines that can help designers to make good design decisions (**MS-Guidelines**); and a defined set of steps to follow when designing a display (**MS-Process**). The **MS-Taxonomy** (light purple) track is the most important part of the thesis and develops a new way of understanding how to display information that is less dependent on each particular sense (visual, auditory, and haptic) but rather relies on how a designer thinks about displaying information in space (spatial metaphors), time (temporal metaphors), and directly to a particular sense (direct metaphors). Because the MS-Taxonomy is so important, it forms the central station, whereas the three senses (visual, auditory, and haptic) overlap with the three approaches to designing displays (spatial, temporal, and direct metaphors).

Red and dark blue tracks:

The final track shown in the map is **Case Study**, the application of the theory developed in the thesis to the perceptual data mining of stock market data. Here, the **MS-Process** was applied and the **MS-Guidelines** used to develop and evaluate a range of new multisensory interfaces, such as the **3D bar chart** and the **bidAsk landscape**. These interfaces were tested in a range of different **Virtual Environments**, such as the **WEDGE**, the **Responsive Workbench**, and the **i-CONE**. Designing such displays is a highly iterative process (that is, parts of the design process have to be repeated multiple times). This is visually encoded at the end of the **Case Study** (red) and **MS-Process** (dark blue) tracks, which together create a loop.

Technique

In 1931, Beck designed the first diagrammatic map of the London Underground. The now-iconic Tube map is an extraordinary example of directional signage, as it renders the complex layout of the London subway system into horizontal, vertical, and diagonal lines with stops spaced at regular intervals. (Beck received a sum of 21 pounds for this milestone work.)

Technically, a metro map is a graph that consists of a set of nodes with a set of edges connecting these nodes and a set of paths that cover all the nodes and edges of the graph. Some vertices and edges may appear in more than one path, but each occurs in at least one path.

Hand-drawn metro maps exhibit a number of specific properties such as

- Each path is as straight as possible.
- Edge crossings are minimized.
- Labels do not overlap.
- Paths are placed horizontally, vertically, or at 45-degree angles.
- Each metro line is drawn with a unique color.
- Space nodes occur evenly on paths.
- Space angles of edges incident to a node appear evenly.

The first version of Nesbitt's *PhD Thesis Map* was drawn by hand (see below); the final map was rendered in PowerPoint. Recently, a number of algorithms have been published that support the semiautomatic layout of graphs in metro map style.

Reference System

The base map of the *PhD Thesis Map* is shown below in gray. Using this base map, diverse tracks of thought and their interconnections can be overlaid.

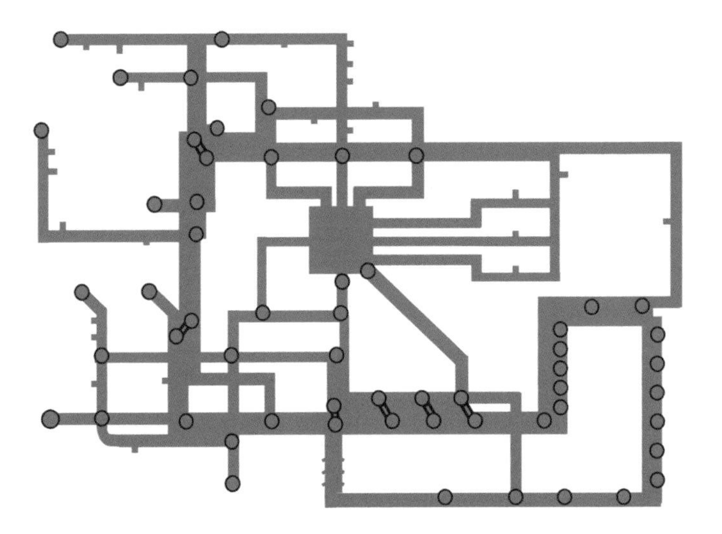

Data Overlays

Readers can be guided through the diverse "tracks," and every "station," as well as its connections to other tracks of abstract thought, can be explained.

For example, the pink track of **Abstract Data** begins in the top left corner of the map. Traveling along this route, one encounters the following stations: **large; multi-attributed; finding patterns; human perceptual tools; virtual abstract worlds; virtual hybrid worlds; stock market data;** and **data characterization**. **Abstract Data** is often **large** and **multi-attributed; finding patterns** in such large, multi-attributed, abstract data sets is a primary motivation for designing multisensory displays—hence the connection to the gray **Data Mining** track. The **virtual abstract worlds** station connects to the light blue **Information Display** track. The **virtual hybrid worlds** describe different types of models, or worlds, that can be built from abstract data using **Virtual Environments**. People can then explore these virtual worlds to find patterns in the abstract data. In his thesis, Nesbitt built a number of real displays, or worlds, using **stock market data**, which were tested in **Case Studies**. The final station on the pink track is **data characterisation**, a key step in the general process of building a virtual world for display.

Details

The metro map layout supports both episodic and semantic memory models. A series of studies showed the efficiency, memorability, and learnability of metro map displays for different tasks. The six maps on the right show details for the **Human Perception** (dark green), **MS-Taxonomy** (light purple), **MS Guidelines** (green), and **Case Study** (red) tracks.

Unique Features

Nesbitt painted the image to the left at the end of his PhD journey. It shows how the metro map was abstracted from three-dimensional objects in the real world into the familiar two-dimensional map of colored lines that we recognize and use today. On the left we see the realistic rendering of a train as it moves through a landscape of hills, rivers, tunnels, and bridges—abstracted on the right into the routes of the actual London metro map.

Nesbitt frames the image within an unusual shape to reflect the importance of the display space when designing a map. It is critical that maps communicate the frame of reference or context, without which they cannot be interpreted. The metro map has a distinct display space: tracks do not run as they do in the real world, and we cannot use the map to estimate the actual distance between stations. Nevertheless, metro maps perfectly preserve the order of the stations as well as every connection between them.

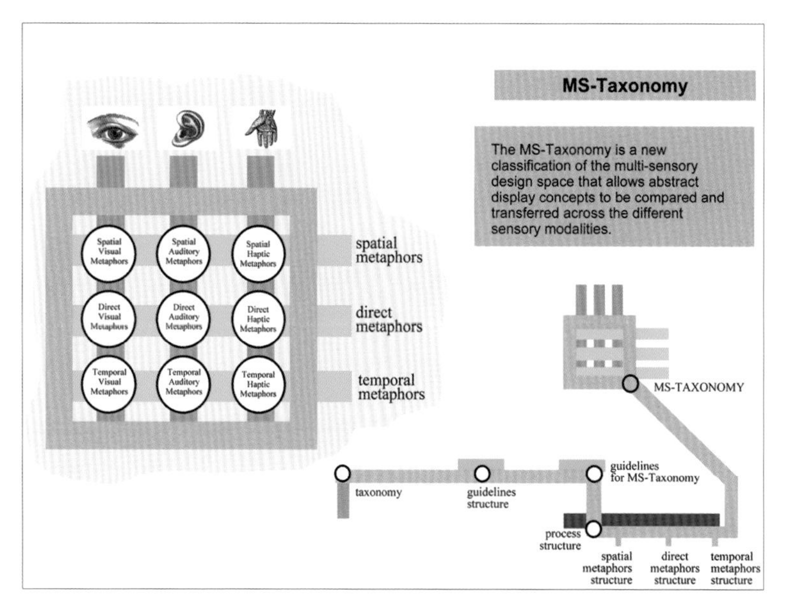

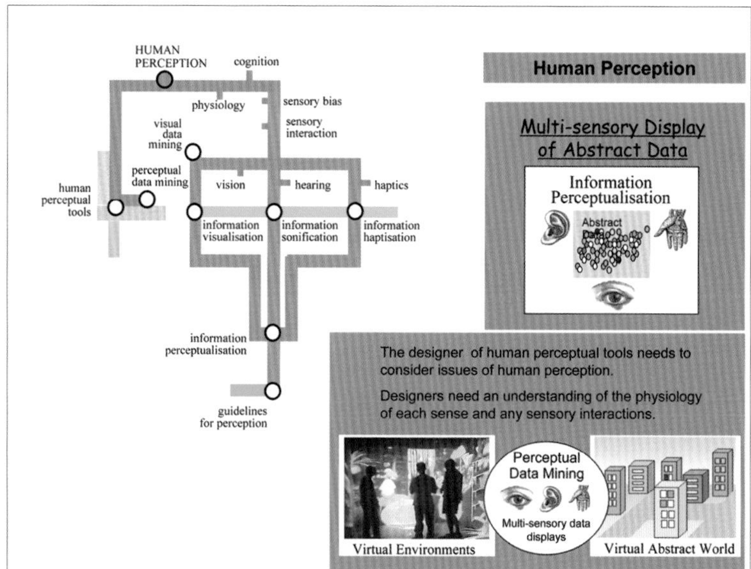

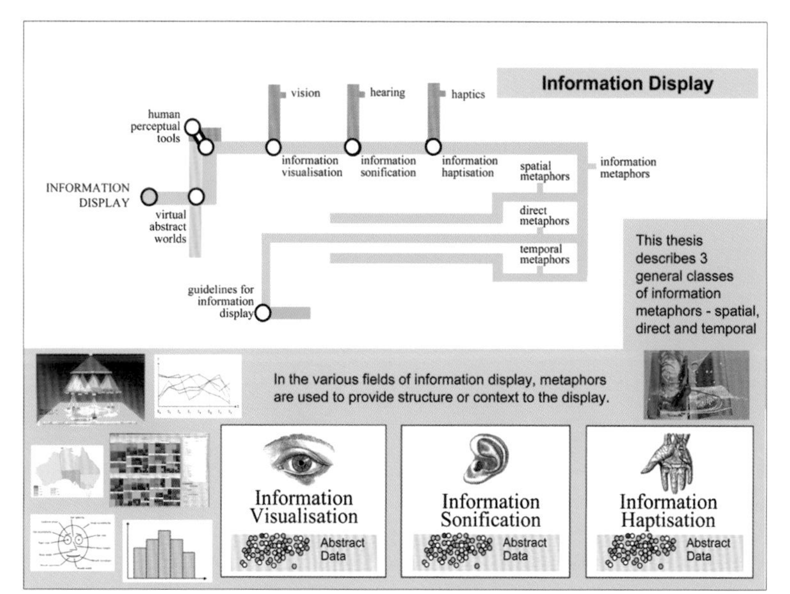

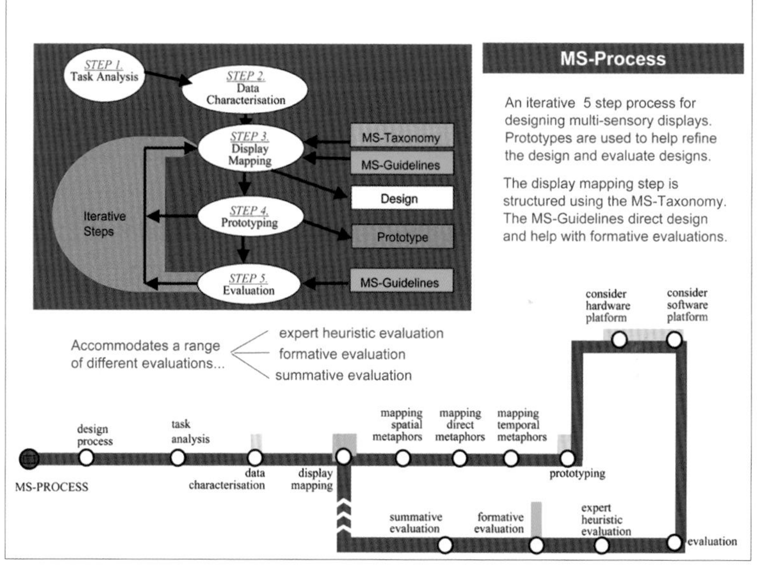

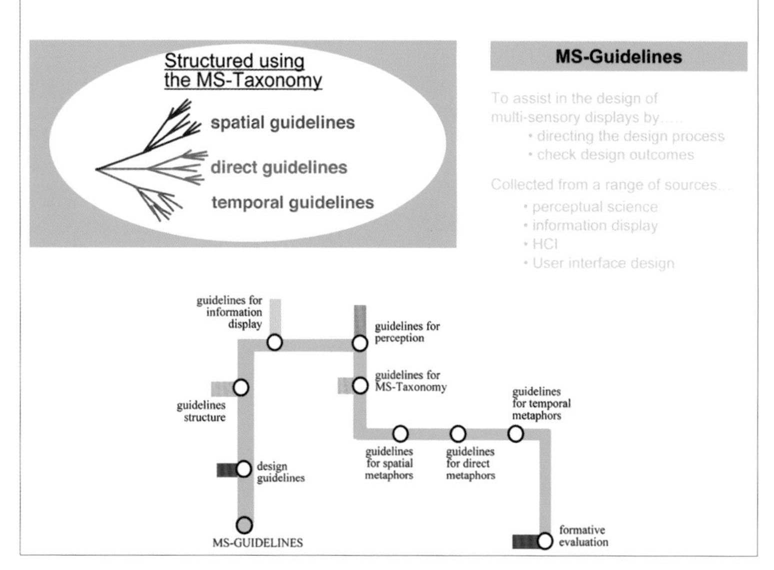

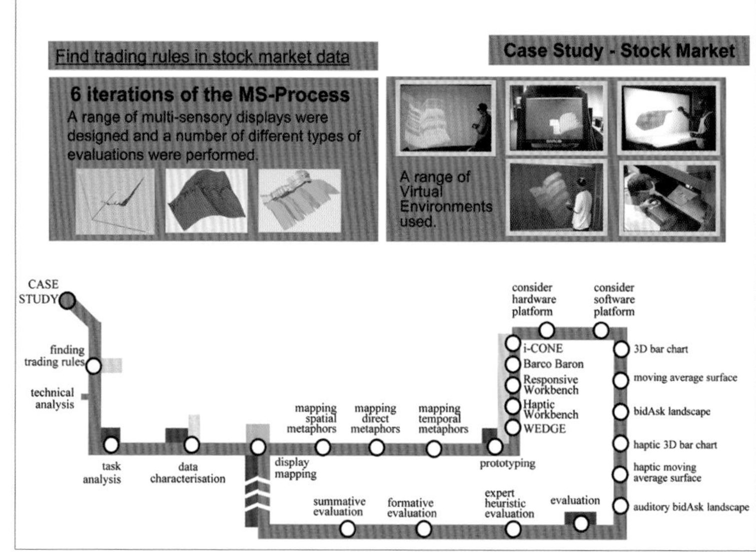

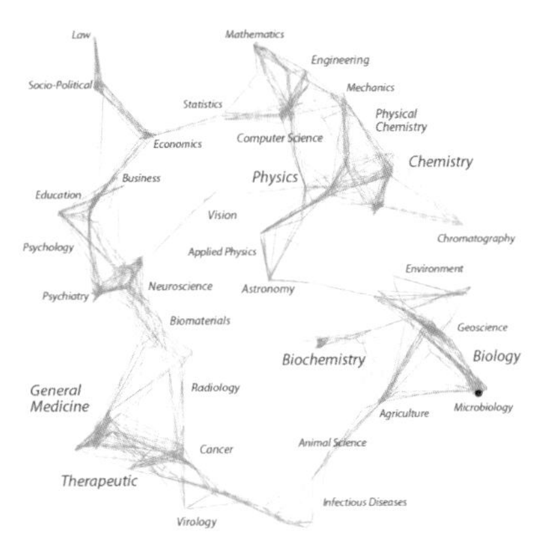

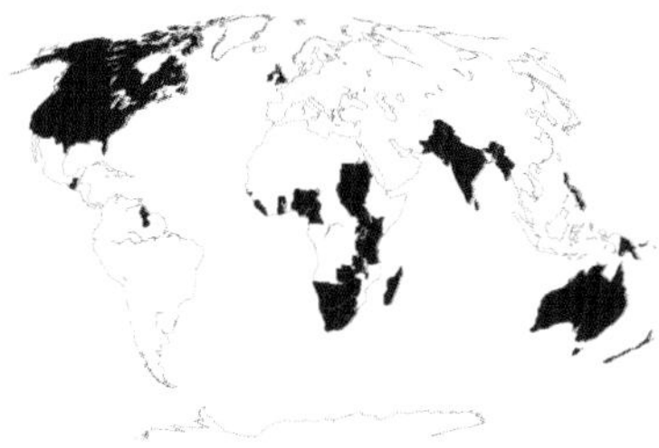

2005

Timeline of 60 Years of Anthrax Research Literature

By Steven A. Morris

STILLWATER, OKLAHOMA, 2005

Courtesy of Steven A. Morris, Oklahoma State University, Stillwater

Aim

Morris's work focuses on the study of research specialties. He is particularly interested in (1) identification and ranking of individual entities, for example, highly cited researchers, "rising stars," centers of excellence, high-impact papers, and key journals; (2) examination of groups and their relations, for example, terms (subtopic vocabularies), papers (research fronts, or papers grouped by subtopic), references (exemplar reference groups and paradigms), paper authors (research teams), reference authors (schools of thought), paper journals (research report libraries), and reference journals (base knowledge libraries); and (3) the study of specialty dynamics, such as trends, growth and decline, obsolescence of knowledge, geographic migration of research activity, discontinuous events, discoveries, external events, and forecasting.

Interpretation

The crossmap at right shows the historical development of subspecialties related to anthrax research. Relevant papers were gathered from Thomson Reuters Science Citation Index. Hierarchical agglomerative clustering, based on common references cited by paper pairs—also called bibliographic coupling—was used to group papers that tend to report on similar research topics.

The papers are then plotted in a year by topic crossmap such that the *x*-axis represents the years 1945–2004 covered by the data set and topics are listed on the *y*-axis. The left margin shows a dendrogram that displays the clustering hierarchy for the 35 clusters. Labels derived from manually browsing the titles of the papers in each cluster for themes appear in the right margin.

Each paper is represented by a circle whose size is proportional to the total number of citations the paper received. The intensity of each circle's red fill color is linearly proportional to the number of times the paper was cited in the final year of the data set. The darkest red denotes 34 citations—the highest citation count for a paper from papers published in the final year of the data set. Assuming the final year of the data set is the current year, this fill technique tends to show papers that are currently of great interest among researchers in the specialty. Manually added annotation indicates trends and major external events such as the anthrax postal bioterror attacks in 2001. The emergence and obsolescence of research topics, historically important papers (large circles), and currently important papers (red-filled circles) can be observed.

Steven A. Morris is an electrical engineer and native of Tulsa, Oklahoma. He earned BSEE and MSEE degrees from Tulsa University in 1983 and 1986, respectively, and worked as a research scientist in the petroleum industry for 15 years. Returning to academic studies in 1999, he conducted research in the mapping of journal and patent literature for technology forecasting and technology management applications. After earning a PhD in electrical engineering from Oklahoma State University in 2005, he developed crossmaps as a means to visually examine and communicate correlations among the diverse scholarly entities and their attributes.

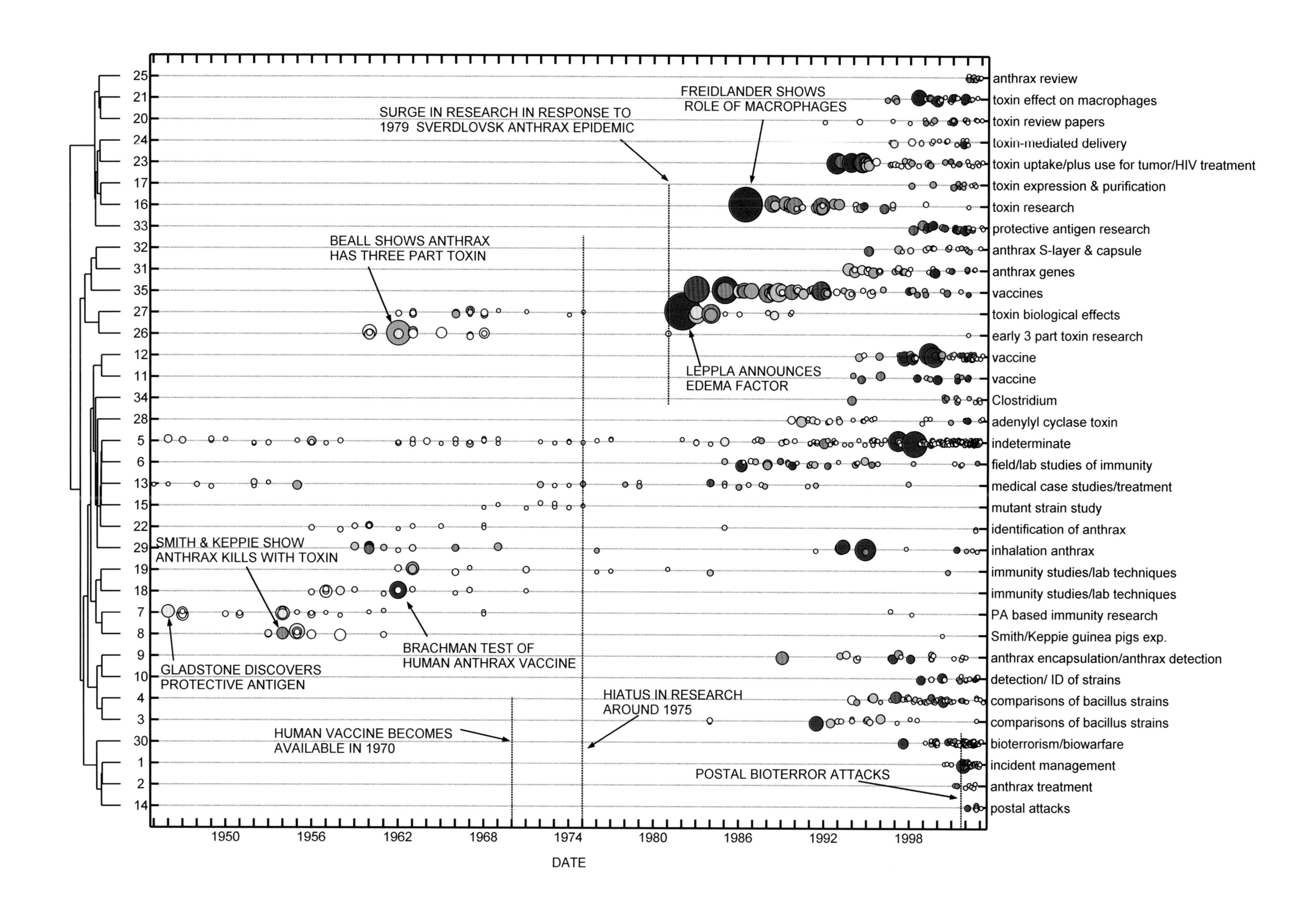

FREIDLANDER SHOWS ROLE OF MACROPHAGES
SURGE IN RESEARCH IN RESPONSE TO 1979 SVERDLOVSK ANTHRAX EPIDEMIC
BEALL SHOWS ANTHRAX HAS THREE PART TOXIN
LEPPLA ANNOUNCES EDEMA FACTOR
SMITH & KEPPIE SHOW ANTHRAX KILLS WITH TOXIN
BRACHMAN TEST OF HUMAN ANTHRAX VACCINE
GLADSTONE DISCOVERS PROTECTIVE ANTIGEN
HIATUS IN RESEARCH AROUND 1975
HUMAN VACCINE BECOMES AVAILABLE IN 1970
POSTAL BIOTERROR ATTACKS
anthrax review
toxin effect on macrophages
toxin review papers
toxin-mediated delivery
toxin uptake/plus use for tumor/HIV treatment
toxin expression & purification
toxin research
protective antigen research
anthrax S-layer & capsule
anthrax genes
vaccines
toxin biological effects
early 3 part toxin research
vaccine
vaccine
Clostridium
adenylyl cyclase toxin
indeterminate
field/lab studies of immunity
medical case studies/treatment
mutant strain study
identification of anthrax
inhalation anthrax
immunity studies/lab techniques
immunity studies/lab techniques
PA based immunity research
Smith/Keppie guinea pigs exp.
anthrax encapsulation/anthrax detection
detection/ ID of strains
comparisons of bacillus strains
comparisons of bacillus strains
bioterrorism/biowarfare
incident management
anthrax treatment
postal attacks
25
21
20
24
23
17
16
33
32
31
35
27
26
12
11
34
28
5
6
13
15
22
29
19
18
7
8
9
10
4
3
30
1
2
14
1950
1956
1962
1968
1974
1980
1986
1992
1998
DATE

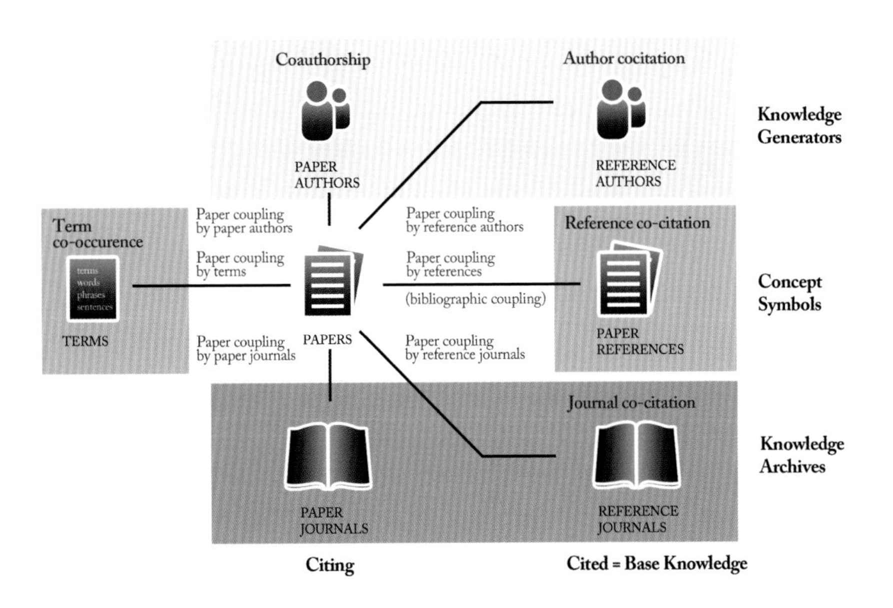

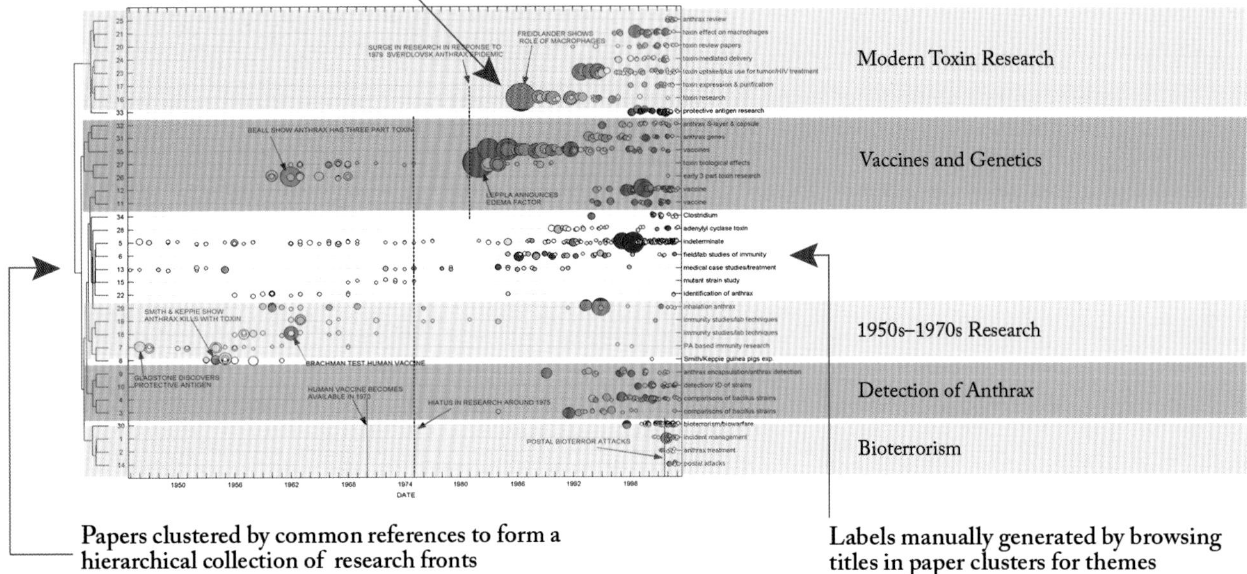

Data

The anthrax dataset consists of 2,472 journal papers covering 60 years of anthrax research. The data was retrieved from Thomson Reuters's Science Citation Index, using the search terms "anthrax OR anthracis." The 2,472 papers have 4,493 unique authors and were published in 605 different journals. They have 25,007 unique references from 7,433 different reference journals and by 16,563 different reference authors.

Technique

Journal papers are assumed to be excellent source material for the mapping of research specialties as they constitute a permanent, public record of scholarly communication and are vetted through peer review. A typical research specialty usually has fewer than 5,000 papers in its corresponding research literature.

Authors of a research specialty tend to study a common research topic, to present their work at the same conferences and in the same journals, to read and cite one another's work, and to belong to the same "invisible colleges."

In his PhD work, Morris maps research specialties through collections of papers that comprehensively sample a specialty's literature (see figure above). The basic model of the collection of papers is of a collection of bibliographic entities of seven different entity types. **PAPERS** are the base entity. Each **PAPER** is associated with its **PAPER AUTHORS**, relevant **TERMS**, the **PAPER JOURNAL** in which it was published, and the **REFERENCES** it cites. Each reference is linked to its associated **REFERENCE AUTHOR** and **REFERENCE JOURNAL**. Authors represent **Knowledge Generators** (figure above, top level), terms and references represent **Concept Symbols** (middle), and paper journals represent **Knowledge Archives** (bottom). Reference authors, references, and reference journals are symbols of the **Base Knowledge** on which authors build their work.

Entity groups are identified using co-occurrence relations. Co-occurrence relations for papers are shown in gray; for all other entities, they are shown in red. Paper networks can be generated by interlinking those papers that share many terms, paper journals or reference journals, paper references, or many reference authors or paper authors. Coauthorship networks consist of author nodes interlinked by edges that represent joint papers. Edges can be weighted by the number of joint papers, received citations, or other joint-associated entities. Network analysis and clustering techniques can be applied to any of the mentioned networks. Attributes and distributions, such as the number of papers and references per year, can be calculated as well.

Reference System

Crossmaps use a two-dimensional reference system to correlate diverse scholarly entities and their features (for example, research topics versus time or authors versus research topics). They require categorical data and meaningful orderings of axis labels (for example, alphabetical sorting of author names or grouping of paper clusters based on their similarity). Dendrograms are used to visually communicate the results of fusions made during agglomerative hierarchical clustering.

The *Timeline of 60 Years of Anthrax Research Literature* incorporates a reference system of research topics (vertical) versus time (horizontal). Bibliographic coupling was used to group papers on similar topics into clusters. A simulated annealing routine was used to arrange the order of leaves in the dendrogram (and corresponding clusters) such that similar subspecialties tend to be close together. Note that the highest topics relate to **Modern Toxin Research**; the ones below to **Vaccines and Genetics**; beyond that to **1950s and 1970s Research**, the **Detection of Anthrax**, and finally to **Bioterrorism** (see image below).

Data Overlays

The reference system was used to plot the 987 papers in the anthrax dataset that pair-wise share (cocite) at least five references. Topic-related papers are plotted in horizontal tracks by publication date. Vertical jittering is applied to ensure that papers with identical topics and publication years are distinguishable. Circle size and color are used to represent the number of citations received and the currency of citations, respectively. Important papers are labeled. External events are shown—note the "birth" of bioterrorism-related topics after the **POSTAL BIOTERROR ATTACKS**. Citations and other linkages, as well as cluster boundaries and other data, could be overlaid but have been omitted for clarity.

Details

Complementary crossmap views depict citation of papers over **Time** (opposite page, top left), the number of papers per **Authors** over **Time** (top right), and organization of **References** or **Index Terms** (bottom row) by **Research Topics**. Rectangular overlays with red labels highlight major authors and topics.

Unique Features

Crossmaps employ diverse gestalt principles to improve readability. The principle of "Continuance" is used to communicate temporal relationships. "Proximity" helps to show what topics are related. "Alignment" is used to group related papers. "Size" communicates the importance of papers. Hierarchical sequencing of information is used to communicate the overall structure of research topics being displayed.

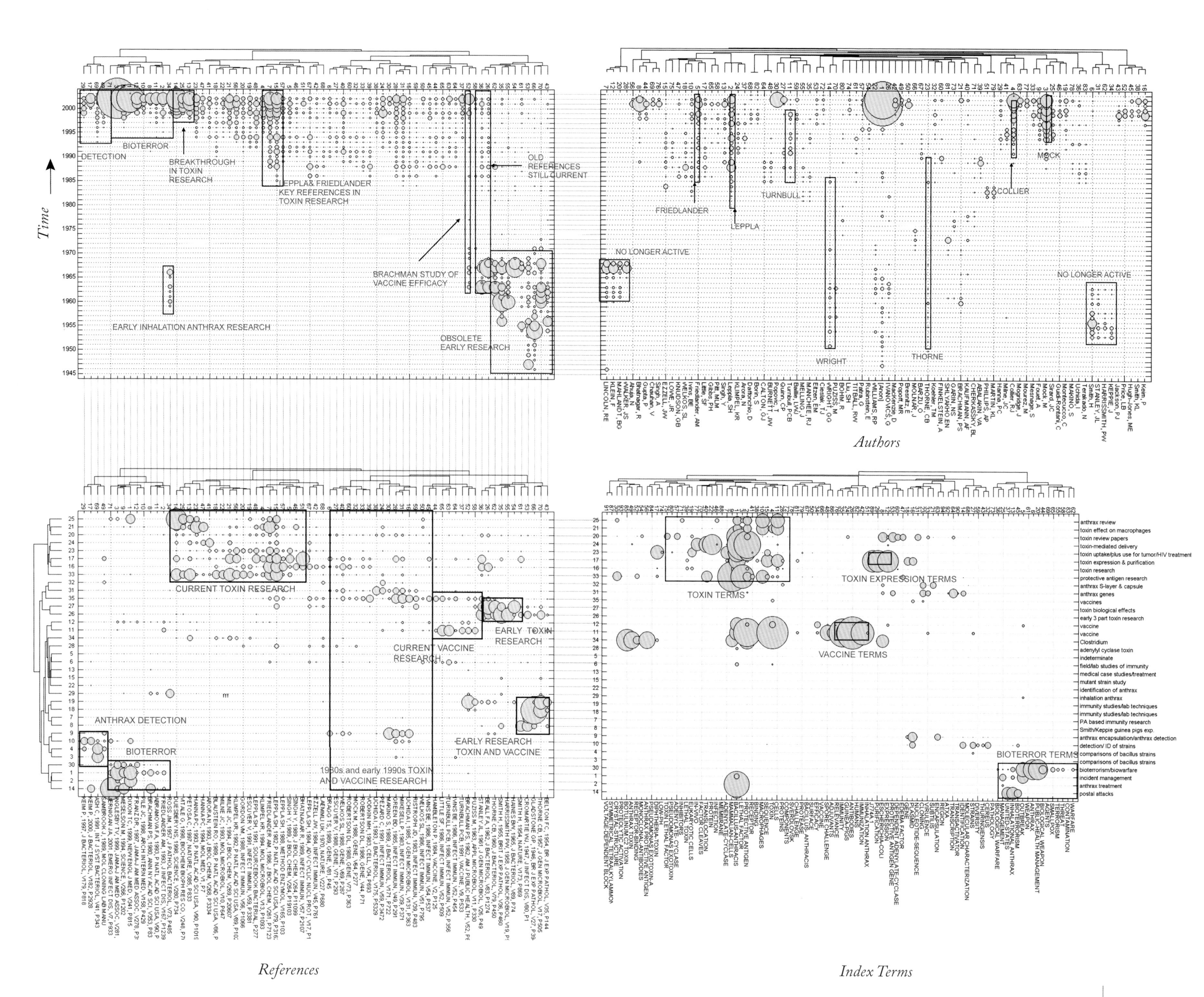
Time
DETECTION
BIOTERROR
BREAKTHROUGH IN TOXIN RESEARCH
LEPPLA& FRIEDLANDER KEY REFERENCES IN TOXIN RESEARCH
OLD REFERENCES STILL CURRENT
BRACHMAN STUDY OF VACCINE EFFICACY
EARLY INHALATION ANTHRAX RESEARCH
OBSOLETE EARLY RESEARCH
FRIEDLANDER
LEPPLA
TURNBULL
MOCK
COLLIER
NO LONGER ACTIVE
WRIGHT
THORNE
Authors
CURRENT TOXIN RESEARCH
CURRENT VACCINE RESEARCH
EARLY TOXIN RESEARCH
ANTHRAX DETECTION
BIOTERROR
EARLY RESEARCH TOXIN AND VACCINE
1980s and early 1990s TOXIN AND VACCINE RESEARCH
References
TOXIN TERMS
TOXIN EXPRESSION TERMS
VACCINE TERMS
BIOTERROR TERMS
Index Terms

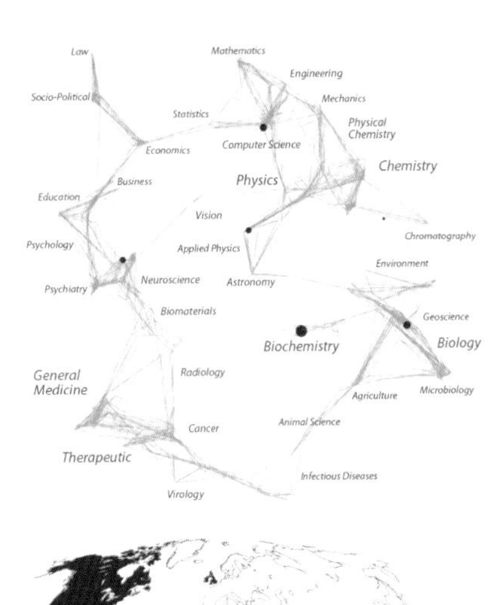

2005

Treemap View of 2004 Usenet Returnees

By Marc A. Smith and Danyel Fisher

REDMOND, WASHINGTON, 2005

Courtesy of the Community Technologies Group, Microsoft Research

Aim

How do social cyberspaces such as e-mail threads, newsgroups, and the blogosphere evolve? Are the social groups that create them different from or similar to face-to-face groups? Netscan was designed to answer these and many other questions. It provided detailed reports about Microsoft (MS) Usenet activity for any newsgroup—at any day, week, month, quarter, or year from October 1999 to May 2007. Interfaces to Netscan included *author profile,* to view Usenet newsgroup usage data for a selected author; *thread view,* to view all the messages in a specified conversation, independent of the thread to which it was posted or from which responses were made; *treemap,* to view the hierarchical relationships of newsgroups in the Usenet; and *cross posts,* to view the relationships between neighboring newsgroups within Usenet. Researchers, practitioners, and educators were able to use Netscan to find newsgroups that share unique interests, monitor the health of newsgroups related to certain interests and pursuits, stay informed on current events and the latest trends, locate sources for technical assistance and information, examine troubling issues and hot topics not covered in Microsoft product documentation, track the contributions of favorite authors across Usenet newsgroups, and gather historical data on the development of social cyberspaces.

Interpretation

The map at right shows the hierarchical relationships between and activity of Microsoft Usenet newsgroups in 2004. The treemap interface was employed to graphically depict 189,144 newsgroups and their 257,442,374 postings. It is based on the treemap layout originally introduced by Ben Shneiderman at the University of Maryland. Each newsgroup is represented by a square. The size of each square corresponds to the number of people who posted at least twice. Color-coding is used to show the increase or decrease in the number of posters compared to the 2003 data: red indicates fewer, green denotes more. Each square is labeled as a literal hierarchy (that is, "**rec.pets**" contains "**rec.pets.cats**"). The growth of certain newsgroups (for example, "**alt.binaries**", at bottom left) and the decline of the "**comp**" groups (at middle right) can be seen at a glance.

Marc A. Smith worked as a sociologist and senior researcher at Microsoft Research, specializing in the social organization of online communities, when he designed this map. He is the coeditor of *Communities in Cyberspace* (Routledge)—a collection of essays exploring the ways identity, interaction, and social order develop in online groups. Smith's research focuses on the ways group dynamics change when they take place in social cyberspaces, and his goal is to visualize these social cyberspaces, mapping and measuring their structures, dynamics, and life cycles. In support of this research, he directed the setup of the Netscan social network analysis infrastructure. Marc received a BS in International Area Studies from Drexel University in Philadelphia in 1988; an MPhil in Social Theory from Cambridge University in 1990; and a PhD in Sociology from the University of California, Los Angeles, in 2001.

Danyel Fisher is a researcher in the VIBE (Visualization and Interaction for Business and Entertainment) group at Microsoft Research. He received an MS from the University of California, Berkeley, in 2000 and a PhD in computer science from the University of California, Irvine, in 2004. His research centers on using qualitative and quantitative social network analysis and visualization to understand the ways that people in online groups interact—via e-mail, Usenet, blogs, and contemporary social software. He uses these results to design the next generation of interfaces for social software, as well as to address broader questions of social structure. His recent research looks at visualizing large data sets created from human activities, including Web server logs.

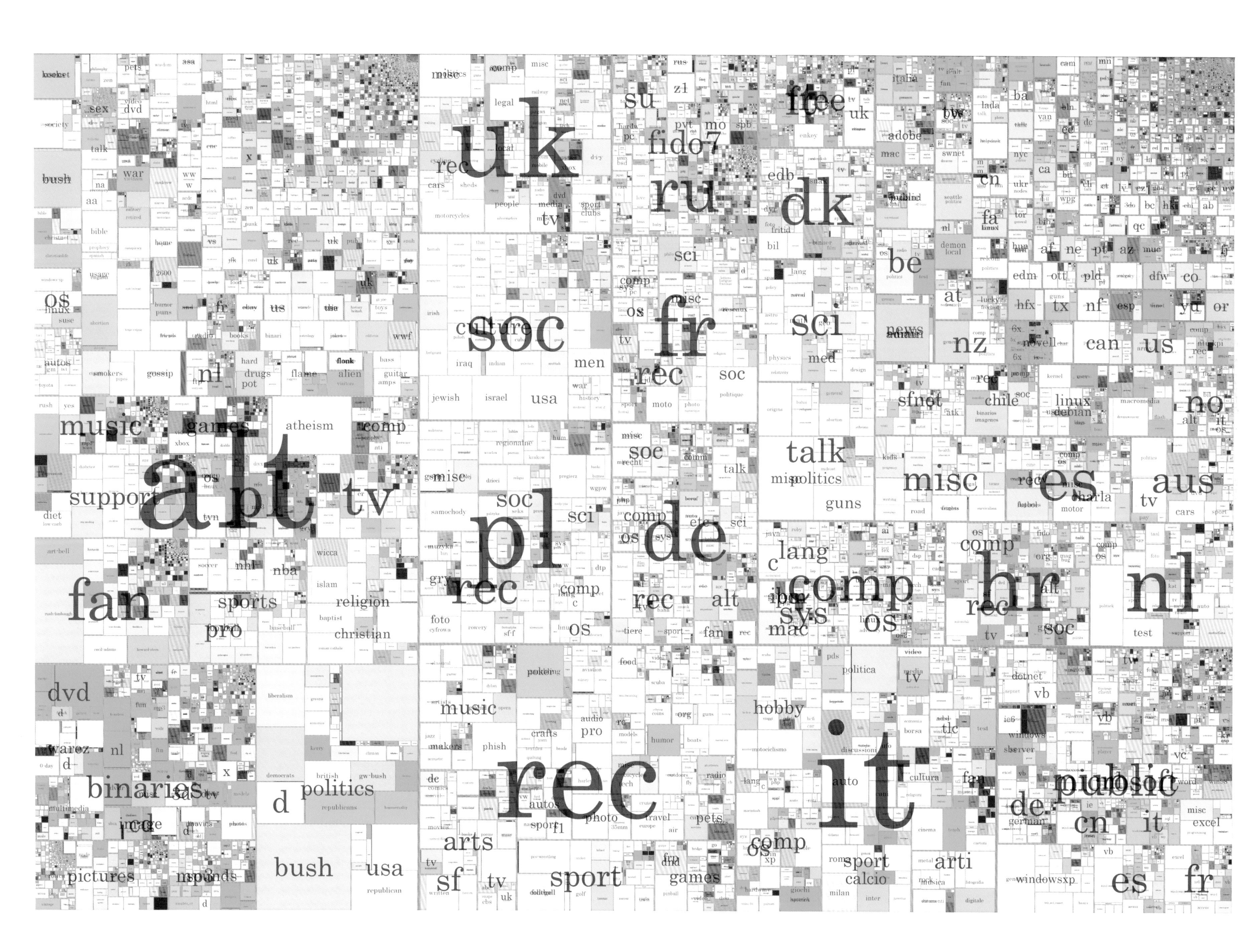
alt
pt
tv
music
games
atheism
comp
support
fan
sports
religion
christian
pro
dvd
binaries
pictures
bush
usa
politics
d
uk
rec
soc
culture
men
usa
pl
soc
sci
rec
music
phish
arts
sf
sport
su
ru
fido7
fr
rec
soc
de
rec
alt
dk
sci
talk
politics
guns
lang
comp
sys
mac
os
it
hobby
sport
calcio
arti
nz
can
us
misc
es
aus
hr
nl
test
public
microsoft
cn
it
es
fr
windowsxp

Data

Usenet is a distributed, threaded conversation, organized into hierarchical topic-oriented newsgroups. Users read individual newsgroups to find the messages in which they are interested. Each newsgroup has a unique name—increasingly specific terms delimited by periods—that dictates its place in the hierarchy. An example is **comp.lang.perl.misc,** where **comp** refers to discussions about computing, the **lang** label within **comp** to discussions of computer languages, the **perl** label within **lang** to the computer language Perl, and the **misc** label within **perl** to miscellaneous topics relating to Perl. An author can post a message to one Usenet group or cross-post to more than one Usenet group; in either case, that post counts as one message. However, if a single person posts the same message under three distinct online names, that message will be counted three times—under three different authors.

Between October 1, 1999, and May 31, 2007, the Netscan engine collected information on 68 million unique participants in more than 175,000 newsgroups, posting more than 2 billion messages in about 250,000 threads.

There is no guarantee that all activity data was captured. By its nature, Usenet functions as a loose, distributed network of servers, and although most messages propagate to most servers, some messages miss many servers. The news feed collected was obtained from the Microsoft Research news server, which had received Usenet data from multiple upstream providers, including the University of Washington; the University of California, Berkeley; and Internet service providers Cable and Wireless and UUNET.

The complete data set has a size of 6 terabytes. A subset of the data has been shared with several universities for research and educational purposes and can be requested by e-mailing the authors.

Reference System

A treemap layout was used to create a base map of the Usenet hierarchy. Treemaps take the root of a tree and a rectangular drawing area defined by upper left and lower right coordinates as input (see image below). The recursive treemap algorithm then starts with the root node as *active node*, the drawing area as active area, and any initial *partitioning direction* (for example, horizontal). In each recursion, it determines the number of outgoing edges from the *active node*. If that number is larger than one, then the *active area* is divided in *partitioning direction,* where the size of the partitions might correspond to an attribute value of the children's nodes (for example, size). The partitioning direction is changed (for example, to the vertical). The recursive treemap algorithm is restarted for each child node, its respective drawing area, and the new partitioning direction. The algorithm stops whenever the active node has no children.

Treemaps utilize 100 percent of the display space, show the nesting of hierarchical levels, can be used to represent node attributes by area size and color, and are scalable to data sets of a million items. They

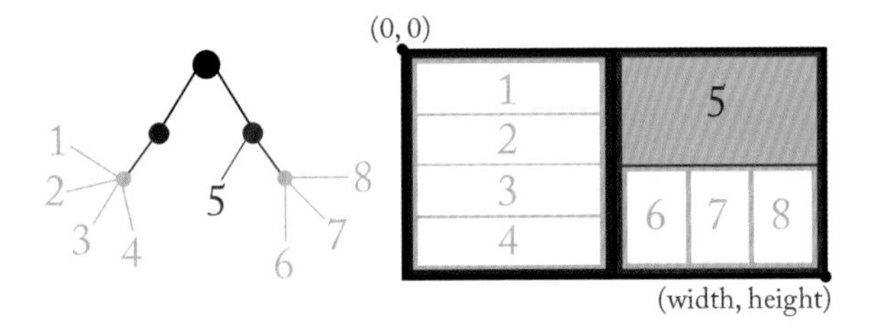

do not easily support size comparisons; labeling might result in clutter when many thin or small areas exist, showing only leaf node attributes.

The original treemap layout algorithm generates long, thin rectangles, which are hard to compare and label. Squarified treemaps optimize the aspect ratio of the rectangles, making them as close to square as possible, yet they cannot guarantee the stability of a rectangle's position over time. Usenet treemaps follow a compromise algorithm between the illegibility of slice-and-dice treemaps and the instability of squarified layouts proposed by Ben Shneiderman and Wattenberg, which fixes certain items in pivotal locations to keep positioning consistent and aspect ratio low.

To aid in gestalt perception of the subtrees and subgroups within the treemap, and to offer a macro/micro reading of the map, Usenet treemaps add space within each rectangle and around and between its children. The size of the inner space is large at the upper levels of the treemap hierarchy and shrinks linearly down the hierarchy, so it is smallest at deep leaf nodes. The outline of each rectangle is drawn with a line thickness proportional to that node's depth in the hierarchy. Thus the outer-level rectangle, which contains all the other rectangles, has the thickest outline, and from there the line thickness decreases linearly down the hierarchy (see image below).

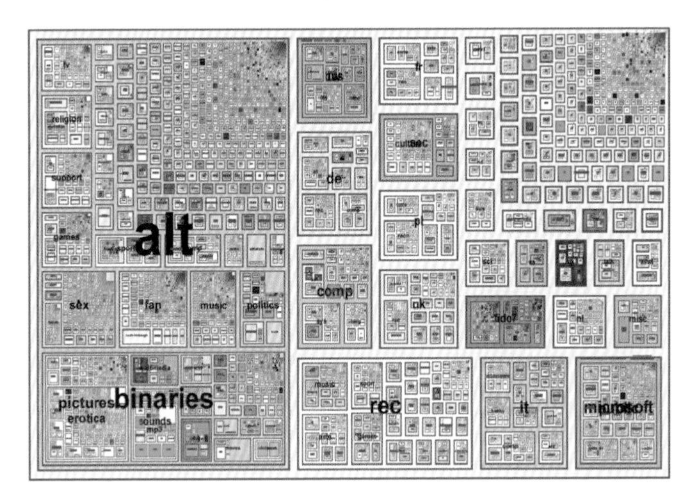

Each newsgroup is represented by a box whose area size is proportional to the number of messages posted in the newsgroup over some period of time. When a newsgroup contains other newsgroups, its size reflects the cumulative number of messages in the newsgroup itself and in all of its children. If a parent has no messages of its own, then its children occupy the entire area.

The boxes nest within one another according to the hierarchical naming system. Using the example given above, "**comp**" contains (among other items) "**lang**", which contains "**perl**", which contains "**misc**", which represents the newsgroup "**comp.lang.perl.misc**".

Note that rectangles are sorted by size; for instance, small rectangles are rendered in the upper right corner of the *Treemap View of 2004 Usenet Returnees* and in the treemaps at right.

Data Overlays

Treemaps easily encode three variables at a time: one in the size of the boxes, one in their color, and one in their hierarchical arrangement. In the *Treemap View of 2004 Usenet Returnees*, color represents the growth or decline of groups. Color-coding (green represents decline and red represents increase) indicates changes in the number of postings. To avoid jarring discontinuities, the color of each parent newsgroup reflects the weighted average of the colors of its children.

Unique Features

Through the online Netscan interface system shown below, diverse statistics could be produced, such as the treemap renderings of the **2005** and **2006 Posters** and their **Posts**, shown opposite. Here area size encodes the number of people that posted to a certain Usenet group and the number of their posts respectively. Note that these maps have a much larger proportion of **alt** posters and postings than the *Treemap View of 2004 Usenet Returnees*—the left half of the 2005/2006 Posters map and about 2/3 of the Posts map is devoted to **alt**. Note that of the "**alt**" postings, about half of all posters post about five-sixths of their postings to "**alt.binaries.***". While the map of returnees shows only those users who posted more than once, many one-time users frequently post to alt, which explains its disproportionately large size.

In sum, the treemaps provide a macro/holistic view of the complete Usenet system. They highlight major structural subcomponents. For instance, alt encompasses the greatest area; national hierarchies and Microsoft are the next largest. Some places are growing more than others—which ones, and why? Some newsgroups are more subdivided than others—what accounts for these differences? Some attract far more people and retain them for longer periods than do their neighbors. When treemaps of different types (that is, posts, returnees, posters) are contrasted, additional differences emerge. For instance, "**alt.binaries**" has lots of traffic (posts) but few replies and fewer returnees. Such contrasting elements highlight the fact that raw volumes do not necessarily indicate social solidarity or community.

Netscan supported the interactive selection, manipulation, and visualization of specific Usenet data sets (see interface snapshot below). Data from a specific Usenet group for a time span could be selected; shown here is a listing of "**rec.music**" for a **Month** starting 10/31/2004. For each Newsgroup, the number of **Posts**, **Posters**, **Returnees**, **Replies**, and **Repliers** is listed.

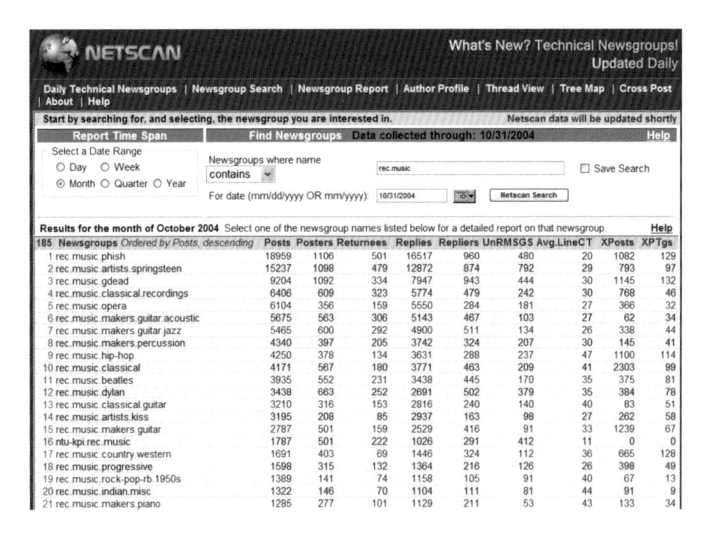

NETSCAN

What's New? Technical Newsgroups! Updated Daily

Daily Technical Newsgroups | Newsgroup Search | Newsgroup Report | Author Profile | Thread View | Tree Map | Cross Post | About | Help

Start by searching for, and selecting, the newsgroup you are interested in. Netscan data will be updated shortly

Report Time Span | Find Newsgroups Data collected through: 10/31/2004 Help

Select a Date Range: Day / Week / Month / Quarter / Year

Newsgroups where name contains rec.music | Save Search

For date (mm/dd/yyyy OR mm/yyyy): 10/31/2004 | Netscan Search

Results for the month of October 2004 Select one of the newsgroup names listed below for a detailed report on that newsgroup. Help

185	Newsgroups *Ordered by Posts, descending*	Posts	Posters	Returnees	Replies	Repliers	UnRMSGS	Avg.LineCT	XPosts	XPTgs
1	rec.music.phish	18959	1106	501	16517	960	480	20	1082	129
2	rec.music.artists.springsteen	15237	1098	479	12872	874	792	29	793	97
3	rec.music.gdead	9204	1092	334	7947	943	444	30	1145	132
4	rec.music.classical.recordings	6406	609	323	5774	479	242	30	768	46
5	rec.music.opera	6104	356	159	5550	284	181	27	366	32
6	rec.music.makers.guitar.acoustic	5675	563	306	5143	467	103	27	62	34
7	rec.music.makers.guitar.jazz	5465	600	292	4900	511	134	26	338	44
8	rec.music.makers.percussion	4340	397	205	3742	324	207	30	145	41
9	rec.music.hip-hop	4250	378	134	3631	288	237	47	1100	114
10	rec.music.classical	4171	567	180	3771	463	209	41	2303	99
11	rec.music.beatles	3935	552	231	3438	445	170	35	375	81
12	rec.music.dylan	3438	663	252	2691	502	379	35	384	78
13	rec.music.classical.guitar	3210	316	153	2816	240	140	40	83	51
14	rec.music.artists.kiss	3195	208	85	2937	163	98	27	262	58
15	rec.music.makers.guitar	2787	501	159	2529	416	91	33	1239	67
16	ntu-kpi.rec.music	1787	501	222	1026	291	412	11	0	0
17	rec.music.country.western	1691	403	69	1446	324	112	36	665	128
18	rec.music.progressive	1598	315	132	1364	216	126	26	398	49
19	rec.music.rock-pop-rb.1950s	1389	141	74	1158	105	91	40	67	13
20	rec.music.indian.misc	1322	146	70	1104	111	81	44	91	9
21	rec.music.makers.piano	1285	277	101	1129	211	53	43	133	34

The top menu selects different views of the data, such as **Author Profile, Thread View,** and **Tree Map**. TreeMap was used to map the entire Usenet dataset (opposite page). The interactive Tree Map view supports zooming for greater focus on objects of interest while minimizing label occlusion.

2005 Posters

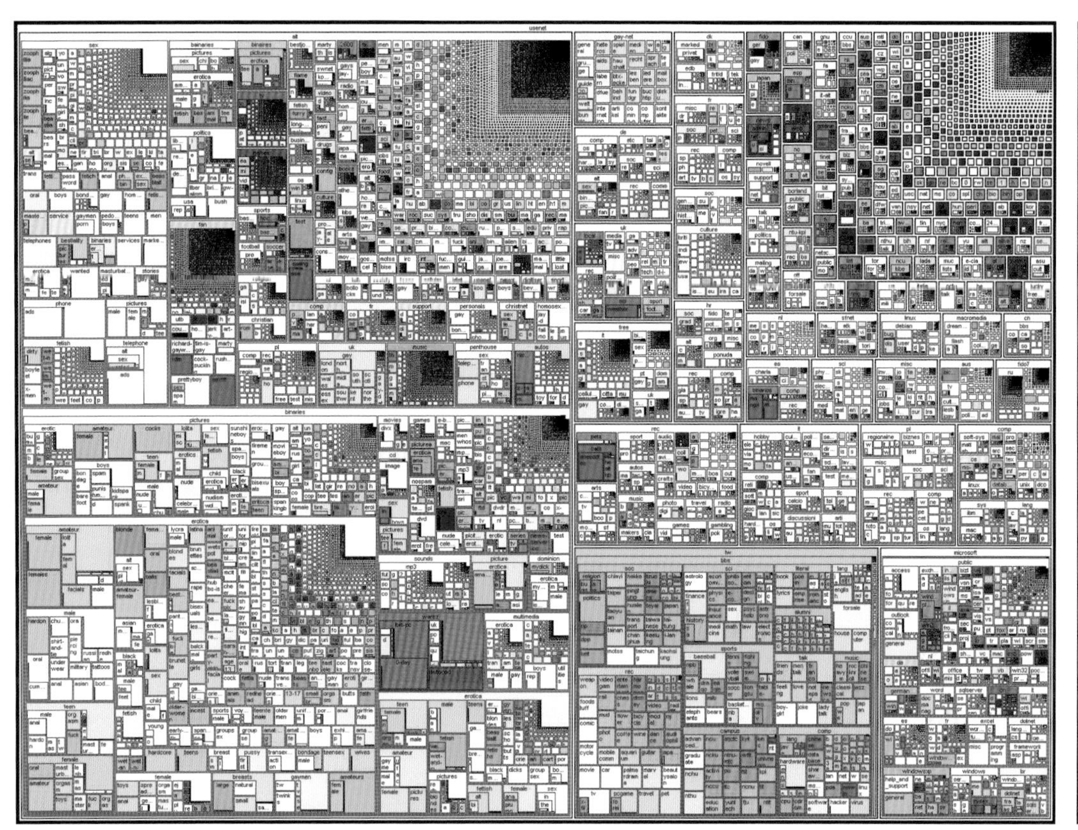

2006 Posters

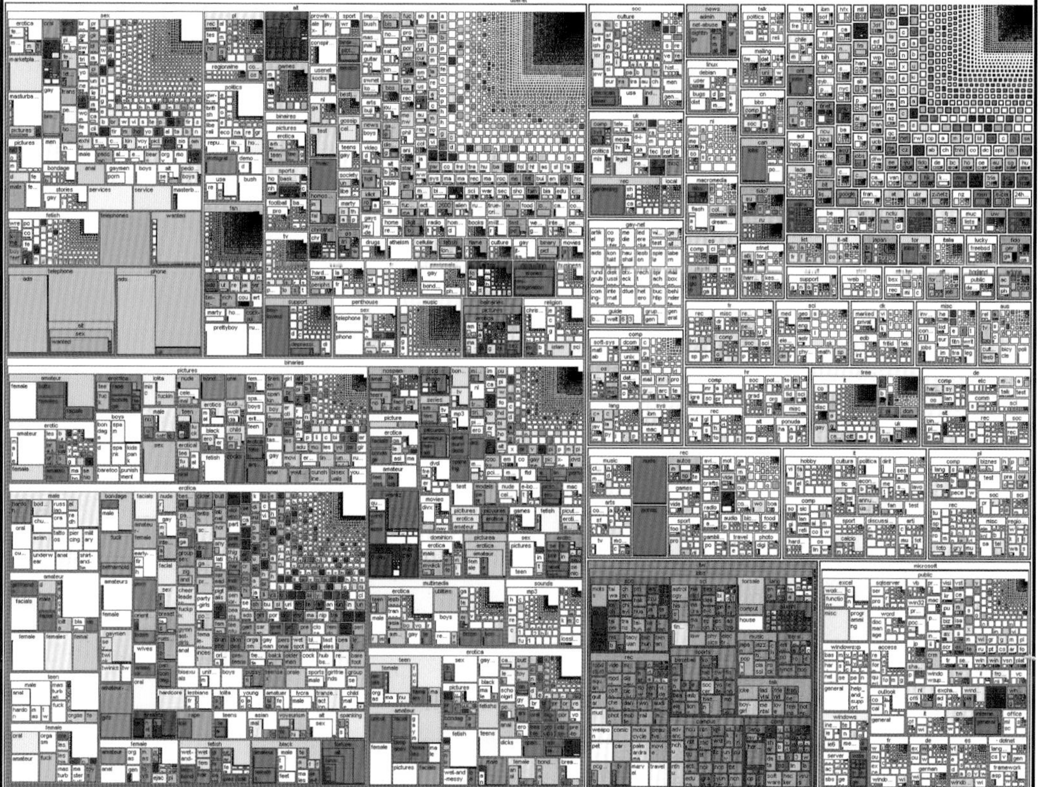

2005 Posts

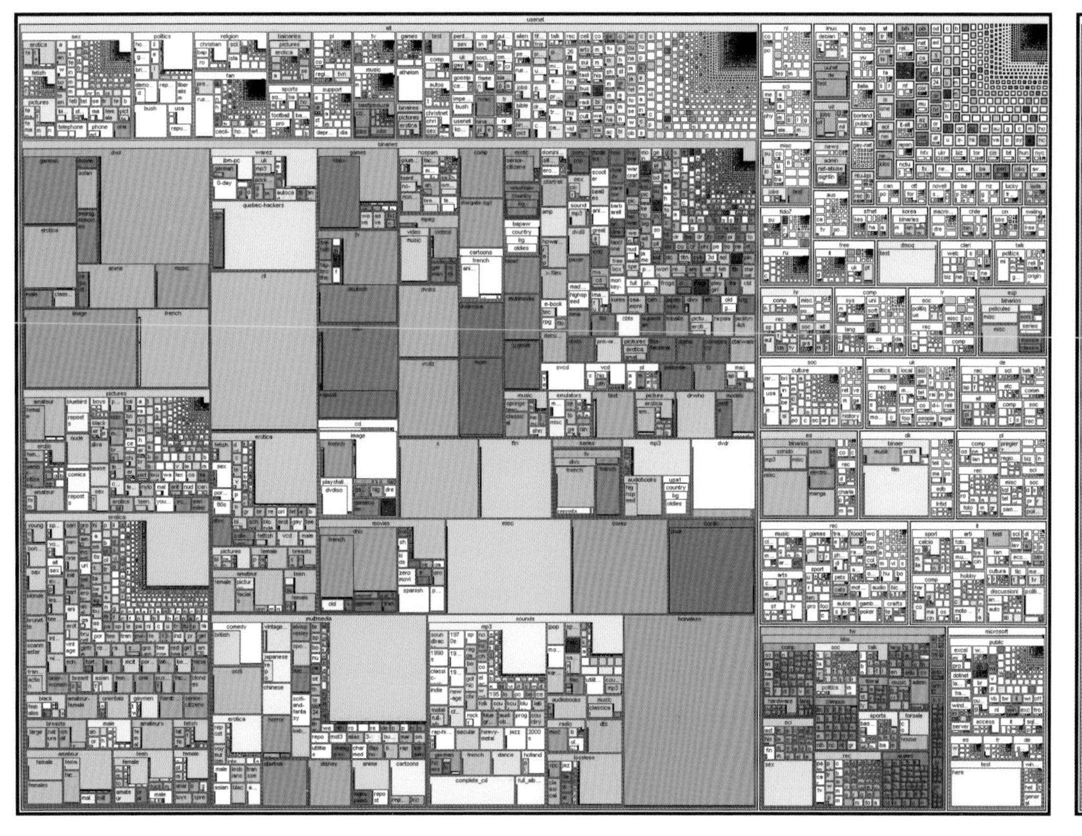

2006 Posts

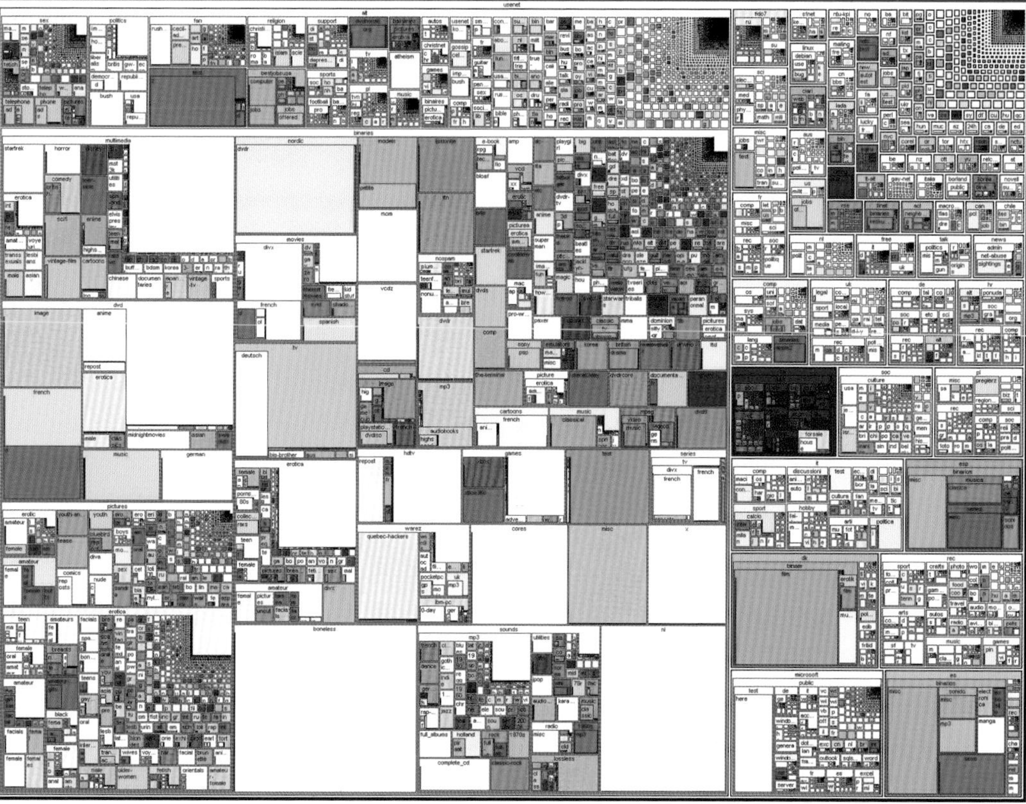

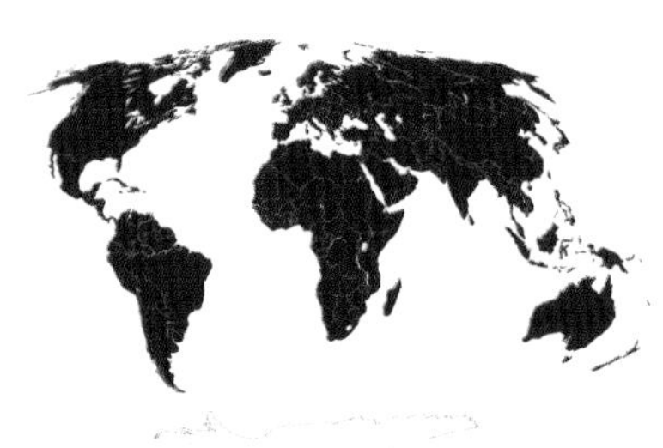

2005

In Terms of Geography

By André Skupin

NEW ORLEANS, LOUISIANA, 2005

Courtesy of André Skupin, San Diego State University, San Diego

Aim

Many areas of research are extensive in terms of data, published works, scholarly expertise, and practice. Social and intellectual interrelations are highly complex. How can we gain access to broad, global structures of whole research domains as well as to finer, regional, and local structures within subdisciplines? How can we understand major areas, existing and nonexisting connections, and the homogeneity of subareas? Skupin seeks to understand how cartographic design and geographic technology can be applied to map not only geographic space but also abstract semantic spaces.

Interpretation

The map at right shows the content coverage of geography research for the 10-year period from 1993 to 2002. It was generated from more than 22,000 abstracts submitted to the annual meetings of the Association of American Geographers. Mountains represent areas of higher topical focus, with word stems serving as labels. **Community** studies are most dominant in the middle of the map. **Soil**, **climate**, **population**, **migration**, **women**, **social**, and **health** are other major areas of study. Valleys represent regions with less topical focus. One can think of these as capturing information sediments from the surrounding mountains, leading to a mixture of topics. For example, **nature** and **management** are valleys surrounded by the major mountains of **water**, **land**, **development**, and **environment**. Likewise, geographers' investigations of **planning** and **transportation** valleys are heavily influenced by those topics that are indicated on neighboring mountains.

Notice how the arrangement of labeled mountains and valleys replicates major global subdivisions of the geographic knowledge domain. The upper left corner contains topics associated with **physical geography**, while the upper right corner—including **GIS**, **model**, **spatial**, and **data**—covers the area now known as **geographic information science**. Much of the remainder of the map reflects various topics investigated within **human geography**, including such further subdivisions as **economic geography** in the lower right corner. Conversely, smaller topical structures within the major mountains are visible, such as the **cover** and **use** regions of the **land** mountain, which reflect documents containing the phrases "land cover" and "land use."

André Skupin is an associate professor of geography at San Diego State University. Previously he held an associate professorship at the University of New Orleans. He received a master's degree in cartography at the Dresden University of Technology, Germany, and a doctoral degree in geography at the State University of New York at Buffalo. During his graduate studies he performed research at the National Center for Geographic Information and Analysis (NCGIA). He has worked in the geographic information systems (GIS) industry in Germany, the United States, and South Africa. Skupin's core research area is the application of geographic metaphors, cartographic principles, and computational methods in the visualization of nongeographic information. His research is strongly interdisciplinary, aimed especially at increased cross-fertilization between geography, information science, and computer science. For example, he has developed new approaches to creating map-like knowledge domain visualizations on the basis of high-dimensional vector space models and artificial neural networks. Other recent work includes novel methods for visualizing individual human movement and demographic change as trajectories in n-dimensional attribute space. His research results have been published in such journals as *IEEE Computer Graphics and Applications, Computing in Science and Engineering*, and the *Proceedings of the National Academy of Sciences*. Skupin has served as coeditor of a book on self-organizing maps, a special issue of *Environment and Planning B: Planning and Design*, and the 2007 U.S. National Report to the International Cartographic Association, which was published as a special issue of *Cartography and Geographic Information Science.*

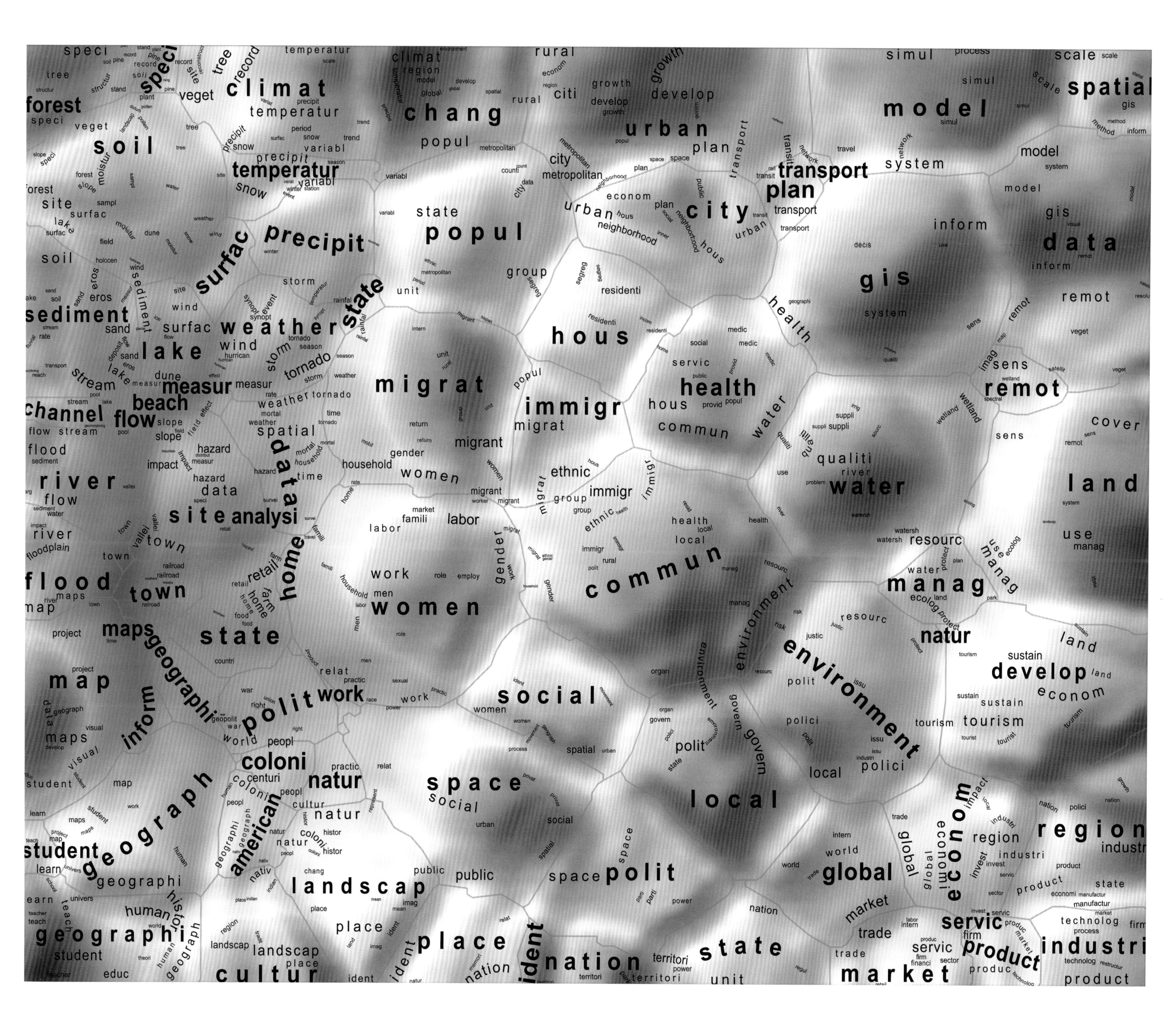

forest
soil
speci
climat
chang
rural
urban
simul
scale
spatial
model
temperatur
precipit
popul
city
plan
transport
gis
data
surfac
sediment
weather
state
hous
health
lake
measur
beach
channel
flow
migrat
immigr
remot
river
data
site analysi
commun
water
land
flood
town
home
women
manag
maps
state
geographi
natur
map
inform
polit
work
social
environment
develop
coloni
natur
space
local
tourism
student
geograph
american
landscap
polit
econom
region
global
geographi
place
ident
nation
state
servic
product
industri
cultur
market

Data

More than 22,000 abstracts submitted to the annual meetings of the Association of American Geographers (AAG) during 1993–2002 were analyzed to generate this map. Data for the early years was provided as RTF and Word documents by the conference organizers, while later abstracts were derived from conference CDs containing PDF documents.

Reference System

The self-organizing map (SOM) method, a special form of artificial neural networks, is used to create a two-dimensional semantic map of documents. Input documents are represented as term vectors, which serve as the layer of input nodes during training of the SOM, while the output layer consists of nodes arranged as a two-dimensional lattice. Each of the nodes in the output space is assigned an n-dimensional weight vector having the same dimensionality as the input vectors (see image below).

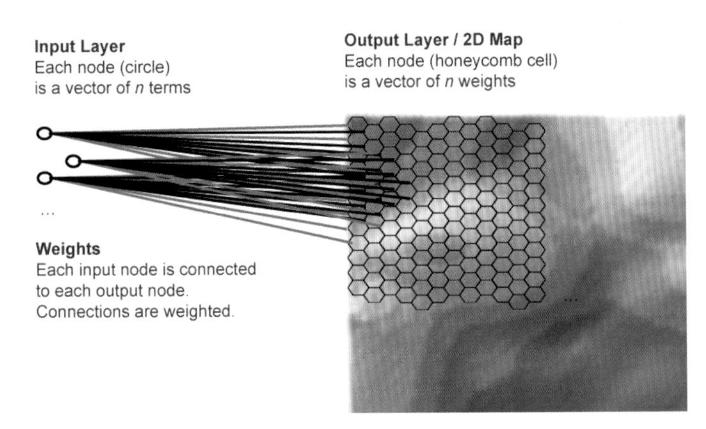

The goal of SOM training is to replicate in the two-dimensional output space major topological structures existing in the n-dimensional input space. This results in a two-dimensional model of the input data. That model itself can then be visualized; *In Terms of Geography*—featured at the 2005 Conference on Spatial Information Theory (COSIT)—was produced this way. Alternatively, one could map the input data or even other data onto this two-dimensional model.

In order to create a two-dimensional map of the geographic knowledge domain, each document is first represented as an n-dimensional vector of term occurrences, where n equals the number of unique terms occurring in any of the 22,000 documents. A total of 10,000 output nodes are arranged as a two-dimensional lattice. With each of the inside nodes having six neighbors, this results in a honeycomb-like spatial configuration, as shown in the image above.

Before the neural network is trained, the n weights associated with each output node are initialized with random values. Documents are then presented in order, and the following steps are performed. For each input document, the similarities/distances to all output nodes are computed. The winning output node j is selected, and its weights are updated toward an even better match with the input document. In addition—and this is the real secret behind spatial self-organization—the two-dimensional neighbors of node j are updated as well. This search for the winning node, as that node and its neighbors are updated, is then repeated for each input document, up to a specified maximum number of iterations. The resulting two-dimensional grid of n-dimensional neuron vectors is the basis for a series of transformations leading up to the map shown here.

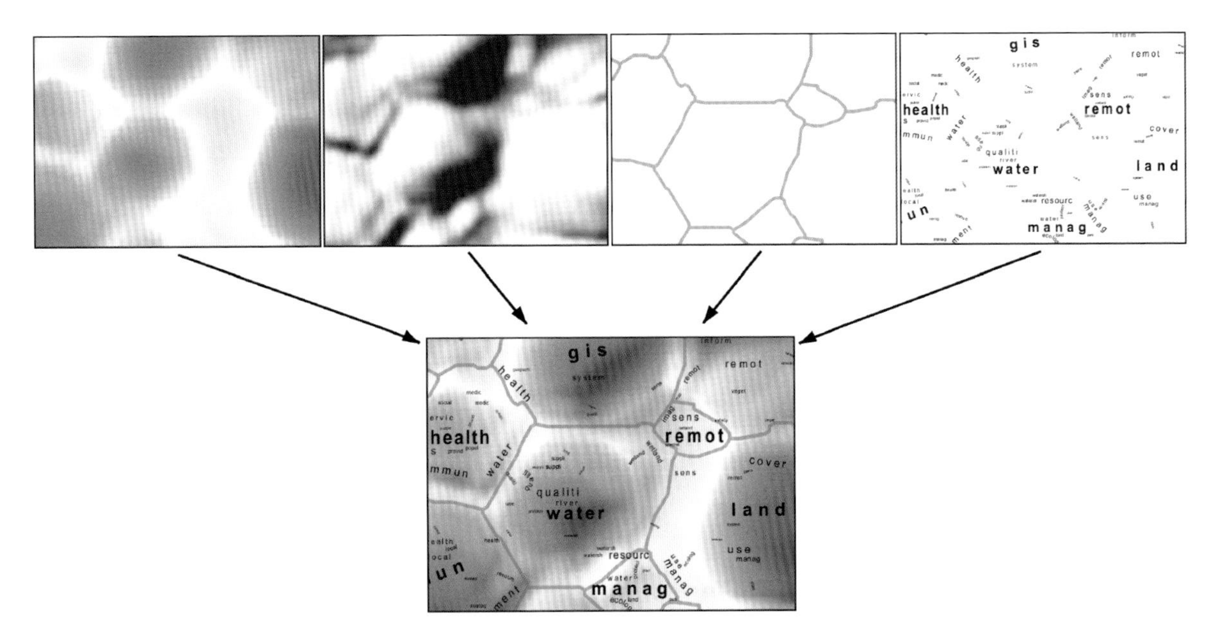

Data Overlays

The trained SOM is transformed into a number of layers, which are then rendered using standard geographic information system (GIS) software (see image above). Landscape features express the degree of topical focus, with elevated areas corresponding to more well-defined topical regions and low-lying areas corresponding to a mingling of various topics. The landscape visualization combines color-coding with hill shading. In addition, labels are computationally generated for regions within this map, at five different scale levels. Broad labels extend over larger regions, while smaller regions are associated with more specific labels. The large display size of the map allows simultaneous overlay of multiple levels of labels, with a visual hierarchy established through label size and GIS software that prevents placement conflicts.

Details

The resulting base map of geography can be used to map any text document that can be represented in the same n-dimensional space. The simplest approach is to find the one neuron that is most similar to the new document vector and to place it into the honeycomb-shaped (hexagonal) area associated with that neuron.

An obvious data set to be mapped onto this base map is the set of documents that were initially used for training. Due to the topology-preserving nature of SOM training, similar documents will now tend to closely congregate within the map. When the 22,000 AAG abstracts are mapped back onto the SOM consisting of 10,000 output nodes/regions, then, on average, every node will be associated with two abstracts. To reduce the resulting overplotting, each abstract is randomly placed inside its assigned honeycomb cell and thus has a unique two-dimensional location.

A portion of the resulting map overlay is shown below, with a close-up of the valley separating the **GIS** and **health** mountains. The region labels are derived from the titles of submitted abstracts, while the placement of abstracts is based on their full text. Hence, there can be some repetition of labels among the peaks and valleys.

Unique Features

A map of presenters and discussants who participated in a workshop on Improved Visualization of Uncertain Information organized by the National Academy of Sciences in March 2005 was created. For each presenter and discussant, Skupin downloaded information from the Web (for example, a CV when available), transformed it into a term vector, and mapped it onto the SOM.

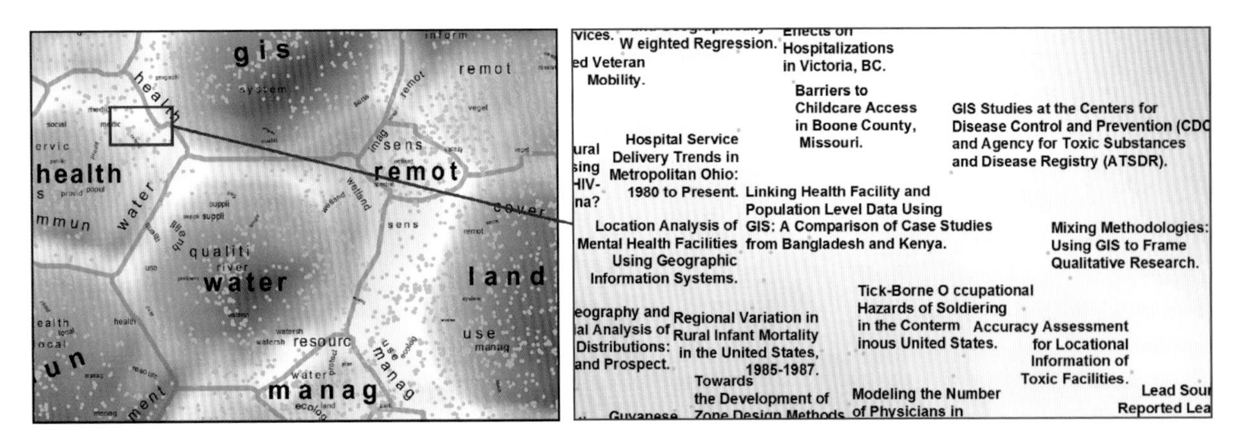

The first session, on "Large, structured data sets with ample time for interpretation," included presentations by David **Scott**, Rice University; Bill **Cleveland**, Purdue University; and Michael **Friendly**, York University (shown as red circles in the middle map to the right). Alex **Szalay**, Johns Hopkins University, acted as discussant (red triangle). The second session, on "Large, unstructured, changing data sets where the relevance, significance, and conceptual links among the data have yet to be discovered," included Ronald **Coifman**, Yale University; André **Skupin**, then at the University of New Orleans; and Stephen **Eick**, University of Illinois at Chicago and SSS Research. Discussant David **Harris**, National Security Agency, offered responses. (These participants are shown in green.) Finally, the workshop schedule included a third session, on "Data sets that are reduced for rapid understanding in time-pressured situations," with presentations by Chris **Wickens**, University of Illinois at Urbana-Champaign; Peter **Fisher**, City University, London; Henry **Rolka**, Centers for Disease Control and Prevention; and Pat **Hanrahan**, Stanford University. Discussion was led by Robert **Frey**, State University of New York, Stony Brook. (These participants are shown in blue.) The resulting overlay revealed one key detail previously overlooked by all the participants: for all three sessions, the discussant was *at home* in a significantly different location than the respective presenters. Notice how far the triangles are from the circles of the same color. Skupin interprets this as discussants having been specifically chosen to provide true perspective. To appreciate the larger image, one must step back—as did the discussants with their particular background and expertise, evidenced in both their responses and publications.

Skupin has also been developing methods to project the trajectories through a knowledge domain followed by people over time. For example, he extracted temporally sliced sections from the list of publications by Michael Goodchild, University of California, Santa Barbara—a leading expert in geographic information science. With a 10-year sliding window and 5-year overlap, each slice can be turned into a temporal vertex. When Goodchild is conceptualized as a discrete object, a linear path can thus be generated, of which a portion is shown (above right). Skupin is also thinking about alternative conceptualizations and their corresponding visualizations. For example, one could conceptualize Goodchild as simultaneously occupying the *entire* geographic knowledge domain, but with different intensity in different locations. For multitemporal data, one could derive from this a change landscape, indicating his increasing or decreasing intensity of topical interest. Identical source data can thus lead to an array of distinct visual expressions.

Technique

The combination of intense computation with geographic metaphors and cartographic design considerations is at the core of Skupin's work. His use of a highly detailed base map—onto which various data can be overlaid—is inspired by the use of topographic maps as the geometric base for many types of thematic maps, such as geological or tourist maps. From a computational perspective, the use of a self-organizing map that contains a very large number of neurons is fairly unique. Skupin aims to explore how far we can go in the design of maplike information visualizations. The use of a range of label sizes (from very large to very small) on a large-format map and the omission of a legend are aimed at challenging traditional notions of interactivity, by encouraging viewers to vary their distance from the map and to engage in discussion.

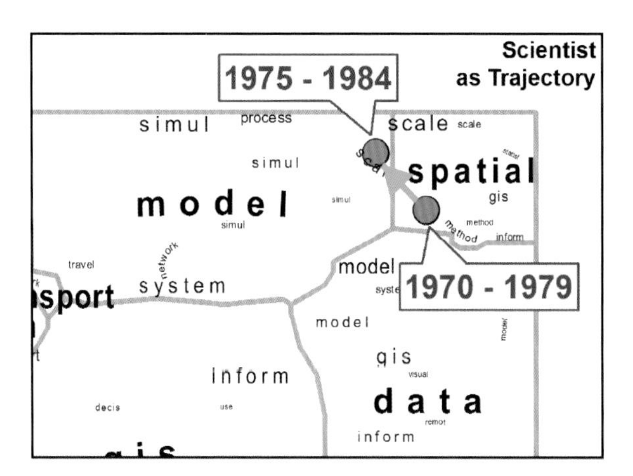

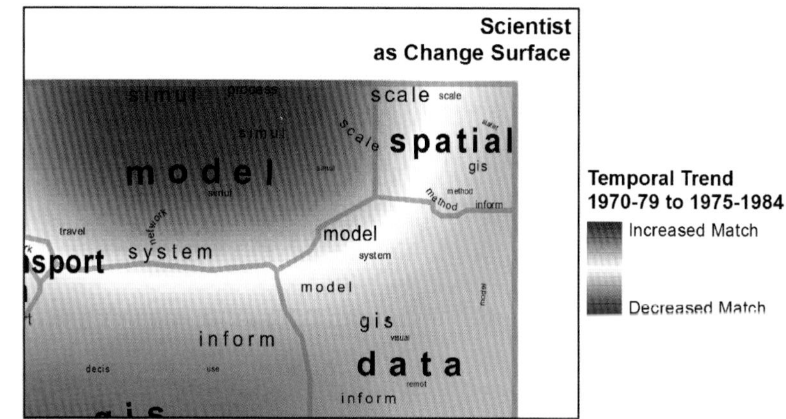

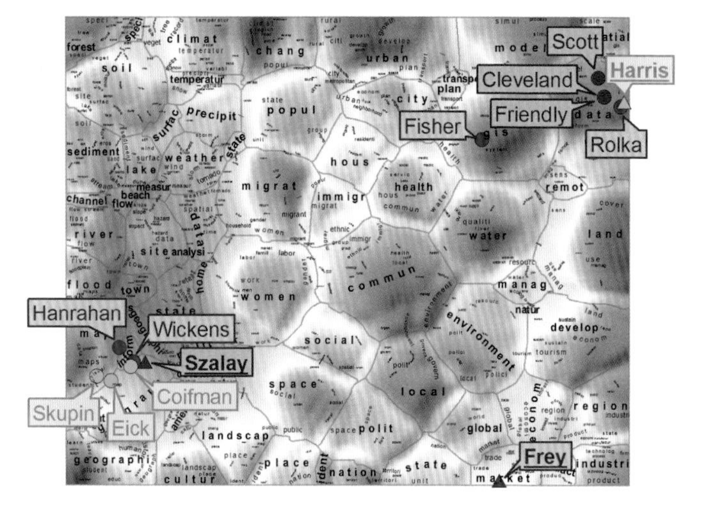

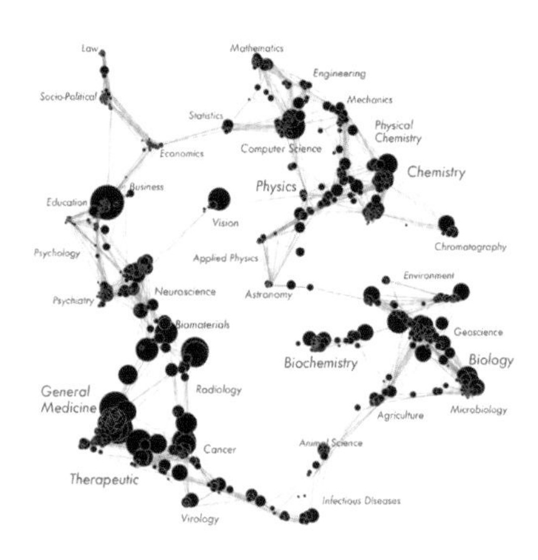

2005

The Structure of Science

By Kevin W. Boyack and Richard Klavans

ALBUQUERQUE, NEW MEXICO, AND BERWYN, PENNSYLVANIA, 2005

Courtesy of Kevin W. Boyack and Richard Klavans, SciTech Strategies, Inc.

Aim

How many scientific fields exist? How large are they? How are they interconnected? Who cites whose research? Can science be visually depicted and made tangible to young and old? How can the processing power of computers be used to make sense of the enormous number of scholarly records available today? What algorithms should be applied in what sequence to arrive at maps that have the highest accuracy both locally and globally? How do we determine the best metaphors for different users?

Interpretation

This map was generated from citation relationships among the scientific papers of the combined 2002 Science and Social Science Citation Indexes. It is actually a superposition of two separate maps, one based on clusters of journals (see **page 12, Toward a Reference System for Science**) and the other based on clusters of the papers themselves. The journal map was calculated first. Using bibliographic coupling at the journal level, nearly 7,300 journals were placed into 671 clusters of journals. A separate calculation was then done to cluster the papers. Again using bibliographic coupling, more than 800,000 papers were placed into 96,500 clusters. These clusters of papers—representing research communities—were then located on the journal map using the journal positions of the papers. A galaxy metaphor was employed to visualize the resulting map of the 96,500 research communities. Each white dot in the galaxy map represents a research community. Research communities with papers from only one journal cluster are located in the same position as that cluster. Inspection of the map shows that there are concentrations of dots in areas where there are many journal clusters. By contrast, research communities with papers from many different journal clusters are located between the clusters. These show up as lines between major areas of science (such as between **Chemistry** and **Biochemistry**) or as dispersed clouds (in the medical areas, for example). These communities reflect the interdisciplinary nature of science. In 2005, this was the most comprehensive and most accurate literature map ever generated. It has been used for many purposes, including as a template on which to show topic distributions and agency funding profiles.

Kevin W. Boyack joined SciTech Strategies, Inc. in 2007 after working at Sandia National Laboratories, where he spent several years in the Computation, Computers, Information and Mathematics Center. He holds a PhD in Chemical Engineering from Brigham Young University. His current interests and work are related to information visualization, knowledge domains, science mapping with associated metrics and indicators, network analysis, and the integration and analysis of multiple data types.

Richard Klavans is the president of SciTech Strategies, Inc. He holds a PhD in management from the Wharton School of the University of Pennsylvania. His current work is related to the generation of highly accurate maps of science, using multiple techniques, such as bibliographic coupling, cocitation, and coword, as well as the associated metrics and indicators that allow government and industry users to make more effective policy decisions. He is interested in semantics, augmented cognition, and the application of mathematical tools to information spaces.

The Structure of Science

5 The Social Sciences are the smallest and most diffuse of all the sciences. Psychology serves as the link between Medical Sciences (Psychiatry) and the Social Sciences. Statistics serves as the link with Computer Science and Mathematics.

1 Mathematics is our starting point, the purest of all sciences. It lies at the outer edge of the map. Computer Science, Electrical Engineering, and Optics are applied sciences that draw upon knowledge in Mathematics and Physics. These three disciplines provide a good example of a linear progression from one pure science (Mathematics) to another (Physics) through multiple disciplines. Although applied, these disciplines are highly concentrated with distinct bands of research communities that link them. Bands indicate interdisciplinary research.

2 Research is highly concentrated in Physics and Chemistry. These disciplines have few, but very distinct, bands of research communities that link them. The thickness of these bands indicates an extensive amount of interdisciplinary research, which suggests that the boundaries between Physics and Chemistry are not as distinct as one might assume.

Law
Socio-Political
Economics
Business
Education
Psychology
Psychiatry
Mathematics
Computer Science
Statistics
Mechanics
Electrical Engineering
Materials
Physical Chemistry
Optics
Physics
Chemistry
Chromatography
Applied Physics
Neuroscience
Astronomy
Environment
GeoScience
Radiology
Biology
Biochemistry
Microbiology
Agriculture
General Medicine
Oncology
Animal Science
Medical Treatments
Immunology
Infectious Disease

4 The Medical Sciences include broad therapeutic studies and targeted areas of Treatment (e.g. central nervous system, cardiology, gastroenterology, etc.) Unlike Physics and Chemistry, the medical disciplines are more spread out, suggesting a more multi-disciplinary approach to research. The transition into Life Sciences (via Animal Science and Biochemistry) is gradual.

3 The Life Sciences, including Biology and Biochemistry, are less concentrated than Chemistry or Physics. Bands of linking research can be seen between the larger areas in the Life Sciences; for instance between Biology and Microbiology, and between Biology and Environmental Science. Biochemistry is very interesting in that it is a large discipline that has visible links to disciplines in many areas of the map, including Biology, Chemistry, Neuroscience, and General Medicine. It is perhaps the most interdisciplinary of the sciences.

We are all familiar with traditional maps that show the relationships between countries, provinces, states, and cities. Similar relationships exist between the various disciplines and research topics in science. This allows us to map the structure of science.

One of the first maps of science was developed at the Institute for Scientific Information over 30 years ago. It identified 41 areas of science from the citation patterns in 17,000 scientific papers. That early map was intriguing, but it didn't cover enough of science to accurately define its structure.

Things are different today. We have enormous computing power and advanced visualization software that make mapping of the structure of science possible. This galaxy-like map of science (left) was generated at Sandia National Laboratories using an advanced graph layout routine (VxOrd) from the citation patterns in 800,000 scientific papers published in 2002. Each dot in the galaxy represents one of the 96,000 research communities active in science in 2002. A research community is a group of papers (9 on average) that are written on the same research topic in a given year. Over time, communities can be born, continue, split, merge, or die.

The map of science can be used as a tool for science strategy. This is the terrain in which organizations and institutions locate their scientific capabilities. Additional information about the scientific and economic impact of each research community allows policy makers to decide which areas to explore, exploit, abandon, or ignore.

We also envision the map as an educational tool. For children, the theoretical relationship between areas of science can be replaced with a concrete map showing how math, physics, chemistry, biology and social studies interact. For advanced students, areas of interest can be located and neighboring areas can be explored.

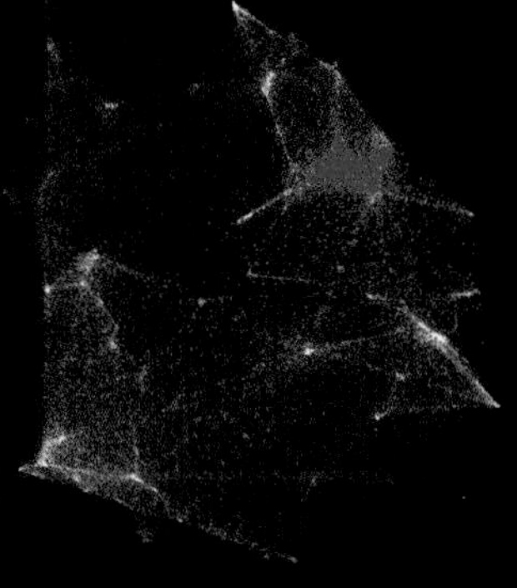

Nanotechnology

Most research communities in nanotechnology are concentrated in Physics, Chemistry, and Materials Science. However, many disciplines in the Life and Medical Sciences also have nanotechnology applications.

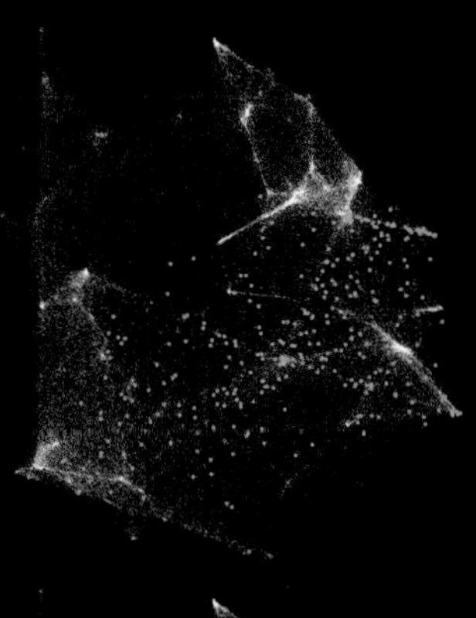

Proteomics

Research communities in proteomics are centered in Biochemistry. In addition, there is a heavy focus in the tools section of chemistry, such as Chromatography. The balance of the proteomics communities are widely dispersed among the Life and Medical Sciences.

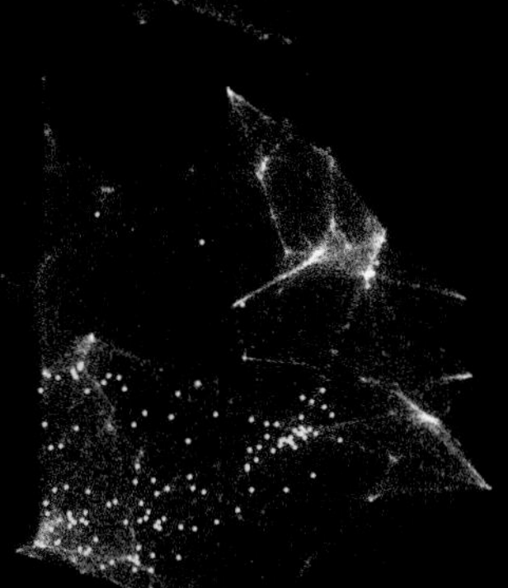

Pharmacogenomics

Pharmacogenomics is a relatively new field with most of its activity in Medicine. It also has many communities in Biochemistry and two communities in the Social Sciences.

Data

This map is based on paper-level and journal-level data from the combined 2002 Science Citation Index and Social Sciences Citation Index by Thomson Reuters. The data set comprises about 1.07 million papers published in 7,300 journals together with their 24.5 million references.

Technique

Two separate maps were superpositioned: a *Map of Disciplines* generated from journal-level data and a *Map of Research Communities* based on paper-level data. The generation of both maps is explained below.

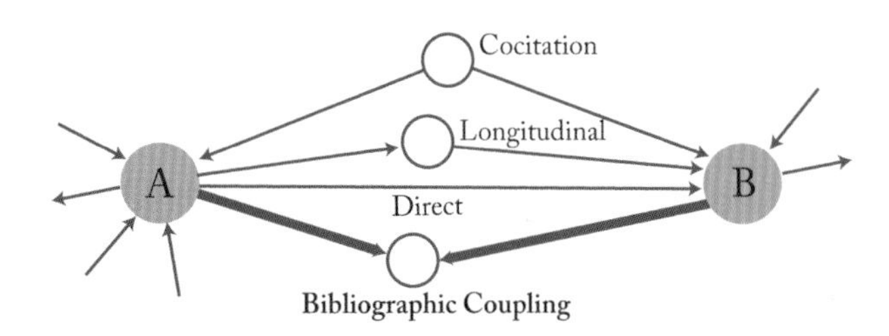

Both maps employ bibliographic coupling and cosine similarity, and they iteratively use VxOrd and a modified single-link algorithm to achieve nested clustering. Bibliographic coupling occurs when two papers (or journals) have common elements in their reference lists or bibliographies (see graph). This process is commonly used to create maps of current science.

The figure below exemplifies the iterative application of spatial layout routines and single linkage clustering to arrive at nested hierarchies of paper or journal clusters. The general process of database search leading to **Aggregate Records**, **Bibliographic Coupling** resulting in a **Coupling Matrix**, **VxOrd** use to generate a **Graph Layout** that can be displayed via VxInsight (see also the section on VxOrd and VxInsight), and further **Cluster Assignment** to identify **Emergent Categories** or major areas of science is also shown (top left on opposite page).

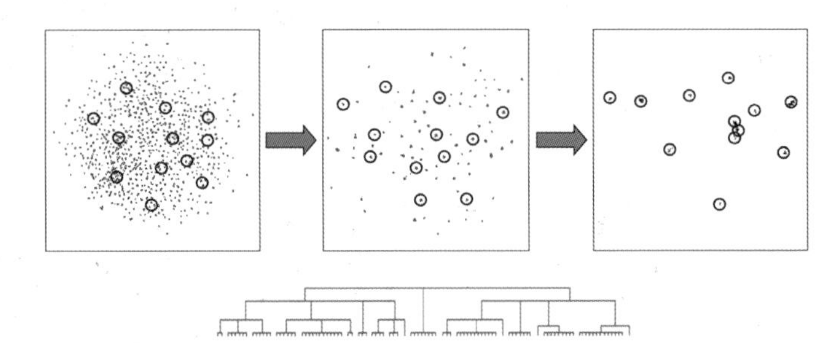

Reference System

Disciplinary Map Based on Journals

The design of the disciplinary map is a multistep process. First, the set of nearly 7,300 unique journals in the 2002 dataset was identified. Next, the similarity between journals based on bibliographic coupling of reference data was calculated. Coupling was done at the paper level, and then aggregated to the journal level. A modified cosine index was then calculated for each journal-journal interaction. This similarity matrix was then fed into the VxOrd layout algorithm (see **Details** section) to determine the *x, y* coordinates for each journal. Single link was then used to assign journals to clusters, and 671 journal clusters emerged.

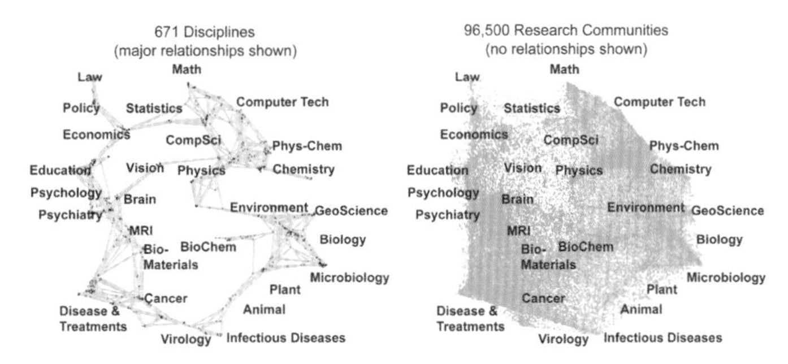

Disciplinary Map Research Community Map

Coupling counts were reaggregated at the journal cluster level and the *x, y* position for each journal cluster was calculated using VxOrd. During this process, 25 multidisciplinary journals—such as *Science*, *Nature*, and *PNAS*—were not allowed to aggregate with other journals but given their own clusters to avoid overaggregation. Major areas were labeled manually.

Research Community Map Based on Papers

In this step, research communities are computed and mapped. Again, this is a multistep process. First, the set of approximately 1,000,000 unique papers is identified. About 730,000 are research papers; the other papers are editorials, book reports, or similar with no citation links. Subsequently, only the research papers are considered. Again using bibliographic coupling, the research papers are placed in 96,500 clusters. Clustering is done using the processing pipeline explained on **page 12, Toward a Reference System for Science**, with VxOrd doing the *x, y* layout and a modified single-link clustering algorithm assigning papers to clusters based on their *x, y* positions. These clusters of papers or research communities are then located on the journal map using the journal positions of the papers.

The resulting ring-shaped map is invariant to mirroring and rotation. The alignment convention chosen was to place mathematics, the most theoretical science, at the top, followed in clockwise direction by computer technology, physics, chemistry, earth sciences, biology, medicine, psychology, education, statistics, and mathematics once again. There is an interesting peninsula drawn from economics to policy and law. Vision connects physics and brain sciences.

The result is a more accurate look at the disciplinary structure of science. The map constitutes a spatial reference system on which to display information at a disciplinary level (for example, funding amounts by discipline/agency) or topic distributions (for example, nanotechnology moving into medicine).

Data Overlays

Each white dot in *The Structure of Science* map represents one of the 96,500 research communities. Color-coded labels were assigned to visually distinguish five major areas of science. The three maps on the right highlight research communities that study **Nanotechnology**, **Proteomics**, and **Pharmacogenomics** in red, green, and yellow, respectively.

Details

This map can be used to identify the core competency (or funding patterns) of different institutions, such as the U.S. Department of Energy (DOE), the National Science Foundation (NSF), and the National Institutes of Health (NIH) (see maps below). Notice the complementarity in funding by the NSF and NIH. The NSF and NIH funding overlays were generated by linking papers with the grants that were likely to have funded the work. Papers and grants were linked by matching the principal investigator on a grant with the first author of a paper. A match between the institutions on the grant and the paper was also required to assign a match. A time lag of three years between funding of the grant and publication of the paper was assumed. The DOE funding overlay was calculated directly using the list of papers published by DOE laboratories (for example, Los Alamos, Lawrence Livermore, Sandia, Argonne, and Oak Ridge).

VxOrd and VxInsight

VxOrd and VxInsight were developed by Sandia National Laboratories. VxOrd employs a highly scalable force-directed placement layout algorithm to lay out networks with up to 10 million nodes and many more edges. Its run time increases linearly with the number of nodes [O(n)], and one needs about three hours on an average personal computer to lay out 1 million nodes and 7 million edges. The resulting layout is displayed as a scatterplot.

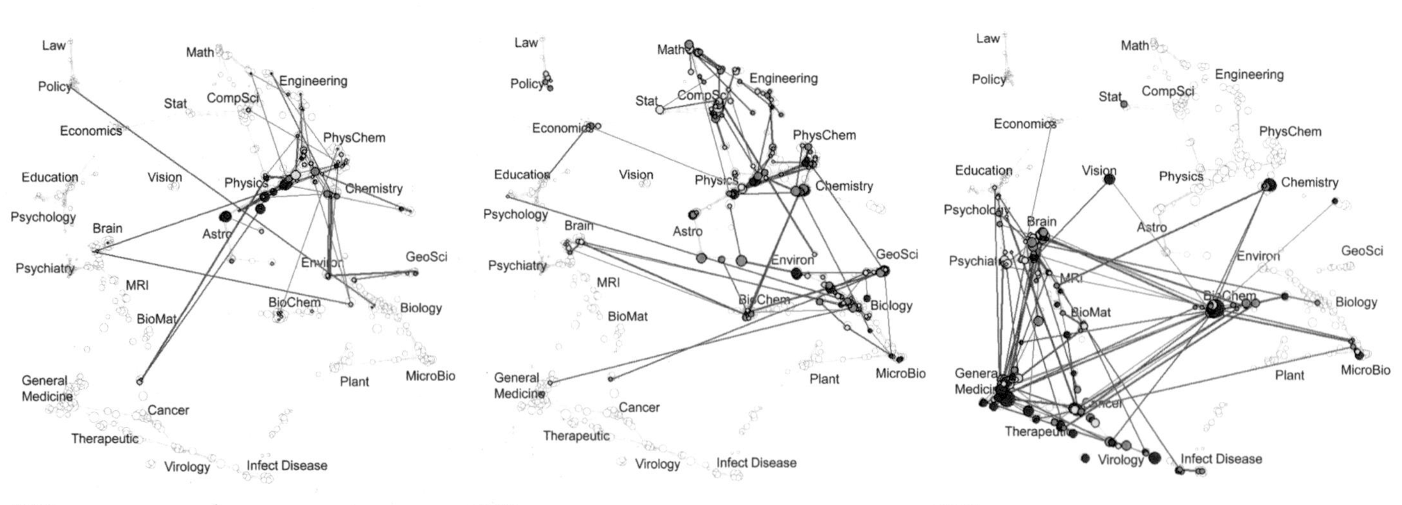

DOE NSF NIH

The algorithm takes a set of nodes and their weighted edges as input. During the incremental layout, each object tries to minimize an energy equation using a solution space exploration algorithm. The algorithm employs boundary jumping and edge cutting (pruning) to separate larger clusters. It uses pruning instead of thresholding to depict the "backbone" of a network—as thresholding (that is, keeping only edges with greater than a given weight) is not an effective strategy for networks with very different local dynamics. For example, physics or biology papers can easily acquire up to 1000 citation linkages in a few years. Papers in the humanities rarely acquire such high citation counts. If a threshold based on the number of citations were applied, most humanities records would disappear. Using pruning, the top-n edges per paper are kept. This strategy avoids the disciplinary bias that would result from thresholding, while preserving edge degree along with the macro-structure of the data set. Common values for n are 10–20.

VxInsight uses VxOrd to lay out and facilitate the interaction with large-scale networks. The resulting layout is displayed using an intuitive terrain or grid metaphor or it is shown as scatterplot. This exposes an implicit structure in large graphs and gives a context for the investigation of subgraphs. VxInsight enables analysts to navigate and explore graph structures at multiple levels of detail through drilldown (see interface and different data views below). Metadata associated with graph objects is shown as labels, and detailed information can be retrieved for each node.

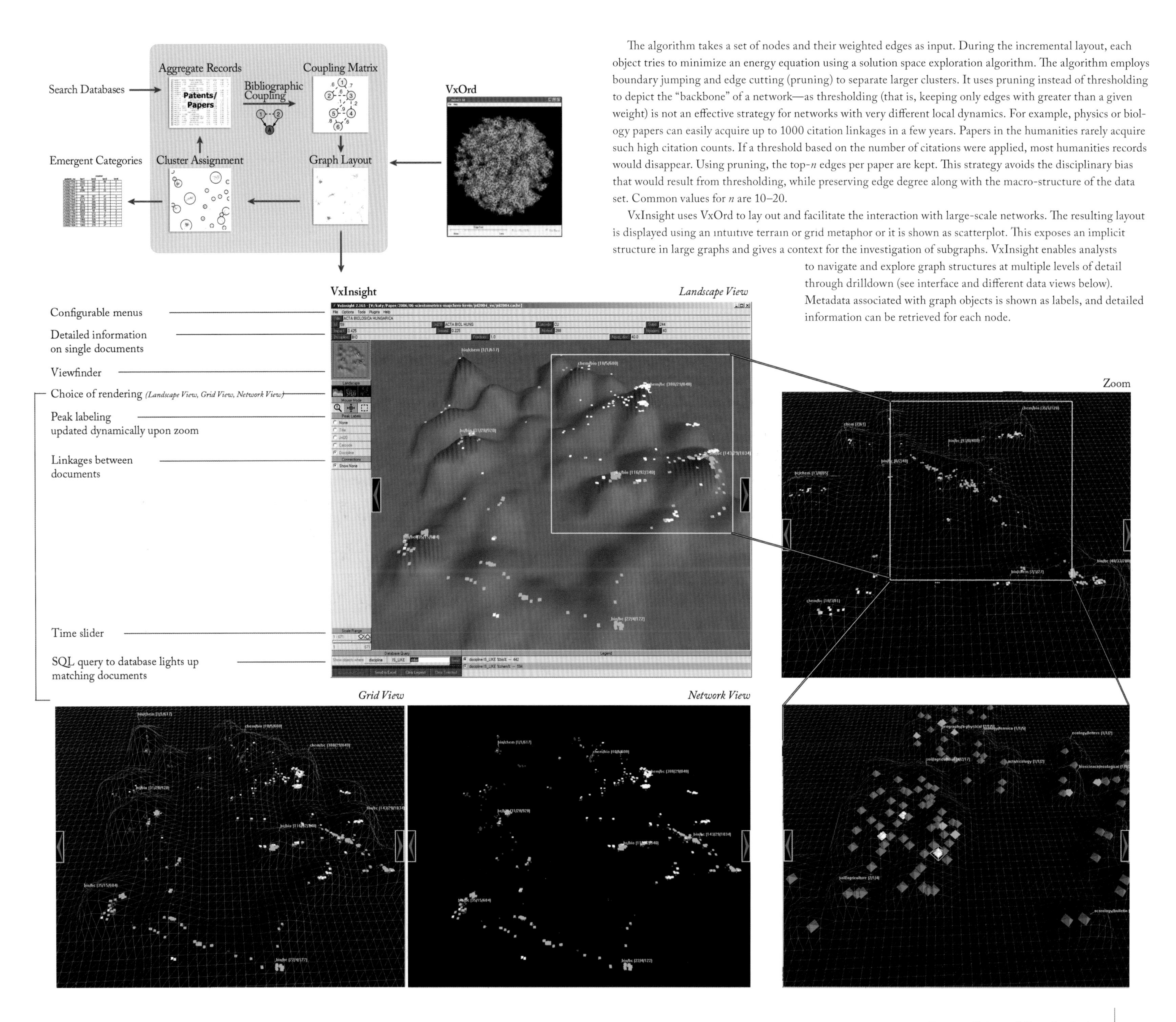

Second Iteration of Exhibit (2006): The Power of Reference Systems

Four Existing Reference Systems
versus
Six Science Reference Systems

This iteration aims to inspire discussion about a common reference system for all existing scholarly knowledge. Throughout history, scientists have battled to agree on standardized reference systems for their respective fields of research. These standards are invaluable for indexing, storing, accessing, and managing scientific data efficiently.

Results include the description of the electromagnetic spectrum, the periodic table of elements, geographic projections, and multilevel celestial reference systems, shown here. Note that the geographic map encapsulates the transition from paper to geographic information systems (GIS). All four maps are designed for public use and consumption.

In comparison to these four existing systems are six potential reference systems for scholarly knowledge. Each reference system—from the one-dimensional timeline and the geographic system to the semantic system—could be used to identify the location of an author, paper, patent, or grant and its trajectory or contribution.

Four Existing Reference Systems

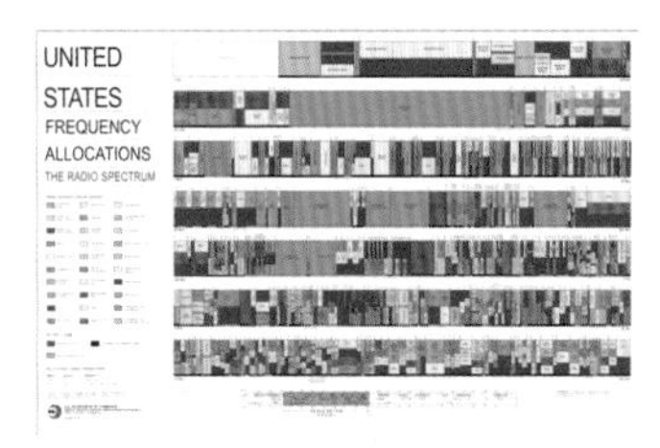

National Telecommunications and Information Administration, U.S. Frequency Allocations Chart, 2003

Murray Robertson and John Emsley, The Visual Elements Periodic Table, 2005

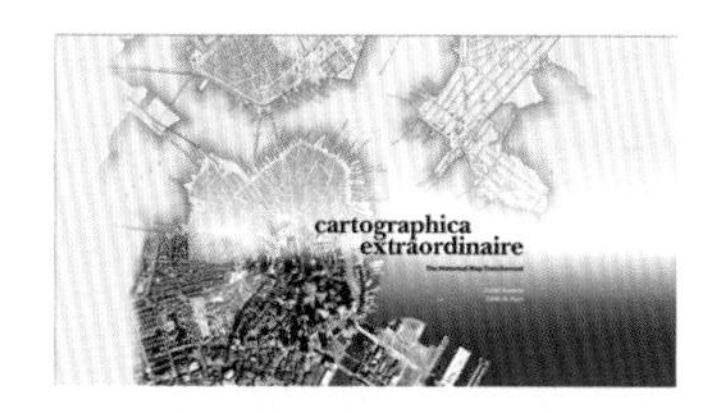

David Rumsey and Edith M. Punt, Cartographica Extraordinaire: The Historical Map Transformed, 2004

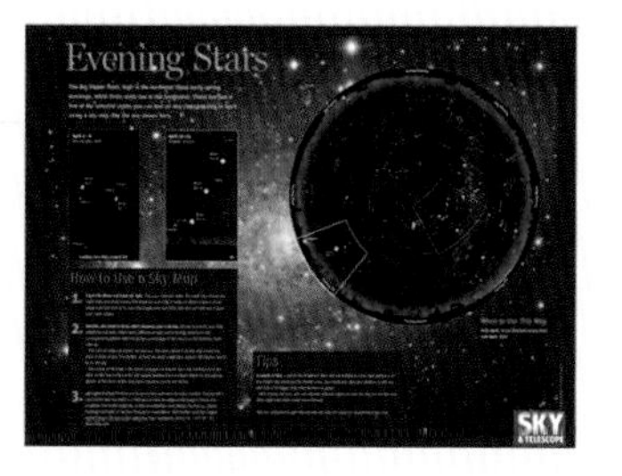

Roger W. Sinnott, Interactive Factory, Sky Chart of New York City in April 2006, 2006

Six Science Reference Systems

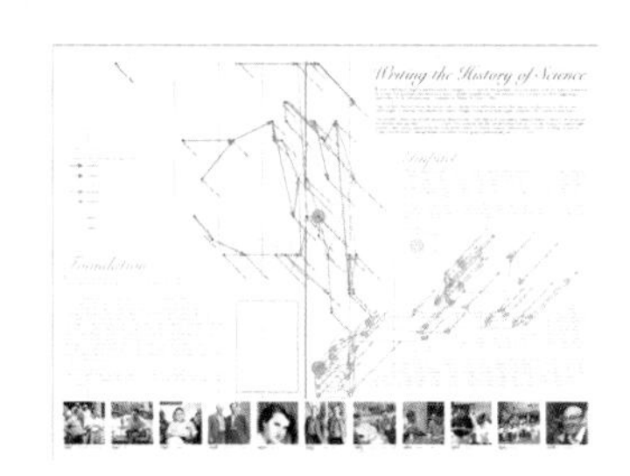

Eugene Garfield, Elisha F. Hardy, Katy Börner, Ludmila Pollock, and Jan Witkowski, HistCite Visualization of DNA Development, 2006

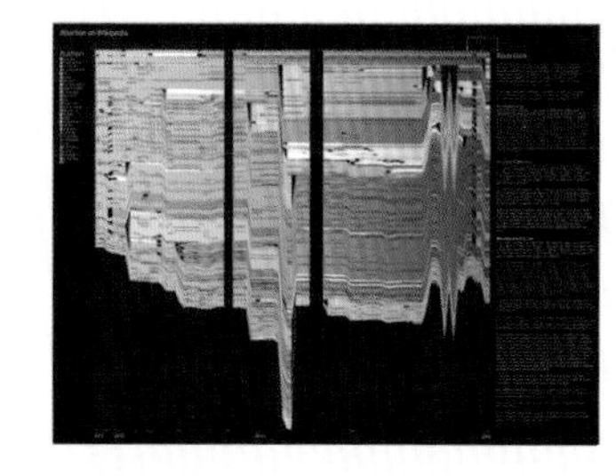

Martin Wattenberg and Fernanda Viégas, History Flow Visualization of the Wikipedia Entry "Abortion," 2006

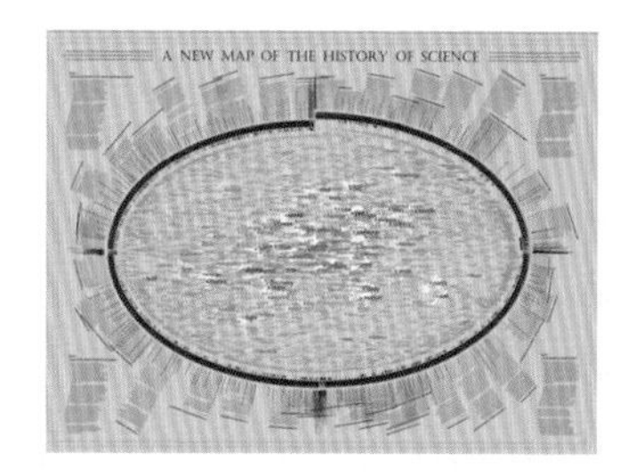

W. Bradford Paley, TextArc Visualization of "The History of Science," 2006

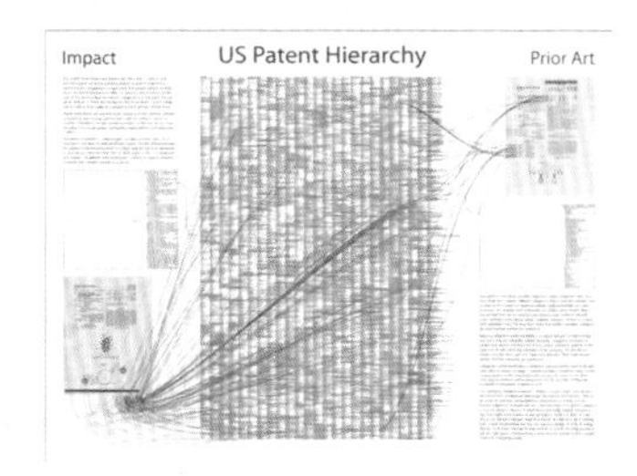

Katy Börner, Elisha F. Hardy, Bruce W. Herr II, Todd M. Holloway, and W. Bradford Paley, Taxonomy Visualization of Patent Data, 2006

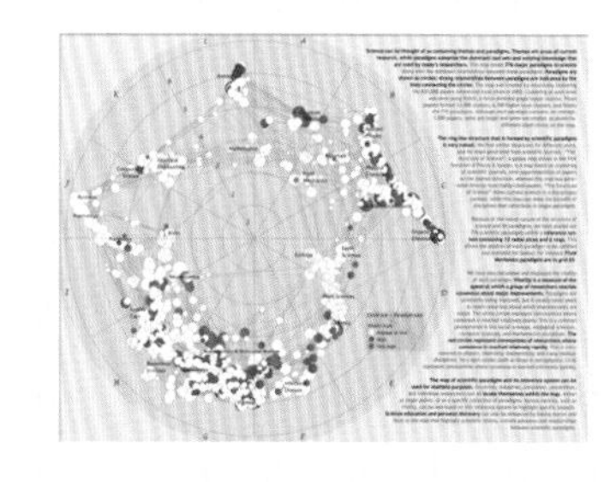

Kevin W. Boyack and Richard Klavans, Map of Scientific Paradigms, 2006

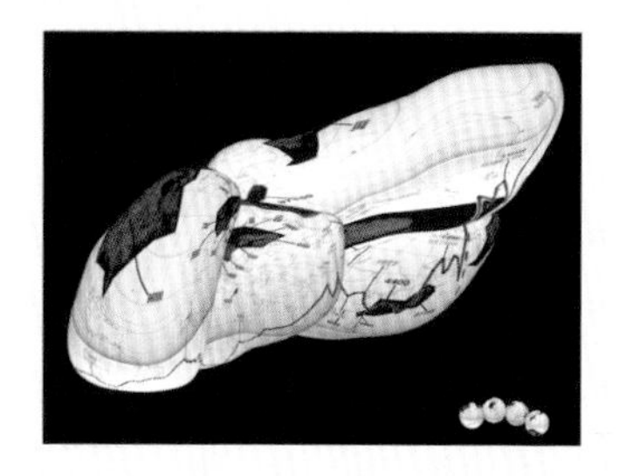

Ingo Günther, Zones of Invention—Patterns of Patents, 2006

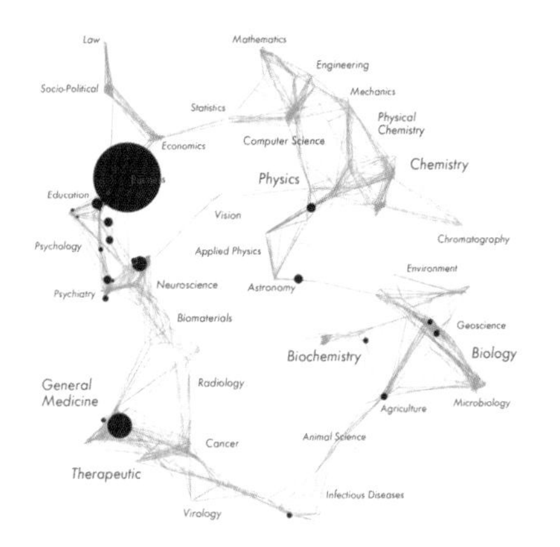

2003

U.S. Frequency Allocations Chart

By the National Telecommunications and Information Administration

WASHINGTON, DC, 2003

Courtesy of the Office of Spectrum Management

The NTIA's Spectrum Chart, also known as the U.S. Frequency Allocations Chart, depicts the radio frequency spectrum allocations to radio services operated within the United States. Published in October 2003, the chart depicts the allocation decisions that were made by the NTIA and the Federal Communications Commission (FCC) up to July 1, 2003; it replaces a similar chart printed by NTIA in 1996. The U.S. domestic spectrum uses may differ from international allocations that comply with international regulations or bilateral agreements.

The chart uses a one-dimensional reference system to graphically partition the radio spectrum–extending from 3 kilohertz (kHz) to 300 gigahertz (GHz)–into more than 450 frequency bands (see zoom into legend and total frequency spectrum below). Because 30 different U.S. radio services are allocated portions of the spectrum in more than 450 separate frequency bands, many allocation issues quickly become quite complex. Through the use of graphics and multiple colors, this chart helps widely diverse audiences gain a general understanding of U.S. domestic spectrum allocations policies. The chart was produced in Adobe PageMaker. Copies of the chart and background information are available at **http://www.ntia.doc.gov/osmhome/allochrt.html**.

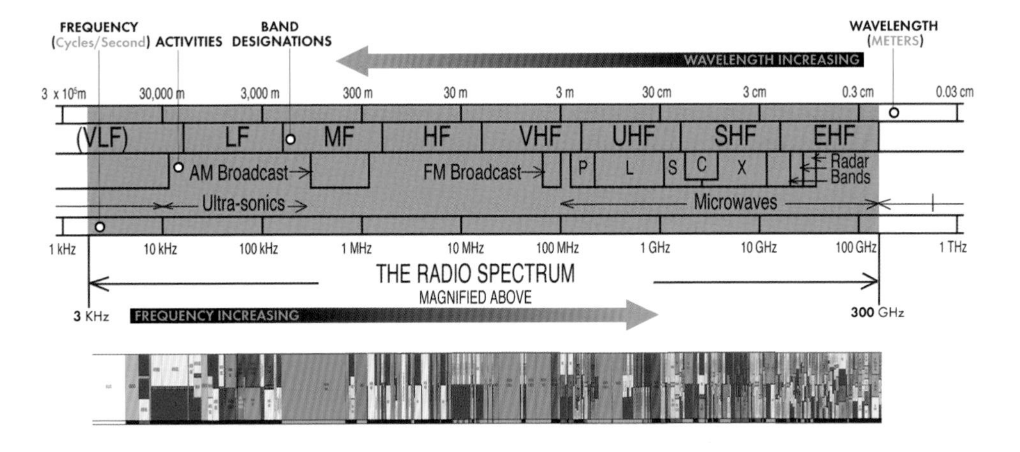

The National Telecommunications and Information Administration (NTIA) is an agency within the U.S. Department of Commerce, established in 1978. Prior to NTIA's inception, the executive branch had established the Office of Telecommunications Policy (OTP) in order to spearhead the administration's communications policy in certain areas, most notably cable television. The NTIA succeeded this unit, combining the responsibilities and mission of the OTP with those of the Commerce Department's Office of Telecommunications. The agency is now the principal advisor to the president on both domestic and international telecommunications and information policy activities. The NTIA also manages the federal use of the spectrum; performs cutting-edge telecommunications research and engineering, such as resolving technical telecommunications issues for the federal government and private sector; administers infrastructure and public telecommunications facilities grants; and operates the research and engineering Institute for Telecommunication Sciences in Colorado.

UNITED STATES FREQUENCY ALLOCATIONS

THE RADIO SPECTRUM

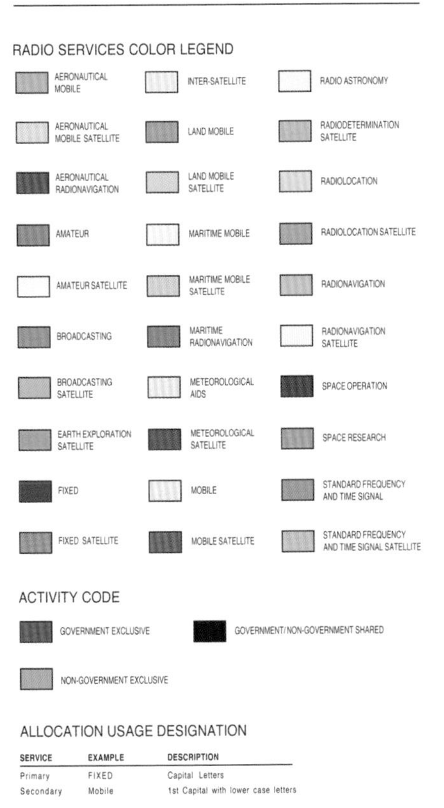

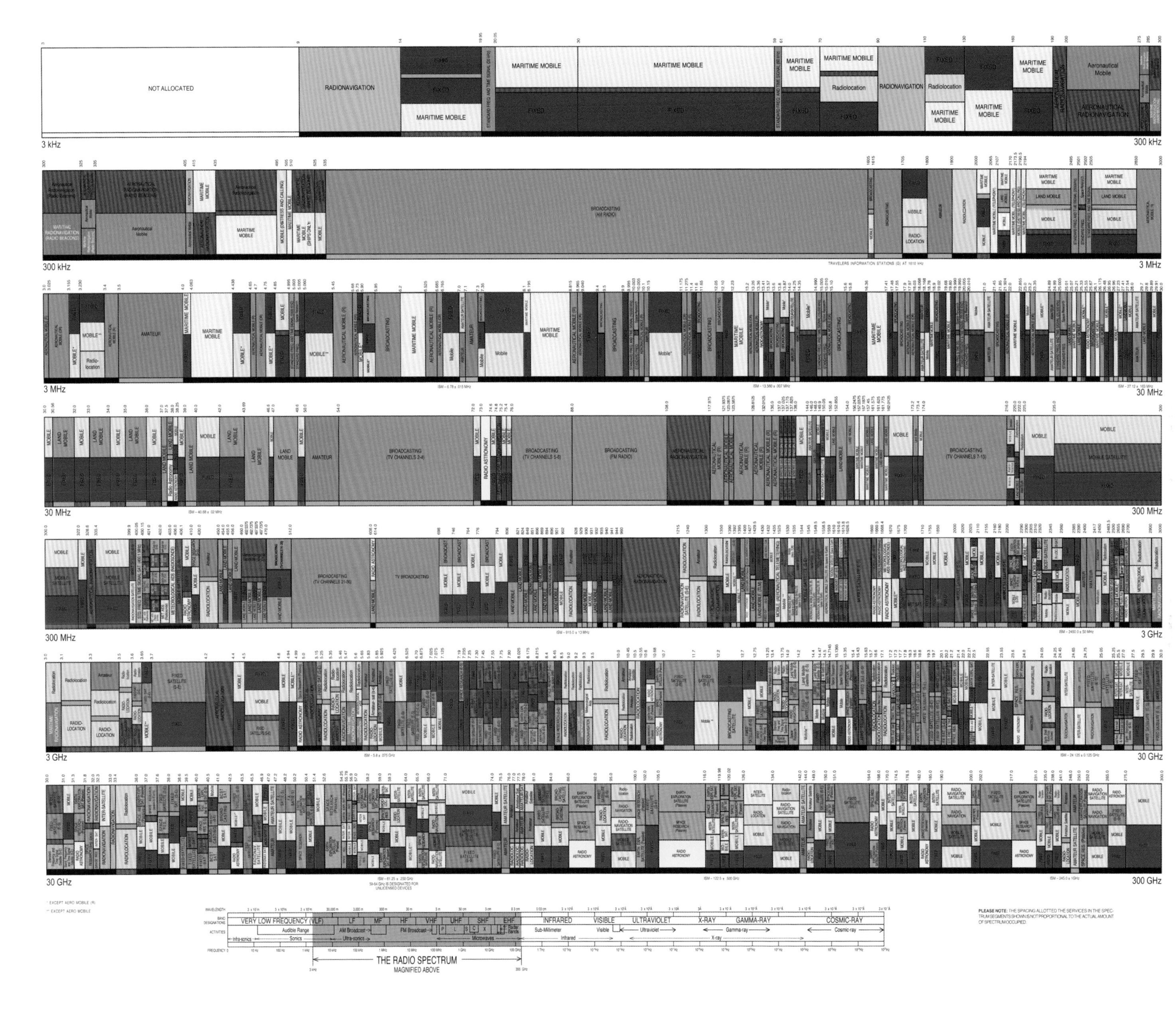

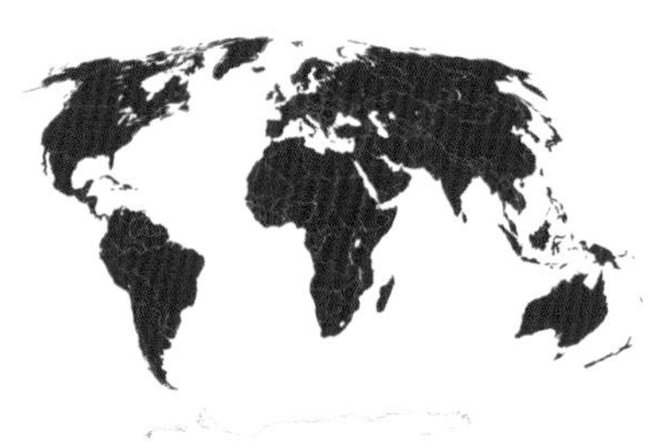

2005

The Visual Elements Periodic Table

By Murray Robertson and John Emsley
LONDON, UNITED KINGDOM, 2005
Courtesy of the Royal Society of Chemistry. Images: copyright 1999–2006. Murray Robertson

No chemistry textbook, classroom, auditorium, or research laboratory is complete without a copy of the periodic table of the elements. Since the earliest days of chemistry, attempts have been made to arrange the known elements in ways that would reveal commonalities between them. Different taxonomies were suggested based on similarities of the elements' properties, like the triads of Johann W. Döbereiner and John A. R. Newlands's octaves.

In 1869, the Russian chemist Dmitri Mendeleev noticed that the properties of the chemical elements are not arbitrary but depend on the structure of the atom and vary according to atomic weight in a periodic way. Mendeleev decided to follow the scheme of Newlands's octaves, which ordered the elements into eight columns according to increasing atomic weight. However, he extended the scheme by moving each element to the location appropriate for its atomic weight according to the periodicity of its properties, leaving "holes" in the table for elements perhaps yet to be discovered. It was later discovered that the atomic number, rather than the atomic weight, defined the periodicity.

Mendeleev also conducted an in-depth study to determine more accurate atomic weights and hence better placement of the elements in the table. He introduced long periods (transition metals) to avoid placing metals below nonmetal elements. "Leaving holes" had the extraordinary effect of facilitating the discovery of new elements. More than 700 versions of the periodic table were produced in the century after Mendeleev.

The table shown here includes images of the elements and investigates the manner in which they affect our daily lives. The images were created by using three-dimensional modeling software to transpose two-dimensional graphs of various chemical properties onto fractal landscapes. An interactive version is available online at **http://www.chemsoc.org/viselements**.

The periodic table of the elements is an example of a two-dimensional tabular reference system. The columns of the table correspond to groups of elements sharing the same chemical and physical properties, defined by their valence and interaction properties with other elements. Within rows, elements of different properties are grouped according to similarity in weight and number of atomic orbitals.

Murray Robertson is an illustrator who has been involved with the visualization of ideas in science since 1998. As an artist and printmaker at Glasgow Print Studio, he exhibits regularly both within the United Kingdom and internationally. Robertson's works of graphic precision draw upon the iconography of signs and symbols inherent to the matrix of both mystical and scientific knowledge systems, echoing diverse cultures and empirical paradigms in an exploration of humanity's eternal and universal quest for knowledge and meaning. Projects have included working with Scottish Television (STV), the Medical Research Council, the UK Clinical Virology Network, and most recently the Electrical Engineering department of Glasgow University, producing a series of images based on current developments in nanotechnology.

John Emsley is a chemist as well as an award-winning popular science writer and broadcaster. His books include *Nature's Building Blocks: An A–Z Guide to the Elements* (Oxford University Press, 2003), *Elements of Murder: A History of Poison* (Oxford University Press, 2005), and *Better Looking, Better Living, Better Loving* (Wiley-VCH, 2007). He has written a comprehensive history of the periodic table of elements, available at **http://www.chemsoc.org/viselements/pages/history**.

The table was drawn by Robertson based on scientific data provided by Emsley. It was commissioned and supported by the Royal Society of Chemistry in the United Kingdom as part of an arts and science collaborative project called Visual Elements. The project aims to produce a new and vibrant visual assessment of the startling diversity of material that constitutes the world in which we live.

The Visual Elements Periodic Table

This chart shows the 111 currently known and officially named elements that comprise the Periodic Table (IUPAC 2004). Each element is represented visually by an image produced for the Visual Elements project.

The Periodic Table is an arrangement of all known elements in order of increasing atomic number. The Periodic Table fits all the elements, with their widely diverse physical and chemical properties, into a logical pattern. There are eighteen vertical columns in the table which divide the elements into groups. Elements within a group have closely related physical properties. Horizontal rows list the elements in order of their increasing mass and are called series or periods. Properties of elements change in a systematic way through a period.

Visual Elements is an arts and science collaborative project supported by the Royal Society of Chemistry which aims to explore and reflect upon the diversity of elements that comprise matter in as unique and innovative manner as possible. All the images displayed here, together with screensavers, postcards and chemical data for each element can be viewed on the Visual Elements web site, hosted by the RSC.

Visit the periodic table on the web at:
www.chemsoc.org/viselements

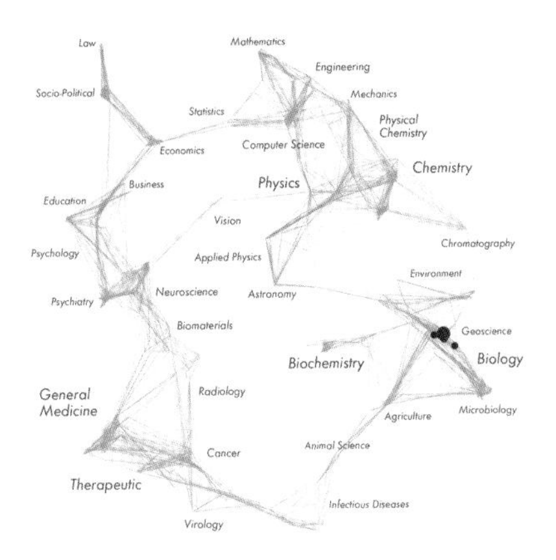

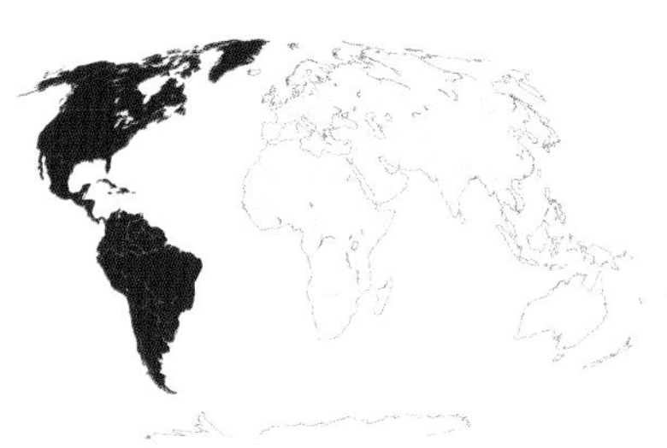

2004

Cartographica Extraordinaire: The Historical Map Transformed

By David Rumsey and Edith M. Punt

SAN FRANCISCO, CALIFORNIA, AND REDLANDS, CALIFORNIA, 2004

Shown here is the cover of *Cartographica Extraordinaire: The Historical Map Transformed* by David Rumsey and Edith M. Punt. The book features stunning reproductions from the renowned David Rumsey Map Collection—one of the largest and most complete collections of its kind. Focused largely on cartography of the Americas from the 18th and 19th centuries, the collection comprises more than 150,000 items: maps, atlases, globes, school geographies, maritime charts, and a variety of pocket, wall, children's, and manuscript maps, as well as contextual supporting documents.

The delicacy and rarity of collections such as Rumsey's necessitate careful storage and restricted-use policies. Driven by an intense desire to make the collection available to a wider public, Rumsey launched a Web site (**http://www.davidrumsey.com**) in March 2000. By the end of 2009, the Web site provided access to more than 20,000 maps. Using sophisticated yet intuitive software, visitors can view maps side by side, zoom in for inspection of the smallest details, and save and print. A comprehensive catalog provides information about each map's cartographic relevance and provenance, author, publisher, and date of publication, as well as other historical and geographic facts. By 2009, 120 of the historical maps were available in the Featured Content layer of Google Earth. The georeferenced maps wrap the virtual globe in their modern spaces, allowing explorations in time and space.

Cartographic maps tell a myriad of distinct, important, and sometimes controversial stories, along two main paths of inquiry: how did a continental wilderness become a civilization and how has the development of cartographic science changed the ways we perceive, describe, study, and use that land?

David Rumsey is president of Cartography Associates, a digital publishing company based in San Francisco, and director of Luna Imaging, a provider of enterprise software for online image collections. He was a founding member of Yale Research Associates in the Arts, a group of artists working with electronic technologies. He subsequently became associate director of the American Society for Eastern Arts in San Francisco. Later, he entered a 20-year career in real estate development and finance, during which he had a long association with the General Atlantic Holding Company of New York and served as president and director of several subsidiaries. General Atlantic eventually became the Atlantic Trust, a Bermuda-based philanthropic foundation that is now one of the world's largest charities. Rumsey retired from real estate in 1995 and founded Cartography Associates, beginning a third career as a digital publisher, online library builder, and software entrepreneur.

Edith M. Punt graduated from McGill University in Montreal with a BSc in physical geography and Northern studies. After an apprenticeship in cartography with Nunavik Graphics in Montreal, she went on to the College of Geographic Sciences in Nova Scotia, where she earned a diploma in cartography, and was recognized by the Nova Scotia House of Assembly for outstanding achievement in cartography. She was the recipient of the Nicholson Cartographic Scholarship in 1995 and the President's Prize for Best Colour Map in 1996—both from the Canadian Cartographic Association. In 1996 she also won the National Geographic Society Award in Cartography and the Intergraph Award in Computer Cartography from the Canadian Institute of Geomatics. She has been a cartographer and writer at ESRI in Redlands, California, since 1996.

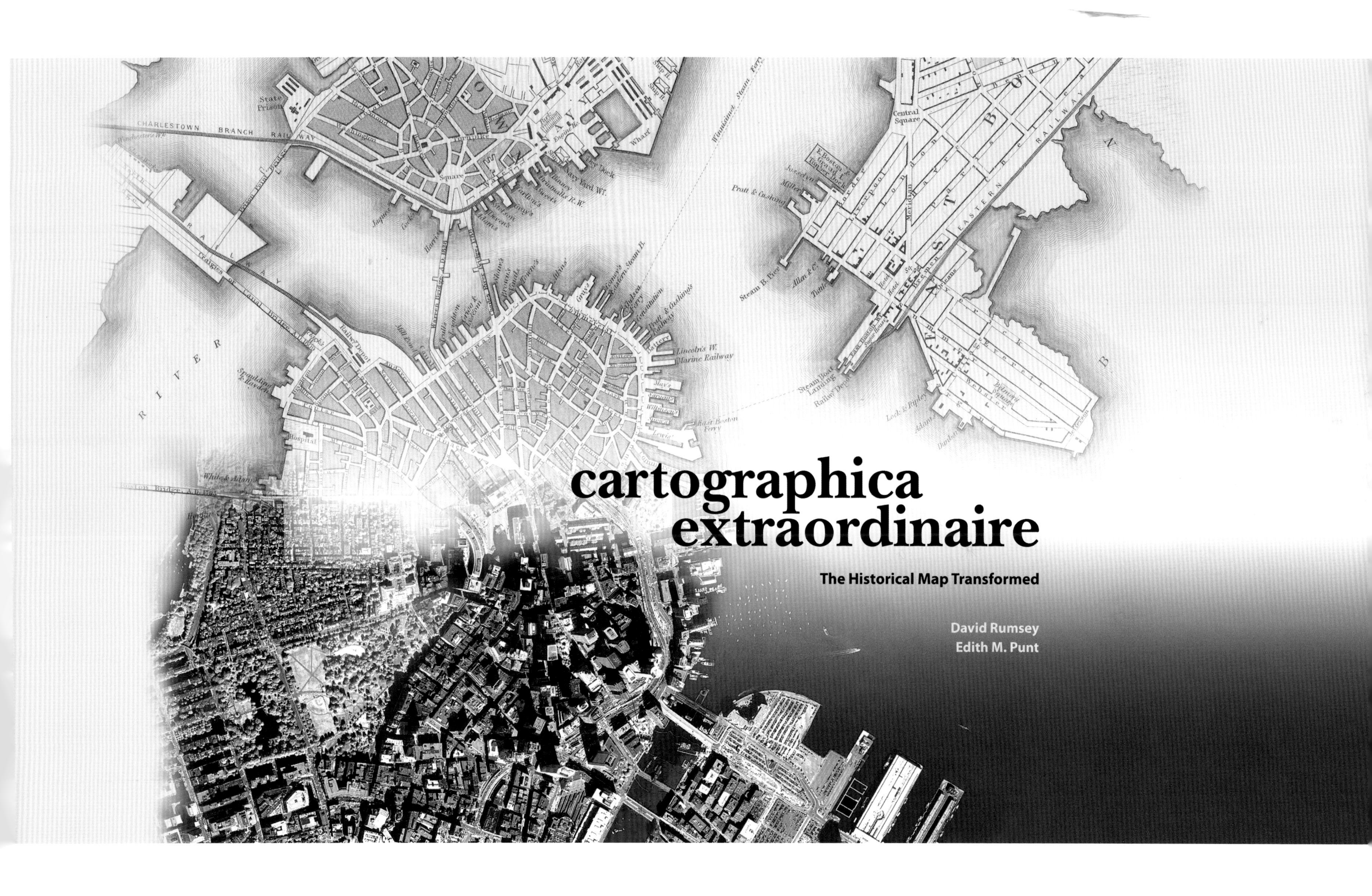

cartographica extraordinaire

The Historical Map Transformed

David Rumsey
Edith M. Punt

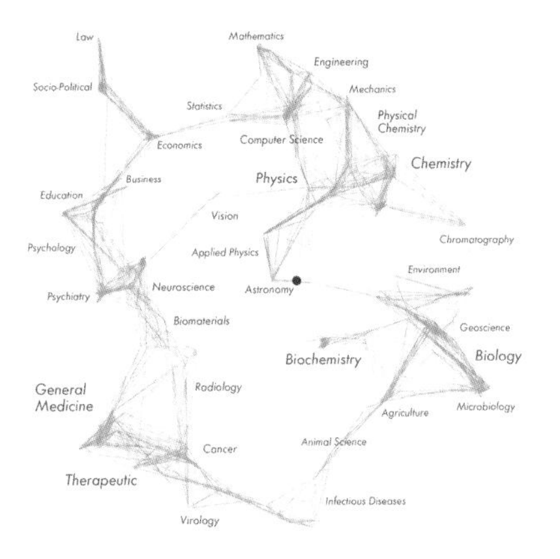

2006

Sky Chart of New York City in April 2006

By Roger W. Sinnott and the Interactive Factory
CAMBRIDGE, MASSACHUSETTS, 2006
Courtesy of Sky & Telescope

This is a printed still of an *Interactive Sky Chart*, designed by Roger W. Sinnott and iFactory for *Sky & Telescope*. Online versions can simulate a naked-eye view of the sky from any location on Earth, at any time of night, on any date from 1600 to 2400. The circle seen here simulates the view of a dome centered over New York City in April 2006. The yellow outlined sky section represents the view looking into the southeastern part of the dome (see zoom below). The purple outlined section is a view into deep space. The stars and planets charted are the ones typically visible without optical aid under clear suburban skies. (They might not be visible in the city.) Deep-sky objects that can be seen through binoculars are also plotted. Observations over many centuries are used to predict which stars and planets will be visible from different areas at various times of the year.

April 12–14
Around 10 p.m.
γ Vir
Moon Apr 12
Spica
Moon Apr 13
10°
Moon Apr 14
α Lib
Jupiter
SE

To specify a point on the Earth or any celestial sphere, geometers use spherical coordinates, such as latitude and longitude on Earth. Astronomers expand the planet's coordinates out into the celestial sphere using coordinates called declination and right ascension. Imagine the lines of latitude and longitude ballooning outward from the Earth and becoming imprinted upon the surrounding sphere of sky. Lines of both right ascension and declination stay fixed with respect to the stars. That is why they can be permanently printed on star maps. However, this also means that the one-to-one connection between right ascension and longitude is broken the moment after the two systems rotate with respect to each other.

Roger W. Sinnott is the editor and author of numerous astronomy books. He coauthored *Sky Catalogue 2000.0*, a comprehensive collection of standard data for astronomical objects. The introduction contains a Messier list, an OB stellar association, and a Local Group and Cluster of Galaxies table—as well as photos of M45, M13, M24, M104, and several non-Messier objects; a comprehensive list of references; and a glossary of selected astronomical names.

The Interactive Factory (iFactory) is a privately held company based in Boston. Its mission is to serve clients through the conception and execution of innovative, intelligent, and usable digital media solutions. IFactory thrives on the challenges of information architecture, content development, usability testing, and innovative, beautiful design. It believes in the fusion of purpose and play: that functional, useful tools should also compel and that engineering knowledge should encompass everything from legacy database systems to the latest tools for content management and graphics.

Sky & Telescope magazine is an institution in the world of astronomy. From its birth in 1941 to the present day, it has managed to retain a high level of quality that has earned a devoted following. A complete history of the magazine is available online.

Evening Stars

The Big Dipper floats high in the northeast these early spring evenings, while Orion sinks low in the southwest. These are just a few of the celestial sights you can find on any clear evening in April using a sky map like the one shown here.

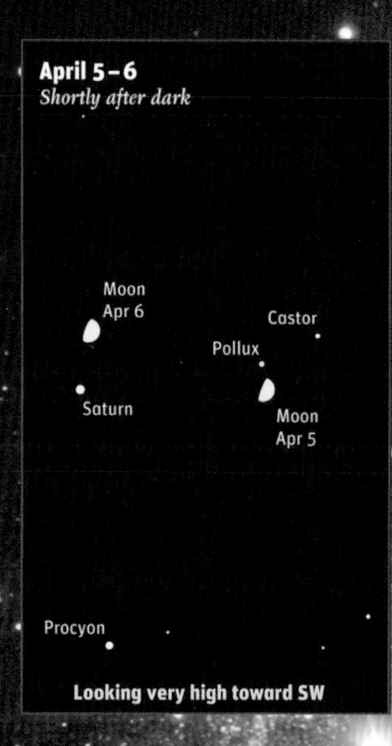

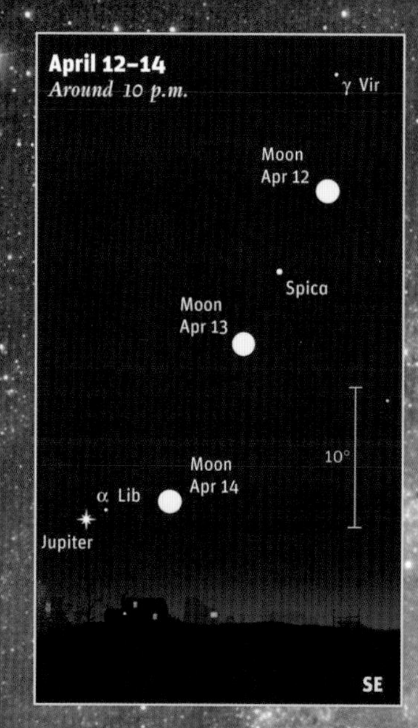

When to Use This Map

Early April: 10 pm (daylight-saving time)

Late April: Dusk

How to Use a Sky Map

1. **Check the dates and times at right.** Take your map out under the night sky around the right time, and bring along a flashlight to read it by. It helps to attach a piece of red paper over the front or to use a flashlight with red LEDs; the dim red light won't spoil your night vision.

2. **Outside, you need to know which direction you're facing.** (If you're unsure, just note where the Sun sets; that's west.) Whichever way you're facing, make sure the corresponding yellow label along the curved edge of the map is at the bottom, right-side up.

This curved edge represents the horizon. The stars above it on the map match the stars in front of you. The farther up from the map's edge they appear, the higher they'll be in the sky.

The center of the map is the zenith (straight overhead). So a star halfway from the edge of the map to the center will appear halfway from straight ahead to straight up. Ignore all the parts of the map above horizons you're not facing.

3. **Let's give it a try!** Pretend you're facing the southwest horizon (labeled "Facing SW"). Just a little way up (that is, a little way in from the edge of the map) is Sirius, the brightest star in the night sky, in the constellation Canis Major. Farther up, nearly halfway overhead, is the star Procyon in Canis Minor. Still farther up is the ringed planet Saturn. Go out at the right time, face southwest, and look up into the sky — there they are!

Tips

A couple of tips: Look for the brightest stars and constellations first; light pollution or moonlight may wash out the fainter ones. And remember that star patterns in the sky will look a lot bigger than they do here on paper.

With a map like this, you can identify celestial sights all over the sky. Go out the next clear night and make some starry friends!

You can customize a night-sky map for any time and place at SkyandTelescope.com.

PHOTO COURTESY OF ROBERT GENDLER

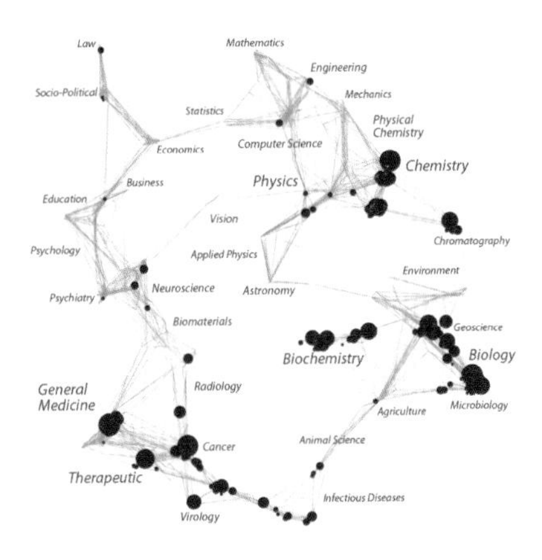

2006

HistCite Visualization of DNA Development

By Eugene Garfield, Elisha F. Hardy, Katy Börner, Ludmila Pollock, and Jan Witkowski

PHILADELPHIA, PENNSYLVANIA; BLOOMINGTON, INDIANA; AND NEW YORK, NEW YORK, 2006

Courtesy of Eugene Garfield, Thomson Reuters, Indiana University, and Cold Spring Harbor Laboratory

Aim

In 1964, Garfield and colleagues published "The Use of Citation Data in Writing the History of Science." The work opens with this question: "Can a computer write the history of science?" It is followed by this response: "Probably not in the sense usually implied." The paper presents the first attempt to write the history of science by automatic means.

Forty years later, Garfield's HistCite tool automatically generates chronological tables and historiographies of topical paper collections. It helps researchers, librarians, and others to identify core papers on a topic in question; to understand the impact of specific authors, papers, and journals; and to make sense of the history of old and new research topics. The HistCite tables support the interactive display and permit sorting of papers chronologically, by authors, journals, and citation scores. In addition, the tool helps to find papers that were missed in the initial search and gives easy access to full citation records.

Interpretation

This map compares, contrasts, and combines a modified version of the manually created citation graph published in 1964 (left) with the HistCite graph automatically generated in 2006 (right). The 1964 graph shows the network of papers and citation linkages that led to the discovery of the DNA structure published by James Watson and Francis Crick in 1953. The automatically generated graph plots major papers that cite the Watson and Crick primary paper. Nodes on the maps represent papers and are displayed in chronological order using a one-dimensional reference system. Node sizes denote the relative number of citations each core paper has received. Edges represent citation linkages. The interactive version and other HistCite examples can be found at **http://www.histcite.com**.

Eugene Garfield is chairman emeritus of the Institute for Scientific Information (ISI) of Thomson Reuters. He received an SB in chemistry and an MS in library science from Columbia University and a PhD in structural linguistics from the University of Pennsylvania. He is a pioneer in information retrieval systems and inventor of Current Contents, Index Chemicus, Science Citation Index, and Arts and Humanities Citation Index. He is an eclectic science communicator, founding publisher/editor of *The Scientist*, author of more than 1,000 articles and books, and developer of HistCite. Garfield serves on the editorial boards of *Scientometrics* and the *Journal of Information Science & Technology* and is past president of the American Society for Information Science and Technology.

Elisha F. Hardy (graphic design) **See page 158.**
Katy Börner (concept and design) **See page 132.**

Ludmila Pollock has been the executive director of Libraries and Archives at the Cold Spring Harbor Laboratory (CSHL) in New York, which is responsible for cutting-edge research in cancer, genetics, bioinformatics, and neurobiology, since 1999. Pollock has been vice president and president of the Medical and Scientific Libraries of Long Island, in 1997 and 1998, respectively, and served on the library advisory committee for the Nature Publishing Group from 2002 to 2005. She is a member of the New York Academy of Sciences, the American Association for the Advancement of Science, the Oral History Association, and the History of Science Society.

Jan Witkowski is the executive director of the Banbury Center at Cold Spring Harbor Laboratory (CSHL) and a professor at the Watson School of Biological Sciences. He obtained an SB at the University of Southampton and a PhD at the University of London. With James Watson, he coauthored the textbook *Recombinant DNA*. He also edited *Illuminating Life and Inspiring Science: Jim Watson and the Age of DNA*. Witkowski is a member of the Scientific Advisory Board of the Baker Veterinary Research Institute at Cornell University, editor in chief of *Trends in Biochemical Sciences*, and on the editorial board of *Encyclopedia of the Life Sciences*.

Writing the History of Science

In their 1964 paper, Eugene Garfield and his colleagues try to answer the question: Can a computer write the history of science? To answer this question, they selected a recent scientific breakthrough – the discovery of a structure for DNA suggesting a mechanism for its self-duplication – published by Watson & Crick in 1953.

They use Isaac Asimov's book *The Genetic Code* to identify forty milestone works that lead to the discovery as well as their interlinkages. In addition, they identify the citation linkages among those forty papers using the *1961 Science Citation Index*.

The detailed comparison of both networks demonstrates a high degree of coincidence between Asimov's account of events and the citation data, see also *Foundation* chart. They conclude that the use of citation data to write the history of science might provide a new modus operandi for the study of the history of science, research administration, and the sociology of science. Today, their HistCite™ tool generates interactive citation graphs automatically, see *Impact* chart.

Impact

Hardly a day goes by when we do not read of the gene for this or that disease, or see DNA fingerprinting on a television crime show. There is so much emphasis on the biological functions of DNA that it is easy to forget that it is a molecule, made of atoms in a particular spatial pattern. Determining the pattern of atoms in DNA was precisely what led to the double helix but the Watson and Crick 1953 paper, and the accompanying papers by Wilkins and Franklin and their colleagues, was not the end of the story. As the chart on the right shows, X-ray crystallographic studies of DNA continued for many years, and a rigorous confirmation of the structure did not come until the 1970s.

Not surprisingly, there were continuing discoveries and some surprises. One was that not all DNA was double stranded. Robert Sinsheimer found that a small bacteriophage – a virus that attacks bacteria – had a single DNA strand. Many years later, this bacteriophage played an important role when techniques were developed to sequence, to determine the order of the bases in DNA.

NODE SIZE
Number of Citations
3,000 300

A number of the papers in this chart deal with an unsuspected form of DNA. The standard structure is a right handed helix, as if the DNA chains are wrapped clockwise around a central core. Alexander Rich found that under certain circumstances and with particular sequences of the bases, DNA can assume a left-handed helical structure. The role of these left-handed sequences is till unclear but they may be involved in the regulation of gene expression.

One of the key features of the double helix was that its structure immediately suggested how the molecule could be duplicated. The two strands separate, and each acts as a template for the synthesis of a new strand, base pairing determining the order bases in the new strand. Arthur Kornberg discovered DNA polymerase, an enzyme that carried out that reaction. This was greeted at the time with great hyperbole – that life had been created in the test tube – but the enzyme plays an essential role much of the research that flows from the double helix. In the early 1970s, methods were developed for manipulating DNA and genes, and that unprecedented control over genetic material – genetic engineering – has led to a new industry, biotechnology, and to the *Human Genome Project* that holds great promise for improving human health.

LINK COLOR

Historical Links (Identified by Isaac Asimov)

→ explicit

---→ implicit

Coincident Citation Links

→ explicit

--→ implicit

Non-Coincident Citation Links

⋯⋯→ explicit

⋯⋯→ implicit

Foundation

Even the most revolutionary of scientific discoveries owes a great debt to what has gone before, and the discovery of the DNA double helix is no exception.

This chart shows major lines of scientific enquiry that contributed to Watson and Crick's insight in 1953. On the top is the line of research on the chemical analysis of proteins. Fischer was one of the great German biochemists who analyzed the basic components of proteins, amino acids. Martin and Synge developed techniques that were used by Chargaff in his analyses of DNA. The sequence of amino acids in proteins was worked out by Fred Sanger, but the impact of his work on the field of molecular genetics came after the double helix. The central line is that of genetics, beginning with Mendel, and the central citation on this line is that of Avery, Macleod and McCarty whose work established that DNA was the hereditary substance. Not shown is work by T. H. Morgan who won the Nobel Prize in 1933. Also missing is the Phage Group, founded by Max Delbruck and Salvador Luria who sought to explore the mysteries of the gene with the intellectual rigor employed by the physicists. The bottom line captures the earliest studies of the chemical nature of DNA and RNA, from the fundamental analyses of Miescher and Kossel, through the speculations of Phoebus Levene to Ernst Chargaff.

Not visible are the social interactions of scientists. Rosalind Franklin's interactions with Maurice Wilkins, Chargaff's disdain for Watson and Crick, and the rivalry between Watson and Crick and Linus Pauling, all contributed to the discovery in ways that papers and their interlinkages simply cannot reveal.

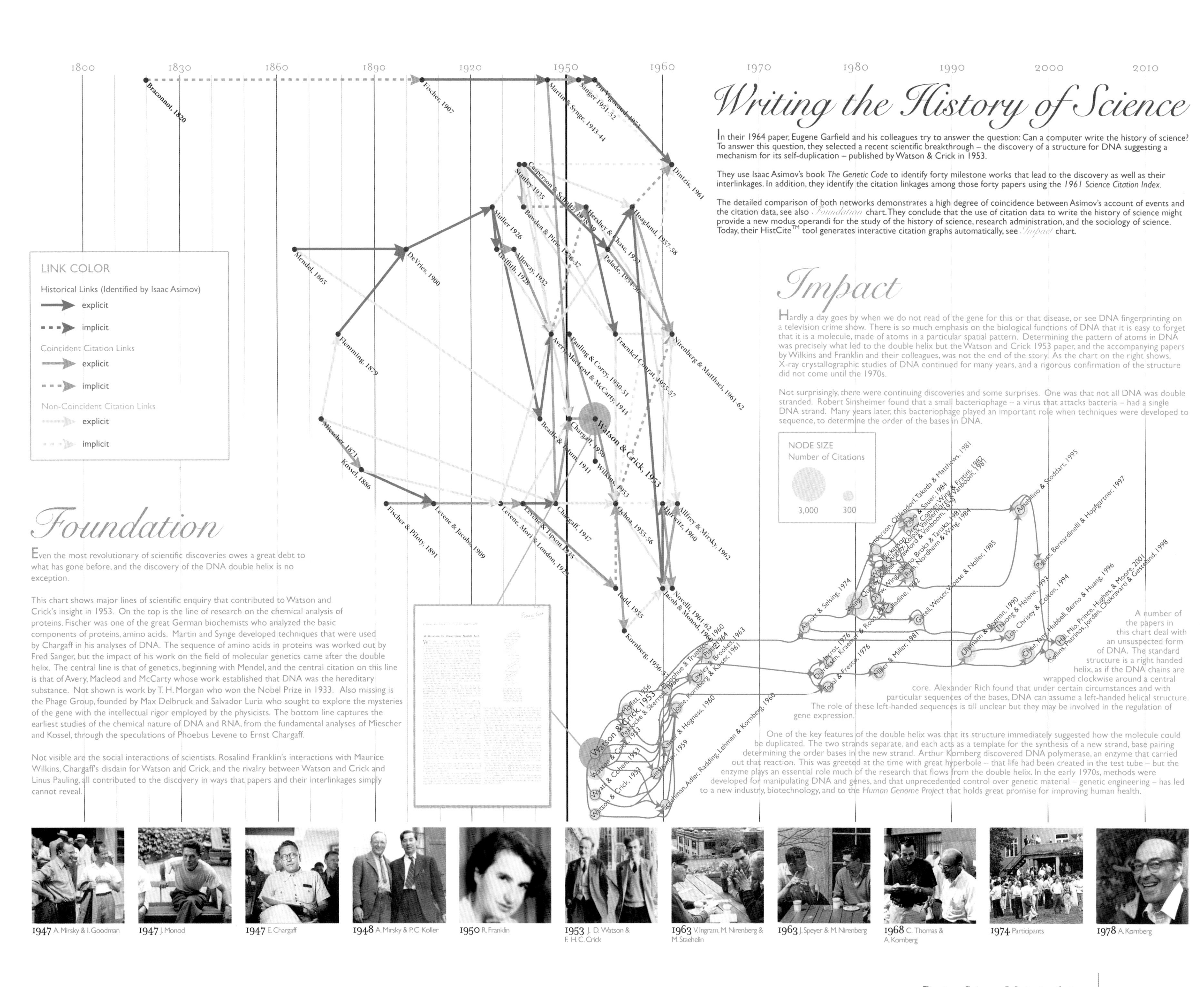

1947 A. Mirsky & I. Goodman 1947 J. Monod 1947 E. Chargaff 1948 A. Mirsky & P.C. Koller 1950 R. Franklin 1953 J. D. Watson & F. H. C. Crick 1963 V. Ingram, M. Nirenberg & M. Staehelin 1963 J. Speyer & M. Nirenberg 1968 C. Thomas & A. Kornberg 1974 Participants 1978 A. Kornberg

Manually Compiled Historiographs

This page shows the historiograph originally published in "The Use of Citation Data in Writing the History of Science" (1964) and explains its generation and modification for inclusion in the *HistCite Visualization of DNA Development* map.

Data

Garfield and colleagues were interested to understand if citation data could be used to write the history of science. They decided to use the then recent discovery of the structure of DNA as an example. Science historian Isaac Asimov's book *The Genetic Code* had just been published, describing the major scientific developments that eventually led to the laboratory duplication of the DNA-controlled process of protein synthesis. Garfield and his collaborators carefully identified the specific papers involved in the discoveries and constructed a topological network diagram of 40 milestone events as described by Asimov. Bibliographic data about the papers involved in the events were extracted from the 1963 Genetics Citation Index and the 1961 Science Citation Index.

Technique

The authors conceptualize the history of science as a chronological sequence of events in which each new discovery is dependent upon earlier discoveries. The history of science is represented as a chronologic map or topological network diagram, also called a historiograph. Garfield and his colleagues strongly argue in their papers and presentations that their work does not diminish the role of the scholar in writing the history of science. Instead, the approach and tool provide the scholar with a new modus operandi which might significantly affect future historiography.

Reference System

A one-dimensional reference system is used to organize papers and events in time, from bottom (oldest) to top (youngest). In the historiograph shown here, paper number **1**, representing **Braconnot, 1820**, is the oldest and paper number **40**, **Nirenberg/Matthaei, 1961–1962**, is the youngest. The layout of nodes with identical time stamps aims to minimize link crossings so that the layout remains readable. Two different time resolutions are possible: by publication year or by month. If a field evolves extremely quickly, then it is desirable to record and map publication data on a monthly basis or even examine dates of manuscript submission and acceptance.

Data Overlays

The networks extracted from Asimov's book and derived from the citation indexes were extensively analyzed and compared to examine the degree of coincidence between the historian's account of events and the citational relationship between these events. Special transparent overlays were prepared to communicate the result of this examination.

First, each paper was represented by a square box arranged in chronological order (see graph on right). Each box contains the nodal number, nodal author named by Asimov, and the publication year for the nodal work. Nodes are shape-coded for broad fields (see legend

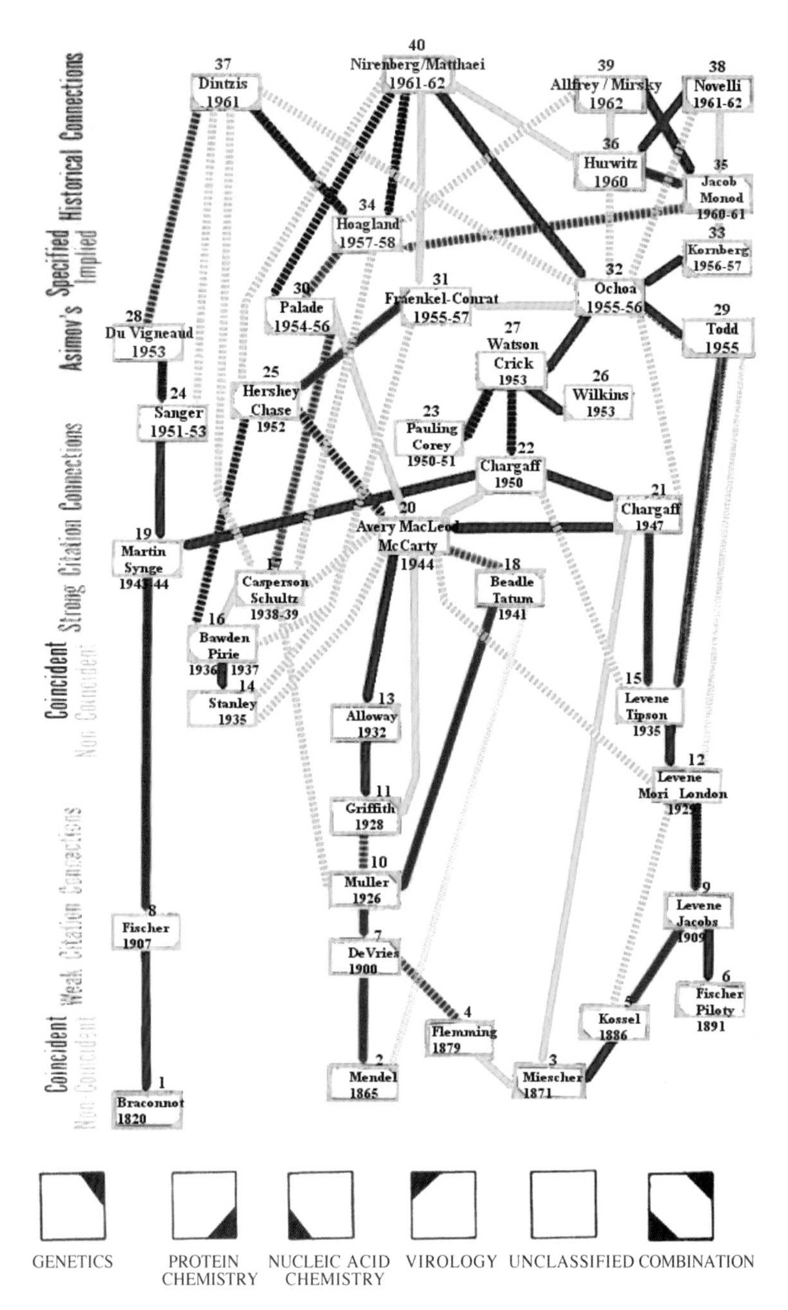

below graph). In some nodes, combinations exist; the **1944** paper by **Avery MacLeod and McCarty** with number **20, for example**, is coded both for **GENETICS** and **NUCLEIC ACID CHEMISTRY**.

Edges in the network are type- and color-coded. Solid lines indicate direct citations or strongly implied citations, that is, citations to closely related works by the same authors of the earlier paper, connection via an intermediate self-citation, or intermediate connections through any other references cited in the later paper. Dashed lines represent implied relationships whereby a nodal author refers to the work of an earlier nodal author by text description or through personal communication but not by explicit citation. Line color is used to denote the origin of linkages—red for Asimov, yellow for citation indexes—and correlations—blue representing citations that occur in Asimov and the citation data compiled for the 65 papers.

Different types of linkages are drawn on separate transparencies. When all the transparent overlays are superimposed, a complete comparative picture of both coincidence and noncoincidence of the Asimov historical network and the citation network can be observed. The edges that were not reinforced by citation connections stand out as pure red lines. The citations that coincide with Asimov's historical connections are purple (a combination of red and blue). Citation connections that are not coincident with Asimov's historical connections stand out as pure yellow lines.

The reader should keep in mind that the citation connections were established almost exclusively on the basis of nodal data, not on the basis of locating citation data from all possible sources.

The authors concluded that the citation network technique provides the scholar with a new modus operandi that can help improve the pace and accuracy of science historiography.

Modifications

The original graph was too low-resolution to be included in the exhibit map. In addition, though it places papers in the correct chronological order, the years are not evenly spaced. In adition, citation linkages start at a citing paper and point to cited reference papers. As well, in the new map, the redesigned map is higher resolution, has evenly spaced years (1800-2010, on top), and citation linkages were inverted to represent the flow of information from cited paper to citing paper. The seminal paper by Watson and Crick was highlighted. Color-coding was changed to fit the color scheme of the map. Paper node labels were redesigned. Descriptions of line color and shape encodings were simplified to meet the needs of a general audience.

Unique Features

Developed over many years, HistCite (**http://www.histcite.com**) is a software tool that automatically generates chronological tables and historiographies from searches of WoS or one of its indexes: Science Citation Index (SCI), Social Sciences Citation Index (SSCI), or Arts and Humanities Citation Index (AHCI). The WoS export files can be created based on subject, author, or journal search; by cited reference or author search; or by retrieving records by institution, countries, or a combination thereof. Export files must contain all cited references.

HistCite can be applied to identify papers imperative to the development of a topic, important papers "missed" by keyword search, most prolific and most cited authors and journals, and other keyword searches that can be used to expand the collection. It supports the analysis of publication productivity and citation rates within a collection of research papers to compar countries and institutions from which authors publish, most prolific and most cited authors within the groups, and citation statistics for groups and subgroups (such as mean and median citation rates of papers or number of authors per paper). HistCite supports the reconstruction of the history and development of a research field by emphasizing highly cited papers, important coauthor relationships, and earlier publications and documents important to the development of the author's work. It displays a timeline of the authors' publications and a graph that shows the key papers and timeline of a research field. See **http://garfield.library.upenn.edu/histcomp** for a listing of more than 300 historiographs.

Automatically Generated Historiographs

This page shows the historiographs automatically generated based on papers citing the famous 1953 Watson and Crick paper "Molecular Structure of Nucleic Acids—A Structure for Deoxyribose Nucleic Acid." The data set was retrieved from Thomson Reuters. It comprises 2,916 papers published between 1953 and 2006, by 4,781 unique authors in 1,059 journals. The papers have 113,789 outer references to papers not included in the set of 2,916 and 5,660 unique words in titles and abstracts.

Several steps are involved in the generation of the historiographs and their modification for inclusion in the *HistCite Visualization of DNA Development* map. HistCite uses the ISI format, which includes authors, titles, journals, volume pages and years, and the list of cited references. It then creates a series of tables and matrices that list all papers. By default, papers are sorted in chronological order and within each year by journal, volume, issue number, and page. The user can also sort papers by authors, journals, and citation frequencies (see tables for the *DNA Dataset*, to the right).

The tool also helps to identify missing references—in separate tables, the closest matching paper in the collection is displayed and changes are suggested.

HistCite computes the local citation score (LCS), that is, the number of times a paper is cited in a given collection, and the global citation score (GCS), that is, the number of cites in the entire SCI/SSCI as given in the times cited (TC) field of the ISI record. Users can select a threshold for LCS or GCS to reduce the number of papers to be mapped. If there are 500 source papers, then a 5 percent selection threshold would produce 25 core papers. These core papers should be of prime interest, especially to a researcher who is not familiar with the subject matter. HistCite was applied to parse the ISI-formatted *DNA Dataset* file and to produce the tables at right, which are also available at **http://garfield.library.upenn.edu/histcomp/watson-nature-1953**.

The historiograph view lays out papers chronologically, interlinked by their citations. The oldest paper is placed at the top, the youngest at the bottom. Annual or monthly (issue) resolution is supported. Using a GCS >= 360 (max GCS=2,822), the citation graph consists of 35 nodes and 64 links. Year/month labels are shown at left. Each paper is represented by a circle that is size-coded according to the number of citations it has received (GCS if a GCS map was requested, and LCS otherwise). Each circle is labeled by a paper identification number, and respective papers are listed in the lower right corner. When viewing the graph online, double-clicking a node brings up its complete source record.

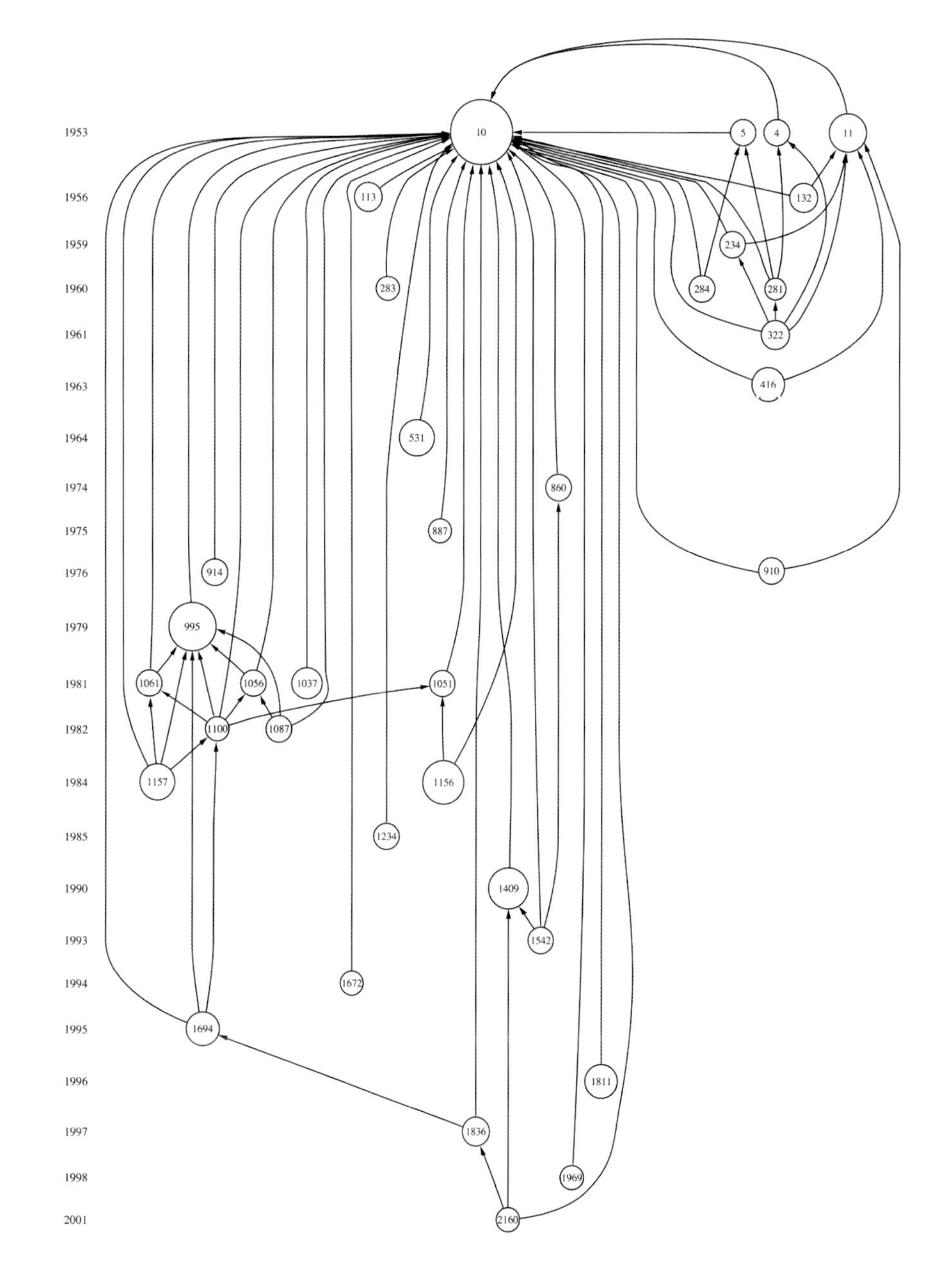

4 WYATT & COHEN, 1953
5 WATSON & CRICK, 1953
10 WATSON & CRICK, 1953
11 WATSON & CRICK, 1953
113 MOFFITT, 1956
132 PEACOCKE & SKERRETT, 1956
234 SINSHEIMER, 1959
281 SCHACHMAN et al., 1960
283 DONOHUE & TRUEBLOOD, 1960
284 KAISER & HOGNESS, 1960
322 JOSSE et al., 1961
416 LAWLEY & BROOKES, 1963
531 PLATT, 1964
860 ARNOTT & SELSING, 1974
887 DIERCKSEN et al., 1975
910 TOPAL & FRESCO, 1976
914 JACROT, 1976
995 WANG et al., 1979
1037 MILLER & MILLER, 1981
1051 ANDERSON et al., 1981
1056 DREW et al., 1981
1061 WANG et al., 1981
1087 CALLADINE, 1982
1100 DICKERSON et al., 1982
1156 PABO & SAUER, 1984
1157 RICH et al., 1984
1234 GUTELL et al., 1985
1409 UHLMANN & PEYMAN, 1990
1542 THUONG & HELENE, 1993
1672 LEE et al., 1994
1694 AMABILINO & STODDART, 1995
1811 CHEE et al., 1996
1836 PIGUET et al., 1997
1969 COLLINS et al., 1998
2160 HILL et al., 2001

Nodes: 2916. Authors: 4781. Journals: 1059. Outer References: 113789. Words: 5660
Collection span: 1953 - 2006
View: Overview. Sorted by **GCS**. Page **1** of 30: [1 2 3 4 5 6 7 8 9 10] 11 21

#	LCR	NCR	Node / Date / Journal / Author	LCS	GCS
1	0	7	10 1953 NATURE 171 (4356): 737-738 WATSON JD; CRICK FHC *MOLECULAR STRUCTURE OF NUCLEIC ACIDS - A STRUCTURE FOR DEOXYRIBOSE NUCLEIC ACID*	2822	2822
2	5	38	995 1979 NATURE 282 (5740): 680-686 WANG AHJ; QUIGLEY GJ; KOLPAK FJ; CRAWFORD JL; VANBOOM JH; et al. *MOLECULAR-STRUCTURE OF A LEFT-HANDED DOUBLE HELICAL DNA FRAGMENT AT ATOMIC RESOLUTION*	136	1544
3	5	96	1156 1984 ANNUAL REVIEW OF BIOCHEMISTRY 53: 293-321 PABO CO; SAUER RT *PROTEIN-DNA RECOGNITION*	4	1321
4	4	456	1409 1990 CHEMICAL REVIEWS 90 (4): 543-584 UHLMANN E; PEYMAN A *ANTISENSE OLIGONUCLEOTIDES - A NEW THERAPEUTIC PRINCIPLE*	29	1109
5	1	11	11 1953 NATURE 171 (4361): 964-967 WATSON JD; CRICK FHC *GENETICAL IMPLICATIONS OF THE STRUCTURE OF DEOXYRIBONUCLEIC ACID*	463	1036
6	9	218	1157 1984 ANNUAL REVIEW OF BIOCHEMISTRY 53: 791-846 RICH A; NORDHEIM A; WANG AHJ *THE CHEMISTRY AND BIOLOGY OF LEFT-HANDED Z-DNA*	26	861
7	1	29	331 1964 SCIENCE 146 (364): 347-& PLATT JR *STRONG INFERENCE - CERTAIN SYSTEMATIC METHODS OF SCIENTIFIC THINKING MAY PRODUCE MUCH MORE RAPID PROGRESS THAN OTHERS*	2	855
8	1	31	1811 1996 SCIENCE 274 (5287): 610-614 Chee M; Yang R; Hubbell E; Berno A; Huang XC; et al. *Accessing genetic information with high-density DNA arrays*	4	804
9	4	33	416 1963 BIOCHEMICAL JOURNAL 89 (1): 127-& LAWLEY PD; BROOKES P *FURTHER STUDIES ON ALKYLATION OF NUCLEIC ACIDS AND THEIR CONSTITUENT NUCLEOTIDES*	7	743
10	19	712	1694 1995 CHEMICAL REVIEWS 95 (8): 2725-2828 Amabilino DB; Stoddart JF *Interlocked and intertwined structures and superstructures*	7	735

Total: 4781, Nodes: 2916
View: Overview. Sorted by **GCS**. Page **1** of 24: [1 2 3 4 5

#	Name	TLCS	TGCS	Pubs
1	WATSON JD	3628	5081	11
2	CRICK FHC	3577	4951	7
3	Rich A	355	4077	22
4	WANG AHJ	197	2903	4
5	VANBOOM JH	181	2301	6
6	KOLPAK FJ	171	2042	2
7	QUIGLEY GJ	171	2042	2
8	VANDERMAREL G	171	2042	2
9	KORNBERG A	135	1579	12

Total: 5660, Nodes: 2916, Word count: 18405, All words count: 27704
View: Overview. Sorted by **number of publications (Pubs)**. Page **1** of 2

#	Word	Pubs	Percent	TLCS	TGCS
1	DNA	553	19.0	1179	18052
2	STRUCTURE	230	7.9	4545	13929
3	MOLECULAR	193	6.6	3268	7366
4	ACID	188	6.4	4395	13773
5	NUCLEIC	141	4.8	3445	9282
6	ACIDS	108	3.7	3320	8195
7	SYNTHESIS	99	3.4	285	4242
8	BASE	98	3.4	354	3111
9	DOUBLE	93	3.2	395	4368
10	GENETIC	92	3.2	54	2252

Total: 1059, Nodes: 2916
View: Overview. Sorted by **number of publications (Pubs)**. Page **1** of 6: [1 2 3 4 5 6]

#	Name	TLCS	TGCS	Pubs
1	NATURE	4125	11385	109
2	JOURNAL OF THE AMERICAN CHEMICAL SOCIETY	311	4661	93
3	JOURNAL OF MOLECULAR BIOLOGY	458	7501	68
4	PROCEEDINGS OF THE NATIONAL ACADEMY OF SCIENCES OF THE UNITED STATES OF AMERICA	448	3763	59
5	BIOCHIMICA ET BIOPHYSICA ACTA	169	1822	49
6	SCIENCE	220	5860	46
7	BIOPOLYMERS	64	880	44
8	JOURNAL OF THEORETICAL BIOLOGY	39	582	36
9	BIOCHEMISTRY	77	1445	33
10	JOURNAL OF BIOLOGICAL CHEMISTRY	141	2431	30

Nodes: 2916
Page **1** of 30: [1 2 3 4 5 6 7 8 9 10] 11 21

cited nodes	LCR	NCR	Nodes	LCS	GCS	citing nodes
10	1	20	1 1953 FRANKLIN RE	57	268	2 40 44 48 51 72 74 136 151 159 161 230 285 333 386 469 548 646 677 678 690 720 722 739 757 810 858 944 976 1004 1041 1062 1089 1144 1193 1261 1270 1276 1334 1349 1373 1390 1423 1431 1510 1769 2104 2192 2341 2397 2403 2477 2547 2594 2664 2677 2714
1 10 12	3	11	2 1953 FRANKLIN RE	14	81	40 44 48 72 136 159 161 162 678 690 939 1025 1062 1482
10	1	7	3 1953 SMITH CL	1	5	51
10	1	35	4 1953 WYATT GR	31	453	32 51 86 90 162 172 187 202 215 262 269 281 290 294 304 309 322 357 363 370 377 419 473 481 485 597 643 699 991 2440 2788
10	1	25	5 1953 WATSON JD	83	452	20 42 43 51 67 69 79 80 87 88 93 104 125 130 136 140 164 181 197 198 201 215 216 248 255 262 265 267 274 281 284 294 298 302 310 328 356 398 457 462 466 473 501 523 566 569 595 610 649 667 710 733 747 748 790 798 799 834 941 1050 1055 1129 1238 1246 1295 1313 1455 1580 1595 1614 1661 1766 1770 1810 1824 2047 2395 2404 2493 2664 2719 2729 2838
10	1	3	6 1953 WYATT GR	4	16	17 20 51 90
10	1	36	7 1953 LARK KG	2	55	20 90
10	1	7	8 1953 STENT GS	4	18	20 42 69 115
10 13	2	3	9 1953 REY LR	0	1	

Modifications

For the exhibit map, the HistCite output was redrawn to arrive at an equal spacing of years. It was turned 90 degrees to fit the chronological reference system. Citation linkages were inverted to represent the flow of information from cited paper to citing paper. Node IDs were replaced by the author names and publication years of the respective papers. Node size was added manually, as it was not supported in the version of HistCite available to us. The central paper by Watson and Crick was highlighted. Color-coding was changed to fit the color scheme of the map.

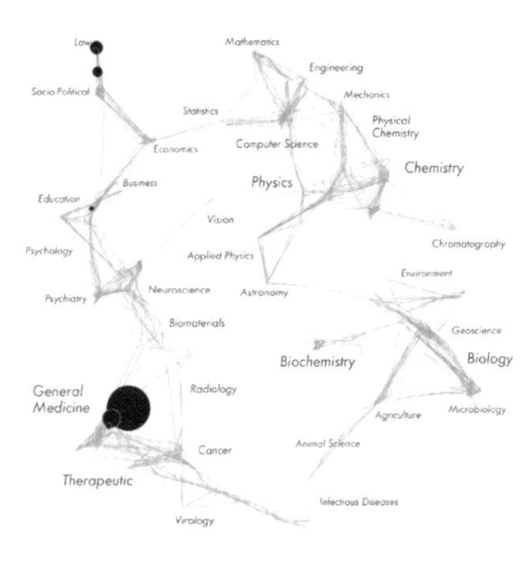

2006

History Flow Visualization of the Wikipedia Entry "Abortion"

By Martin Wattenberg and Fernanda B. Viégas

CAMBRIDGE, MASSACHUSETTS, 2006

Courtesy of Martin Wattenberg, Fernanda B. Viégas, and IBM Research

Aim

The History Flow visualization technique aims to chart the evolution of a document as it is edited by many people. It shows broad trends in revision histories while preserving details for closer examination. It is here applied to show the evolution of Wikipedia entries created by people all over the world. It might also be useful to study other collaborative situations, such as patterns of communication, conflict, and contributions in large-scale software development.

Interpretation

This map shows the edit history of the Wikipedia entry "Abortion." The left column lists all authors who contributed to the entry with their assigned color codes. The right column shows the final version of the entry as of April 20, 2003, at 5:32 pm. The text is color-coded according to the author of the final edit. The middle column gives the History Flow visualization. Each vertical line represents a version of the entry—from December 2001 to June 2003. The total length of the line reflects the length of the entry. Line color-coding indicates which author has edited which part(s) of the text. White to gray represent the contributions of anonymous authors. Entry versions are sorted in time, from left to right. As can be seen, the page has gone through many changes over time. Note that the entry survived two complete deletions that happened in December 2002 and in February 2003.

Martin Wattenberg

received an SM in mathematics from Stanford University in 1992 and a PhD in mathematics from the University of California, Berkeley, in 1996. Martin was the director of Research and Development at SmartMoney.com, where he designed Internet-based financial software. In 2002, he joined IBM Watson Research Center in Cambridge, Massachusetts, where he now leads the Visual Communication Lab. Wattenberg is known both for his scientific and applied work in the field of information visualization and for his information-based digital art. Wattenberg's work has been exhibited at Ars Electronica, Linz; the Institute of Contemporary Arts, London; the New Museum, New York; the Whitney Museum of American Art, New York; and at galleries and festivals internationally. Commissions include work created for Ars Electronica, the NASA Art Program, New Radio and Performing Arts, the Smithsonian National Museum of American History, the Walker Art Center, and the Whitney Museum of American Art. *Technology Review* named Wattenberg "one of the world's 100 top young innovators."

Fernanda B. Viégas

received master's and PhD degrees in Media Arts and Sciences from MIT. She joined the Visual Communication Lab at IBM in 2005. Her research focuses on the social side of visualization, exploring representations of online communities to support online identity, collective memory, and storytelling. Her current research interests include visualizing social dynamics in open-source communities, exploring collaborative uses of visualization applications, and investigating online privacy as it applies to visualization. Her projects have been exhibited at the Institute for Contemporary Art in Boston and in galleries in New York and Los Angeles.

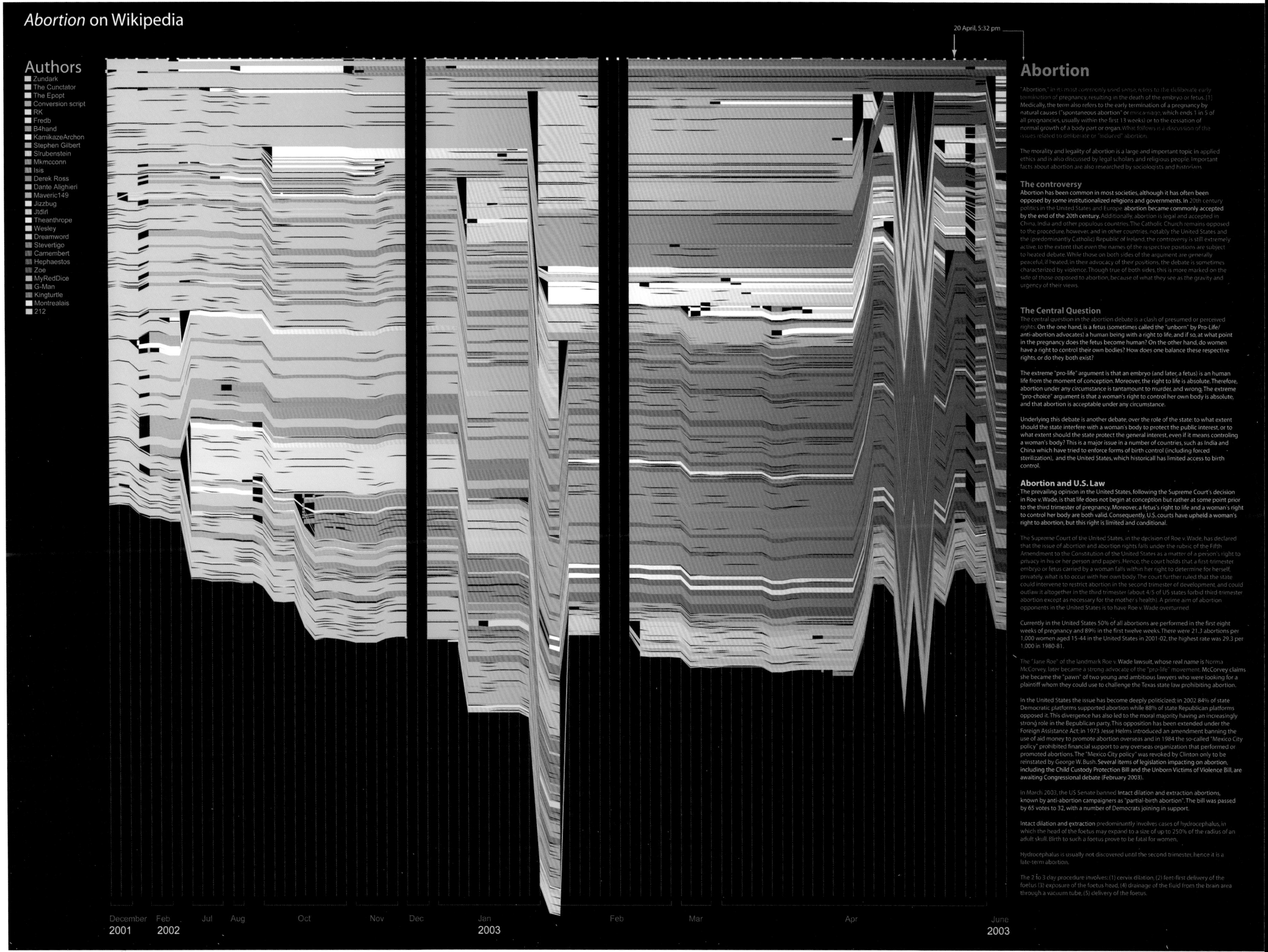

Abortion on Wikipedia

Authors
- Zundark
- The Cunctator
- The Epopt
- Conversion script
- RK
- Fredb
- B4hand
- KamikazeArchon
- Stephen Gilbert
- Slrubenstein
- Mkmcconn
- Isis
- Derek Ross
- Dante Alighieri
- Maveric149
- Jizzbug
- Jtdirl
- Theanthrope
- Wesley
- Dreamword
- Stevertigo
- Camembert
- Hephaestos
- Zoe
- MyRedDice
- G-Man
- Kingturtle
- Montrealais
- 212

20 April, 5:32 pm

December 2001 | Feb 2002 | Jul | Aug | Oct | Nov | Dec | Jan 2003 | Feb | Mar | Apr | June 2003

Abortion

"Abortion," in its most commonly used sense, refers to the deliberate early termination of pregnancy, resulting in the death of the embryo or fetus.[1] Medically, the term also refers to the early termination of a pregnancy by natural causes ("spontaneous abortion" or miscarriage, which ends 1 in 5 of all pregnancies, usually within the first 13 weeks) or to the cessation of normal growth of a body part or organ. What follows is a discussion of the issues related to deliberate or "induced" abortion.

The morality and legality of abortion is a large and important topic in applied ethics and is also discussed by legal scholars and religious people. Important facts about abortion are also researched by sociologists and historians.

The controversy

Abortion has been common in most societies, although it has often been opposed by some institutionalized religions and governments. In 20th century politics in the United States and Europe, abortion became commonly accepted by the end of the 20th century. Additionally, abortion is legal and accepted in China, India and other populous countries. The Catholic Church remains opposed to the procedure, however, and in other countries, notably the United States and the (predominantly Catholic) Republic of Ireland, the controversy is still extremely active, to the extent that even the names of the respective positions are subject to heated debate. While those on both sides of the argument are generally peaceful, if heated, in their advocacy of their positions, the debate is sometimes characterized by violence. Though true of both sides, this is more marked on the side of those opposed to abortion, because of what they see as the gravity and urgency of their views.

The Central Question

The central question in the abortion debate is a clash of presumed or perceived rights. On the one hand, is a fetus (sometimes called the "unborn" by Pro-Life/anti-abortion advocates) a human being with a right to life, and if so, at what point in the pregnancy does the fetus become human? On the other hand, do women have a right to control their own bodies? How does one balance these respective rights, or do they both exist?

The extreme "pro-life" argument is that an embryo (and later, a fetus) is an human life from the moment of conception. Moreover, the right to life is absolute. Therefore, abortion under any circumstance is tantamount to murder, and wrong. The extreme "pro-choice" argument is that a woman's right to control her own body is absolute, and that abortion is acceptable under any circumstance.

Underlying this debate is another debate, over the role of the state: to what extent should the state interfere with a woman's body to protect the public interest, or to what extent should the state protect the general interest, even if it means controling a woman's body? This is a major issue in a number of countries, such as India and China which have tried to enforce forms of birth control (including forced sterilization), and the United States, which historicall has limited access to birth control.

Abortion and U.S. Law

The prevailing opinion in the United States, following the Supreme Court's decision in Roe v. Wade, is that life does not begin at conception but rather at some point prior to the third trimester of pregnancy. Moreover, a fetus's right to life and a woman's right to control her body are both valid. Consequently, U.S. courts have upheld a woman's right to abortion, but this right is limited and conditional.

The Supreme Court of the United States, in the decision of Roe v. Wade, has declared that the issue of abortion and abortion rights falls under the rubric of the Fifth Amendment to the Constitution of the United States as a matter of a person's right to privacy in his or her person and papers. Hence, the court holds that a first-trimester embryo or fetus carried by a woman falls within her right to determine for herself, privately, what is to occur with her own body. The court further ruled that the state could intervene to restrict abortion in the second trimester of development, and could outlaw it altogether in the third trimester (about 4/5 of US states forbid third-trimester abortion except as necessary for the mother's health). A prime aim of abortion opponents in the United States is to have Roe v. Wade overturned.

Currently in the United States 50% of all abortions are performed in the first eight weeks of pregnancy and 89% in the first twelve weeks. There were 21.3 abortions per 1,000 women aged 15-44 in the United States in 2001-02, the highest rate was 29.3 per 1,000 in 1980-81.

The "Jane Roe" of the landmark Roe v. Wade lawsuit, whose real name is Norma McCorvey, later became a strong advocate of the "pro-life" movement. McCorvey claims she became the "pawn" of two young and ambitious lawyers who were looking for a plaintiff whom they could use to challenge the Texas state law prohibiting abortion.

In the United States the issue has become deeply politicized; in 2002 84% of state Democratic platforms supported abortion while 88% of state Republican platforms opposed it. This divergence has also led to the moral majority having an increasingly strong role in the Republican party. This opposition has been extended under the Foreign Assistance Act; in 1973 Jesse Helms introduced an amendment banning the use of aid money to promote abortion overseas and in 1984 the so-called "Mexico City policy" prohibited financial support to any overseas organization that performed or promoted abortions. The "Mexico City policy" was revoked by Clinton only to be reinstated by George W. Bush. Several items of legislation impacting on abortion, including the Child Custody Protection Bill and the Unborn Victims of Violence Bill, are awaiting Congressional debate (February 2003).

In March 2003, the US Senate banned Intact dilation and extraction abortions, known by anti-abortion campaigners as "partial-birth abortion". The bill was passed by 65 votes to 32, with a number of Democrats joining in support.

Intact dilation and extraction predominantly involves cases of hydrocephalus, in which the head of the foetus may expand to a size of up to 250% of the radius of an adult skull. Birth to such a foetus prove to be fatal for women.

Hydrocephalus is usually not discovered until the second trimester, hence it is a late-term abortion.

The 2 to 3 day procedure involves: (1) cervix dilation, (2) feet-first delivery of the foetus (3) exposure of the foetus head, (4) drainage of the fluid from the brain area through a vacuum tube, (5) delivery of the foetus.

Data

The Wikimedia Foundation generously makes public a complete copy of all Wikimedia wikis at **http://download.wikipedia.org.** The data used here was downloaded in May 2003. It comprises 130,596 content pages from the English Wikipedia together with their revision histories, spanning an editing time of two years—from January 15, 2001 (the launch date of Wikipedia.org) to May 2003. Note that each revision entry is stored with a time stamp and the username of its author (or the author's IP address if the author has chosen to remain anonymous).

Reference System

The history flow visualization uses a two-dimensional reference system of time versus number of "sentences" (see below). Time runs from left to right. Each modification to a content page is represented by a vertical bar. Content page length runs top to bottom and corresponds to the number of "sentences" in a page, defined as pieces of text delimited by periods or HTML tags. Note that tiny changes, such as the correction of one letter or the insertion of a comma, show up as a change to the entire sentence.

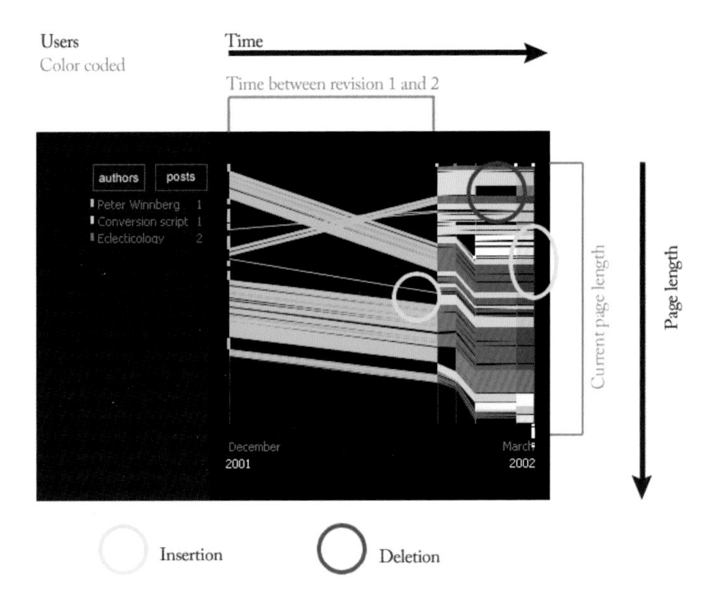

Data Overlays

The example above shows all eight versions of the Wikipedia page "IBM" from December 2001 to March 2002. The page had been collaboratively edited by three authors in eight different instances. Each author is represented by a different color. The color scheme used is shown on the left side together with the number of posts per author. White to gray represent the contributions of anonymous authors. The length of each colored line indicates the amount of text that the respective author has written or revised. Text sections that have been kept the same between consecutive revisions are visually interlinked by connections—essentially shaded areas in the colors assigned to authors. The width of a connection corresponds to the length of the text attributed to that author. Insertions (yellow ovals) and deletions (red oval) correspond to unconnected line segments–"gaps" in the visualization that can be easily spotted. Movements of text blocks cause crossing line segments. As can be seen, most of the activity happened in February and March 2002.

Details

Finding matching sections of two document revisions is a well-studied problem in computer science, with many possible solutions. History flow uses a simple algorithm by Heckel to match up text tokens; this produces adequate results with fair efficiency.

Unique Features

The interactive history flow tool is written in Java 1.4 and runs as a standalone program. It has four main visualization modes that can be selected at the top of the tool (see below). (1) **Group** view shows all contributions from different authors, color-coding the text to indicate the author of each sentence. (2) **Individual** view highlights the contributions of a single author over time. (3) **Text changes** view shows the new content in each version of the wiki page, independent of authorship. Portions of the text that have been edited the most can be easily identified. (4) **Text age** view uses a grayscale gradient from white (brand-new contribution) to dark gray (very old contribution) to represent the persistence of different contributions. In addition, a user can toggle between **SPACING** by **date** or by **versions**. The latter places revisions in an equidistant fashion.

Growth

History flow visualizations show that almost all pages in Wikipedia change continually over time. Some grow gradually while other pages develop in bursts (see Wikipedia page **"Iraq"** below). Note that the mass deletion in May 2003 was not repaired for many days.

Vandalism

The history flow visualization can be applied to understand the frequency and timing of acts of vandalism, which are rather common in publicly editable sites such as Wikipedia.

Interestingly, malicious edits are often repaired immediately, such that most users never see them. The two visualizations in the top row, opposite page, show a different view of the Wikipedia page **"Abortion."** On the left, revisions are plotted to be equidistant. The vertical black interruptions indicate times when an author deleted most of the page. On the right, the horizontal spacing of revisions corresponds to the time a revision was made. As the vandalism was repaired almost instantaneously, malicious and correcting edits show up on top of each other and cannot be detected at the zoom level shown on the right. The visualizations can be used to study to what extent anonymity affects the likelihood of vandalism.

Persistence

The overall stability of a page in terms of size and content is important in assessing the reliability of group-authored Web sites. History flow visualizations show that the text of some Wikipedia pages persists for a long time while other pages are revised frequently. Similarly, text contributed by specific authors can be short or long lived. One example is the Wikipedia page "Islam," in which white portions of the visualization represent text contributed by a single author, named RX (see opposite page, bottom row). Note that the revisions are spaced equidistant by "versions."

"Iraq": Growth

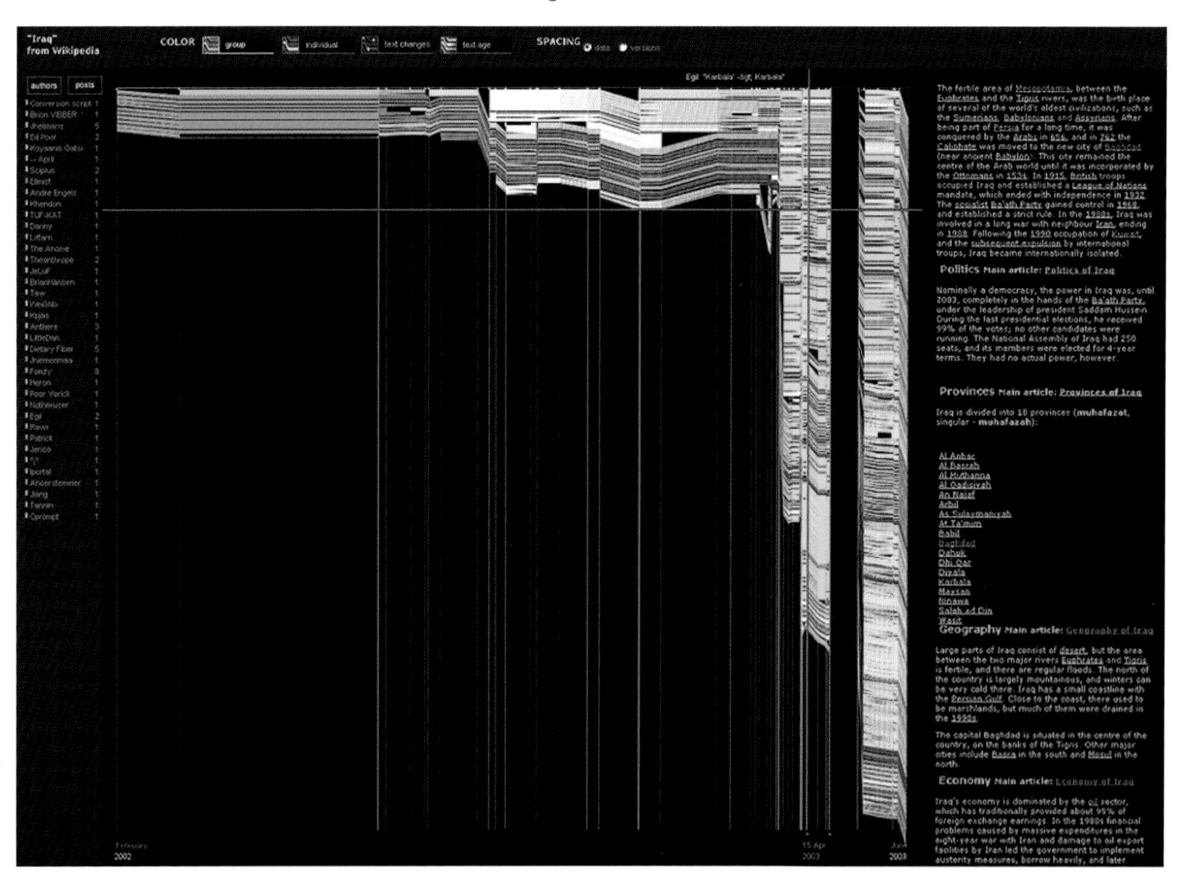

"Abortion": Vandalism

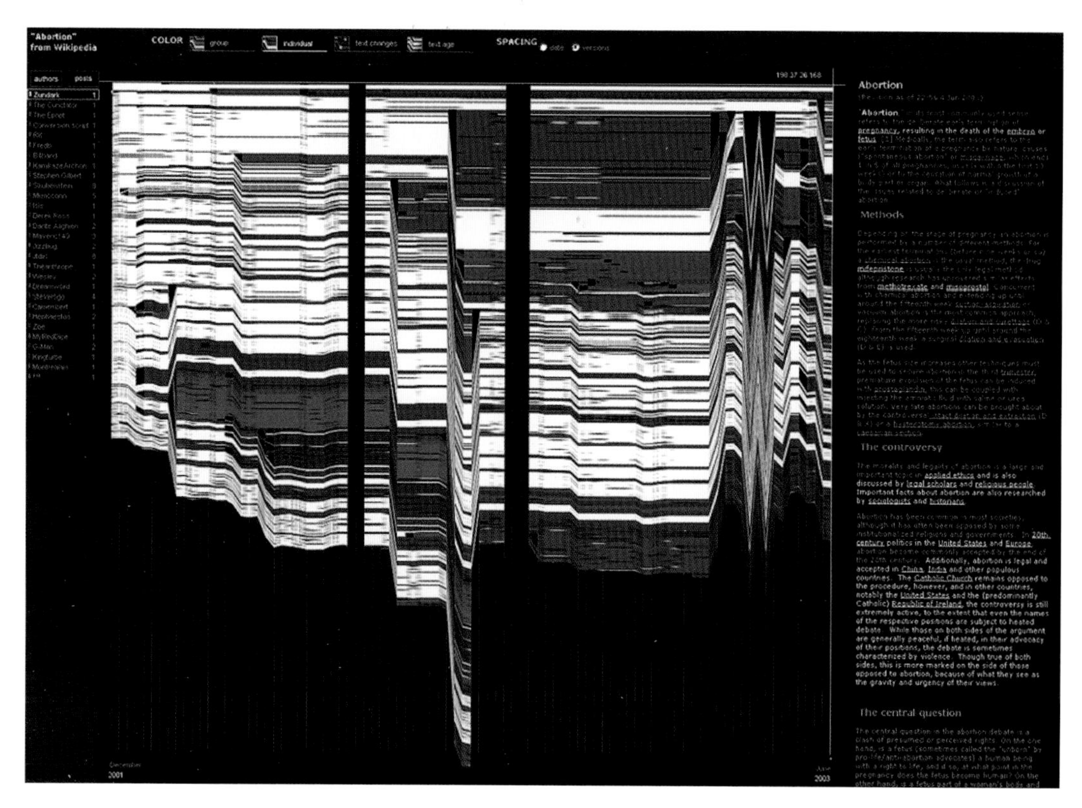

Equal Distance Spacing

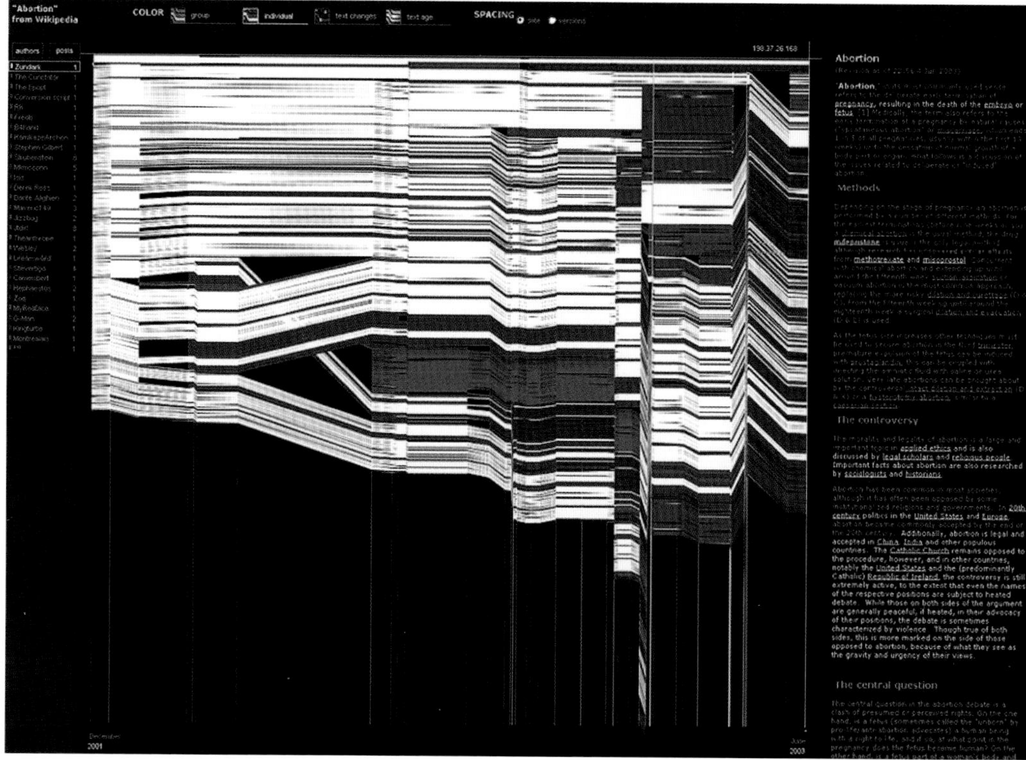

Date Spacing

"Islam": Persistence

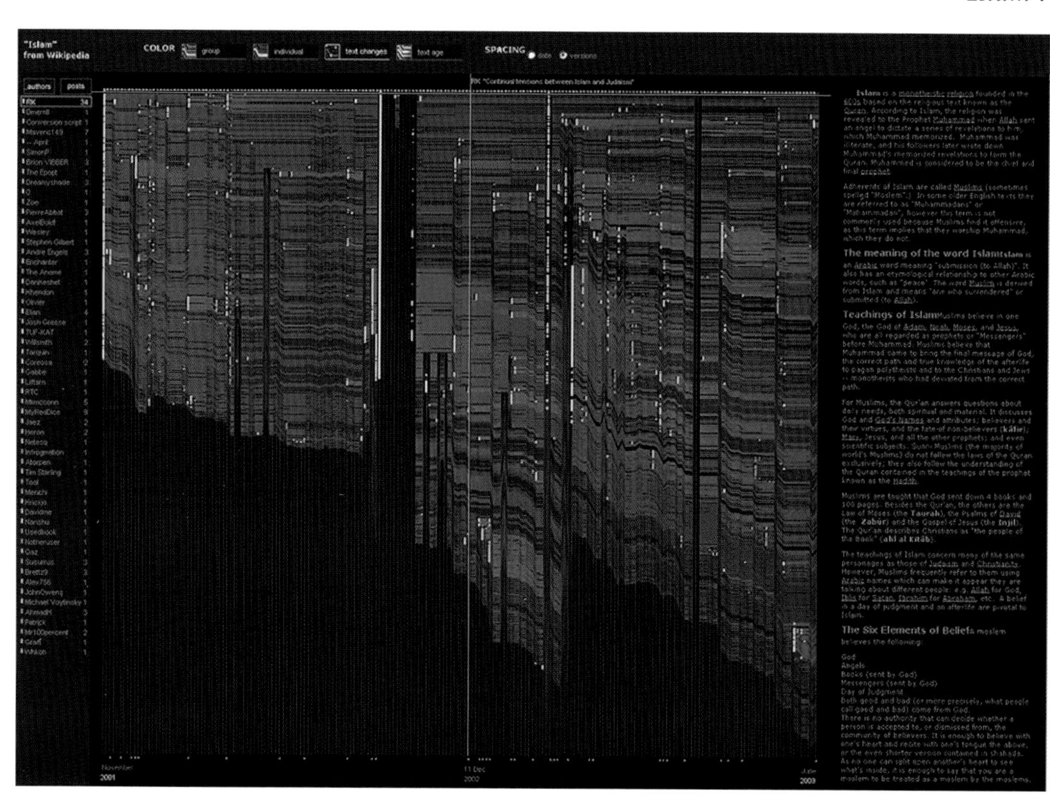

User Color Coded

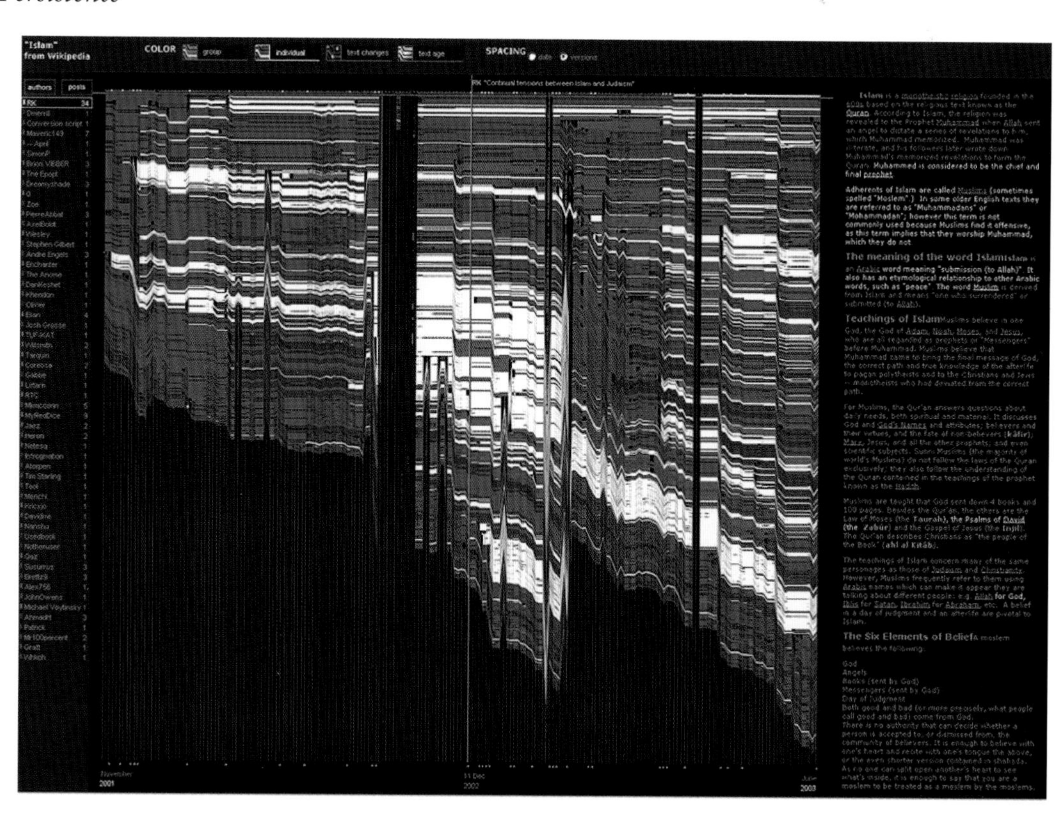

User RX Highlighted

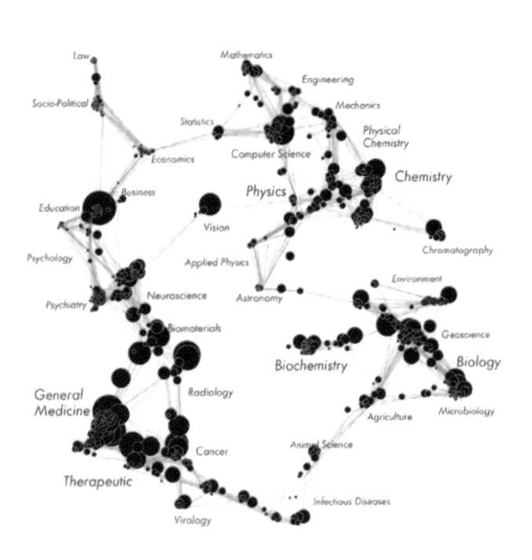

2006

TextArc Visualization of The History of Science

By W. Bradford Paley
NEW YORK, NEW YORK 2006
Courtesy of W. Bradford Paley

Aim

Imagine you are given a manuscript with thousands of lines, and you need to make sense of it in very little time. Is there any way that you could possibly understand the main characters that drive the story, their relationships and interactions, and the topics covered without having sufficient time to read the book? Is there any way to visually analyze the structure of a text and gauge the professionalism of the writer? TextArc was designed to address not only these needs, but also to map the history of science.

Interpretation

This map shows a TextArc visualization of four volumes of a book by Henry Smith Williams, *A History of Science*. The intelligent organization of the books and the culturally recognized way of organizing science quickly become apparent. The history's first two volumes are organized in a strictly chronological fashion, so as the book wraps from 12:00 to 6:00 around the right side of the ellipse, it is organized as a timeline. The next two volumes distinguish two major domains—making two timelines—of more recent scientific exploration: the physical sciences from 6:00 to 9:00 and the life sciences from 9:00 back to 12:00. Since the scattered words are pulled toward the places where they are used in the text (see the map itself for a better description of the layout), a particular kind of structure emerges: names of individuals that are mentioned but once or twice appear along the outer boundaries, while frequently cited concepts that are common to science of all eras—for example, **system**, **theory**, and **experiment**—are drawn to the center. Even more interesting is that the main subjects of focus for certain areas are neither near the specific edges nor the general center, but in a local, topical band between the two: **mind**, **knowledge**, and **conception** during the philosophic beginnings of science; **moon**, **earth**, **sun**, and **stars** at a later time; **electricity**, **light**, and **natural forces** in the recent physical sciences; and **animals**, **disease**, **development**, and **brain** in the recent life sciences, for example.

W. Bradford Paley uses computers to create visual displays with the goal of making readable, clear, and engaging expressions of complex data. His visual representations are inspired by the calm, richly layered information in natural scenes. He applies rendering methods used by fine artists and graphic artists, informed by their possible underpinnings in human perception, and uses them to create narrowly scoped, almost idiosyncratic representations whose visual semantics are often driven by the real-world metaphors of the experts who know the domains best.

Paley rendered his first computer graphics in 1973; graduated Phi Beta Kappa from the University of California, Berkeley, in 1981; founded Digital Image Design Incorporated (**http://www.didi.com**) in 1982; and started creating financial and statistical data visualization in 1986. He has exhibited at the Museum of Modern Art, is in the Artport permanent collection of the Whitney Museum of American Art, and has received multiple grants and awards for both art and design. He recently finished redesigning the interface at the center of the New York Stock Exchange trading floor. He is an adjunct associate professor at Columbia University and is director of Information Esthetics (**http://www.informationesthetics.org**), a fledgling interdisciplinary group exploring the creation and interpretation of data representations that are both readable and aesthetically satisfying.

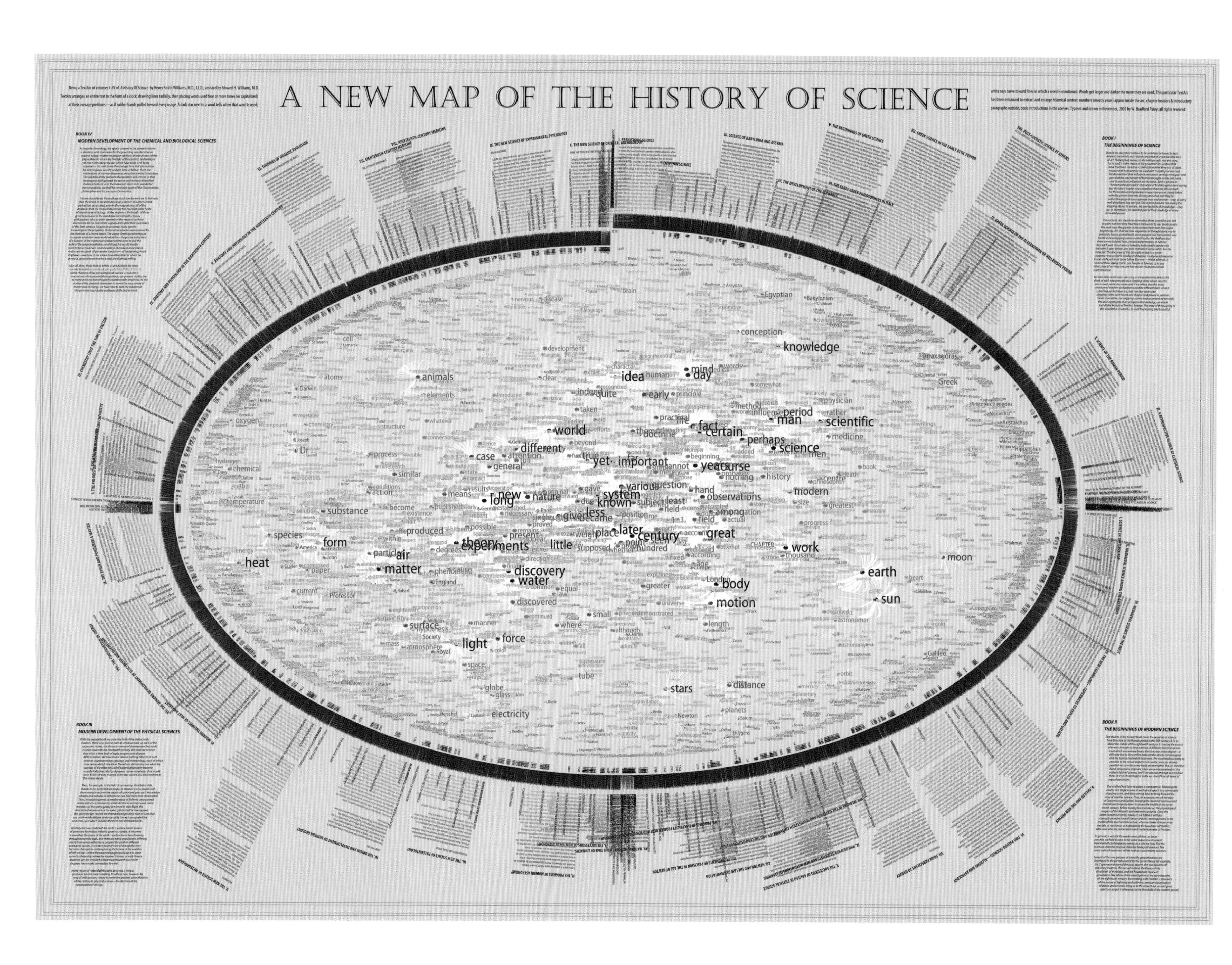

Being a TextArc of volumes I–IV of A History Of Science by Henry Smith Williams, M.D., LL.D.; assisted by Edward H. Williams, M.D. TextArc arranges an entire text in the form of a clock: drawing lines radially, then placing words used four or more times (or capitalized) at their average positions—as if rubber bands pulled toward every usage. A dark star next to a word tells when that word is used;
A NEW MAP OF THE HISTORY OF SCIENCE
white rays curve toward lines in which a word is mentioned. Words get larger and darker the more they are used. This particular TextArc has been enhanced to extract and enlarge historical context: numbers (mostly years) appear inside the arc, chapter headers & introductory paragraphs outside, book introductions in the corners. Typeset and drawn in November, 2005 by W. Bradford Paley; all rights reserved
BOOK IV
MODERN DEVELOPMENT OF THE CHEMICAL AND BIOLOGICAL SCIENCES
BOOK I
THE BEGINNINGS OF SCIENCE
BOOK III
MODERN DEVELOPMENT OF THE PHYSICAL SCIENCES
BOOK II
THE BEGINNINGS OF MODERN SCIENCE
knowledge
idea
mind
day
early
world
different
fact
certain
man
period
scientific
perhaps
science
yet
important
course
year
case
general
new
long
nature
system
known
various
observations
modern
less
great
later
century
work
theory
experiments
little
form
air
matter
heat
substance
species
discovery
water
body
motion
earth
sun
moon
light
force
surface
stars
distance
electricity
animals
conception

Data

The e-text of *A History of Science* volumes 1–4 by Henry Smith Williams was downloaded from Project Gutenberg (**http://www.gutenberg.org**; see **page 18, Michael S. Hart**). The volumes are entitled *The Beginnings of Science*, *The Beginnings of Modern Science*, *Modern Development of the Physical Sciences*, and *Modern Development of the Chemical and Biological Sciences* and each has about 80,000 words.

Technique

Paley engages with his primarily expert clients to obtain a deep understanding of their needs and wants. Equipped with comprehensive knowledge about their workflow and priorities, he designs software that harnesses the tremendous capacity of today's computers—their ability to quickly and interactively render large amounts of data—to augment his clients' intellect. TextArc is one of Paley's many creations. It renders large amounts of text to function as an index, concordance, and summary combined. The TextArc applet (at **http://textarc.org**) quickly engages the viewer's eyes and mind in a journey to uncover meaning.

TextArc connects directly to Project Gutenberg, creating a unique interface to works such as *Alice in Wonderland*, *Hamlet*, and thousands of others. Not all texts are equally legible in TextArc. Smaller texts lay out particularly nicely (one is advised to search by number of lines in the text: simply enter "1000-2000" in the "Line Count" field). The stemming of words (the reduction of *playing*, *playful*, or *player* to *play*, for example) and the listing of common (and therefore dispensable) words or phrases are available only for English. Lewis Carroll's *Alice in Wonderland,* shown with black background at right, will be used to explain TextArc's interactive capabilities, which are not preserved in the static map of *The History of Science*.

Reference System

TextArc represents each line in a text as arranged in an arc, stepping clockwise, starting and ending at 12:00. The text lines might be drawn horizontally, as in the *Alice in Wonderland* visualization to the right, or radially, as in the map of *The History of Science*. The arc creates a circular reference system in which words can be placed according to their occurrence in the original text.

Data Overlays

TextArc uses statistical natural language processing and computational linguistics approaches to stem, stop-word, and analyze text for easy human interpretation. The result is rendered in a way that helps people discover patterns and concepts in any text by leveraging a powerful, underused resource: human visual processing. For example, more frequent words are rendered in brighter colors, leveraging our preattentive processing capabilities. Note that key characters like Alice, the Mad Hatter, the King and Queen of Hearts, and the Gryphon stand out in the spiral text arc, as do other words evocative of the story, such as **poor, dear, door,** and **little.** Important typographic features, like the mouse-tail shape of a poem at about 2:00 can be seen because the tiny lines retain their formatting.

All words that appear more than once in the text are drawn at their average position inside the spirals. Imagine each word attached to its place in the spiral by a tiny rubber band (drawn in yellow in the images below); if the word appears in two places, two rubber bands are attached. The net result of this rubber band tug-of-war is that a word will appear closer to places where it is used more often. Hence, words draw attention to where they appear in the document.

Distribution information is revealed when the analyst points at a word: its "rubber band" rays become visible, linking it to every place it appears in the text. On the left below, we see that **Alice** appears evenly throughout the story, whereas **Gryphon** appears in the last quarter—at 9:00 (see images to the right).

Clicking a word selects it, leaving its rays visible when the cursor moves away. A text view can show every line that uses that word. Selecting one word and pointing at another shows where words are collocated and how they interleave.

A curved line connects the words in the order in which they appear in the text, showing how the TextArc space relates to the original linear space of the text.

Coupled windows ensure that **Rabbit,** referring to the White Rabbit, is highlighted in the arc and an overlay full-text window. Lines containing **Rabbit** are drawn in green around the arc, in the text window, and even in the scrollbar.

A concordance shows how many times each word is used. Words can be looked up in a thesaurus and drawn in red.

The online interactive version of the *Alice in Wonderland* visualization (**http://textarc.org/Alice.html**) reads the entire text to the viewer. Whenever a major word is read, its major relationships to other words are activated and animated.

Details

The TextArc visualization of *A History of Science* is a static rendering of four book volumes. A rendering of the complete text produces a black circle. Numbers appearing in the text are plotted along the inner side of the circle, providing a chronological ordering and a sense of passing time. Outside the circle, the prefaces of all the books are given beside the respective texts, filling each of the four corners. Their wavy text alignment indicates their similarity in function. The title and first paragraph of each chapter are plotted in concentric lines parallel to the outer spiral. Subheadings radiate outward like sunbeams. The beginning and end of each volume has a higher number of subheadings, emphasizing the 12:00, 3:00, 6:00, and 9:00 division of the circle. The inner workings of the circle are most interesting. Just like in the *Alice in Wonderland* rendering, the position of words reflects their usage in the text. The type size denotes the number of occurrences. Words that start with an uppercase letter—mostly names of people and places—are shown in red. White feather patterns indicate the number of times each word occurs and approximate placement of words in the original text. A closer examination reveals which topics of science were studied by whom, where, and at what time. It also highlights the topics that are used throughout the text, which are central to all scientific endeavors, such as **experiment**, **discovery**, and **fact**.

The TextArc visualization also works on a macro level: topics observed from a distance look like a sacred land of wisdom or unreachable in clouds that merely suggest their shape.

Unique Features

Intermediate renderings of an earlier version of the TextArc visualization of *A History of Science* are shown on the right, revealing the sequence in which the postscript file was composed.

Alice in Wonderland

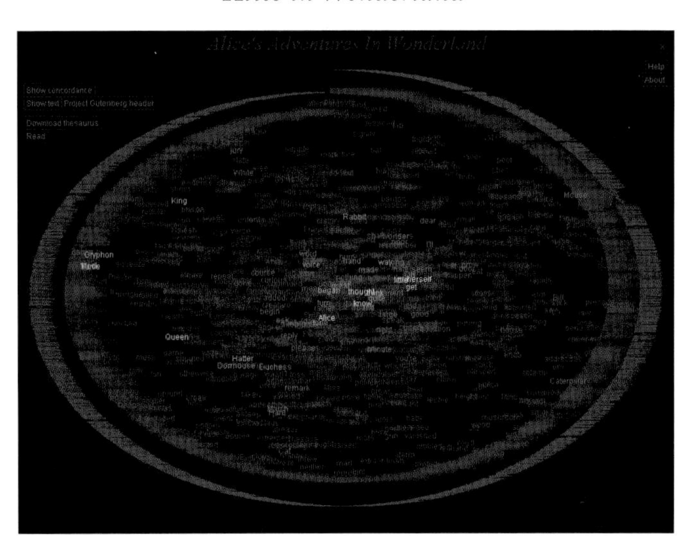

Size and color coding of words

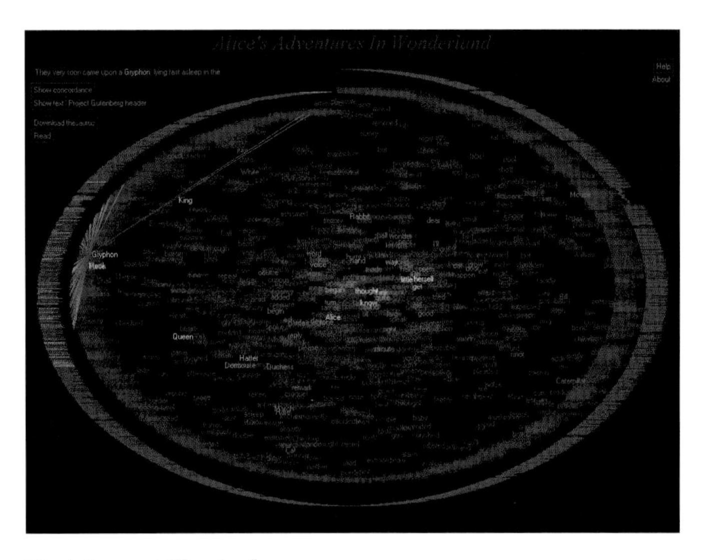

Word placement: "Gryphon"

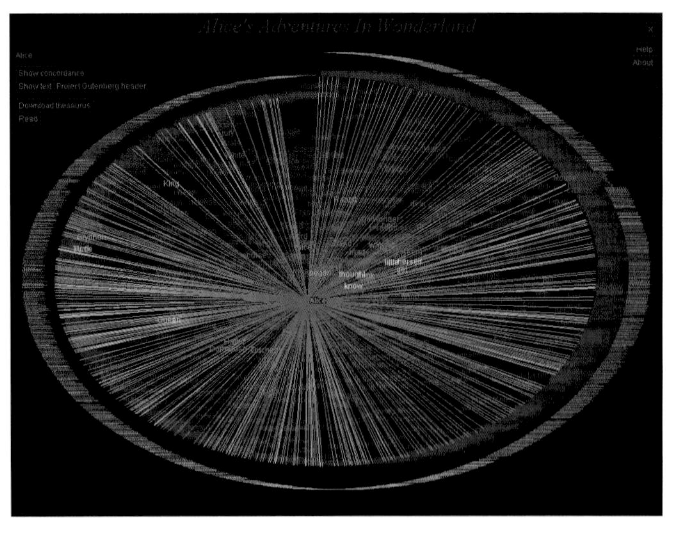

Word placement: "Alice"

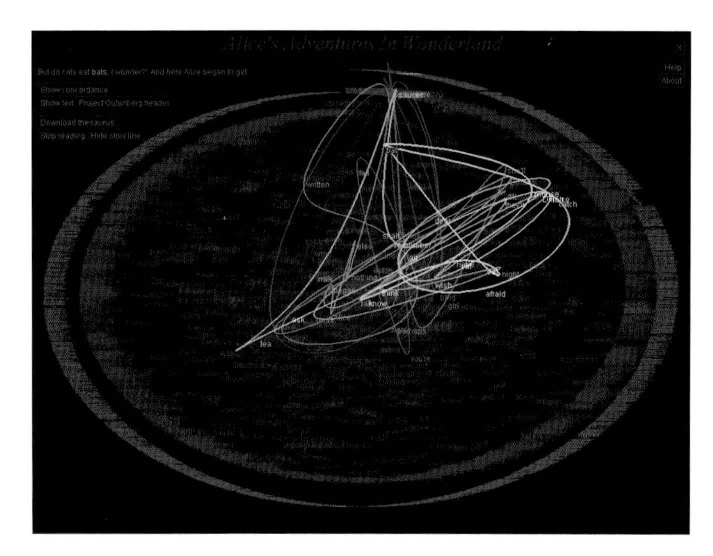

Word sequence indicated by curved line

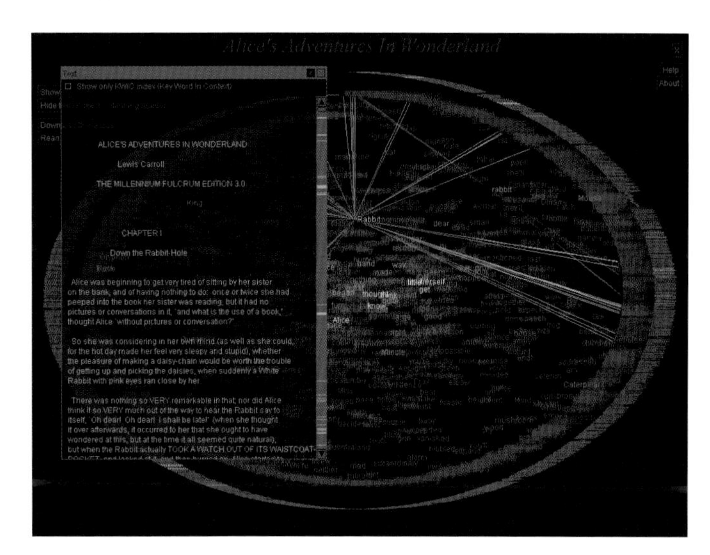

Coupled windows showing "Rabbit"

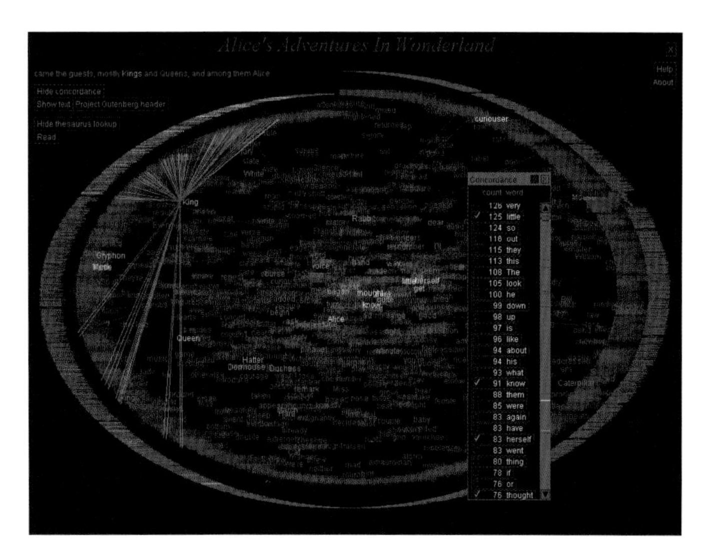

Concordance with word counts

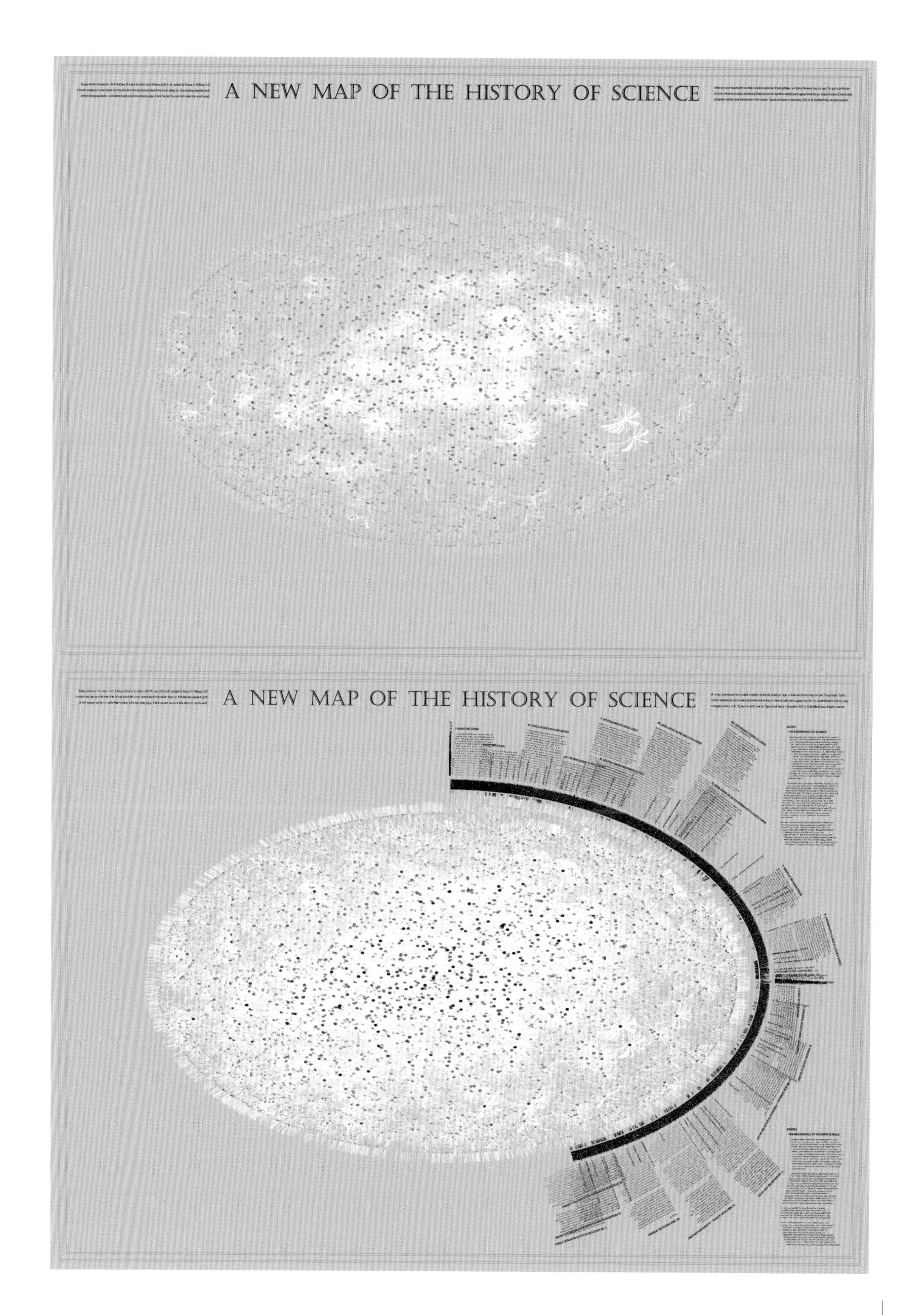

2006

Taxonomy Visualization of Patent Data

By Katy Börner, Elisha F. Hardy, Bruce W. Herr II, Todd M. Holloway, and W. Bradford Paley

BLOOMINGTON, INDIANA, AND NEW YORK, NEW YORK, 2006

Courtesy of Indiana University and W. Bradford Paley

Aim

Taxonomies, classification hierarchies, ontologies, and controlled vocabularies help us deal with the information flood. We use directory structures to organize our files and organizational hierarchies to structure our work environment. We hold on to those manmade structures of highly abstract knowledge spaces when conquering the sea of corporate and private data. However, because of their size and complexity, the design and continuous update of organizational schemas often proves too challenging. The taxonomy visualization and validation tool (TV) supports the semiautomatic validation and optimization of organization schemas imposed on a data set as a means of ordering, structuring, and naming. By showing the "goodness of fit" of a schema potentially capable of organizing millions of items, the TV eases the identification and reclassification of misclassified information entities, the identification of classes that grew overproportionally, the evaluation of the size and homogeneity of existing classes, and the examination of the "well-formedness" of an organization schema.

Interpretation

The TV was applied to visualize the U.S. Patent and Trademark Office (USPTO) patent classification. The fabriclike pattern shows a listing of the first three levels of the patent hierarchy in 25 columns. It can be seen as a 1.5-dimensional reference system that captures the main structure of this complex information space. Next to each patent class are bar graphs that show the goodness of fit of all patents organized into this class.

Exemplarily overlaid are two patents. GORE-TEX—the lightweight, durable synthetic fiber used as tissue filler in cosmetic implants, waterproof clothing, and many other products—shows the impact a patent might have. The Gold Nanoshell patent was selected to show the prior art of a patent. The cover pages of both patents and their position in the classification hierarchy are shown. Line overlays represent citation linkages. Red lines denote 182 citations of the GORE-TEX patent; they are sorted in time with dark red indicating older and bright red younger citations. Blue lines represent the 16 prior art references of the Gold Nanoshell patent to the classes of the cited patents.

Katy Börner (concept and design) holds an SM in engineering in electronics from the University of Technology, Leipzig, and a PhD in computer science from the University of Kaiserslautern, Germany. In 1999, she joined Indiana University, where she is Victor H. Yngve Professor of Information Science at the School of Library and Information Science, adjunct professor in the School of Informatics and Computing and at the Department of Statistics in the College of Arts and Sciences, core faculty of Cognitive Science, research affiliate of the Biocomplexity Institute, fellow of the Center for Research on Learning and Technology, member of the Advanced Visualization Laboratory, and founding director of the Cyberinfrastructure for Network Science Center. She coedited *Visual Interfaces to Digital Libraries* (Springer-Verlag) and a special *PNAS* issue, *Mapping Knowledge Domains*. She has authored more than 130 papers. Börner directs diverse projects related to the large-scale analysis, modeling, and visualization of scientific data sets that are funded by NSF, NIH, and private foundations.

Elisha F. Hardy (design) See page 158.

Bruce W. Herr II (programming) See page 166.

Todd M. Holloway (programming) was a PhD candidate in computer science at Indiana University. He holds a BA from Grinnell College and an SM from Indiana University, both in computer science. His research interests include data mining, case-based reasoning, information retrieval, personalization, and information visualization.

W. Bradford Paley (concept and design) See page 128.

The US Patent Hierarchy

Impact

The United States Patent and Trademark Office does scientists and industry a great service by granting patents to protect inventions. Inventions are categorized in a taxonomy that groups patents by industry or use, proximate function, effect or product, and structure. At the time of this writing there are 160,523 categories in a hierarchy that goes 15 levels deep. We display the first three levels (13,529 categories) at right in what might be considered a textual map of inventions.

Patent applications are required to be unique and non-obvious, partially by revealing any previous patents that might be similar in nature or provide a foundation for the current invention. In this way we can trace the impact of a single patent, seeing how many patents and categories it affects.

The patent on Goretex—a lightweight, durable synthetic fiber—is an example of one that has had significant impact. The box below enlarges the section of the hierarchy where it is filed, and the red lines (arranged to start along a time line from 1981 to 2006) point to the 130 categories that contain 182 patents, from waterproof clothing to surgical cosmetic implants, that mention Goretex as "prior art."

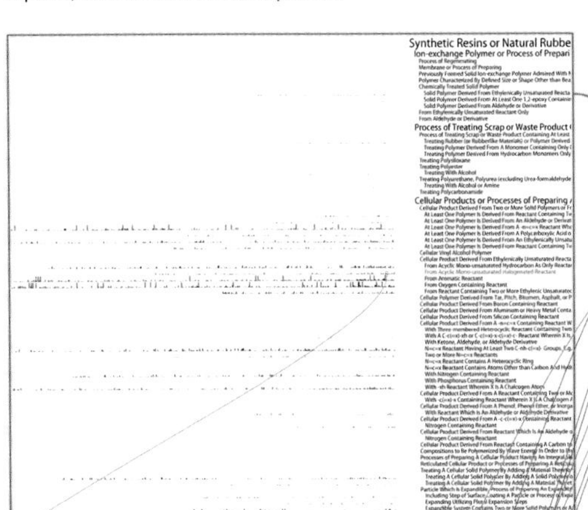

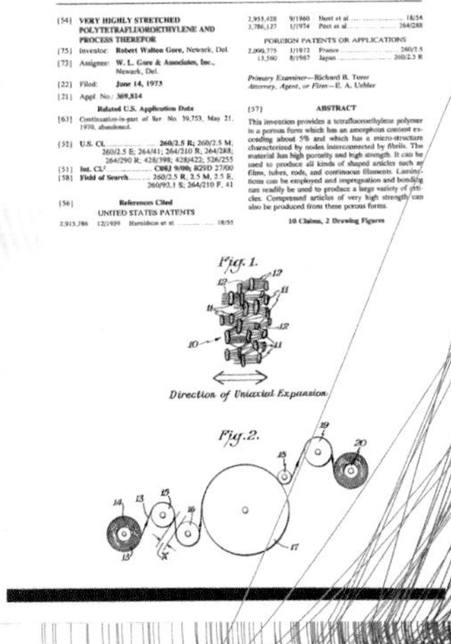

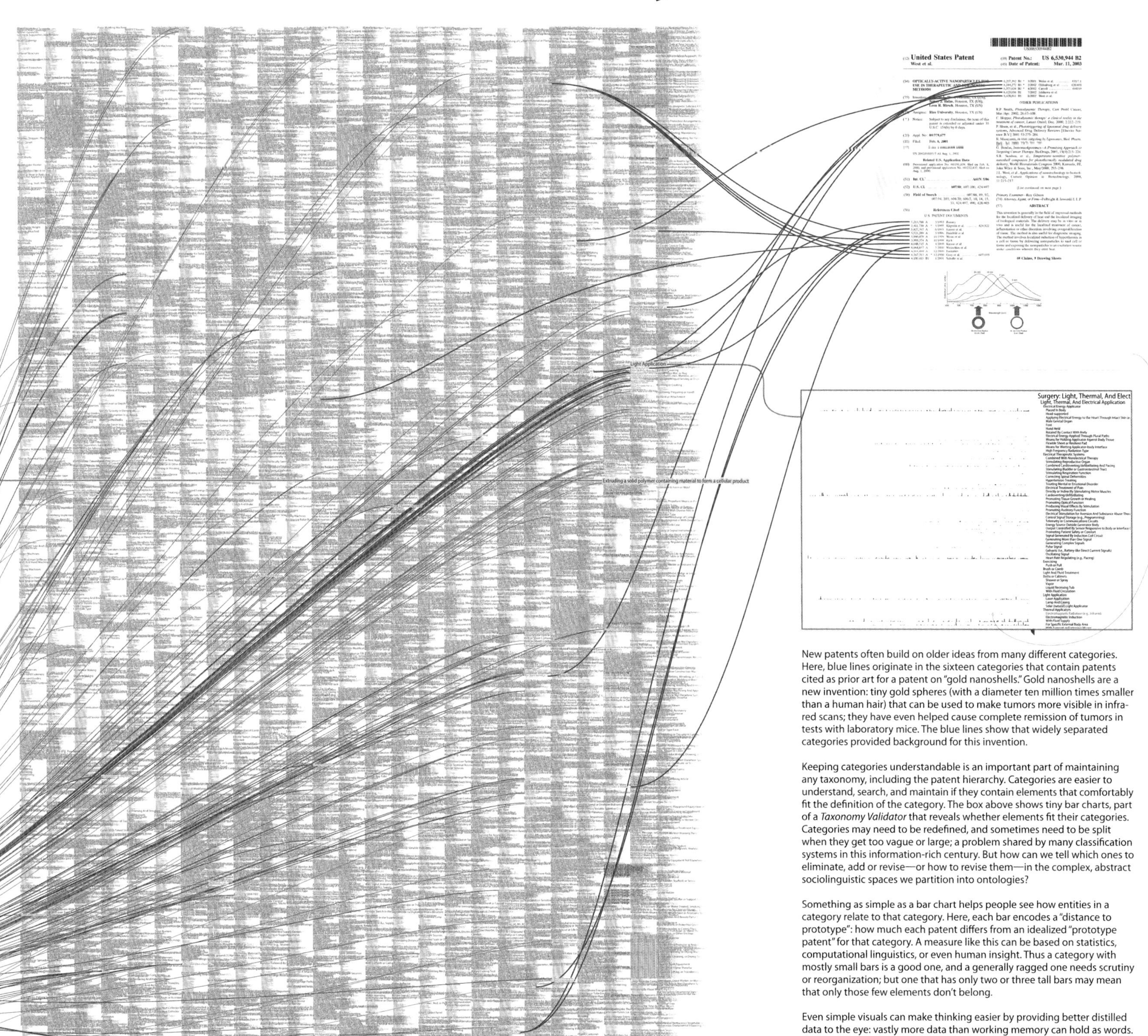

Prior Art

New patents often build on older ideas from many different categories. Here, blue lines originate in the sixteen categories that contain patents cited as prior art for a patent on "gold nanoshells." Gold nanoshells are a new invention: tiny gold spheres (with a diameter ten million times smaller than a human hair) that can be used to make tumors more visible in infrared scans; they have even helped cause complete remission of tumors in tests with laboratory mice. The blue lines show that widely separated categories provided background for this invention.

Keeping categories understandable is an important part of maintaining any taxonomy, including the patent hierarchy. Categories are easier to understand, search, and maintain if they contain elements that comfortably fit the definition of the category. The box above shows tiny bar charts, part of a *Taxonomy Validator* that reveals whether elements fit their categories. Categories may need to be redefined, and sometimes need to be split when they get too vague or large; a problem shared by many classification systems in this information-rich century. But how can we tell which ones to eliminate, add or revise—or how to revise them—in the complex, abstract sociolinguistic spaces we partition into ontologies?

Something as simple as a bar chart helps people see how entities in a category relate to that category. Here, each bar encodes a "distance to prototype": how much each patent differs from an idealized "prototype patent" for that category. A measure like this can be based on statistics, computational linguistics, or even human insight. Thus a category with mostly small bars is a good one, and a generally ragged one needs scrutiny or reorganization; but one that has only two or three tall bars may mean that only those few elements don't belong.

Even simple visuals can make thinking easier by providing better distilled data to the eye: vastly more data than working memory can hold as words. They focus people on exactly the right issues, and support them with the comprehensive overviews they need to make more informed judgements.

Data

Patent data was downloaded from the USPTO and imported into a PostgreSQL database—which consists of patents issued from January 1, 1976, to January 1, 2005, including patent abstracts, classifications, authors, and prior art references, among other information. The data set comprises 3,173,537 patents and 160,523 distinct patent classes, organized into a hierarchy up to 15 levels deep.

Reference System

An organizational schema could be rendered as a tree or as an indented list. The latter is analogous to a table of contents or a file directory structure, where node depth in the hierarchy is indicated by the amount of indenting. In the figure below, circles represent organizational nodes, rectangles organizational labels. Black-filled circles and rectangles indicate the root nodes. Gray- and white-filled nodes denote intermediate and leaf nodes, respectively. Both representations quickly reveal the structure of a hierarchy. However, the labels in the tree representation occlude each other—particularly if many nodes share the same level. Node labels are easy to read in the indented list representation, and this representation is used in the TV.

Tree Structure *Indented List Representation*

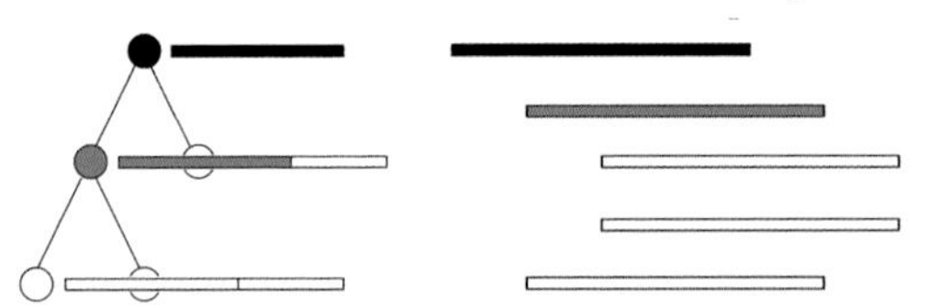

The original plan was to print the complete hierarchy—all 160,523 classes, up to 15 levels deep. However, printing only 100,000 classes using 6-point type and 1-point line spacing (7 points or 4 millimeters space per line) results in a list of 400,000 millimeters (1,312 feet) length. Hence, only the first three levels of the hierarchy—13,529 categories—were plotted, using rather small type. Specifically, 7-point type is used for level 1, 3.5-point type indented by 1.5 points for level 2, and 1-point type indented by 3 points for level 3. To render all these categories, 25 columns were needed. The result is the fabric-like pattern shown. The area can be seen as a 1.5-dimensional reference system that captures the main structure of this complex information space.

Data Overlays

Given a reference system, diverse data elements can be made visible. For example, all classes in which a certain inventor, company, or country has patents can be highlighted. Item interrelations (for example, patent citations) and class interrelations (for example, similarity according to number of shared classifications) can be denoted by line overlays. The number and goodness of fit of all patents in each patent class can be visually represented (using bar graphs, for example; see figure below and **Zoom into Bar Graphs** to the right).

Bar Graphs *Indented List Representation*

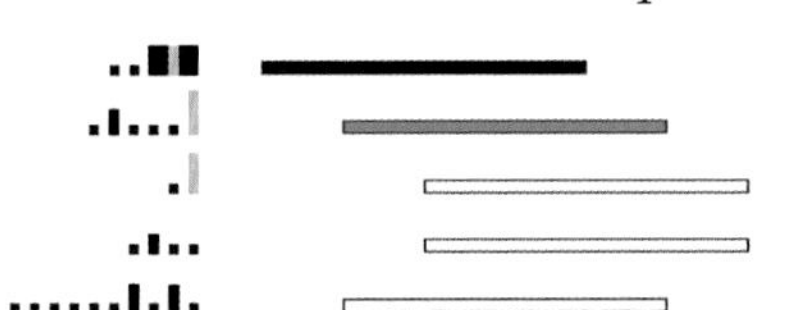

Zoom into Bar Graphs

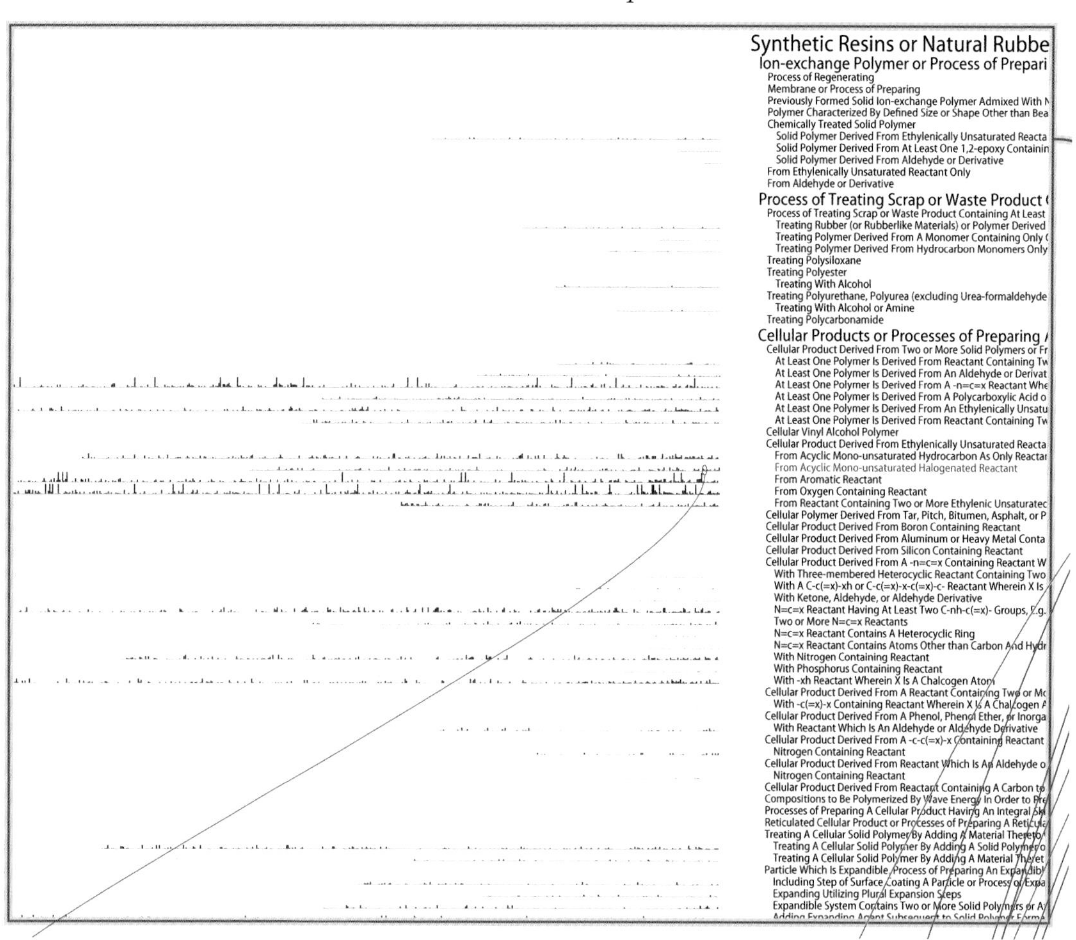

The TV depicts each patent, represented as a bar on the left side of the class to which it belongs. The resulting bar graphs can be sorted by item properties such as age, citation count, or distance to class prototype. The height of the bars can be used to represent item properties.

In the *Taxonomy Visualization of Patent Data* map, the height of each patent bar represents its similarity to the patent prototype of its patent class. The prototype of a class (PTc) is defined to be a feature vector of length equal to the number of unique references occurring across all patents in the class. Each feature corresponds to the prior probability of a reference occurring in that class, that is, the number of patents that have this reference divided by the total number of patents in this class. A patent is represented as a feature vector *P* of the same length as the prototype of its class. It is given binary features indicating the presence or absence of a reference. For each patent feature vector, the similarity between it and its class prototype is computed using the cosine similarity measure that equals the dot product of the two vectors divided by the product of the magnitudes of the two vectors.

To ensure that outliers would be easily visible (stick out from the crowd), similarity is converted into distance and patents that do not fit well in a class (that is, they have a large distance from their prototype) are represented by a high bar. Patents with a small distance get a low bar. Final bar heights are linearly scaled to fit between 0 and 4 millimeters.

Details

The reference system was used to show the impact and prior art of two patents.

The patent for GORE-TEX was selected to show the impact a patent might have. The Gold Nanoshell patent was selected to show the prior art of a patent. Gold nanoshells are a new type of optically tunable nanoparticles. Their ability to "tune" to a desired wavelength is critical to in-vivo therapeutic applications such as thermal tumor destruction, wound closure, tissue repair, and disease diagnosis.

For both patents, close-ups of their position in the 25-column classification hierarchy are shown on the opposite page. The zoom above shows the bar graphs next to each class that indicate how many patents are in that class and the similarities they share. Collectively, the bars show the "goodness of fit" between the hierarchy and the patents it organizes.

Blue line overlays interlink 16 prior art references of the Gold Nanoshell patent to the classes of the cited patents (see zoom on opposite page). Red lines denote 182 citations of the GORE-TEX patent. The latter are sorted in time, with dark red indicating older and bright red younger citations.

In the interactive mode, the TV can be queried for item or class properties. Matching entities (for example, all new items or all classes that contain items of a certain type) are highlighted. Bars can be selected to retrieve more item details on demand.

Unique Features

Some of the TV analysis, display, and interaction techniques are newly developed; others had been combined in an unusual and unique way. The TV is unique in its usage of (1) class nodes to apply a divide-and-conquer strategy during the analysis and visualization of potentially very-large-scale data sets, (2) bar graphs to display properties of class nodes (for example, size) and entities (for example, similarity or age), and (3) a static (yet interactively navigable) "base map" of organizational schema(s), bar graphs, and static or dynamic "link overlays."

Original Map

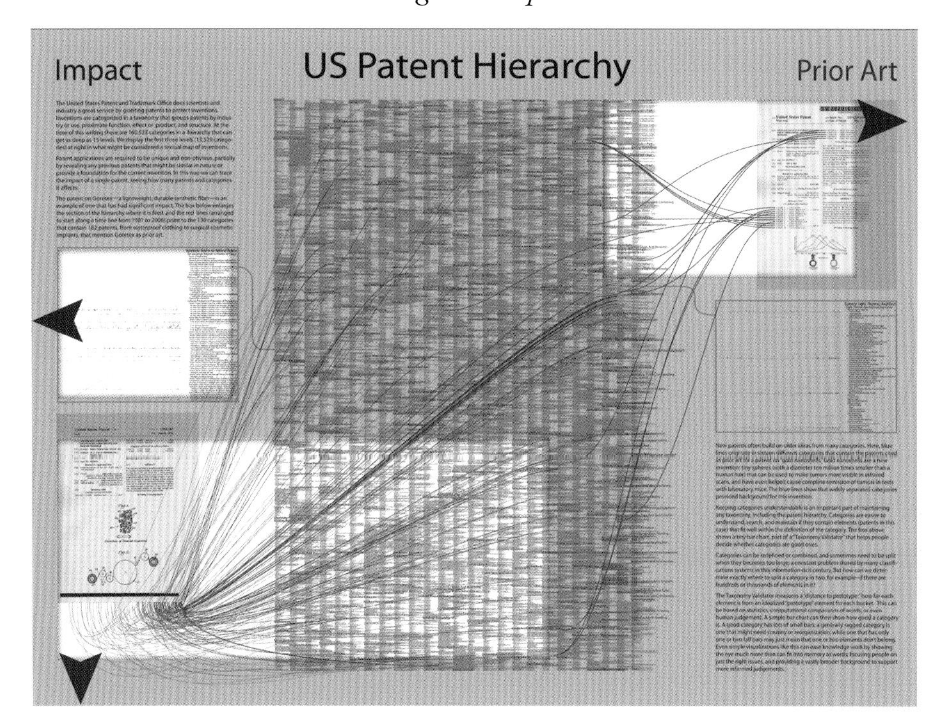

Zoom into References cited by Gold Nanoshell Patent: "Prior Art"

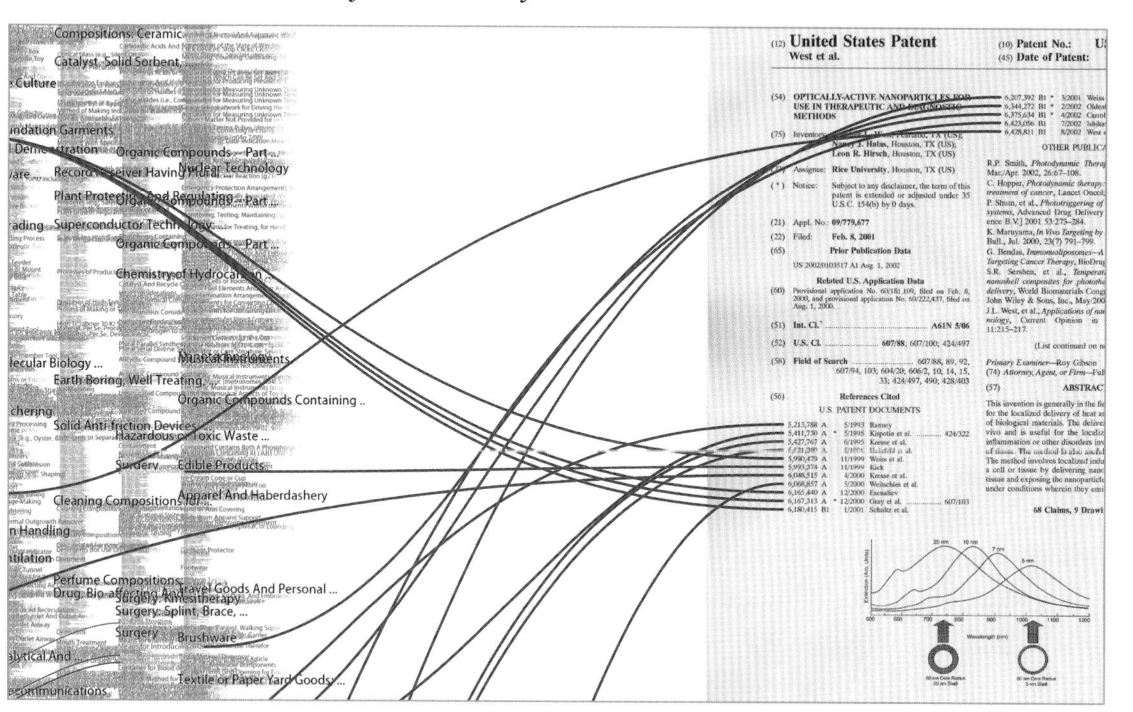

Zoom into Citations GORE-TEX Patent: "Impact"

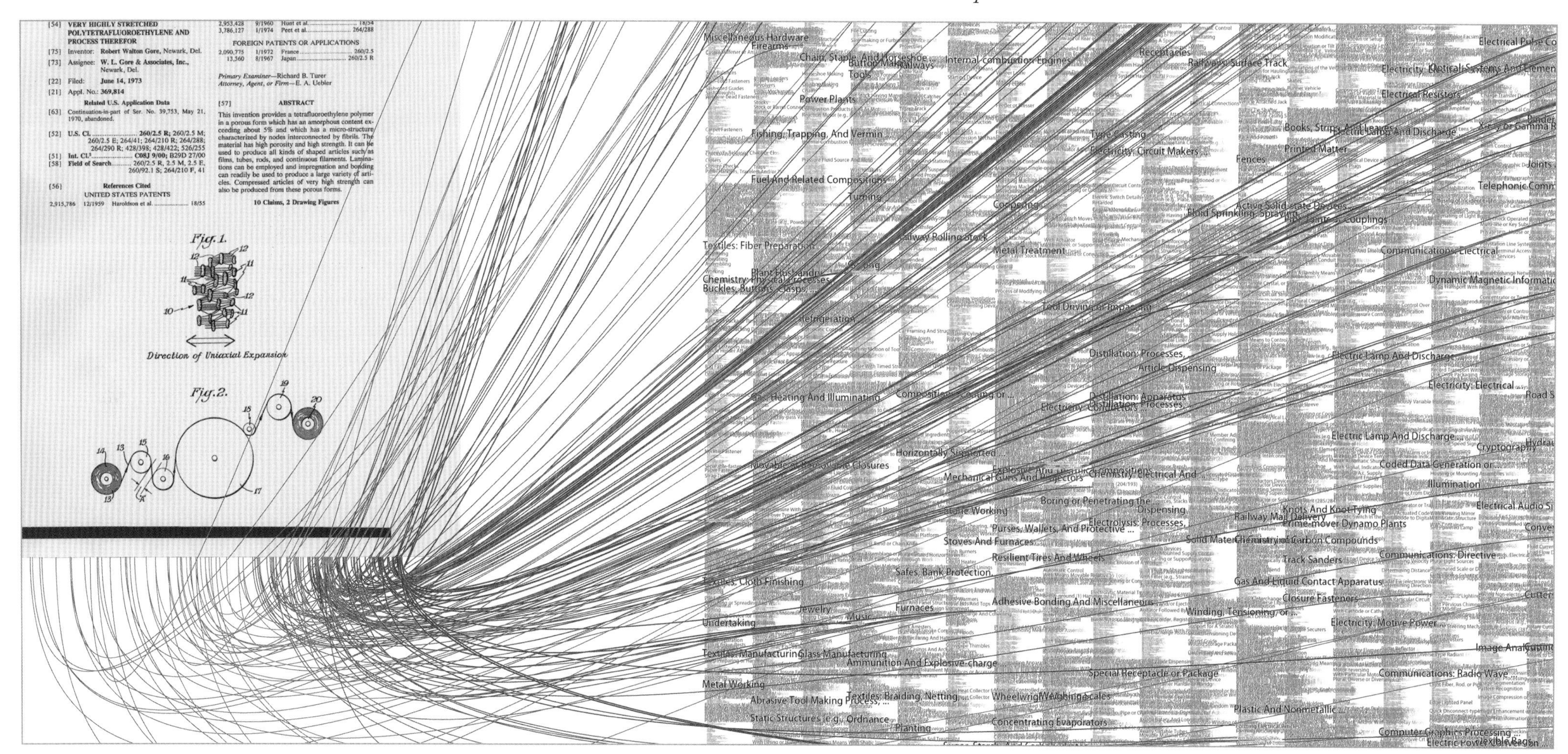

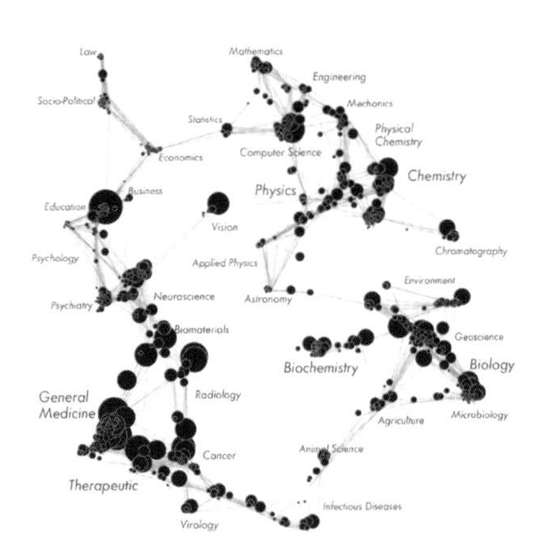

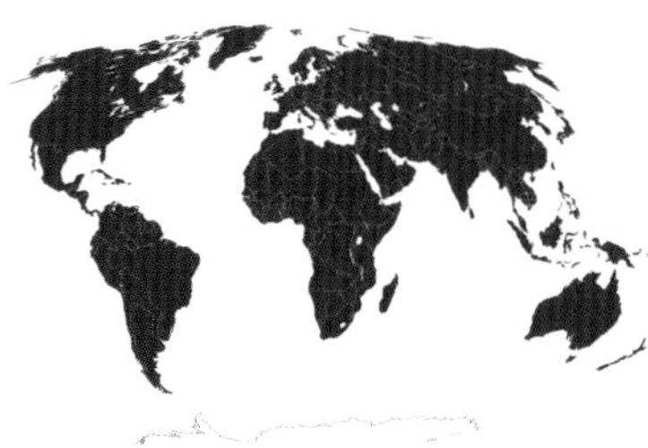

2006

Map of Scientific Paradigms

By Kevin W. Boyack and Richard Klavans

ALBUQUERQUE, NEW MEXICO, AND BERWYN, PENNSYLVANIA, 2006

Courtesy of Kevin W. Boyack and Richard Klavans, SciTech Strategies, Inc.

Aim

Science can be thought of as containing themes and paradigms; themes are current areas of research, while paradigms comprise the dominant tool sets and existing knowledge that are used by current researchers. What would a paradigm map of science look like? How many paradigms are currently active? How large and how vital are they?

Interpretation

This map was generated by recursively clustering the 820,000 most important papers referenced in 2003 using the processing pipeline described on **page 12, Toward a Reference System for Science**. The result is a map of 776 paradigms, which are shown as circles on the map. Although each paradigm contains an average of 1,000 papers, they range in sizes, as shown by the variously sized circles on the map. The most dominant relationships between paradigms were also calculated and are shown as lines between paradigms. A reference system was added for means of navigation and communication.

Color-coding indicates the vitality of a research topic—the darker the red, the younger the average reference age and the more vital and faster moving the topic. The white circles represent paradigms where consensus is reached relatively slowly. This is a common phenomenon in the social sciences, ecological sciences, computer sciences, and mathematics disciplines. The red circles represent communities of researchers where consensus is reached relatively rapidly. This is more common in physics, chemistry, biochemistry, and many medical disciplines. Very dark circles (such as those in quantum physics) represent communities where consensus is reached most quickly.

Countries, industries, companies, and individual researchers can all locate themselves within the map, either as single points or as a specific collection of paradigms. Science education and discovery can also be enhanced by linking to the map stories and facts that highlight content and relationships between scientific paradigms.

Kevin W. Boyack joined SciTech Strategies, Inc. in 2007 after working at Sandia National Laboratories, where he spent several years in the Computation, Computers, Information and Mathematics Center. He holds a PhD in chemical engineering from Brigham Young University. His current interests and work are related to information visualization, knowledge domains, science mapping with associated metrics and indicators, network analysis, and the integration and analysis of multiple data types.

Richard Klavans is the president of SciTech Strategies, Inc. He holds a PhD in management from the Wharton School of the University of Pennsylvania. His current work is related to the generation of highly accurate maps of science using multiple techniques, such as bibliographic coupling, cocitation, and coword, as well as the associated metrics and indicators that allow government and industry users to make more effective policy decisions. He is interested in semantics, augmented cognition, and the application of mathematical tools to information spaces.

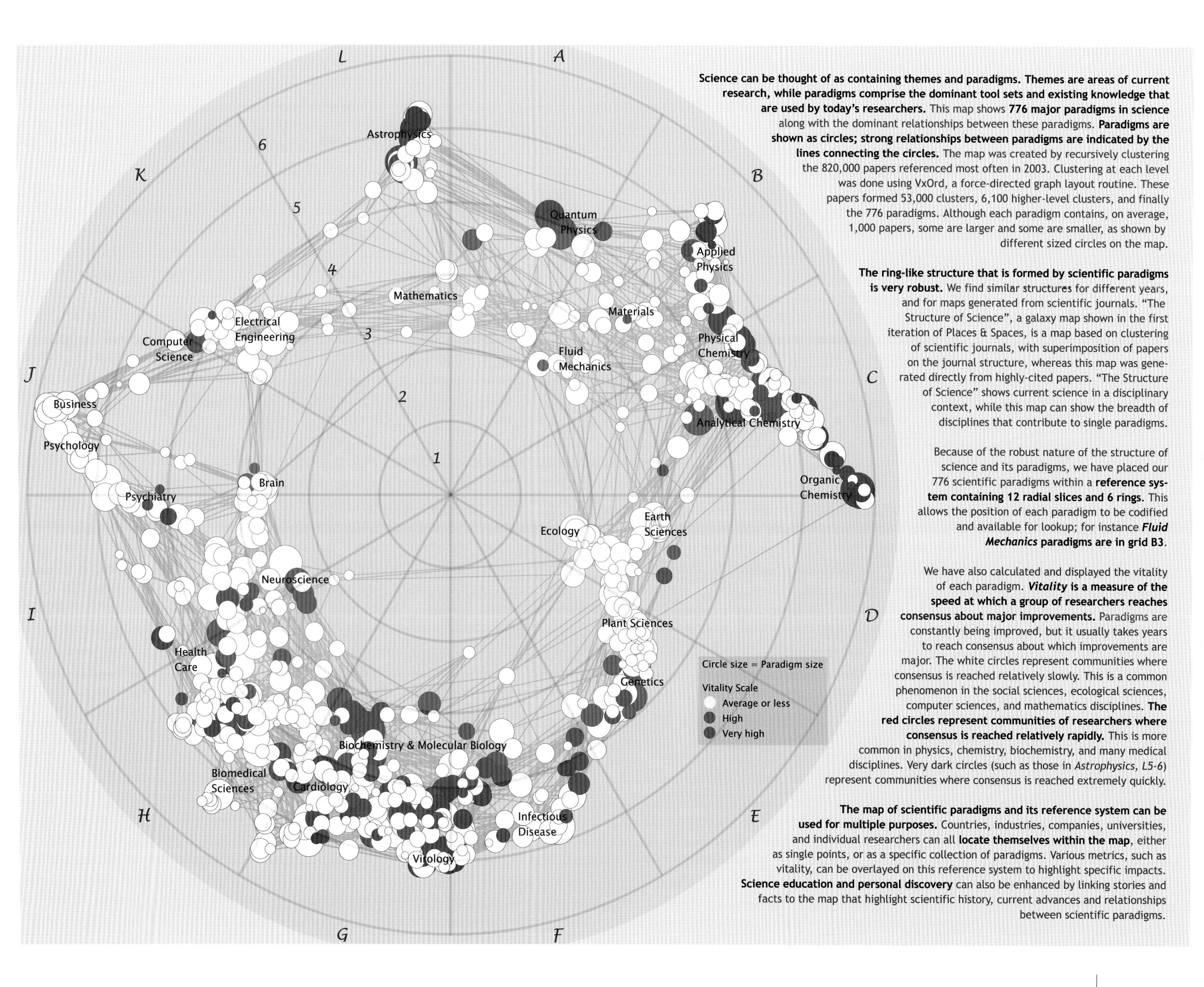

Science can be thought of as containing themes and paradigms. Themes are areas of current research, while paradigms comprise the dominant tool sets and existing knowledge that are used by today's researchers. This map shows **776 major paradigms in science** along with the dominant relationships between these paradigms. **Paradigms are shown as circles; strong relationships between paradigms are indicated by the lines connecting the circles.** The map was created by recursively clustering the 820,000 papers referenced most often in 2003. Clustering at each level was done using VxOrd, a force-directed graph layout routine. These papers formed 53,000 clusters, 6,100 higher-level clusters, and finally the 776 paradigms. Although each paradigm contains, on average, 1,000 papers, some are larger and some are smaller, as shown by different sized circles on the map.

The ring-like structure that is formed by scientific paradigms is very robust. We find similar structures for different years, and for maps generated from scientific journals. "The Structure of Science", a galaxy map shown in the first iteration of Places & Spaces, is a map based on clustering of scientific journals, with superimposition of papers on the journal structure, whereas this map was generated directly from highly-cited papers. "The Structure of Science" shows current science in a disciplinary context, while this map can show the breadth of disciplines that contribute to single paradigms.

Because of the robust nature of the structure of science and its paradigms, we have placed our 776 scientific paradigms within a **reference system containing 12 radial slices and 6 rings.** This allows the position of each paradigm to be codified and available for lookup; for instance ***Fluid Mechanics*** **paradigms are in grid B3.**

We have also calculated and displayed the vitality of each paradigm. ***Vitality*** **is a measure of the speed at which a group of researchers reaches consensus about major improvements.** Paradigms are constantly being improved, but it usually takes years to reach consensus about which improvements are major. The white circles represent communities where consensus is reached relatively slowly. This is a common phenomenon in the social sciences, ecological sciences, computer sciences, and mathematics disciplines. **The red circles represent communities of researchers where consensus is reached relatively rapidly.** This is more common in physics, chemistry, biochemistry, and many medical disciplines. Very dark circles (such as those in *Astrophysics, L5-6*) represent communities where consensus is reached extremely quickly.

The map of scientific paradigms and its reference system can be used for multiple purposes. Countries, industries, companies, universities, and individual researchers can all **locate themselves within the map**, either as single points, or as a specific collection of paradigms. Various metrics, such as vitality, can be overlayed on this reference system to highlight specific impacts. **Science education and personal discovery** can also be enhanced by linking stories and facts to the map that highlight scientific history, current advances and relationships between scientific paradigms.

Data

This map is based on paper-level data from the combined 2003 Science Citation Index and Social Sciences Citation Index by Thomson Reuters. The data set comprises about 820,000 highly cited reference papers.

Reference System

Three levels of clustering were used to group the 820,000 reference papers into (1) 53,000 communities, (2) 6,100 first-level clusters, and (3) 776 paradigms. Each clustering involves layout via VxOrd (see **page 106, The Structure of Science**). A fourth layout was done for the resulting 776 paradigms. Current papers (about 760,000) were later assigned to the paradigms and links between paradigms based on the distribution of their references.

Cocitation was used to determine the modified cosine similarity of papers. Cocitation occurs when two papers A and B are cited by a subsequent paper (see below). The more often papers A and B are cited together the higher their cocitation similarity. The result is an implied network with weighted edges.

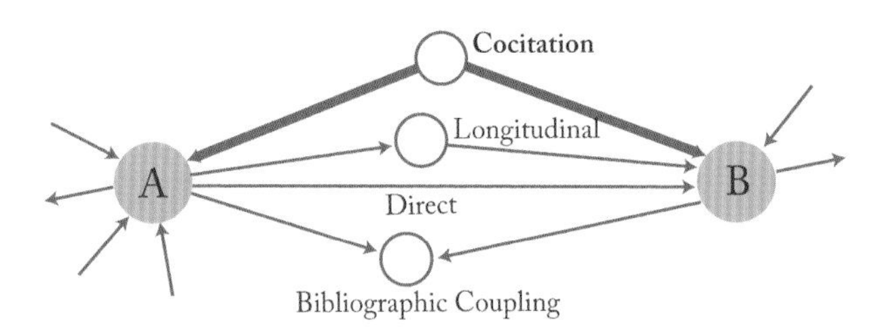

In each clustering step, the cocitation counts were reaggregated. The fourth iteration gave the *x, y* layout of the 776 paradigm clusters—a look at the paradigms or sets of tools used in current science.

The ringlike structure that is formed by scientific paradigms is very robust. The authors have found similar structures for different years and for maps generated from scientific journals (see **page 106, The Structure of Science**, which is based on 2002 data). A comparison of this paradigm map with the journal-based 2002 map is very instructive; although fields of science have roughly the same order around the ring, the distributional differences are important. For example, mathematics occupies a small discrete location at the top of the 2002 map. By contrast, papers from mathematics journals are spread throughout the entire upper half of the paradigm map, indicating that mathematics is an enabling science that is used widely in computer science, engineering, and the physical sciences. The paradigm map can show the breadth of disciplines that contribute to single paradigms.

Patent Yield

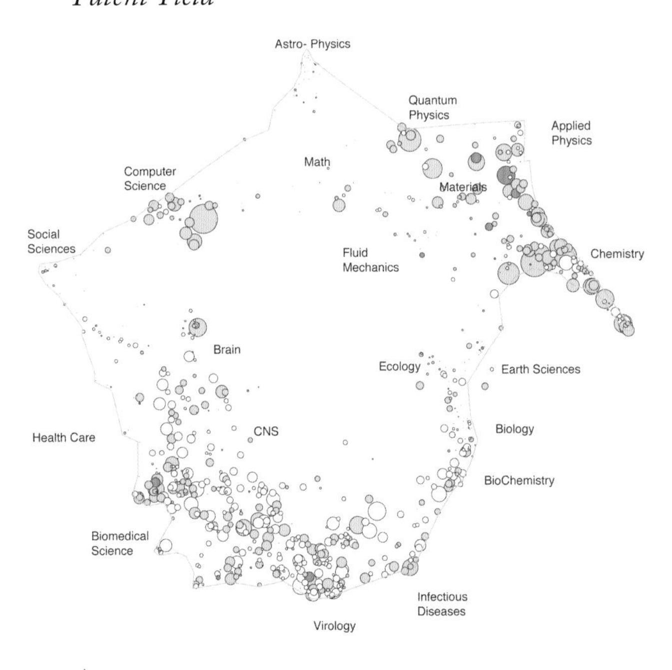

Country Profile

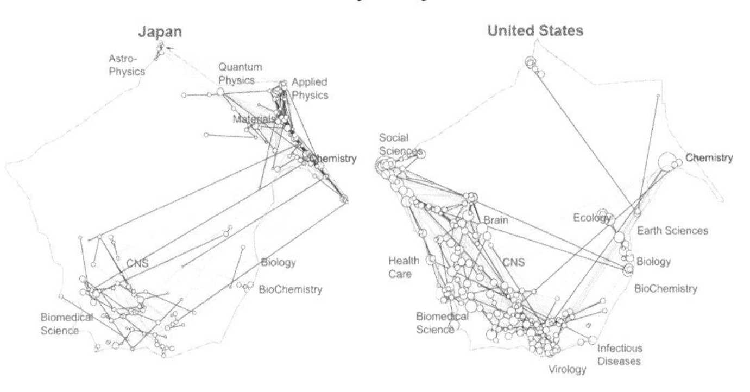

Data Overlays

Each paradigm is represented by a circle. Edges show strongest communication patterns between paradigms. The size of the circle represents the number of papers per paradigm (1,000 papers on average). The color of the circle indicates the vitality of a paradigm. *Vitality* is defined as a measure of the speed at which a group of researchers reaches consensus. It is calculated from the reference ages of a community of papers. The darker the red, the younger the average reference age and the more vital and faster-moving the topic, the quicker the communities build upon more recent work, and the faster the pace at which recent research findings are incorporated. Publishing mechanisms such as e-prints, common in astronomy and physics, support this fast pace. Communities of researchers that use books reach consensus relatively slowly. Detailed labels (bigrams) for each node were extracted from titles of papers but are not shown in the map. High-level labels were assigned by hand.

Details

The size and color of the circles, as well as their linkages, can also be used to overlay other data, such as patent yield or country or industry profiles.

Patent Yield

The map below identifies scientific communities that yield patents and thus contribute directly to the economy. In this patent/publication ratios map, circle size represents the number of patents divided by the number of papers for all paradigms with at least 100 papers. Color indicates the patent quality, measured by the number of citations minus expected number of citations on a per-patent basis.

Country Profiles

The United States is strong in the medical sciences (at the lower left), but underpublishes in the area of physical sciences and is in danger of losing its competitive position in physical sciences and engineering. Japan is strong in physics, as shown by the nodes at the upper right, but weak in the social sciences.

Institutional Strengths

Institutional (for example, national, regional, sector, or university) strengths, as indicated by relative overpublishing (using the entire world as a basis), can also be displayed on the map. Examples for eight industries are given at right.

Institutional Strengths

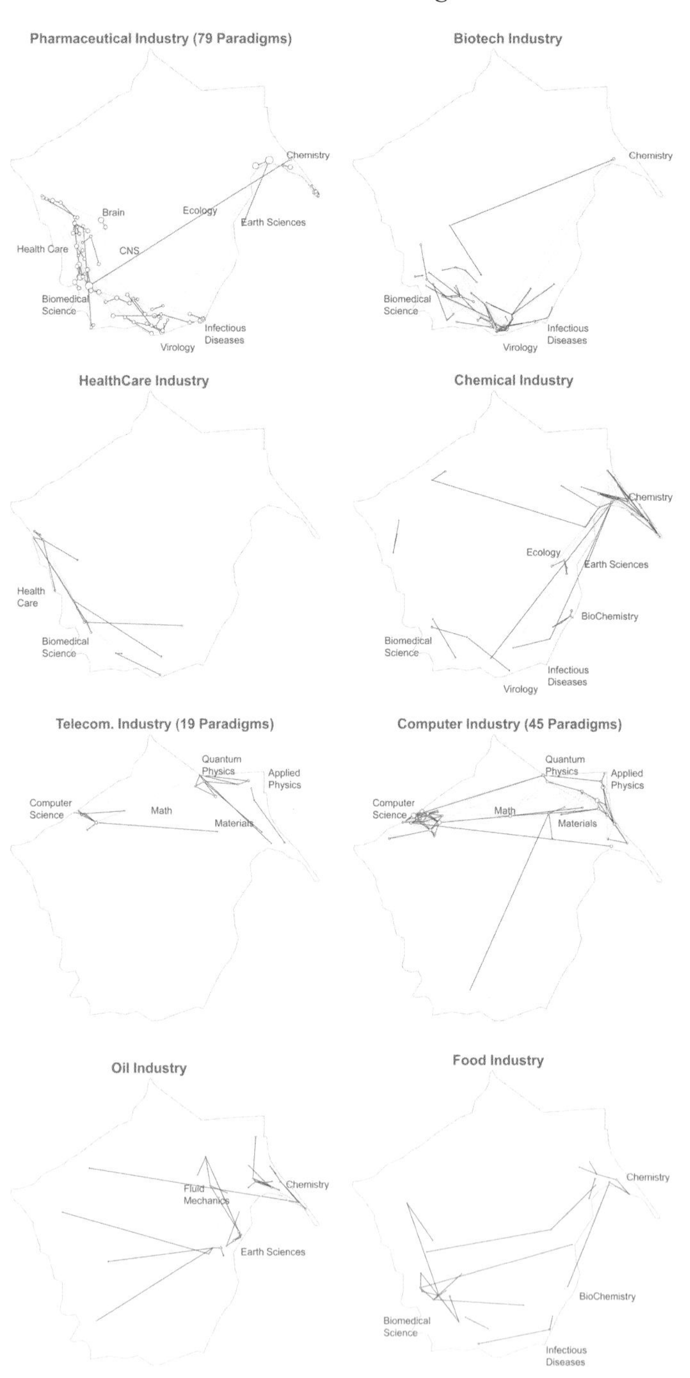

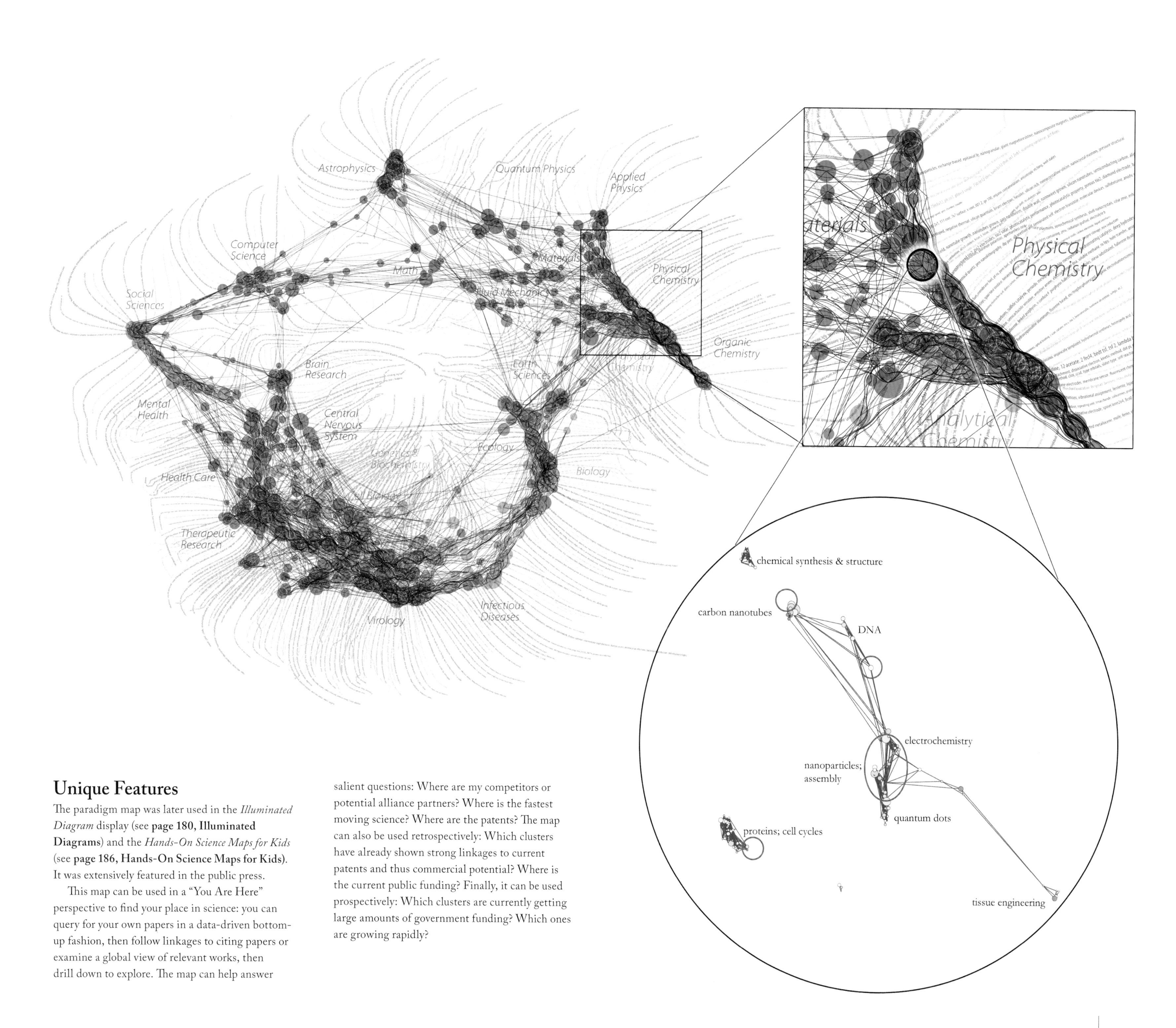

Unique Features

The paradigm map was later used in the *Illuminated Diagram* display (see **page 180, Illuminated Diagrams**) and the *Hands-On Science Maps for Kids* (see **page 186, Hands-On Science Maps for Kids**). It was extensively featured in the public press.

This map can be used in a "You Are Here" perspective to find your place in science: you can query for your own papers in a data-driven bottom-up fashion, then follow linkages to citing papers or examine a global view of relevant works, then drill down to explore. The map can help answer salient questions: Where are my competitors or potential alliance partners? Where is the fastest moving science? Where are the patents? The map can also be used retrospectively: Which clusters have already shown strong linkages to current patents and thus commercial potential? Where is the current public funding? Finally, it can be used prospectively: Which clusters are currently getting large amounts of government funding? Which ones are growing rapidly?

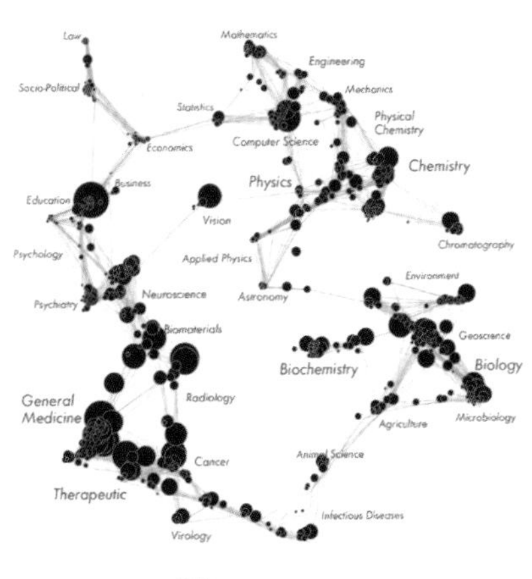

2006

WorldProcessor: Zones of Invention—Patterns of Patents

By Ingo Günther

NEW YORK, NEW YORK, 2006

Courtesy of Ingo Günther

Aim

Globes are used to chart continents, seas, and countries—as well as political borders that separate them. They can also be used to depict socioeconomic data and the social, cultural, and political conflicts arising from them. WorldProcessor globes (**http://www.worldprocessor.com**) depict a broad spectrum of global data sets of political conflicts; socioeconomic studies; environmental data; technological developments; and the spread of people, knowledge, and disease. Over the last 20 years, Günther has mapped data on globes as navigational guides in a globalized world. WorldProcessor is one of the first projects that introduced the notion of information mapping to the art world.

Interpretation

The WorldProcessor globe of *Patterns of Patents & Zones of Invention* (WorldProcessor #286) plots the total number of patents granted worldwide, beginning with nearly 50,000 in 1883, reaching 650,000 in 1993 (near the North Pole), and rapidly approaching 1 million in 2002 (in the southern hemisphere). Geographic regions where countries offer environments conducive to fostering innovation are represented by topology. Additionally, nations and countries that have an average of 500 or more U.S. patents per year granted to their residents or companies are called out in red by their respective averages in the years after 2000. Günther sculpted a three-dimensional distortion of the physical globe in such a way that would recover the original shape of the data graph (see also **page 192, WorldProcessor Globes**).

Ingo Günther, a sculptor and media artist, is a visiting professor at the Tokyo University of the Arts. He studied ethnology and cultural anthropology at the Johann Wolfgang Goethe University (Frankfurt am Main) and art at the Düsseldorf Academy of Fine Arts, Germany. In the 1970s, his travels took him to Northern Africa, North and Central America, and Asia. Günther's early sculptural works with video led him toward more journalistic projects, which he pursued in TV, print, and art. Based in New York, he played a crucial role in the evaluation and interpretation of satellite data gathered from political and military crisis zones; the results were distributed internationally through print media and TV news. The goal was to make previously inaccessible military and ecological information available to the public in order to have a direct impact on political processes. Since 1988, Günther has used globes as a medium for his artistic and journalistic interests. In 1989, nine months before the unification of Germany, he founded the first independent TV station in Eastern Europe, Kanal X (Channel X), in Leipzig, in order to contribute to the establishment of a free media landscape. The interviews and research he conducted during a journey of several months through the Cambodian refugee camps in Thailand became the basis of a series of articles published in the German newspaper *Die Tageszeitung* (*taz*). This journey, and later travels to refugee camps around the world, became the foundation for Günther's concept of the Refugee Republic, on which he has been working ever since.

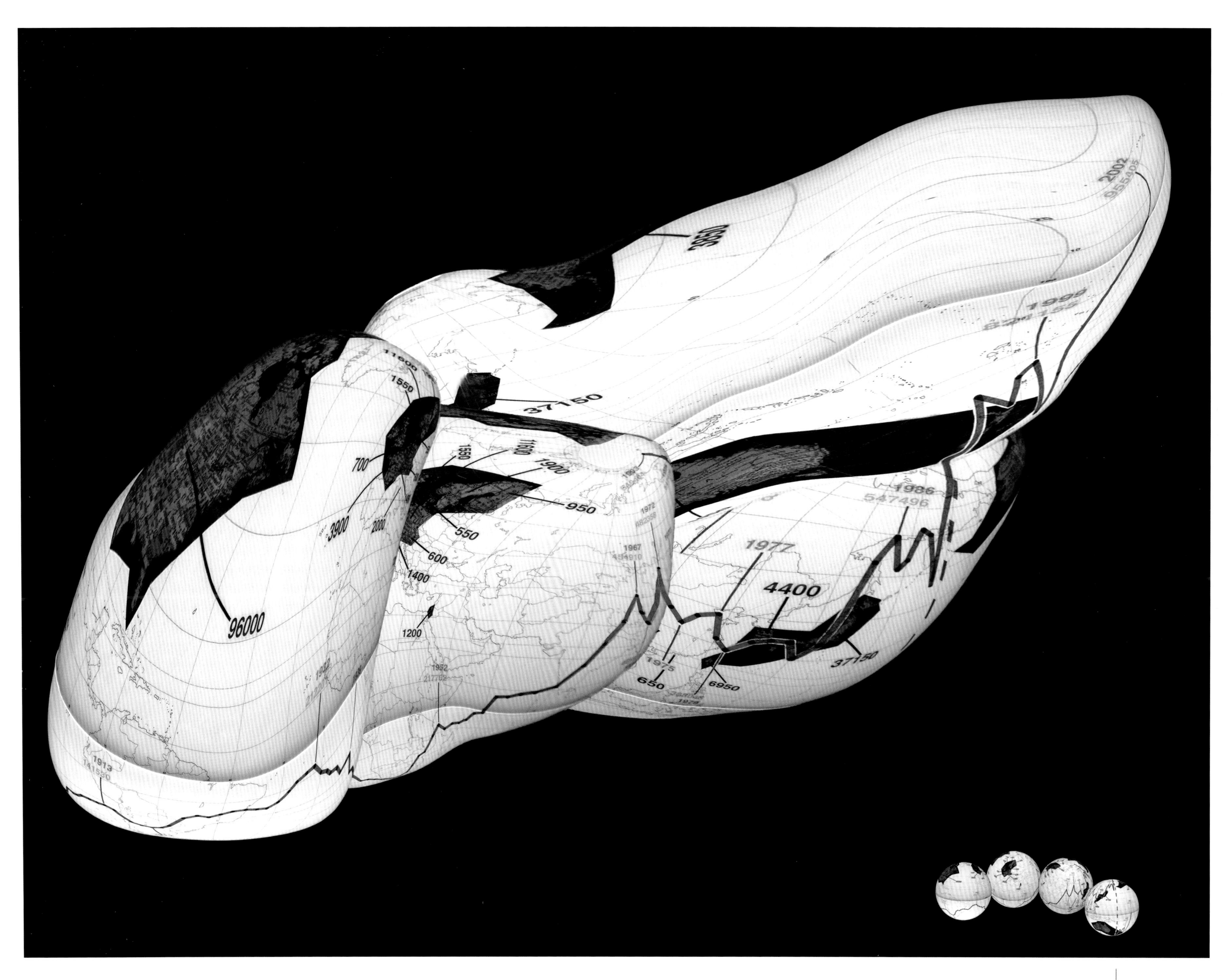

96000
3900
700
1550
2000
1400
600
550
950
1900
37150
1200
4400
1977
1986
547496
2002
955405
1999
650
6950
1975
37150
1967
1952
1913

Data

The patents shown here were retrieved from the database of the World Intellectual Property Organization (WIPO), as well as from the U.S. Patent and Trademark Office (USPTO).

Technique

The WorldProcessor project began in 1988. It is intended to be "as instant an interface to the world as possible, documenting ongoing information studies on civilization." So far, Günther has mapped 350 topics on more than 1000 plastic globes. For many topics, alternative graphic renderings are generated in order to balance the connotational/emotional values that are inherent to color and visual representations. His biggest challenge is to keep the globes current, as the underlying data statistics are changing. He has considered adding expiration dates to the titles and captions to avoid potential misunderstandings and misuse.

Reference System

All WorldProcessor globes use Earth as a reference system. The globes have a 12-inch (about 30 centimeters) diameter and represent the Earth at a scale of 1:42 million. While the topography of land and water masses does not change, political borders do as countries are united, divided, and renamed.

Data Overlays

In the absence of a "visual Esperanto," complex data sets have to be rendered into something that is easy to understand and remember. To map the enormous increase in the number of patents, WorldProcessor #286 shows a graph of the total number of patents granted worldwide, beginning with nearly 50,000 in 1883, reaching 650,000 in 1993 (near the North Pole), and rapidly approaching 1 million in 2002 (in the southern hemisphere). The impact of the economic and political systems on innovation and patent generation is indicated by rendering the respective geographic regions and nations by topography. All other regions have been omitted. All nations where residents are granted an average of 500 or more U.S. patents per year are called out in red by their respective averages in the years after 2000.

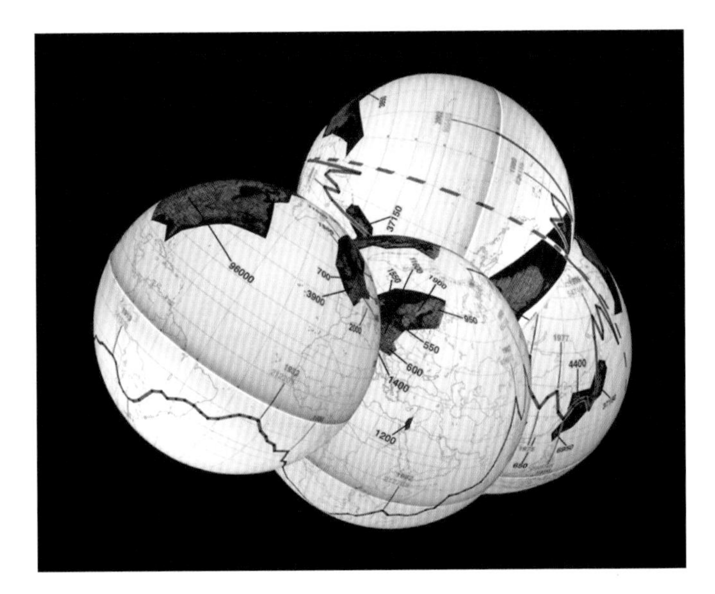

Unique Features

Data sets for other globes, such as the ones at right, come from publicly available sources, such as institutions, governments, fact books, newspapers, and magazines. Data sets are often incompatible and noncomparable, as they were acquired in different time spans, for different purposes, and using different statistical methods. This needs to be addressed in the formatting and rendering of the data. A healthy skepticism is advisable when reconciling data sets.

The globes on the opposite page are numbered in chronological sequence. Version numbers are given as well. **[1-4] TV Ownership 1990** is the fourth version of the first globe created. **[31] Fertilizer Pollution 1988** uses dark red circles to indicate oil spills and gray-shaded areas to point out sea pollution and land pollution from chemical fertilizers. Shown is only a small part of the entire pollution spectrum in 1988. At an average of every three months this globe becomes obsolete due to yet another major oil spill. **[155-10] Company vs. Country** reveals that some company's yearly gross income is larger than the entire GNP of a given country. This is the 10th update and revision of this globe. **[327] Mobile Teledensity** is a recent globe that shows the world wide penetration of mobile phones. Sparse black lines represent less than 50% ownership, semi-concentrated lines 50% to 100%, and most dense indicate more phones than inhabitants. Countries with more than double the world's average growth rate of 23% from 2002 - 2007 are encircled with bright dots. There are 4 billion mobile phones, 3 times the number of landlines.

To walk through an exhibit of nearly 100 globes is a remarkable experience. A typical display is shown below. Other exhibits have been set up using globes placed in a grid with some empty spaces that let viewers rest and digest. The globes seem to invite interaction. Despite the "Do not touch" signs, Günther has seen—via personal observation and hidden cameras—many men and a few women spinning the globes, perhaps feeling for a brief moment to have the world's data at their fingertips.

Viewers need to mentally complete the image and decipher and relate the information in order to create intellectual ownership of the experience. Therefore information is layered with increasing complexity in order to present yet another challenge to the audience while supplying visual intrigue and satisfaction.

Presented just below eye level, the head-size globes seem to invite viewers into a kind of personal communication. The proportions of the globes upon poles (or occasionally on tripods) further underscore their anthropomorphic appearance. Because of the three-dimensional experience, viewers find themselves facing one another and therefore easily involved in discussion—perhaps more so than they would be standing side by side before a two-dimensional work (see also **page 192, WorldProcessor Globes**).

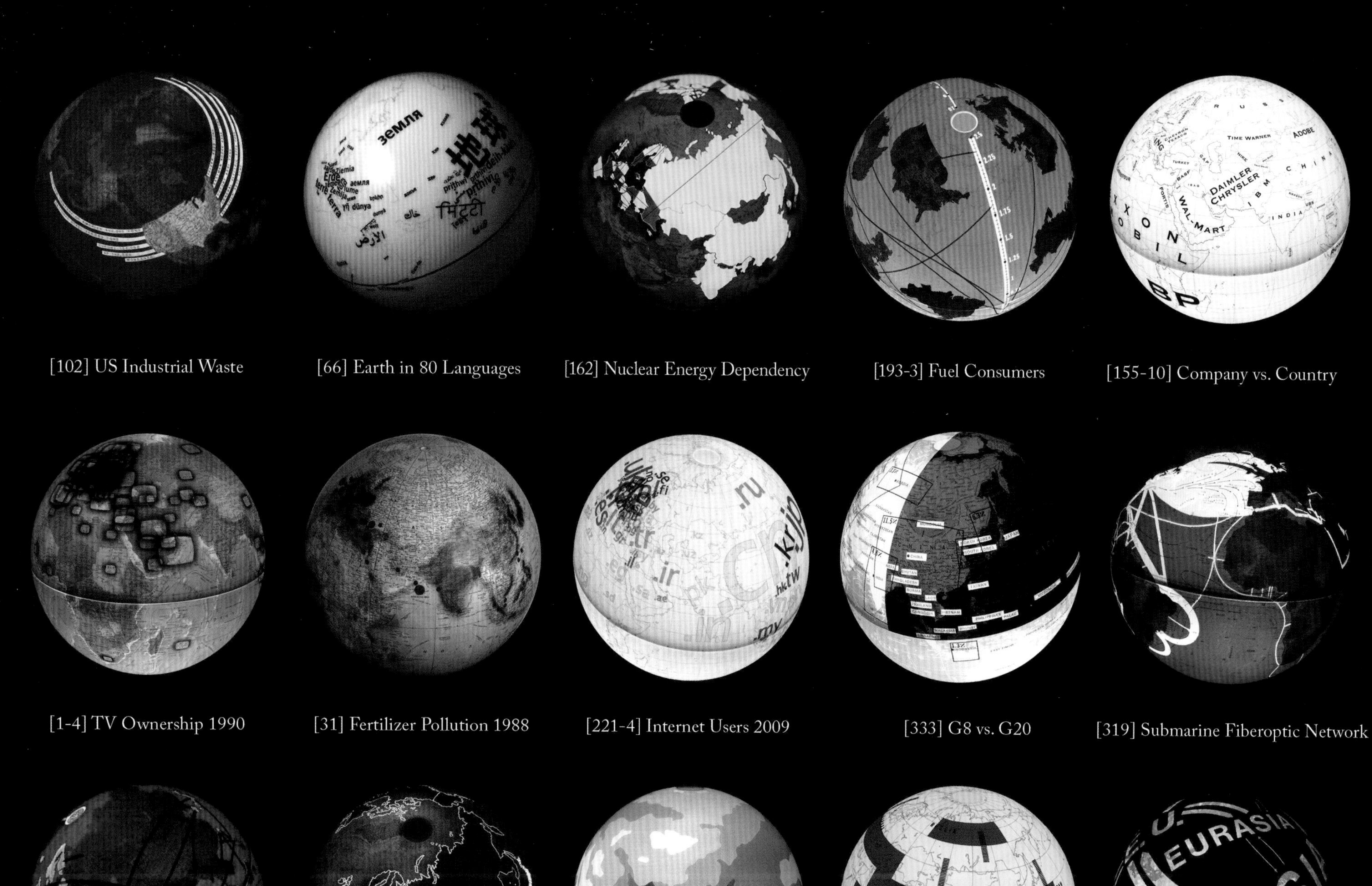

[102] US Industrial Waste

[66] Earth in 80 Languages

[162] Nuclear Energy Dependency

[193-3] Fuel Consumers

[155-10] Company vs. Country

[327] Mobile Teledensity

[1-4] TV Ownership 1990

[31] Fertilizer Pollution 1988

[221-4] Internet Users 2009

[333] G8 vs. G20

[319] Submarine Fiberoptic Network

[30] CO2 Spiral

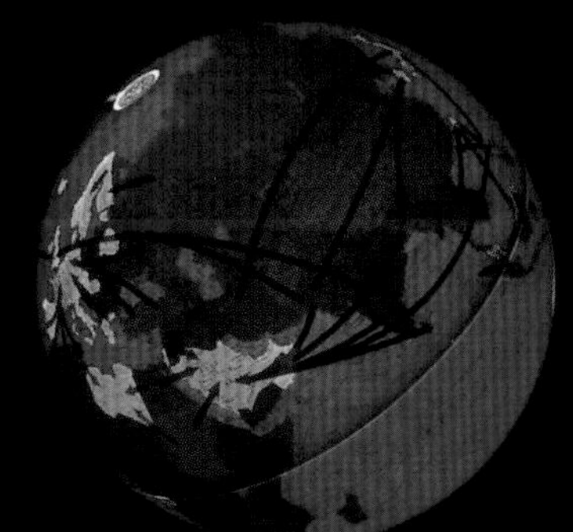

[166] Labor Migration

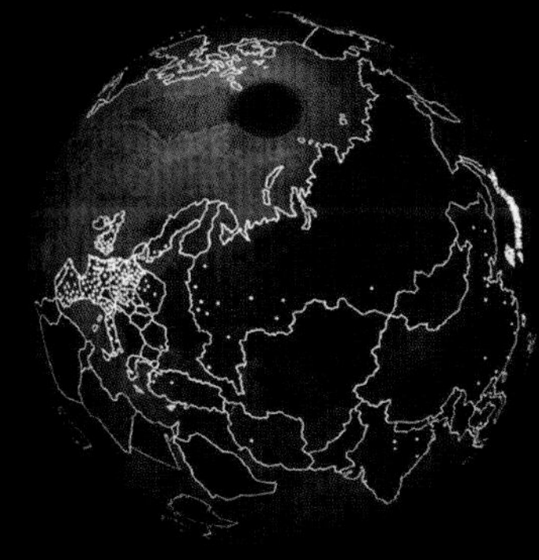

[202] Car Populations

[233] Rainfall

[331] Energy Consumption vs.GDP

[156] Global Trade Currents

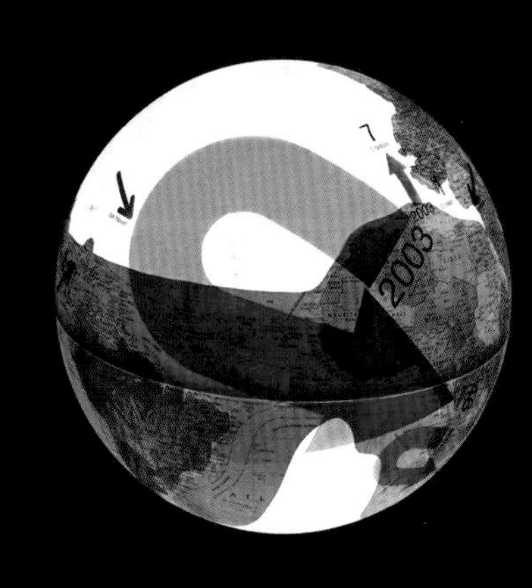

[294] South - South Investment

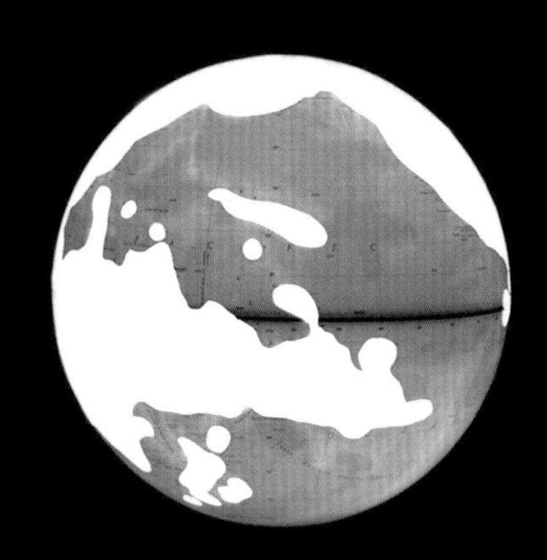

[107] Extended Exclusive Economic Zones

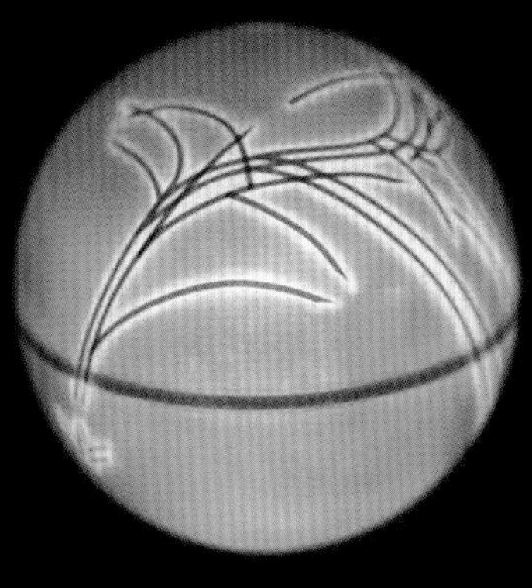

[231] DNA Traces

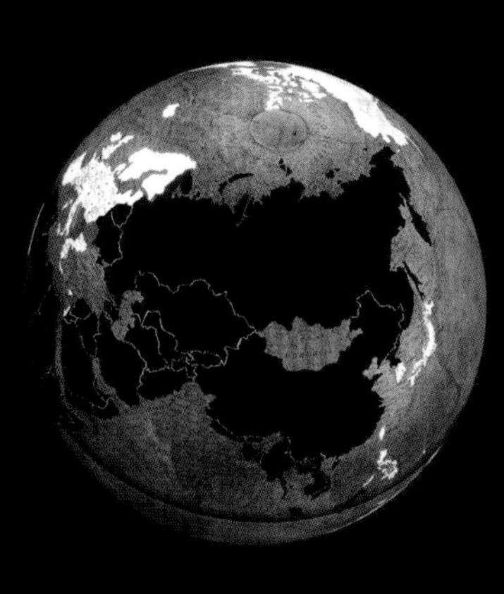

[250-2] Freedom of the Press

[177-2] Fuel Consumption and Prices

[8-4] Life Expectancy 2002

[312] Time Zone Conflicts

Third Iteration of Exhibit (2007): The Power of Forecasts

Four Existing Forecasts
versus
Six Science Forecasts

The third iteration of the exhibit compares and contrasts seismic hazard, economic, resource depletion, and epidemic forecast maps with maps forecasting the structure and evolution of science.

Real-time weather forecasts are served by the National Oceanic and Atmospheric Administration (NOAA) or the National Aeronautics and Space Administration (NASA). Computational models of the movements of tectonic plates help reduce losses due to earthquakes, volcanic activity, and tsunamis. Epidemic models make us understand how interconnected we all are and how actions far away affect us right here. Economic models let us simulate catastrophic and sustainable futures for mankind.

Daily science and technology forecasts would show science maps with overlays of top experts/institutions/countries, major activity bursts, or emerging research frontiers, augmenting our knowledge and decision-making. Why are they not available on TV, in the press, and online?

Four Existing Forecasts

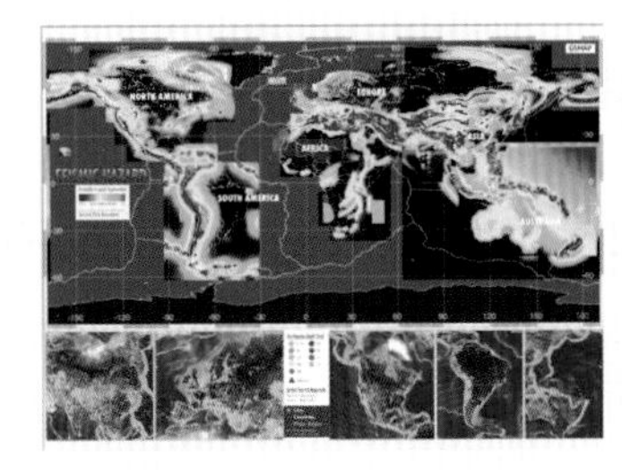

Michael W. Hamburger, Charles Meertens, and Elisha F. Hardy, Tectonic Movements and Earthquake Hazard Predictions, 2007

Rob Bracken, Dave Menninger, Michael Poremba, and Richard Katz, The Oil Age: World Oil Production 1859 to 2050, 2006

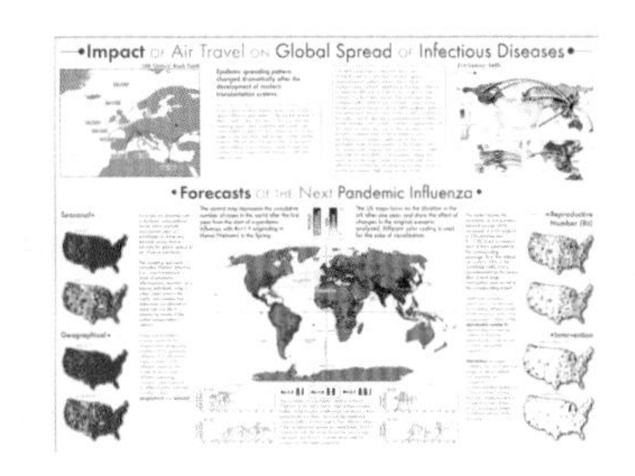

Vittoria Colizza, Alessandro Vespignani, and Elisha F. Hardy, Impact of Air Travel on Global Spread of Infectious Diseases, 2007

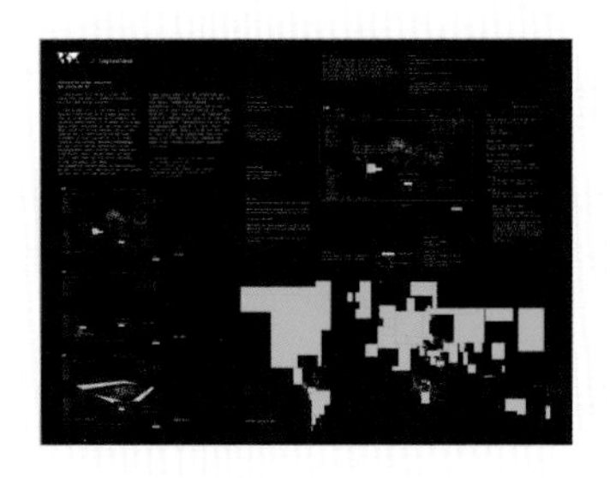

Michael Aschauer, Maia Gusberti, Nik Thoenen, and Sepp Deinhofer, [./logicaland] Participative Global Simulation, 2007

Six Science Forecasts

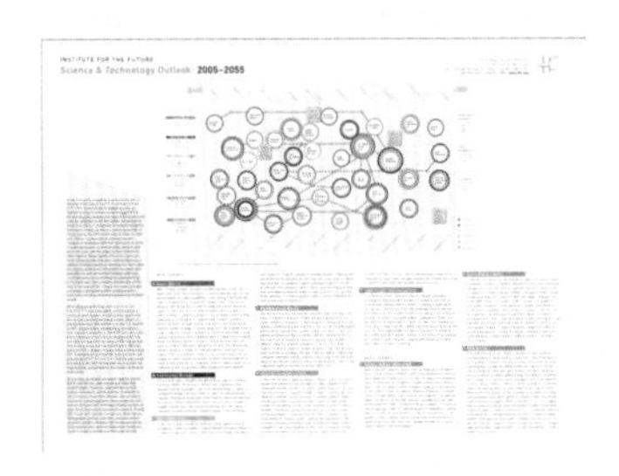

Marina Gorbis, Jean Hagan, Alex Soojung-Kim Pang, and David Pescovitz, Science & Technology Outlook: 2005-2055, 2006

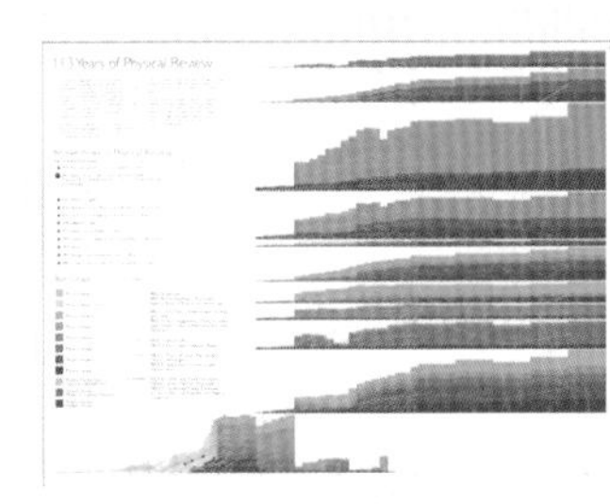

Bruce W. Herr II, Russell J. Duhon, Katy Börner, Elisha F. Hardy, and Shashikant Penumarthy, 113 Years of Physical Review, 2007

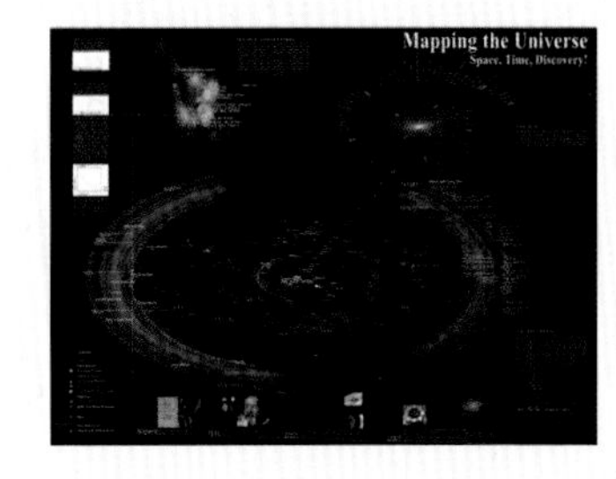

Chaomei Chen, Jian Zhang, Michael S. Vogeley, J. Richard Gott III, Mario Juric, and Lisa Kershner, Mapping the Universe: Space, Time, and Discovery! 2007

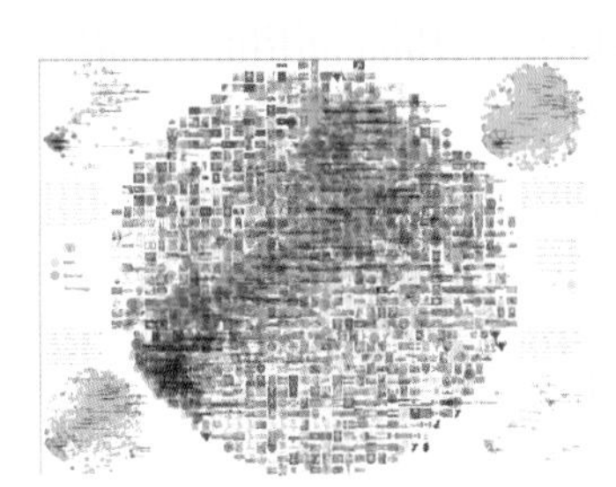

Bruce W. Herr II, Todd M. Holloway, Katy Börner, Elisha F. Hardy, and Kevin Boyack, Science-Related Wikipedian Activity, 2007

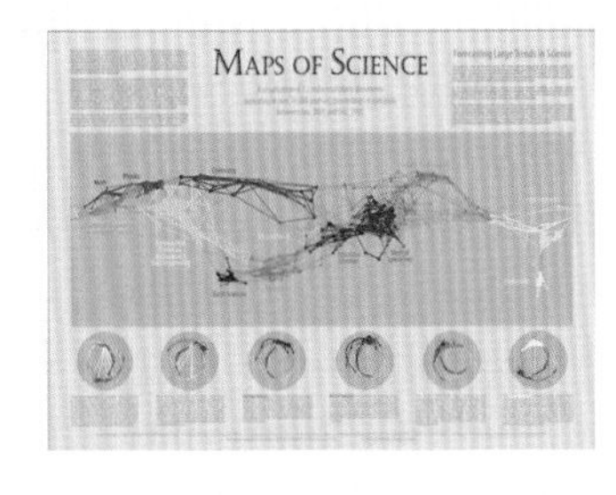

Richard Klavans and Kevin Boyack, Maps of Science: Forecasting Large Trends in Science, 2007

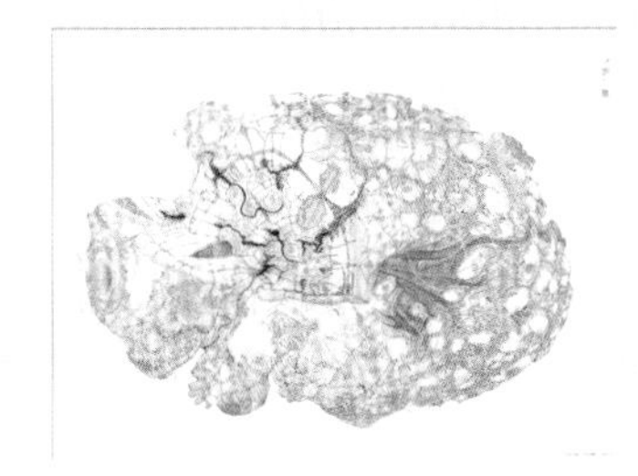

Daniel Zeller, Hypothetical Model of the Evolution and Structure of Science, 2007

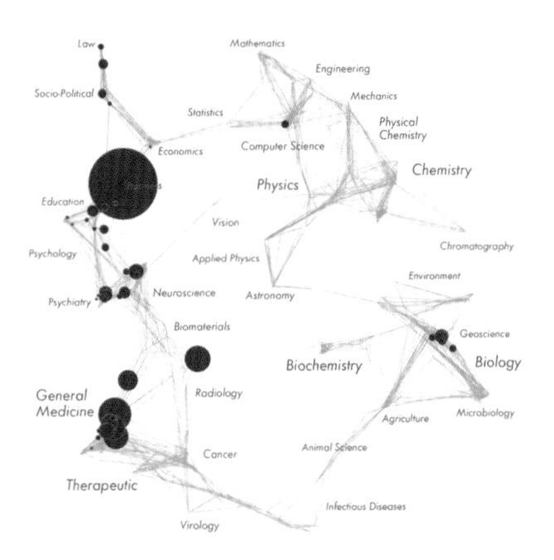

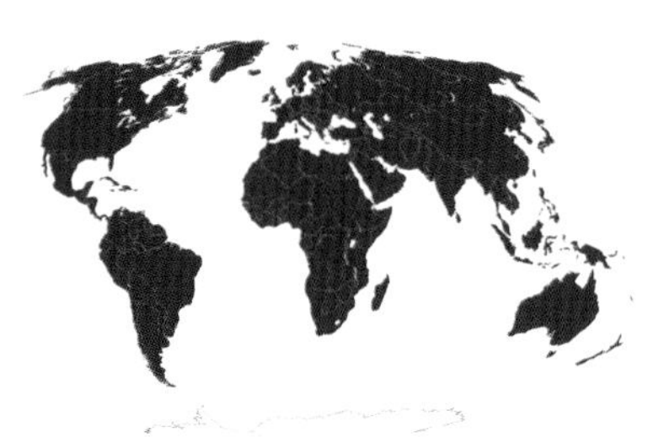

2007

Tectonic Movements and Earthquake Hazard Predictions

By Michael W. Hamburger, Charles Meertens, and Elisha F. Hardy

BLOOMINGTON, INDIANA, AND BOULDER, COLORADO, 2007

Courtesy of Indiana University

Scientists from the UNAVCO Consortium in Boulder, Colorado, and at Indiana University in Bloomington, Indiana, created Jules Verne Voyager, a precision interactive map tool for the virtual exploration of Earth and other worlds. The tool provides users with an extraordinary array of geological, geographic, and geophysical databases and novel means to understand the interrelationships of geophysical and geologic processes, structures, and measurements using high-precision global positioning systems (GPS) data and solutions.

The online browser interface at **http://jules.unavco.org** allows users to create "maps on demand" using a wide range of base maps, geophysical overlays, and geographical information. A "junior" version of the map tool provides educational users with rapid and user-friendly access to research-quality Earth visualization tools (**http://jules.unavco.org/VoyagerJr**).

The seismic hazard map shown here was derived from the International Lithosphere Program and the Global Seismic Hazard Assessment Program, using a model built from historical seismicity catalogs and geologic and geodetic data to predict where earthquakes might occur, how large they might be, and how often they occur. The source data has a resolution of 10 pixels per degree.

Seismic hazard is represented in a probabilistic fashion: as the peak ground acceleration (in meters/second^2) with a 10 percent chance of exceedance in a 50-year period. The prediction maximum (shown in red) was set to the equivalent of Earth's gravitational acceleration (9.8 meters/second^2).

The inset maps show the topographic, seismological, volcanic, and tectonic data for several of the major seismically active plate boundaries that make up the "Ring of Fire" surrounding Asia, Europe, North and South America, and the western Pacific. Arrows indicate the inferred direction of motion of the Earth's crust with respect to an arbitrarily "fixed" plate at the center of each map. Note the intense concentration of earthquake and volcanic activity near the boundaries of these tectonic plates—as well as in a few anomalous zones that occur within the interior of the Earth's tectonic plates. Earthquake depths can be used to chart the consumption of tectonic plates as they are submerged into the Earth's mantle along convergent plate boundaries.

Jules Verne Voyager works equally well for visualizing other planets and moons. Data sets for most major bodies of the solar system will soon become available.

Michael W. Hamburger (data and visualization) is professor of geological sciences at Indiana University. His research interests center on the relation of earthquakes to global geological processes, earthquake hazards, and volcanic activity. His research in seismology and volcanology has included field investigations in Alaska, the Philippines, the South Pacific, Central Asia, and the central United States. Recent research involves the application of new satellite surveying techniques to studies of deformation of the Earth's crust in the U.S. midcontinent and in active volcanoes in the Philippines. He has also taken on a leading role in creating the U.S. Educational Seismology Network (USESN), which brings seismology research instruments into hundreds of schools across the country. Hamburger has been involved in organizing campus and community response to major natural disasters, including the Indian Ocean tsunami of 2004 and the Pakistan earthquake of 2005.

Charles Meertens (data and visualization) is the facility director and interim president of UNAVCO in Boulder, Colorado. UNAVCO provides GPS project support, technology development, and data management and archiving services to the UNAVCO research community—supported by the National Science Foundation (NSF)—and to NASA's Global GPS Network (GGN). He is also one of the coprincipal investigators on GEON (the Geosciences Network), which began as a collaborative research project funded under the NSF Information Technology Research (ITR) program. GEON is conducting fundamental research toward developing a cyberinfrastructure for the earth sciences. Meertens is actively engaged in education and outreach activities at UNAVCO, particularly in the areas of visualization and data access. He received a PhD in geophysics from the University of Colorado in 1987 and subsequently conducted GPS research in earthquake, volcanic, and tectonic processes there, at the University of Utah, and at UCAR/UNAVCO (while the UNAVCO facility was still part of the University Corporation for Atmospheric Research).

Elisha F. Hardy (graphic design) **See page 158.**

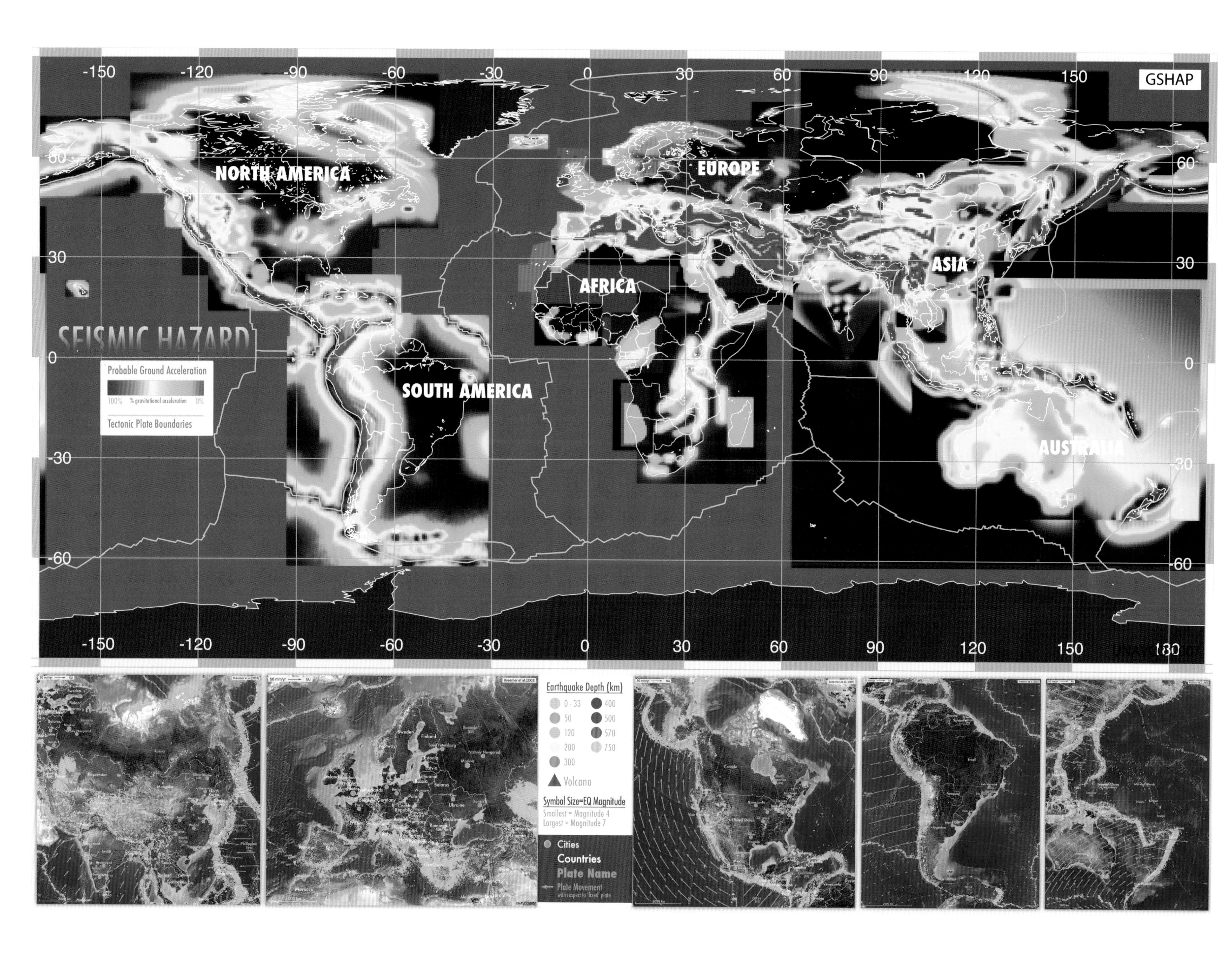
GSHAP
NORTH AMERICA
EUROPE
ASIA
AFRICA
SOUTH AMERICA
AUSTRALIA
SEISMIC HAZARD
Probable Ground Acceleration
Tectonic Plate Boundaries
Earthquake Depth (km)
Volcano
Symbol Size=EQ Magnitude
Smallest = Magnitude 4
Largest = Magnitude 7
Cities
Countries
Plate Name
Plate Movement
with respect to 'fixed' plate

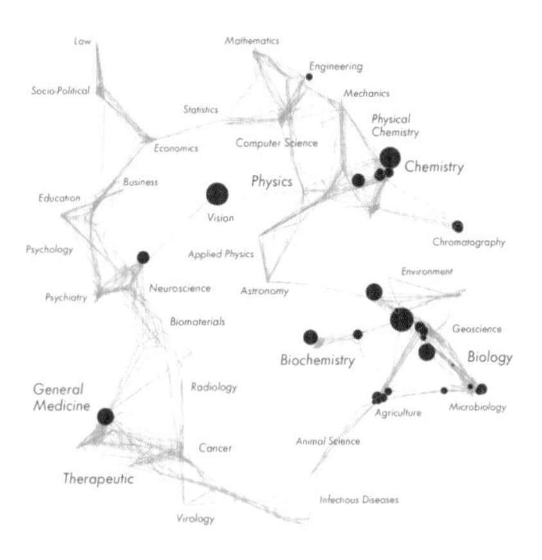

2006

The Oil Age: World Oil Production 1859 to 2050

By Rob Bracken, Dave Menninger, Michael Poremba, and Richard Katz
SAN FRANCISCO, CALIFORNIA, 2006
Courtesy of San Francisco Informatics

No understanding of industrial civilization would be complete without taking into account the central role of fossil fuels—oil in particular. Virtually everything we consider modern—from cars to air travel to plastics—would vanish without the empowering force of petroleum, the most energy-dense and versatile substance known to man. It's no exaggeration to say that we live in the age of oil.

It took millions of years to create the world's original endowment of some two trillion barrels of oil; yet in just 100 years we have pumped nearly half of it. This means that the world has arrived at the point of maximum production, after which oil becomes harder and more expensive to extract, and its flow inexorably diminishes. That turning point is called peak oil, and the world's leading geologists say that we have reached it, or will very soon.

The Oil Age illuminates the history of oil from critical angles, charting its steady growth in production, mapping its geographical sources, and showing its deep connection to sociopolitical events and modern technologies. The poster draws on a wide range of sources, including government statistics and the work of leading experts, including Colin J. Campbell, whose oil depletion model provides the central image covering most of *The Oil Age* (1859-2050). By displaying future projections of oil production, the poster poses a difficult question: How will mankind deal with the inexorable depletion of its most valuable resource?

Copies of *The Oil Age* are available at **http://www.oilposter.org**. The poster has been distributed to every member of the U.S. Congress and donated to more than 2,500 teachers across the United States.

Rob Bracken (writer) is a writer and communications manager for a major software company. He holds a master's degree in English from the University of Missouri. Rob became interested in oil and energy issues several years ago, after reading *The Party's Over* by Richard Heinberg. Rob lives in San Francisco with his wife, Deidra, and their cat.

Dave Menninger (graphic artist) received a bachelor's degree in visual design from Stanford University. Since then he has worked as a graphic and user interface designer in the San Francisco Bay Area for various computer startup companies and America Online. In 2006, he presented to the Midwest Model United Nations on peak oil and its implications. He lives with his fiancée, Monica, and their cat in Kiev, Ukraine.

Michael Poremba (statistician) works as a statistician and data architect, designing and constructing information systems. He holds bachelor's degrees in computer science and philosophy with additional studies in art history and design. Following the U.S. invasion of Iraq in 2003, Poremba began studying resource wars, energy depletion, and related issues. He lives with his wife, Anna, in San Francisco.

Richard Katz (catalyst) holds a degree in cinematography from New York University, and by invitation did advanced studies in cinematography at the American Film Institute. He has worked as a catalyst and secular evangelist at several major software companies including Intuit. Richard became interested in resource depletion in the early 1970s. He lives with his wife, Ann, their cat, and their dog in San Francisco.

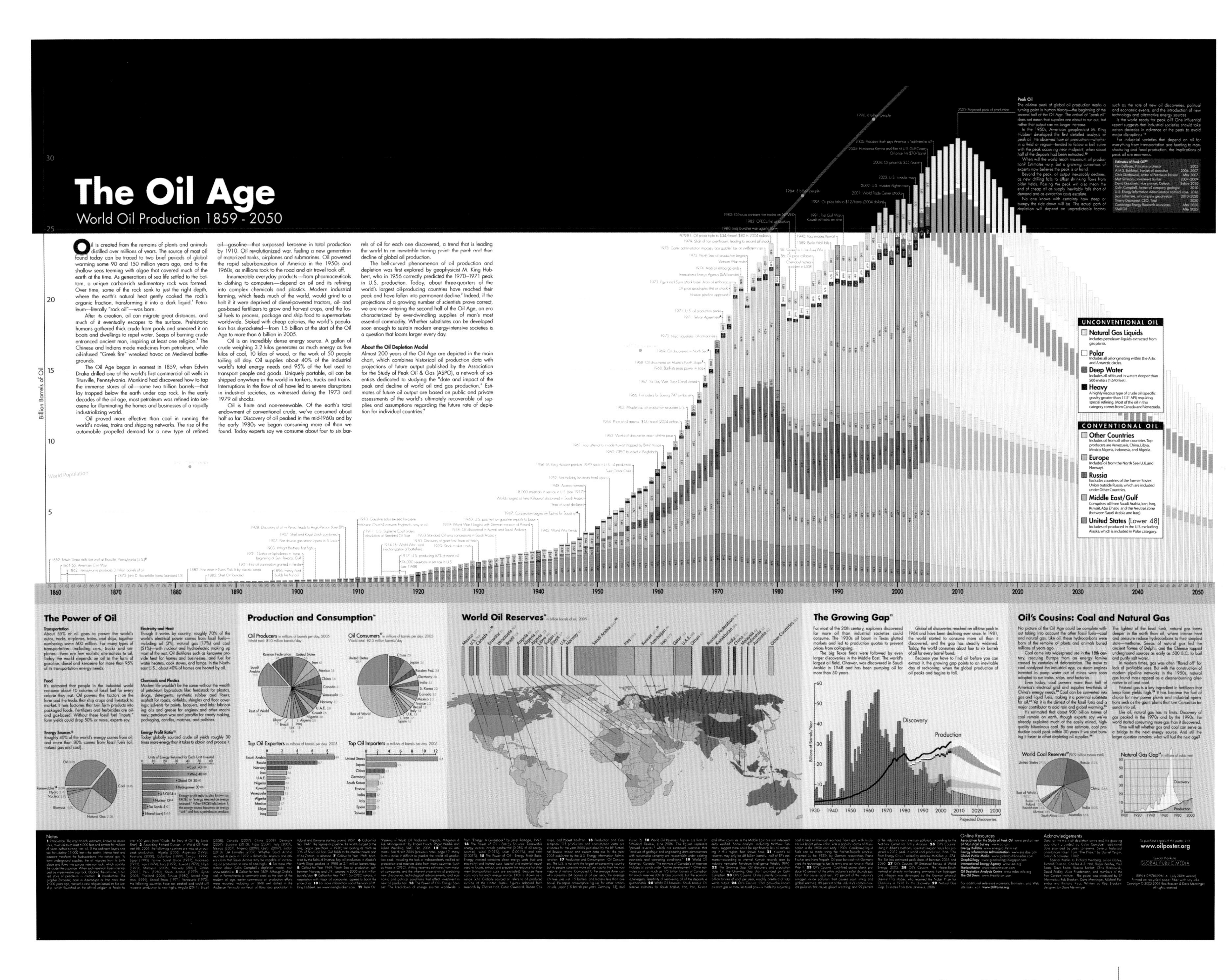

The Oil Age
World Oil Production 1859 - 2050
Billion Barrels of Oil
World Population
UNCONVENTIONAL OIL
Natural Gas Liquids
Polar
Deep Water
Heavy
CONVENTIONAL OIL
Other Countries
Europe
Russia
Middle East/Gulf
United States (Lower 48)
Peak Oil
About the Oil Depletion Model
The Power of Oil
Production and Consumption
World Oil Reserves
The Growing Gap
Oil's Cousins: Coal and Natural Gas
Oil Producers
Oil Consumers
Top Oil Exporters
Top Oil Importers
Discovery
Production
Notes
Online Resources
Acknowledgements
www.oilposter.org

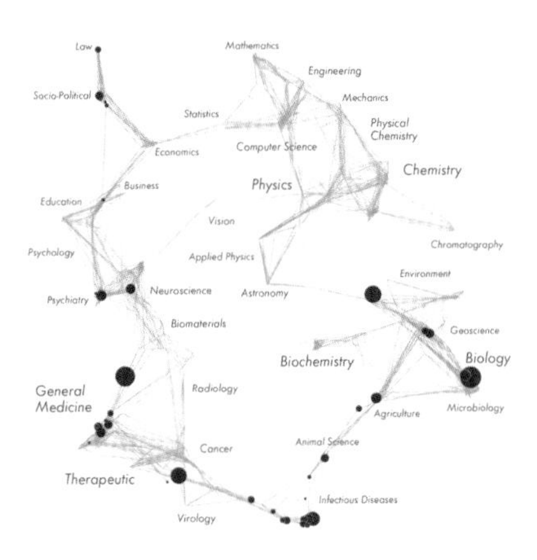

2007

Impact of Air Travel on Global Spread of Infectious Diseases

By Vittoria Colizza, Alessandro Vespignani, and Elisha F. Hardy
BLOOMINGTON, INDIANA, 2007
Courtesy of Indiana University

This map shows the impact of air travel on the global spread of epidemics. In the 14th century, infectious diseases like the Black Death spread in smoothly diffuse waves over long time periods (see contour lines at upper left of map). Modern transportation systems cause a networklike spread of epidemics; for example, the global transmission of severe acute respiratory syndrome (SARS) in 2002–2003 happened within a few weeks as the epidemic spread rapidly from airport to airport (see upper right of map).

Detailed knowledge of the worldwide population distribution and movement patterns of individuals can be used to build large-scale, stochastic, spatial-transmission models. These models augment the study of epidemic spread patterns, which helps to predict and contain future epidemics. The effect of travel patterns, origin of outbreak, seasonality, virus transmissibility, and different medical interventions can all be modeled. Modeling results are communicated via maps that need to be interpreted and understood by different stakeholders to be truly useful in the identification, design, and implementation of appropriate intervention strategies.

Forecast maps in the lower part show the spreading of the next pandemic influenza using different model parameters: the redder a geographical area, the greater the infection rate.

U.S. maps show the effect of different initial outbreak conditions, for example, seasonal (starting in spring or fall) or geographical (starting in Chicago or Bucharest). They also show specific disease parameters, such as the average number of infections generated by a sick person (R_0) or different intervention strategies. Even a limited worldwide sharing of antiviral drugs, for example, by the United States and Western countries, leads to fewer infections, thereby benefiting both drug donors and recipients.

The central map shows the cumulative number of worldwide influenza cases for model parameters: Spring, Hanoi, infection rate R_0=1.9, and no intervention. The graphs below plot the simulated fraction of infected individuals for selected countries and cities over time.

Vittoria Colizza (research and data) is a research scientist at the Institute for Scientific Interchange (ISI Foundation, Turin, Italy) where she leads the Computational Epidemiology Laboratory. Vittoria obtained her PhD in physics at the International School for Advanced Studies (SISSA) in Trieste, Italy, in 2004. After holding a research position at the Indiana University School of Informatics in Bloomington, Indiana, she spent a year as visiting assistant professor at Indiana University (time of the creation of this map), and she joined the ISI Foundation in Turin in 2007. She was recently awarded a Starting Independent Career Grant in Life Sciences by the European Research Council Ideas Program.

Alessandro Vespignani (research) is James H. Rudy Professor of Informatics and Computing and adjunct professor of physics and statistics at Indiana University, where he is also the director of the Center for Complex Networks and Systems Research (CNetS) and associate director of the Pervasive Technology Institute. He obtained his PhD at the University of Rome "La Sapienza." After holding research positions at Yale University and Leiden University, he was a member of the condensed matter research group at the International Center for Theoretical Physics (UNESCO) in Trieste. Before joining Indiana University, Vespignani was faculty of the Laboratoire de Physique Théorique at the University of Paris-Sud, working for the French National Council for Scientific Research (CNRS), of which he is still a member at large. Vespignani is an elected fellow of the American Physical Society and is serving on the board/leadership of a variety of professional associations and journals and the Institute for Scientific Interchange Foundation in Turin, Italy.

Elisha F. Hardy (graphic design) **See page 158**.

Impact OF Air Travel ON Global Spread OF Infectious Diseases

14th Century: Black Death

Dec. 1350
June 1350
Dec. 1349
June 1349
Dec. 1348
June 1348
Dec. 1347
June 1347
Atlantic Ocean

Epidemic spreading pattern changed dramatically after the development of modern transportation systems.

In pre-industrial times disease spread was mainly a spatial diffusion phenomenon. During the spread of Black Death in the 14th century Europe, only few traveling means were available and typical trips were limited to relatively short distances on the time scale of one day. Historical studies confirm that the disease diffused smoothly generating an epidemic front traveling as a continuous wave through the continent at an approximate velocity of 200-400 miles per year.

The SARS outbreak on the other hand was characterized by a patched and heterogeneous spatio-temporal pattern mainly due to the air transportation network identified as the major channel of epidemic diffusion and ability to connect far apart regions in a short time period. The SARS maps are obtained with a data-driven stochastic computational model aimed at the study of the SARS epidemic pattern and analysis of the accuracy of the model's predictions. Simulation results describe a spatio-temporal evolution of the disease (color coded countries) in agreement with the historical data. Analysis on the robustness of the model's forecasts leads to the emergence and identification of epidemic pathways as the most probable routes of propagation of the disease. Only few preferential channels are selected (arrows; width indicates the probability of propagation along that path) out of the huge number of possible paths the infection could take by following the complex nature of airline connections (light grey, source: IATA).

21st Century: SARS

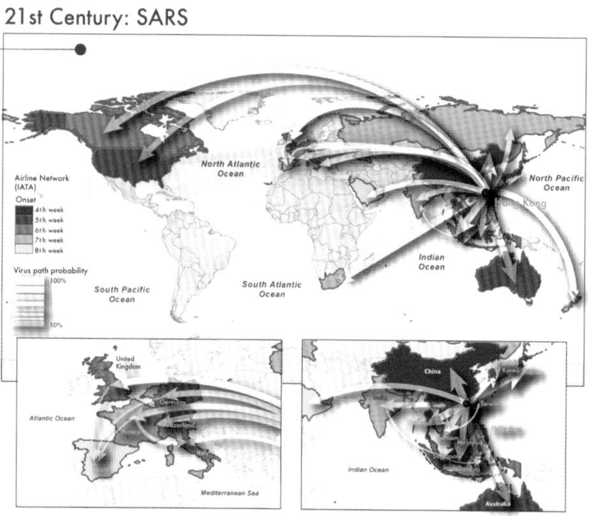

Forecasts OF THE Next Pandemic Influenza

Seasonal

SPRING

FALL

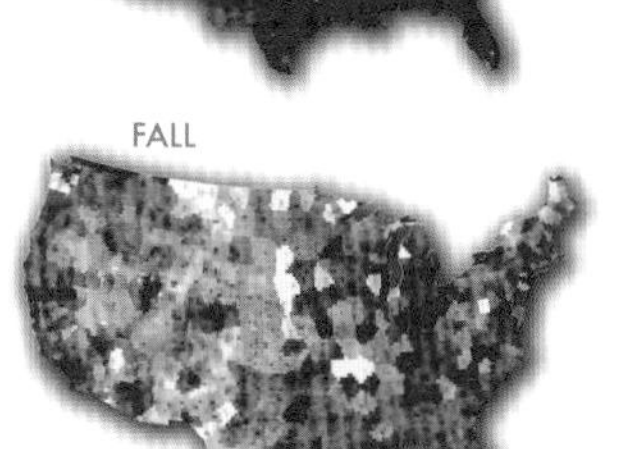

Geographical

CHICAGO

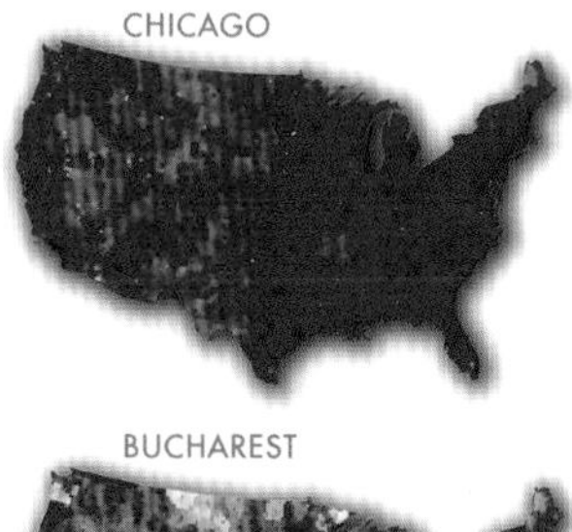

BUCHAREST

Forecasts are obtained with a stochastic computational model which explicitly incorporates data on worldwide air travel and detailed census data to simulate the global spread of an influenza pandemic.

The modeling approach considers infection dynamics (i.e., virus transmission, onset of symptoms, infectiousness, recovery, etc.) among individuals living in urban areas around the world, and assumes that individuals are allowed to travel fom one city to another by means of the airline transportation network.

Numerical simulations provide results for the temporal and geographic evolution of the pandemic influenza in 3,100 urban areas located in 220 different countries. The model allows to study different spreading scenarios, characterized by different initial outbreak conditions, both **geographical** and **seasonal**.

The central map represents the cumulative number of cases in the world after the first year from the start of a pandemic influenza with R0=1.9 originating in Hanoi (Vietnam) in the Spring.

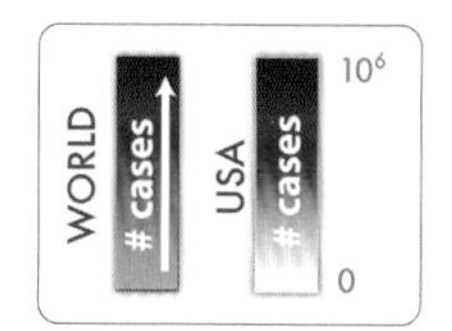

The US maps focus on the situation in the US after one year, and show the effect of changes in the original scenario analyzed. Different color coding is used for the sake of visualization.

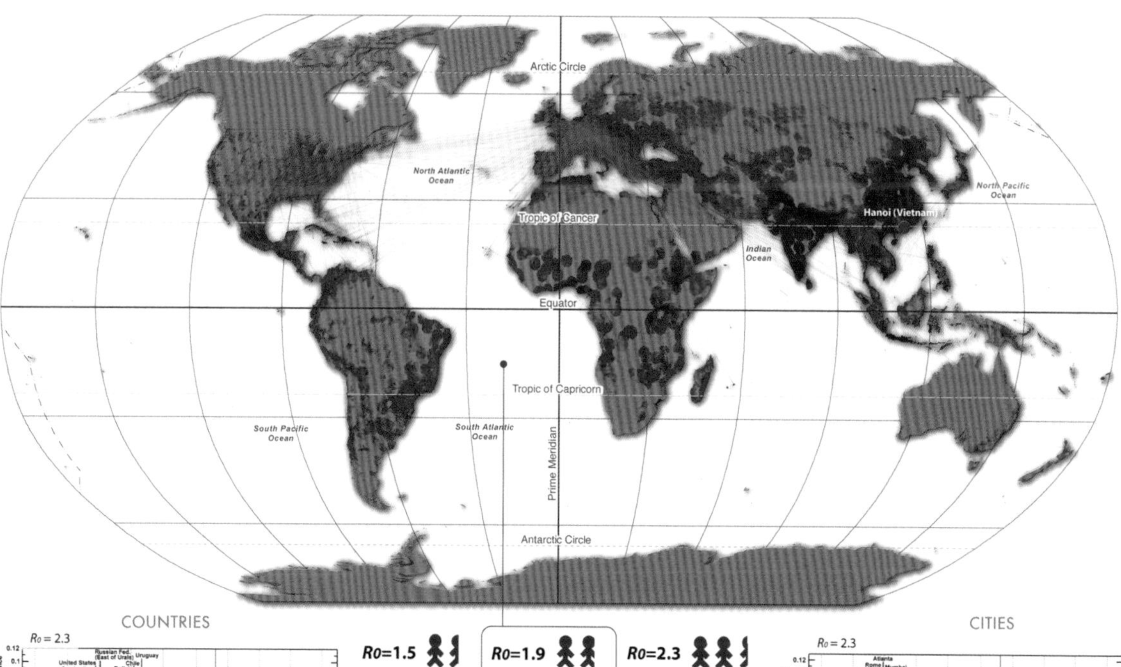

Time evolution of a pandemic starting in Hanoi (Vietnam) in the Fall in the no intervention scenario. Profiles of the fraction of infectious individuals in time (prevalence) are shown for some representative countries (left) and cities (right). Two different values of the reproductive number are considered: R0=1.5, consistently with the values shown for the US map (top right), and R0=2.3, in order to provide the comparison with faster spreading.

The model inlcudes the worldwide air transportation network (source: IATA) composed of 3,100 airports in 220 countries and E=17,182 direct connections, each of them associated to the corresponding passenger flow. This dataset accounts for 99% of the worldwide traffic and is complemented by the census data of each large metropolitan area served by the corresponding airport.

Additional spreading scenarios can be obtained by modeling different levels of infectiousness of the virus, as expressed in terms of the **reproductive number** R0, representing the average number of infections generated by a sick person in a fully susceptible population.

Intervention strategies modeling the use of antiviral drugs can be considered. Two scenarios are compared: an uncooperative strategy in which countries only use their own stockpiles, and a cooperative intervention which envisions a limited worldwide sharing of the resources.

Reproductive Number (R0)

1.7

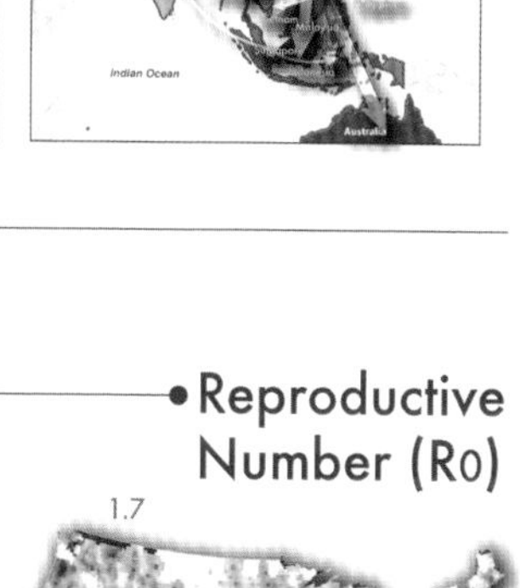

1.5

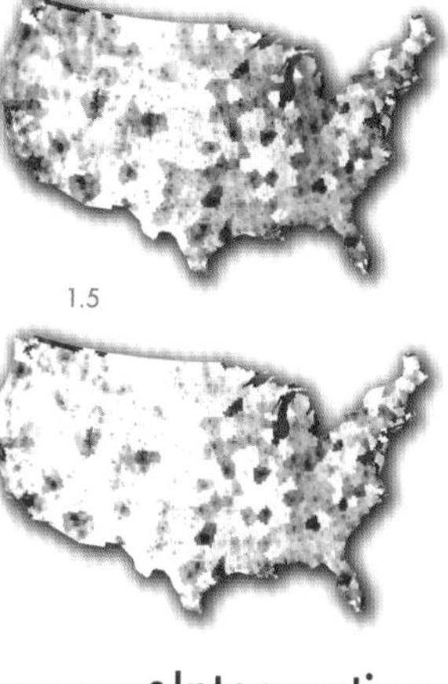

Intervention

UNCOOPERATIVE

COOPERATIVE

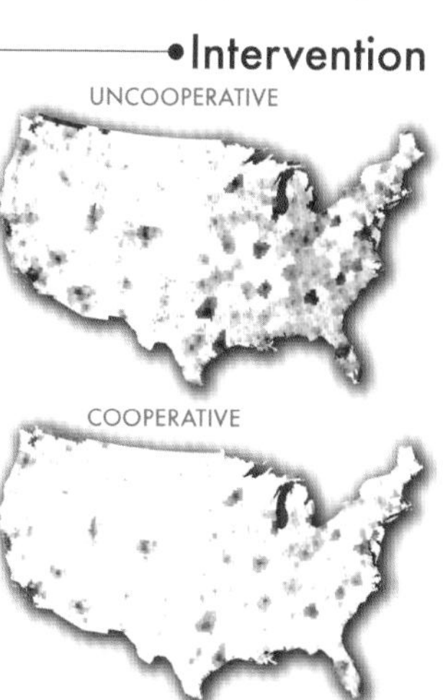

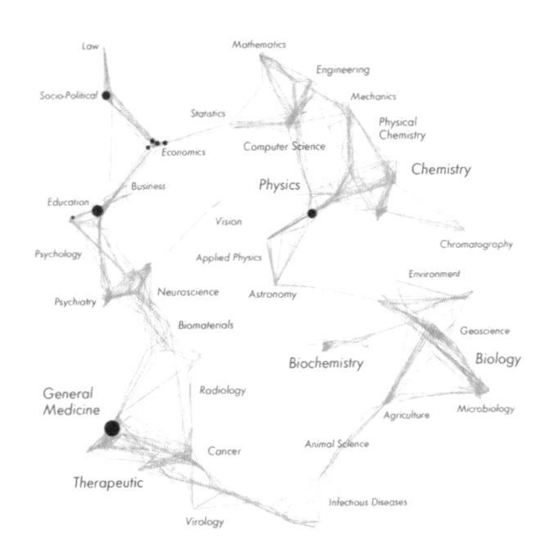

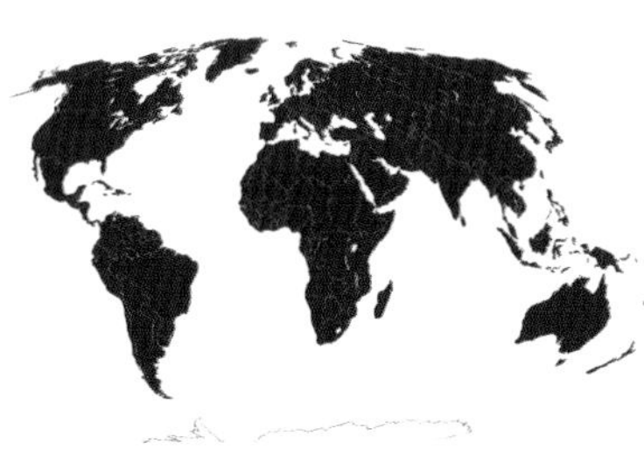

2007

[./logicaland] Participative Global Simulation

By Michael Aschauer, Maia Gusberti, and Nik Thoenen, in collaboration with Sepp Deinhofer

VIENNA, AUSTRIA, 2007

Courtesy of Michael Aschauer, Maia Gusberti, and Nik Thoenen, in collaboration with Sepp Deinhofer; re-p.org

The map shows different screenshots of the Web-based participative global simulation [./logicaland], including the geographic origin of its online participants. Financial and natural resource endowments of 185 nations can be manipulated for this visual project study. The simulation aims to raise awareness and understanding of our world's complex economic, political, and social systems.

The digital simulation is based on a dynamic global world model called Regionalized World III (RW-3). The RW-3 difference equation model was designed by Frederick Kile and Arnold Rabehl in Wisconsin in the mid-1970s.

The original model of [./logicaland] simulated regional areas and was modified to include simulation of the interacting financial and natural resources of specific countries. The simulation is initilized with "real" values from the year 2001, derived from the statistics in the CIA's *World Factbook*.

Anyone with Internet access can log on to **http://logicaland.net** and participate in rounds of play lasting up to 22 hours. Input parameters include target investment distribution in such sectors as industry, agriculture, and technology; adjustments can be made in the percentage distributions of development aid—food and utilities surplus—and can be changed for one or more countries. The parameter changes made by participants become "votes" that are polled by the server and fed back into the simulation so that possible effects can be examined. However, a single user's influence is minimal as it is a fraction of all participants' actions. Major change requires coordinated collective action.

Geographically, [./logicaland] encompasses the entire world. More importantly, it explicitly reveals how economics, demographics, politics, and the environment are interlinked through processes whose effects cross national borders. Further detailed study of these processes itself crosses disciplinary boundaries. The simulation visualizes and makes palpable the changes in complex economic, political, and social systems that occur when a community actively intervenes. Finally, in contrast to most tools in science that are neither participatory nor public, [./logicaland] allows for public participation in all the various levels of global simulation.

The source code for [./logicaland], licensed under the GNU General Public License (GNU GPL), is available at **http://logicaland.net/download.html**.

Michael Aschauer earned a master's degree in digital arts/visual media design from the University of Applied Arts in Vienna. Prior to that, he studied philosophy at the University of Vienna and informatics at the Technical University of Vienna. He lives and works as an artist and freelance technical and artistic services provider in Vienna. His works have been shown at numerous international festivals and exhibitions.

Maia Gusberti studies graphic design in Biel and media arts at the University of Applied Arts in Vienna. In 1999, she cofounded the design cooperative re-p.org—an independent unit for research, concepts, experiments, and output in the fields of visual design and art. Since 2007, she has worked as an artist and communications designer in Switzerland. Gusberti has received awards in Net Vision at Prix Ars Electronica (Linz, Austria) and a Premiere Grant from the University of Applied Arts. She has held residencies in Sofia, Bulgaria; Paris, France; and Cairo, Egypt.

Nik Thoenen is a graduate of the School for Graphic Design in Biel. He is a cofounder of the design cooperative re-p.org in Vienna. Since 2007, he has worked as an artist as well as a graphic and type designer in Switzerland.

Sepp Deinhofer studied informatics and computer science at the Vienna Technical University, as well as visual media design at the University of Applied Arts in Vienna. He lives and works in Vienna as a freelance software engineer, and is closely involved with the design cooperative re-p.org.

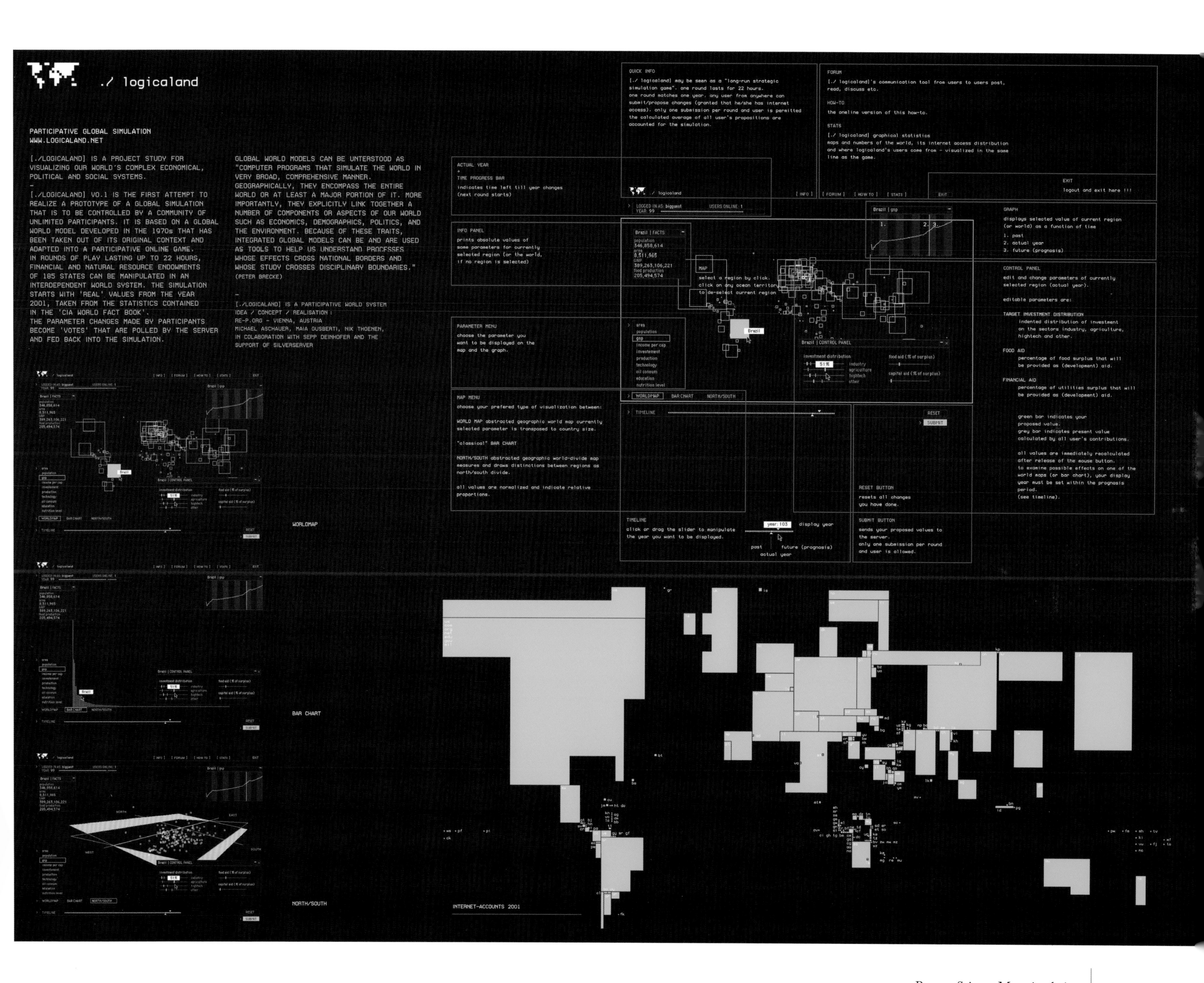

./ logicaland
PARTICIPATIVE GLOBAL SIMULATION
WWW.LOGICALAND.NET
[./LOGICALAND] IS A PROJECT STUDY FOR VISUALIZING OUR WORLD'S COMPLEX ECONOMICAL, POLITICAL AND SOCIAL SYSTEMS.
-
[./LOGICALAND] VO.1 IS THE FIRST ATTEMPT TO REALIZE A PROTOTYPE OF A GLOBAL SIMULATION THAT IS TO BE CONTROLLED BY A COMMUNITY OF UNLIMITED PARTICIPANTS. IT IS BASED ON A GLOBAL WORLD MODEL DEVELOPED IN THE 1970s THAT HAS BEEN TAKEN OUT OF ITS ORIGINAL CONTEXT AND ADAPTED INTO A PARTICIPATIVE ONLINE GAME.
IN ROUNDS OF PLAY LASTING UP TO 22 HOURS, FINANCIAL AND NATURAL RESOURCE ENDOWMENTS OF 185 STATES CAN BE MANIPULATED IN AN INTERDEPENDENT WORLD SYSTEM. THE SIMULATION STARTS WITH 'REAL' VALUES FROM THE YEAR 2001, TAKEN FROM THE STATISTICS CONTAINED IN THE 'CIA WORLD FACT BOOK'.
THE PARAMETER CHANGES MADE BY PARTICIPANTS BECOME 'VOTES' THAT ARE POLLED BY THE SERVER AND FED BACK INTO THE SIMULATION.
GLOBAL WORLD MODELS CAN BE UNTERSTOOD AS "COMPUTER PROGRAMS THAT SIMULATE THE WORLD IN VERY BROAD, COMPREHENSIVE MANNER. GEOGRAPHICALLY, THEY ENCOMPASS THE ENTIRE WORLD OR AT LEAST A MAJOR PORTION OF IT. MORE IMPORTANTLY, THEY EXPLICITLY LINK TOGETHER A NUMBER OF COMPONENTS OR ASPECTS OF OUR WORLD SUCH AS ECONOMICS, DEMOGRAPHICS, POLITICS, AND THE ENVIRONMENT. BECAUSE OF THESE TRAITS, INTEGRATED GLOBAL MODELS CAN BE AND ARE USED AS TOOLS TO HELP US UNDERSTAND PROCESSES WHOSE EFFECTS CROSS NATIONAL BORDERS AND WHOSE STUDY CROSSES DISCIPLINARY BOUNDARIES."
(PETER BRECKE)
-
[./LOGICALAND] IS A PARTICIPATIVE WORLD SYSTEM
IDEA / CONCEPT / REALISATION :
RE-P.ORG - VIENNA, AUSTRIA
MICHAEL ASCHAUER, MAIA GUSBERTI, NIK THOENEN,
IN COLABORATION WITH SEPP DEINHOFER AND THE SUPPORT OF SILVERSERVER
QUICK INFO
[./ logicaland] may be seen as a "long-run strategic simulation game". one round lasts for 22 hours. one round matches one year. any user from anywhere can submit/propose changes (granted that he/she has internet access). only one submission per round and user is permited the calculated average of all user's propositions are accounted for the simulation.
FORUM
[./ logicaland]'s communication tool from users to users post, read, discuss etc.
HOW-TO
the oneline version of this how-to.
STATS
[./ logicaland] graphical statistics maps and numbers of the world, its internet access distribution and where logicaland's users come from - visualized in the same line as the game.
EXIT
logout and exit here !!!
ACTUAL YEAR
+
TIME PROGRESS BAR
indicates time left till year changes (next round starts)
LOGGED IN AS: bigguest
USERS ONLINE: 1
YEAR: 99
[INFO]
[FORUM]
[HOW TO]
[STATS]
EXIT
INFO PANEL
prints absolute values of some parameters for currently selected region (or the world, if no region is selected)
Brazil | FACTS
population
346,858,614
area
8,511,965
GNP
389,263,106,221
food production
205,494,574
MAP
select a region by click.
click on any ocean territory to de-select current region
Brazil | gnp
1.
2.
3.
GRAPH
displays selected value of current region (or world) as a function of time
1. past
2. actual year
3. future (prognosis)
PARAMETER MENU
choose the parameter you want to be displayed on the map and the graph.
area
population
gnp
income per cap
investement
production
technology
oil consum
education
nutrition level
Brazil
Brazil | CONTROL PANEL
investment distribution
51%
industry
agriculture
hightech
other
food aid (% of surplus)
capital aid (% of surplus)
CONTROL PANEL
edit and change parameters of currently selected region (actual year).
editable parameters are:
TARGET INVESTMENT DISTRIBUTION
indented distribution of investment on the sectors industry, agriculture, hightech and other.
FOOD AID
percentage of food surplus that will be provided as (development) aid.
FINANCIAL AID
percentage of utilities surplus that will be provided as (development) aid.
green bar indicates your proposed value.
grey bar indicates present value calculated by all user's contributions.
all values are immediately recalculated after release of the mouse button.
to examine possible effects on one of the world maps (or bar chart), your display year must be set within the prognosis period.
(see timeline).
MAP MENU
choose your prefered type of visualization between:
WORLD MAP abstracted geographic world map currently selected parameter is transposed to country size.
"classical" BAR CHART
NORTH/SOUTH abstracted geographic world-divide map measures and draws distinctions between regions as north/south divide.
all values are normalized and indicate relative proportions.
WORLDMAP
BAR CHART
NORTH/SOUTH
TIMELINE
RESET
SUBMIT
RESET BUTTON
resets all changes you have done.
TIMELINE
click or drag the slider to manipulate the year you want to be displayed.
year: 103
display year
past
future (prognosis)
actual year
SUBMIT BUTTON
sends your proposed values to the server.
only one submission per round and user is allowed.
WORLDMAP
BAR CHART
NORTH/SOUTH
NORTH
SOUTH
EAST
WEST
INTERNET-ACCOUNTS 2001

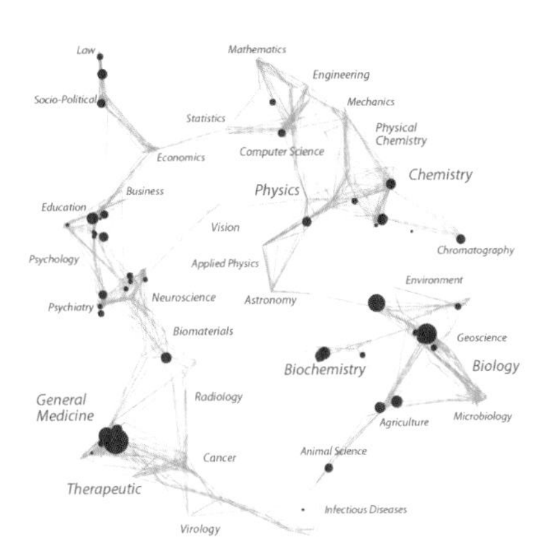

2006

Science & Technology Outlook: 2005–2055

By Marina Gorbis, Jean Hagan, Alex Soojung-Kim Pang, and David Pescovitz

PALO ALTO, CALIFORNIA, 2006

Courtesy of The Institute for the Future

Aim

In March 2005, the U.K. Government Office for Science (GO Science) established the Horizon Scanning Centre to help inform cross-governmental and departmental priority setting and strategy formation.

The *Science & Technology Outlook: 2005–2055* map was developed by the Institute for the Future (IFTF) for the Horizon Scanning Centre as part of a study of future trends in science and technology. It is an internally consistent, plausible view of the future based on the best expertise available. Rather than being a prediction, it aims to provide the reader with a contextualized understanding of the intricacies and interdependencies between current and future trends. An outlook is neither a statement of policy nor an indication of preference. Rather, it is meant to provide a context for planning and policy.

This outlook map represents a convergence of two formerly distinct practices in futures work. IFTF has a history of producing visual maps in its meetings and brainstorming sessions, but it has traditionally published text-intensive white papers. The map is one of many IFTF maps that both summarizes the collective wisdom of its experts and findings of its researchers and supports facilitated processes that turn that wisdom into strategy, policy, and action.

Interpretation

Science & Technology Outlook: 2005–2055 shows key developments suggested by workshop participants associated with major research trends. Key developments are organized in a timeline—starting with 2005 on the left and ending with 2055 on the right. The colored circles surrounding them are keyed to nine map themes and metathemes identified by IFTF researchers as major drivers. Map themes include **Small World**, **Intentional Biology**, **Extended Self**, **Mathematical World**, **Sensory Transformation**, and **Lightweight Infrastructure**. The three metathemes relate to the structure and geography of science and technology: democratized innovation, transdisciplinarity, and emergence. The time scale and event dates are intentionally vague, reflecting the inherent uncertainty of the enterprise. On the top and bottom of the main timeline are additional challenges, such as **population growth**, **climate changes**, or the **rise of China**, which participants suggested would shape the direction of science and technology. A series of perspectives—each focusing on one of the nine trends—accompanies the map.

Marina Gorbis is the executive director of the Institute for the Future (IFTF) and director of the Technology Horizons Program, which focuses on innovations at the intersection of new technologies and social organization. She holds a Master of Public Policy from the University of California, Berkeley; a certificate in international business from the University of London; and a BA in industrial psychology from the University of California, Berkeley. A native of Odessa, Ukraine, Gorbis is particularly suited to see things from a global perspective. She has directed international programs and led international development projects for SRI International in China, Japan, Vietnam, India, and Eastern Europe. Gorbis has also authored publications on international business and economics, with an emphasis on regional innovation and competitiveness.

Jean Hagan is a creative director at IFTF and designer of *Science & Technology Outlook: 2005–2055*.

Alex Soojung-Kim Pang is a research director at IFTF. He holds a PhD in History and Sociology of Science from the University of Pennsylvania. Pang brings strong historical and theoretical perspectives to his work on the future of pervasive computing, the end of cyberspace, and the coevolution of technology and society. He is also the founding editor of Future Now, IFTF's blog on emerging technologies, and is involved in other IFTF experiments involving social networking media, new publishing platforms, and futures research.

David Pescovitz is a research director at IFTF; a writer-in-residence for the College of Engineering, including Electrical Engineering and Computer Sciences (EECS), at the University of California, Berkeley; and coeditor of the popular blog Boing Boing. Pescovitz holds a BFA in electronic media from the University of Cincinnati and an MS in journalism from the University of California, Berkeley. Pescovitz has also written for *Scientific American*, *Popular Science*, *Business 2.0*, the *New York Times*, the *Washington Post*, *Salon*, the *Los Angeles Times*, *IEEE Spectrum*, *The Industry Standard*, *Spin*, MTV (online), *Interior Design*, the Discovery Channel (online), *Flash Art*, *Small Times*, Britannica Online, and *New Scientist*, among many other publications and Web sites.

INSTITUTE FOR THE FUTURE

Science & Technology Outlook: 2005–2055

Technology Horizons Program
Institute for the Future
124 University Avenue, 2nd Floor, Palo Alto, CA 94301
t 650.854.6322 f 650.854.7850 www.iftf.org

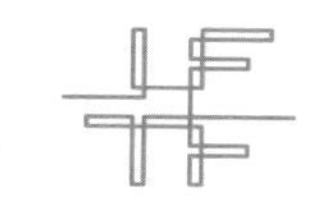

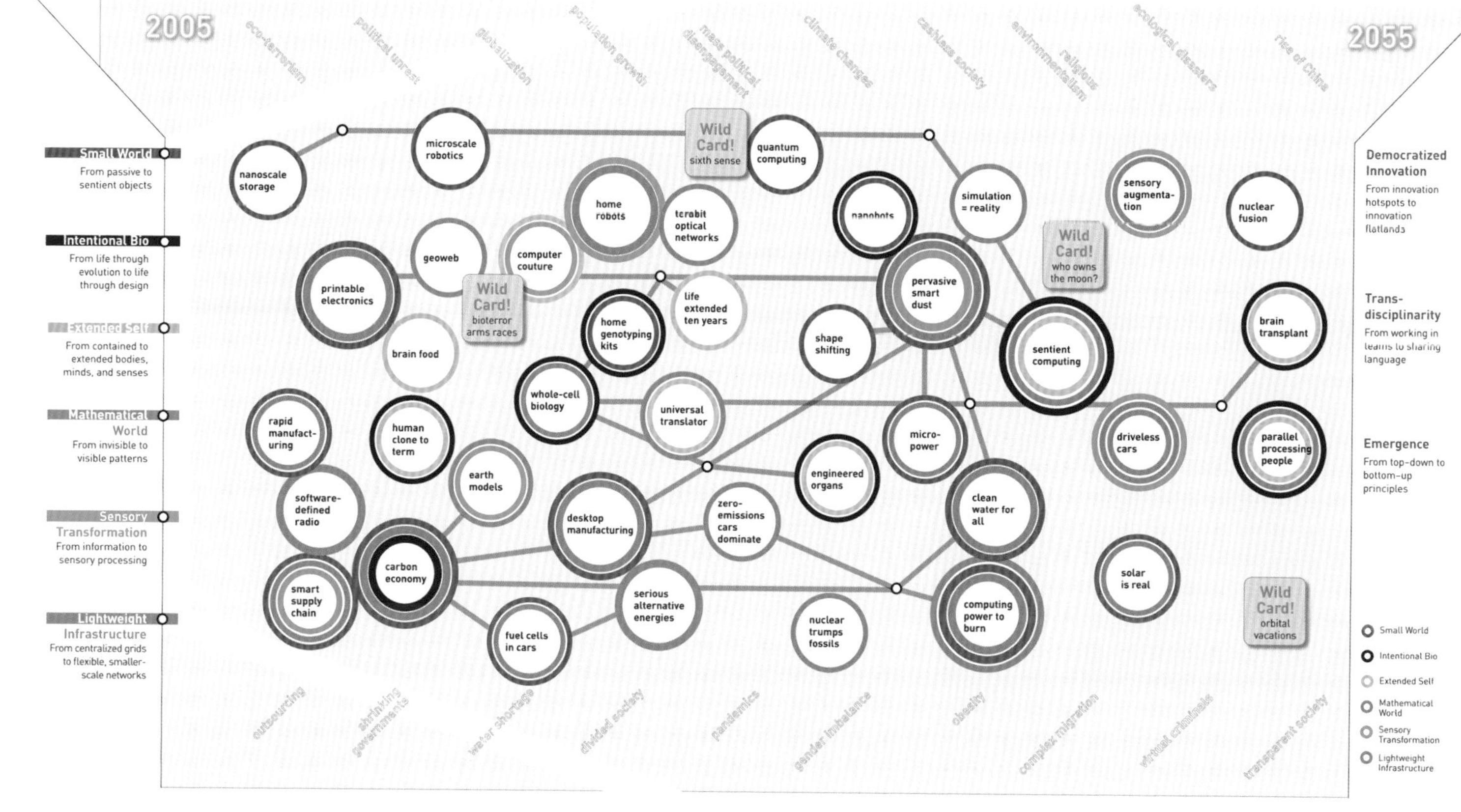

A map is a tool for navigating an unknown terrain. In the case of this map, Science & Technology Outlook: 2005–2055, the terrain we're navigating is the uncharted territory of science and technology (S&T) in the next 50 years. However, the map of the future is not a tool for prediction or, for that matter, the product of predictions. Nor is it comparable to modern navigation techniques in which we rely on a shrinking number of strong signals, like GPS coordinates, to show the right path. Rather, it's more akin to classical low-tech navigational techniques with their reliance on an array of weak signals such as wind direction, the look and feel of the water, and the shape of cloud formations. Taken together, these signals often prove more useful for navigation than high-tech methods because, in addition to aiding travelers in selecting the "right" path, the signals contextualize information and reveal interdependencies and connections between seemingly unrelated events, thus enriching our understanding of the landscape. That's precisely the intention of this map of the future of S&T—to give the reader a deeper contextual understanding of the landscape and to point to the intricacies and interdependencies between trends.

While developing the map, the Institute for the Future (IFTF) team listened for and connected a variety of weak signals, including those generated during interviews and workshop conversations involving more than 100 eminent U.K. and U.S. experts in S&T—academicians, policymakers, journalists, and corporate researchers. The IFTF team also compiled a database of outlooks on developments that are likely to impact the full range of S&T disciplines and practice areas over the next 50 years. We also relied on IFTF's 40 years of experience in forecasting S&T developments to create the map and an accompanying set of S&T Perspectives that discuss issues emerging on the S&T horizon and are important for organizations, policymakers, and society-at-large to understand.

On this map, six themes are woven together across the 50-year horizon, often resulting in important breakthroughs. These are supported by key technolgies, innovations, and discoveries. In addition to the six themes, three meta-themes—democratized innovation, transdisciplinarity, and emergence—will overlay the future S&T landscape influencing how we think about, learn about, and practice science. Finally, S&T trends won't operate in a vacuum. Wider social, demographic, political, economic, and environmental trends will both influence S&T trends and will be influenced by them. Some of these wider trends surround the map to remind us of the larger picture.

MAP THEMES

Small World

After 20 years of basic research and development at the 100-nanometer scale, the importance of nanotechnology as a source of innovations and new capabilities in everything from materials science to medicine is already well-understood. Three trends, however, will define how nanotechnology will unfold, and what impacts it will have. First, nanotechnology is not a single field with a coherent intellectual program; it's an opportunistic hybrid, shaped by a combination of fundamental research questions, promising technical applications, and venture and state capital. Second, nanotechnology is moving away from the original vision of small-scale mechanical engineering—in which assemblers build mechanical systems from individual atoms—toward one in which molecular biology and biochemistry contribute essential tools (such as proteins that build nanowires). Finally, nanotechnology will also serve as a model for transdisciplinary science. It will support both fundamental research and commercially oriented innovation; and it will be conducted not within the boundaries of conventional academic or corporate research departments, but in institutional and social milieux that emphasize heterogeneity.

Intentional Biology

For 3.6 billion years, evolution has governed biology on this planet. But today, Mother Nature has a collaborator. Inexpensive tools to read and rewrite the genetic code of life will bootstrap our ability to manipulate biology from the bottom up. We'll not only genetically reengineer existing life but actually create new life forms with purpose. Still, we will not be blind to what nature has to teach us. Evolution's elegant engineering at the smallest scales will be a rich source of inspiration as we build the bio-nanotechnology of the next 50 years.

Extended Self

In the next 50 years, we will be faced with broad opportunities to remake our minds and bodies in profoundly different ways. Advances in biotechnology, brain science, information technology, and robotics will result in an array of methods to dramatically alter, enhance, and extend the mental and physical hand that nature has dealt us. Wielding these tools on ourselves, humans will begin to define a variety of different "transhumanist" paths—that is, ways of being and living that extend beyond what we today consider natural for our species. In the very long term, following these paths could someday lead to an evolutionary leap for humanity.

Mathematical World

The ability to process, manipulate, and ultimately understand patterns in enormous amounts of data will allow decoding of previously mysterious processes in everything from biological to social systems. Scientists are learning that at the core of many biological phenomena—reproduction, growth, repair, and others—are computational processes that can be decoded and simulated. Using techniques of combinatorial science to uncover such patterns—whether these are physical, biological, or social—will likely occupy an increasing share of computing cycles in the next 50 years. Such massive computation will also make simulation widespread. Computer simulation will be used not only to help make decisions about large complex scientific and social problems but also to help individuals make better choices in their daily lives.

Sensory Transformation

In the next ten years, physical objects, places, and even human beings themselves will increasingly become embedded with computational devices that can sense, understand, and act upon their environment. They will be able to react to contextual clues about the physical, social, and even emotional state of people and things in their surroundings. As a result, increasing demands will be placed on our visual, auditory, and other sensory abilities. Information previously encoded as text and numbers will be displayed in richer sensory formats—as graphics, pictures, patterns, sounds, smells, and tactile experiences. This enriched sensory environment will coincide with major breakthroughs in our understanding of the brain—in how we process sensory information and connect various sensory functions. Humans will become much more sophisticated in their ability to understand, create, and manage sensory information and ability to perform such tasks will become keys to success.

Lightweight Infrastructure

A confluence of new materials and distributed intelligence is pointing the way toward a new kind of infrastructure that will dramatically reshape the economics of moving people, goods, energy, and information. From the molecular level to the macroeconomic level, these new infrastructure designs will emphasize smaller, smarter, more independent components. These components will be organized into more efficient, more flexible, and more secure ways than the capital-intensive networks of the 20th century. These lightweight infrastructures have the potential to boost emerging economies, improve social connectivity, mitigate the environmental impacts of rapid global urbanization, and offer new future paths in energy.

META-THEMES

Democratized Innovation

Before the 20th century, many of the greatest scientific discoveries and technical inventions were made by amateur scientists and independent inventors. In the last 100 years, a professional class of scientists and engineers, supported by universities, industry, and the state, pushed amateurs aside as a creative force. At the national scale, the capital-intensive character of scientific research made world-class research the property of prosperous advanced nations. In the new century, a number of trends and technologies will lower the barriers to participation in science and technology again, both for individuals and for emerging countries. The result with be a renaissance of the serious amateurs, the growth of new scientific and technical centres of excellence in developing countries, and a more global distribution of world-class scientists and technologists.

Transdisciplinarity

In the last two centuries, natural philosophy and natural history fractured into the now-familiar disciplines of physics, chemistry, biology, and so on. The sciences evolved into their current form in response to intellectual and professional opportunities, philanthropic priorities, and economic and state needs. Through most of the 20th century, the growth of the sciences, and academic and career pressures, encouraged ever-greater specialization. In the coming decades, transdisciplinary research will become an imperative. According to Howard Rheingold, a prominent forecaster and author, "transdisciplinarity goes beyond bringing together researchers from different disciplines to work in multidisciplinary teams. It means educating researchers who can speak languages of multiple disciplines—biologists who have understanding of mathematics, mathematicians who understand biology."

Emergence

The phenomenon of self-organizing swarms that generate complex behavior by following simple rules—will likely become an important research area, and an important model for understanding how the natural world works and how artificial worlds can be designed. Emergent phenomena have been observed across a variety of natural phenomena, from physics to biology to sociology. The concept has broad appeal due to the diversity of fields and problems to which it can be applied. It is proving useful for making sense of a very wide range of phenomena. Meanwhile, emergence can be modeled using relatively simple computational tools, although those models often require substantial processing power. More generally, it is a richly suggestive as a way of thinking about designing complex, robust technological systems. Finally, emergence is an accessible and vivid a metaphor for understanding nature. Just as classical physics profited from popular treatments of Newtonian mechanics, so too will scientific study and technical reproductions of emergent phenomena likely draw benefits from the popularization of its underlying concepts.

Data

The Horizon Scanning Centre is engaged in identifying future issues of potentially significant impact or opportunity over the next 10, 20, and 50 years. This Delta Scan database contains the core science and technology information used by the Horizon Scanning Centre to meet this objective. The bulk of the database was created by IFTF in 2005 and early 2006.

To design the database, IFTF conducted interviews and a series of six workshops involving more than 100 eminent U.K. and U.S. experts in science and technology. Participants included scientists, journalists, venture capitalists, research and development managers, policy makers, graduate students, and postdoctoral students.

The workshops took place in the United Kingdom and the United States. In each country, there was one workshop held for scientists under the age of 40.

Participants were asked to record (1) potential discoveries of great importance, (2) events they expected to happen, (3) wild cards, and (4) major issues that needed to be addressed.

These items were written on large Post-it notes and placed on a timeline of science and technology spanning 50 years into the future. The outlook draws upon those six workshop maps.

Experts' suggestions were collected in a database of outlooks on science and technology that are likely to impact the full range of science and technology disciplines and practice areas over the next 50 years. Utilizing IFTF's 40 years of expertise in forecasting science and technology developments, institute researchers identified six key themes and three metathemes that drive these developments. Wider social, demographic, political, economic, and environmental challenges that will both influence and be influenced by science and technology trends were also extracted.

The resulting data can be accessed via Delta Scan (**http://www.deltascan.org**), which contains a hundred outlook pages covering a wide range of scientific disciplines and technologies.

Technique

Road mapping originated in the 1970s and has been improved continuously. Road maps are widely used to support innovation, strategy, and policy development. A recent summit on road mapping extracted more than 450 examples of graphic road map representations from more than 900 public-domain road map documents.

Road maps help improve communication during the design process, as participants have to agree on the items to be included and the visual language used to represent them. They are also helpful in later stages by working to align the goals and strategies of different entities and organizations. Well-designed road maps are an efficient means to communicate complex ideas using visual metaphors and language. Each road map is a functional rather than ornamental object—a tool to stimulate discussion in strategy and planning sessions, a static object supporting a dynamic process.

Road mapping is the collaborative effort of major stakeholders. The goal is to create a map that visually captures and communicates the collective goal as well as the "road" toward achieving it. For example, a product road map links product, technology, and resource plans with market opportunities and business strategies.

Reference System

A one-dimensional chronological reference system, as well as color-coded trends, is used to map and interlink key developments. Six color-coded themes are used to structure the map.

Data Overlays

Deciding on the type and quantity of information needed to display imagery or text often proves difficult. The outlook map represents the interwoven networklike trends and developments as a map. The elevator-speech overview, introduction, details, and interviews about trends are given as text in the accompanying *Science & Technology Outlook: 2005–2055* report. Pages about Small World and Emergence have been reproduced as examples on the right.

Details

The Delta Scan forum supports the interactive exploration and search for key developments, trends, and challenges. One can click on an item in the outlook map list to see what was written about it. The "Find Pages" option supports a keyword-based search. The "Editorial Guidelines" explain how to comment on the contents of the database. The "Project" link leads to details about the project.

As a result, macro trends can be linked to local issues in such areas as technology and society, health and healthcare, and global business trends.

Unique Features

Currently, one of the most important users of Delta Scan is the U.K. government. With the recent decision by the HM Treasury that science and technology would be one of the areas on which the government would be focusing its five-year plans, the demand among midlevel policy makers for the kind of information that the database offers has increased considerably.

David King, the chief scientist of the U.K. government, argues that horizon scanning will have a powerful influence on policy making—and not only in Whitehall. He says that the scans "are not 'predicting' the future, [but] rather setting out a broad range of different possibilities and challenging assumptions. … [The] government can't just sit back and wait for it to happen … government has to identify opportunities and risks at least 5 to 10 years ahead when making policy. It can then make decisions that might move us from an unfavourable to a favourable scenario." He adds that "Although [Delta Scan] was designed as a tool for government, I believe it will also have a broader use across the private sector."

While still in the development stage, the horizon scans have already started to influence policy making. They have, for example, aided the Health and Safety Executive (HSE) in planning for the future of workplace health and safety, and the HM Treasury in writing its report "Opportunities and Challenges for the UK: Analysis for the 2007 Comprehensive Spending Review."

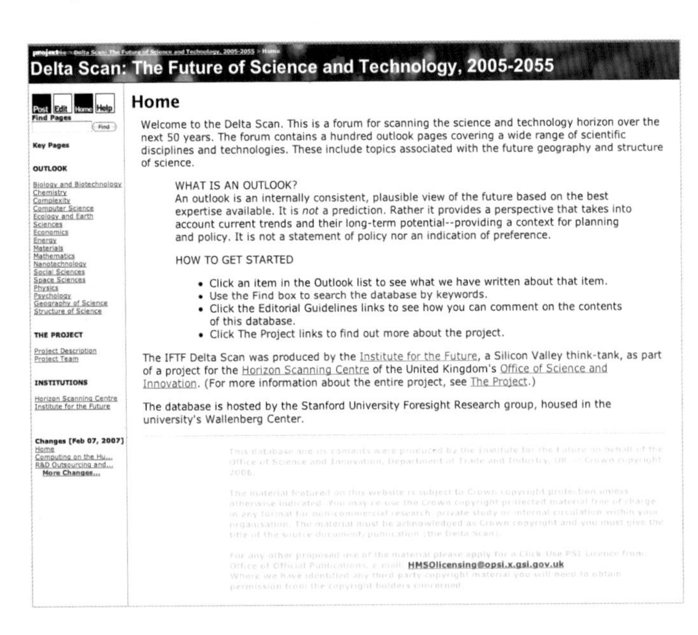

SMALL WORLD

From passive to sentient objects

What's new to say about nanotechnology? Every forecaster sees it as the "Next Big Thing." First described in the mid-1980s, nanotechnology has moved steadily from the fringes to the center of science and seems poised to transform everything from materials to computing. However, as it has moved into the mainstream, it's also changed in some important ways. We think that there are three key points to understand regarding the future of nanotechnology.

It's Not About Nanotechnology, It's About the Small World

First of all, nanotechnology gets the attention, but it's really the combination of nanotechnology and micro-electro-mechanical systems (MEMS) that promises to be transformational. MEMS is nanotechnology's quiet older brother, and in the last decade has found its way into everything from automobiles (as sensors in engines and airbag accelerometers) to medical instruments. MEMS are, as the Institute for the Future's Paul Saffo puts it, "training wheels for nano."

First, MEMS have proven the value of mixing sensing and reactive capability into once passive objects and products. Companies and designers who have become skilled at MEMS integration are well poised to exploit nanotechnologies. Second, MEMS are manufactured using the same lithographic and etching processes that turn out cheap microprocessors by the billions. As those processes move into the sub-100nm range, they open the possibility of applying existing familiar manufacturing processes.

It's Not About Machines, It's About Efficiency and Sentience

Bill Joy, Michael Crichton, and Ray Kurzweil have created a public image of nanotechnology as being primarily concerned with creating tiny robots. So far, however, nanotechnology and MEMS have been applied in more familiar areas.

While there is the possibility that robotics will be an important part of the future of small-world technology, the big story isn't about creating new forms of autonomous intelligence; it's about weaving sentience and responsiveness into existing things.

MEMS is a key technology for creating sensors that allow computers to sample and react to their surroundings. They also serve as a bridge connecting built objects—everything from cell phones to highways—to the digital world.

Likewise, while nanotechnology's main uses have been to enhance materials and biomedical products, in the future the cutting edge of nanotechnology will shift to creating quantum computational devices and nanoscale sensors. At the start of the last century, we energized and illuminated our cities; in this century, we're going to make them intelligent.

It's Not About Machines, It's About Hybrids

The original language of nanotechnology spoke of foundries and factories, molecule-sized gears and levers, atoms as switches. But some of the most interesting nanoscale devices and processes owe as much to biology as to mechanical engineering. On the production side, scientists are using proteins and prions to spin nano-wires, or creating conditions in which amino acids self-assemble into nanotubes. Not only do small-world devices sometimes spring from biological roots, they also appropriate tools from biology. Lab-on-a-chip developers, for example, have experimented with harnessing flagellates as tiny motors and pumps. In other words, scientists working in the small world are less likely to generate an inventory of mechanical systems, than a menagerie of hybrids and chimeras.

DETAILS

Making existing things more sentient

A 2005 study revealed that the largest target industries for nanotechnology were biomedical/life sciences, materials, consumer products, and chemicals. In these industries, nanotechnology has mainly served to improve the performance and efficiency of existing products, processes, and materials. Carbon nanotubes, for example, have been used to strengthen composite materials, while finely milled zinc nanoparticles are used in sunscreens.

Nanotechnology—not a discipline but an opportunity

Global spending on nanotechnology has been increasing rapidly. The spending is distributed among both leading scientific nations and up-and-coming countries like South Korea, India, and China.

Another notable characteristic of this spending is that much of it has been devoted to investing in new laboratories and other infrastructure or awarded through national initiatives designed to build fundamental capabilities in nanotechnology research. While there is an expectation of substantial commercial payoffs, virtually every country funding nanotechnology research is taking a long-term, strategic view.

The number of patents granted that deal with nanotechnology has also grown quickly. There has been a particulary sharp increase in the number of patents—and by implication, level of research activity—since 1995.

1 TARGET INDUSTRIES FOR NANOTECHNOLOGY

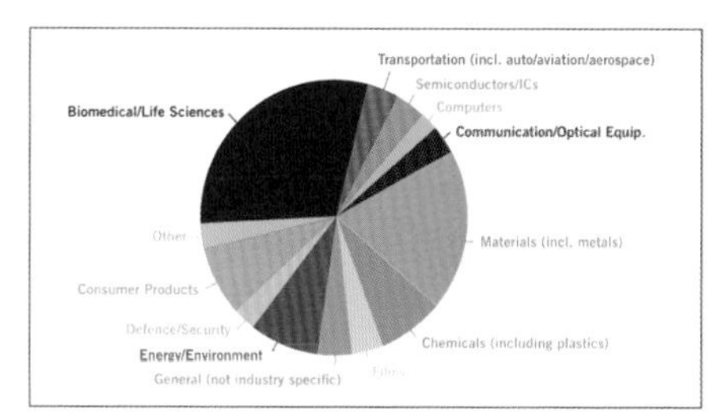

Source: President's Council of Advisors on Science and Technology, 2005. *The National Nanotechnology Initiative at Five Years: Assessment and Recommendations of the National Nanotechnology Advisory Panel.*

2 NANOTECHNOLOGY SPENDING IS INCREASING RAPIDLY

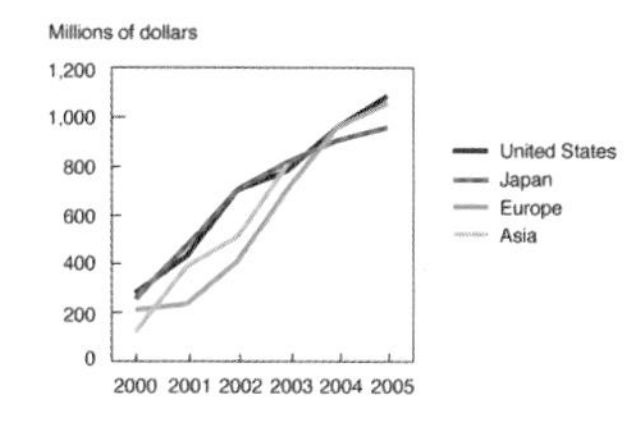

Source: National Science Foundation; President's Committee of Advisors on Science and Technology

Looking below the numbers, however, we see that small-world science breaks new sociological ground. Traditionally, as Thomas Kuhn argued in *The Structure of Scientific Revolutions*, scientific fields have been united by a few big questions, a deeply held epistemological viewpoint, and workhorse instruments and laboratory techniques. Disciplines have tended toward ever-greater specialization and fruitful isolation. MEMS and nanotechnology, in contrast, are less about a coherent world view and intellectual program, than a set of opportunities (commercial as well as scientific) and a trading zone of techniques and theories. Small-world science is the first great non-paradigmatic science, less a discipline than a permanent revolution.

It's about hybrids

Ian Pearson, Futurologist at BT, had this to say about the role of hybrids in nanotechnology: "After 2020, we'll use genetic modification to start harnessing the engines of propagation that exist in nature—using proteins to assemble things just as nature uses proteins to assemble things. And, of course, the control system you've got for doing that is DNA. This will ultimately have the effect of creating new kinds of intelligence-producing or even intelligent species. If we modify DNA in situ in a bacterium to assemble electronic circuits, which has already been demonstrated in principle today, then we can, in principle, make smart bacteria."

3 NANOTECH PATENTS ON THE RISE

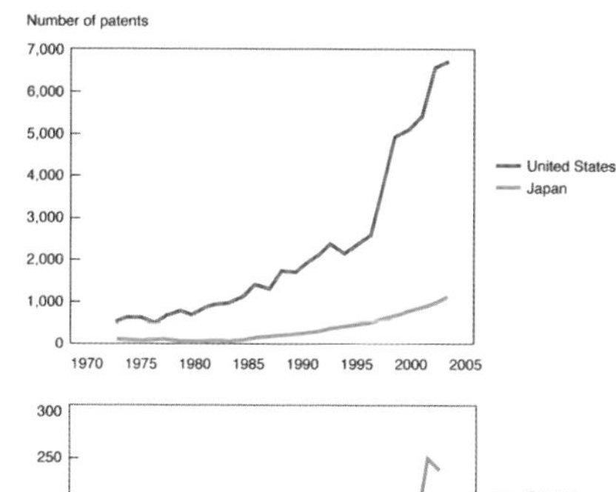

Source: Zan Huang, Hsinchun Chen, Alan Yip, Gavin Ng, Fei Guo, Zhi-Kai Chen, and Mihail C. Roco, Longitudinal patent analysis for nanoscale science and engineering: Country, institution and technology field, *Journal of Nanoparticle Research* 5: 333–363, 2003.

EMERGENCE

From top–down to bottom–up thinking

The wide diffusion of sensors and computing capabilities into everyday environments will make emergence not only a major paradigm for thinking but also for designing technological and social systems. At the core of emergence is the idea that patterns of complex behavior emerge when sets of independent agents follow simple rules—think, "dumb ants, smart behaviors." In the next few decades, unlimited computational resources will allow researchers increasingly to understand patterns of complex emergent behavior and, using this knowledge, simulate and design complex emergent systems in everything from life sciences and economics to trading and marketing. This means that many of these disciplines and organizational domains will be re-invented around the principles of emergence.

De-Coding Emergence

Emergent phenomena have been observed across a variety of natural phenomena, from physics to biology and sociology. An emergent behavior or emergent property can appear when a number of simple entities (agents) operate in an environment, forming more complex behaviors as a collective. The behaviors themselves are often unpredictable and unprecedented, and represent a new level of the system's evolution. In other words, such behaviors or properties are not characteristic of any single entity, nor can they easily be predicted or deduced from the behavior of the individual entities. The shape and behavior of a flock of birds or school of fish are good examples. Flocking is not a quality of any individual bird; it only emerges as a property of a group of birds. The appeal of the concept of emergence has several sources. It is proving useful for making sense of a very wide range of phenomena that are difficult to explain using simple causality. Emergence can also be modeled using relatively simple computational tools, but those models often require substantial processing power. With access to virtually unlimited computational power widening from a few research labs and large organizations to many, the ability to decode and understand emergent patterns in a variety of new domains will grow.

Designing for Emergence

Greater understanding of emergent systems and processes will lead to re-thinking of design principles in many domains outside biology and natural systems. Recent marketplace experiments—from eBay to Wikipedia to multiple open-source projects—are early indicators. These experiments are showing new structures for production, new webs of exchanges, and new processes for value creation that together point to an alternative "emergent" framework for organizing economic life. Many companies are beginning to use emergence as a design principle for marketing, viewing their customers not through the top–down lens of segments (hierarchical fixed categories), but through the lens of dynamic swarms and networks. As such, they're trying create conditions to encourage consumer swarming to their products and services. In technology areas, from robotics to nanotechnology, scientists are increasingly using principles of emergence to build complex autonomous systems. Social scientists, including economists, sociologists, and psychologists, are increasingly likely to use complexity theory and simulations both as an explanatory framework and design paradigm.

Emergence as a Mindset and Toolkit Shift

Emergence represents a mindset shift in sciences from focus on causality to focus on simulation; from looking at homogeneous independent variables to identifying complex patterns among heterogeneous and interdependent events; from traditional statistics to power laws with their long tails. This mindset shift requires new tools and new skill sets. Just like the discipline of economics is slowly being re-shaped by application of the principles of complexity and emergence, social scientists will need to increasingly understand new methodologies and tools, and apply them to understanding social phenomena.

DETAILS

Wikipedia: emergent knowledge creation

Wikipedia is a free online encyclopedia, written collectively from the bottom up using a wiki—an open, public, writable Web page that anyone can edit and change. It is a public knowledge base, created by aggregating individual contributions. Rather than relying on experts to create and edit individual entries, Wikipedia relies on scores of volunteers, experts, and other passionate amateurs to create content. Thus far, Wikipedia has amassed 750,000 articles in English, making it several times larger than *Encyclopædia Britannica.* It has a total of 2.2 million articles in 30 different languages, making it the largest global encyclopedia. Unlike any other encyclopedia which costs millions of dollars to produce, Wikipedia is produced at virtually no cost—it operates on small donations and is run primarily by volunteers.

According to Jimmy Wales, Wikipedia's founder, the goal is to "give every single person on the planet free access to human knowledge." Wikipedia uses a very simple set of rules to assign roles for administering and monitoring content areas and rights to copy, modify, and distribute Wikipedia's content, providing a good example of what emergent knowledge production might look like in the future. In addition to its online encyclopedia, the Wikimedia Foundation is now involved in a number of other knowledge-base projects using the same principles of emergent content creation, among them:

- **Wikinews**—a free online news source where any site visitor can add or edit stories
- **Wikibooks**—a collection of open-content textbooks that anyone can edit
- **Wikimedia Commons**—a repository of free content images, sound and other multimedia files
- **Wiktionary**—a free wiki dictionary, including thesaurus and lexicon

1 WIKIPEDIA, THE FREE ONLINE ENCYCLOPEDIA WRITTEN BY VOLUNTEERS

Source: Wikipedia

2 WIKIMEDIA FOUNDATION'S GOAL IS TO DEVELOP AND MAINTAIN OPEN CONTENT

Source: Wikimedia Foundation

Emergent organizational structures: peer-to-peer markets

eBay is a quintessential example of a platform that creates a playing field for ad hoc trading. eBay doesn't sell merchandise. It provides the environment, the tools, and social accounting mechanisms for trading, in other words, a set of simple rules and tools that enable the emergence of an active marketplace. One of the key mechanisms is the reputation system for rating sellers and buyers that ultimately creates the trust needed to sustain the auction market. According to traditional economics, a one-time exchange with a stranger typically ends up with the sucker's payoff—you send your money and the Tiffany lamp doesn't arrive. At eBay, individual buyers contribute a small amount of time and energy rating sellers. The ratings are aggregated into scores that indicate to other buyers who are highly reputable sellers—the so called Power Sellers. The value of the Power Seller reputation marker is translated into higher prices for the seller and reliable service for the buyer.

A number of other online markets now employ the same basic model for bringing together potential buyers and sellers. eBay has also spawned related services such as Picture it Sold that sells your items for you on eBay for a fee, and Intuit's method for using eBay auction prices to determine the fair market value of goods donated to charities for the purpose of accurate tax deductions.

Emergent air transport: from scheduled air traffic to air taxis

Re-orienting disciplines and inquiries around emergence will require a completely different set of tools and processes for analysis and design. For example, agent-based modeling (ABM) is a widely used technique for understanding emergent phenomenon. Instead of looking for causality, ABM relies on simulating a system composed of "behavioral entities"—traffic jams, shopping, voting, organizational dynamics. Instead of asking, "Can you explain it?" researchers using ABM techniques ask, "Can you grow it?"

DayJet is a new airline that plans to revolutionize business travel by utilizing ABM to simulate regional business travel. Using a few questions—where do you want to go? when do you need to be there? how early can you leave? how many seats do you need?—DayJet software will, within five seconds, map a best-case route and generate a detailed price quote. DayJet's goal is that 85% of its offers to potential customers are unlikely to be refused. On the surface, this may seem like a basic routing problem, but the Dayjet solution reflects a vast increase in complexity over anything the aviation industry has tried to date.

3 EBAY POWER SELLER DESIGNATION INSPIRES TRUST

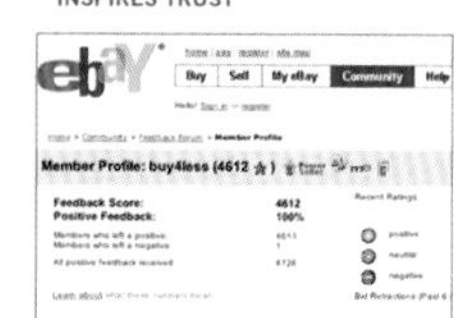

Source: www.eBay.com

4 DAYJET APPLIES EMERGENCE TO BUSINESS TRAVEL

Source: DayJet

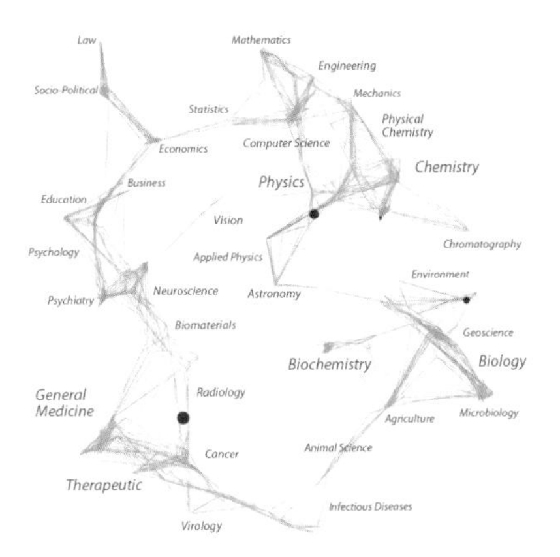

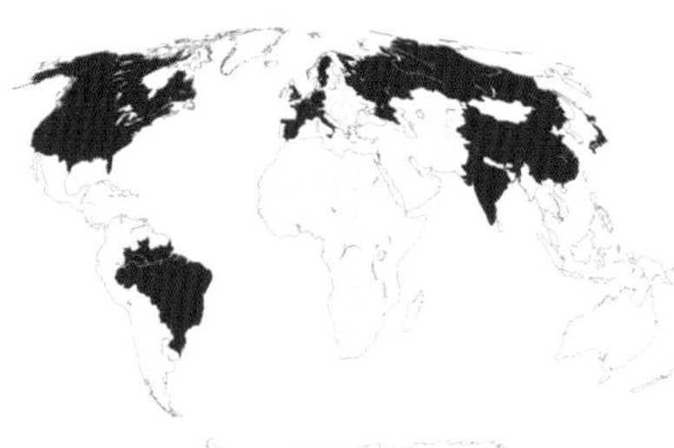

2007

113 Years of Physical Review

By Bruce W. Herr II, Russell J. Duhon, Elisha F. Hardy, Shashikant Penumarthy, and Katy Börner

BLOOMINGTON, INDIANA, 2007

Courtesy of Indiana University

Aim

How did the field of physics evolve over the last 100 years? When did the many different subfields of physics emerge, die, split, and merge? How are these subfields connected via permanent citation linkages woven by thousands of physicists over the many decades? Can the web of papers their authors interlinked via coauthor and paper-citation linkages be used to identify high-impact papers? Can it be mined to predict the future, or at least the next Nobel laureate?

Interpretation

This is the very first map of a 113-year scholarly data set that captures the structure and evolution of the entire field of physics. The visualization aggregates 389,899 papers published in 720 volumes of 11 journals between 1893 and 2005. Time runs horizontally. In 1975, the *Physical Review* introduced the Physics and Astronomy Classification Scheme (PACS) codes. In this visualization, the top-level PACS codes run vertically and are labeled from **PACS 0 General** to **PACS 9 Geophysics, Astronomy and Astrophysics** on the right. The 91,762 papers published from 1893 to 1976 take up the left third of the map. The 217,503 papers published from 1977 to 2000, for which there are references but no citation data, occupy the middle third of the map. The 80,634 papers from 2001 to 2005, for which citation data is available, fill the last third of the map. Each annual bar is further subdivided vertically into journals, and each journal is further subdivided horizontally into the volumes of the journal. The size of each journal-by-volume area is proportional to the number of papers published. Overlaid on this two-dimensional base map are all citations from every *Physical Review* paper published in 2005.

Each year, Thomson Reuters predicts three Nobel Prize awardees in physics based on data from its ISI Web of Knowledge, including citation counts, high-impact papers, and discoveries or themes worthy of special recognition. The map uses small Nobel Prize medals to indicate all Nobel prize–winning papers. Correct predictions are highlighted.

Bruce W. Herr II (data mining and visualization) **See page 166.**

Russell J. Duhon (data mining and visualization) designs, programs, parses, researches, and visualizes at the Cyberinfrastructure for Network Science Center at Indiana University. He likes making the little algorithms that fit between the big ones. His areas of interest include economics-inspired modeling of scientific activity, statistical methods for understanding data, and unusual data sets. He received a bachelor's degree in computer science from the School of Informatics at Indiana University in Bloomington, Indiana.

Elisha F. Hardy (graphic design) received a BA in fine arts at Indiana University and is currently a graduate student in human–computer interaction design at IU's School of Informatics and Computing. As a member of the Cyberinfrastructure for Network Science Center, led by Katy Börner, she has had the opportunity to work not only on this atlas but also on a range of other designs including logos, handouts, Web pages, and the maps of science featured here. Hardy's other artistic interests include ceramics and drawing. She is a Mac fanatic, having used them for more than 10 years.

Shashikant Penumarthy (data preparation) is a PhD student at the School of Library and Information Science, Indiana University. He holds a master's degree in computer science from Indiana University. His research pertains to enabling interoperability among information visualization tools by using ontologies to describe and process visualization concepts in an automated manner. As part of this line of research he developed the Virtual World Toolkit (VWTk), which is freely available via SourceForge. He also works as an independent consultant on the design and development of visualization-centric modeling and analysis tools and user interfaces.

Katy Börner (concept) **See page 132.**

113 Years of Physical Review

This visualization aggregates 389,899 articles published in 720 volumes of 11 journals between 1893 and 2005. The 91,762 articles published from 1893 to 1976 take up the left third on the map. In 1977, the Physical Review introduced the Physics and Astronomy Classification Scheme (PACS) codes, and the visualization subdivides into the top-level PACS codes. The 217,503 articles from 1977 to 2000, for which good citation data is not available, occupy the middle third on the map. The 80,634 articles from 2001 to 2005, for which good citation data is available, fill the last third of the map.

Each vertical bar is subdivided vertically into the journals that appear in it with height proportional to the number of papers, and each journal is subdivided horizontally into the volumes of the journal appearing in the column.

On top of this base map, all citations from the papers in every top-level PACS code in 2005 are overlaid and then drawn from the source area to the individual volumes containing papers cited.

The small Nobel Prize medals indicate the 24 volumes containing the 26 papers appearing in Physical Review for 11 Nobel prizes between 1990 and 2005. Each year, Thomson ISI predicts three Nobel Prize awardees in physics based on citation counts, high impact papers, and discoveries or themes worthy of special recognition. Correct predictions by Thomson ISI are highlighted.

Nobel Prizes in Physical Review

Year of Nobel Prize Winners Publication Year(s) (indicated by Nobel Prize medals on the right)

- 2005 Roy J. Glauber, John L. Hall, and Theodor W. Hänsch 1963, 1971
- 2004 David J. Gross, H. David Politzer, and Frank Wilczek 1973
  Thomson ISI successfully predicted a winner in this year, with the following paper.
  Gross D, Wilczek F. Ultraviolet Behavior of Non-Abelian Gauge Theories. Physical Review Letters 30: 1343 & 1973
- 2003 Anthony J. Leggett 1970
- 2002 Raymond Davis Jr., Masatoshi Koshiba, and Riccardo Giacconi 1962, 1968, 1987
- 2001 Eric A. Cornell, Wolfgang Ketterle, and Carl E. Wieman 1995, 1996
- 1998 Robert B. Laughlin 1982, 1983
- 1997 Steven Chu and William D. Phillips 1985, 1986, 1988
- 1996 David M. Lee, Douglas D. Osheroff, and Robert C. Richardson 1972
- 1995 Martin L. Perl 1959, 1975
- 1994 Bertram N. Brockhouse and Clifford G. Shull 1955, 1958
- 1990 Jerome I. Friedman, Henry W. Kendall, and Richard E. Taylor 1969

Bar Graph

- Physical Review
- Physical Review Series I
- Physical Review A
- Physical Review B
- Physical Review C
- Physical Review D
- Physical Review E
- Physical Review L
- Physical Review Special Topics Accelerated Beams
- Physical Review Physics Educational Research
- Physical Review Modern Physics

Lines

- PACS 0 General
  PACS 8 Interdisciplinary Physics and Related Areas of Science and Technology
- PACS 1 The Physics of Elementary Particles and Fields
  PACS 4 Electromagnetism, Optics, Acoustics, Heat Transfer, Classical Mechanics, and Fluid Dynamics
- PACS 2 Nuclear Physics
  PACS 3 Atomic and Molecular Physics
- PACS 5 Physics of Gases, Plasmas, and Electric Discharges
  PACS 9 Geophysics, Astronomy, and Astrophysics
- PACS 6 Condensed Matter: Structure, Mechanical and Thermal Properties
  PACS 7 Condensed Matter: Electronic Structure, Electrical, Magnetic, and Optical Properties

PACS
9
8
7
6
5
4
3
2
1
0
None

1893 1900 1910 1920 1930 1940 1950 1960 1970 1980 1990 2000 2005

Data

Physical Review Data Set

The *Physical Review* data set was provided by the American Physical Society (APS). It includes optical character recognition (OCR) of the full text, SGML/XML (Extensible Markup Language) of the full text, and bibliographic metadata in XML format for 398,005 papers (see table below).

Journal Name	# Papers	Start Date	End Date
Physical Review Series I	1,469	01/01/1893	01/01/1912
Physical Review Series II	47,941	01/01/1913	01/01/1970
Physical Review A	45,021	01/01/1970	01/01/2006
Physical Review B	117,229	01/01/1970	01/01/2006
Physical Review C	26,501	01/01/1970	01/01/2006
Physical Review D	47,183	01/01/1970	01/01/2006
Physical Review E	27,366	01/01/1993	01/01/2006
Physical Review Letters	81,775	01/01/1958	01/01/2006
Physical Review Special Topics: Accelerators and Beams	716	01/01/1998	01/01/2006
Physical Review Special Topics: Physics Education Research	12	01/01/2005	01/01/2006
Reviews of Modern Physics	2,792	01/01/1929	01/01/2006
Total	398,005		

XSLT (Extensible Stylesheet Language Transformations) was used to parse the many files that reported the bibliographic metadata in XML format. The resulting data was used to generate *113 Years of Physical Review*. Among other things, the *PNAS* dataset provides a unique Digital Object Identifier (DOI) for each paper; a listing of its authors; date(s) when the paper was received, revised, corrected, and published; PACS code(s); and citation references was extracted from the new data.

PACS codes have been in use since 1975. They are assigned by the author in consultation with reviewers and editors. The average number of PACS codes per paper since then is about 1.5. There is no "main" PACS code for a paper.

Nobel Laureates

The list of Nobel laureates in physics was obtained from the Nobel Foundation Web site (**http://nobelprize.org**). Using citation graphs of the most highly cited paper and other statistics from the ISI Web of Knowledge, the core paper that earned the Nobel prize was manually determined. The Nobel Prize is awarded to one or multiple persons for a distinct piece of work. On average, nearly 30 years pass from the time the work is first conducted until the Nobel Prize is awarded.

Nobel Predictions

Since 1989, Thomson Reuters has used a selection process to predict the next Nobel Prize winners in medicine, chemistry, physics, and economics. The predicted winners are named Thomson Reuters Scientific Laureates in recognition of the significant contribution their citations make to navigation within the ISI Web of Science. The selection process considers citation counts from the past 30 years, the quantity of high-impact papers, and themes worthy of special recognition, to ensure the individual chosen has made a fundamental contribution to the discovery in question. Predictor prizes and other recognitions are also used in the selection process.

The names of Thomson Reuters's Scientific Laureates were downloaded from the Thomson Reuters Scientific Hall of Laureates on July 14, 2006. The prediction for the 2004 Nobel Prize for David J. Gross and Frank Wilczek was made in 1990 by Angela Martello. Thomson Reuters Scientific Laureates who did, in fact, win the Nobel Prize include Michael B. Green, Shuji Nakamura, John H. Schwarz, Yoshinori Tokura, and Edward Witten.

Reference System

How to fit 389,899 papers published over 113 years on an 18.5-inch x 25-inch (about 47 centimeters x 63 centimeters) canvas at 360 dots per inch? Using the temporal and topical PACS organization of papers, a simple two-dimensional reference system is used. Time runs from left to right, and the PACS codes increase from bottom to top, with the lowest line of bar graphs having no PACS codes assigned. The first 83 years take up the left third of the page, the next 25 years occupy about 1 centimeter (less than half an inch) each in the middle third, and the last 5 years (which contain about 20 percent of all *Physical Review* publications) are plotted on the right third of the map.

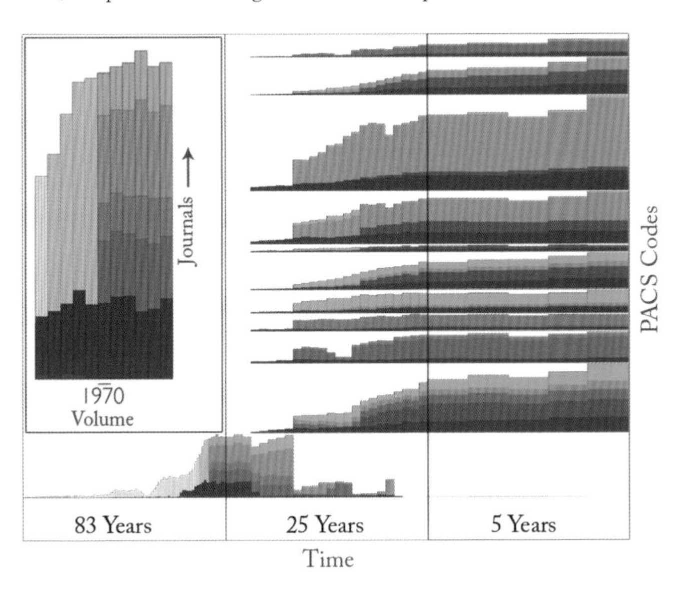

But how should the papers be laid out within each PACS-by-year box? The original idea was to group papers published in the same year by bibliographic coupling. That is, the more references two papers share, the more similar they are. Hierarchical agglomerative clustering would then be applied to compute a partition of the paper set that had high within-cluster similarity and low between-cluster similarity. Clusters would then be size-coded by the number of papers they contained and laid out using self-organizing maps or a Dorling cartogram.

We found that a layout using circles, however, would not be most space efficient. Additionally, the same journals and volumes would be positioned in different places in the 113 years-by-10 PACS boxes, as the layout algorithms are invariant to rotation and mirroring. Hence, we decided to group and display papers using their common "containers": journals and journal volumes.

The space was divided into horizontal slices for each journal, proportional in height to the quantity of papers from that journal containing that particular PACS. Those slices were then subdivided vertically for each volume in that journal, proportional in width to the quantity of papers.

Of the 261,190 papers with PACS codes, 110,978 were found to contain more than one top-level PACS code. Each paper was placed in all its PACS areas, thereby increasing not only the size of the respective "journal-by-volume" areas but also contributing their citations (yellow lines) to the set of citations among the different PACS.

Data Overlays

Papers

The different journals are color-coded (see legend), and they always appear in the same sequence, improving the readability of the map considerably. As mentioned previously, the height of each journal bar corresponds to the number of papers published by the journal in that year. Journal volume subdivisions reflect the percentages of papers published in them.

Citation Linkages

Interested in illustrating the cross-fertilization of different areas of research, we tried to find a way to show the citation linkages among the 389,899 papers. Metaphorically and visually, citation linkages might resemble "blood vessels" that interconnect and feed different areas of science.

For 2005 alone, there were 958,913 citation linkages. Exactly 42,516 *Physical Review* papers published in 2005 cited 145,665 papers within the data set and 176,473 papers outside the set. Aggregated at the level of journal volumes, there were 17,343 links (shown as tendrils) from 10 top-level PACS codes in 2005, linked to 2,691 journal volumes.

To create a flow map layout, papers were aggregated into journal volumes, and volumes were clustered into a tree according to their spatial arrangement. Lines were then drawn from the root (top-level PACS code) to each volume, remaining merged into single lines until the bounding box for the upcoming cluster was approached, then splitting into lines proportional to the amounts that "flow" to the respective volumes of the cluster. For this map, the flow map layout was independently applied to all 2005 citations from each top-level PACS code into other *Physical Review* papers. Spatially, each journal volume is represented by a constructed node in the center of the area it covers on the map. (The opposite page provides more information on flow maps.)

Nobel Laureates

To overlay the Nobel laureates whose Nobel-winning papers were published in *Physical Review*, the winning papers were crosschecked with the *Physical Review* database. If the Nobel-winning papers were found, then a Nobel icon was placed where the paper belongs on the base map, determined by looking at its year, PACS code, and journal. If the Nobel-winning paper was predicted by Thomson Reuters, then a larger Nobel icon was used (see image below).

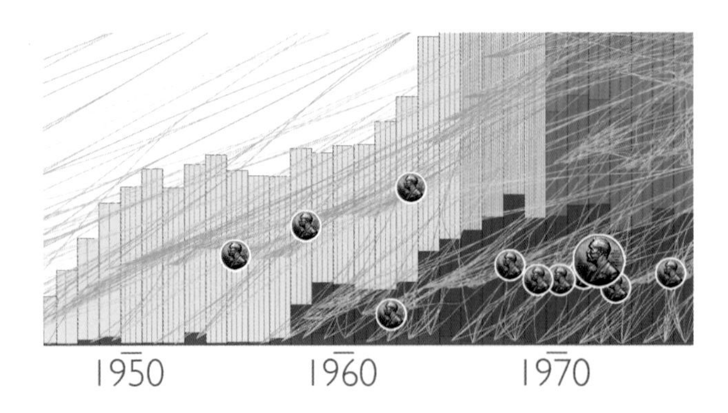

With the final reference system and data overlays, the growth and decline of the number of papers in different PACS codes and journals can be seen. The number and density of the overlaid 2005 citation links show that physics draws extensively on old and new knowledge.

Unique Features

Given a base map of physics, other data can be overlaid. Examples include trajectories of single physicists through the topic space; genealogies of generations of mentors and students; the effects of geographic movement (for instance, from Europe to the United States) on coauthorship networks; and scientific revolutions.

Flow Maps

Flow maps (also called movement maps) help to communicate the movement of tangible objects (for example, people, bank notes, or goods) in geographic space, as well as that of intangible objects (for example, energy, ideas, or reputation) in digital space. In *113 Years of Physical Review*, the maps are used to show the flow of knowledge via citation linkages.

Some of the first flow maps were drawn by hand by Henry Drury Harness, Alphonse Belpaire, and Charles Joseph Minard. A close-up of one of Minard's maps that illustrates the approximate amount of cotton imported by Europe in 1864 is shown below.

Flow maps are generated from tables that represent the amount of flow occurring between pairs of places, for example, migration tables produced by the U.S. Census Bureau. Typically, the greatest amount of movement is between spatially near places, modulated by the size of the places. The amount of traffic follows a power law; that is, there are few pairs of places that have enormous traffic and many pairs with minimal traffic.

The *113 Years of Physical Review* map uses a flow map layout algorithm, developed and made freely available by Doantam Phan and colleagues at Stanford University. The algorithm uses hierarchical clustering to generate a flow map tree for a particular root, given a set of nodes, positions, and flow data between the nodes. During layout, the flow map tree is used to merge edges that run in the same direction, thus minimizing edge crossings and improving readability.

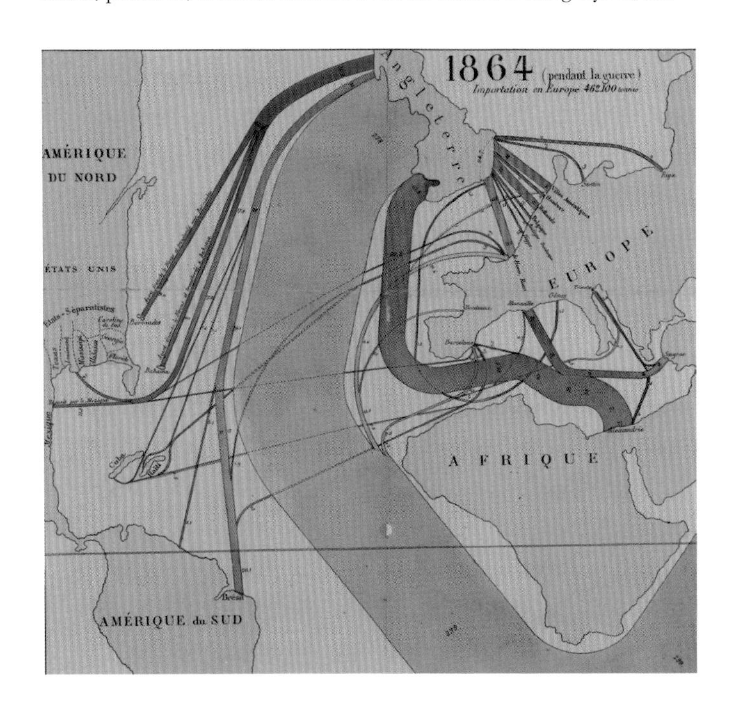

There are two types of flow maps: continuous and discrete.

Continuous flow maps

Continuous flow maps use vector fields or streamlines to show continuous flow patterns. As an example, the maps by Waldo Tobler below show estimated state-to-state net migration in the United States, depicted as a vector field with scalar potential, contour lines, and estimated trajectories.

Gaining and losing states, based on the marginals (geographically adjacent) of a 48-by-48 state migration table, see below.

The pressure to move in the United States, based on a continuous spatial gravity model, see below.

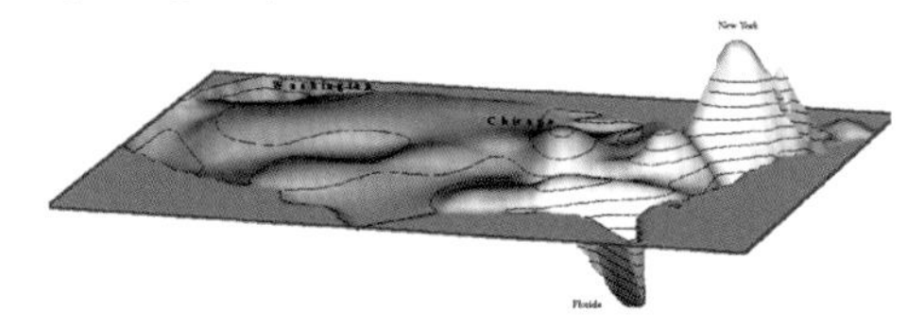

Migration potentials and gradients with the potentials shown as contours are given below.

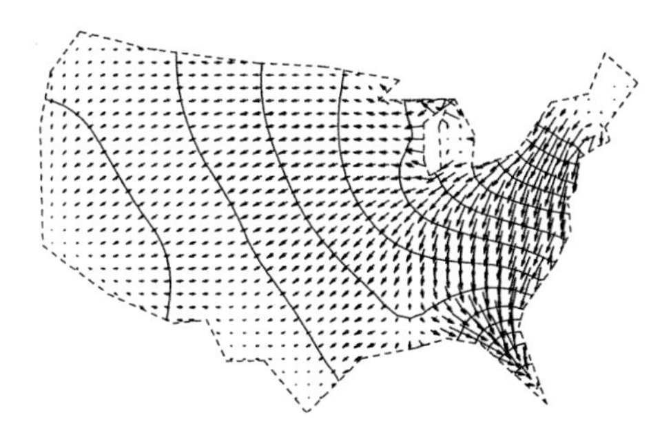

Migration potentials and streaklines with the gradient vectors connected to form streaklines are shown below.

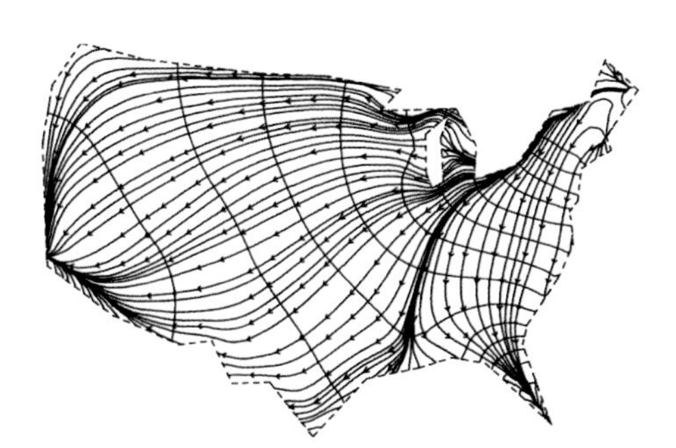

Discrete flow maps

Discrete flow maps use bands or arrows whose width is proportional to the volume moved. One of the first computer-generated flow maps (1959) showed a "Cartographatron display of 9,931,000 desire line traces of personal trips in Chicago."

In 1987, Tobler designed the Flow Mapper program. An updated version is freely available via the Center for Spatially Integrated Social Science. The program was used to generate the flow map of the 1995–2000 net migration shown below, left. It has also been used to map journal inter-citations using data by Clyde H. Coombs and colleagues. In that case, both citation directions are shown using half-arrows (see below, right).

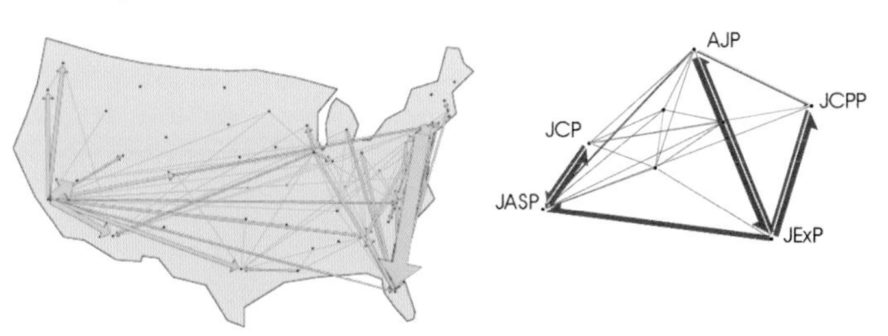

A flow map of migration from California to all other U.S. states by Phan et al. for 1995–2000 is shown below.

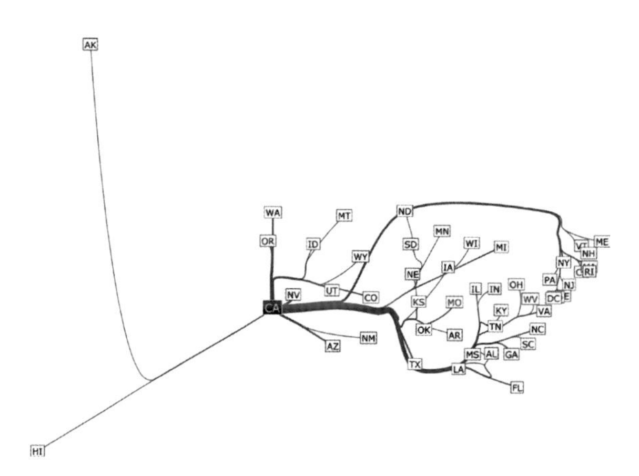

The algorithm also supports the layering of multiple flows. Shown here are the top 10 states from which residents migrate to California and to New York. The spatial pattern reveals that New York tends to attract people from the East Coast, while California residents come from diverse geographic regions of the United States.

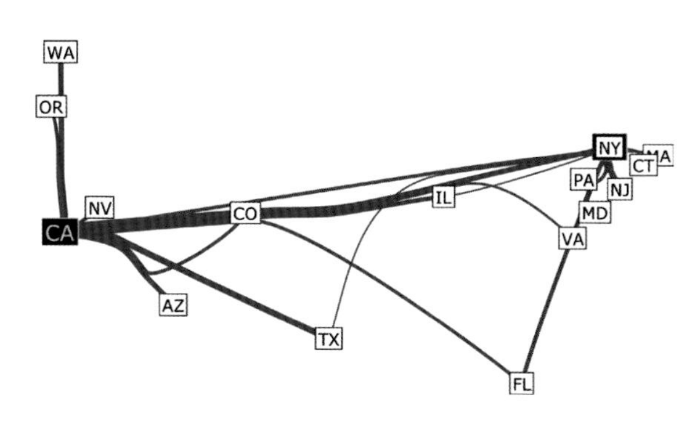

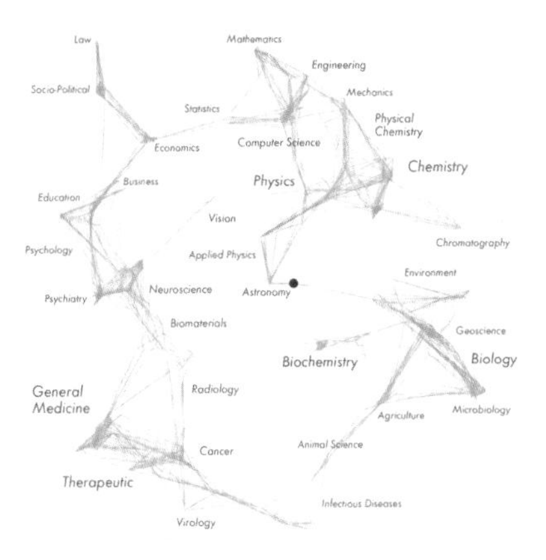

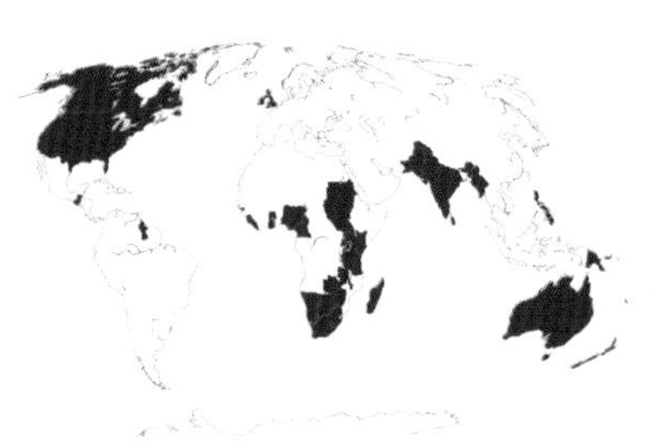

2007

Mapping the Universe: Space, Time, and Discovery!

By Chaomei Chen, Jian Zhang, Lisa Kershner, Michael S. Vogeley, J. Richard Gott III, and Mario Juric

PHILADELPHIA, PENNSYLVANIA, AND PRINCETON, NEW JERSEY, 2007

Courtesy of Drexel University and Princeton University

Aim

People have always been fascinated by the stars and their relationship to cradle Earth. Astronomy was one of the first sciences practiced. Many children were inspired to choose scientific careers by the first man in space, the first man on the moon, or the Mars Exploration Rover Mission. But how can we best communicate the immense size and complexity of the physical data, scholarly activities, and resulting scientific theories?

This map aims to communicate the structure of the universe and discoveries made by the Sloan Digital Sky Survey (SDSS). Relevant scientific literature, SDSS data, and imagery are used to show how theory and tool developments have influenced progress in astronomy.

Interpretation

This map represents space, time, and our discoveries of phenomena in both. Space is shown as a large circular **Map of the Universe**. Planet Earth—the starting point of mankind's discoveries—occupies the center. More than 600,000 astronomical objects, including some of the most distant quasars discovered by the SDSS, were positioned according to their correct ascension and the natural logarithm of their distance from Earth. The map also shows major discovery dates and the durations of accelerated citation growth (continuous bursts of citations). Time is captured as a **Time Spiral** at the top right, which plots the sequence of newly emergent themes over time. Themes were extracted from astronomical literature relevant to SDSS's work. Discoveries and their many interlinkages are shown as an evolving **Network of Scientific Literature**. Yellow lines cross-reference concepts, citation hubs, and the paths leading to discoveries; they also highlight past and current hot spots. Short-term predictions of research trends can be made by linear extrapolation of the current average citation acceleration rate in the SDSS literature of 3.17 years with a standard deviation of 1.8 years. Candidates for points of growth in the near future are suggested in the network and in the time spiral.

Chaomei Chen is an associate professor at the College of Information Science and Technology, Drexel University. His research interests include mapping scientific frontiers and visualization of emerging trends and patterns of scientific literature. He is the editor in chief of *Information Visualization*, and the author of *Information Visualization: Beyond the Horizon* (Springer-Verlag, 2004; 2nd edition, 2006) and *Mapping Scientific Frontiers: The Quest for Knowledge Visualization* (Springer-Verlag, 2003). He is the creator of CiteSpace, a free Java application for detecting and visualizing intellectual trends in scientific literature.

Jian Zhang is a doctoral student at the College of Information Science and Technology, Drexel University. He received a master's degree in science and technology journalism from Texas A&M University.

Lisa Kershner is a graphic designer at the College of Information Science and Technology, Drexel University.

Michael S. Vogeley is an associate professor of physics at Drexel University. His research interests include observational cosmology, formation and evolution of galaxies and active galactic nuclei, and statistical analysis of large data sets. He received a PhD in astronomy from Harvard University and has held positions at Johns Hopkins University, the Space Telescope Science Institute, and Princeton University.

J. Richard Gott III is a professor of astrophysics at Princeton University, working on cosmology and general relativity. He discovered an exact solution to Einstein's field equations for two moving cosmic strings—a solution of particular interest as it would allow time travel to the past. He has studied the topology of large-scale structure in the universe, and with Mario Juric coauthored the paper "Map of the Universe."

Mario Juric is a member of the Institute for Advanced Study in Princeton, New Jersey. He holds a PhD in astrophysical sciences from Princeton University. Juric is a theorist working on a diverse set of topics, ranging from long-term dynamics of extrasolar planets and the structure and evolution of the Milky Way to studies of the properties of asteroids in the solar system.

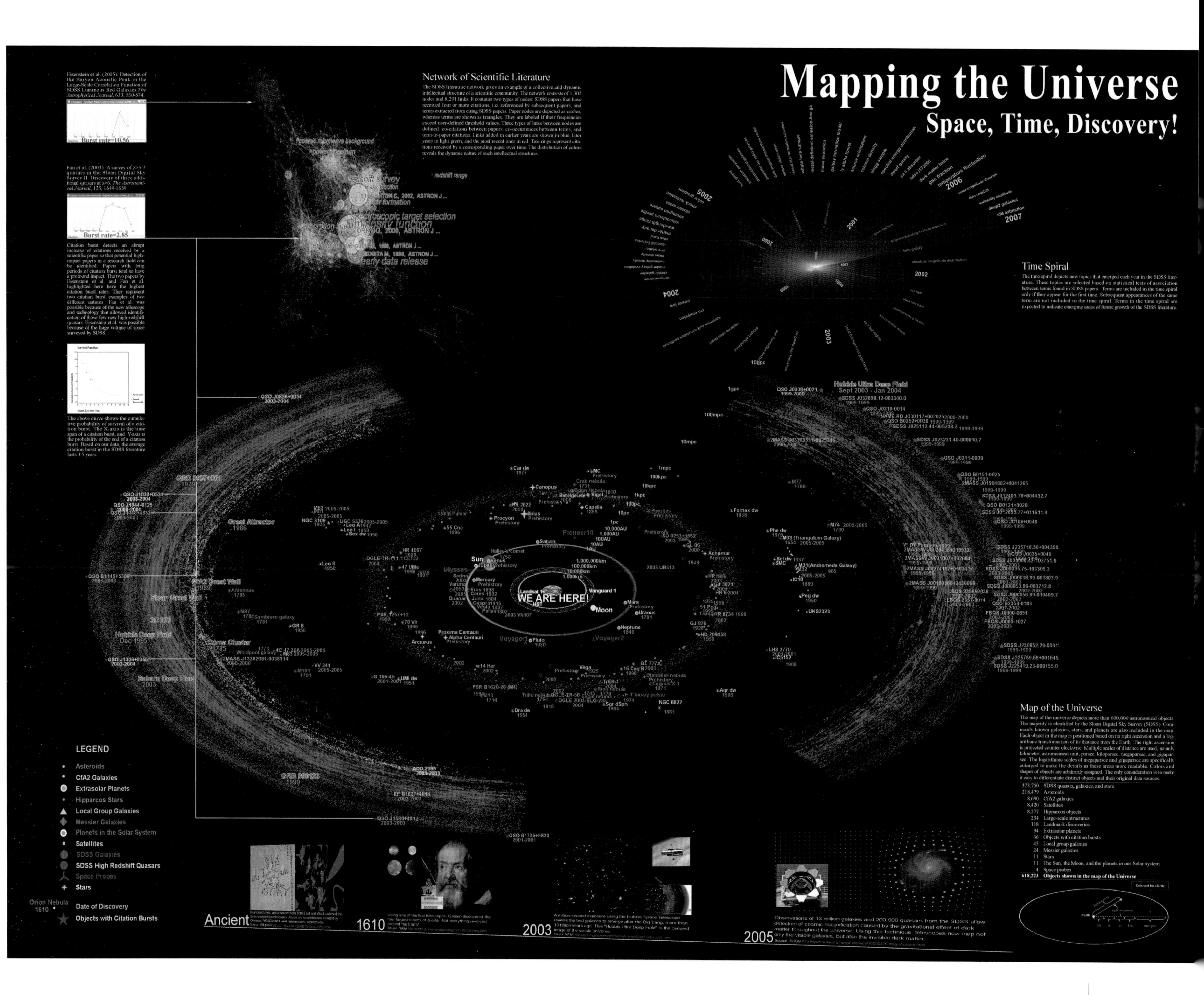

Mapping the Universe
Space, Time, Discovery!
Network of Scientific Literature
The SDSS literature network gives an example of a collective and dynamic intellectual structure of a scientific community. The network consists of 1,302 nodes and 8,291 links. It contains two types of nodes: SDSS papers that have received four or more citations, i.e. referenced by subsequent papers, and terms extracted from citing SDSS papers. Paper nodes are depicted as circles, whereas terms are shown as triangles. They are labeled if their frequencies exceed user-defined threshold values. Three types of links between nodes are defined: co-citations between papers, co-occurrences between terms, and term-to-paper citations. Links added in earlier years are shown in blue, later years in light green, and the most recent ones in red. Tree rings represent citations received by a corresponding paper over time. The distribution of colors reveals the dynamic nature of such intellectual structures.
Eisenstein et al. (2005). Detection of the Baryon Acoustic Peak in the Large-Scale Correlation Function of SDSS Luminous Red Galaxies. The Astrophysical Journal, 633, 560-574.
Burst rate=10.56
Fan et al. (2003). A survey of z>5.7 quasars in the Sloan Digital Sky Survey II: Discovery of three additional quasars at z>6. The Astronomical Journal, 125, 1649-1659.
Burst rate=2.85
Citation burst detects an abrupt increase of citations received by a scientific paper so that potential high-impact papers in a research field can be identified. Papers with long periods of citation burst tend to have a profound impact. The two papers by Eisenstein et al. and Fan et al. highlighted here have the highest citation burst rates. They represent two citation burst examples of two different natures. Fan et al. was possible because of the new telescope and technology that allowed identification of those few new high-redshift quasars. Eisenstein et al. was possible because of the huge volume of space surveyed by SDSS.
Survival Function
The above curve shows the cumulative probability of survival of a citation burst. The X-axis is the time span of a citation burst, and Y-axis is the probability of the end of a citation burst. Based on our data, the average citation burst in the SDSS literature lasts 3.5 years.
Time Spiral
The time spiral depicts new topics that emerged each year in the SDSS literature. These topics are selected based on statistical tests of association between terms found in SDSS papers. Terms are included in the time spiral only if they appear for the first time. Subsequent appearances of the same term are not included in the time spiral. Terms in the time spiral are expected to indicate emerging areas of future growth of the SDSS literature.
2001
2002
2003
2004
2005
2006
2007
Map of the Universe
The map of the universe depicts more than 600,000 astronomical objects. The majority is identified by the Sloan Digital Sky Survey (SDSS). Commonly known galaxies, stars, and planets are also included in the map. Each object in the map is positioned based on its right ascension and a logarithmic transformation of its distance from the Earth. The right ascension is projected counter clockwise. Multiple scales of distance are used, namely kilometer, astronomical unit, parsec, kiloparsec, megaparsec, and gigaparsec. The logarithmic scales of megaparsec and gigaparsec are specifically enlarged to make the details in these areas more readable. Colors and shapes of objects are arbitrarily assigned. The only consideration is to make it easy to differentiate distinct objects and their original data sources.
373,750 SDSS quasars, galaxies, and stars
218,479 Asteroids
8,690 CfA2 galaxies
8,420 Satellites
8,277 Hipparcos objects
234 Large-scale structures
118 Landmark discoveries
94 Extrasolar planets
66 Objects with citation bursts
45 Local group galaxies
24 Messier galaxies
11 Stars
11 The Sun, the Moon, and the planets in our Solar system
4 Space probes
618,223 Objects shown in the map of the Universe
Enlarged for clarity
Earth
LEGEND
Asteroids
CfA2 Galaxies
Extrasolar Planets
Hipparcos Stars
Local Group Galaxies
Messier Galaxies
Planets in the Solar System
Satellites
SDSS Galaxies
SDSS High Redshift Quasars
Space Probes
Stars
Orion Nebula 1610
Date of Discovery
Objects with Citation Bursts
WE ARE HERE!
Sun
Moon
Hubble Ultra Deep Field
Sept 2003 - Jan 2004
Great Attractor
1986
CfA2 Great Wall
Sloan Great Wall
Hubble Deep Field
Coma Cluster
Subaru Deep Field
M33 (Triangulum Galaxy)
M31(Andromeda Galaxy)
Ancient
1610
Using one of the first telescopes, Galileo discovered the four largest moons of Jupiter. Not everything revolved around the Earth!
2003
A million-second exposure using the Hubble Space Telescope reveals the first galaxies to emerge after the Big Bang, more than 13 billion years ago. This "Hubble Ultra Deep Field" is the deepest image of the visible universe.
2005
Observations of 13 million galaxies and 200,000 quasars from the SDSS allow detection of cosmic magnification caused by the gravitational effect of dark matter throughout the universe. Using this technique, telescopes now map not only the visible galaxies, but also the invisible dark matter.

Data

Two data sets were used to produce the map: astronomical data made available via the Sloan Digital Sky Survey (SDSS) and publication data from the Web of Science (WoS).

SDSS is an international effort to capture high-resolution optical images of more than a quarter of the sky using a dedicated 2.5-meter telescope at Apache Point, New Mexico. The goal is to create a three-dimensional map of about a million galaxies and quasars. The telescope's 120-megapixel camera can image 1.5 square degrees of sky at a time, about 8 times the area of the full moon. A pair of spectrographs measures the spectra of more than 600 galaxies and quasars in a single observation, and their distances can be computed from that data. In June 2005, SDSS had imaged more than 8,000 square degrees of the sky in 5 bandpasses, detecting nearly 200 million celestial objects and measuring the spectra of more than 675,000 galaxies, 90,000 quasars, and 185,000 stars. SDSS data is released to the scientific community and the general public in annual increments. It is used in studies concerning topics ranging from asteroids and nearby stars to the large-scale structure of the universe. For this map, SDSS Data Release 1 was used.

Publication data was downloaded from the Web of Science (WoS) using a topic search for SDSS, resulting in 1,254 papers published between 2000 and 2007.

Technique

The **Network of Scientific Literature** depicts the accumulated view of the dynamic structure of the research field. It identifies papers with citation bursts and high centrality as indicators of significant discoveries. Astronomical positions of these discoveries are marked in the **Map of the Universe** with corresponding durations of citation bursts. The distribution of objects associated with citation bursts indicates the current focal or priority areas of SDSS research on the backdrop of the universe. The **Time Spiral** depicts the emergence of new concepts over time.

Map of the Universe

The circular **Map of the Universe** in the center of the chart uses a Cartesian reference system to lay out 618,223 astronomical objects, such as planets, stars, asteroids, galaxies, and quasars, as well as manmade objects, such as satellites and space probes. Each object is positioned according to its correct ascension and a logarithmic transformation of its distance from Earth (see explanatory figure at lower right of the exhibit map and below). Different scales of distance are shown: kilometer, astronomical unit, parsec, kiloparsec, megaparsec, and gigaparsec. Different types of objects are assigned different colors and shapes (see legend).

The circular map was inspired by "The Map of the Universe" by Gott and Juric. Both maps depict the entire visible universe, and both cover the

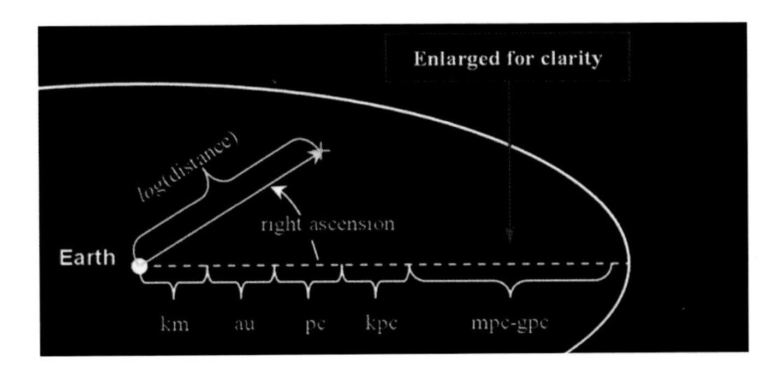

same set of multiple scales, from the surface of the Earth to the objects gigaparsecs distant. The circular shape is designed to make the map more intuitive, as if the viewer were standing on Earth and looking around.

Overlaid on the map are significant discoveries, including SDSS discoveries and discoveries made by other projects (for example, the Hubble Ultra Deep Field). The next layer is the predictive overlay. Here the most highly cited and highly bursty publications were identified and associated with SDSS discoveries and objects marked by a red star. The yellow label next to the star indicates the duration of the corresponding citation burst.

Time Spiral

Time is represented as a timeline spiral, starting with 1990 at the center and spiraling outward to 2007. This reference system is used to order and display emerging topics, presumed to be areas of future growth in the SDSS literature. The terms were extracted using feature selection algorithms based on log-likelihood ratio tests. The algorithm estimates the strength of association between low-frequency and high-frequency terms.

Specifically, the **Time Spiral** uses the following process to filter the noun phrases:
(1) take the extracted noun phrases and measure statistical association strengths between the terms, (2) select terms that have statistically significant strengths, (3) identify the first appearance of these terms and discard subsequent occurrences, and (4) plot the first appearances of selected terms on the time spiral. Since emerging terms tend to appear with low frequencies in their early stages, the time spiral depicts fresh terms that have statistically significant connections to the established ones.

Literature Network

The WoS literature data set comprises 1,254 papers with 11,968 unique references (cited papers) and 11,345 unique terms. Exactly 1,147 papers in the set received more than 4 citations within the data set and they are shown in the network together with the top 155 terms. Paper nodes are connected by cocitation links that have normalized weights above 0.4. Term nodes are connected by co-occurrence links if their normalized weights are above 0.4. A paper is linked to a term if the term appears in the paper. The final network consists of 1,302 nodes and 8,291 links. Major term and paper nodes are labeled. The top five highly cited papers are given in the table below.

The five terms used most often are "luminosity function" (215 counts), "early data release" (131), "redshift survey" (126), "spectroscopic target selection" (115), and "absorption line" (94). Two terms used at greater count were excluded from the survey: "Sloan Digital Sky Survey" (611; part of the search query used to retrieve the literature data set) and "data release" (235; the central SDSS activity).

CiteSpace (see opposite page) was used to analyze and lay out the literature network. The procedure was as follows:

1. Set analysis window from 2000 to 2007 in single-year slices.
2. Extract noun phrases from the title and abstract fields of each paper, as well as keywords listed in the fields of descriptors and identifiers. Note that the phrase extraction is limited to phrases containing two, three, or four words.
3. Set threshold values for times cited (c), times cocited (cc), and normalized link weight (ccv), to generate the network to be visualized. Here, c=4, cc=3, and ccv=40 was selected for all time slices (see top screenshot, opposite page). That is, a paper will only be shown if it was cited four times or more in at least one time slice. A cocitation link between two papers is selected if the two papers are either cocited three or more times in that time slice or the normalized link weight is greater than or equal to 0.4. Normalization is performed by dividing the number of times two papers are cocited by the square root of the product of the citation counts for these two papers.
4. Papers and terms in each slice form a network. Eight yearly networks are merged by CiteSpace. The resulting network depicts the earliest connection found in the sequence (see bottom screenshot, but also exhibit map). Note that the network layout is nondeterministic.
5. Adjust display threshold values and select nodes to be labeled.

Independent of their type, links added in earlier years are shown in blue or light blue (2000–2002), later years in green (2003–2005), and the most recent ones in red (2006–2007). The color of a paper node is rendered according to the time it was cited. Earlier citations result in a blue node color. Citations received within one year add an outer ring to a node. Nodes that are cited over multiple years have many "tree rings" and can be easily spotted. Highly cited papers are from earlier years, including York et al., 2000 and Stoughton et al., 2002.

Times Cited	Authors	Year	Title	Journal	Volume	First Page
585	York, Donald G. and 143 coauthors	2000	The Sloan Digital Sky Survey: Technical Summary	ASTRON J	V120	P1579
424	Stoughton, Chris. and 191 coauthors	2002	Sloan Digital Sky Survey: Early Data Release	ASTRON J	V123	P485
412	Fukugita, M., Ichikawa, T., Gunn, J. E., Doi, M., Shimasaku, K., Schneider, D. P.	1996	The Sloan Digital Sky Survey Photometric System	ASTRON J	V111	P1748
345	Gunn, J. E. and 39 coauthors	1998	The Sloan Digital Sky Survey Photometric Camera	ASTRON J	V116	P3040
314	Schlegel, David J., Finkbeiner, Douglas P., Davis, Marc	1998	Maps of Dust Infrared Emission for Use in Estimation of Reddening and Cosmic Microwave Background Radiation Foregrounds	ASTROPHYS J 1	V500	P525

CiteSpace

Chen aims to develop a generic approach to detecting and visualizing emerging trends and transient patterns in scientific literature. His particular interest is the design of so-called progressive knowledge domain visualizations, which show how research fronts and intellectual bases of a field change over time.

The concept of a *research front* refers to the approximately 40 to 50 papers in a field that scientists actively cite. As scientists tend to cite new work, the papers in a research front are generally recent and change as time passes. Here, a research front is conceptualized as cluster of cocited papers (see image opposite page, top left).

The *intellectual base* of a research front is its citation and cocitation footprint in scientific literature—an evolving network of scientific publications cited by research front papers.

On the opposite page, we explain the data analysis algorithms and visualization metaphors used in CiteSpace.

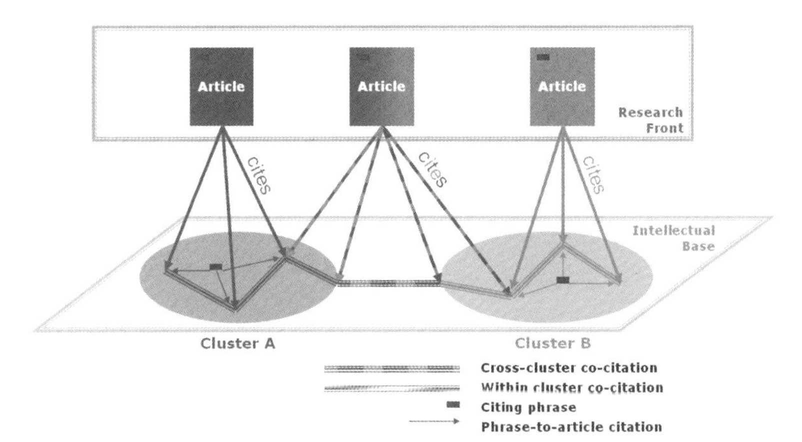

Analysis

Research front terms are detected from the source papers based on their burstiness. Users can choose to have these terms extracted from any combination of the title, abstract, descriptor, and identifier fields.

Emergent *research front concepts* are identified using Kleinberg's burst detection algorithm. The algorithm employs a probabilistic automaton whose states correspond to the frequencies of individual words. State transitions correspond to points in time around which the frequency of the word changes significantly. The algorithm returns a ranked list of the most significant word bursts in the paper stream, together with the intervals of time in which they occurred. This serves as a means of identifying topics or concepts that rose to prominence over the course of the stream, were discussed actively for a period of time, and then faded away.

Pivotal points of paradigm shift over time are computed using a betweenness centrality metric. For each node x, this metric computes the number of shortest geodesic paths from any node in the network to any other node that pass though x. Nodes that occur on many shortest paths between other nodes have higher betweenness.

Cocitation networks are often very dense, their display cluttered and unreadable. Pathfinder network scaling is applied to reduce the number of links shown at a time to improve readability. It relies on a triangle inequality test to determine whether a particular link should be preserved or eliminated. The selection criterion is that the weight of a single-link path should not exceed that of alternative paths of multiple links.

Visualization

CiteSpace supports two complementary views: cluster views and time-zone views. The latter is shown below. A screenshot of the main interface and the visualization window are given to the right.

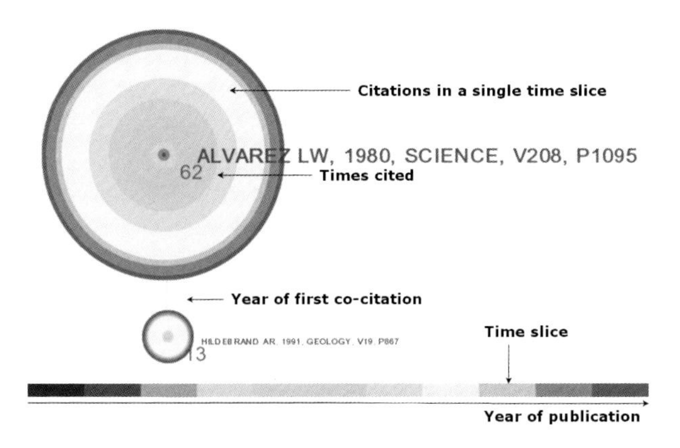

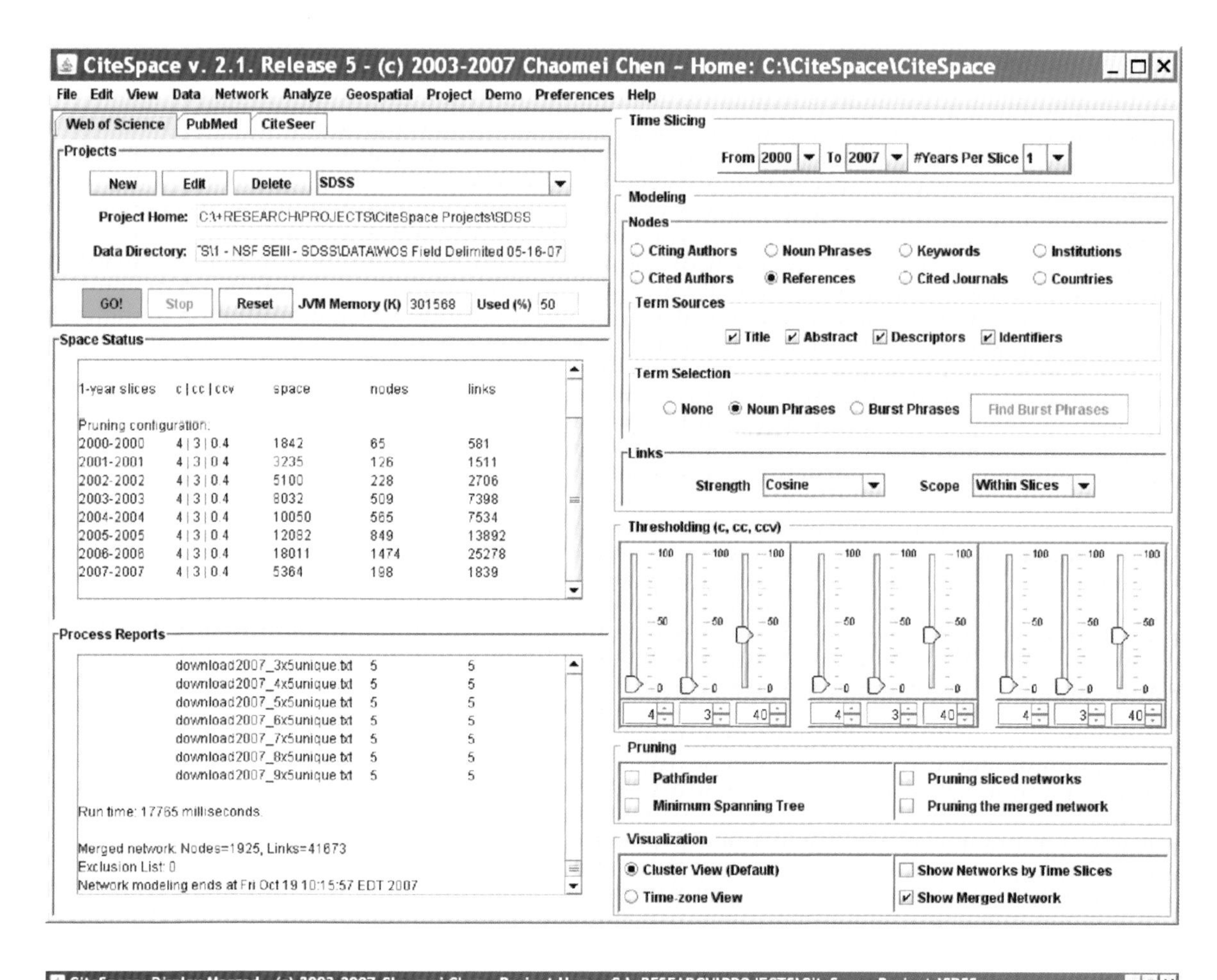

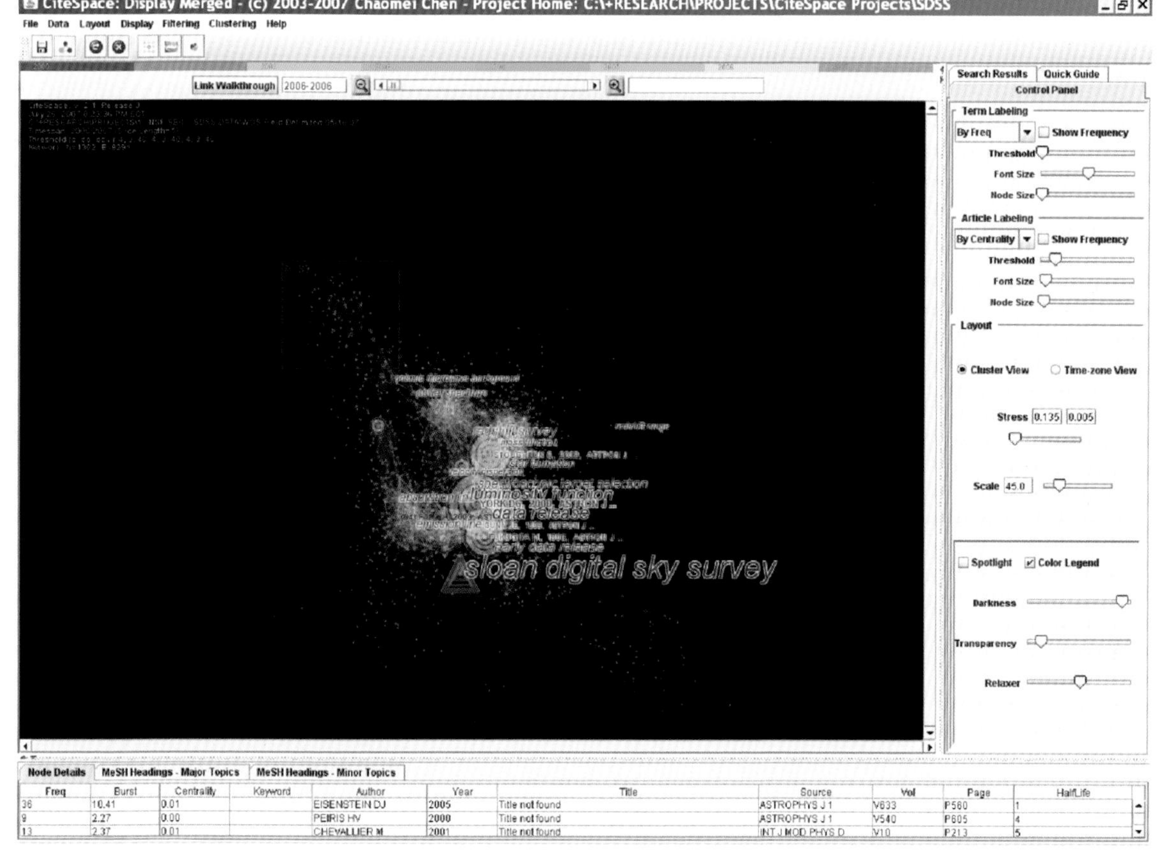

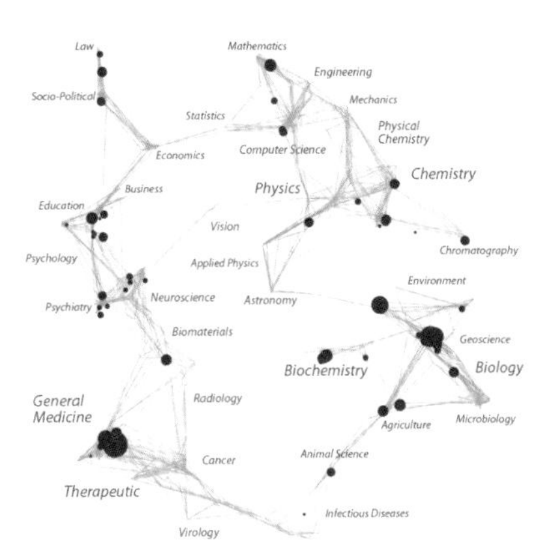

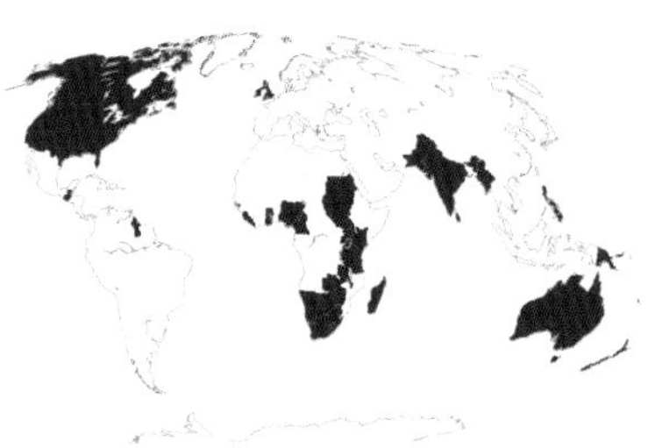

2007

Science-Related Wikipedian Activity

By Bruce W. Herr II, Todd M. Holloway, Elisha F. Hardy, Kevin W. Boyack, and Katy Börner

BLOOMINGTON, INDIANA, 2007

Courtesy of Indiana University

Aim

Wikipedia (**http://wikipedia.org**), created in 2001 by Jimmy Wales, is growing fast. In 2007, it included 7.4 million articles in more than 250 languages. About 9,700 new articles were added every day. The English Wikipedia had more than 276,000 contributors. According to the Alexa Web-ranking service, Wikipedia was the 37th most visited Web site. (See **page 18, Encyclopedias** for 2009 numbers). While the structure of editorially controlled encyclopedias like Encyclopaedia Britannica or MSN Encarta is well known, nobody has ever seen the evolving structure of Wikipedia. What major areas does it cover? How much of the content relates to math, science, and technology? How are those areas interlinked? Which topics are growing, declining, merging, or splitting, and how fast? The activity of the thousands of Wikipedians might be even more interesting: What are the hot, most discussed articles? Are today's bursts of editing and linking activity an indicator for tomorrow's growth?

Interpretation

This map shows the structure and activity patterns of 659,388 articles in the English Wikipedia, based on a network constructed in early 2005 and full history data from April 2, 2007. In the middle is a base map of all English Wikipedia articles, each represented by a small gray circle. A 37 x 37 half-inch (about 1.2 centimeters) grid was overlaid, and a relevant image was downloaded for each grid area and rendered underneath the network of articles. The data overlay consists of articles tagged as **Math**, **Science**, and **Technology**—drawn as blue, green, and yellow dots respectively. The sizes of the dots represent the certainty that these articles are in fact related to one of the three categories. The top 150 math, science, and technology articles are labeled by title.

The four corners show smaller versions of the base map with articles size-coded according to the **Article Edit Activity** (top left), number of **Major Edits** made in 2007 (top right), **Number of Bursts** in edit activity (bottom right), and the **Article Popularity,** measured by the number of times other articles link to it (bottom left). These visualizations serve to highlight current trends, predict future editing activity, and estimate potential increase in Wikipedia articles related to math, science, and technology.

Bruce W. Herr II (data mining and visualization) was a full-time software developer at the Cyberinfrastructure for Network Science Center at Indiana University when he designed this map. Bruce received an SB in Computer Science from Indiana University. He enjoys making beautiful, extensible, usable, and maintainable software. His research interests include information visualization, human–computer interaction, taking advantage of cognitive processing in software, software design, aesthetics in visualization, and extensible software. He made major contributions to the Taxonomy Validator, InfoVis Cyberinfrastructure, Cyberinfrastructure Shell (CIShell), Network Workbench, and SciMaps.org.

Todd M. Holloway (data mining) **See page 132.**

Elisha F. Hardy (graphic design) **See page 158.**

Kevin W. Boyack (graph layout) **See page 106.**

Katy Börner (concept) **See page 132.**

The Information Visualization Lab at Indiana University promotes and supports research, practice, and education toward highly usable and useful information visualizations. **http://ivl.slis.indiana.edu**

The Cyberinfrastructure for Network Science Center at Indiana University provides data sets, software, computing resources, and expertise for the study of social, biological, and technological networks. **http://cns.slis.indiana.edu**

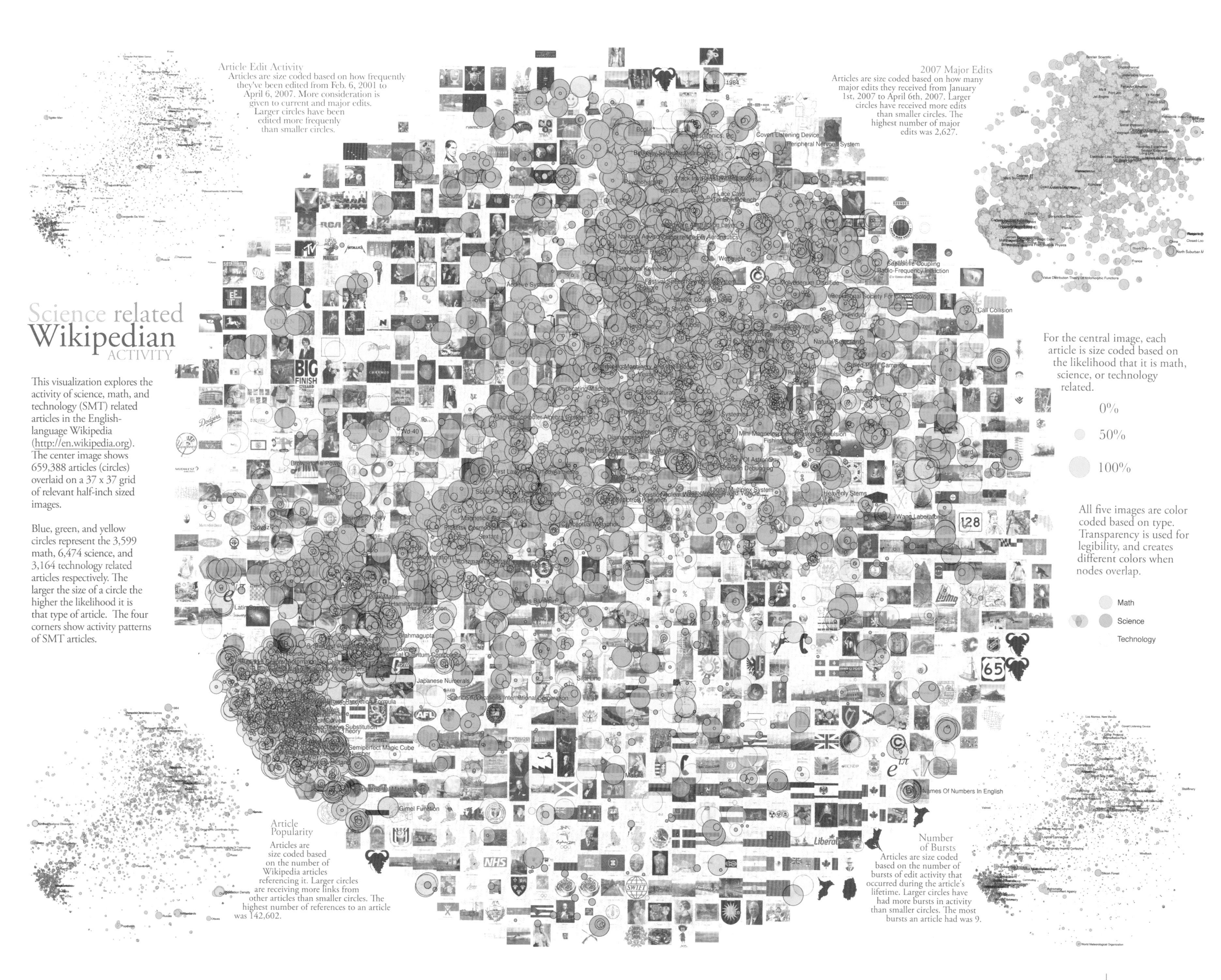
Science related
Wikipedian
ACTIVITY
This visualization explores the activity of science, math, and technology (SMT) related articles in the English-language Wikipedia (http://en.wikipedia.org). The center image shows 659,388 articles (circles) overlaid on a 37 x 37 grid of relevant half-inch sized images.
Blue, green, and yellow circles represent the 3,599 math, 6,474 science, and 3,164 technology related articles respectively. The larger the size of a circle the higher the likelihood it is that type of article. The four corners show activity patterns of SMT articles.
Article Edit Activity
Articles are size coded based on how frequently they've been edited from Feb. 6, 2001 to April 6, 2007. More consideration is given to current and major edits. Larger circles have been edited more frequenly than smaller circles.
2007 Major Edits
Articles are size coded based on how many major edits they received from January 1st, 2007 to April 6th, 2007. Larger circles have received more edits than smaller circles. The highest number of major edits was 2,627.
For the central image, each article is size coded based on the likelihood that it is math, science, or technology related.
0%
50%
100%
All five images are color coded based on type. Transparency is used for legibility, and creates different colors when nodes overlap.
Math
Science
Technology
Article Popularity
Articles are size coded based on the number of Wikipedia articles referencing it. Larger circles are receiving more links from other articles than smaller circles. The highest number of references to an article was 142,602.
Number of Bursts
Articles are size coded based on the number of bursts of edit activity that occurred during the article's lifetime. Larger circles have had more bursts in activity than smaller circles. The most bursts an article had was 9.

Data

The "wiki" technology was invented by Ward Cunningham in 1953. The defining feature of wikis is that each page has an "edit this page" link that takes users to an editing view of the page's content. A user can make and submit changes to the text, which immediately replace the previous version of the text. Hence, readers can easily become authors of the page. Users can register to create and retain a user profile or decide to remain anonymous. When anonymous users make changes, their IP addresses are logged. Each wiki page also has a "page history" link that provides access to previous versions of the page, as well as a "recent changes" link that lists the most recent edits and helps track changes.

The largest public wiki is Wikipedia, a free "encyclopedia of everything," started by Jimmy Wales and Larry Sanger in 2001. On November 1, 2005, there were 10 different Wikipedias—in English, German, French, Japanese, Polish, Italian, Swedish, Dutch, Portuguese, and Spanish—each containing more than 50,000 articles. Less than five years after its creation, it included more than 2,700,000 articles, written by about 90,000 different contributors, in 195 languages. The English Wikipedia was largest, with 800,342 articles and 78,977 categories. Three racks of servers process 60 million Wikipedia hits per day. (See **page 18, Encyclopedias** for 2009 numbers).

Wikipedias in different languages are loosely interlinked; otherwise the Wikipedias exist independently of one another. The Wikimedia Foundation generously makes complete data dumps of all current articles and past revisions freely available at **http://download.wikimedia.org**. For this visualization, a network constructed from the English Wikipedia in early 2005 and a full history dump from April 2, 2007, is used. There are three data sources of interest for this study: (1) page links, containing the links between articles; (2) category links, including the articles' category, subcategory, and supercategory membership relations; and (3) stub-meta-history, containing the full history of each article in Wikipedia.

The dumps were parsed using custom Python, Perl, and Bash scripts. The data comprises 659,388 articles, their 16,582,425 links, and 52,300,922 edit records. In this data set, the mean number of links per page is 25 and the mean number of edits is 79. A graph of the number of **New Articles** and the number of **Edits** over time is shown above right.

Out of all 659,388 articles, exactly 3,599 are tagged with the category math, 6,474 with science, and 3,164 with technology. Of these, 8,181 articles are in one category; 2,348 in two; and 73 in three categories—which produces interesting color mixtures in all five maps in *Science-Related Wikipedian Activity*.

Reference System

To create a base map for the visualization, a cocitation metric was computed based on the 14.5 million interlinkages of the 659,388 Wikipedia articles. Using this metric, it is possible to determine the level of similarities between two articles. In this instance, that similarity information was given to a spatial layout algorithm, VxOrd, to generate coordinates for each article, such that more related articles are placed more closely in space. As a result, we see clusters of highly related articles throughout the map. Two candidate layouts created by VxOrd are shown at bottom. The layout on the left was chosen for its round shape, which not only signifies wholeness or completion, but also, having greater surface area, allows for more images to be displayed.

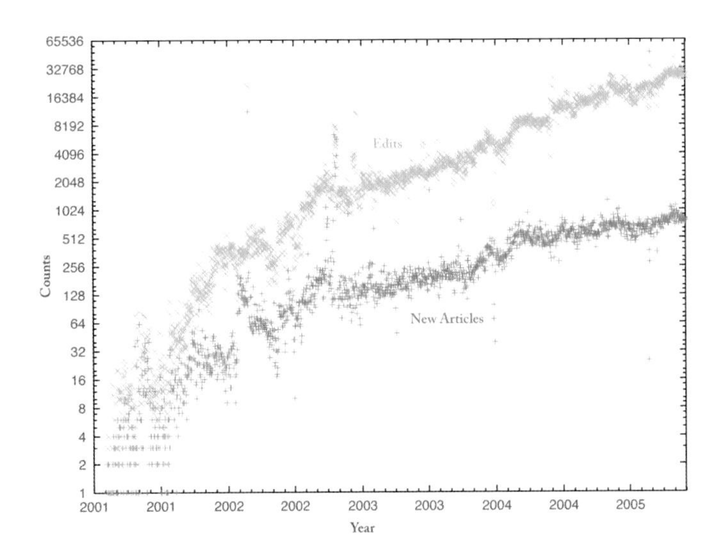

Data Overlays

Five different data overlays—all using the same reference system—are shown.

The largest map, in the center, aims to communicate the main structure of articles related to math, science, and technology in Wikipedia. To facilitate this, each article is color-coded by category and sized by the probability that it pertains to math, science, or technology. Labels identify the 150 highest probability articles. Finally, images are used to help clarify the map. The initial idea, to place science-relevant images around clusters of nodes like dandelion tendrils, would have caused major occlusion problems. Instead, an invisible 37 x 37 half-inch (about 1.2 centimeters) grid is drawn underneath the network of articles and filled with relevant images. If there is a popular math, science, or technology article in a grid cell, the main image illustrating the article is used as the underlay. Otherwise, an image from the most popular article, regardless of category, is used.

The four corner maps show different metrics to help predict the future growth of Wikipedia articles related to math, science, and technology. Possible metrics include article edit activity; date of creation; last major edit and last edit (major or minor); number of major edits and number of edits (major or minor); number of major edits in 2007 (as of April 2, 2007) and number of edits in 2007 (as of April 2, 2007; major or minor); 2005–2006 major edit trajectory (number of edits in 2006/number of edits in 2005) and 2005–2006 edit trajectory (number of edits in 2006/number of edits in 2005; major or minor); in-degree (the number of links an article receives from other articles); the number of bursts in edit activity; and the variance of an article's burst activity.

The final decision was to use edit activity (top left), number of 2007 major edits (top right), number of bursts in edit activity (bottom right), and the article's popularity (measured by the number of times other articles link to that entry; bottom left), as those links are presumed to have the most predictive value. For each of the chosen metrics, an overlay was created that size-coded the articles by the metric chosen and color-coded them according to their category, if applicable; otherwise articles were left gray.

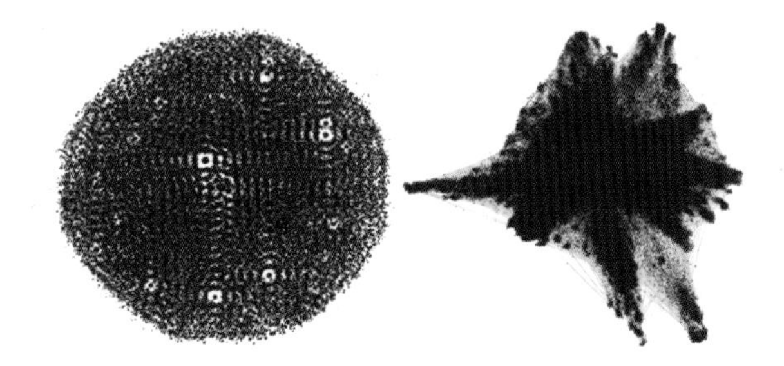

Details

Spatially, most of the math-related articles are located in a tight cluster in the lower left portion of the base map. This tight grouping is a direct result of the fact that, for the most part, articles about math reference other math articles. Science occupies a wide swath of Wikipedia, intermingling with subjects such as history, technology, and popular culture. Technology is located throughout most of Wikipedia because of its wide applicability.

The images also have some stories to tell. Given the space limitations, about 1000 half-inch (about 1.2 centimeters) images were used, which represent only about 1/10th of the math, science, and technology articles and 1/600th of all the articles. Even with these relatively few images, however, clusters of articles can be easily spotted just by exploring the images.

Looking at the edit activity layer (map on previous page, top left), corresponding to how often an article has been edited, reveals that much of what is hotly edited is not hard-core science but is more tied to common knowledge. However, **Evolution** (see zoom below), the most hotly contested and hence largest circle, is different. Its prominence is likely due to the divisive nature of evolution, driving edit wars between those on both sides of the evolution debate.

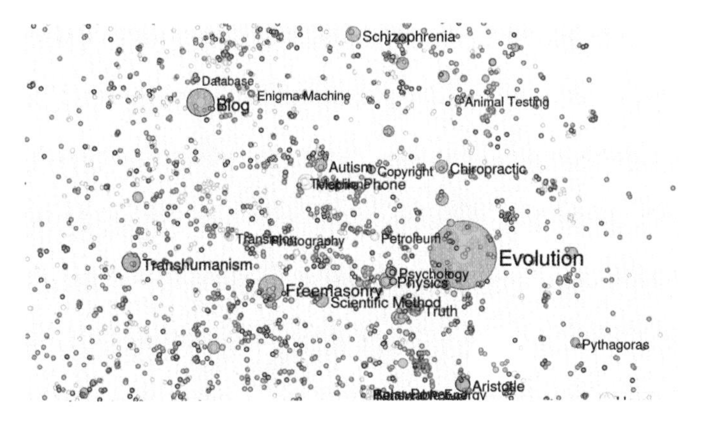

The article popularity layer (bottom left), corresponding to the number of articles linked to a particular article, shows that the math, science, and technology that many other articles link to cover more general topics.

The 2007 major edits layer (top right), corresponding to the number of edits an article received from January 1 to April 2, 2007, reveals that most of the math, science, and technology articles were still being actively edited as of 2007.

The number of bursts layer (bottom right), corresponding to the number of bursts or sudden increases in edit activity for an article, shows that a number of math, science, and technology articles feature recurring periods of intense editing activity. These articles appear to be fairly obscure, perhaps revealing that there are not many who work on those articles; but when they do work, they do so often.

Unique Features

The very same data set and base map was used to map *An Emergent Mosaic of Wikipedian Activity* on a 5-foot x 5-foot (approximately 1.52 meters x 1.52 meters) canvas. (See close-up of the map, below, and online interactive Google map at **http://scimaps.org/maps/wikipedia**.) Here, each of the 659,388 entries is represented by a yellow circle. Circles are size- and color-coded by the number of times the article was revised or edited, with priority given to newer edits. This way, the map emphasizes entries that might be controversial, the subject of considerable vandalism, or merely a topic whose content frequently changes.

To keep the map readable, only the largest nodes are labeled. A 120 x 120 half-inch (about 1.2 centimeters) grid was underlaid, and exactly 10,324 images were placed in those grid cells that contained at least one article with an image that was not a Wikipedia icon. If there were multiple images in a grid cell, then the image from the article with most links was selected.

A closer examination reveals idiosyncratic priorities of Wikipedians. The 10 most frequently revised articles as of April 2, 2007, concern Jesus, Adolf Hitler, October 2003, Nintendo revolution, Hurricane Katrina, India, RuneScape, anarchism, Britney Spears, and PlayStation 3.

Some think *An Emergent Mosaic of Wikipedian Activity* is similar to Piet Mondrian's drawings. Others see it as a proof that there truly is a "global brain" of human intellect evolving on Earth.

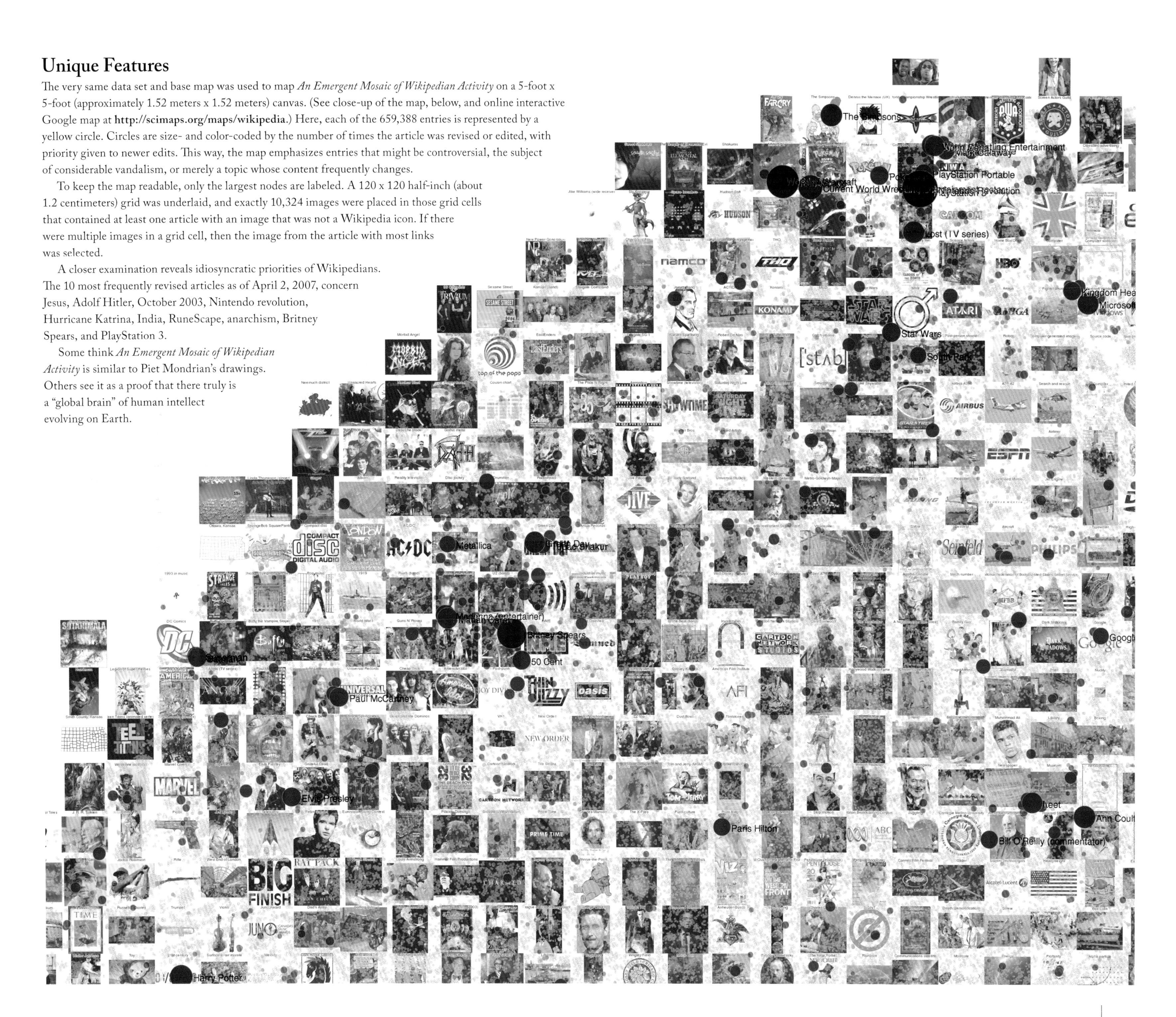

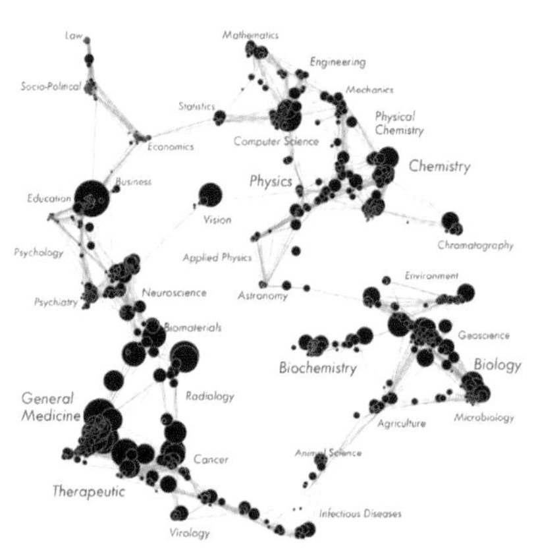

2007

Maps of Science: Forecasting Large Trends in Science

By Kevin W. Boyack and Richard Klavans

BERWYN, PENNSYLVANIA, AND ALBUQUERQUE, NEW MEXICO, 2007

Courtesy of Richard Klavans, SciTech Strategies, Inc.

Aim

All previous large-scale maps of science were generated using data from the Science Citation Index (SCI) and Social Sciences Citation Index (SSCI). How would the map of science change if data from the Arts and Humanities Citation Index (AHCI) were added? Would that create a second continent, or would arts and humanities constitute a peninsula? Which discipline is bridging the gap between the sciences, arts, and humanities? What might happen if Scopus data were folded in as well? Scopus covers only the last 10 years but has twice as many titles as SCI and SSCI combined. Will the global structure of science change with the addition of all this new data? Do we now have enough data to predict future changes in the structure of science based on year-to-year changes in a five-year time window?

Interpretation

This most recent map of science, also called the UCSD Map of Science (see **page 12, Science Maps and Their Makers**), is based on the largest set of scientific literature yet mapped—about 7.2 million papers published in more than 16,000 separate journals, proceedings, and series over a five-year period (2001–2005) retrieved from WoS and Scopus databases. A three-dimensional layout places disciplines—groups of journals—on a sphere. This overcomes problems with previous maps of science that had imposed borders and avoids potential boundary effects. Using this spherical projection to understand scientific disciplines as topography upon a globe, viewers can now explore science in all directions without "falling off the map." Using a Mercator projection, the spherical layout was flattened onto a two-dimensional map to ease navigation and exploration.

A forecast of how the structure of science may evolve in the near future was generated by evaluating the changes in the connectedness of various regions of the map between 2001 and 2005. In that time frame, the rate of change has been stable, and it will likely continue to be in the near future. This map and variations on it are used daily by their makers for planning, evaluation, and education at national, corporate, and personal levels. These maps serve as tools to determine which areas of science are most closely connected, which are most or least intellectually vital, and which produce the most patents.

Kevin W. Boyack joined SciTech Strategies, Inc. in 2007 after working at Sandia National Laboratories, where he spent several years in the Computation, Computers, Information and Mathematics Center. He holds a PhD in chemical engineering from Brigham Young University. His current interests and work are related to information visualization, knowledge domains, science mapping with associated metrics and indicators, network analysis, and the integration and analysis of multiple data types.

Richard Klavans is the president of SciTech Strategies, Inc. He holds a PhD in management from the Wharton School of the University of Pennsylvania. His current work is related to the generation of highly accurate maps of science using multiple techniques, such as bibliographic coupling, cocitation, and coword, as well as the associated metrics and indicators that allow government and industry users to make more effective policy decisions. He is interested in semantics, augmented cognition, and the application of mathematical tools to information spaces.

MAPS OF SCIENCE

A visualization of 7.2 million scholarly documents appearing in over 16,000 journals, proceedings or symposia between Jan, 2001 and Dec, 2005

This map of science was constructed by sorting more than 16,000 journals into disciplines. Disciplines, represented as circles, are sets of journals that cite a common literature; links (the lines between disciplines) are pairs of disciplines that share a common literature. A three-dimensional model was used to determine the position of each discipline on the surface of a sphere based on the linkages between disciplines. The model treats links like rubber bands attempting to bring two disciplines close to each other. Pairs of disciplines without links tend to end up on different sides of the map.

The spherical map, which is not shown here, was unrolled in a mercator projection (the same one used to show the continents of the earth on a two-dimensional map) to give the large map shown below. This projection allows inspection of the entire map of science at once. Note that the disciplines tend to string along the middle of the map - if this were a map of the earth it would be like a single continent undulating along the equator. There are no disciplines at the top (north pole) or the bottom (south pole). Mercator projections also introduce distortions. We tend to forget that the left side is connected to the right side, and assume that the middle is most important. In this map, the social sciences (yellow) on the right connect with the computer sciences (pink) on the left in one continuous swath.

The six map projections shown at the bottom are images of what one would see if looking directly down at the south pole of the map, at six different rotations. When viewed this way, the map looks like a wheel with an inner ring and outer ring. This wheel of science corresponds very closely with the two-dimensional maps we have previously produced.

Forecasting Large Trends in Science

Calculations were performed using the large colored groupings of disciplines (fields) to determine if any of them were likely to cause large scale changes in the structure of science over time. Connectedness coefficients between fields were calculated for each individual year, 2001-2005. A simple regression analysis was conducted to see if there were significant changes in these connectedness coefficients from year-to-year.

If the structure of science shown below is moving toward stability, we would expect connectedness between neighboring fields to increase, and connectedness between distant fields to decrease. We found the opposite, suggesting that the underlying structure is unstable and likely to change dramatically over the next decade.

Six stories, representing how the structure is likely to change, are provided below. Maps with white arrows represent instances of distant fields that are likely to be pulled closer to each other in the future. Maps with dark arrows represent fields that are currently close-knit, that are likely to become more dispersed. We expect that future maps of science will show changes in stucture corresponding to these observations. Medicine will disperse slightly, while the physical sciences will tighten and draw closer to the medical fields.

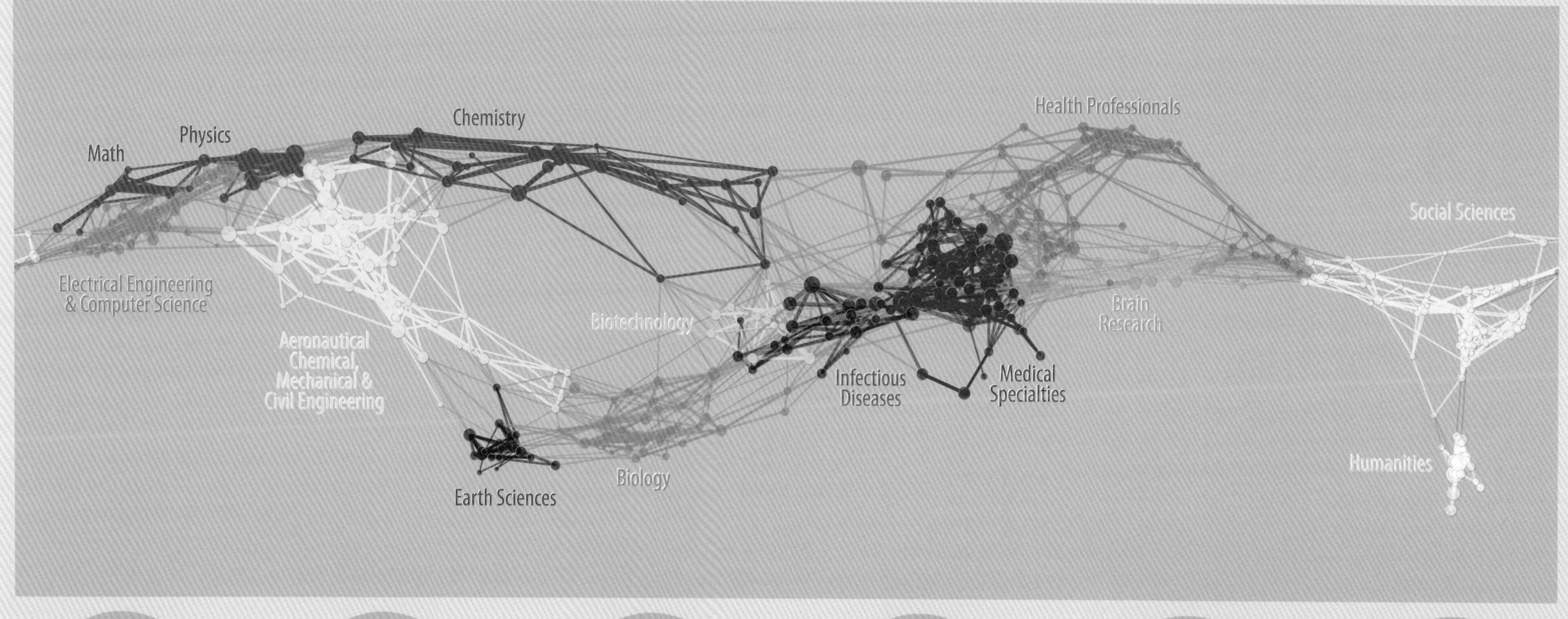

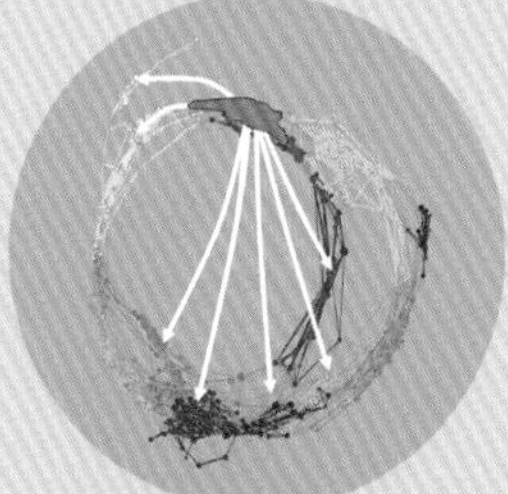

Electrical Engineering & Computer Science (EE/CS), indicated by the pink shape in the view above, is a field whose connectedness has been increasing much more quickly (15%) than expected. Connectedness has increased between EE/CS and all other fields from 2001-2005. The connections with the largest annual increases (>10%) are shown by white arrows. Over time, these stronger connections will distort the map, and may bring EE/CS into a more central position.

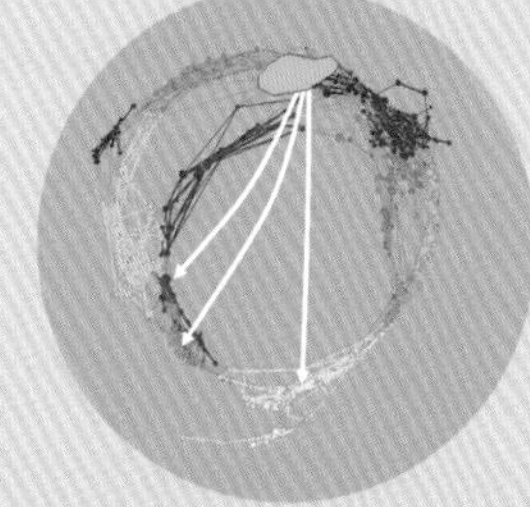

Biotechnology, indicated by the light green shape above, has the largest overall increase in connectedness with other fields (16%). It has relatively few connections with the EE/CS, Math & Physics, and Social Sciences fields, but these three connections had the largest fractional increase. The connection with EE/CS, which had the single largest growth rate (91%) of any connection, reflects recent growth in the area of bioinformatics.

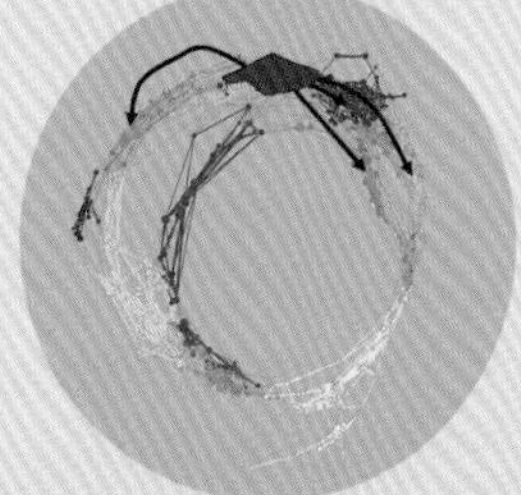

Infectious Diseases, indicated by the dark red shape above, has an overall decrease in connectedness (2%) with other fields. Decreases in connection strength between this field and the fields of Biology, Medical Specialties, Health Professionals and Brain Research (all >3%) are shown as black arrows, and will drive a slow dispersion of the medical fields compared to the current structure.

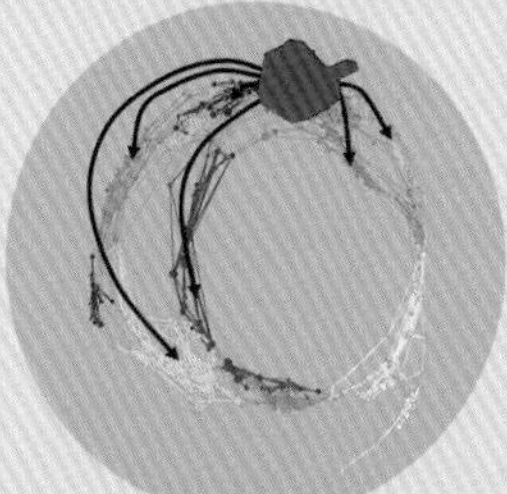

Medical Specialties, indicated by the red shape above, has an overall decrease in connectedness (2%) with other fields. This is dominated by decreasing connection strength to the other medical fields and biology, as shown by the black arrows. The only connection increasing in strength is the one to EE/CS, which is not shown here, but was shown as a white arrow in the first story.

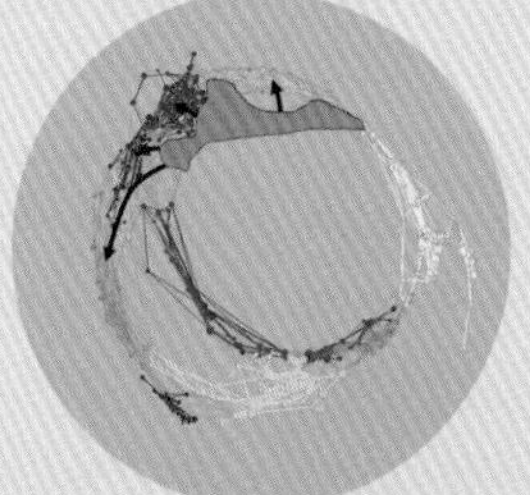

The Health Professionals field, indicated by the orange shape above, has the largest overall decrease in connectedness (4%) to other fields. As with the other medical fields, its connection strength with medicine and biology is decreasing in all cases, as shown by the black arrows. With the decreasing connection strengths throughout medicine, we expect the map structure in these areas to relax slightly over time.

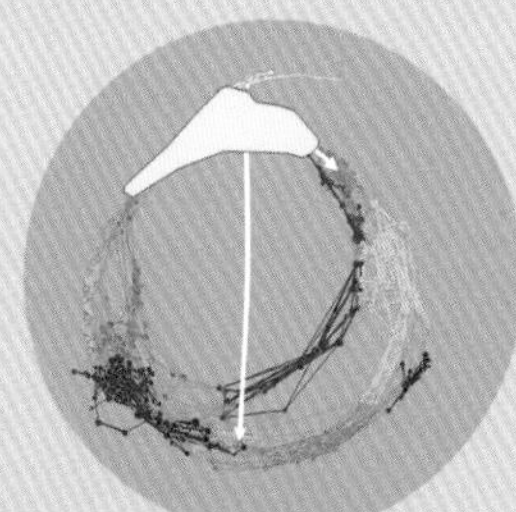

The Social Sciences, indicated by the yellow shape above, had an overall increase in connectedness (9%) with other fields. Although its greatest connectedness gains were with EE/CS and Biotechnology (see white arrows), it also had consistent connection increases with nearly all the other fields. In general the fields of EE/CS, Biotechnology, and the Social Sciences are become more connected, and are pulling on the physical sciences as well.

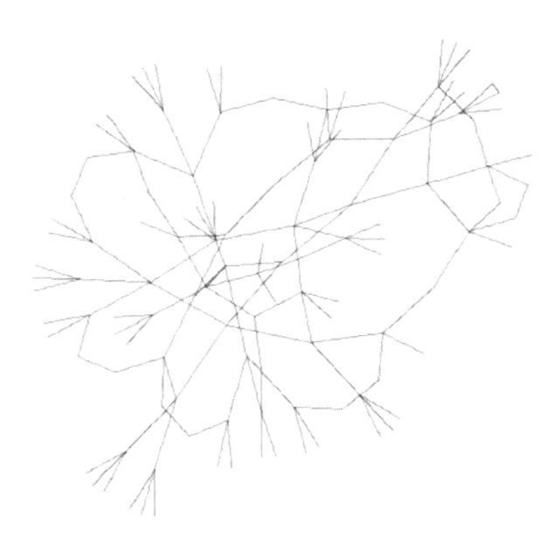

Two-Dimensional
Euclidian Space (R^2)

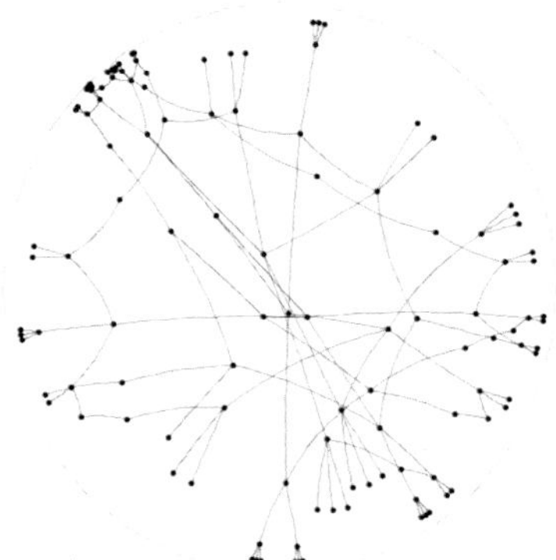

Two-Dimensional
Hyperbolic Space (H^2)

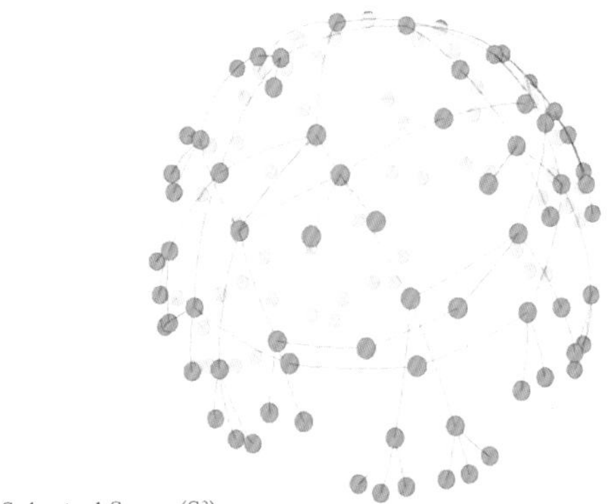

Spherical Space (S^2)

Data

This map is based on paper-level data from Web of Science (WoS) by Thomson Reuters for the years 2001–2005 and publication data from Elsevier's Scopus covering 2001–2005. In total, this data set captures about 7.2 million papers published in over 16,000 separate journals, proceedings, and series.

Technique

As with **The Structure of Science** map (**page 106**) and the **Map of Scientific Paradigms** (**page 136**) by the same two authors, multiple layout and clustering steps were required to produce this map.

Reference System

Bibliographic coupling, using both highly cited references and keywords, was applied to determine the similarity of journals. The utilization of keywords supported the placement of about 1,500 MEDLINE journals for which neither database had reference lists. The references were grouped by applying a distance measure and an average link-clustering algorithm.

Using a hierarchical clustering procedure, journal clusters were clustered again. No information was deleted at each subsequent level of clustering—the original cocitation counts were aggregated to the appropriate clusters and levels. Hierarchical clustering was stopped when there were fewer than 1000 nodes; here there are 554. This higher level of aggregation represents paradigms. Multidisciplinary journals were parsed into multiple clusters to avoid overaggregation.

The third iteration used a three-dimensional layout of the disciplines (groups of journals), placing those disciplines on a sphere using a 3-D Fruchterman-Reingold layout with a forced unit radius of 1.0. A comparison of the layout of the same graph in **two-dimensional Euclidian space (R^2)**, **two-dimensional hyperbolic space (H^2)**, and **spherical space (S^2)** is shown to the left.

The **Spherical Layout** was then flattened using a **Mercator Projection** (below) to create a two-dimensional version of the map. Each journal has a specific *x*, *y* position and can be "science located" on the map.

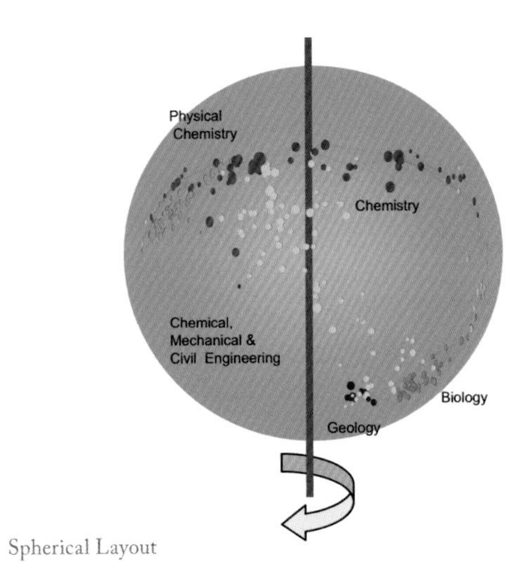

Spherical Layout

Mercator Projection

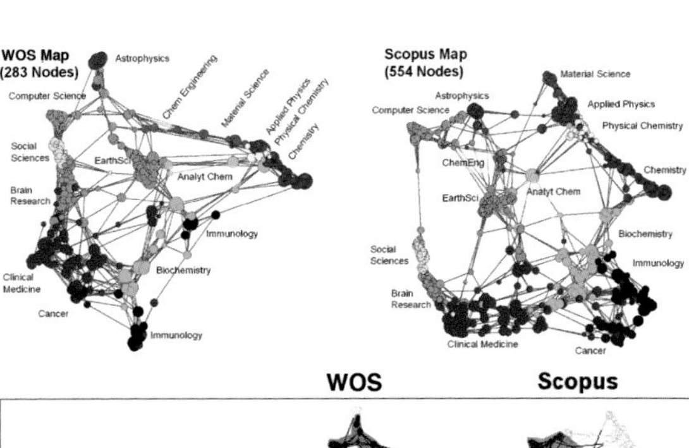

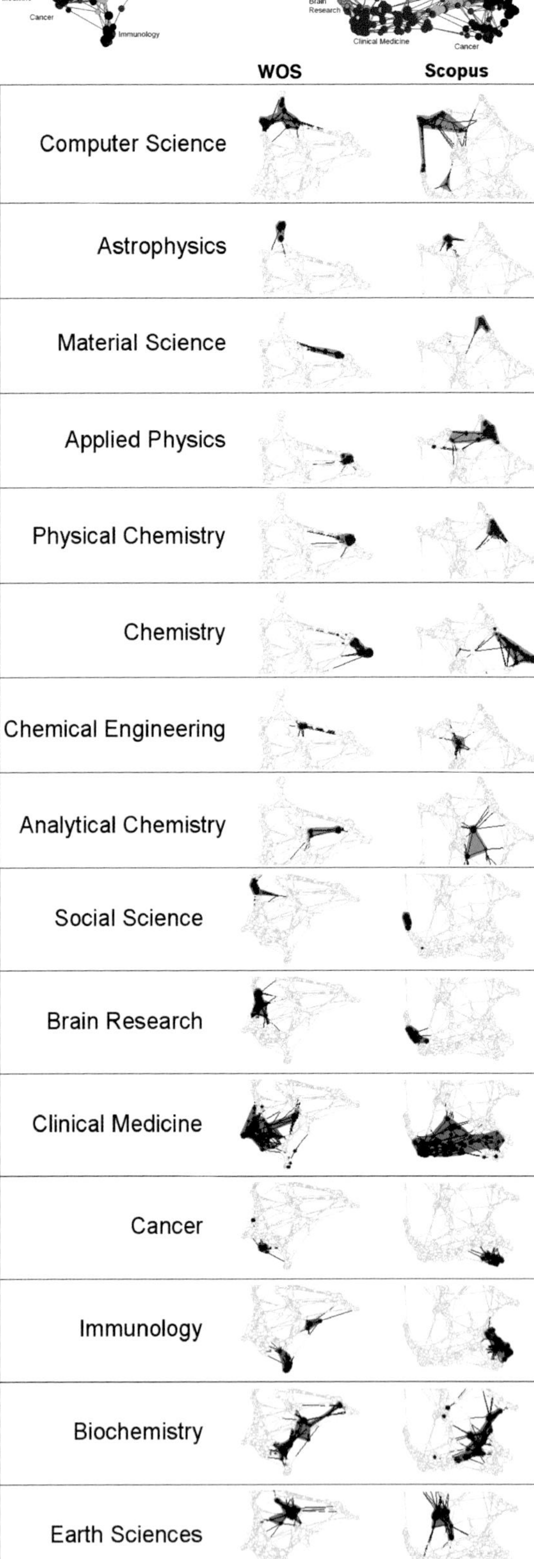

Data Overlays

The resulting map was labeled manually, with each discipline given a separate label based on journal membership. High-level color groupings are based on discipline labels. Detailed metadata (key phrases) for each node was extracted from titles of papers. The result is a map that provides the most comprehensive look at the disciplinary structure of science. As five years of data are used, the map is robust in structure, allowing long-term use of the map as a science reference system.

Details

Creation of the map of 16,000 journals followed closely on the heels of a study comparing the 2004 WoS and Scopus databases. After being mapped separately, the databases were split into multiple categories, so that the category sizes, shapes, and connections could be visually compared (see maps to the right). Each map was then divided into 15 different areas, using a manual process of examining the paradigms, their dominant journal constituents, and the distribution of journals from current papers assigned to the paradigms.

This study had three major benefits. First, a quantitative comparison of the WoS and Scopus databases for 15 areas of science was provided. Second, the influence of the resulting differences on the map of science became clear. Third, the study showed that these differences do not affect the fundamental shape and structure of science; rather, they create local differentiation and improve our understanding of local relationships.

Unique Features

Patent and grant data can be mapped to disciplines using detailed metadata matching via extracted keywords. Detailed drilldowns can be generated for each discipline in support of a variety of disciplinary and policy studies. The online interface (at **http://mapofscience.com)** offers multiple means to utilize this new base map of science.

Different areas of science can be selected; funding of different directorates of the National Science Foundation (NSF) and various branches of the National Institutes of Health (NIH) can be overlaid (see overlay of the NSF's Office of Cyberinfrastructure in the screenshot below); and the core competency of U.S. institutions, such as the Department of Agriculture, Department of Transportation, Department of Energy, and NASA can be requested. Diverse drilldowns are available as well (see 12 maps on the right).

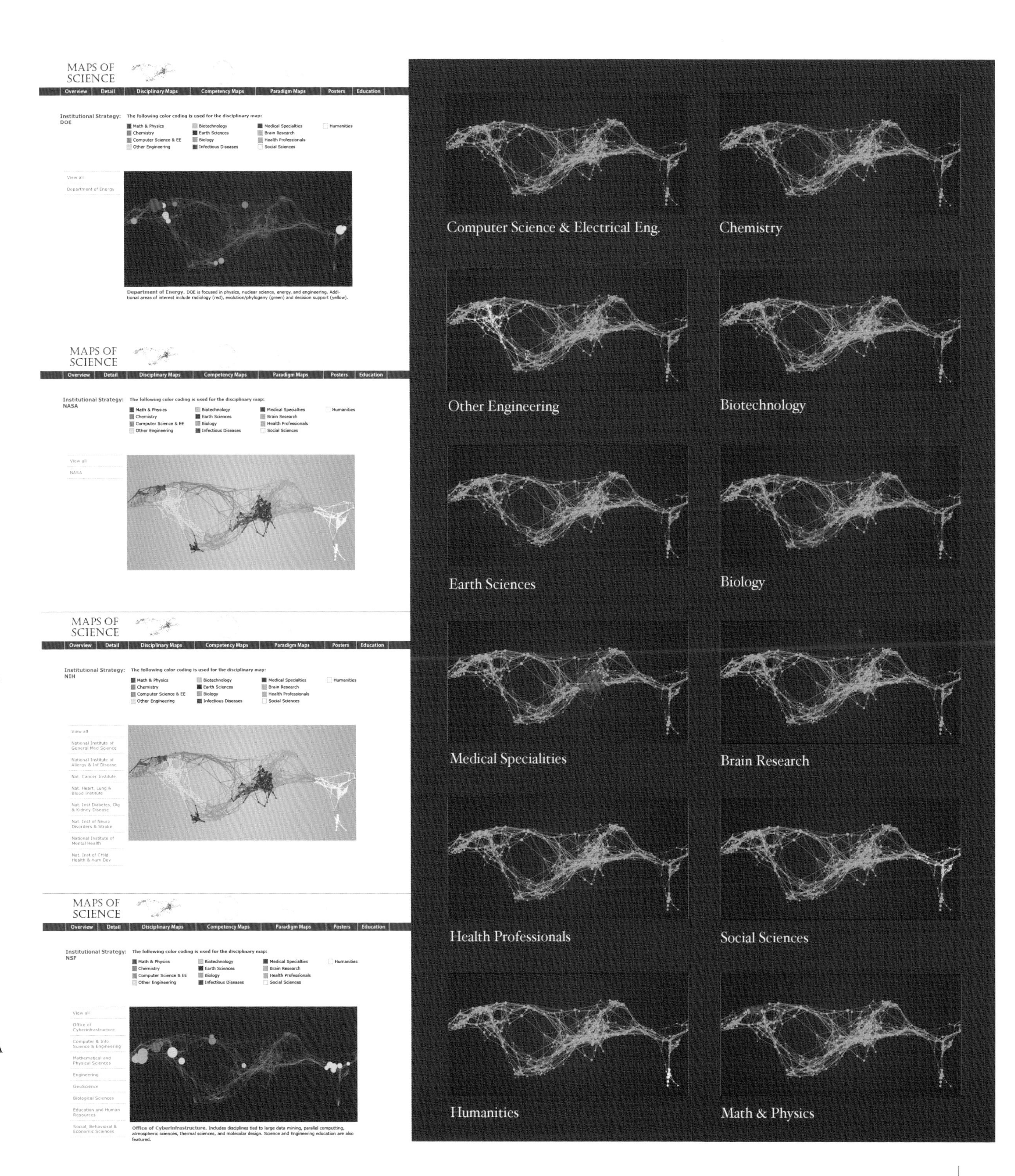

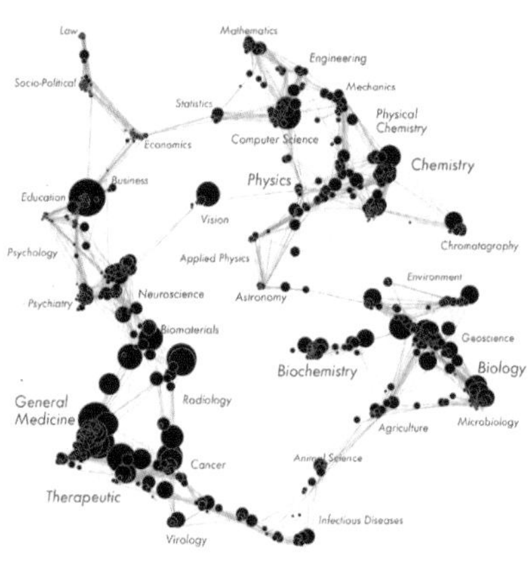

2007

Hypothetical Model of the Evolution and Structure of Science

By Daniel Zeller
NEW YORK, NEW YORK, 2007
Courtesy of Daniel Zeller

Aim

How can the structure and evolution of science be conceptually represented and visually mapped? What metaphors might work best to depict established and emerging fields, the impact of funding, the continuous and often desperate search for novelty and reputation? Will there be monsters that guard undiscovered lands of knowledge?

Interpretation

This drawing conceptualizes science as layers of interconnected scientific fields. Starting with the original scientific thought, science grows outward in all directions. Each year, another layer is added to the meteor-shaped manifestation of knowledge. New fields (blue) emerge and established fields (brown) merge, split, or die. The cutout reveals a layering of fat years that produce many new papers and slim years when few papers are added. Each research field corresponds to a tube-shaped object. Some have rapid growth patterns, due to electronic papers that are interlinked within days. Other fields communicate knowledge via books—in which case, years may pass before the first citation bridge is established. Blue tentacles could symbolize the search for opportunities and resources or activity bursts due to hype and trends. The injection of money (yellow) has a major impact on how science grows. There are voids in our knowledge that may take the shape of monsters. The trajectories of scientists who consume money, write papers, interlink papers via citation bridges, and fight battles on the front lines of research could be overlaid. Yet, scientists are mortal. All they leave behind are the knowledge structures on which future generations can build.

Daniel Zeller holds a BFA and an MFA in sculpture from the University of Connecticut, Storrs, and the University of Massachusetts, Amherst, respectively. He lives and works in New York as a professional artist. Daniel Zeller's drawings depict abstract spaces with beauty and inspiring complexity. The visual language he developed stimulates people's intellect and emotions in unexpected ways.

Zeller's works can be seen as travel documents—charts or aerial views that reflect a wide range of sources, from satellite imagery to electron micrographs, anatomical diagrams to road maps. For Zeller, drawing, like travel, is a fluid action, a continuous response to itself, an endless effort to be present where the mark is being made. It is also a discipline in which past experience and study inform the mark, guiding it in ways that can become increasingly predictable. It thus becomes imperative to combine learned patterns into something else, to alter the predictable and discover that which is different yet somehow familiar. It is exploration, an active desire to find what is uncharted or hidden. On a geographic level, people still engage in this activity as individuals. But collectively they have reached a certain level of saturation—most of the frontiers have met one another and disappeared. Maps and charts show people where to explore and how to get to places that others have already seen.

Zeller was awarded a Pollock-Krasner Foundation grant in 2001. His work is featured in the collections of the Museum of Modern Art, the Arkansas Art Center, the Los Angeles County Museum of Art, and the Whitney Museum of American Art. Recent solo exhibitions include *Drawings* (2001), *Empirical Cryptosis* (2003), *Cyclical Redundance* (2005), and *Geomorphical Fluxitosis* (2008) at Pierogi Gallery (Brooklyn, New York); *Recent Drawings* (2004) and *New Drawings* (2007) at Daniel Weinberg Gallery (Los Angeles, California); *Biodegradable* (2004) and *Erreur Infinie* (2006) at g-module (Paris, France); and *Drawings* (2008) at MichelSoskine Inc. (Madrid, Spain).

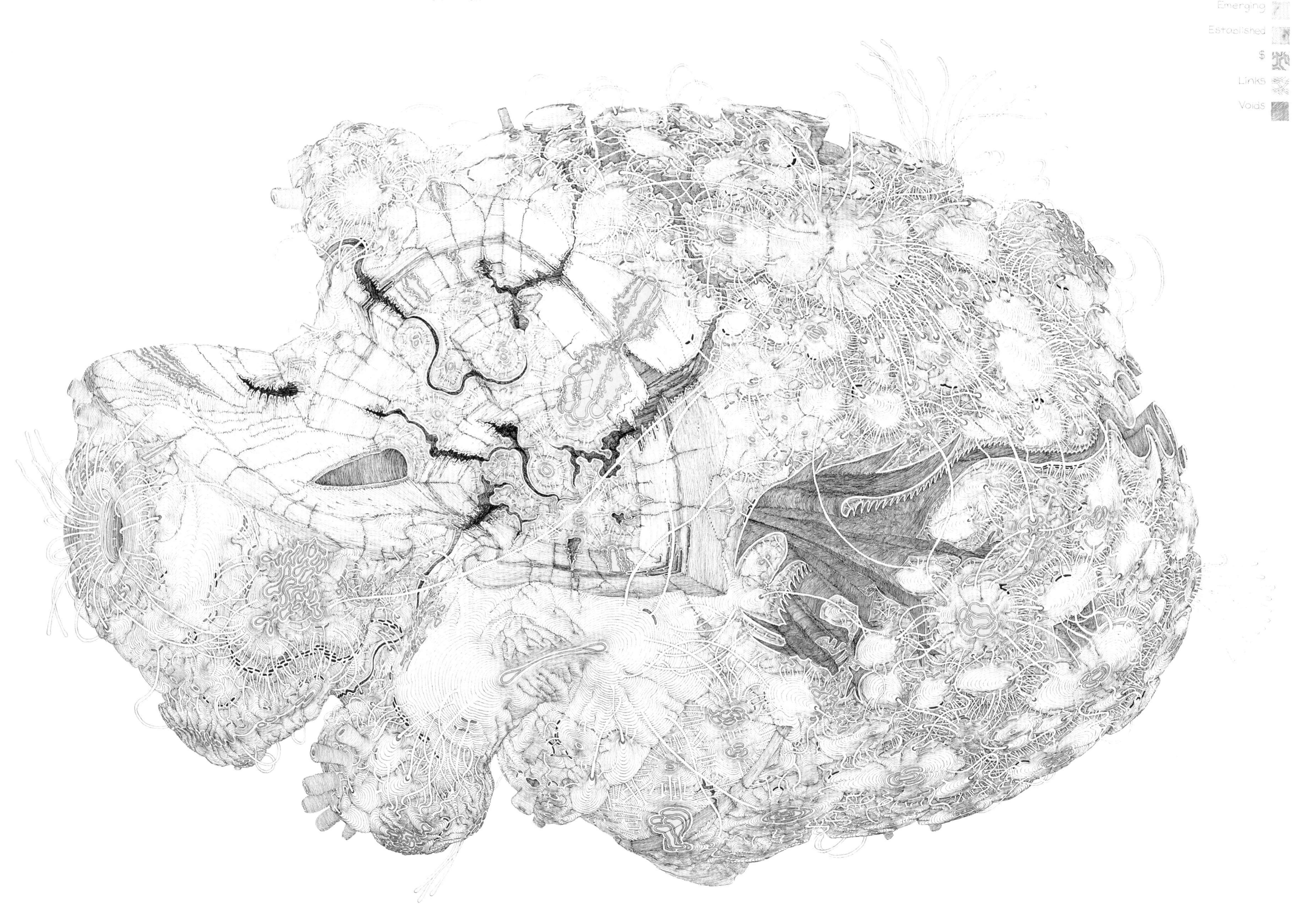
HYPOTHETICAL MODEL of the EVOLUTION and STRUCTURE of SCIENCE
Emerging
Established
$
Links
Voids
One of Many Possible Interpretations
Daniel Zeller 2007

Data

This drawing was inspired by a workshop on "Forecasting Science" in October 2006. On the initial drawings, Katy Börner provided acknowledgments, a description of the conceptualization of scientific growth, and e-mail and phone feedback. This is an excerpt from Börner's correspondence, on the conceptualization of scientific structure and growth:

"Dan,

"Enclosed are text and imagery that aim to communicate my understanding of how science is structured and grows. When selecting one of your many images as a starting point for designing a visual pattern language of scientific growth, I was definitely drawn toward larger, high-resolution versions. When looking at a science map, I expect to see a realistic area size (or a zoomed-in portion of it) and a certain complexity. Images like **Superficial Inquiry, 2005** are gorgeous but seem to represent very detailed area of research—not a map of all of science.

"In **Microbial Interaction, 2005**, I see 'normal areas' of science: small world networks (red areas) with many local and only few 'weak' linkages (red pipes/bridges) to other areas. The yellow part could be an interdisciplinary boundary/wall/gulf that is hard to bridge. Maybe somebody poured (yellow) money into this area in support of interdisciplinary collaboration. Bridges are slowly growing into it. The most (visually) amazing artifact is the blue furuncle. It could be a very focused new area of research that draws from two domains, sucking out their lifeblood/money. The layering and coloring might represent time. New layers of knowledge/expertise networks grow on top of old ones, rendering older ones obsolete while 'standing on their shoulders' at the same time. The fine red/blue shading might be useful to render time and diverse dynamics in more detail.

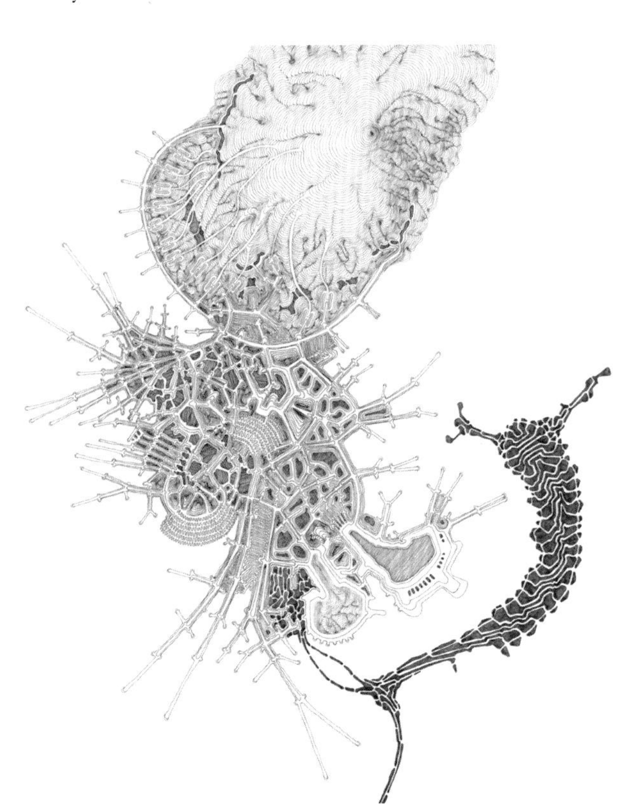

Superficial Inquiry, 2005

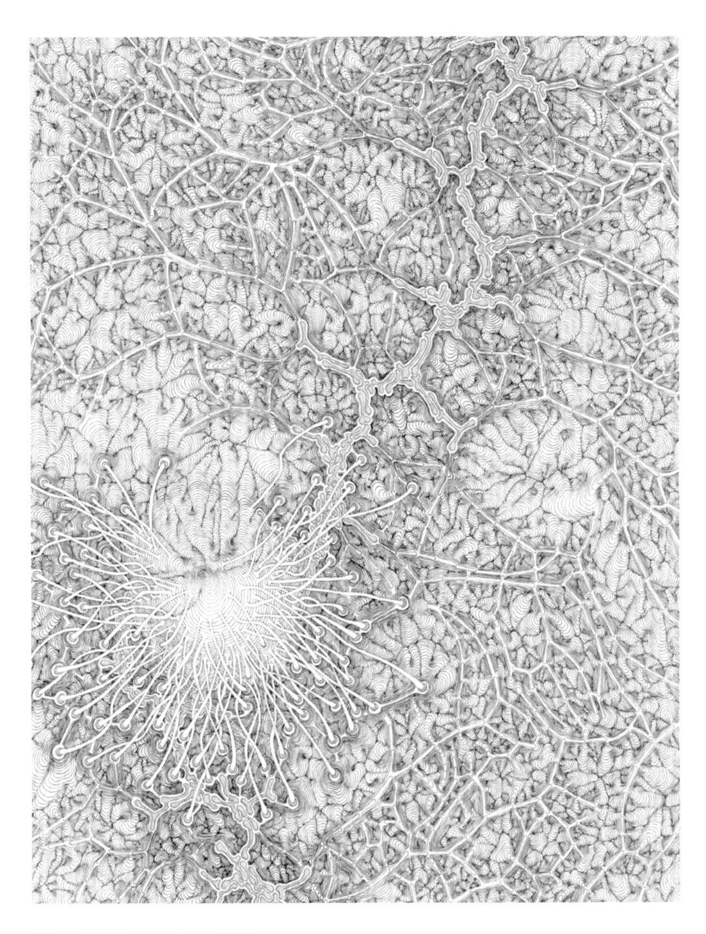

Microbial Interaction, 2005

"Science grows from the first initial scientific idea outwards in all directions, forming a meteorite-shaped object. It might look a little bit like your **Permeable Unit, 2006**. The initial scientific idea is at the center. Each year, new papers are added that are interlinked to previous papers, creating tree-ring-like layers. Tubelike elements start anywhere between the center and the outer crust (today's knowledge) of science. They represent scientific fields that grow and shrink in size over time. Tubes can merge and split, representing the merge and split of fields. Tubes can be weakly or strongly coupled by bridges representing fields that coevolve. Tubes might be terminated, for example, by scientific revolutions.

"There are good years (with lots of funding) and bad years; see de Solla Price's graph of number of scientific papers and the 50 percent decrease in paper production during WWI and WWII [the graph appears on **page 2, Knowledge Equals Power**]. The resulting shape of science is not perfectly round, as different areas of science have very different dynamics/growth patterns. Some interlink new ideas/papers published in e-prints within days. The humanities uses books—it takes years to get the first citation. Medical research requires major funding; $100,000 does not last very long. A social scientist can jumpstart a highly productive research career with the very same amount.

"There are bursts of activity at the outer crust, such as hypes, fashions, etc. They reflect the interest level in different areas of research according to press coverage, money, etc. They are like sunbursts. No papers/new knowledge might be added, but there definitely is interest and lots of scientists might consider moving into this area.

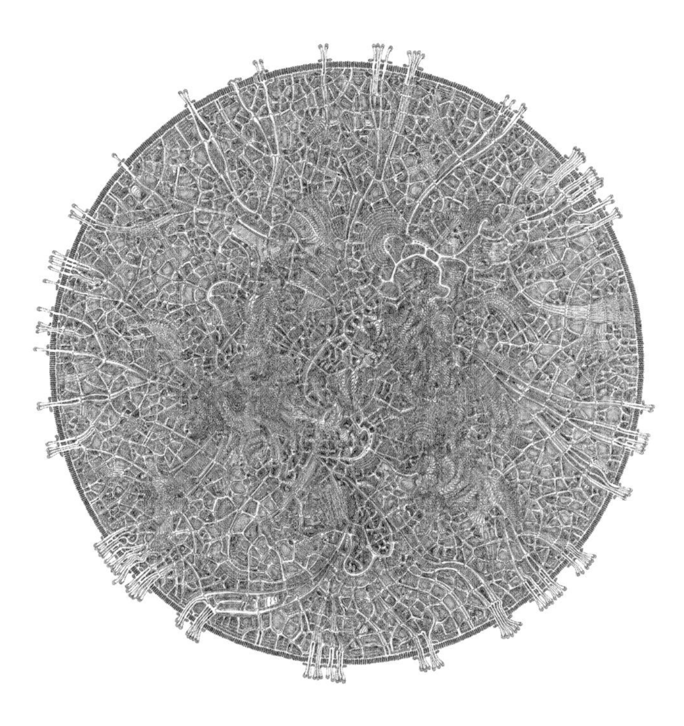

Permeable Unit, 2006

"Still, there lingers danger. Just like there are oceans on our planet, there are empty spaces, and unexplored lands inhabited by monsters on the map of science.

"I believe cutouts, labeling, and context-and-detail techniques from medical drawings would work well. We will need a legend that explains the different entities shown. The labels can be very small—almost part of the pattern, so only very curious viewers will see them. All others will simply enjoy the beauty of science.

"I wonder how to create a very rich pattern of very different dynamics. Huge differences in the sizes and dynamics of fields/areas would be realistic. In fact, the outer shape might very much look like a meteorite.There could be a larger ocean or river of money-flowing from one mountaintop to the next-down the food chain. You could have cool long tentacles on a few active areas only. Areas do differ in terms of what bridges they (can) create. Tree/science rings of different thicknesses are realistic. You could even decide to paint only 60 percent of everything and leave some parts as sketch—making the point that we do not have data for all areas. Now, I want to zoom in to explore details."

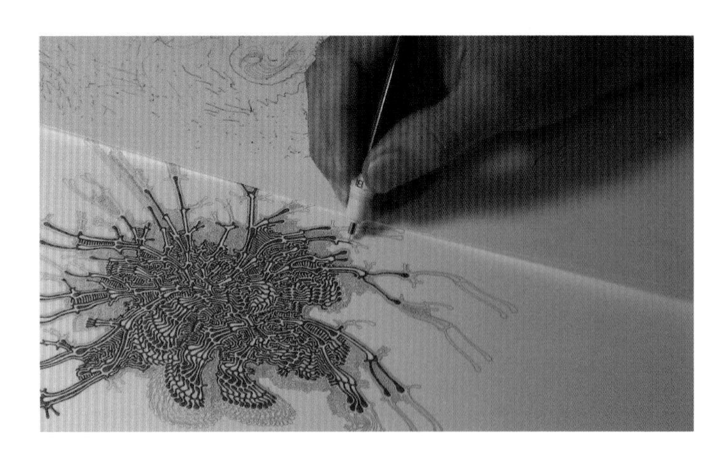

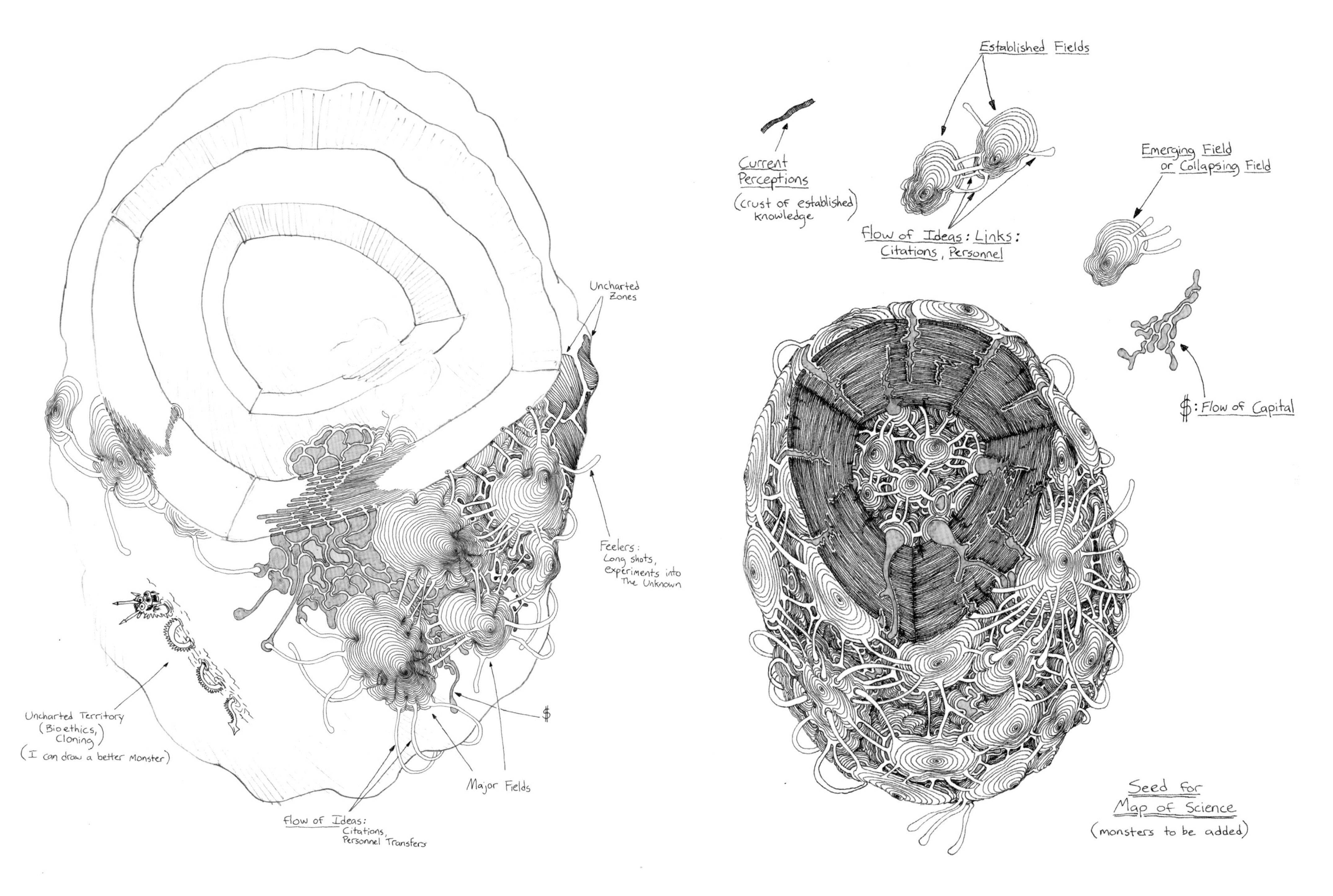

Technique

The original drawing was rendered by hand in roughly 80 hours of pure drawing time. Normally there would be no preliminary sketch, but for this drawing, the general shape and cutout area were initially laid out in pencil, providing a loose structure for the details to be built upon. The ink was then applied line by line, with no detailed global plan, but following certain local rules or patterns, which ultimately accrued to form this network of hypothetical connections.

Zeller did not see **Maps of Science: Forecasting Large Trends in Science (page 170)** before completing his drawing, yet both maps have arrived at the same conceptualization: current science happens on the crust or surface of a round object. This way, one can never fall off the surface of science—going straight in one direction on the outer crust of science leads one back to the initial position.

Reference System

A spherical reference system is used. Events are organized chronologically from the center of the sphere outward. The placement of scientific fields on the sphere minimizes edge crossings.

Data Overlays

Colors indicate emerging fields of science (blue), established fields (brown), money (yellow), links (blue), and voids (black). Many major elements of science are represented by the shapes of objects: clusters, tentacles, and monsters.

Details

Initial design sketches for the *Hypothetical Model of the Evolution and Structure of Science* are above. They show an outline of the final shape and cutout areas of science, funding as gold arteries, tentacles searching for topic areas with high scientific yield, the very first monster on a map of science, and different types of legend design. While the pencil outline may be changed, the ink interior is permanent.

Additional Elements of the Exhibit

Certainly science maps and data graphs work to engage viewers intellectually—but can they also capture the imagination, as did the early maps of the world? Is it possible to involve viewers in a more dynamic way that heightens both their awareness and appreciation of data, information, and knowledge? What can be learned from theater, movies, and art exhibits—as well as science displays—to improve the ability of science maps to entertain while educating, to inspire while being true to facts, and to change the way we see our world, empowering viewers to engage in science in novel ways?

Additional exhibit elements allow viewers to become participants—to touch and interact with science. The **Illuminated Diagrams** (ID) display combines the exceptional high data density of two large prints—a current map of the world and a map of today's science—with the flexibility of an interactive program that drives a touch panel display and two projectors that illuminate the maps. Using the touch panel display, the viewer can study both: all the locations where papers on any given topic are being published and which topics are being studied in any given geographic location.

The *Hands-On Science Maps for Kids* invite children to see, explore, and understand science from above. The maps use the ID base maps overlaid with watercolor drawings. Children and adults alike are invited to help solve a puzzle by placing images of major inventions/discoveries and inventors/scientists in their appropriate places on the maps. The **WorldProcessor Globes** show the support of various countries for patenting activity and the resulting distribution of patents. **Shape of Science** is the first attempt to sculpt the structure of science. The **Video of the Exhibition** documents the exhibit display at the New York Public Library (NYPL) and features interviews with curators, mapmakers, and NYPL officials, who discuss their motivations and ambitions to map science.

CANAD
APAN

Illuminated Diagrams

By W. Bradford Paley, Kevin W. Boyack, Richard Klavans, John Burgoon, and Peter Kennard
ALBUQUERQUE, NEW MEXICO, BERWYN, PENNSYLVANIA, AND NEW YORK, NEW YORK, 2006

Illuminated Diagrams

The setup of the Illuminated Diagrams display adds the flexibility of an interactive program to the exceptional high-data density of a print. This technique is generally useful when the data is relatively stable but there is too much pertinent data to be displayed on a screen.

The large-scale high-resolution maps show a topic map of science (at left and also **page 182**) and a current map of the world (at right and also **page 183**). Projectors are used to illuminate parts of the maps, directing the viewers' eyes to important features. By acting as smart spotlights, they help to give a radarlike "grand tour" of science, while also animating stories in static data (for example, the influence and dissemination of ideas), and highlighting query results.

Topic Map: How Scientific Paradigms Relate

The word *science* covers a vast range of topics, including astronomy, biology, chemistry, mathematics, physics, and psychology, as well as scientific approaches in the arts and humanities. This map illustrates how distinct areas of study are defined and related.

The map represents more than 1.5 million scientific papers (760,000 papers published in 2004 and their 820,000 highly cited reference papers) as white dots. Each *scientific paradigm* (represented by a red circle) contains papers that were often cited together. Some paradigms have few papers, others many, as denoted by circle sizes. The word filaments—or flowing labels—are made up of common words unique to each paradigm, thus revealing the actual language used by the scientists who work in that area. Curved lines show how paradigms are related—the stronger the relationship between paradigms, the thicker and darker those lines.

Lines are curved to facilitate visibility, so the eye can travel them with ease. We show 4,370 lines here, leaving thousands of fainter ones undrawn. A layout of the 776 paradigms was made using the VxOrd algorithm originally developed at Sandia National Laboratories. The resulting circular structure is no accident, nor is it arbitrarily imposed on the data—it comes from the structure of science itself. If you imagine that every link is a rubber band (stronger when darker) and that every node has a small force field around it that pulls it toward similar nodes and pushes it away from all other nodes, this dynamic balance of forces automatically creates the layout. Thus the map shows how **Physics** (at about 1:00) is related to **Astrophysics** and **Astronomy** (12:30) as well as to **Chemistry** (at about 2:00). The jutting peninsula of **Organic Chemistry** at 3:00 has only few connections to the thicket of **Medicine** between 5:30 and 7:00. Instead, it connects to **Medicine** through **Analytical Chemistry** (the large node at the base of the peninsula) with its tool base including **Spectroscopy** and **Proteomics**, which are commonly used in **Medicine** (see map **D** on **page 185** for area labels).

Geographic Map: Where Science Gets Done

Here, the same papers are arranged on a more familiar map of the world. Each white dot on the map represents not a city, but a set of 10 or fewer papers. Paper sets are scattered around the exact location of their authors for visibility. The size of the labeled brown circles is proportional to the number of papers authored in that location. The circle shape is distorted because of the projection chosen. City names and postal codes are used instead of word filaments. Otherwise, the design of the geographic map resembles that of the science map.

Interactive Exploration

The displays are interconnected via "brushing and linking." Viewers can interactively examine all the locations where papers on any given topic are being published and which topics are being studied in any given geographic location using a touch panel screen in a lectern. With the brush of a finger over the geographic map, the selected location will glow on the lectern's touch screen. Additionally, any topic studied in that location will be illuminated on the front or rear projected printed maps. The brighter a topic glows, the more papers there are in that area on that topic of choice. Conversely, selecting a topic node via the touch screen allows one to see all the locations in the world where that topic is being studied.

Buttons can be pushed to highlight interdisciplinary areas such as **Nanotechnology**. In addition, the influence of several major inventors and scientists (such as **Albert Einstein**) can be simultaneously made visible on both the map of science and the map of the world. Three levels of influence are highlighted in this order: 1) the topic areas and workplaces of all papers of the selected individual, 2) the topic areas and workplaces of those papers that cite the selected individual's papers, and 3) the topic areas and workplaces of those who have cited the citing papers from level 2. The exhibit video at **http://scimaps.org** shows an animation of the display.

W. Bradford Paley (typography, graphics, and interaction design) **See page 128**.

Kevin W. Boyack (scientometrics and data shaping) **See page 106**.

Richard Klavans (scientometrics and node layout) **See page 106**.

John Burgoon (geographic mapmaking) was a master's degree student in information science at the School of Library and Information Science, Indiana University, and a research associate in the Information Visualization Lab, led by Katy Börner, when designing this map. His focus is on information visualization, graph theory, and geographic information systems as practically applied to business. Having returned to school after 15 years in industry, he revels in collaborative projects involving complex systems of information, people, and software.

Peter Kennard (system design and programming) began his work with digital media more than 25 years ago. After an education in architecture and five years of professional experience in photography, lighting, and industrial design, he became involved in the field of print graphics in 1980. As a graphic designer for the Mansfield Stock Chart Service (New York), he oversaw its first wave of digital distribution of financial graphics. Later he developed high-performance graphic arts tools for Caligari Corporation and Autodesk. Today, he is a partner of the aRt&D lab (San Francisco), in the field of digital entertainment media. He lives in New York and is involved in many projects in the area of architecturally integrated media and high-performance digitally mediated interactive visualization.

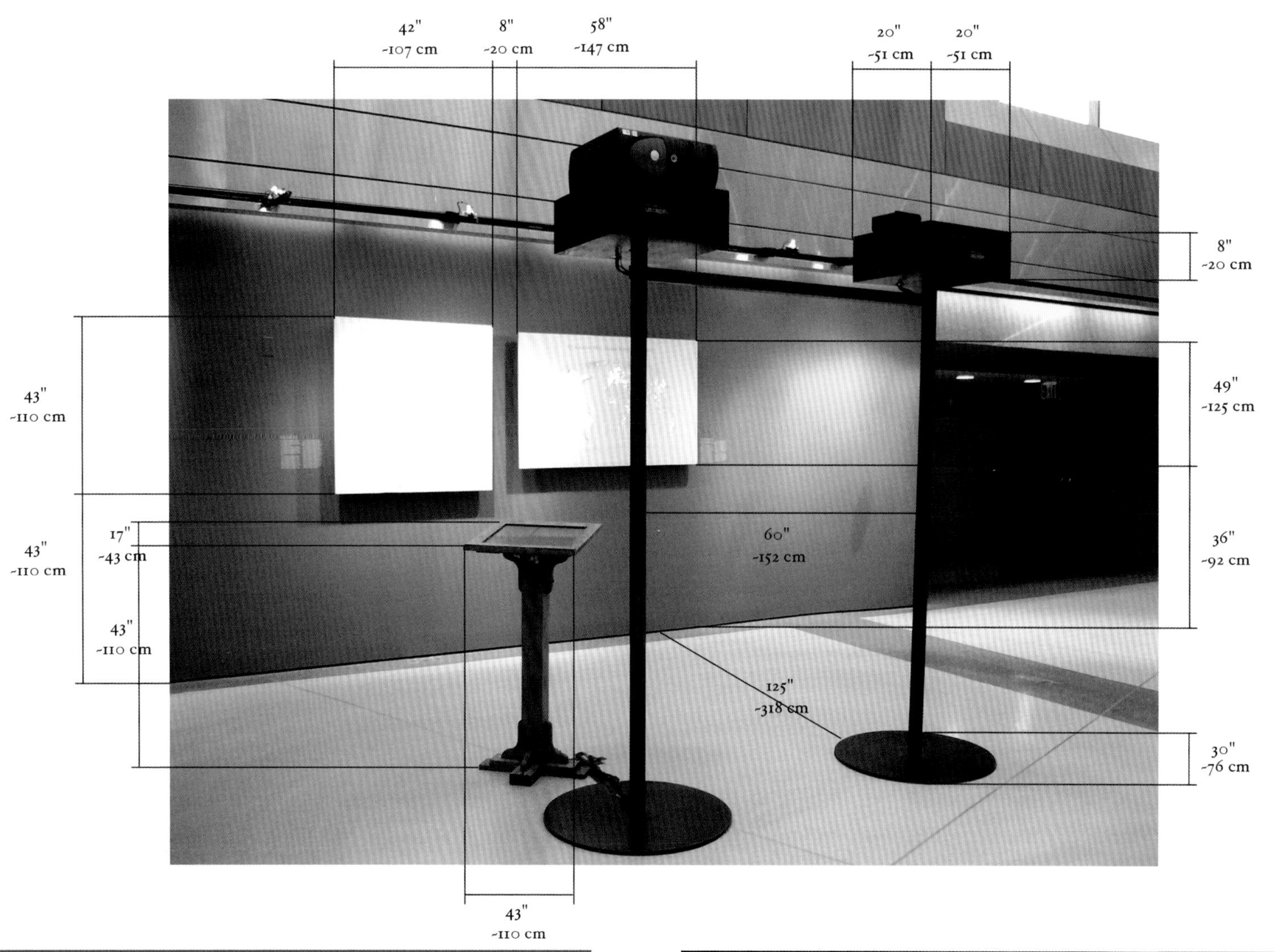
42"
~107 cm
8"
~20 cm
58"
~147 cm
20"
~51 cm
20"
~51 cm
8"
~20 cm
49"
~125 cm
36"
~92 cm
30"
~76 cm
43"
~110 cm
43"
~110 cm
17"
~43 cm
43"
~110 cm
60"
~152 cm
125"
~318 cm
43"
~110 cm

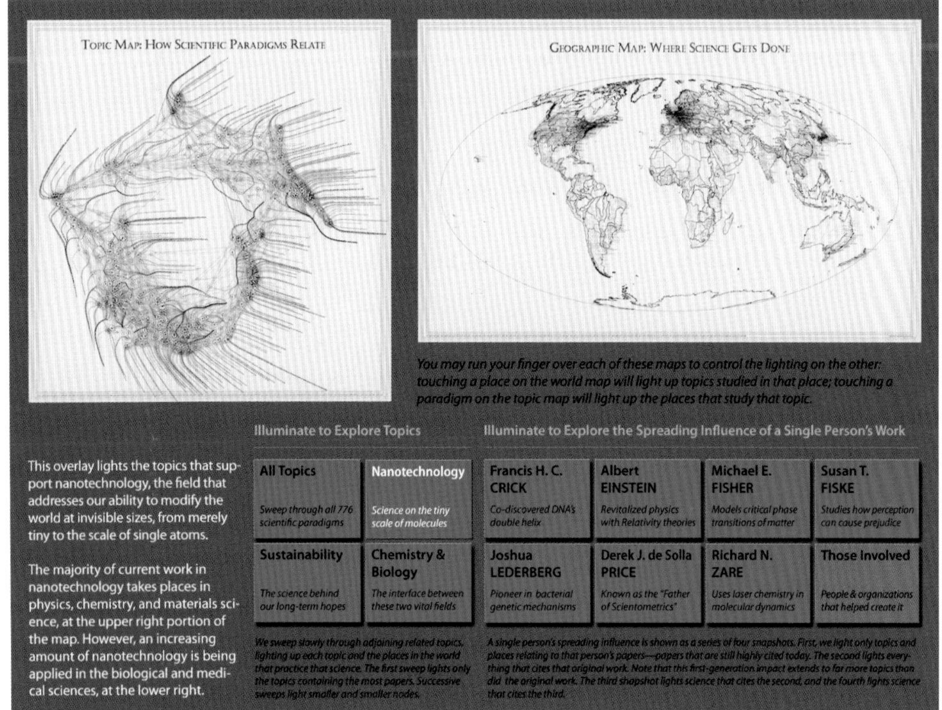
Topic Map: How Scientific Paradigms Relate
Geographic Map: Where Science Gets Done
You may run your finger over each of these maps to control the lighting on the other: touching a place on the world map will light up topics studied in that place; touching a paradigm on the topic map will light up the places that study that topic.
Illuminate to Explore Topics
Illuminate to Explore the Spreading Influence of a Single Person's Work
This overlay lights the topics that support nanotechnology, the field that addresses our ability to modify the world at invisible sizes, from merely tiny to the scale of single atoms.
The majority of current work in nanotechnology takes places in physics, chemistry, and materials science, at the upper right portion of the map. However, an increasing amount of nanotechnology is being applied in the biological and medical sciences, at the lower right.
All Topics
Sweep through all 776 scientific paradigms
Nanotechnology
Science on the tiny scale of molecules
Sustainability
The science behind our long-term hopes
Chemistry & Biology
The interface between these two vital fields
Francis H. C. CRICK
Co-discovered DNA's double helix
Albert EINSTEIN
Revitalized physics with Relativity theories
Michael E. FISHER
Models critical phase transitions of matter
Susan T. FISKE
Studies how perception can cause prejudice
Joshua LEDERBERG
Pioneer in bacterial genetic mechanisms
Derek J. de Solla PRICE
Known as the "Father of Scientometrics"
Richard N. ZARE
Uses laser chemistry in molecular dynamics
Those Involved
People & organizations that helped create it
We sweep slowly through adjoining related topics, lighting up each topic and the places in the world that practice that science. The first sweep lights only the topics containing the most papers. Successive sweeps light smaller and smaller nodes.
A single person's spreading influence is shown as a series of four snapshots. First, we light only topics and places relating to that person's papers—papers that are still highly cited today. The second lights everything that cites that original work. Note that this first-generation impact extends to far more topics than did the original work. The third shapshot lights science that cites the second, and the fourth lights science that cites the third.

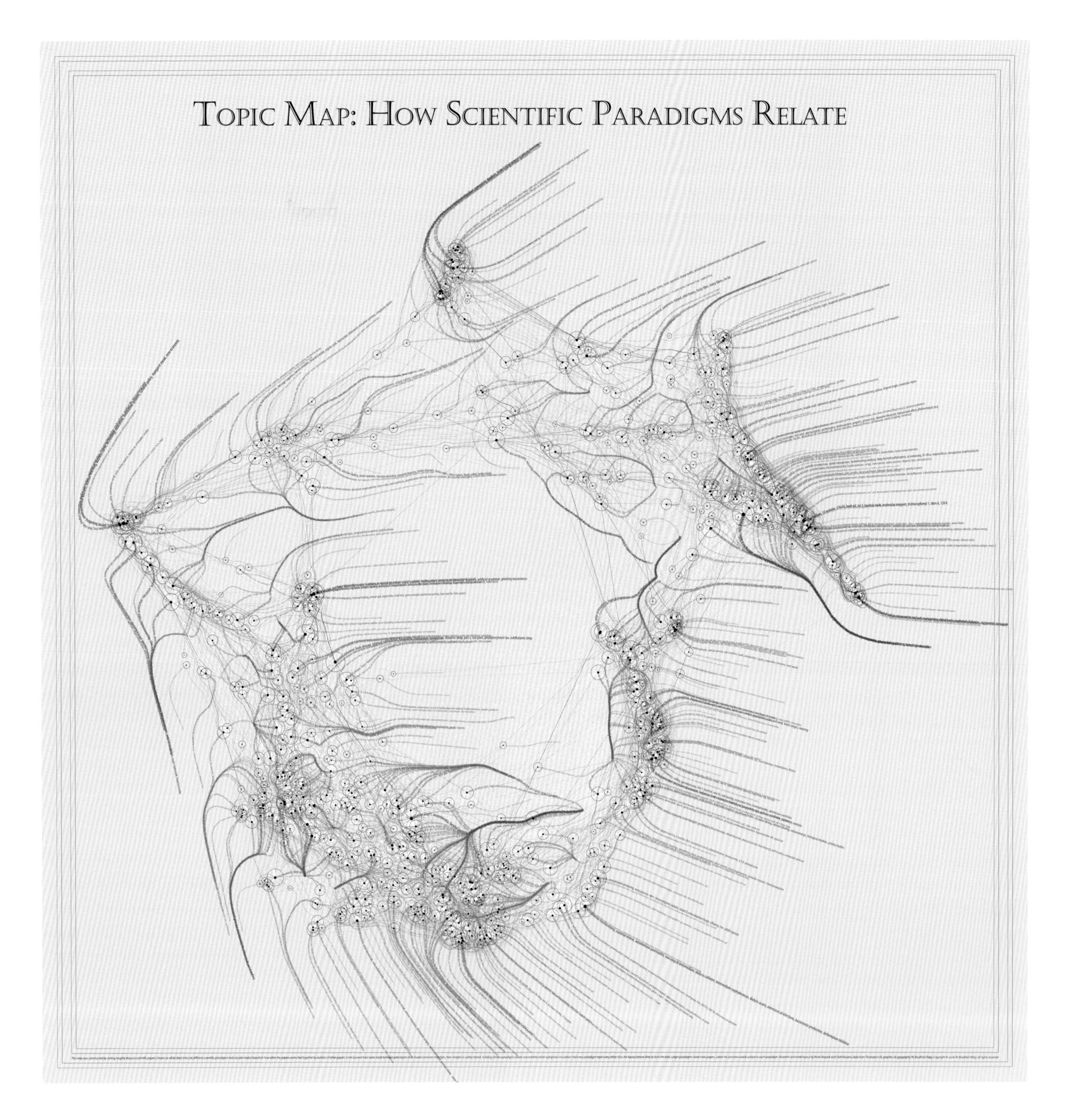
Topic Map: How Scientific Paradigms Relate

GEOGRAPHIC MAP: WHERE SCIENCE GETS DONE
33411: Tokyo, Japan

Reference System

Shown below are four maps: the **(A) Map of Scientific Paradigms** featured in the second iteration of the exhibit (see **page 136, Map of Scientific Paradigms**, by Kevin W. Boyack and Richard Klavans), the **(B) ID Topic Map**, and the two other maps that appeared in magazines: the **(C)** ***Nature*** **"Brilliant Display" Map**, and the **(D)** ***SEED*** **Magazine Map**. All four maps use the very same reference system for node layout.

Data Encoding

In all four maps, node area size encodes the number of papers produced per paradigm. However, the maps differ considerably in the visual rendering of node color and shape; color, width, and shape of linkages; and the labeling of nodes and scientific disciplines. Consequently, the four maps provide different insights.

Map **A** uses node color to encode the vitality of the 776 paradigms. Nodes in map **D** are color coded according to their membership in a discipline of science (for example, paradigm nodes that are mostly biology are given in green, blue is used for chemistry, brown for earth sciences, yellow for the social sciences, and red for medical; see also **page 13, UCSD Map of Science**).

Links between paradigms that share papers are represented by lines, see close-up on **page 139, Map of Scientific Paradigms**. In all four maps, node proximity and darker or thicker lines indicate the strength of the relationship between two paradigms. Lines in map **A** are straight. Those in maps **B–D** are curved to avoid overlap in support of legibility.

Maps **A** and **D** have area and discipline labels, respectively, that were assigned by hand. The labels in map **A** are more specific than the labels in map **D**. CNS is the abbreviation for Central Nervous System.

Flowing Labels

Maps **B–D** have flowing node labels made up of sequences of common words unique to each paradigm. The layout of the labels can be best described using a pictorial description provided by Paley. He imagined the labels as flowing downhill away from the nodes. To create the terrain, he envisioned the base map as a tabletop and each node as a tall spike (with height proportional to node size) arising from the table. Then he convolved these spikes with Gaussian bumps having a large diameter of approximately one-quarter the size of the image. Later, he put in a smaller, steeper Gaussian structure on top, one-twentieth the width of the image, to help bend the label paths more strongly away from nearby nodes in clusters. With this setup, a label path is started at a distance that is approximately the width of the letter "M" from the node and follows the descending gradient. The path is stopped when reaching a local minimum. The result is the topic map layout in the ID setup (see **B**, below). In Paley's words:

"Antique maps carefully wrapped text along features in the geography. This saves space—if the features were spatially separated the labels stayed in the vicinity of the feature, not complicating the labeling of other features. It also helped tie the label more firmly to the geographic feature. I'd suggest that the cognitive binding of perceptually-related objects—especially objects next to one another and with similar shapes—is stronger than that of labels that are merely nearby or point at the feature. Note that for this image the features are only dots and are so tightly packed that I couldn't do what I think optimal: track the feature shape as a river name tracks the river in antique maps, so I fall back at pointing at the node."

In map **B**, many flowing labels overlap and are unreadable in the densest areas (where one presumably most wants to read them). Tweaking the algorithm parameters led to map **C**. Paley summarizes the strengths and weaknesses of the improved label layout as follows:

"Strengths:

- Better (more even) use of space for labels than simple linear placement, especially in dense areas.
- No "callout lines" necessary: the text itself points toward the node.
- Dual representation for each node: large circle allows for greater visual range to express the data's variability; small central dot anchors the node to space, overlaps less and provides a visual anchor or the text (being roughly x-height and baseline-aligned)—while retaining the easy ability to connect the large circle with the small dot—people seem to have evolved to finding pupils in irises, splashes in ripples, etc.
- The more relaxed paths of the nodes may elicit a useful intellectual frame of reference for certain kinds of data: it resonates with historical geographic labeling conventions and generally feels more organic: good for organically grown or flow-based data sets.

Weaknesses:

- Much too organic and "hairy" looking: the overall effect draws too much attention to the labeling algorithm and away from the data.
- Hard to read upside-down lettering—ideally they would be reversed when that happens.
- Too many overlaps and text paths that bump over earlier (larger) ones.
- The current paths are too choppy, even if we allow the complex paths—more powerful software and computers are required to manage all the moving points (thousands per path) at a finer spatial resolution."

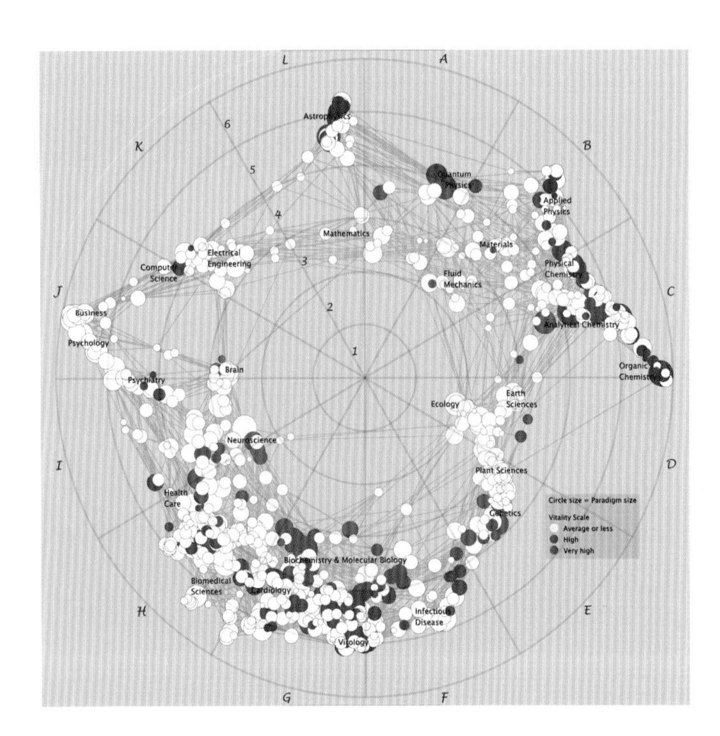

(A) Map of Scientific Paradigms

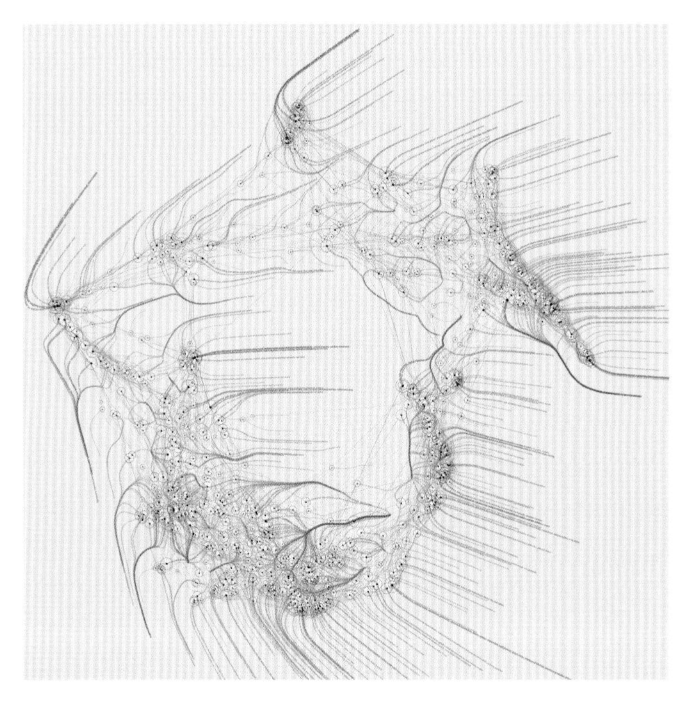

(B) Illuminated Diagram Topic Map

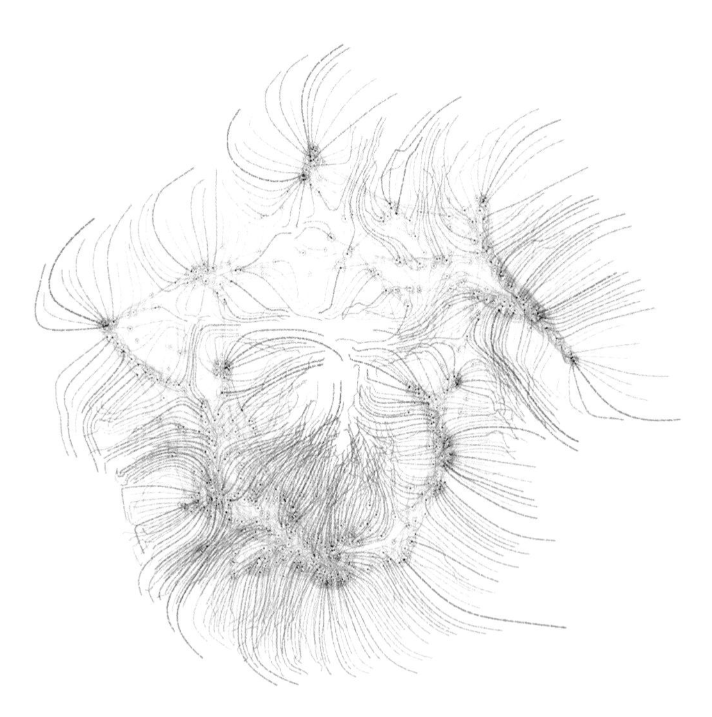

(C) *Nature* "Brilliant Display" Map

Map **D** prevents overlap by starting with the most important (here, biggest) nodes on the blank page. As each node's label path is laid out, a very small and steep Gaussian curve (about the diameter of twice the font height) is convolved with the whole path so that later paths do not meet earlier paths. This results in slightly distracting label paths with some turned upside down; however, each label is readable. The map was later recolored (see photographs on right).

Maps **B–D** have inspired many naming attempts, including "a beast that needs a haircut" (Paley), a "feather boa" (Barbara Tversky), "a filamentous microorganism you might see under a microscope" (press release from Sandia National Laboratories), and a "flying spaghetti monster" (Slashdot comment).

Unique Features

Since May 2007, the *Illuminated Diagrams* have been available as a rear projection. Instead of projectors that cast illumination patterns on large-scale mounted maps (see **page 180**), two large-scale display screens, such as the 42-inch (approximately 107 centimeters) TVs shown to the right, are used to illuminate adhesive prints of the world and science maps. The software has been reimplemented to drive multiple walls, so that both the world and science can be explored at multiple levels of resolution.

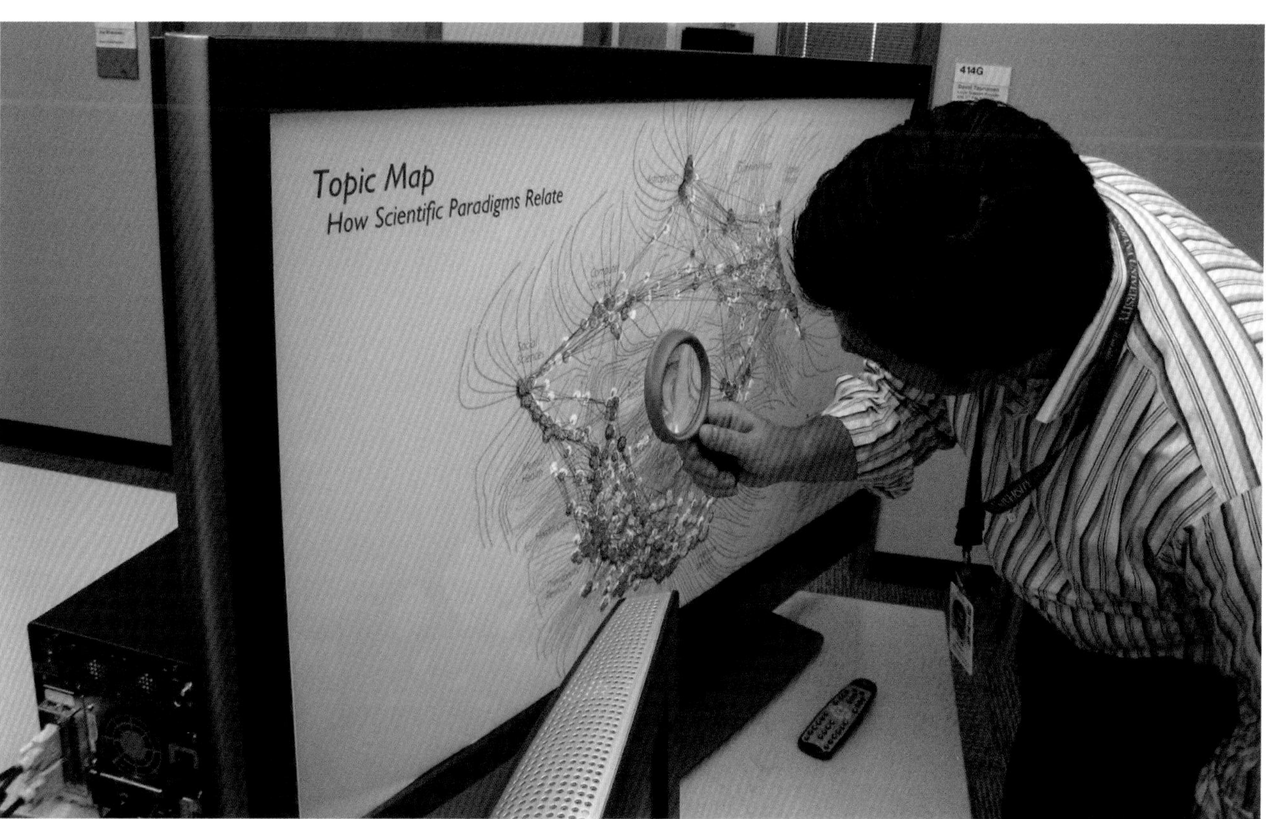

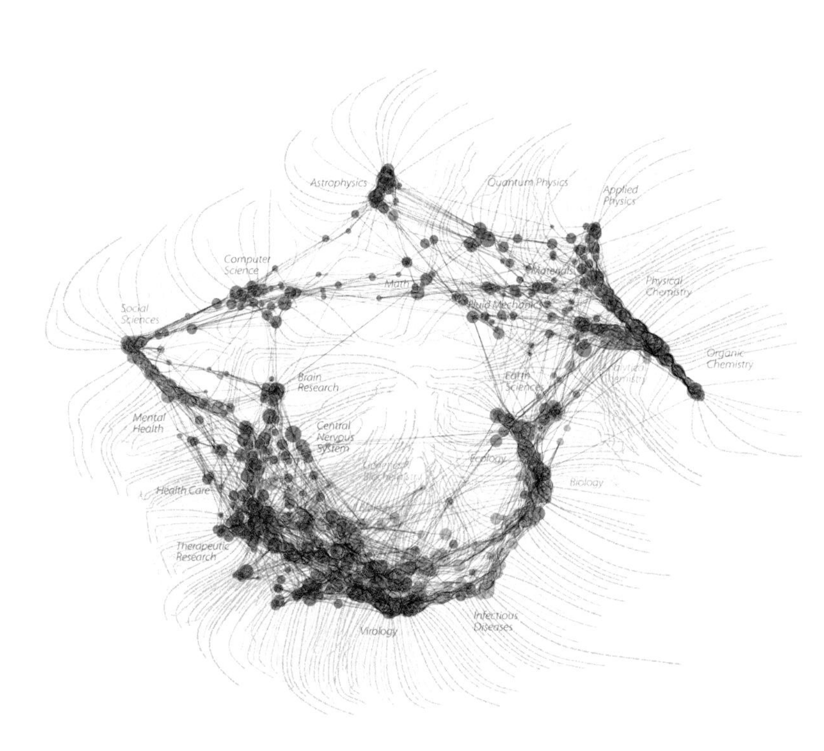

(D) *SEED* Magazine Map

Hands-On Science Maps for Kids

By Fileve Palmer, Julie M. Davis, Elisha F. Hardy, and Katy Börner

BLOOMINGTON, INDIANA, 2006

The base map is taken from the Illuminated Diagrams display by Kevin Boyack, Richard Klavans, and W. Bradford Paley. We would like to thank Stephen Miles Uzzo, director of technology, and Michael Lane, director of exhibit services, at the New York Hall of Science for manufacturing the physical maps.

Courtesy of Indiana University, 2006

Aim and Interpretation

It is interesting to note that textbooks and encyclopedias in the United States feature mostly U.S. inventions/discoveries and inventors/scientists. The same types of books in Europe feature mostly European inventors/scientists and achievements. Perhaps this holds true for other regions. In truth, of course, major inventions/discoveries and inventors/scientists exist all over the world—and across all areas of science. The *Hands-On Science Maps for Kids* invite children to see, explore, and understand science from above. They aim to help children and adults alike to learn where major inventions/discoveries and inventors/scientists can be found on a map of the world and on a map of science.

One map shows the world and the geographic locations where science is practiced and researched. The other map shows major areas of science and their complex interrelationships. Watercolor paintings were digitally added to bring a tangible quality to both continents and areas of science. On the maps are empty places for 18 puzzle pieces; on those pieces are images of major inventions/discoveries on the front and major inventors/scientists on the back. Participants are invited to solve the puzzle by placing these images in their appropriate places. They are given these instructions: "Start by selecting either of the two maps. Decide if you want to place famous inventors or major inventions first. Turn the map over when you are done and start over. Look for the many hints hidden in the drawings to find the perfect place for each puzzle piece. What other inventors and inventions do you know? Where would your favorite science teachers and science experiments go? Which area of science do you want to explore next?"

Fileve Palmer (painting) was an East Harlem high school history teacher before embarking on graduate studies in anthropology at Indiana University. In her former life, she also studied visual arts and showed paintings in New York galleries. When working on the map, Palmer worked as artistic and ethnographic consultant at Katy Börner's Information Visualization Lab. She helped design and paint the images for the Hands-On Science Maps for Kids. As a former educator, an artist, and now a student, she looks forward to promoting learning in her cross-disciplinary endeavors.

Julie M. Davis (data acquisition) has a bachelor's degree in English and anthropology from Indiana University. Her undergraduate thesis work focused on developing better interpretations of remote sensing data in archaeological fieldwork. She has a strong interest in seeing the knowledge generated in research institutions and universities made better available to the general public in more universally comprehensible forms. She has explored this interest in the past while working with local museums, creating presentations of history and archaeological research. She served as curator of the *Places & Spaces: Mapping Science* exhibit from September 2006 to August 2007.

Elisha F. Hardy (graphic design) See page 158.

Katy Börner (concept) See page 132.

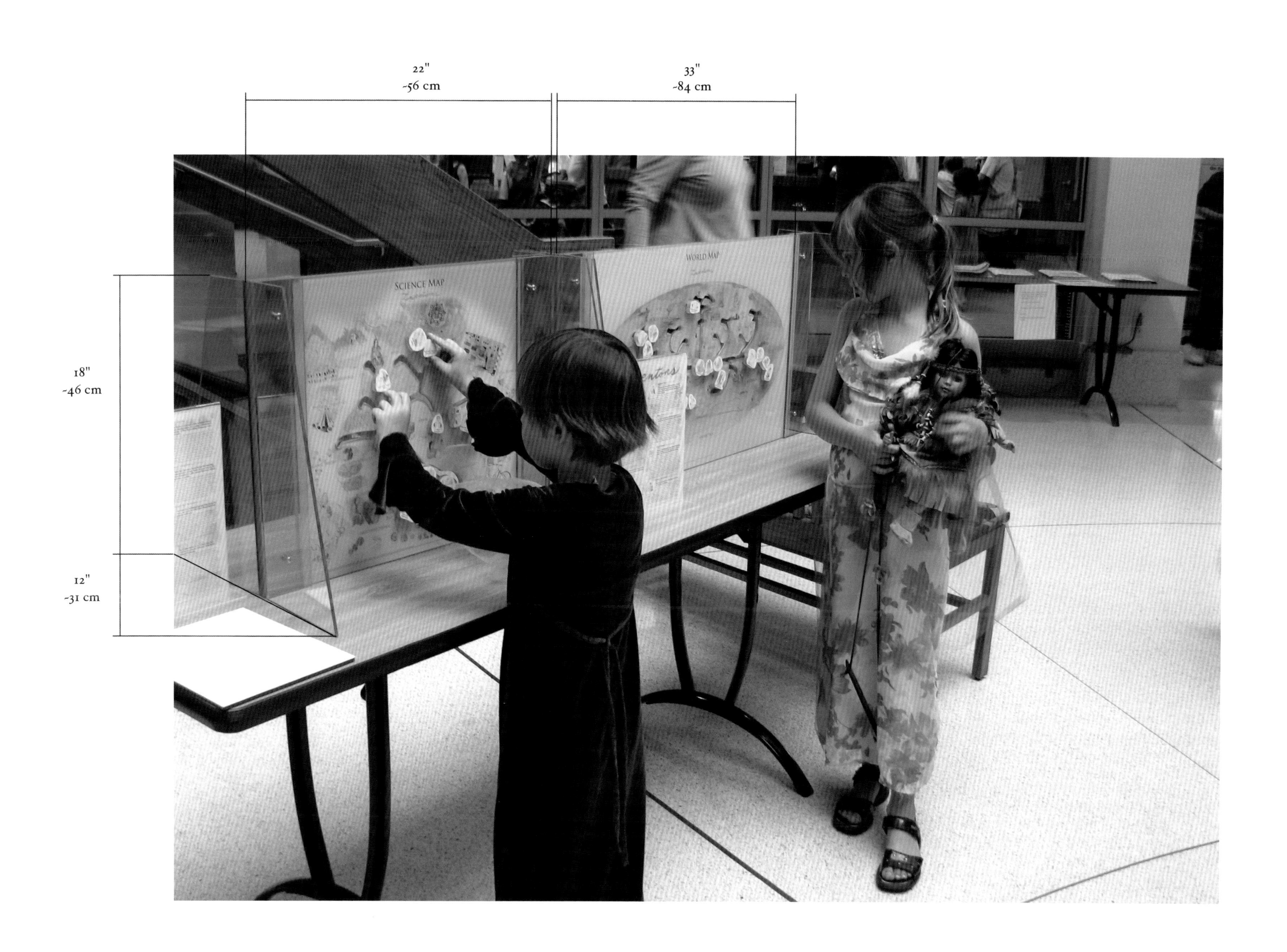
22"
~56 cm
33"
~84 cm
18"
~46 cm
12"
~31 cm
Science Map
World Map

World Map

Inventions

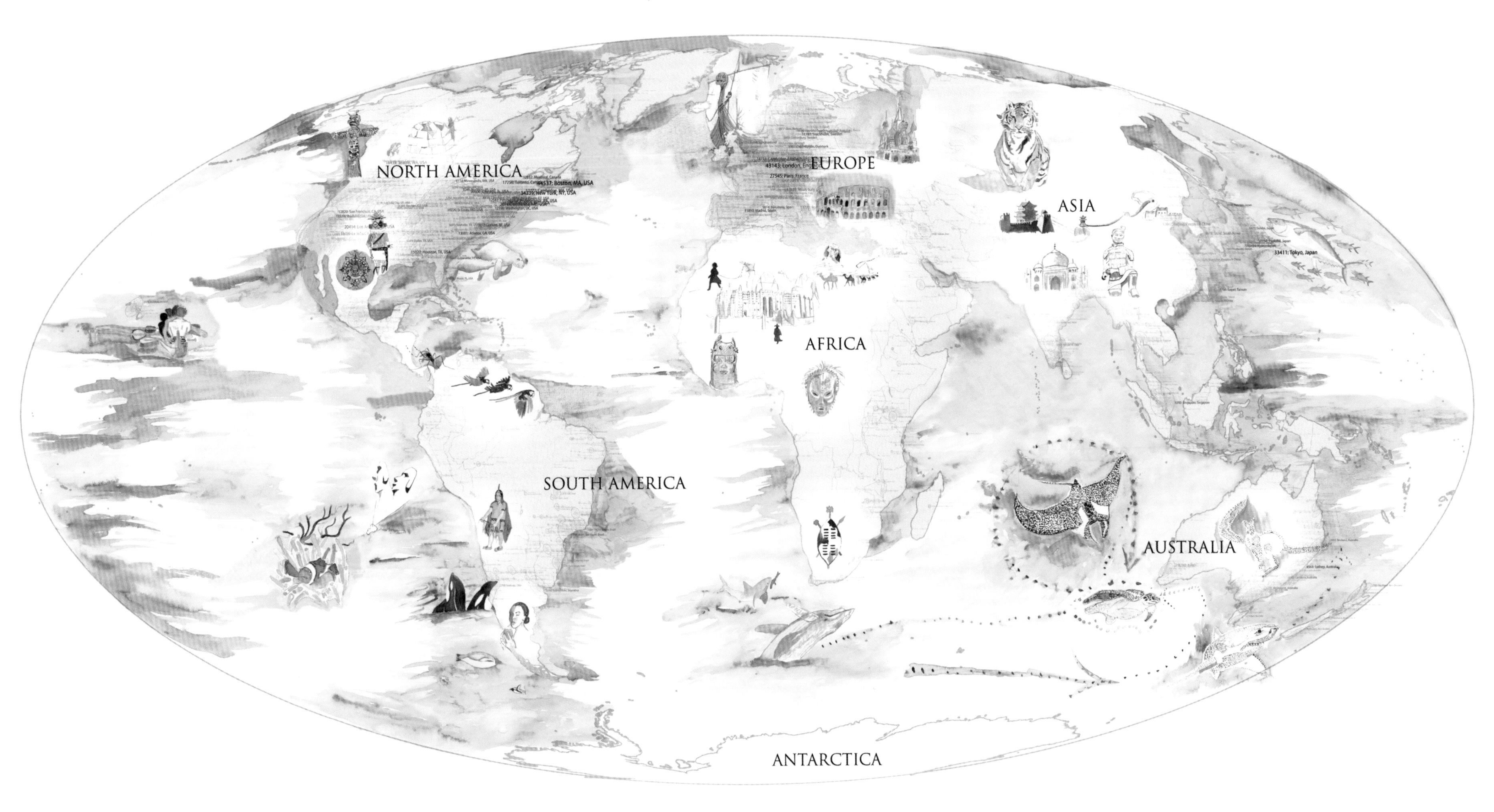

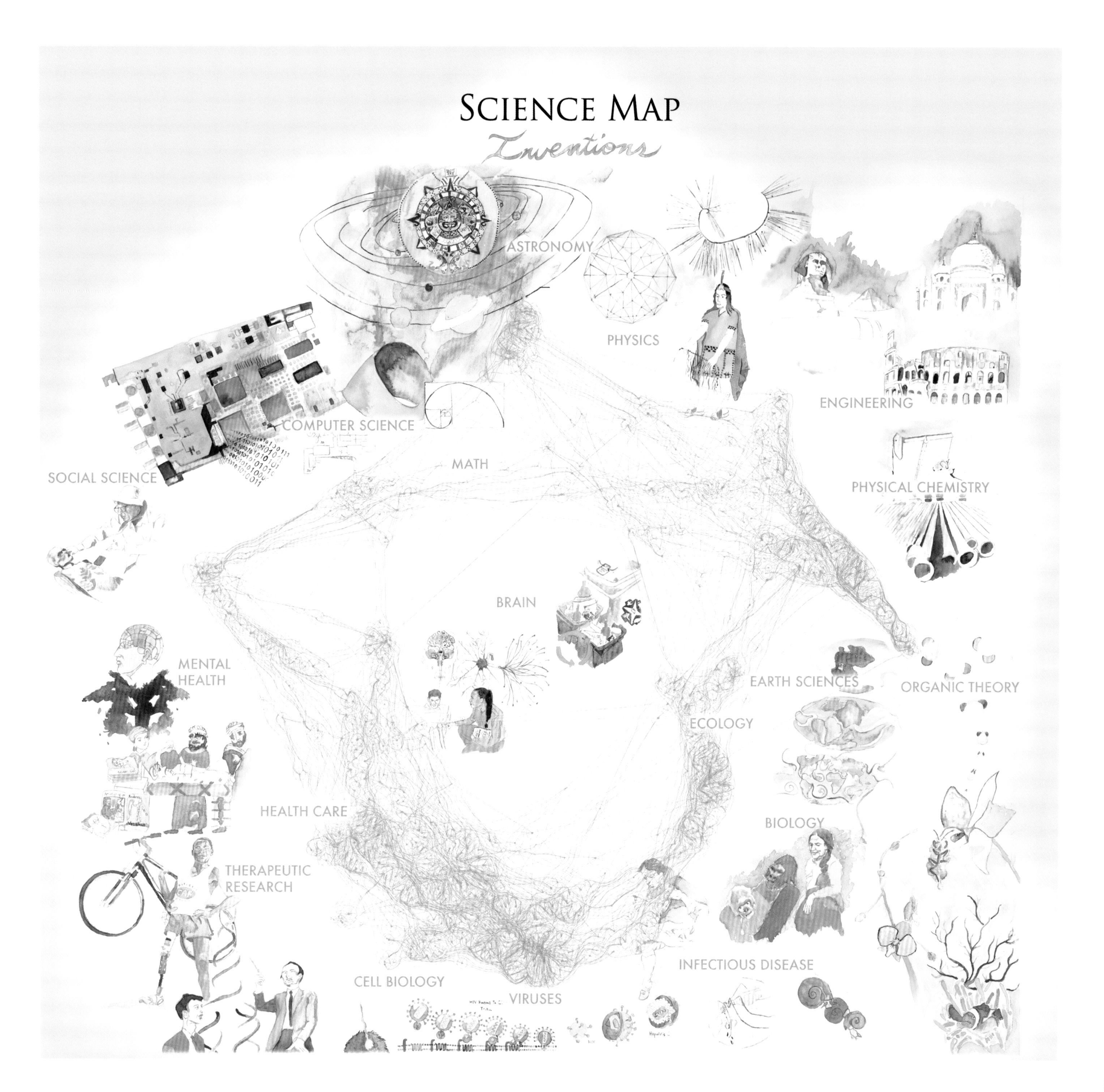
SCIENCE MAP
Inventions
ASTRONOMY
PHYSICS
ENGINEERING
COMPUTER SCIENCE
MATH
SOCIAL SCIENCE
PHYSICAL CHEMISTRY
BRAIN
MENTAL HEALTH
EARTH SCIENCES
ORGANIC THEORY
ECOLOGY
HEALTH CARE
BIOLOGY
THERAPEUTIC RESEARCH
INFECTIOUS DISEASE
CELL BIOLOGY
VIRUSES

Data

The reference world map and science map were taken from the **Illuminated Diagrams (page 180)** display. Inventions/discoveries and inventors/scientists were selected based on the study of diverse science encyclopedias, Web resources, and expert suggestions.

Technique

Both maps show the raw data, allowing viewers to imagine the quantity of records shown. Watercolor drawings were added to bring the geographic places and scientific spaces to life. Labels in the science map were modified so that children can understand them more easily (the original title VIROLOGY, for example, was replaced by VIRUSES). Different designs for the layering and the puzzle pieces were explored. Below are images of one cardboard puzzle piece, the first tracks of the science map (made to work on both sides), and the final puzzle map design.

Learning Objectives

The maps are intended to provide a global view of the geographic and scientific origins of major inventions/discoveries and inventors/scientists. Major contributions from all areas of the world and across all areas of science are used (see timeline below for a list of all the inventions/discoveries and inventors/scientists used in the two maps).

While the base map of our world is taught extensively in school, the base map of science is less well known or understood. For those educated in the western system, it may be easy to correctly place the U.S. and European inventions/discoveries and inventors/scientists; the placement of puzzle pieces corresponding to other parts of the world, however, may be more challenging.

The puzzle maps offer a hands-on experience. They serve as a tangible exercise that allows participants to use their spatial and motor skills to explore the structure of science and to remember where puzzle pieces go. After playing with the puzzle, children and adults hopefully walk away with a broader and more global understanding of both world geography and the world of science. They learn that inventions/discoveries and inventors/scientists exist all over the world and across all areas of science. Ideally, this will help children find their own places in science.

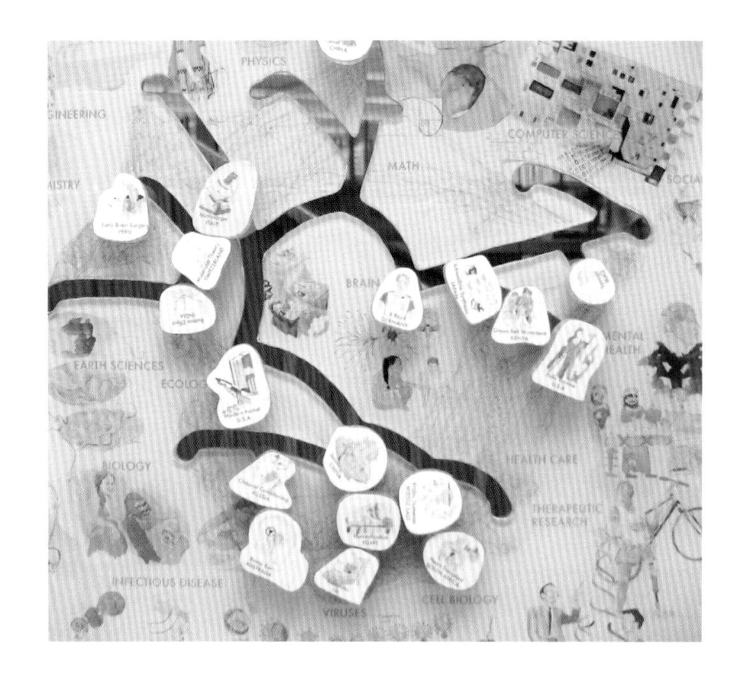

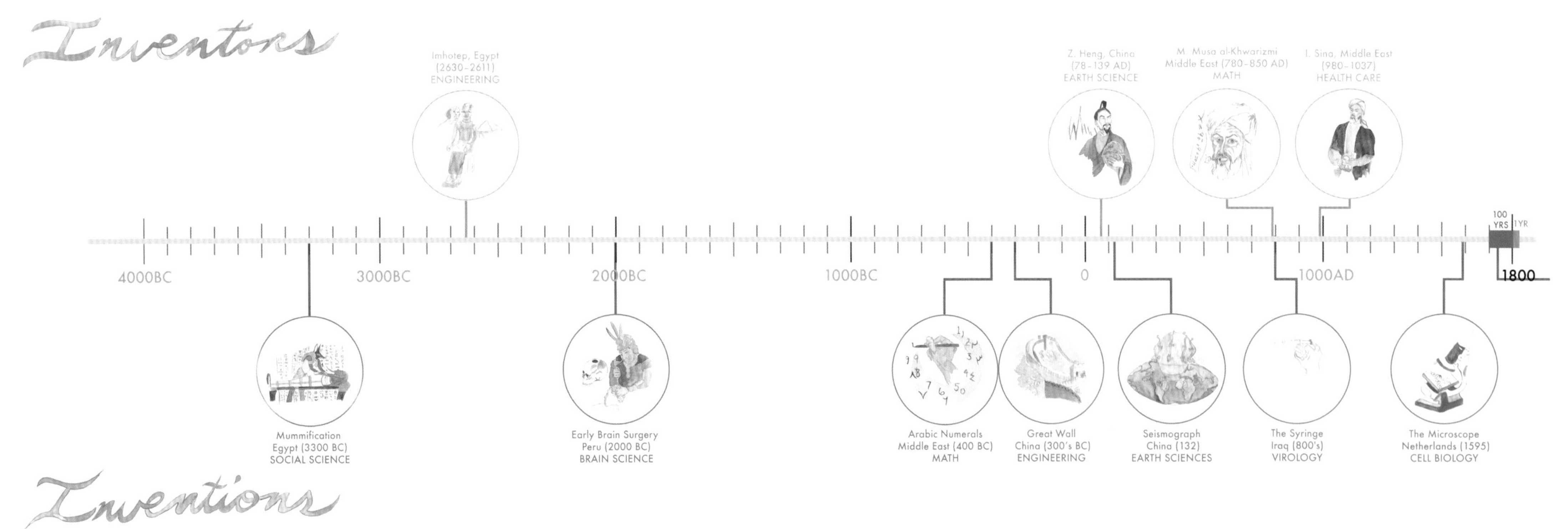

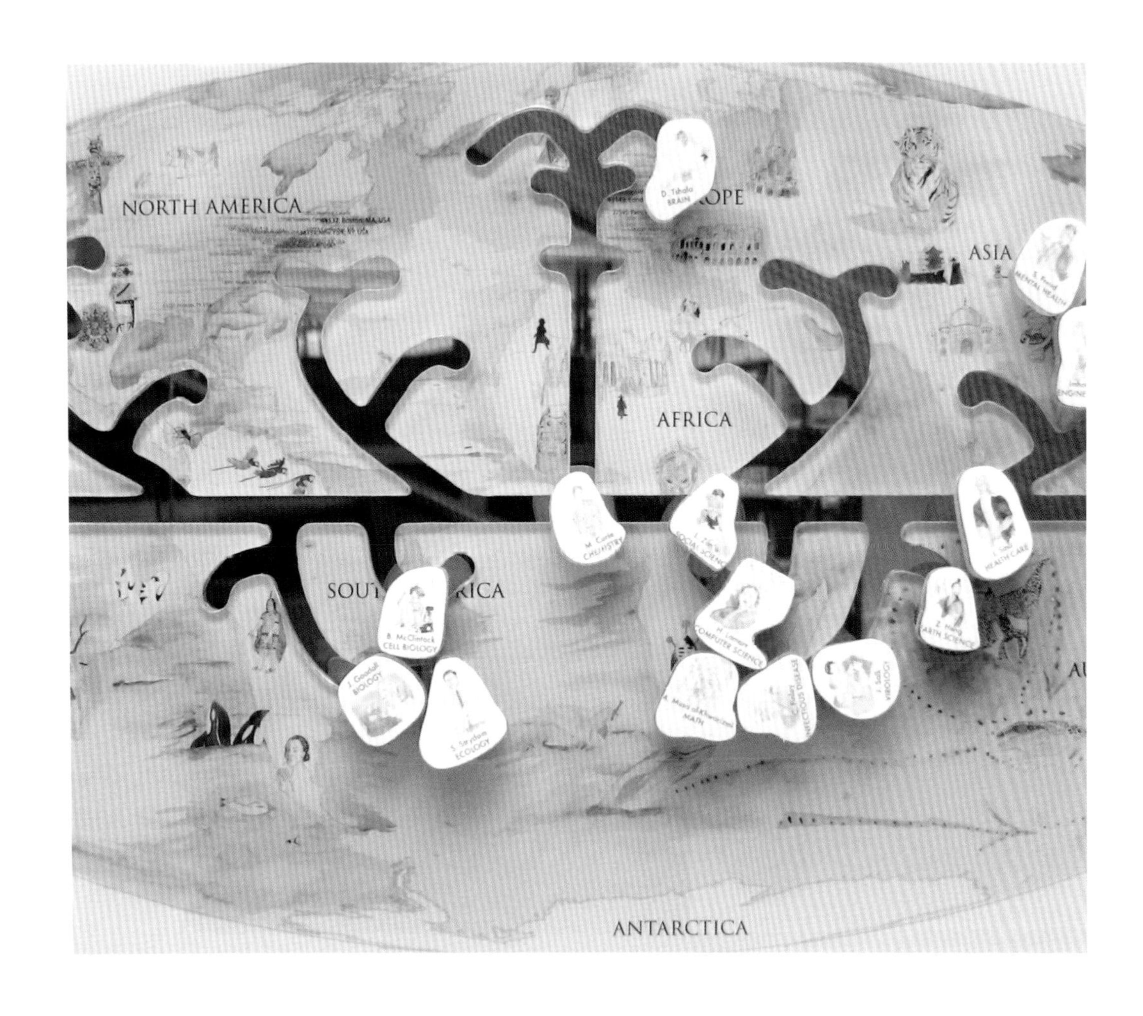
NORTH AMERICA
ASIA
AFRICA
ANTARCTICA

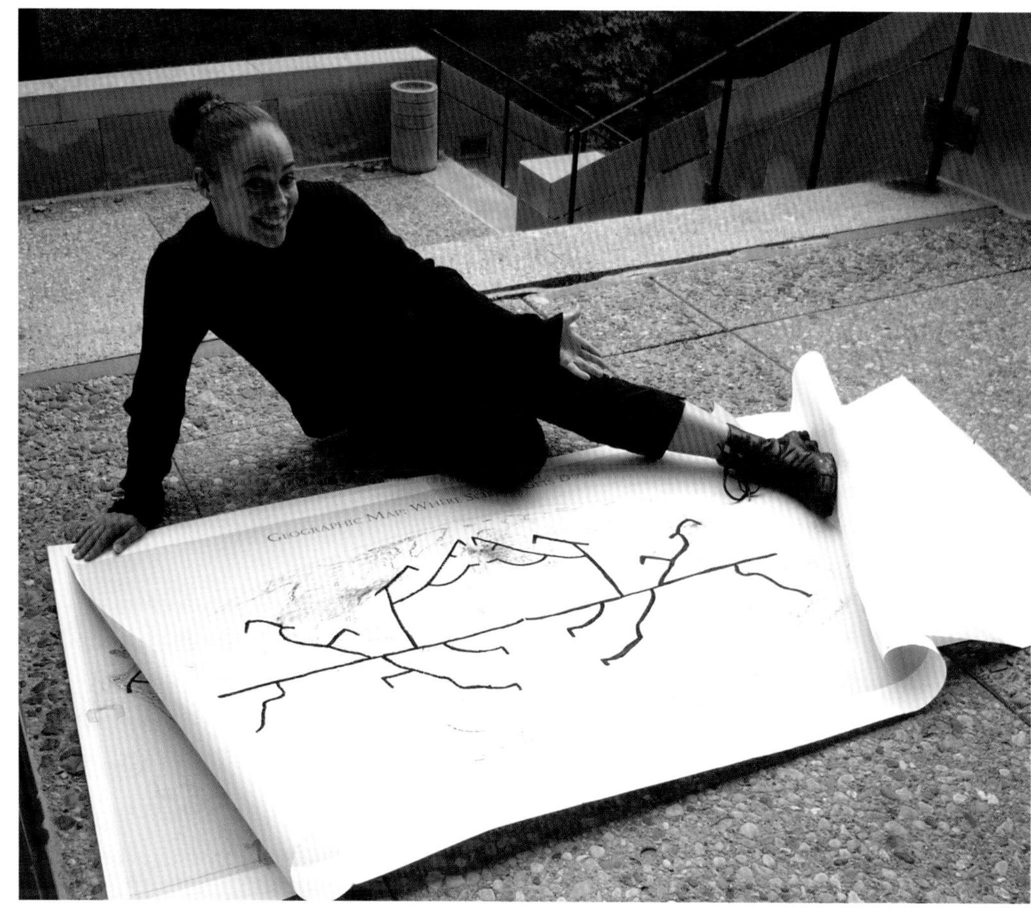

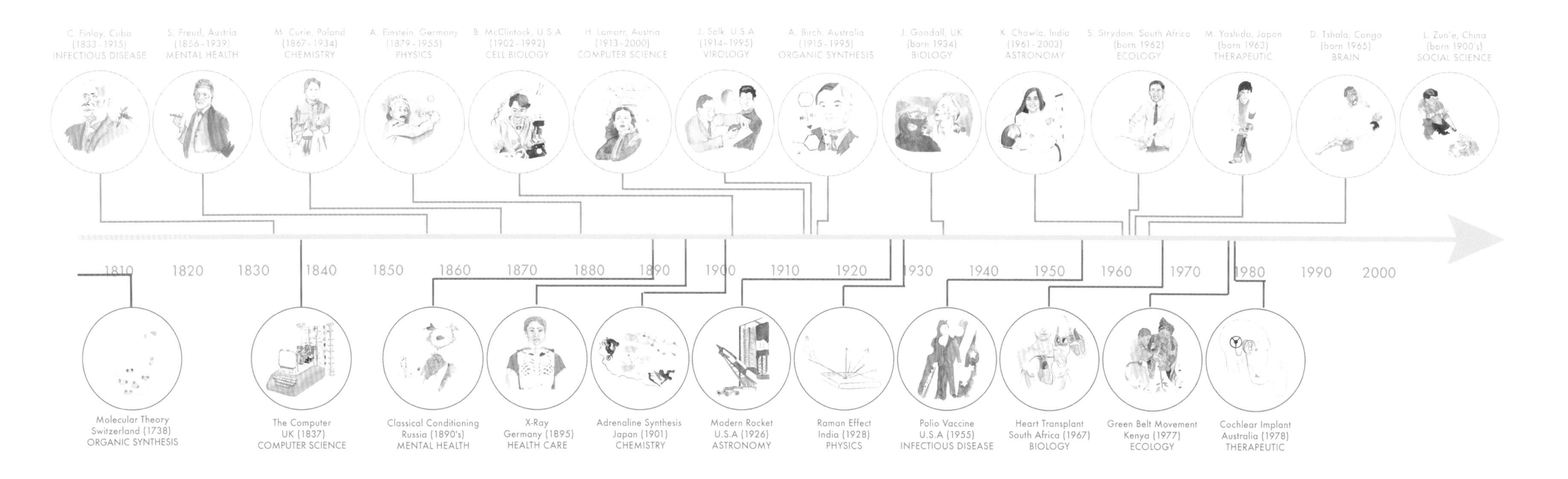
C. Finlay, Cuba (1833-1915) INFECTIOUS DISEASE
S. Freud, Austria (1856-1939) MENTAL HEALTH
M. Curie, Poland (1867-1934) CHEMISTRY
A. Einstein, Germany (1879-1955) PHYSICS
B. McClintock, U.S.A (1902-1992) CELL BIOLOGY
H. Lamarr, Austria (1913-2000) COMPUTER SCIENCE
J. Salk, U.S.A (1914-1995) VIROLOGY
A. Birch, Australia (1915-1995) ORGANIC SYNTHESIS
J. Goodall, UK (born 1934) BIOLOGY
K. Chawla, India (1961-2003) ASTRONOMY
S. Strydom, South Africa (born 1962) ECOLOGY
M. Yoshida, Japan (born 1963) THERAPEUTIC
D. Tshala, Congo (born 1965) BRAIN
L. Zun'e, China (born 1900's) SOCIAL SCIENCE
1810
1820
1830
1840
1850
1860
1870
1880
1890
1900
1910
1920
1930
1940
1950
1960
1970
1980
1990
2000
Molecular Theory Switzerland (1738) ORGANIC SYNTHESIS
The Computer UK (1837) COMPUTER SCIENCE
Classical Conditioning Russia (1890's) MENTAL HEALTH
X-Ray Germany (1895) HEALTH CARE
Adrenaline Synthesis Japan (1901) CHEMISTRY
Modern Rocket U.S.A (1926) ASTRONOMY
Raman Effect India (1928) PHYSICS
Polio Vaccine U.S.A (1955) INFECTIOUS DISEASE
Heart Transplant South Africa (1967) BIOLOGY
Green Belt Movement Kenya (1977) ECOLOGY
Cochlear Implant Australia (1978) THERAPEUTIC

WorldProcessor Globes

By Ingo Günther, John Burgoon, Stephen C. Oh, and Dongxia Monika Zhu

The second iteration also features three WorldProcessor Globes: Foreign US Patent Holders (WorldProcessor #284), Patterns of Patents & Zones of Invention (WorldProcessor #286), and Shape of Science.

Aim and Interpretation

Foreign U.S. Patent Holders ***(WorldProcessor #284)***

This globe represents half of all patents in the United States—which are registered to foreign holders. Countries with more than 1,000 patents registered in the United States are indicated by name, with the point size of the representative text scaled according to the square root of the total number of U.S. patents held. Were the number of domestically held U.S. patents to be indicated according to this methodology, the entire surface of the globe would be covered.

Patterns of Patents and Zones of Invention ***(WorldProcessor #286)***

This globe plots the total number of patents granted worldwide, beginning with nearly 50,000 in 1883, reaching 650,000 in 1993 (near the North Pole), and rapidly approaching 1 million in 2002 (note that the scale shifted to the southern hemisphere). Geographic regions where countries offer environments conducive to fostering innovation are represented by topology. Additionally, nations and countries that have an average of 500 or more U.S.-registered patents per year granted to their residents or companies are highlighted in red by their respective averages in the years after 2000.

Shape of Science

This sculpture of the *Shape of Science* spatializes the quantified connectivities and relative flows of inquiry within the world of science. It gives the map by Klavans and Boyack (see **page 136, Map of Scientific Paradigms**) a physical form that can be touched and explored in new ways.

Ingo Günther (concept and design) **See page 140.**

John Burgoon (geocoding) **See page 180.**

Stephen C. Oh holds a BFA from the Rhode Island School of Design. Previously, he studied at the Design Academy Eindhoven (Netherlands). He has assisted Ingo Günther in the continuation of the WorldProcessor project and in the development of *Shape of Science*. He lives and works in Brooklyn, New York.

Dongxia Monika Zhu graduated from the Zhejiang Art School (now the Zhejiang Art Academy) of the National Art Academy. She has worked for the *Zhejiang Daily* newspaper as a reporter and for the *Qiang Jiang Evening News* as a fashion columnist. Her first book, *Yao Tiao*, was published in 1990. From 1990 to 1993, Zhu attended the Toronto School of Art. Since her move to New York in 1994, her illustrations and designs have been published in the *New York Times*, the *Daily News*, and other commercial and editorial publications. She continues to write for the *Qiang Jiang Evening News* with her own column, "Monika's New York." Zhu has also contributed recent articles and photographs to *Contemporary Art*, the Chinese edition of *National Geographic Traveler*, and *Soho Xiaobao*, a Beijing-based publication. Throughout her career she has collaborated with other artists on projects such as the controversial 1989 work Dialog (phone booth shooting). Zhu has contributed to Günther's WorldProcessor project since 2001.

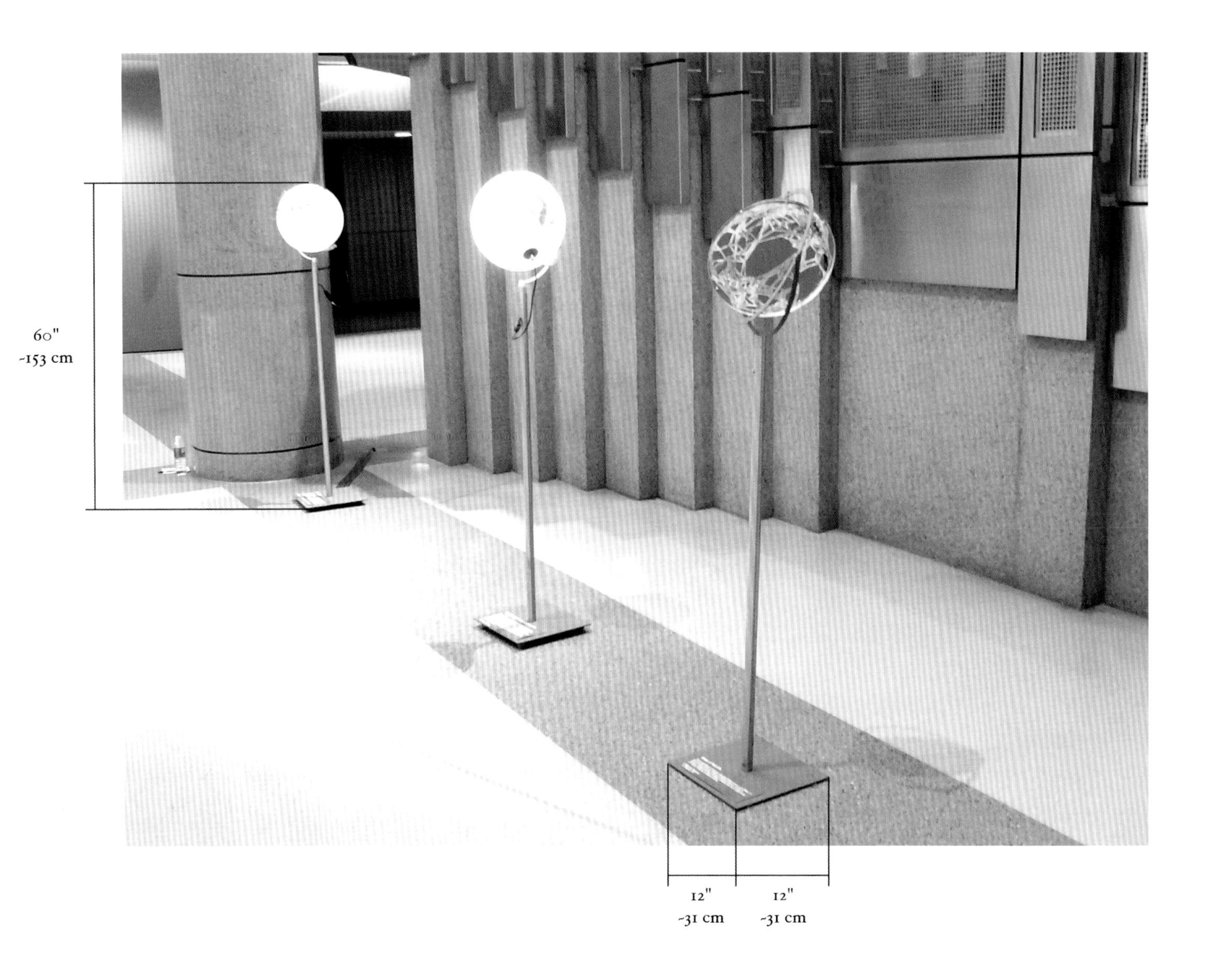
60"
-153 cm
12"
-31 cm
12"
-31 cm

1900
11600
1550
700
3900
96000

TAIWAN
HONG KONG
AUSTRALIA

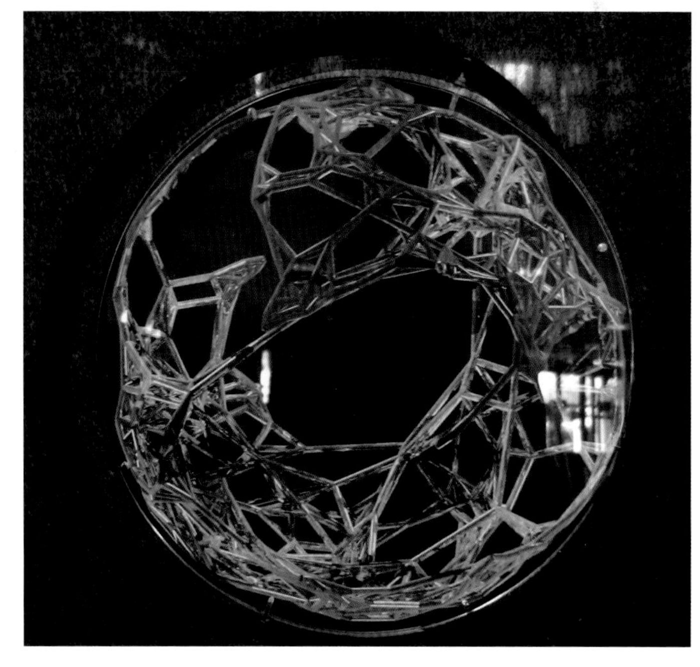

Video of the Exhibition

By Chad Redmon and Aaron Raskin, Harbinger Media, Inc., Katy Börner, Ingo Günther and Stephen Oh, Elisha F. Hardy, and Monika Herzig

We would like to thank Anne Prieto, Indiana University, for providing the brain images; Nicole (Nikki) Roberg for designing and sharing the science maps for kids. Classroom pictures are by the 2005/2006 Love/Creek class at Rogers Elementary School, Bloomington, IN.
Courtesy of Indiana University, 2006

Aim and Interpretation

An exhibit is a unique event that may last from several hours to many years. The first and second iterations of *Places & Spaces: Mapping Science* were on display from April 3 to August 31, 2006 at the Science, Industry and Business Library (SIBL) of the New York Public Library (NYPL).

The *Places & Spaces: Mapping Science* video captures this public display. Curators Deborah MacPherson and Katy Börner walk the viewer through the exhibit. John Ganly (SIBL) comments on the match of the exhibit with the responsibility of libraries to make scientific resources easily accessible. Mapmakers present at the opening reception were interviewed about their motivations for mapping science. The DVD also contains a PDF file with 150 dots per inch versions of all exhibit maps, together with the biographies and stories of their makers.

We hope that scientists will continue to be inspired, students and teachers encouraged, and the general public fascinated by this multilayered, accessible approach to the worlds of modern scientific thought. The video (in English and with French and Chinese subtitles) can be ordered via the exhibit Web site (**http://scimaps.org**).

Chad Redmon and Aaron Raskin, Harbinger Media, Inc. (producers). Harbinger Media, Inc. (**http://harbingerpro.com**) is a decentralized media production company with operatives worldwide and offices in New York, Baghdad, Johannesburg, and Shanghai. Using digital technology and exposing unique stories found nowhere else, Harbinger has successfully branded itself as a hip, socially conscious, international media think tank. Recent clients include Nike, Hugo Boss, and Google.

Katy Börner (director) **See page 132**.

Ingo Günther and Stephen Oh (cover design) **See page 192**.

Elisha F. Hardy (graphic design) **See page 158**.

Monika Herzig (music composition and performance) completed her doctorate in music education and jazz studies at Indiana University, where she now teaches in the Arts Administration program. As a touring jazz artist and Owl Studios recording artist, she has performed at many prestigious jazz clubs and festivals. Groups under her leadership have toured Germany and opened for acts such as Tower of Power, Sting, the Dixie Dregs, Yes, and more. As a recipient of the 1994 DownBeat Magazine Award for her composition "Let's Fool One" and with several big band arrangements published with the University of Northern Colorado Press, Herzig has also garnered international recognition for her writing skills. More information on CD releases, sound clips, videos, and performance dates can be found at **http://monikaherzig.com**.

"What we wanted to achieve is that we introduce people to maps of science in a way that reminds them of their usage of cartographic maps ... [We wanted them to understand that maps of science can be used] to see what kind of emerging research frontiers exist, what major researchers and papers and patents and grants exist in a certain area of research, and in general to get a more global view of what we collectively know."
Katy Börner

"My interest is ... how regular people, just, you know, the general public, a teacher, could use maps of science. You know you've got all these people that make maps and there's interest in how maps can help the National Science Foundation and other people with their policies in understanding how science goes together. What I want to do is be able to make them so that regular people understand how to dial into them, how to move through them."
Deborah MacPherson

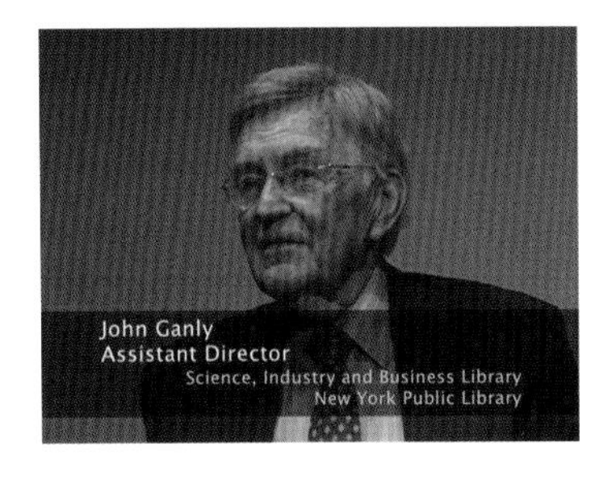

"The exhibit that we're doing is in keeping with our responsibility at the New York Public Library for being the central focus for science materials within the library."
John Ganly

"This is the first iteration of the Places & Spaces exhibit. It compares and contrasts the first maps of our planet to first maps of sciences. The first maps of our planet were not perfectly accurate, as people had just started to explore all the different continents. ... Similarly, in today's maps of science we know we don't have the complete coverage and the high coverage of the data that we would like to have, and therefore the maps of sciences which we create today cannot be complete and as correct as they should be. The second iteration of the exhibit compares and contrasts reference systems. We are interested to create a reference system of sciences which then can be used to overlay other types of data, such as the research portfolios of different institutions or countries."
Katy Börner

"Our first map ... shows the structure of science in kind of a galaxy-type form. One thing that's interesting is that this map showing science in a disciplinary fashion allows you to look at the distribution of a particular topic over the entire map of sciences. So, for instance, this little cutout here shows nanotechnology and it shows that it's centered in the physical chemistry and physics and materials area."

"This is our next iteration that appears here in the [...] second iteration of the Places & Spaces exhibit. ... What you see here are about eight hundred different circles that show sets of papers rather than sets of journals, and this map is something that we're now using to overlay other information. For instance, we can generate networks that show national strengths and national science strategies. We can generate funding profiles in different agencies. ... A funding profile for the National Science Foundation would show lots of money going into physics and chemistry."
Kevin W. Boyack

"My name is Brad Paley and I am a designer who specializes in massive amounts of information. What we have here, fitting in with the rest of the show, is a map of science showing how the different areas of science relate to the different areas of the world as well as to one another. It's using a relatively innovative display technology called an illuminated diagram, which is an old-fashioned print and a projector that's controlled by a computer, projecting light onto the print to help direct your attention to specific issues. This particular print is showing a map of seven hundred and seventy-six different specific areas in science. The reason you would map science or any abstract concept in general is to take advantage of your innate ability to deal with space as a foundation, and then on top of that foundation, do your analysis."
W. Bradford Paley

"My interest really is trying to understand the structure of knowledge presented in a visual way that corresponds to the way knowledge actually is structured. The idea of the physical analogy, especially to something that responds to our eyes visually, I think is very important ... [and enables us] to take extremely complex information, and yet present it in a way, so that the unimportant details go into the back and the really important, structural elements come to the front."
Richard Klavans

"The problem is that there's an ever-increasing number of outlets for scientific publication and just the number of publications themselves as well, and so it's increasingly hard for scientists to keep track of the latest developments and in particular with respect to the growing interdisciplinary nature of science. ... We need new tools to gain, first of all, an overview of the structure of a scientific domain, like geography or geology or mathematics, as well as the structure of science as a whole. And then, we can also use visualization to, for instance, find information, so we can use visualization in terms of search and navigation, as well as for the exploration of science as a whole or parts of science. As a geographer, one thing that I'm very interested in/excited about is that we can maybe use visualizations of science in science education K–12, and above, and below. Where science feels today very nebulous, very tough, hard to grasp, for children and students at all levels, and if we make science more exciting by making these maps of them, we could, for instance, induce what you might call a sense of place in science, that students see, Ahh, this is the structure of science, and there are these places, very real places in science, things that people do, and I might find a home in that place or in that place. So that's one of the promising aspects of it as well."
André Skupin

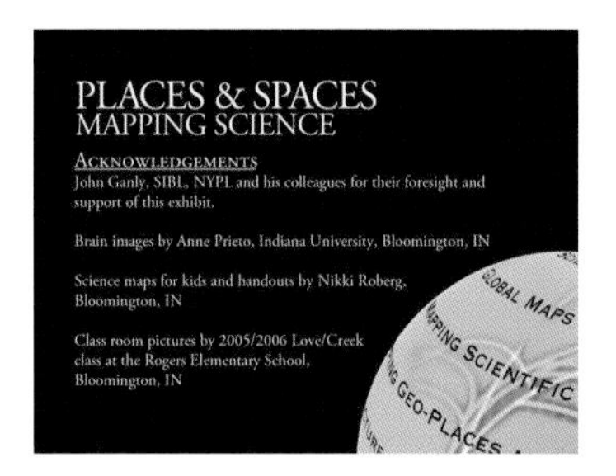

"I think you have to understand the makers of science maps are very much in the early stages of their evolution. There are no accepted standards; there are no equivalent Mercator projections for science or any of the common things that you see in geography. And you also have to understand that mapping science is not necessarily a commonsense sort of notion, because fields of science don't necessarily have a location. The locations that we assign to different fields can be completely artificial in the minds of some people, or they can be concrete, so mapping science is a more of a mental map than it is a physical map. And the whole objective here, in my opinion, is to see if we can meaningfully understand the structure of science."
Henry G. Small

6304: Boulder, CO, USA
5806: Denver, CO, USA
12820: San Francisco, CA, USA
20414: Los An…, USA
7919: San Diego, CA, USA
6349: Dallas, TX, USA
15020: Houston, TX, USA
3868: San Antonio, TX, USA
10226: St Louis, MO, USA
5915: Columbus, OH
20744: Bethesda, MD, USA
12390: Washington, DC, USA
3141: Lexington, KY
4569: Charlottesville, VA, USA
2615: Richmond, VA, USA
5681: Nashville, TN, USA
9019: Durham, NC, USA
3558: Memphis, TN, USA
1784: Little Rock, AR, USA
13001: Atlanta, GA, USA
4413: Birmingham, AL, USA
2217: Charleston, SC, USA
3616: New Orleans, LA, USA
6047: Gainesville, FL, USA
2396: Tampa, FL, USA
3733: Miami, FL, USA
5012: Mexico City, Mexico
3089: Rio De Janeiro, Brazil
4691: Sao Paulo, Brazil
SOUTH AMERICA

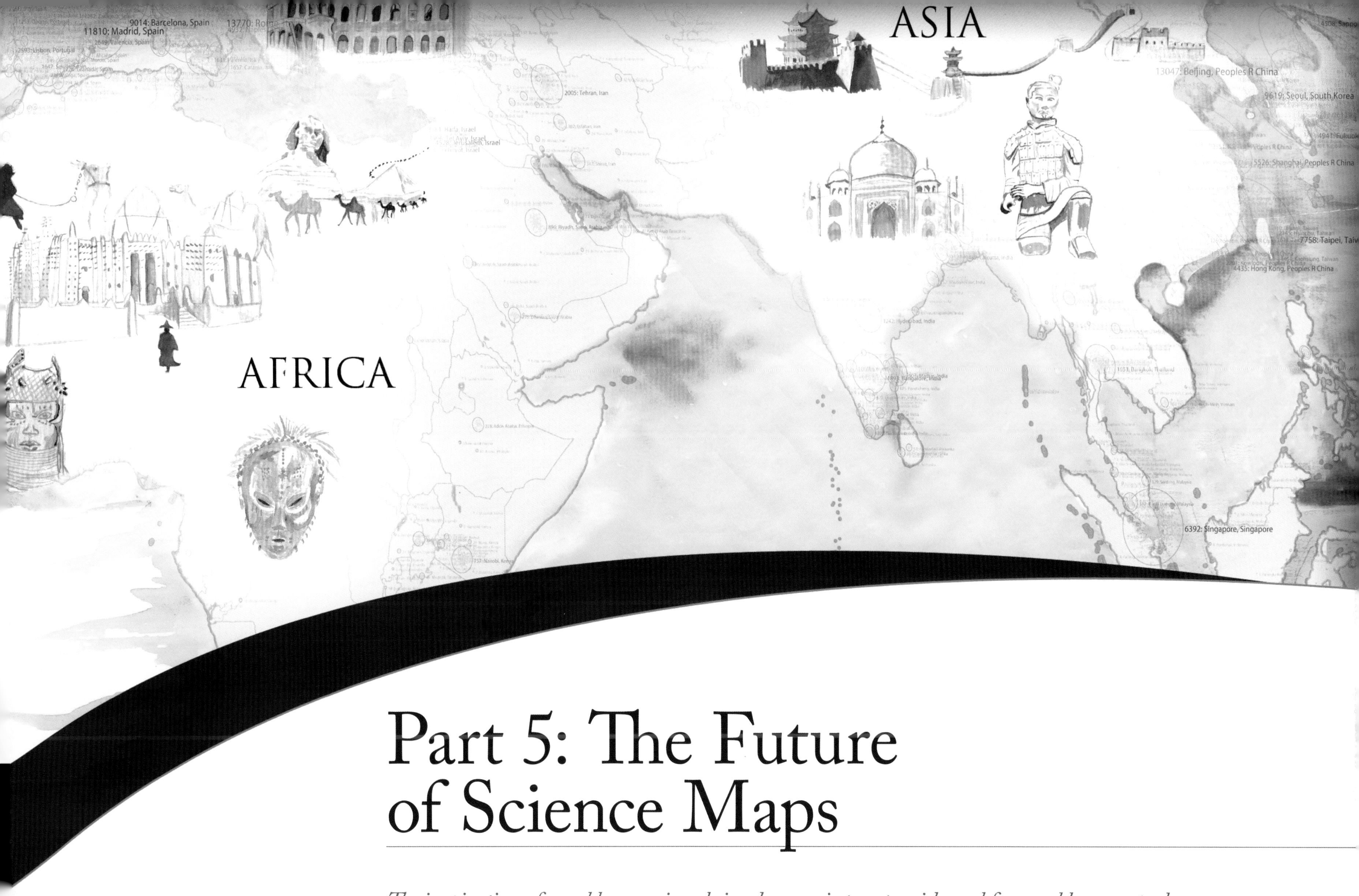

Part 5: The Future of Science Maps

The inspiration of a noble cause involving human interests wide and far, enables men to do things they did not dream themselves capable of before, and which they were not capable of alone. The consciousness of belonging, vitally, to something beyond individuality; of being part of a personality that reaches we know not where, in space and time, greatens the heart to the limit of the soul's ideal, and builds out the supreme of character.

Joshua L. Chamberlain

Science Maps as Visual Interfaces to Scholarly Knowledge

Libraries have service, research, and educational missions (see **page 8, Knowledge Needs and Desires**). They help us to compose, reproduce, distribute, catalog, critique, collect, and codify knowledge. Here we review current socio-technical developments and argue for the exploitation of spatial metaphors and social navigation of nonlinear network structures in support of collective knowledge production, navigation, and consumption to improve the utility, usability, and value of knowledge access tools. We then review existing and envision future visual interfaces to digital libraries.

A library of mankind's knowledge available online could be the greatest achievement of our generation.
Brewster Kahle

Identification and Preservation of Valuable Knowledge

Laura Campbell, director of the U.S. Library of Congress National Digital Library Program (NDLP) estimated in 2007 that 161 exabytes of digital content would be generated in 2007, while 988 exabytes would be produced by 2010. The sheer immensity of raw instrument or simulation data precludes complete collection and preservation.

Citation counts, usage data, and expert ratings have an important role to play when selecting which scientific data and scholarly works to preserve. The expertise of trained librarians will need to be harnessed toward analyzing the activity patterns of potentially billions of users through advanced data mining techniques. The visualization and communication of analysis results is expected to provide further insights into clusters, patterns, trends, and activity bursts on knowledge usage and growth.

Adding Value: Ratings and Context

Librarians and data providers use citation counts and journal impact factors by Thomson Reuters, Scopus, and other publishers, as well as reshelving data and usage data to optimize their collections.

In addition, there are diverse services that rank Web resources by the number of times other resources link to them. Among them are Traffic Detail pages of Alexa Internet, Bing (formerly Live Search, Windows Live Search, MSN Search), BlogRolling!, Delicio.us, Digg, IceRocket, Google's PageRank, Reddit, Spurl, Technorati, Google's webmaster tools (for searching site links including backlinks), Yahoo! Services, or WhoLinksToMe.com, which provides aggregated statistics of in-links and is also available as an RSS feed).

Finally, libraries aim to provide rich contextual information, description, and metadata to promote deeper analysis and understanding of collections and materials. They develop and implement tools and services to promote enhanced interpretation, context, and understanding—all of which are invaluable for navigating and managing large-scale resources.

Collaborative Knowledge Production and Consumption

There is no human being who could read, process, and understand the entirety of our collective scholarly knowledge. Hence, the problem is not how one person can access all knowledge but how we can access and manage all of humanity's knowledge as a productive team.

Web 2.0 applications merge blogs, social networking sites, and Web bulletin boards to let people freely share their knowledge. Anybody can select from the pool of best ideas and technologies. Reputation systems keep track of who provided the most valuable or most used content. Free online reference management systems let anybody save links to favorite articles, references, Web sites, and other online resources so that they can be easily accessed. Tags can be added to further ease retrieval. Some systems automatically extract bibliographic information from library services or full text records. Many systems support "social bookmarking" so that people can share and benefit from one another's collections to discover new, interesting content. The resulting social networks are based on mutual interest, rather than on prior acquaintance. For example, MyBlogLog lets individuals connect and benefit from the suggestions of fellow readers; Citation Machine helps students, teachers, and independent researchers in formatting citation references in different styles; BibSonomy supports the sharing of bookmarks and lists of literature; LibraryThing lets individuals catalog their books online; CiteULike allows for scholarly papers to be stored, organized, and shared; and Connotea offers the ability to import and export references to and from desktop reference management software and supports standards such as DOIs and OpenURLs used by academic publishers. PsychLinker offers annotated links to psychological resources. Biolicious supports the browsing of literature and the ranking of scientists. MemeStreams helps scientists, lawyers, artists, entrepreneurs, educators, and others build social networks. Connexions is an environment for collaboratively developing, freely sharing, and rapidly publishing scholarly content on the Web; its Connexions Consortium contains general educational materials, organized in small modules that easily tap into larger collections or educational courses.

Supporting Social Navigation

Typically, information search is a lonely activity. Two friends may be searching for identical terms on the same search engine at precisely the same moment. Through that process, however, they have no way of connecting to one other in order to discuss and compare results. The Answer Garden systems from Mark S. Ackerman and Thomas W. Malone were among the first to facilitate information finding through social networks. They were initially designed to facilitate informal data flows and capture of information. Later versions provided an explicit expertise-location engine.

Today's systems generate recommendations that are tailored to each user's need. Amazon's item-to-item collaborative filtering exploits records of past customer behavior to generate high-quality recommendations; the more customers search or buy, the more refined the service's recommendations. Some systems such as MyLibrary@LANL support the creation of multiple personae, which allows one individual to pursue a range of different topics. Other systems, such as Noshir Contractor's IKNOW, help users locate knowledgeable experts on any topic, based on coauthor relationships, text content analysis, or chains of personal connections.

Combining Top-Down and Bottom-Up Knowledge Organization

Critical issues in knowledge navigation, management, and utilization are posed by many factors, including the tremendous increase in and diversification of databases and their holdings, the shrinking grain size of knowledge nuggets (for example, from books to papers to memos), and the growing complexity of scholarly results (data sets, software, and workflows).

Existing organizational schemas, developed by librarians and domain experts, can help bring structure and navigational guidance. The Dewey Decimal Classification and the Library of Congress Classification are among the most commonly used systems. In addition, there are subject headings such as MeSH terms (used to describe more than 19 million MEDLINE papers) or the USPTO classification hierarchy (which organizes about 3 million U.S. patents into about 180,000 distinct patent classes). The structures are mostly hierarchical and can be readily converted into a map: the root node can be seen as the "planet," its children as "continents," their children as "countries," children of countries as "counties," and so on. The resulting reference system can be used to map data sets of different origins, such as Wikipedia and MEDLINE data.

Online community sites use tag clouds to label and retrieve records of interest. Ideally, top-down expert-designed structure would be augmented by bottom-up user-driven tagging.

Visual Interfaces to Digital Libraries

Most of today's digital libraries support searching, alerting, linking, and exporting data. Some also offer visual interfaces to their holdings. Originally, these visual depictions of a library's content and structure were generated by hand (see Harold Johann Thomas Ellingham's map of natural science and technology, opposite top left).

Today, science maps are generated automatically using computers. Visual interfaces first provide a database overview and then support zoom and filtering, as well as the retrieval of details. An example is the MEDLINE browser, developed by Antarctica System, Inc., which supported the interactive explo-

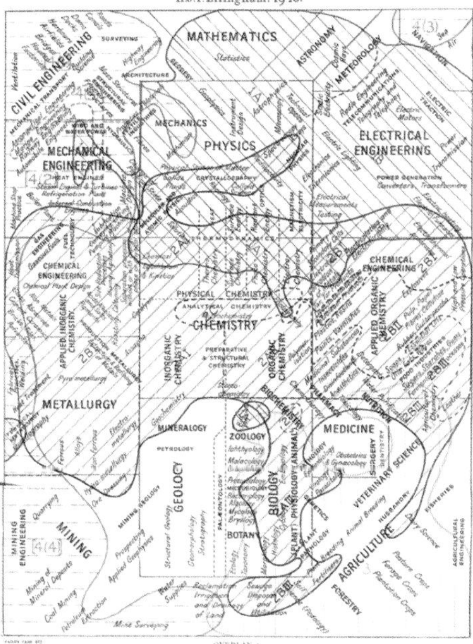

ration of the National Library of Medicine's PubMed database. The main interface and a close-up of the **Anatomy/Body Regions** section are shown below.

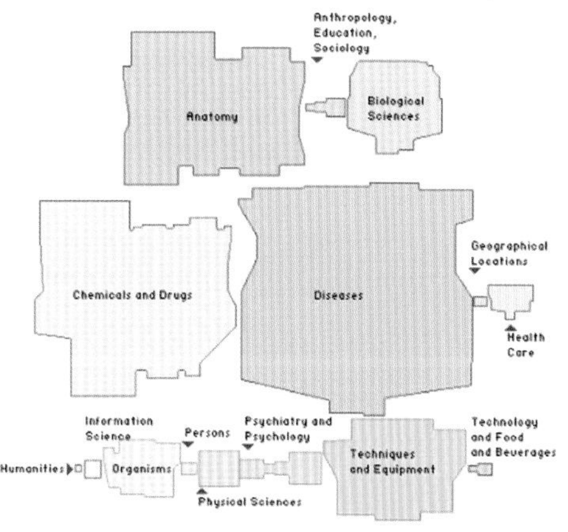

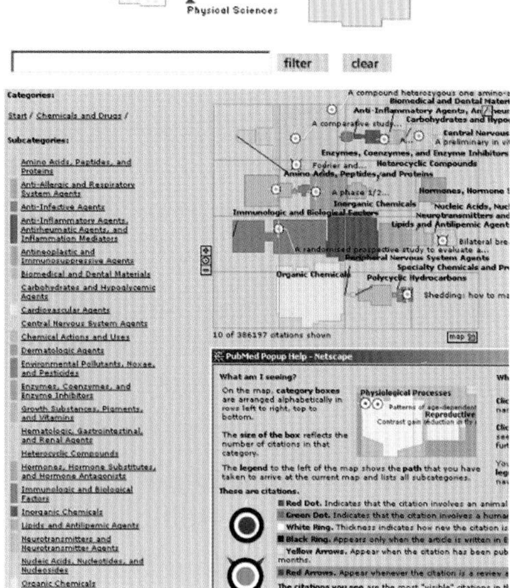

The main interface shows the top-level Medical Subject Headings (MeSH) categories arranged alphabetically in rows from left to right, top to bottom (for example, **Anatomy and Anthropology**). The colored areas represent the papers filed under each top-level category, with the size indicating the number of citations. The shape of the areas is arbitrary but supports recognition. Users of this interface gain an immediate overview of available categories and the number of papers these categories contain.

They can click on an area of interest to zoom in and see it enlarged (bottom left). Users can filter papers of interest by entering a keyword in the search window. Matching papers will be marked on the map to facilitate visual browsing and Boolean search. Close-ups are further subdivided into subcategories, if any. The subcategories are listed in the legend on the left, and labeled in bold on the map. On the map, individual citations are represented using target graphics and nonbold titles. The target graphics indicates how new the citation is; if the article or paper was written in English; if it involves human or nonhuman subjects; if it is a review paper; or if it has been published within the last three months.

Placing the mouse over a target displays the paper's title, author, date of publication, and PubMed ID number. Clicking on the paper's title retrieves its summary from the PubMed database.

Other interfaces generate the coauthorship network environment of an author (see the CiNii Researchers Link Viewer, below). Here the focal author, shown in the middle of a circle, is surrounded by his or her coauthors, who are further surrounded by their coauthors. The size of the circles is used to encode additional attributes.

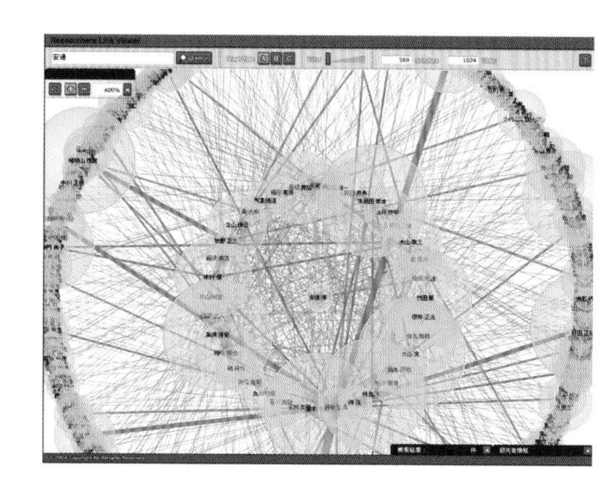

TextArc, discussed in **TextArc Visualization of "The History of Science," page 128**, visually communicates the content and structure of a book. Tools such as HistCite and CiteSpace (see **page 120, HistCite Visualization of DNA Development**, and **page 162, Mapping the Universe: Space, Time, and Discovery!**, for a detailed explanation of both) accept common library output formats and generate paper-citation graphs or visualizations of emerging research frontiers.

Memory Palaces and Mirror Gardens

Memory palaces were developed in classical Greek culture to help people manage and recite great quantities of information. They are a highly evolved mnemonic structures that are responsive to the user's position in an imagined space.

Spatial Memory Organization

Spatial metaphors constitute a fundamental part of human cognition. In daily life, we organize data spatially by filing our papers or shelving our books. Spatial organization of digital papers allows for easier access, management, and relocation of content. While time and labor are needed to reorganize physical libraries, electronic information spaces can be resorted and restructured by author, date of publication, or relevance to a query at the press of a button.

Virtual Online Spaces

Advances in computer and networking technologies fuel the rapid growth of three-dimensional browser systems that enable the creation of compelling, multimodal, navigable, and collaborative virtual online environments. These virtual environments are inhabited by avatars—graphical icons acting as placeholders for human users in a cyberspace system—and provide means for interacting with the objects in the environment, with embedded information sources and services, or with other users and visitors of the environment. They can be seen as a local social information structure, a communication medium, a tool to improve local democracy and participation, a practical resource to organize everyday life, or a space in which to experience and experiment with cyberspace.

Librarea was created as a virtual world in Active Worlds, in which real librarians could create functional, information-rich environments, meet with other librarians from around the world, create works of art, or teach (see **page 9, Children**, and **page 186, Hands-On Science Maps for Kids** for educational applications of this technology).

Collaborative Memory Palaces

Active Worlds technology has been applied to design a collaborative Memory Palace that provided more effective means to collaboratively access, manage, and utilize scholarly data, algorithms, papers, and expertise. The scholarly items were laid out as a semantic space. Anybody online could access them.

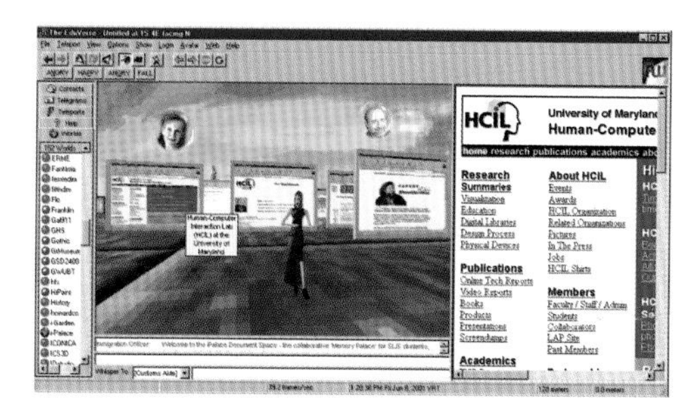

When accessing items, users automatically organize themselves by interest and expertise—for example, biology students, faculty, and practitioners met in the biology area and educators grouped in the education space—thereby creating communities of expertise and practice.

Evolving Mirror Gardens

A second world, Mirror Garden, was used to visualize user interaction data such as navigation, manipulation, chatting, and Web access activity recorded in the Memory Palace world. It provided a means to guide users, to evaluate the effectiveness and usability of the world, to optimize design properties, and to examine the evolving user community. The term "Mirror Garden" was derived from a merging of the term "Mirror World," as envisioned by David H. Gelernter, and "PeopleGarden," coined by Rebecca Xiong and Judith Donath.

Implementation

Prototypes of both worlds were implemented at the School of Library and Information Science (SLIS) at Indiana University in 2000. The Memory Palace world was seeded with about 8,000 links to online records such as text, images, video, and software demonstrations—all collected from course pages and bookmark lists of SLIS faculty. About 300 students in Bloomington; 200 students in Indianapolis; faculty, prospective students, and alumni in both cities had access to the Memory Palace world. Several individuals used this space to learn about SLIS, contact experts, or simply to keep up with continually evolving research topics.

Mapping Intellectual Landscapes for Economic Decision-Making

The information age requires knowing and serving customers' needs effectively and efficiently. Companies need to exploit social networks to succeed in existing and new markets. They have to absorb knowledge and creativity to design, manufacture, and sell competitive products. Business managers of small, midsize, and large companies need to identify pockets of innovation, to initiate and sustain synergistic collaborations, and to promote the flow of ideas into products. Effective visualizations of economic, scientific, political, and other data streams have the potential to vastly improve competitiveness.

It is being realized with a thud that the world is probably going to be ruled by those who know how, in the fullest sense, to apply science.
Vannevar Bush

Monitoring Customer Activity

Customers' responses to qualitatively novel products such as eBay and YouTube are generally unpredictable. However, running user studies is considered low priority, as all companies strive to "be first." Instead, the aim is to track customer behavior extensively and increase the production of "bestselling" products immediately. Today, Google Analytics provides multiple means to monitor online spaces. Other tools show not only the number of customers per site but also their origin and final destination in real time. "Information scent" studies let you model and predict the flow of customers through a Web page before it is even designed.

As video cameras, radio-frequency identification (RFID) tags, geolocation devices, and other forms of tracking real-world behavior become widespread, similar analyses and optimization can also be employed for offline customer behavior.

Exploiting Social Networks

Today, project-oriented, loosely formed, interdisciplinary teams replace the concept of lifelong employment. One must now keep track of the continually changing work landscape as new colleagues appear, roles shift, and different projects take shape. Salespeople, managers, and researchers need to monitor their networks more closely. Tools like LinkedIn and Facebook, as well as open-source software such

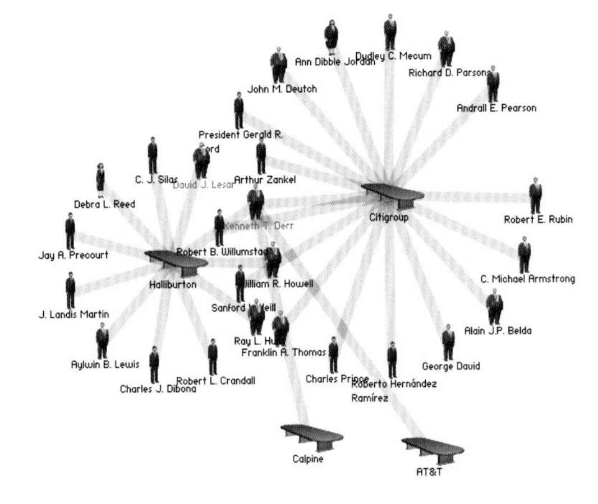

as GenIsis, help people visualize, analyze, and evaluate their social, organizational, and value networks.

Mark Lombardi has portrayed the extensive network structure of political and financial misdeeds based on his handwritten database of 12,000 index cards (see **page 32, Milestones in Mapping Science**).

The Web site They Rule lets anybody map the boards of some of the most powerful U.S. companies. Many of these companies share directors. A network of corporations (table icons) interlinked to directors (person icons) is shown above.

Identifying Experts and Innovations

Companies that grow interdisciplinary "gene pools" of innovators to effectively "absorb" new expertise and knowledge and then convert it into strategic intelligence have a higher chance of surviving. Research in economics shows the importance of bidirectional flows of knowledge between universities and industry: Academic research programs generate "knowledge spillovers," such as personnel, licenses on inventions/patents, and tacit knowledge of value for industry. Industry offers technological challenges, technologies, and financial resources for researchers that help align research topics to industry needs. Faculty—acting as advisors and collaborators—help shape new technologies. They develop graduate programs that incorporate new innovations. The academia-industry networks turn out to be invaluable for the diffusion of new technology. Alliances and collaborations with preeminent researchers, consortia of companies, and workshops are also central to the development and promotion of standards. Maps of science can help identify major players, trends, and bursts of activity in the coevolving academic and intellectual property realms.

Communicating Intelligently and Effectively

As the complexity and dynamics of business processes increase, and as project teams become more interdisciplinary and internationally distributed, the need to communicate across language, cultural, and disciplinary/departmental boundaries increases as well.

Dialogue maps capture what was said in a discussion. They record arguments for or against a certain decision in an easy to edit and understand hypertext diagram on a screen. A special visual grammar encodes questions, ideas, and pros and cons (see CogNexis screenshot below). Subnetworks can be collapsed to avoid clutter. Networks can be traversed at the local and global levels.

As a meeting unfolds, the map too develops on screen before all the participants. It serves as a "group memory," helping members focus on important points, reduce redundant arguments, understand how comments relate to one another, and recognize which points are still open. As the map

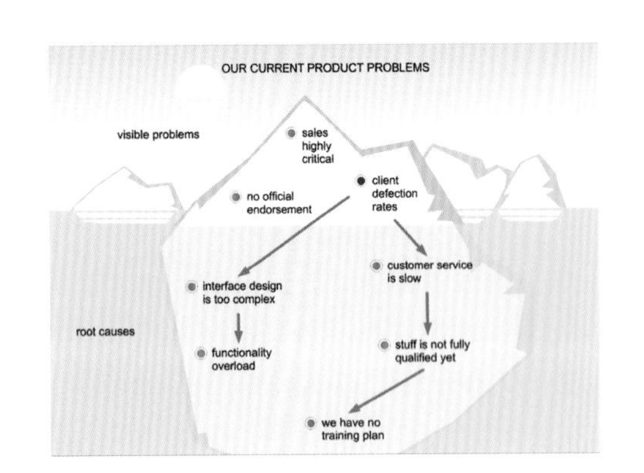

is created and owned collectively, it engenders a shared understanding about the current problems, potential solutions, possible interpretations, and the roles and responsibilities of those involved. The map also provides an effective means to capture and communicate the meeting results.

Mapping the Future

Road mapping techniques are widely used in business strategy and policy development to support innovation, strategy, and policy development and dissemination. Graphic depictions of road maps use metaphors to render complex interdependencies, challenges and opportunities, or next steps in a visual, easy to digest format. They are an effective means for discussion and consensus finding. Graphic road maps may be printed in large format and hung in public places within a company or be served in annotatable online formats.

Mapping the Market

Visualizations such as the *Map of the Market* by SmartMoney provide up-to-date information on the trading volume and trading value of more than 500 stocks simultaneously. The maps are updated every 15 minutes based on stock data provided by ComStock Partners, Inc. (with a 20 minute delay), historical prices and fundamental data by Hemscott, Inc., earning estimates by Zacks Investment Research, and insider trading data by the financial division of Thomson Reuters.

This treemap layout (see also **page 98, Treemap View of 2004 Usenet Returnees**) represents each stock value with a rectangle. Stocks are grouped by industry. The size of a rectangle (individual stock) represents its market capitalization. Color gradation depicts the level of losses (bright red is -6 percent) or gains (bright green is +6 percent). Hovering the mouse over a rectangle brings up the company's name and advises whether its stock price is going up or down. Clicking on a rectangle provides more detailed information.

Mapping the Economic Landscape

As economic markets go global and accelerate in pace, it becomes ever more important to gain an overview of major developments happening at different scales and with different dynamics. Economic, scientific, political, ecological, and other data streams have to be harvested, correlated, and understood to arrive at intelligent decisions.

Maps have been used to show the intake of imported products from around the globe for a specific country (see the **U.K. 2006 Global Ecological Footprint** shown on **page 43, Milestones in Mapping Science**).

Here we discuss *The Product Space,* a map of 775 industrial products exported in tandem during the period of 1962–2000, by Cesar A. Hidalgo, Bailey Klinger, Albert-László Barabási, and Ricardo Hausmann. The map is based on world trade flows as researched by Robert C. Feenstra and colleagues and published by the National Bureau of Economic Research (NBER). A maximum spanning tree algorithm was used to reduce the complete coexport matrix to less than 1 percent of the links. The resulting network of product nodes and their coexport linkages was laid out using a force-directed layout algorithm. Node sizes represent the value of traded products in thousands of U.S. dollars. Their color corresponds to 10 product groups identified using the Leamer classification. Each product class is labeled by an icon. Link width and color indicates the frequency of joint exports. The visual representation of this data set supports simultaneous observations of the structure of the product space, the classification of the products, the strength of their connections, and their participation in world trade.

The network has a core-periphery structure with higher-value product classes (for example, metallurgy, machinery, and chemicals) in the core and lower-quality classes (for example, fishing and garments) in the periphery. Products at the core of the network are highly interconnected while those at the periphery are less diversified (that is, countries operating in the core of the product space have capabilities to develop and manufacture a wide range of products, while countries operating at the periphery have fewer opportunities for diversification).

Countries can be characterized by the differences in the number and quality of goods they export. Developing countries tend to specialize in lower-quality products while developed countries cover primarily high-quality goods at the core of the network. Countries can improve their export portfolio by including products that use similar production expertise, methodology, and resources. Successful development strategies for countries need to take into account their current positions in the product space.

Using the product space as a base map, the export portfolios of countries can be plotted over time. As an example, three time slices of data on the industrial evolution of Malaysia are shown on top. From 1980 to 2000, Malaysia successfully expanded its product offerings into the electronics sector, indicated by the increased size and number of black squares covering the light blue nodes at the bottom of the network.

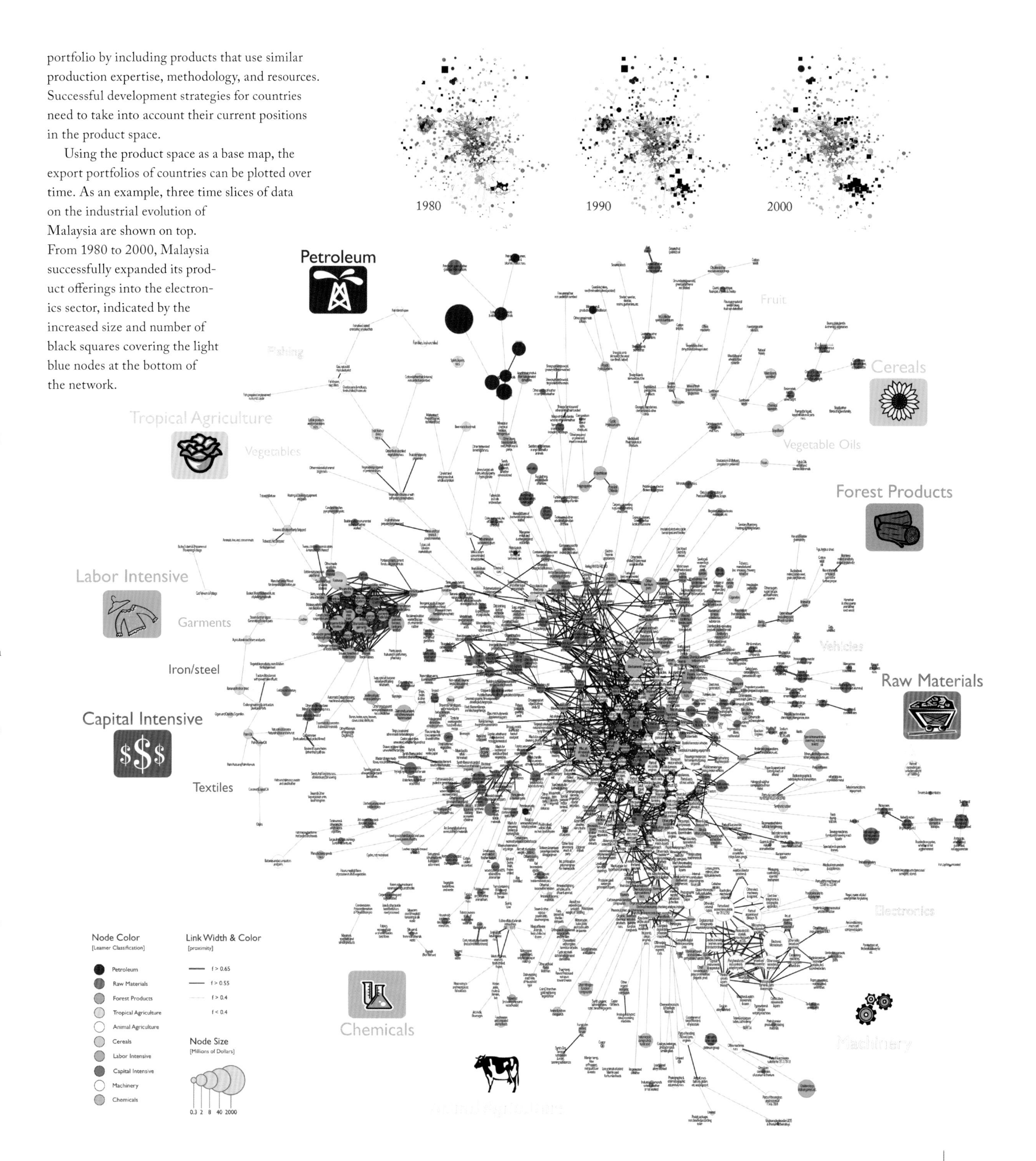

Science of Science Policy Maps for Government Agencies

The globalization of science and associated "brain drains," the demand for accountability, and the increasing ratio of high-quality research proposals to limited budgets demand a theoretically grounded and practically validated science policy. The promoted "science of science" should be seen as a theoretical framework with associated tools that augment the capabilities of policy makers, program directors, and reviewers.

Wide consensus also exists on the importance of federally funded science to our nation's long-term economic competitiveness. ... Neither this Administration nor any future one can escape the urgent demands of 21st-century realities. ... Globalization is bringing the problems of countries around the world to our doorstep.

John H. Marburger III

Toward a "Science of Science"

A scientific field of "knowledge about science" was first proposed in 1923, and it was renamed the "science of science" in 1928. The publication of John Desmond Bernal's *Social Function of Science* (1939) was a key turning point, but the field then lay dormant until after World War II, when Derek J. de Solla Price published *Science Since Babylon* (1961) and *Little Science, Big Science* (1963). Eugene Garfield connected the works of Bernal and de Solla Price in "Citation Indexes for Science" (1955) (see **page 120, Eugene Garfield**). In 1969, the term *bibliometrics*, used to mean "the application of mathematical and statistical methods to books and other media of communication" was introduced by Alan Pritchard. In the same year, Vasily Nalimov and Z. M. Mulchenko coined the term *scientometrics* to describe "the application of those quantitative methods which are dealing with the analysis of science viewed as an information process." However, it is only today that we have the data and the algorithmic and computational resources to study the structure and evolution of science on a large scale.

Funding Science

Each country needs to decide what percentage of its gross national product (GNP) it should spend in support of education, science, and technology (see **page 5, Science and Society in Equilibrium**). Decisions are typically based on previous budgets, national priorities, and budget requests from funding agencies. In 2006, Japan had 704,949 full-time researchers, the United States 1,394,682, the Organisation for Economic Co-operation and Development (OECD) 3,891,123, and the EU-27 1,301,022. The gross domestic expenditure on research and development (GERD) as a percentage of gross domestic product (GDP) for Japan, the United States, the OECD, and the EU-27 for 1988–2006 as compiled by OECD is shown above, right.

Countries must then allocate research and development (R&D) spending to different agencies. Funding agencies have different priorities and focus areas. For example, the U.S. Department of Energy (DOE) focuses on energy-related research, while the National Institutes of Health (NIH) funds mostly biomedical research (see funding profiles in **page 108, The Structure of Science**). Trends for major U.S. agencies between 1976–2009 and general U.S. spending trends in federal research by discipline for 1970–2007, as compiled by the American Association for the Advancement of Science (AAAS), are shown at bottom right.

Subsequently, each funding agency/directorate needs to decide which solicitations will attract the most promising and transformative research proposals while keeping in mind the funding portfolios of other agencies. Project proposals need to be matched with funding programs and reviewer groups. Scientific opportunity must be balanced with national need when awarding funding.

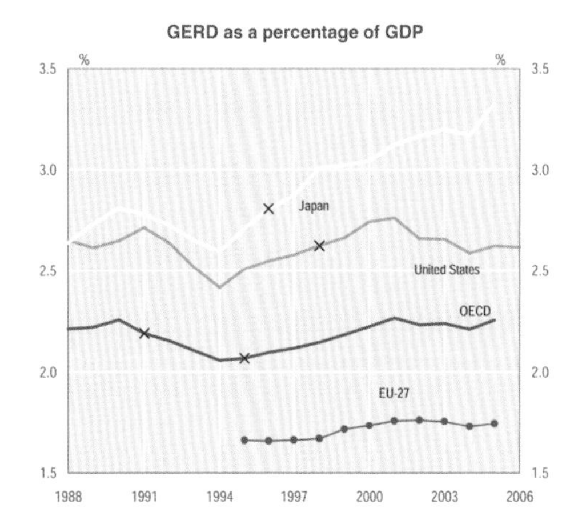

Judging Research Quality

Funding decisions must be based on the quality of proposed, ongoing, and completed research of an investigator team in the context of all other work. Research in bibliometrics and scientometrics has produced a staggering variety of measures of research excellence. The U.K. Royal Academy of Engineering recently identified a comprehensive set of research excellence measures comprising academic/conference/trade publications; peer recognition; consultation; evidence of partnerships, industrial support, and partnering activities; patents and practical research outputs; independence indicators; core and support capability indicators; involvement with society indicators; and strategic program and resource planning indicators. The measures are used alone or in combination to rank order persons, institutions, or nations.

Citation rankings of authors, journals, universities, and countries can be accessed online through services provided by Thomson Reuters, Scopus, SCImago, Eigenfactor, and many others. Rankings can be requested for subject areas or other categories. For instance, the *h*-index, total papers, total references, total cites, and self-citations can be compared and contrasted and thresholds can be applied.

However, one number may not be able to fully represent a person, institution, or nation and the complex positive and negative feedback cycles in which they are involved (see **page 58, Conceptualizing Science: Science Dynamics**). In such instances, time-series, geographic, and network analyses with corresponding visualizations come into play. They provide objective and effective means of accessing and correlating primary data sources; compare and contrast different types of comprehensive evidence in support of or against resource allocation decisions and policy changes; and help communicate insights and decisions.

Computing Funding Impact

Funding agencies are interested to know the impact of their programs on the development of science and technology. A common measure is cost-effectiveness, calculated as the number of citations per dollar spent. Given that there is nearly a four-year lag before papers emerge from funded research, and an additional multi-year period before these papers reach their citation peak, it is typical to compare expenditure figures for each year with citation counts four

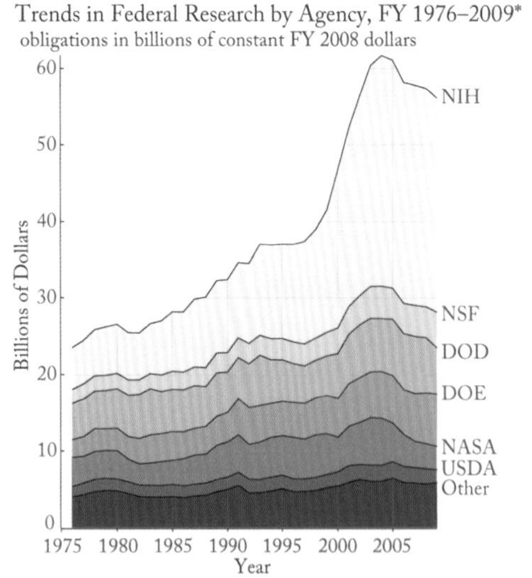

* FY 2009 figures are latest AAAS estimates of FY 2009 request. Research includes basic research and applied research. 1976–1994 figures are NSF data on obligations in the Federal Funds survey.

Source: AAAS analyses of R&D in annual AAAS R&D reports. MARCH '08 REVISED © 2008 AAAS

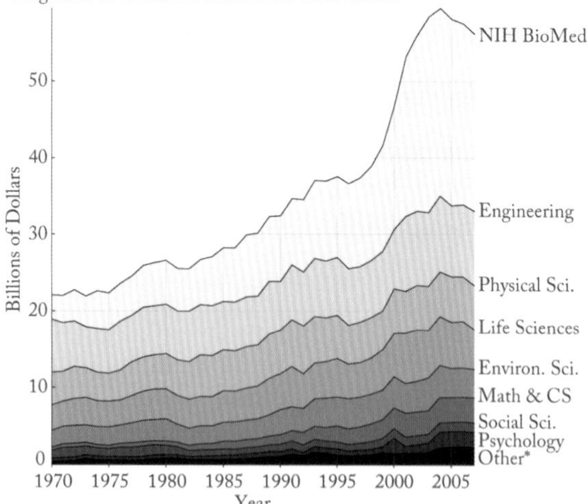

* Other includes research not classified (includes basic research and applied research; excludes development and R&D facilities)
Life sciences–split into NIH support for biomedical research and all other agencies' support for life sciences.

Source: National Science Foundation, Federal Funds for Research and Development FY 2005, 2006, 2007, 2008. FY 2006 and 2007 data are preliminary. Constant-dollar conversions based on OMB's GDP deflators. FEB. '08 © 2008 AAAS

to six years later and to track this over time. Work by Jonathan G. Lewison and Robert M. May examined the cost-effectiveness of G7 countries for 1986–1991 and showed that for every million U.K. pounds spent on scientific research, publications in the United States acquire approximately 150–200 citations per year; publications in the United Kingdom 150 citations (with an upward trend); publications in Canada 140 citations; and publications in Japan, Germany, France, and Italy 40–60 citations.

Other indicators are media citations per budget dollar and Web traffic per budget dollar. In 2006, the most cost-effective think tank received 3.9 cites for every $10,000 spent.

The impact of funding on the diffusion of knowledge can be calculated by linking awards to papers and their citations. Shown below are author-supplied linkage patterns (blue lines) from one NIH grant on aging research to all resulting publications. Citations are shown as green lines.

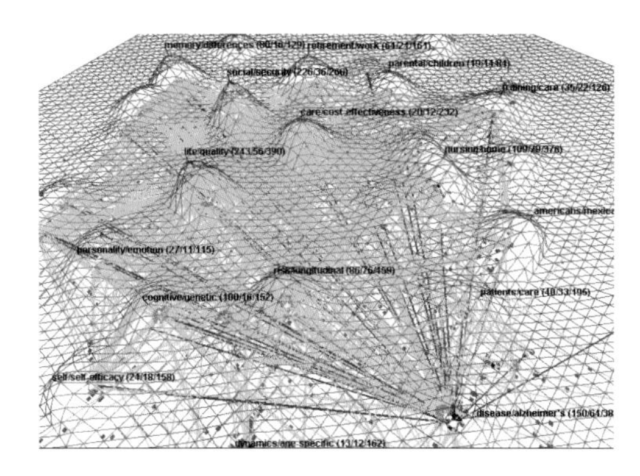

The **Illuminated Diagrams (page 180)** extend this network traversal outward by recursively showing citations to these papers in geographic space and science space. The geographic and scientific impact of a person, country, or any other basic scholarly unit can be portrayed in a similar manner.

Geography of Science

Geographic depictions of publications and citation profiles provide additional information with respect to research performance (see the four maps in **Knowledge Equals Power, page 2**). The geographic distribution of the most highly cited scientists and their institutions as rendered by Michael Batty is shown at the top of the next column.

Research by Caroline S. Wagner and Loet Leydesdorff shows that the number of papers that have coauthors from different countries has increased significantly over the last decade.

Science Dynamics

The term *brain drain* refers to the emigration of talented scientists, engineers, and other bright

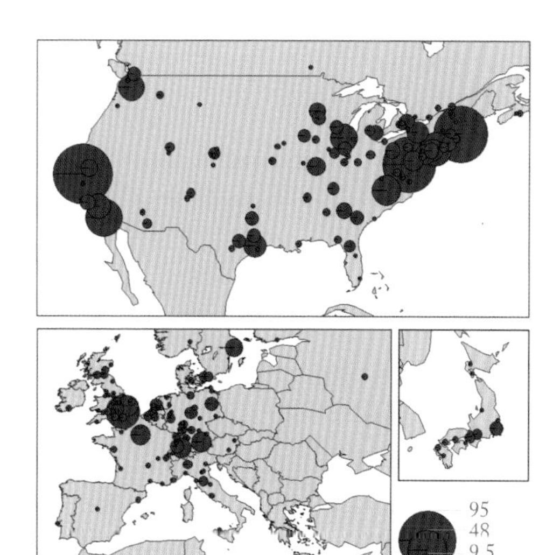

minds to highly developed countries for the opportunities provided by leading universities, dynamic companies, free economic and social environments, and a high standard of living. Today, the term *reverse brain drain* reflects changes in the direction of emigration streams caused by the continuously changing economic and political landscape. It is interesting to study where the greatest talents are trained, which environments foster creativity and innovation, and where that innovation is most effectively converted into profit. Additionally, it is important to understand how a scientific field has evolved over several decades, the impact of external events, and the interplay of that field with other areas of science. Shown below is the area of chemistry research, highlighted in a global map of science (left), and diffusion flows among its 14 major subdisciplines (right) in 2004. The original work shows an animation of the evolving structure, topical composition, and knowledge diffusion via citations over a 30-year time span.

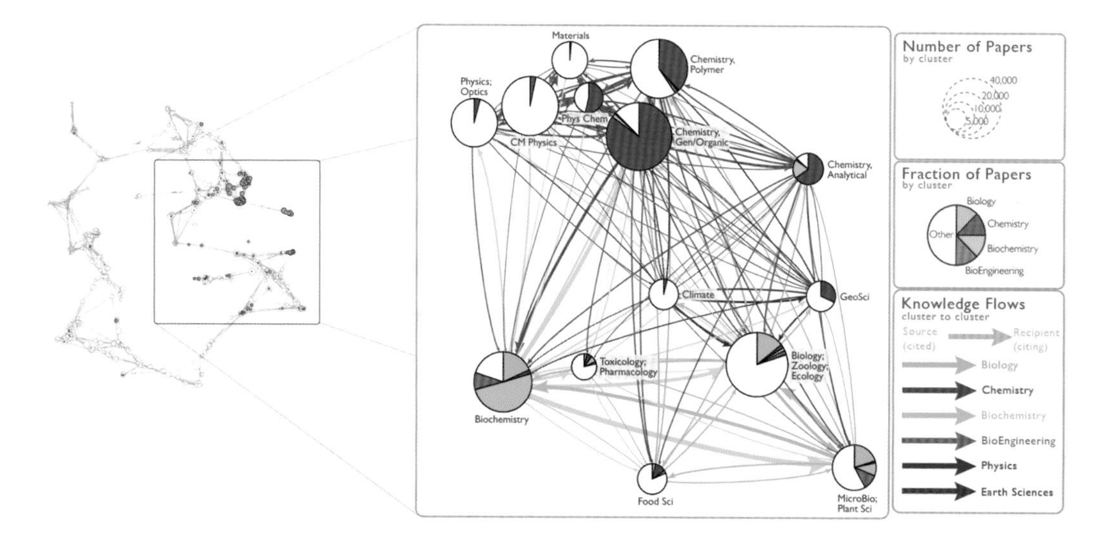

The Scientific Wealth of Nations

In a globally connected world, the comparison of the scientific output of economically competing countries becomes an important factor for science policy decisions. Public funding must focus on preserving and attracting talent. The ease with which scientific results are converted into profitable products and technology is another concern. It is of little use to have many highly cited papers if the economical value and societal benefits for these results are created abroad.

In "The Scientific Wealth of Nations" (1997), May compared and contrasted the scientific output of different nations according to the quantity and quality of their contribution to the world's total number of publications and their share of major international science prizes. In 2004, David A. King's article "The Scientific Impact of Nations" compared economic and scientific wealth and disciplinary strengths of major nations. In 2006, Ping Zhou and Loet Leydesdorff discussed the emergence of China as a leading nation in science. In 2007, Masatsura Igami and Ayaka Saka published "Capturing the Evolving Nature of Science," and Hugo Horta and Francisco M. Veloso compared EU and U.S. scientific output by scientific field.

Recent work by Richard Klavans, Kevin W. Boyack, and W. Bradford Paley overlays the scientific strength of 10 nations on a science base map—the *Map of Scientific Paradigms* discussed in **Map of Scientific Paradigms (page 138)**. Shown at right are overlays for the United States, Germany, and Japan. Nodes and edges are highlighted if a nation publishes significantly more in that area than expected. As can be seen, the United States tends to publish a high number of papers in the social sciences (yellow), the medical sciences (red), and areas of agricultural science (brown). The exceptionally large node in the chemistry area (green) reflects the work that interprets basic research in chemistry for use in the more applied areas of medicine. Germany and Japan excel in physics (purple) and chemistry. Nations that emphasize interdisciplinary work have more weight on long edges, linking topics that are more distant from one another. See **References and Credits (page 246)** for additional overlays and more extensive explanations.

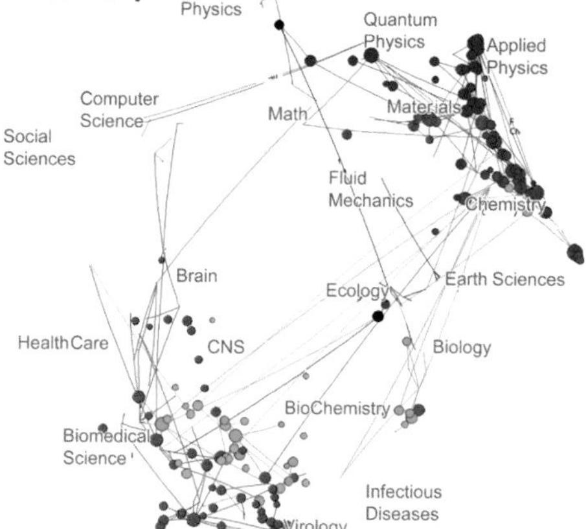

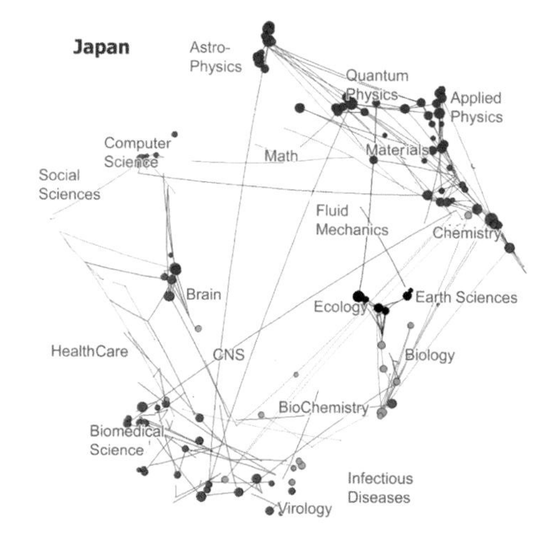

Professional Knowledge Management Tools for Scholars

As mentioned in **Knowledge Needs and Desires (page 8)**, scholars have multiple responsibilities. They are researchers, authors, editors, reviewers, teachers, inventors, investigators, team leaders, and science administrators. In fulfilling each of these roles, scholars benefit from a deeper understanding of and more global perspective on science. As science becomes more interdisciplinary and international, scholars must become "sciencetrotters" as well as globetrotters, crossing existing boundaries with ease. They must be fluent in multiple science languages and cultures to harvest the best theory, tools, and expertise, independent of their origins. Maps of science can guide scholars in knowledge access and management activities, as well as in their evaluation of what knowledge is novel and relevant.

Scholars try to maximize the yield from their foraging activities by incorporating (ingesting) fresh ideas and insights culled from their natural habitats. Academic reputations are based, in part, on the accumulation of certifiable novelty.
Blaise Cronin and Carol A. Hert

Scholars as Researchers and Authors

As researchers and authors, scholars need to know their place and value in science. They are nodes in the networks of support (as they are acknowledged in papers and social networks) and influence (as their works and theories are disseminated based on a work's merit or author's reputation). Their selection of research topics, students, collaborators, and publication venues (for example, journals or conferences, book publishers, and funding agencies) influence and shape their place in these networks and ultimately their reputations (see **page 59, Symbolic Capital and the Cycle of Credibility**).

A commonly used measure of an author's reputation is citation counts, and there are many different tools to support the retrieval and analysis of citations. For instance, the freely available software tool Publish or Perish (see **page 45** in **Milestones in Mapping Science**) retrieves raw citations via Google Scholar and generates statistics such as total number of papers/citations; average number of citations per paper, citations per author, papers per author, and number of citations per year; Hirsch's *h*-index and related parameters; Leo Egghe's *g*-index; contemporary *h*-index; age-weighted citation rate; two variations of individual *h*-indices; and number of authors per paper. The SCImago Journal and Country Rank (SJR) generates "country profiles" based on Scopus data, which are presented as tables, citation networks, and bubble charts. Many more science analysis tools and services can be found on **pages 26-47, Milestones in Mapping Science**.

Global science maps show the context of basic units of science. The reputation and connectivity of scholars, hot and cold research topics, core competencies, high-impact journals, bursts of activity, and funding per research area, as well as their potential correlations, can all be examined. Current and proposed work can be evaluated in a global context, opportunities and vulnerabilities can be checked, and future trends can be seen. The maps can be explored at a global, local, or individual level to strategize and optimize efforts that increase reputation and keep scientists engaged in positive feedback loops.

Scholars as Editors

Editors are the gatekeepers of science. For each journal, they determine the editorial board, assign paper reviews, and ultimately accept or reject papers. Editors need to know the position of their journals in the evolving world of science. They need to advertise their journals appropriately and attract high-quality submissions, which will in turn increase the journal's reputation and lead to higher-quality submissions. They need to ensure that the editorial board and reviewers properly cover the range of submitted material. They must also write editorials describing the interrelationship of papers in the current issue to those in prior issues as well as to other areas of research.

Scholars as Reviewers

Many scholars serve as reviewers. They read, critique, and suggest changes to help improve the quality of papers and funding proposals. They reject papers and proposals that do not merit publication or funding. This peer-review system is at the core of scholarly activity. Because of the increasing specialization of science, a single scholar's area of expertise now tends to be rather small. Maps of science can help identify reviewers, paper duplications, and related works that should be cited.

Scholars as Teachers

Scholars teach classes, train doctoral students, and supervise postdoctoral researchers. They are part of a scholarly genealogy (see Lenoir's *Scholarly Genealogies* in **Milestones in Mapping Science, page 38**).

The reputation and ranking of both scholars and the institutions at which they are based has a major impact on the quality of students attracted to those institutions. THE–QS World University Rankings, published by Times Higher Education (THE) in London, evaluate universities from around the world based on Scopus data. The Webometrics Ranking, produced by the Cybermetrics Lab of the National Research Council (CSIS) in Spain, computes bibliometric rankings for more than 4,000 universities and 1,000 research labs worldwide based on the volume, visibility, and impact of Web pages. In the United States, the *Chronicle of Higher Education*, *U.S. News and World Report*, and other periodicals provide extensive information about college and university rankings, read by thousands of students and parents each year.

The utility of science maps in teaching is discussed on **page 186, Hands-On Science Maps for Kids**.

Scholars as Inventors

Some scholars are inventors. They create intellectual property and obtain patents, thus needing to navigate and make sense of research spaces as well as intellectual property spaces. Maps of patents (see **page 8**, figure **Claiming Intellectual Property Rights via Patents**) or intellectual property (see *SparkCluster Map*, **page 45, Milestones in Mapping Science**) help inventors make informed decisions.

Scholars as Investigators

Supporting students, hiring staff, purchasing equipment, and attending conferences all cost money. While some areas of research require very little funding to produce outstanding results, others require considerable amounts to even get started. Matching research interests and proposals to existing federal and commercial funding opportunities, finding industry collaborators and sponsors, and acquiring hardware donations are nontrivial tasks. Superimposing funding opportunities on a map of science dimensionalizes the nearly 25,000 funding opportunities in the United States, worth more than $33 billion, as listed on the Community of Science Web site in 2010.

Scholars as Team Leaders and Science Administrators

Many scholars direct multiple research projects simultaneously. Some have full-time staff, research scientists, and technicians in their laboratories and centers. Science administration easily consumes a major part of their time. Tools like the one discussed below, on the opposite page, and on **page 202, Science of Science Policy Maps for Government Agencies**, can help them to make informed decisions while advancing their research.

Effective Knowledge Management Tools for Scholars

The excellence of any scholarly team or project depends strongly on the quantity and quality of supporting resources and networks; likewise, the resources and networks produced by that team or project define its level of excellence. Resources include *people*, *research projects*, *teaching opportunities*, *publications*, *presentations*, and co-organized or attended *calls and events* (for example, conferences and workshops), in addition to used or developed *data sets*, *software* packages (including their versions and dependencies), owned and licensed *hardware*, and associated *funding sources*. Resources are highly interlinked; for example, one person may work on many research projects and publish many papers. Similarly, one source of funding typically supports the generation of many scholarly products.

The effective operation of a scholarly team or project requires a new generation of knowledge management (KM) tools that support the entry, interlinkage, access, and management of major scholarly resources, products, and networks. Here, we introduce the IVL system, which supports information access, analysis, and visualization in support of research management. The example data shown represents activities of Börner's Information Visualization Lab (IVL) at Indiana University.

Storing and Accessing Information

The IVL system provides data input forms to store major scholarly resources, their attributes, and inter-linkages in a database. All data pertinent to a lab's operation and online presentation can be served via Web pages, see example at **http://ivl.slis.indiana.edu**. In this way, it becomes easy to find out which papers were published in what year, when PhD students become available for hire, or if an expert has cycles left to serve as a collaborator. Using the IVL system, one can retrieve papers, presentation slides, and information on authors, as well as related data sets and software. In addition, one has the ability to identify the projects that cite, use, or are cited by the various papers, data sets, and software.

Coauthor Network

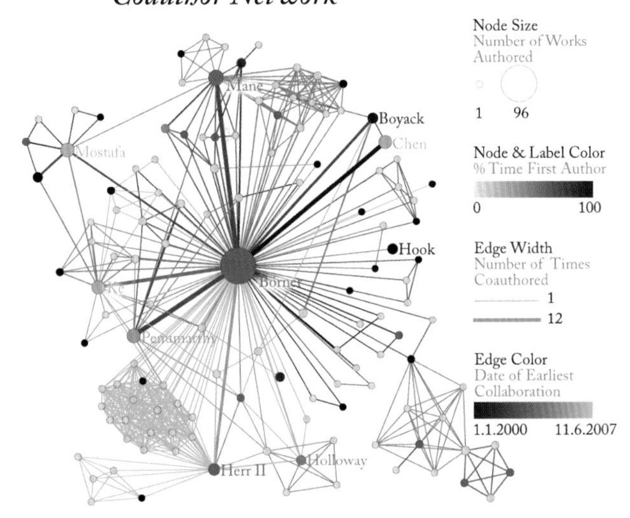

Investigator-Project Network

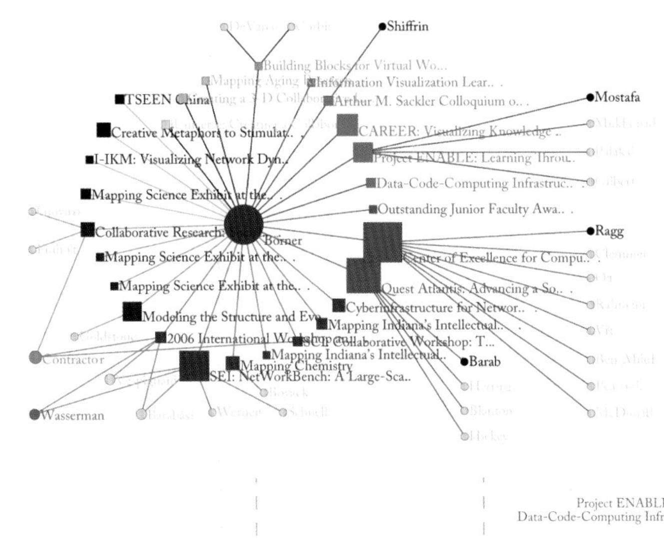

Providing References

Upon entering or selecting the name of a person, complete and up-to-date information can be retrieved for writing letters of reference or for giving phone references.

Generating Project Reports

The IVL system supports the generation of progress reports for funding agencies. When one selects a time frame and a project, all associated data is listed and grouped by record type (for example, team members, publications, and presentations). The result can be formatted in a way that is appropriate for a specific funding agency.

Monitoring Progress

Lists containing the number of people, publications, presentations, and funding per project or in total can be compiled to quickly obtain an overview of general progress and activity. Dollars spent per paper and other derivative data can be calculated. The **Project Timeline** below shows project titles, durations, and amounts for all funding received between 2000–2007. The horizontal bars denote hardware grants, and the green-line graph represents the total dollar amount available per day.

Travels

Mapping Geographic Knowledge Diffusion

Space matters—even in the Internet age. *People*, *presentations*, and *events* have geolocations and can be mapped over time, adding a spatio-temporal dimension to scholarly activity. Shown on the top right, as an example, are **Travels** by lab members for the period 2000–2007, overlaid on a map of the world. Visits by researchers from around the globe can be visualized analogously as a proxy of the diffusion of information via personal contacts.

Mapping Expertise

Coauthor, coprincipal investigator, student-advisor, and other networks can be extracted from the data. The **Coauthor Network** for 2000–2007 is shown on the top left. Node area size represents the number of papers written; the colors correspond to the percentage of times papers were written by a first author, from 0 percent (white) to 100 percent (black). Author names are provided for the 10 nodes with the most papers. Edge width indicates the number of times two authors have collaborated. Edges are colored by date of first collaboration, from 2000 (black) to 2007 (green).

Multipartite Networks

Multiple entity types—such as people and their papers, investigators and projects, or projects and funding—can be correlated. *People* and *project funding* records for 2000–2007 are shown in the **Investigator Project Network** below left. Circular nodes represent investigators, who are coded by area size (based on the number of awards within the network) and by color (based on the percentage of times they served as principal investigator, from 0 percent [white] to 100 percent [black]). Squares represent funded projects, which are coded by area size (based on the total award amount) and by color (based on starting date). Edges between people and funding are color-coded by starting date as well. The evolution of all networks can be animated over time. Networks can be downloaded for further analysis and exploration.

Research Consumption and Production

The knowledge consumption of artifacts by others (for example, the citation of papers, the hiring of students, the number of invited talks) and their production, such as the geographic (national or international) and topical (within domain of research or interdisciplinary) spread of scholarly artifacts over time can be mapped as well.

Project Timeline

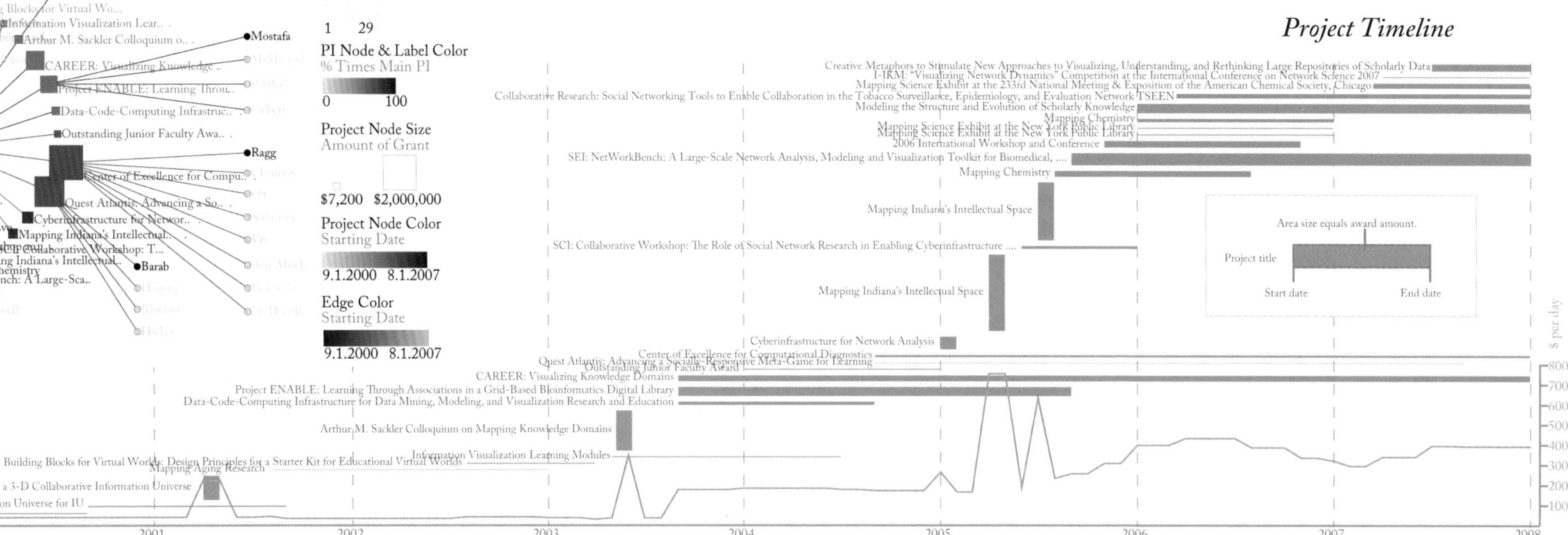

Science Maps for Kids

Children study mathematics, physics, and biology, among many other subjects in school. The subjects are typically taught in isolation by different teachers. However, science is highly interdisciplinary and interconnected—particularly today. Nearly all of the world's major challenges require a close collaboration of scientists from different disciplines. The notion of the lonely genius or visionary driven to exhaustion in search for scientific truth is no longer viable. Breakthrough research and inventions cannot be produced *ex nihilo*.

Cutting-edge science involves large-scale data sets, advanced computational infrastructures, and a close collaboration with computer science and engineering. Though children currently learn about microscopes and telescopes, in an ideal world they would also learn to operate macroscopes, which help identify patterns, trends, and outliers in large-scale data sets. In time, they may even design their own tools to solve problems not known today. Last but not least, children should be engaged in the true work of scientists, feel the excitement of their discoveries, and find their own "place" in science.

Learning to live together in mutual respect and with the definite aim to further the happiness of all, without privilege for any, is a clear duty for mankind, and it is imperative that education shall be brought onto this plane.
Kees Boeke

Science from Above

Children need to learn about and value what the different sciences have to contribute and how they interrelate. They need to understand the very diverse cultures, research approaches, and languages of science. They will have to be able to "speak" more than one science in order to collaborate across scientific boundaries.

Most cognitive scientists believe learning best begins with a big picture, a schema, a holistic cognitive structure, which should be included in the lesson material—often in the text. If a big picture resides in the text, the designers' task becomes one of emphasizing it. If this big picture does not exist, the designers' task is to develop a big picture and emphasize it.
Charles K. West, James A. Farmer, and Philip M. Wolff

Concept maps are widely used in education. In first grade, they may be used to relay the daily schedule. Later, mind maps and argument maps are valuable means of communicating complex systems. Software tools such as Inspiration Software, Compendium Blogware, let's focus, and Rationale help us to visualize our collective knowledge creation, access, sharing, discussion, and utilization. The maps augment and enhance human intellect, ultimately leading to improved decision-making (see **page 202, Mapping Intellectual Landscapes for Economic Decision-Making**). A concept map is composed of four core elements: *nodes;* which are connected by *links;* both of which are described by *words;* and finally resulting in *patterns,* such as a hierarchical or circular ordering of the nodes.

Science maps can be seen as special forms of concept maps. They facilitate a spatial understanding of things, concepts, conditions, processes, or events in the human world. While concept maps tend to be local in scope, science maps are able to convey the structure of all of science (see below and **page 186, Hands-On Science Maps for Kids**).

Mapping Inventions and Inventors

Science maps can be used to place major inventors and inventions in time and space or to plot the lives of major scientists. Consider Einstein, for example. Where would he go on a map of science? Given that he made major contributions in several areas of science, should we divide his image into pieces? How many? Should we place him in the area of his most significant contribution? Which area would that be? Or should we place him in the center of all his contributions, potentially a no man's land? It may be more important that children ask these questions than arrive at any definite conclusions. Where do all the other inventors and inventions go? How can children find their own "place" in science?

Since 2006, the *Places & Spaces: Mapping Science* exhibit has invited children (ages 4 to 14) to draw their favorite scientists and science experiments. More than 138 drawings and related stories have been submitted, including two contest winners: "A Portrait of Albert Einstein" by 11-year-old Luke Layton and "Nature—Science Is All Around Us" by 12-year-old JoHanna Sanders (featured above). Out of all the submissions, the three most popular scientists were Albert Einstein, Thomas Edison, and Steve Irwin; the top three science experiments were "What Is Science?" "Magnets," and "Plants."

Visual Interfaces to Educational Resources

Science maps can also be used as visual interfaces to educational resources (see **page 198, Visual Interfaces to Digital Libraries**)—including national and international science museums; online and offline resources such as books, teaching materials, and exercises; and learning partners. All of these resources can be mapped and explored. Specific examples of these types of educational resources include Open Educational Resources (OER) Commons, a teaching and learning network of shared materials, and the National Science Digital Library (NSDL), a national network of learning environments, resources, and partnerships. Visual interfaces to these resources would raise awareness of their coverage while supporting navigation and utilization.

Science Maps in Teaching

Given a specific topic, such as "sustainability," a teacher must identify which areas, experts, and readings to cover and which sequence of units or pathways to follow in order to best communicate the structure and evolution of the subject over time. Teachers may decide to use the topic coverage and sequence given in a textbook. If no textbook exists, however, teachers are often left to their own devices. As teaching topics quickly evolve and become ever more interdisciplinary, that becomes a daunting, if not impossible, task. Science maps can help by delineating a domain. For example, a query for "sustainability" would highlight relevant research and existing educational material. As a result, review articles, major books, age-appropriate resources, and other context and technical resources could be easily identified. In addition, the structure and evolution of a research topic can be animated and demonstrated to the class (see **page 9, Mapping the Evolution of Coauthor Networks**).

Designing Macroscopes

By using Google Earth, anybody can see Earth from space and then zoom in right into one's own driveway. In the movie *Over the Hedge*, the squirrel Hammy gets an energy drink that makes him move at light speed to activate an extermination system while the world and time seem to stand still.

Children need to be trained to operate tools that speed up and slow down time and help manipulate space over many magnitudes. Just as they are trained in the operation of microscopes and telescopes, they need to be educated in the use of novel macroscope tools, which can help them make sense of large amounts of data. Initially, they may wish to analyze their social networks recorded via e-mail and phone devices or credits collected in gaming environments. Soon enough, they will be measuring and gaining insight from data never considered before. More importantly, they may build their own macroscope tools with functionality superior to those that exist today.

Child Scientists

Children love to do hands-on science, get messy, and see the fruit of their labors: They have been observed to eagerly count and report birds in their backyards to see the collective result via BirdSource. They bury seismic sensors beside their schools to collect, analyze, and interpret scientific data and to share that data with the broader research community. They engage in professional archeology to uncover and interpret the past via primary sources, both textual and artifactual. They also participate in geocaching and geomapping activities, making use of Sensor Webs to measure and study environmental data.

"Edutaining" Knowledge Webs and Games

The enormous increase in our collective knowledge (see **page 4, The Rise of Science and Technology**) and the urgency with which it must be applied to solve the world's challenges requires novel approaches to education. In **Visionary Approaches (page 22 and 25)** we reviewed the works of Kees Boeke, H. G. Wells, Buckminster Fuller, and James Burke. All four developed pioneering ways of teaching and learning. The latter two specifically highlighted the importance of games in understanding existing and imagined complex systems, such as living organisms, ecosystems, technological infrastructures, and political systems.

Retrieval by Association

Both scholarly works and textbooks are generally organized in a linear fashion: one topic follows another. Our minds, however, work by association—we learn by connecting new concepts to existing ones. The network of learned concepts can then be traversed and understood in a non-linear manner. Jimmy Wales's Wikipedia and James Burke's Knowledge Web (K-Web) (see page 18 and 25 in Visionary Approaches) are excellent examples of knowledge structures that support nonlinear traversal and discovery. Visualizations of network structures support the identification of major interconnections or "backbones" and the discovery of highly interconnected knowledge items or clusters. They provide a bird's-eye view of the density and layout of the knowledge network in support of navigation and sensemaking.

Exploring the Web of Knowledge

Books like *Exploring Time* by Gillian Chapman and Pam Robson tell the history of inventions using simple, everyday objects to explain complex phenomena. Readers and listeners can choose between alternative paths, but they rarely get to see their trail from above. While brilliant authors give their readers a sense of the overall picture while explaining unique details, science is too vast and too complex to be completely understood via narratives alone.

James Burke's *Connections* TV series (see **page 25, James Burke**) was intended to illustrate the web of knowledge by interlinking scientific inventions and inventors via nearness in time, geographical space, or topic matter. Burke wanted to communicate the importance of the context in which inventors flourish and inventions are made—to show that inventors share certain values and characteristics and that discoveries tend to follow patterns. The development of the modern ballistic missile resembles the development of the cannonball, for example, and the introduction of the telephone is similar to the creation of medieval church postal services. Scientific development happens not in a linear fashion but within a network of existing knowledge and expertise. Many major inventions (like the Internet) prove to be effective and influential as a direct result of their interaction with other technologies as well as nontechnological factors.

When studying major inventors and discoverers of the past, Burke realized that they did not respect temporal, spatial, or topical boundaries but instead took information and ideas fluidly from different areas to create innovative solutions. This stands in strong contrast to most educational systems, which group information into separate, isolated categories, such as those called biology, history, and social science; these are often further subdivided by time, geography, or topics.

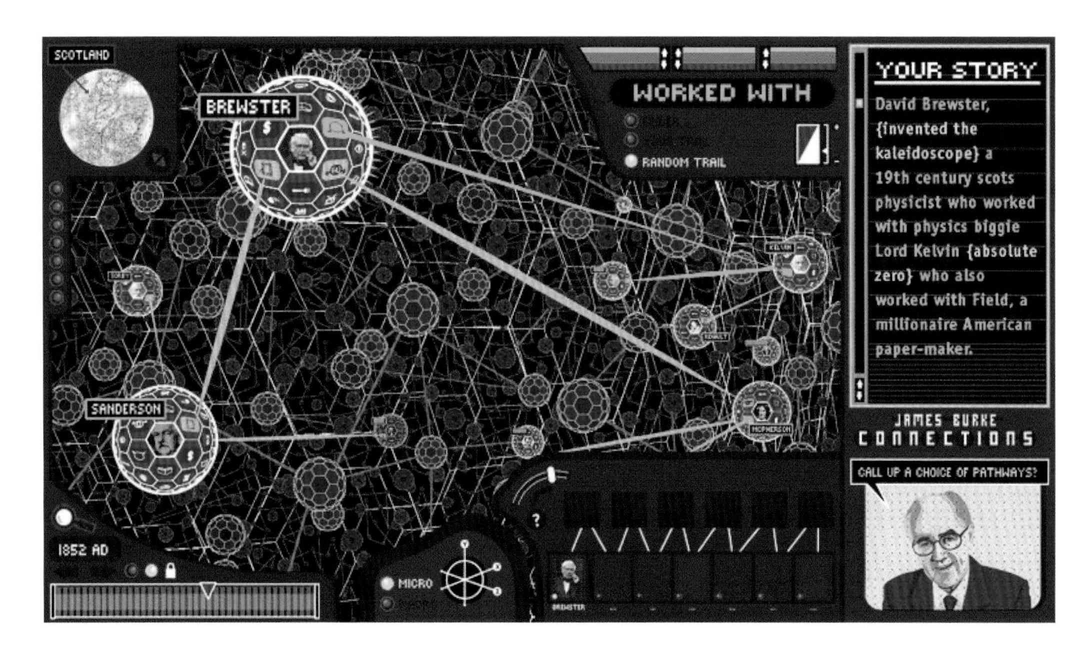

Burke's K-Web software is an online interactive learning tool that enables the exploration of history and the creation of ideas. Users journey through history to encounter thousands of key people, places, and inventions—interconnected in thousands of ways—and to discover the remarkable serendipity behind such tireless commitment to invention and innovation. The subject nodes and their connections are positioned in a three-dimensional dynamic time construct (see figures to the right). Centuries correspond to time-sorted onion skins, representing the ancient past at center and the present in the outer layer.

In *The Knowledge Web* (1999), Burke writes about 142 exemplary gateways to traverse the K-Web. In Burke's words, his work takes us on a "journey across the vast, interconnected web of knowledge to offer a glimpse of what a learning experience might be like in the twenty-first century."

For example, a journey into the history of bioethics and the matter of human cloning might start in the **19th Century** when **Charles Babbage** produced the first computer with the help of mathematician and aristocrat Ada Lovelace (see **page 16, Visionary Approaches**). Interestingly, Lovelace was the daughter of **Lord Byron**, the great Romantic poet and friend of Percy Bysshe and **Mary Shelley**. It was in Lord Byron's Swiss villa that Mary Shelley wrote **Frankenstein**, about a scientist who creates a living being (see image at top right).

In the **20th Century**, the discovery of DNA and genetics led to **The cloning of Dolly** the sheep (see image at middle right).

In the **21st Century** came the completion of the **Human Genome Project**, which raised new **bioethics** questions. The K-Web shows how history links to the cutting edge of modern life, suggesting that if we understand we are the product of history, we can then better grasp and control our destiny.

Computer Games

Games such as Will Wright's SimCity and Spore and Sid Meier's Civilization inspire and engage players of all ages to create the ideal family, city, or civilization. Educational environments such as PowerUp enable exploration of alternative energy resources. Award-winning education software by Tom Snyder Productions is used in more than 400,000 classrooms. Games help players feel like they are part of an alternative world in which they can live a different reality; as a result, an increasing number of players feel "at home" in one or more virtual worlds.

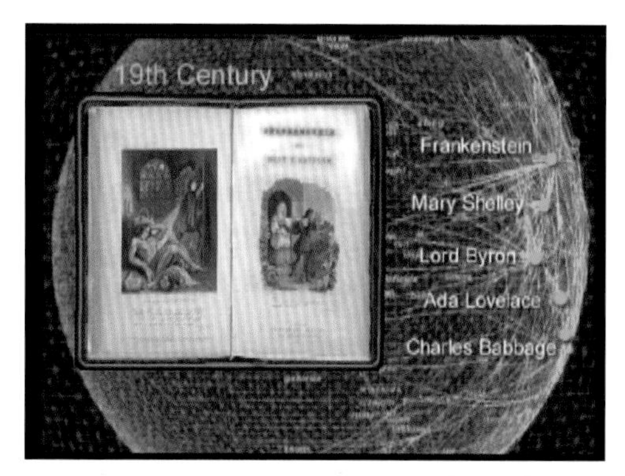

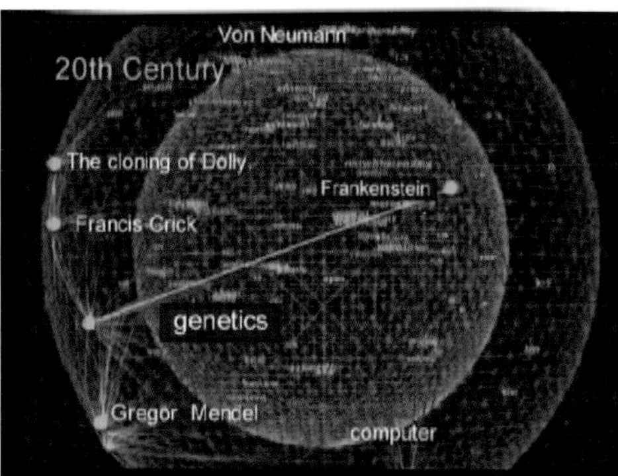

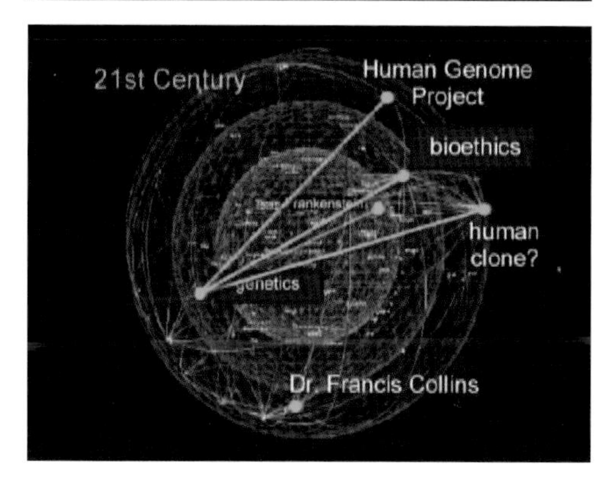

Eduverse is an entire Active Worlds universe dedicated to exploring the educational applications of the Active Worlds technology. It was launched in 1997 and hosts more than 80 educational worlds. Active Worlds technology also powers projects such as Quest Atlantis, which builds on strategies from commercial role-playing games with lessons from educational research on learning and motivation. In all these environments, visitors create virtual personae, talk with other users and mentors, travel to virtual places to perform educational activities or "quests," and engage in curricular tasks that are designed to be both entertaining and educational.

Children growing up today with access to these environments will expect to have tools of comparable sophistication when, as adults, they begin to monitor and support scientific and technological progress and its impact on family life and civilization.

Daily Science Forecasts

Just as meteorologists have learned to measure, map, predict, and dramatize the weather, scientists must also learn how to measure, map, predict, and dramatize science and technology. Weather forecasts have an immediate impact on the ways we dress and the activities we plan. Science and technology forecasts would have a more delayed but longer-term impact. Whether or not current inventions have a direct bearing on today's stock market, they are more than likely to lead to the future development of valuable products, tools, and cures. Science and technology maps and forecasts have another important role to play: the communication of knowledge beyond all boundaries, enabling access by anyone, anywhere. Having access to the best scientific results, industries can strategize product development, teachers can educate with up-to-date information, and researchers can avoid duplication of work as they identify synergies across national and disciplinary boundaries.

The ease with which information spreads is critical to the rate at which change occurs.
James Burke

Everyday Science and Technology for Everybody

Western culture is deeply entrenched in the perspective that science is pure and separate from the profane activities of daily life—that it exists in an ivory tower, remote and inaccessible to all but the select few. Consequently, the quality of scientists who have dedicated themselves to education and outreach is frequently questioned: "If their work is truly exceptional, why don't they publish papers in major journals? Why are they reaching out to us rather than to the best minds, the readers of those journals?" The answer is simple: only a small minority of the total population reads "the major journals." To have a strong impact, science education must reach all individuals in all communities, beyond predetermined expectations and boundaries (see **page 24, Global Brain**). This would seem to be only fair, as many scientists' salaries are drawn out of taxpayers' money.

MarketSite Map

Real-time analyses and depictions of financial data are commonplace online and on TV. Visualizations created by Oculus show data graphs and explanations of the NASDAQ MarketSite (see top of next column) and other financial data.

These visualizations analyze enterprise risk models and communicate results to senior management. The three-dimensional models show a combination of pop-up graphs, charts, and maps

on one screen. They allow the user to input "what if" scenarios, provide analysis of cash flows at risk across regions and assets, and use probability distributions to build up a company-wide picture of portfolio value at risk.

Industry Pull and Science Push

Science and technology maps and forecasts are likely to influence decision-making in both industry and academia. Public information on needed technologies is expected to cause an "industry pull" of research toward much-needed results. Easy to access research results are likely to cause a "research push" toward new technologies and products.

As of 2008, Japan enjoyed effective academic-industry partnerships. In the United States, however, academia focuses on invention while industry pursues commercialization, with few bridges in between.

Easy access to patent data and intellectual property can make a difference. For example, SparkIP provides access to more than three million U.S. patents, dating back to the late 1960s, as well as to new licensable technologies, entered on a daily basis by universities, research labs, corporations, and individual inventors. Search results are displayed in a traditional list format or as SparkCluster Maps (see **page 45, Milestones in Mapping Science**). SparkClusters with green "halos" indicate there are technology listings available, and the option "Browse Listings" leads to all the technology listings categorized in the SparkIP ontology.

Science and Technology Forecast News Online and on TV

Forecasts of scientific developments online and on TV may be the realization of Buckminster Fuller's dream of educating all citizens in an efficient, secure, and inclusive manner (see **page 25, Buckminster Fuller**).

The envisioned science forecasts would resemble weather forecasts. A map of the world or a map of science would be shown with overlays of activity bursts, emerging research areas, major money flows, and discoveries. Much like weather forecasts, the interactive maps would be interpreted by professional commentators. Close-ups of science areas would be augmented by interviews with famous scientists, investors, and beneficiaries. The result might be akin to Bloomberg Television. Moreover, analogous to stock market tickers in TV news, science tickers would show the number of dollars awarded to different sciences, the strength of different nations, and indexes of different scientists.

Diverse data streams could be used to populate the initial science forecasts. They might include science-related stories and RSS feeds by news providers; daily science updates by *The Scientist;* news of hot papers and research fronts; interviews with highly cited scientists; state and national indicators; and newsletters such as "Science Watch"—all available from Thomson Reuters. They would also involve many local efforts to bring science to the public, such as the "Science and the City" webzine of the New York Academy of Sciences. In addition, there

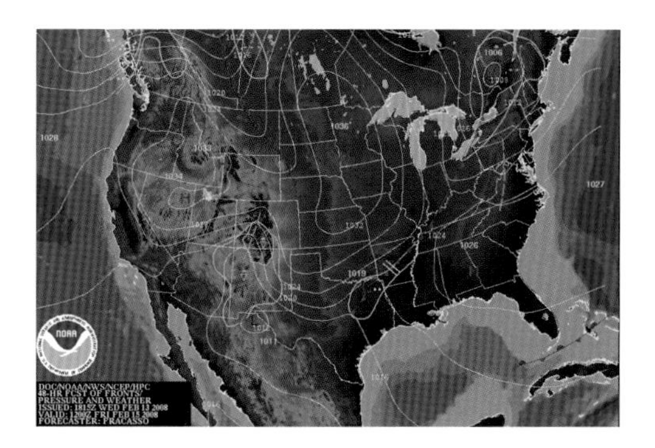

would be many science-relevant materials available via *NOVA scienceNOW,* Science@NASA, YouTube, iTunes, and educational Web sites. Last but not least, stock market data, patent information, and other economic data feeds would be included.

An online science forecast interface might resemble the mashup of NOAA's National Digital Forecast Database (NDFD) Extensible Markup Language (XML) service and Google Maps available via WeatherMole.

People would tune in to the science forecasts regularly to learn about the most recent scholarly advances, see interviews with their favorite scientists, learn how much of their tax money was devoted to any given research field, or simply experience the beauty and wonders of science. It would be the worldwide broadcast of science *Gesamtwetterlage* (total weather conditions), communicated in an easy-to-understand format to all those who might be affected.

Anatomy of Science and Technology Forecasts

Just as weather forecasts require a global network of experts, sensors, and large-scale simulations, science forecasts require harvesting, analyzing, modeling, mapping, and serving science and technology data from around the globe in real time.

Learning from Weather Forecasts

In 1922, Lewis F. Richardson proposed the numerical prediction of weather. He estimated that a regular numerical weather forecast would require a factory of 64,000 mathematicians equipped with calculators. It was not until 1950 that relatively successful predictions were calculated in the United States. Faster computers, higher-quality data, better coverage, and more sophisticated models all helped to quickly improve the accuracy of the predictions. By 1955, weather predictions were performed on a regular basis.

As everything in the atmosphere is connected, successful weather prediction requires close international collaboration. In 1951, the World Meteorological Organization (WMO) was founded—and in 1961, U.S. President John F. Kennedy called

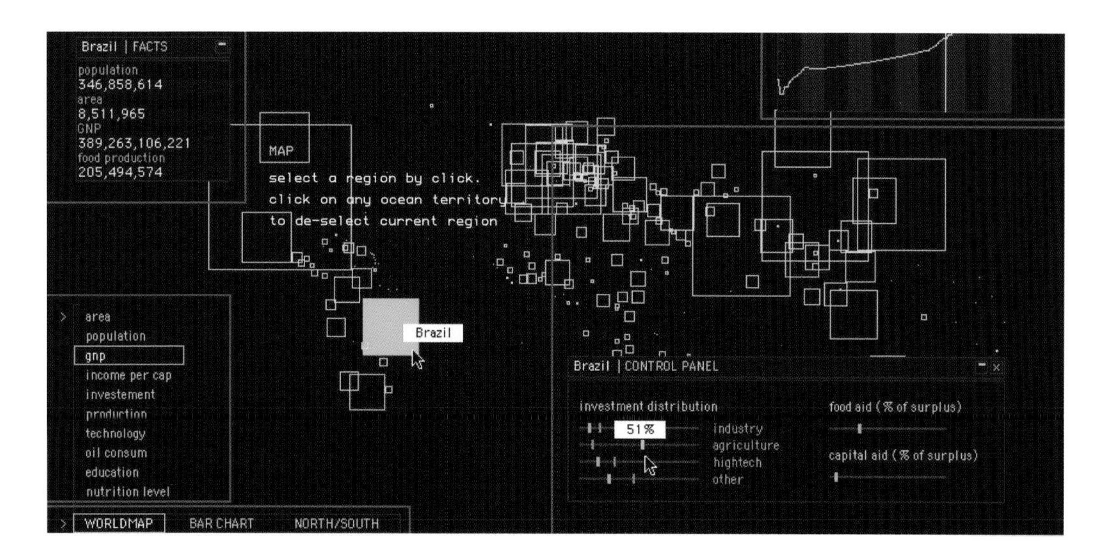

for the development of an international weather prediction program. At the height of the Cold War, 150 countries, including the Soviet Union, joined this effort under the auspices of the WMO.

In 2008, the World Weather Watch global meteorological system collected data and observations from a series of polar-orbiting weather satellites, 12,000 land stations, 7,000 ships and oil rigs, 700 upper-air observation stations, and many commercial airliners. As part of its system, data is transmitted via teletype and radio links to regional and national centers. It is then passed on via high-speed connections to the three WMO centers in Melbourne, Moscow, and Washington, DC. The centers compile global synaptic maps and weather forecasts that are distributed to national weather services every six hours.

Many thousands of meteorologists are employed worldwide. China's weather service alone numbers approximately 65,000 people. The technical infrastructure of millions of interconnected sensors and high-performance computing resources is expensive. However, the utility of storm, flood, and tornado warnings by far exceeds the expenditures—in the United States alone, approximately 10,000 violent thunderstorms, 5,000 floods, and 1,000 tornadoes are recorded each year.

Patchwork of Multiple Forecasts

Analogous to globally collected meteorological or seismic data (see image above right and **page 147, Tectonic Movements and Earthquake Hazard Predictions**, for explanation), science forecast maps will have to be patched together from different data sets that vary in format, aggregation level, and licenses, depending on country and area of research. Some countries and research communities have access to expertise and computing facilities that enable the creation of highly accurate models of science. Others simply collect basic data and have no modeling capability. The availability of a global science forecast, however, would improve the accuracy of local forecasts. Hence, different data streams have to be merged and analyzed, while results need to be redistributed in an intelligent fashion and "translated" to address local needs.

Designing a Scalable Infrastructure

A shared map of our collective scholarly knowledge should be based on the best data sources, algorithms, and expertise. It is preferable that all software applied be open source so that it can be examined, understood, trusted, and advanced by the best minds around the globe. There needs to be a means to share data, algorithms, and interpretations through an environment resembling Wikipedia, Flickr, or YouTube. The envisioned infrastructure would empower everyone to share scholarly knowledge, while providing novel tools to predict and make sense of it (see also **page 68, Scholarly Marketplaces**).

Fragments of this infrastructure already exist. For instance, arXiv supports the efficient sharing of preprints, and MEDLINE provides free and easy access to 19 million biomedical research papers. Many patent and funding award data sets are also freely available online. Connotea, BibSonomy, and CiteULike are examples of sites that provide free online reference management. SparkIP makes it easy to trade intellectual property. Swivel and Many Eyes allow users to upload and view data sets. Resources like nanoHUB and other sites based on its technology ease the sharing of algorithms. SourceForge is a hub for open-source software distribution and services. The site myExperiment supports the sharing of scientific workflows and other research objects; it also helps to build communities. SciVee invites scientists to upload videos and "pubcasts" (podcasts about publications) as a means to make their research known to a wider audience; it also supports virtual commentary, community formation, and blogging features. Ning is a programmable platform for creating customized social networking sites. Many more e-science spaces and cyberinfrastructures are in existence today, all built for diverse scientific communities.

The technical setup of such a highly scalable infrastructure is nontrivial. Many different streams of scholarly data—in many different languages—will need to be federated and advanced algorithms pipelines will have to be applied to analyze this data in order to make predictions and meet requests.

Most likely, science forecasts will be deployed as large-scale Web services. Algorithms will need to be parallelized. Scalability and fault tolerance will be achieved by replicating services across many different machines (see also **page 68, Sociotechnical Cloud Design**). Automatic failure-detection mechanisms will need to be implemented to handle different types of threats and malfunctions. Note that the amount of data Google processes—estimated at several tens of terabytes of uncompressed data in 2008—is much larger than the amount of scholarly data in existence today. Yet, science forecasts will require more advanced data analysis, modeling, and visualization.

Valuing Collective Interpretation and Prediction

The interpretation of science maps and forecasts requires complete transparency in which data sets, algorithms, and visual mappings are used—yet another reason for open-source data and code. Online science simulations that resemble [./logicaland] (see above image and **page 152, [./logicaland] Participative Global Simulation**) or SimCity would empower many to pose or solve new scientific quests or to design and validate science models with the potential to improve our understanding of the inner structure and dynamics of science.

Science and Technology Forecast Politics

Different data sets come with different price tags and access rights. An organization similar to the WMO might have to be founded to coordinate the worldwide acquisition, computation, and distribution of science forecasts. The price tag for an infrastructure and its administration can be justified by the impact of daily science communication on research, education, and product development. We can only imagine how science maps and predictions might change the way in which science is supported and evaluated—transparently and collectively.

Science and Technology Forecast Advantage

The nation to first embrace and implement science forecasts will have an enormous science and technology advantage, as the diffusion of scientific results from academia to industry, bench to bedside, and lab to classroom will tend to increase dramatically. Work by Luís M. A. Bettencourt showed that a doubling of the contact rate between authors publishing on the subject of Bird Flu (H5N1) leads to discovery time savings of approximately one decade. Increased industry buy-in, practical evaluation and appreciation of research, and advances in education are likely to further accelerate innovation and product development. Ultimately, global science forecasts will help us to understand and utilize our collective science and technology knowledge and expertise on a global scale, crossing national and disciplinary boundaries. Successful science and technology cooperation might even help to resolve political conflicts and cultural divides.

Growing a "Global Brain and Heart"

Over the past 50 years, the human population has experienced an exponential increase from about 2.5 billion to 6.6 billion (see graph on **pages 2-3, Knowledge Equals Power**). At the same time, the standard of living in developed countries has also increased considerably, and with it the demand for fresh water, food, materials, and energy, causing substantial and largely irreversible losses in the diversity of life on Earth. High-gloss magazines, television, and online resources extensively feature the lives of the rich and famous and growing numbers of people wish and seek to have more luxury. However, the resources of our planet are limited. It will take the creativity, expertise, empathy, and collaboration of billions of human minds and hearts to develop sustainable ways of living. It is my hope that our collective "global brain and heart" is ready for action before apocalyptic scenarios of humanity's future become reality.

Western civilizations these days place great importance on filling the human 'brain' with knowledge, but no one seems to care about filling the human 'heart' with compassion.
Dalai Lama

Facing Mankind's Global Challenges

Today, we do not know if we have already destroyed vital parts of our planet needed for the survival of our species. We are in desperate need of tools that can help us to make collective sense of existing and future data and knowledge.

Many organizations, such as the United Nations and the World Bank, have set clear goals for reducing poverty and other sources of human deprivation and for promoting sustainable development. However, plans to eradicate famine and disease worldwide cannot be accomplished amid the enormity of recurring environmental damage.

The greatest challenge of the 21st century is to provide every human being on the planet with a long, healthy, and fulfilling life, free of poverty and full of opportunities to participate in the life of their community.
World Bank

The World Resources Institute, *The World Factbook*, and the International Union for Conservation and Nature, among other sources, provide essential statistics for today's environmental data, including energy production, consumption, and trade. However, tables and statistics will need to be translated into easier-to-understand formats to reach a larger percentage of the population intellectually and emotionally. Ideally, people would be able to experience different scenarios of the future, for instance, in online gaming environments to gain extensive knowledge and first-hand expertise of different decisions.

Aligning Science and Technology with Global Challenges

Today's maps of science show how much funding goes into which areas—for instance, well-funded areas appear larger as they attract more researchers that "publish or perish." Given the extremity and urgency of world conditions today, it would be wise to align our science and technology efforts with the most current global challenges. It is reasonable to ask scientists to tag their research with officially compiled lists of challenges, so that tomorrow's maps of science may show the contributions that different areas can make.

Candidate lists of global challenges include the "Millennium Development Goals" compiled by the World Bank or "World Problems" published by the Union of International Associations (UIA). The Millennium Development Goals aim to: (1) eradicate extreme poverty and hunger, (2) achieve universal primary education, (3) promote gender equality and empower women, (4) reduce child mortality, (5) improve maternal health, (6) combat HIV/AIDS, malaria, and other diseases, (7) ensure environmental sustainability, and (8) nurture a global partnership for development. The UIA listing comprises world problems (Issues: 56,564 profiles; 276,791 links) and global strategies (Solutions: 32,547 profiles; 284,382 links) among other subjects.

Certainly, the importance and impact of research results is not always known in advance. Some areas, such as mathematics or computer science, provide the theory and computational infrastructure needed by almost all other sciences; as such, their indirect contributions should be calculated and acknowledged in an objective and scientific way.

Bridging Disciplinary Boundaries

Recent research shows that science today is not driven by single prolific experts but by effectively collaborating teams. A more interdisciplinary, globally connected science is evolving.

The future belongs to a very different kind of person with a very different kind of mind—creators and empathizers, pattern recognizers and meaning makers. These people—artists, inventors, designers, storytellers, caregivers, big picture thinkers—will now reap society's richest rewards and share its greatest joys.
Daniel Pink

Some of the greatest works were accomplished by those who truly embraced science and art, philosophy and technology—such as Leonardo da Vinci, Michelangelo, and Martin Luther—also called "Renaissance men." Today, the concept of "Renaissance teams"—as first proposed by Donna J. Cox—helps to blur the lines between disciplines by bringing together artists, scientists, and technologists to advance science and technology.

Our pre-technological beliefs place philosophy and art at the centre of man's existence and science and technology on the periphery—the former lead the later follow. Yet, the reverse is true. S&T have immeasurably enriched our material lives. If we want to realize the immense potential of a society living in harmony with the systems and artefacts it has created, we must learn—and learn soon—to use S&T to enrich our intellectual lives.
James Burke

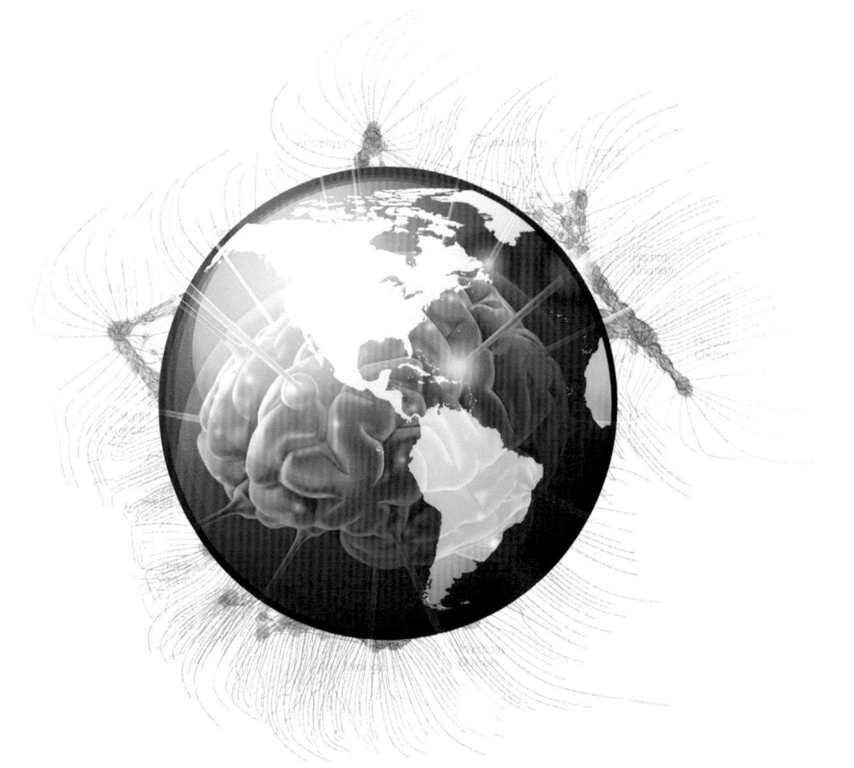

Growing a Global Brain
The subtle human mind faces a flood of information that demands to be processed, managed, and used. There are diverse indicators that a "global brain" is emerging on this planet. While available today primarily in biological wetware, this global brain might very soon be dominated by digital implants that help to effectively merge biological creativity with digital resources and processing speed. This global brain may serve as a means to pool and focus our collective resources to sustain our home planet, Earth, and our species.

Growing a Global Heart

Most of us process scientific data, facts, and results at a rational level. It is rare that we are touched or moved by scientific studies. However, the many global challenges we face require major qualitative changes in the ways we think, behave, and care for one other. Rational decision-making will need to be complemented by compassion in order to align our needs and desires with sustainable ways of living.

The highest education is that which does not merely give us information but makes us live in harmony with all existence.
Rabindranath Tagore

The World Is Flat

Increases in transportation speed and volume seem to shrink our planet (see **page 6, Addictive Intelligence Amplifiers**). In 2006 alone, 4.4 billion passengers were transported via airlines internationally. Innovation in science and technology is occurring worldwide. Innovations and new technologies, as well as epidemics, are able to spread like wildfire. Companies quickly convert their inventions into products, manufacture those products in specific locations, and then distribute them widely around the globe. Historical, cultural, and geographical divisions are thus becoming increasingly blurred.

Despite enormous income disparities—the income of the 225 richest people in the world is equal to that of the poorest 40 percent (2.7 billion people)—we are all interconnected. Global thinking and action are required and often pay off; for example, using a country's vaccines globally can help save more lives locally (see **page 150, Impact of Air Travel on Global Spread of Infectious Diseases**).

Collective Survival of the Fittest

Over tens of thousands of years, evolution favored those who performed most successfully in finding prey and mates while avoiding predators. Today, managing the complexity of knowledge and expertise networks in our world requires a completely different skill set. Even the most extensive training in knowledge management and leadership will not lead to rapid change in perceptual and cognitive abilities. We are optimized for local decision-making, yet we are causing critical planetary change at a global scale.

Growing a Global Heart
Our lives and destinies today are more closely interlinked than ever. An epidemic outbreak in any part of the world has the potential to spread rapidly via transportation networks. Pollution, oil spills, and radioactive leakage would inevitably affect us all. We must care for all living beings as if they were part of our own families. With both mind and heart, we must tend to the Earth, our unique home in space.

Competition and Cooperation

Historically, it is has been commonly accepted that highly competitive environments in which only the best survive lead to superior individuals. Recent observations of animal behavior suggest the contrary is true, as the best collaborating teams are shown to outperform the best individuals. Studies of the egg-laying habits of chickens reveal that breeding programs that focus on selecting the best-performing group (cage of chickens) result in happy hens with higher egg production and a noticeable drop in aggression. In one study, the annual egg production increased 160 percent in only six generations. In a second setting, the best-performing individuals per cage were selected. Here, genes of the most competitive individuals, which used pecking, bullying, and threatening to decrease the egg production of others, were preserved. If these chickens were kept together in one cage, this competitive behavior resulted in bloody fights and fewer eggs. Some scholars of highly ranking institutions might identify themselves and their colleagues as results of the second selection mechanism.

Cooperation is also beneficial in the construction of new levels of organization such as those of multicellular organisms, social insects, and human society. It overwrites "selfish" interests in support of the common good. Five mechanisms seem to underlie the evolution of cooperation: kin selection (donor and recipient are genetic relatives), direct reciprocity (repeated encounters happen between the same individuals), indirect reciprocity (helpful individuals are more likely to receive help), network reciprocity (clusters of cooperators outcompete defectors), and group selection (competition exists also between groups).

Today, both competition and cooperation among researchers, institutes, and nations are vital for local and global success.

The development of the scientific method and the construction of the edifice of science are the greatest group achievements of mankind. Art, music, and poetry might involve greater achievements but they are individual efforts and should be compared with the discovery of Newton's laws or of Einstein's relativity theory. … Science is the only group activity that seems capable, at present, of indefinite improvement and advancement, because it builds on a provable base.

Henry W. Menard

Earth as Common-Pool Resource

The Earth should be seen as our common-pool resource (CPR), consisting of invaluable yet finite natural and manmade resources. Like all other CPRs—such as irrigation systems, fishing grounds, and forests—the Earth itself is in danger of exploitation and misuse. We must find means to avoid the "tragedy of the commons" introduced in 1968 by Garrett Hardin. The term refers to the dilemma of multiple individuals acting independently in their own self-interest, thereby destroying a shared limited resource needed by every one of them. Hardin concluded that no technical solution exists for this kind of tragedy and only strict management of global common goods via increased government involvement or international regulatory bodies is necessary to prevent it. Elinor Ostrom's work shows that local management of common property—as opposed to private or state management—is effective in managing CPRs and can help avoid the tragedy of the commons.

Purely regulatory or intellectual solutions will not suffice to identify, communicate, and protect the planet's core resources to allow for a sustainable future. Rather, the work of the heart—of our human values and morals—must come forward to guide volunteering activities, recycling efforts, resource usage restrictions, and other such local decisions and approaches, all of which can ultimately have profound global impact.

Only through a rite of passage will humanity shift from the love of power to the power of love. This initiation will uproot and transform every aspect of human civilization. It will demand of humankind a new myth, one that insists on cooperation rather than competition, cocreation rather than procreation, networks rather than markets, and sustainability rather than exploitation.
Anodea Judith

Mind-Heart Symbiosis

A global brain with no heart or a global heart with no brain would have little chance of success. When we are able to interconnect and inspire billions of minds and hearts—to care for and to work toward our common good—it will then be possible to truly master the challenges faced by the world today.

Only when generosity is born, a soul is born.
Douglas Hofstadter

We cannot defer the challenges to our children. We cannot pretend they do not exist. We must act now.

References & Credits

This section lists 1,650 citation references, more than 580 image credits, 80 data credits, and 60 software credits. More than 150 scholars provided input on the material presented in the atlas, and their contributions are acknowledged here.

As some spreads have up to 80 references and adding 80 parenthetical references or four-digit numbers to the page layout would considerably hurt readability, the references and credits are not given in the text. Instead, they are listed here by section. References are ordered alphabetically except for those in the **Timeline, pages 26-47,** which are ordered chronologically.

The Web site for the atlas (**http://scimaps.org/atlas**) supports pinpoint citations (that is, references and credits are associated with the specific text they support). In addition, the site will make available EndNote and bibtex files containing all the references.

vi Contents

Image Credits

Group photos courtesy of Katy Börner.

Extracted from: Cosmographia by Ptolemy: Courtesy of the James Ford Bell Library, University of Minnesota, Minneapolis, MN.

Extracted from: Paley, W. Bradford. 2006. TextArc Visualization of "The History of Science." New York, NY. Courtesy of W. Bradford Paley.

Extracted from: Zeller, Daniel. 2007. Hypothetical Model of the Evolution and Structure of Science. New York, NY. Courtesy of Daniel Zeller.

Extracted from: Skupin, André. 2005. In Terms of Geography. New Orleans, LA. In Katy Börner & Deborah MacPherson (Eds.), 1st Iteration (2005): *The Power of Maps, Places & Spaces: Mapping Science* **http://scimaps.org** (accessed May 4, 2009).

Extracted from: Palmer, Fileve, Julie M. Smith, Elisha F. Hardy, and Katy Börner. 2007. "Hands-On Science Maps for Kids." *Places & Spaces: Mapping Science.* **http://scimaps.org/flat/exhibit_info/#Kids** (accessed March 23, 2010).

Extracted from: *Scientific American*. New York, May 27, 1911.

Acknowledgments

References

Alcraft, Rob, and Jon Adams. 1998. *Zoom City.* Oxford: Heinemann Educational Books, Library Division.

Berghaus, Heinrich Karl Wilhelm. 1852. *Physikalischer Atlas: Geographisches Jahrbuch zur Mittheilung aller wichtigern neüen Enforschungen.* Gotha, Germany: Justus Perthes.

Edmonds, William, and Helen Marsden. 1994. *Big Book of Time.* Pleasantville, NY: Reader's Digest Kids.

McCloud, Scott. 1993. *Understanding Comics.* Northampton, MA: Kitchen Sink Press.

McCloud, Scott. 2000. *Reinventing Comics: How Imagination and Technology are Revolutionizing an Art Form.* New York: HarperCollins.

McCloud, Scott. 2006. *Making Comics: Storytelling Secrets of Comics, Manga and Graphic Novels.* New York: HarperCollins.

Schwartz, David M., and Steven Kellogg. 2004. *Millions to Measure.* 3rd ed. New York: HarperCollins.

Tufte, Edward R. 1983. *The Visual Display of Quantitative Information.* Cheshire, CT: Graphics Press.

Tufte, Edward R. 1990. *Envisioning Information.* Cheshire, CT: Graphics Press.

Tufte, Edward R. 1997. *Visual Explanations: Images and Quantities, Evidence and Narrative.* Cheshire, CT: Graphics Press.

Image Credits

Group photos courtesy of Katy Börner.

1 Part 1: Introduction

References

Kurzweil, Ray, and Chris Meyer. 2003. "Understanding the Accelerated Rate of Change." The Cap Gemini Ernst & Young Center for Business Innovation. Published on KurzweilAI.net with permission. **http://www.kurzweilai.net/articles/art0563.html** (accessed May 19, 2008). [Quotation]

Image Credits

Ptolemy (woodcut) from the collection of the James Ford Bell Library, University of Minnesota, Minneapolis, MN.

2 Knowledge Equals Power

Software Credits

Hunter, John. 2010. Matplotlib Home Page. **http://matplotlib.sourceforge.net** (accessed March 15, 2010).

Matplotlib (sponsored by ENThought). 2008. "Cookbook/Matplotlib/Maps." **http://www.scipy.org/Cookbook/Matplotlib/Maps** (accessed March 30, 2010).

Contributors

Elisha F. Hardy, Bryan J. Hook, Peter A. Hook, Luís M. A. Bettencourt, and Stephen M. Uzzo.

Population Growth

References

About, Inc. 2008. "RADAR and Doppler RADAR Invention and History." About.com. **http://inventors.about.com/library/inventors/blradar.htm** (accessed January 1, 2008).

de Solla Price, Derek J. 1965. *Little Science, Big Science.* New York: Columbia University Press.

Encyclopaedia Britannica, Inc. 1991. *The New Encyclopaedia Britannica*, 15th Ed., Phillip W. Goetz (Ed. In Chief). Chicago: Encyclopædia Britannica, Inc., s.v. "Altamira"; "Black Death"; "Cartography"; "Cities"; "Domestication"; "Egypt, Ancient"; "Evolution"; "Greek and Roman Civilization"; "Homo Sapiens"; "Italy"; "Mathematics"; "Papermaking"; "Plague"; "Renaissance"; "Rosetta Stone"; "United States Bill of Rights"; "Universities"; "World War I"; "World War II"; "Writing."

Garfield, Eugene. 1955. "Citation Indexes for Science: A New Dimension in Documentation Through Association of Ideas." *Science* 122: 108–111.

Hawkes, Jacquetta. 1976. *The Atlas of Early Man.* New York: St. Martin's Press.

Headrick, Daniel R. 2000. *When Information Came of Age: Technologies of Knowledge in the Age of Reason and Revolution, 1700–1850.* New York: Oxford University Press.

Kremer, Michael. 1993. "Population Growth and Technological Change: One Million B.C. to 1990." *The Quarterly Journal of Economics* 108, 3: 681–716. **http://www.jstor.org/view/00335533/di976343/97p0125n/0** (accessed October 30, 2007).

Shapiro, Fred R. 1992. "Origins of Bibliometrics, Citation Indexing, and Citation Analysis: The Neglected Legal Literature." *Journal of the American Society for Information Science* 43, 5: 337–339.

United Nations Population Division. 1999. "The World at Six Billion." **http://www.un.org/esa/population/publications/sixbillion/sixbilpart1.pdf** (accessed October 30, 2007).

Weinberg, Bella Hass. 1997. "The Earliest Hebrew Citation Indexes." *Journal of the American Society for Information Science* 48, 4: 318–330.

Wikimedia Foundation. 2009. "University of Al-Karaouine." *Wikipedia, the Free Encyclopedia.* **http://en.wikipedia.org/wiki/University_of_Al-Karaouine** (accessed November 18, 2009).

Williams, Robert V. 2002. "Chronology of Information Science and Technology." **http://www.libsci.sc.edu/bob/istchron/ISCNET/ISCHRON.htm** (accessed October 30, 2007).

Image Credits

Timeline by Russell J. Duhon (data preparation), Elisha F. Hardy (design), Katy Börner (concept), Indiana University.

Data Credits

U.S. Census Bureau. 2007. "Total Midyear Population for the World: 1950–2050." **http://www.census.gov/ipc/www/idb/worldpopinfo.php** (accessed February 10, 2010).

World Population Data, *see* Kremer 1993.

Knowledge and Technology Overload

References

Lesk, Michael. 1997. "How Much Information Is There In the World?" **http://www.lesk.com/mlesk/ksg97/ksg.html** (accessed October 30, 2007).

Lyman, Peter, Hal R. Varian, James Dunn, Aleksey Strygin, and Kirsten Swearingen. 2000. "How Much Information 2000." **http://www.sims.berkeley.edu/how-much-info** (accessed June 10, 2007).

Lyman, Peter, Hal R. Varian, James Dunn, Aleksey Strygin, and Kirsten Swearingen. 2003. "How Much Information 2003." **http://www2.sims.berkeley.edu/research/projects/how-much-info-2003** (accessed October 30, 2007).

Regents of the University of California. 2000. "How Much Information?" **http://www2.sims.berkeley.edu/research/projects/how-much-info/datapowers.html** (accessed April 2, 2010).

TechTarget Corporation. 2010. "What Is: How Many Bytes for …?" **http://searchstorage.techtarget.com/sDefinition/0,,sid5_gci944596,00.html** (accessed March 30, 2010).

Image Credits

How Much Information graph by Russell J. Duhon (data preparation), Elisha F. Hardy (design), Katy Börner (concept), Indiana University.

Us from Above

References

BGP Expert. 2008. "Overview of the Number of IP Addresses …" **http://www.bgpexpert.com/addressespercountry.php** (accessed June 10, 2008).

Boyack, Kevin W., Richard Klavans, W. Bradford Paley, and Katy Börner. 2007. "Mapping, Illuminating, and Interacting with Science." Paper presented at *SIGGRAPH 2007: The 34th International Conference and Exhibition on Computer Graphics and Interactive Techniques* in San Diego, CA, August 5–9.

Lesk, Michael. 1997. "How Much Information Is There in the World?" **http://www.lesk.com/mlesk/ksg97/ksg.html** (accessed October 30, 2007).

Regents of the University of California. 2000. "How Much Information?" **http://www2.sims.berkeley.edu/research/projects/how-much-info/datapowers.html** (accessed April 2, 2010).

Yook, Soon-Hyung, Hawoong Jeong, and Albert-László Barabási. 2002. "Modeling the Internet's Large-Scale Topology." *PNAS* 99, 21: 13382–13386.

Image Credits

2005 World Population, 2003 Scientific Productivity, and 2007 IP Address Ownership by Russell J. Duhon (data preparation), Elisha F. Hardy (design), Katy Börner (concept), Indiana University.

City Lights at Night by Robert Nemiroff and Jerry Bonnell. 2007. "Astronomy Picture of the Day, November 27, 2000." **http://apod.nasa.gov/apod/ap001127.html** (accessed October 30, 2007).

Data Credits

Boyack, Kevin W., John Burgoon, Peter Kennard, Richard Klavans, and W. Bradford Paley. 2006. "Scientific Productivity: Data from Part 4: Illuminated Diagram." GeoMap.

Trustees of Columbia University in the City of New York. 1997–2008. "Socioeconomic Data and Applications Center (SEDAC): Gridded Population of the World and the Global Rural-Urban Mapping Project." Population Grid Data selected, ascii format, year 2005, dowloadable by log-in (file as GL-gpwfe_pcount_05_ascii_quar.zip). http://sedac.ciesin.columbia.edu/gpw/global.jsp (accessed August 17. 2009).

MaxMind, Inc. 2010. "GeoLite Country." http://maxmind.com; http://www.maxmind.com/app/geoip_country (accessed March 15, 2010).

Software Credits

Hunter, John. Matplotlib Basemap (Class) Library. http://matplotlib.sourceforge.net/classdocs.html (accessed October 30, 2007).

The Web of Knowledge

References

Wouters, Paul. 1999. "The Citation Culture." PhD diss., University of Amsterdam. http://garfield.library.upenn.edu/wouters/wouters.pdf (accessed October 30, 2007).

4 The Rise of Science and Technology

The Rise of the Creative Class

References

de Solla Price, Derek J. 1965. *Little Science, Big Science*. New York: Columbia University Press.

Florida, Richard. 2002. *The Rise of the Creative Class: And How It's Transforming Work, Leisure, Community and Everyday Life*. New York: Basic Books.

Menard, Henry W. 1971. *Science: Growth and Change*. Cambridge, MA: Harvard University Press.

Image Credits

People by Russell J. Duhon (data preparation), Elisha F. Hardy (design), Katy Börner (concept), Indiana University.

Data Credits

Today's Percentage of Creative Class, *see* Florida 2002.

U.S. Labor Data, *see* Menard 1971, page 59.

U.S. Census Bureau. 2000. "United States Census 2000." http://www.census.gov/main/www/cen2000.html (accessed June 10, 2008).

National Science Board. 2006. *Science and Engineering Indicators 2006*. 2 vols. Arlington, VA: National Science Foundation (vol. 1, NSB 06-01; vol. 2, NSB 06-01A).

Growth of Science

Books

References

Carter, Susan B., Scott Sigmund Gartner, Michael R. Haines, Alan L. Olmstead, Richard Sutch, and Gavin Wright. 2006. *Historical Statistics of the United States: Millennial Edition*. Online edition. Cambridge: Cambridge University Press. http://hsus.cambridge.org (accessed October 30, 2007).

Internet Archive. 2001. "Million Book Project." http://www.archive.org/details/millionbooks (accessed October 30, 2007).

Lebert, Marie. 2005. "Project Gutenberg, from 1971 to 2005." http://www.etudes-francaises.net/dossiers/gutenberg_eng.htm (accessed October 30, 2007).

Project Gutenberg. 2009. "Main Page." http://www.gutenberg.org/wiki/Main_Page (accessed March 16, 2010).

Image Credits

Books: Russell J. Duhon (data preparation), Elisha F. Hardy (design), Katy Börner (concept), Indiana University.

Data Credits

Bookman data from 1880–2000, *see* Carter et al. 2006.

Project Gutenberg. Search returns (quantity) from 1971–2007, *see* Project Gutenberg, Main Page.

The Million Book Project. "Search returns (quantity) from 1662–2002." *See* http://www.archive.org/about/terms.php (accessed October 30, 2007).

Chemical Abstracts Service. 2007. "CAS Statistical Summary 1907–2006." Columbus, OH: ACS. http://www.cas.org/ASSETS/836E3804111B49BFA28B95BD1B40CD0F/casstats.pdf (accessed October 30, 2007).

OCLC. Worldcat Books. "Search returns (quantity) from 1650–2010." Retrieved September 21–22, 2007 and October 12, 2007. Data taken from OCLC's WorldCat® database, used with OCLC's permission: WorldCat® is a registered trademark of OCLC Online Computer Library Center, Inc. http://www.worldcat.org (accessed October 30, 2007).

Contributors

Michael S. Hart (Project Gutenberg data), Ann Haynes, and Bryan J. Hook.

Papers and Journals

References

de Solla Price, Derek. 1962. *Science Since Babylon*. New Haven: Yale University Press.

Efthimiadis, Efthimios N. 1990. "The Growth of the OPAC Literature." *Journal of the American Society for Information Science* 44, 5: 342–347.

Ginsparg, Paul. 2006. "As We May Read." *The Journal of Neuroscience*, 26, 38: 9606–9608. http://www.jneurosci.org/cgi/content/full/26/38/9606 (accessed October 30, 2007).

Harrison, Chris. 2010. Chris Harrison Home Page. Human-Computer Interaction Institute, Carnegie Mellon University. http://www.chrisharrison.net (accessed March 11, 2010).

Lawrence, Steve, C. Lee Giles, and K. Bollacker. 1999. "Digital Libraries and Autonomous Citation Indexing." *IEEE Computer*, 32, 6: 67–71. http://clgiles.ist.psu.edu/papers/IEEE.Computer.DL-ACI.pdf (accessed October 30, 2007).

May, Robert M. 1997. "The Scientific Wealth of Nations." *Science* 275, 5301: 793–795.

Meadows, Jack. 2000. "The Growth of Journal Literature: A Historical Perspective." *The Web of Knowledge: A Festschrift in Honor of Eugene Garfield*, edited by Helen Barsky Atkins and Blaise Cronin, 87–108. Medford, NJ: Information Today.

Menard, Henry W. 1971. *Science: Growth and Change*. Cambridge, MA: Harvard University Press.

National Federation of Abstracting and Information Services. 1990. "Member Service Statistics 1957–1990." *NFAIS Newsletter* 32, June.

Image Credits

de Solla Price, Derek. 1962. *Science Since Babylon*. New Haven: Yale University Press. Used by permission of Yale University Press, all rights reserved.

Papers and Wikipedia Entries by Russell J. Duhon (data preparation), Elisha F. Hardy (design), Katy Börner (concept), Indiana University.

Growth of Scientific Records, Journals, *see* Menard 1971.

Data Credits

Science growth rates data, *see* Menard 1971.

Ley, Michael. 2008. "The DBLP Computer Science Bibliography." Universität Trier. http://www.informatik.uni-trier.de/~ley/db (accessed June 10, 2008).

NFAIS abstract data set, 1957–1990, *see* National Federation of Abstracting and Information Services 1990.

Web of Science going back to 1981 provided by Henry G. Small.

Chris Harrison (preparer), Brian Amento and Mike Yang (compilers at AT&T Labs). "Royal Society Data: Visualizing the Royal Society Archive." 1665–2005. *Philosophical Transactions of the Royal Society A* (1665–2005); *Philosophical Transactions of the Royal Society B* (1887–2005); *Proceedings of the Royal Society A* (1800–2005); *Proceedings of the Royal Society B* (1905–2005). http://www.chrisharrison.net/projects/royalsociety; data sets available at: http://www.chrisharrison.net/projects/royalsociety/AuthorDistribution.zip; http://www.chrisharrison.net/projects/royalsociety/WordDistribution.zip; http://www.chrisharrison.net/projects/royalsociety/CombinedDistribution.zip. (accessed March 30, 2010.)

CiteSeer data. Produced with data from http://dblp.uni-trier.de (accessed October 30, 2007).

Physical Review Data, 1887–2007.

Chemical Abstracts Service. 2007. "CAS Statistical Summary 1907–2006." Columbus, OH: ACS. http://www.cas.org/ASSETS/836E3804111B49BFA28B95BD1B40CD0F/casstats.pdf (accessed October 30, 2007).

JSTOR data set. 2007. "Search returns (quantity) for journal publications by date from 1665–2007." http://fsearch-sandbox.jstor.org/search (accessed October 30, 2007).

MEDLINE data set. "Statistical reports on MEDLINE®/PubMed® baseline data [Internet]. Bethesda (MD): National Library of Medicine (US), Bibliographic Services Division." Available from: http://www.nlm.nih.gov/bsd/licensee/baselinestats.html (accessed October 30, 2007).

OCLC. "Worldcat Papers: search returns (quantity) from 1650-2010." Retrieved September 21–22, 2007 and October 12, 2007. Data taken from OCLC's WorldCat® database, used with OCLC's permission: WorldCat® is a registered trademark of OCLC Online Computer Library Center, Inc. http://www.worldcat.org (accessed October 30, 2007).

Scopus Coverage, *see* Scopus. 2007. *Content Coverage*. http://www.info.scopus.com/docs/content_coverage.pdf (accessed October 30, 2007).

Scopus data. "Search returns (quantity) for Scopus from January 2001 to December 2005." http://www.scopus.com/scopus/home.url (accessed October 30, 2007).

WoS coverage, *see* Thomson Reuters. 2004. Web of Science 7.0: Science Citation Index Expanded, Social Sciences Citation Index, Arts and Humanities Citation Index. http://www.science.thomsonreuters.com/mjl/wos_ahci_a5020_final.pdf (accessed March 31, 2010).

Wikimedia Foundation. 2007. Article edit data from January 1, 2007 to April 6, 2007.

Contributors

Chris Harrison prepared Royal Society data; Jill ONeill, NFAIS data; Henry G. Small, Web of Science data; Helen de H. J. Mooji and Jaco J. Zijlstra, Scopus data; Jane L. Rosov, MEDLINE data; Michael Krot, John Burns, and Ronald Snyder, JSTOR data; C. Lee Giles, CiteSeer data; Russell J. Duhon, Bryan J. Hook, and Lokman Meho, Books and E-mail data; and Eugene Garfield suggested books and data sources.

Patents

References

eMedia Asia Ltd. 2007. "Bird's Eye View of Patent Playground." *EE Times Asia*. http://www.eetasia.com/ART_8800489316_480100_NT_ceb53150.HTM (accessed March 30, 2010).

Schippel, Helmut. 2001. "Die Anfänge des Erfinderschutzes in Venedig." In Lindgren, Uta (Hrsg). Europäische Technik im Mittelalter. 800 bis 1400. *Tradition und Innovation* 4. Auflage, Berlin: S.539-550.

World Intellectual Property Organization. 2007. "WIPO Patent Report: Statistics on Worldwide Patent Activity." http://www.wipo.int/ipstats/en/statistics/patents/pub_archives/patent_report_2007.html (accessed March 30, 2007).

Contributors

Colin Webb, OECD; and Masatsura Igami, NISTEP.

Image Credits

Patents by Russell J. Duhon (data preparation), Elisha F. Hardy (design), Katy Börner (concept), Indiana University.

Data Credits

World Intellectual Property Organization. 2008. "Patent Applications by Office (1883 to 2006)." *The WIPO Patent Report, 2007 Edition*. http://www.wipo.int/freepublications/en/patents/931/wipo_pub_931.pdf (accessed March 30, 2010).

Top seven companies, *see* eMedia Asia Ltd. 2007.

Investing in the Future

References

HM Treasury. 2004. *Science & Innovation Investment Framework 2004–2014*. http://www.hm-treasury.gov.uk/spending_sr04_science.htm (accessed 1/25/2010).

Menard, Henry W. 1971. *Science: Growth and Change*. Cambridge, MA: Harvard University Press.

Radford, Giles. 2006. "Knowledge Optimization in Research Funding." *Presentation at the Knowledge in Service to Health: Leveraging Knowledge for Modern Science Management Symposium*. Bethesda, MD, February 6. http://grants.nih.gov/grants/KM/OERRM/OER_KM_events (accessed October 30, 2007).

Image Credits

U.S. R&D Expenditures graph by Russell J. Duhon (data preparation), Elisha F. Hardy (design), Katy Börner (concept), Indiana University.

Data Credits

National Science Board. 2006. *Science and Engineering Indicators 2006*. 2 vols. Arlington, VA: National Science Foundation (vol. 1, NSB 06-01; vol. 2, NSB 06-01A).

Welcome Trust data, *see* Radford 2006.

R&D spending data, *see* HM Treasury 2004.

Contributors

Matthew Probus and Mike Pollard, DiscoveryLogic.

Science and Society in Equilibrium

References

de Solla Price, Derek J. 1965. *Little Science, Big Science*. New York: Columbia University Press.

Martino, Joseph P. 1969. "Science and Society in Equilibrium." *Science* 165, 3895: 769–772.

United States Central Intelligence Agency. 2007. "The World Factbook: United States." https://www.cia.gov/library/publications/the-world-factbook/geos/us.html (accessed March 30, 2010).

Image Credits

Martino, Joseph P. 1969. "Science and Society in Equilibrium." *Science* 165, 3895: 769–772, fig. 1. Reprinted with permission from AAAS.

Data Credits

U.S. Population and GDP, *see* United States Central Intelligence Agency 2007.

6 Addictive Intelligence Amplifiers

Shrinking Planet

References

About, Inc. 2008. "RADAR and Doppler RADAR Invention and History." About.com. http://inventors.about.com/library/inventors/blradar.htm (accessed January 1, 2008).

Carter, Susan B., Scott Sigmund Gartner, Michael R. Haines, Alan L. Olmstead, Richard Sutch, and Gavin Wright. 2006. *Historical Statistics of the United States: Millennial Edition*. Online edition. Cambridge: Cambridge University Press. http://hsus.cambridge.org (accessed October 30, 2007).

Lyman, Peter, Hal R. Varian, James Dunn, Aleksey Strygin, and Kirsten Swearingen. 2000. "How Much Information." http://www.sims.berkeley.edu/how-much-info (accessed June 10, 2007).

Pingdom AB. 2008. Pingdom Blog: *Internet 2008 in Numbers*. http://royal.pingdom.com/2009/01/22/internet-2008-in-numbers (accessed January 25, 2010).

Encyclopaedia Britannica, Inc. 1991. *The New Encyclopaedia Britannica*, 15th Ed., Phillip W. Goetz (Ed. In Chief). Chicago: Encyclopædia Britannica, Inc., s.v. "Facsimile"; "Telecommunications"; "Transportation"; "Writing."

Image Credits

"Shrinking of Our Planet…" courtesy of the estate of R. Buckminster Fuller.

"U.S. Transportation and Communications" graphs by Russell J. Duhon (data preparation), Elisha F. Hardy (design), Katy Börner (concept), Indiana University.

Data Credits

Carter et al. 2006.

Internet Systems Consortium, Inc. 2007. "Internet Hosts from 1981 to 2007." http://www.isc.org/index.pl?/ops/ds/host-count-history.php (accessed October 30, 2007).

Contributors

Bonnie DeVarco suggested the Fuller Map.

From Little Boxes to Big Boxes

References

Wellman, Barry. 2002. "Little Boxes, Glocalization, and Networked Individualism." In *Digital Cities II: Computational and Sociological Approaches. Lecture Notes in Computer Science*, edited by Makoto Tanabe, Peter van den Besselaar, and Toru Ishida, 10. Berlin: Springer-Verlag. [Quotation]

Accelerating the Rate of Change

References

Burke, James. 1978. *Connections*. Boston: Little, Brown and Company.

Kurzweil, Ray. 1992. *The Age of Intelligent Machines*. Cambridge, MA: MIT Press.

Kurzweil, Ray. 1998. *The Age of Spiritual Machines: When Computers Exceed Human Intelligence*. New York: The Penguin Group.

Kurzweil, Ray. 2005. *The Singularity is Near: When Humans Transcend Biology*. New York: The Penguin Group.

On-the-Fly Assembly

References

Gardner, Howard. 2007. *Five Minds for the Future*. Boston: Harvard Business School Press.

Goleman, Daniel. 1997. *Emotional Intelligence: Why It Can Matter More Than IQ*. New York: Bantam.

Goleman, Daniel. 2006. *Social Intelligence: The New Science of Human Relationships*. New York: Bantam.

Urban Species

References

Whitehouse, David. 2005. "Half of Humanity Set to Go Urban." BBC News, May 19. http://news.bbc.co.uk/2/hi/science/nature/4561183.stm (accessed August 12, 2009).

Zipf, George K. 1949. *Human Behavior and the Principle of Least Effort*. Cambridge, MA: Addison-Wesley.

Image Credits

See Zipf 1949.

Natural-Born Cyborgs

References

Clark, Andy. 2003. "Natural Born Cyborgs?" In *The New Humanists*, edited by John Brockman, 70. New York: Barnes and Noble.

Moths to the Flame

References

Csikszentmihalyi, Mihaly. 1991. *Flow: The Psychology of Optimal Experience*. New York: HarperCollins.

Rawlins, Gregory J. E. 1997. *Moths to the Flame: The Seductions of Computer Technology*, 87. Cambridge, MA: MIT Press.

Rawlins, Gregory J. E. 1997. *Slaves of the Machine: The Quickening of Computer Technology*. Cambridge, MA: MIT Press. [Quotation, pp. 125-126]

Rheingold, Howard. 2000. *Tools for Thought: The History and Future of Mind-Expanding Technology*. Cambridge, MA: MIT Press.

Man vs. Machine

References

Krauss, Lawrence M., and Glenn D. Starkman. 2004. "Universal Limits on Computation." http://arxiv.org/PS_cache/astro-ph/pdf/0404/0404510v2.pdf (accessed October 30, 2007).

Levitin, Lev B., and Tommaso Toffoli. 2009. "Fundamental Limit on the Rate of Quantum Dynamics: The Unified Bound is Tight." *Physical Review Letters* 103, 16: 160502-160506.

Licklider, Joseph Carl Robnett. 1960. "Man-Computer Symbiosis." *IRE Transactions on Human Factors in Electronics* HFE-1: 4–11. http://groups.csail.mit.edu/medg/people/psz/Licklider.html (accessed October 30, 2007).

Lloyd, S. 2000. "Ultimate Physical Limits to Computation." *Nature* 406: 1047–1054.

Moravec, Hans. 1990. *Mind Children: The Future of Robot and Human Intelligence*. Cambridge, MA: Harvard University Press.

Moravec, Hans. 1998. "When Will Computer Hardware Match the Human Brain?" *Journal of Evolution and Technology* 1. http://www.transhumanist.com/volume1/moravec.htm (accessed October 30, 2007).

Moravec, Hans. 2000. *Robot: Mere Machine to Transcendent Mind*, 70. New York: Oxford University Press.

8 Knowledge Needs and Desires

References

Herbert Simon quoted in Davenport, Thomas H., and John C. Beck. 2001. *The Attention Economy: Understanding the New Currency of Business*, 11. Cambridge, MA: Harvard Business Press. [Quotation, p. 11]

Data Providers and Librarians

References

Börner, Katy, and Chaomei Chen. 2002. "Visual Interfaces to Digital Libraries: Motivation, Utilization and Socio-Technical Challenges." In *Lecture Notes in Computer Science: Visual Interfaces to Digital Libraries*, edited by Katy Börner and Chaomei Chen, vol. 2539: 1–9. Berlin: Springer-Verlag.

Chen, Chaomei, and Katy Börner. 2005. "The Spatial-Semantic Impact of a Collaborative Information Virtual Environment on Group Dynamics." Special Issue (Collaborative Information Visualization Environments), *Presence: Teleoperators and Virtual Environments* 14, 1: 81–103. Cambridge, MA: MIT Press.

Szigeti, Helen. 2001. "The ISI Web of Knowledge Platform: Current and Future Directions." Published online by ISI, Inc., Philadelphia, PA.

Thomson Reuters. 2004. "Web of Science 7.0: Science Citation Index Expanded, Social Sciences Citation Index, Arts and Humanities Citation Index." http://www.science.thomsonreuters.com/mjl/wos_ahci_a5020_final.pdf (accessed March 31, 2010).

Image Credits

Collaborative Visual Interfaces to Digital Libraries, *see* Börner and Chen 2002.

Data Credits

WoS coverage, *see* Thomson Reuters 2004.

Industry

References

Kutz, Daniel. 2004. "Examining the Evolution and Distribution of Patent Classifications." *Proceedings of the 8th International Conference on Information Visualisation (IV'04)*, edited by Ebad Banissi, 983–988. Los Alamitos, CA: IEEE Computer Society.

Image Credits

Claiming Intellectual Property Rights via Patents courtesy of Daniel Kutz.

Science Policy Makers

References

Boyack, Kevin W., Katy Börner, and Richard Klavans. 2007. "Mapping the Structure and Evolution of Chemistry Research." *Proceedings of the 11th International Conference of the International Society for Scientometrics and Informetrics,* edited by Daniel Torres-Salinas and Henk F. Moed, 112–123. Madrid: CSIC.

Radford, Giles. 2006. "Knowledge Optimization in Research Funding." *Presentation at the Knowledge in Service to Health: Leveraging Knowledge for Modern Science Management Symposium.* Bethesda, MD, February 6. http://grants.nih.gov/grants/KM/OERRM/OER_KM_events (accessed October 30, 2007).

Image Credits

Funding Profiles of NIH and NSF by Russell J. Duhon (data preparation), Elisha F. Hardy (design), Katy Börner (concept), Indiana University; the base map of science created by, and used with the permission of, Kevin W. Boyack and Richard Klavans, SciTech Strategies, http://www.mapofscience.com (accessed June 10, 2008).

Data Credits

Welcome Trust data, *see* Radford 2006.

Researchers

References

Ke, Weimao, Katy Börner, and Lalitha Viswanath. 2004. "Analysis and Visualization of the IV 2004 Contest Dataset." IEEE Information Visualization Conference 2004 Poster Compendium, 49–50. http://conferences.computer.org/InfoVis/files/compendium2004.pdf and http://iv.slis.indiana.edu/ref/iv04contest/Ke-Borner-Viswanath.gif (animated version) (accessed October 30, 2007).

Shneiderman, Ben. 1996. "The Eyes Have It: A Task by Data Type Taxonomy for Information Visualizations." *Proceedings of the IEEE Symposium on Visual Languages*, 336–343. Washington, DC: IEEE Computer Society.

Image Credits

Mapping the Evolution of Coauthorship Networks, *see* Ke et al. 2004.

Children

References

Börner, Katy, Fileve Palmer, Julie M. Davis, Elisha F. Hardy, Stephen M. Uzzo, and Bryan J. Hook. 2009. "Teaching Children the Structure of Science." *Proceedings of SPIE: Conference on Visualization and Data Analysis*, 7243, 1: 1-14. Bellingham, WA: SPIE.

Image Credits

"Hands-On Science Maps for Kids," *see* Börner et al. 2009.

Contributors

Fileve Palmer, Katy Börner, Elisha F. Hardy, Julie Smith, Stephen Miles Uzzo (director of technology, New York Hall of Science), and Michael Lane (director of exhibit services, New York Hall of Science).

Society

Image Credits

Places & Spaces: Mapping Science exhibit image courtesy of Katy Börner.

10 The Power of Maps

References

Brown, Lloyd A. 1980. *The Story of Maps*. Mineola, NY: Dover Publications.

Harmon, Katharine. 2004. *You Are Here: Personal Geographies and Other Maps of the Imagination.* New York: Princeton Architectural Press. [Quotation, Introduction, p. 10]

Winchester, Simon. 2002. *The Map that Changed the World: William Smith and the Birth of Modern Geology*. New York: Harper Perennial.

The Power of Stories

References

Belloc, Hilaire. 1913. *The Book of the Bayeux Tapestry: Presenting the Complete Work in a Series of Colour Facsimiles*. London: Chatto & Windus.

Rud, Mogens. 1992. *The Bayeux Tapestry and the Battle of Hastings 1066*. Copenhagen: Christian Eilers Publishers.

Wilson, David M. 1985. *The Bayeux Tapestry*. London: Thames and Hudson.

Image Credits

Bayeux Tapestry, *see* Belloc 1913.

Early Mapmaking

References

Bunbury, E. H. 1883. *A History of Ancient Geography Among the Greeks and Romans from the Earliest Ages Till the Fall of the Roman Empire*, 148–149, fig. 11. London: John Murray.

Harmon, Katharine. 2004. *You Are Here: Personal Geographies and Other Maps of the Imagination*, 10. New York: Princeton Architectural Press.

Miller, Konrad. 1887–1888. "Die Weltkarte des Castorius genannt die Peutingersche Tafel." In *Den Farben des Originals* herausgegeben und eingeleitet von Konrad Miller. Ravensburg, Germany: Maier.

Image Credits

Hecataeus map, *see* Bunbury 1883.

Peutinger map, *see* Miller 1887–1888.

Toward a Geographic Reference System

References

Crone, G. R. 1953. *Maps and Their Makers*. London: Hutchinson University Library.

Skelton, R. A. 1963. "Bibliographical Note." Reproduction of Claudius Ptolemaus, *Cosmographia*, Ulm, 1482. Amsterdam: Theatrum Orbis Terrarum.

Thomson, J. O. 1948. *History of Ancient Geography*. Cambridge: Cambridge University Press.

Turnbull, David. 1989. *Maps Are Territories: Science is an Atlas*. Geelong, Australia: Deakin University Press.

Wood, Denis. 1992. *The Power of Maps*. New York: The Guilford Press.

Scientific Mapmaking

References

Cassini, Giovanni Domenico. 1668. *Ephemerides Bononienses Mediceorum syderum/ex hypothesibus, et tabulis Io*. Typis Emilij Mariæ & Fratrum de Manolessijs.

Headrick, Daniel R. 2000. *When Information Came of Age: Technologies of Knowledge in the Age of Reason and Revolution, 1700–1850*. New York: Oxford University Press.

Hook, Peter A. 2007. "Domain Maps: Purposes, History, Parallels with Cartography, and Applications." *Proceedings of the 11th International Conference on Information Visualisation (IV'07)*, 442-446. Zürich, Switzerland, July 4-6.

Morrison, Philip, and Phylis Morrison. 1987. *The Ring of Truth*, 127. New York: Random House.

Walker, James Thomas, Charles Thomas Haig, James Palladio Basevi, Willam James Heaviside, William Maxwell Cambell, and Sir Sidney Gerald Burrard. 1870–1910. *Account of the Operations of the Great Trigonometrical Survey of India*. Dehra Dun, India: Office of the Trigonometrical Branch, Survey of India.

Wikimedia Foundation. 2009. "World Geodetic System." *Wikipedia, the Free Encyclopedia*. http://en.wikipedia.org/wiki/World_Geodetic_System (accessed November 16, 2009).

Image Credits

Cassini's world map image courtesy of the Houghton Library, Harvard College Library. IC6.C2735.668c.

Mapping the Highest Mountain, *see* Walker et al. 1870–1910.

Thematic Map Making, Statistical Graphics, and Data Visualization

References

Garland, K. 1994. *Mr. Beck's Underground Map*, 20. Middlesex, UK: Capitol Transport Publishing.

Koch, Tom. 2005. *Cartographies of Disease: Maps, Mapping, and Medicine*. Redlands, CA: ESRI Press.

Stamp, L. Dudley. 1964. *A Geography of Life and Death*. Ithaca, NY: Cornell University Press.

Data Charts and Network Visualizations

References

Bernal, John D. 1939. *The Social Function of Science*. London: Routledge & Kegan Ltd.

Bertin, J. 1981. *Graphics and Graphic Information Processing*. New York: Walter de Gruyter.

Burkhard, Remo. 2005. "Knowledge Visualization—The Use of Complementary Visual Representations for the Transfer of Knowledge. A Model, a Framework, and Four New Approaches." PhD diss., Swiss Federal Institute of Technology (ETH).

Card, Stuart, Jock Mackinlay, and Ben Shneiderman, eds. 1999. *Readings in Information Visualization: Using Vision to Think*. San Francisco: Morgan Kaufmann.

Dodge, Martin, and Rob Kitchin. 2002. *Atlas of Cyberspace*. Upper Saddle River, NJ: Pearson Education.

Eppler, Martin J., and Remo Burkhard. 2005. "Knowledge Visualization." In *Encyclopedia of Knowledge Management*, edited by David E. Schwartz. Hershey, PA: Idea Press.

Friendly, Michael, and Daniel J. Denis. 2007. "Milestones in the History of Thematic Cartography, Statistical Graphics, and Data Visualization: An Illustrated Chronology of Innovations." http://www.math.yorku.ca/SCS/Gallery/milestone (accessed October 30, 2007).

Holmes, Nigel. 1993. *The Best in Diagrammatic Graphics*. London: Quarto Publishing.

Kahn, Paul, and Krzysztof Lenk. 2001. *Mapping Websites: Digital Media Design*. Beverly, MA: Rockport Publishers.

Moreno, Jacob L. 1933. "Emotions Mapped by New Geography." *New York Times*, April 3: 17.

Moreno, Jacob L. 1934. *Who Shall Survive?* Washington, DC: Nervous and Mental Disease Publishing Company.

Playfair, William. 1786. *The Commercial and Political Atlas: Representing, by Means of Stained Copper-Plate Charts, the Progress of the Commerce, Revenues, Expenditure and Debts of England during the Whole of the Eighteenth Century*. London: Debrett, Robinson, and Sewell.

Playfair, William. 2005. *The Commercial and Political Atlas and Statistical Breviary, Introduction by Howard Wainer and Ian Spence.* Cambridge: Cambridge University Press.

Slocum, Terry A. 1998. *Thematic Cartography and Visualization*. Upper Saddle River, NJ: Prentice-Hall.

Tufte, Edward R. 1990. *Envisioning Information*. Cheshire, CT: Graphics Press.

Tufte, Edward R. 1997. *Visual Explanations: Images and Quantities, Evidence and Narrative*. Cheshire, CT: Graphics Press.

Tufte, Edward R. 2001. *The Visual Display of Quantitative Information*. 2nd ed. Cheshire, CT: Graphics Press.

Tufte, Edward R. 2006. *Beautiful Evidence*. Cheshire, CT: Graphics Press.

Wainer, Howard. 1997. *Visual Revelations: Graphical Tales of Fate and Deception from Napoleon Bonaparte to Ross Perot*. Mahwah, NJ: Lawrence Erlbaum.

Image Credits

Map of Science, *see* Bernal 1939.

12 Science Maps and Their Makers

References

Atkins, Daniel E., Kelvin K. Droegemeier, Stuart I. Feldman, Hector Garcia-Molina, Michael L. Klein, David G. Messerschmitt, Paul Messina, Jeremiah P. Ostriker, and Margaret H. Wright. 2003. "Revolutionizing Science and Engineering Through Cyberinfrastructure: Report of the National Science Foundation Blue-Ribbon Advisory Panel on Cyberinfrastructure." National Science Foundation. http://www.nsf.gov/od/oci/reports/atkins.pdf (accessed March 25, 2008).

de Solla Price, Derek J. 1965. *Little Science, Big Science*. New York: Columbia University Press.

Emmott, Stephen, Stuart Rison, Microsoft et al. 2005. "Towards 2020 Science." http://research.microsoft.com/towards2020science (accessed October 30, 2007).

Kimerling, A. Jon, Phillip C. Muehrcke, and Juliana O. Muehrcke. 2005. *Map Use: Reading, Analysis, Interpretation*, 5th ed., Madison, WI: JP Publications. [Quotation, p. 520]

Contributions

Computational scientometrics is a term coined by C. Lee Giles.

Scientograph term by George Vladutz, *see* Garfield 1986.

Mapping Science

References

Börner, Katy, Chaomei Chen, and Kevin W. Boyack. 2003. "Visualizing Knowledge Domains." In *Annual Review of Information Science and Technology*, edited by Blaise Cronin, vol. 37: 179–255. Medford, NJ: Information Today/American Society for Information Science and Technology.

Callon, Michel, John Law, and Arie Rip, eds. 1986. *Mapping the Dynamics of Science and Technology: Sociology of Science in the Real World*. London: The Macmillan Press.

Chen, Chaomei. 2002. *Mapping Scientific Frontiers*. London: Springer-Verlag.

Garfield, Eugene. 1986. *Essays of an Information Scientist: Towards Scientography*, vol. 9. Philadelphia: ISI Press.

Garfield, Eugene. 1994. "Scientography: Mapping the Tracks of Science." Reprinted from *Current Contents: Social and Behavioural Sciences* 7: 5–10.

Narin, Francis, and Joy K. Moll. 1977. "Bibliometrics." In *Annual Review of Information Science and Technology*, edited by M. E. Williams, 12: 35–58. White Plains, NY: Knowledge Industry Publications.

Shiffrin, Richard M., and Katy Börner. 2004. "Mapping Knowledge Domains." *PNAS* 101, Suppl. 1: 5183–5185.

Skupin, André. 2002. "On Geometry and Transformation in Map-like Information Visualization." *Visual Interfaces to Digital Libraries: Lecture Notes in Computer Science*, edited by Katy Börner and Chaomei Chen, vol. 2539: 161–170. Berlin: Springer-Verlag.

Skupin, André. 2004. "The World of Geography: Visualizing a Knowledge Domain with Cartographic Means." *PNAS* 101, Suppl. 1: 5274–5278.

Skupin, André, and Sara Irina Fabrikant. 2003. "Spatialization Methods: A Cartographic Research Agenda for Non-Geographic Information Visualization." *Cartography and Geographic Information Science* 30, 2: 99–119.

White, Howard D. 2001. "Author-Centered Bibliometrics through CAMEOs: Characterizations Automatically Made and Edited Online." *Scientometrics* 51, 3: 607–637.

White, Howard D., and Katherine W. McCain. 1997. "Visualization of Literatures." In *Annual Review of Information Science and Technology*, vol. 32: 99–168.

Wilson, Concepción S. 1999. "Informetrics." *Annual Review of Information Science and Technology*, vol. 34: 107–247.

The Utility of a Science Reference System

No references or credits.

Toward A Reference System for Science

References

Boyack, Kevin W., Richard Klavans, and Katy Börner. 2005. "Mapping the Backbone of Science." *Scientometrics* 64, 3: 351–374.

Boyack, Kevin W., Katy Börner, and Richard Klavans. 2009. "Mapping the Structure and Evolution of Chemistry Research." *Scientometrics* 79, 1: 45–60.

Braam, Robert R., Henk F. Moed, and Anthony F. J. van Raan. 1991. "Mapping of Science by Combined Co-Citation and Word Analysis I: Structural Aspects"; "Mapping of Science by Combined Co-Citation and Word Analysis II: Dynamical Aspects." *Journal of the American Society for Information Science* 42, 4: 233–251, 252–266.

Davidson, George S., Brian N. Wylie, and Kevin W. Boyack. 2001. "Cluster Stability and the Use of Noise in Interpretation of Clustering." *Proceedings of IEEE Information Visualization (IV'01)*, 23–30. Los Alamitos, CA: IEEE.

Fabrikant, Sara Irina, Daniel R. Montello, and David M. Mark. 2006. "The Distance-Similarity Metaphor in Region-Display Spatializations." *Special Issue on Exploring Geovisualization: IEEE Computer Graphics & Application*, edited by T. M. Rhyne, A. MacEachren, and J. Dykes, vol. 26, 4: 34–44. Los Alamitos, CA: IEEE Computer Society Press.

Fabrikant, Sara Irina, Daniel R. Montello, Marco Ruocco, and Richard S. Middleton. 2004. "The Distance-Similarity Metaphor in Network-Display Spatializations." *Cartography and Geographic Information Science* 31, 4: 237–252.

Kessler, Michael M. 1963. "Bibliographic Coupling Between Scientific Papers." *American Documentation* 14, 1: 10–25.

Klavans, Richard, and Kevin W. Boyack. 2006. "Identifying a Better Measure of Relatedness for Mapping Science." *Journal of the American Society for Information Science and Technology* 57, 2: 251–263.

Klavans, Richard, and Kevin W. Boyack. 2006. "Quantitative Evaluation of Large Maps of Science." *Scientometrics* 68, 3: 475–499.

Klavans, Richard, and Kevin W. Boyack. 2007. "Is There a Convergent Structure to Science?" *Proceedings of the 11th International Conference of the International Society for Scientometrics and Informetrics*, edited by Daniel Torres-Salinas and Henk F. Moed, 437–448. Madrid: CSIC.

Klavans, Richard, and Kevin W. Boyack. 2008. "Thought Leadership: A New Indicator for National and Institutional Comparison." *Scientometrics* 75, 2: 239–250.

Klavans, Richard, and Kevin W. Boyack. 2009. "Toward a Consensus Map of Science." *Journal of the American Society for Information Science and Technology* 60, 3: 455–476.

Leydesdorff, Loet. 1987. "Various Methods for the Mapping of Science." *Scientometrics* 11: 291–320.

Leydesdorff, Loet. 2007. "Mapping Interdisciplinarity at the Interfaces between the Science Citation Index and the Social Sciences Citation Index." *Scientometrics* 71, 3: 391–405.

Marshakova, Irina V. 1973. "A System of Document Connections Based on References." *Scientific and Technical Information Serial of VINITI* 6: 3–8.

Small, Henry. 1973. "Co-citation in Scientific Literature: A New Measure of the Relationship Between Publications." *Journal of the American Society for Information Science* 24: 265–269.

Image Credits

Backbone of Science, *see* Boyack et al. 2005.

2002 Base Map, *see* Boyack et al. 2009.

Paradigm Map, *see* Klavans and Boyack 2008.

UCSD Map of Science courtesy of Richard Klavans and Kevin W. Boyack, SciTech Strategies, Inc. http://www.mapofscience.com (accessed June 10, 2008).

Mapmakers of Science

No references or credits.

15 Part 2: The History of Science Maps

References

Santayana, George. 1905. *The Life of Reason*. New York: C. Scribner's Sons. [Quotation, p. 284]

Image Credits

Paley, W. Bradford. 2006. TextArc Visualization of "The History of Science." New York, NY. Courtesy of W. Bradford Paley.

16 Visionary Approaches

Image Credits

Visionaries Timeline by Elisha F. Hardy (design), Katy Börner (concept), Indiana University.

18 Knowledge Collection

References

Amazon.com. 2008. "Kindle." http://amazon.com/kindle (accessed November 1, 2009).

Giles, C. Lee, Kurt D. Bollacker, and Steve Lawrence. 1998. "CiteSeer: An Automatic Citation Indexing System." *Proceedings of the 3rd ACM Conference on Digital Libraries (DL'98)*, edited by Ian Witten, Rob Akscyn, and Frank M. Shipman III, 89–98. New York: ACM Press. http://clgiles.ist.psu.edu/papers/DL-1998-citeseer.pdf (accessed October 30, 2007).

Ginsparg, Paul. 2006. "As We May Read." *The Journal of Neuroscience*, 26, 38: 9606–9608. http://www.jneurosci.org/cgi/content/full/26/38/9606 (accessed October 30, 2007).

Lawrence, Steve, C. Lee Giles, and K. Bollacker. 1999. "Digital Libraries and Autonomous Citation Indexing." *IEEE Computer*, 32, 6: 67–71. http://clgiles.ist.psu.edu/papers/IEEE.Computer.DL-ACI.pdf (accessed October 30, 2007).

Wright, Alex. 2007. *Glut: Mastering Information Through The Ages*. Ithaca, NY and London: Cornell University Press.

Michael S. Hart (b. 1947): Project Gutenberg, 1971

References

Ibiblio.org. 2008. "Notes from the Lab." http://www.ibiblio.org/ibiblog (accessed March 12, 2010).

Lebert, Marie. 2005. "Project Gutenberg, from 1971 to 2005." http://www.etudes-francaises.net/dossiers/gutenberg_eng.htm (accessed October 30, 2007).

Project Gutenberg. 2010. "Project Gutenberg News." http://www.gutenbergnews.org (accessed January 25, 2010).

Thomas, Jeffrey. 2007. "Project Gutenberg Digital Library Seeks to Spur Legacy." Bureau of International Information Programs, U.S. Department of State: America.gov. http://www.america.gov/st/washfile-english/2007/July/200707201511311CJsamohT0.6146356.html (accessed November 12, 2009). [Quotation, Michael Hart, interview]

Image Credits

Hart portrait courtesy of Michael Hart.

Contributors

Michael Hart.

Brewster Kahle (b. 1960): Internet Archive, 1996

References

Internet Archive. 2001. "Bios." http://www.archive.org/about/bios.php (accessed October 30, 2007).

Internet Archive. 2001. "Wayback Machine." http://www.archive.org/web/web.php (accessed October 30, 2007).

Kahle, Brewster. 2006. Keynote Address. Presented at *Wikimania 2006*. Cambridge, MA, August 4.

Kahle, Brewster. 2007. "Universal Access to All Knowledge Is Within Our Grasp." *Presentation at Microsoft Faculty Summit 2007*. Redmond, WA, July 16. Previously available at http://research.microsoft.com/workshops/FS2007/presentations/Kahle_Brewster_Faculty_Summit_071607.ppt (accessed January 1, 2008). [Quotation]

Image Credits

Kahle portrait is an unattributed public domain image.

Stevan Harnad (b. 1945): American Scientist Open Access Forum, 1998

References

Berners-Lee, Tim, Dave De Roure, Stevan Harnad, and Nigel Shadbolt. 2005. "Journal Publishing and Author Self-Archiving: Peaceful Co-Existence and Fruitful Collaboration." Unpublished manuscript. http://eprints.ecs.soton.ac.uk/11160 (accessed November 30, 2007).

Brody, Tim, Simon Kampa, Stevan Harnad, Les Carr, and Steve Hitchcock. 2003. "Digitometric Services for Open Archives Environments." In *Research and Advanced Technology for Digital Libraries (ECDL'03)*, edited by T. Koch and I. Sølvberg, 207–220. Berlin: Springer-Verlag. http://eprints.ecs.soton.ac.uk/7503 (accessed October 30, 2007).

Harnad, Stevan. 1990. "Scholarly Skywriting and the Prepublication Continuum of Scientific Inquiry." *Psychological Science* 1: 342–343. Reprinted in *Current Contents* 45, November 11, 1991: 9–13. http://cogprints.org/1581 (accessed November 30, 2007).

Harnad, Stevan. 2003. "Online Archives for Peer-Reviewed Journal Publications." In *International Encyclopedia of Library and Information Science*, edited by John Feather and Paul Sturges. Andover, UK: Routledge. http://users.ecs.soton.ac.uk/harnad/Temp/archives.htm (accessed February 16, 2010). [Quotation]

Harnad, Stevan. 2008. "Harnad E-Print Archives and Psycoloquy and BBS Journal Archives." http://users.ecs.soton.ac.uk/harnad (accessed October 30, 2007).

Harnad, Stevan, and L. Carr. 2000. "Integrating, Navigating and Analyzing Eprint Archives Through Open Citation Linking (the OpCit Project)." Special Issue (Scientometrics: Tribute to Eugene Garfield), *Current Science* 79, 5: 629–638. http://www.iisc.ernet.in/~currsci/sep102000/629.pdf (accessed November 30, 2007).

Lawrence, Steve. 2001. "Online or Invisible?" http://citeseer.ist.psu.edu/online-nature01 (accessed November 30, 2007).

School of Electronics and Computer Science at the University of Southampton. 2007. Cogprints Home Page. http://cogprints.org (accessed October 30, 2007).

Image Credits

Harnad portrait courtesy of Stevan Harnad.

18 Encyclopedias

References

Perseus Digital Library. 2007. "Pliny the Elder." From *The Natural History*, edited by John Bostock and H. T. Riley. http://www.perseus.tufts.edu/hopper/text?doc=Perseus:text:1999.02.0137 (accessed March 30, 2010).

Thayer, Bill. 2006. "Pliny the Elder: the Natural History." http://penelope.uchicago.edu/Thayer/E/Roman/Texts/Pliny_the_Elder/home.html (accessed November 30, 2007).

Wikimedia Foundation. 2010. "Encyclopedia." *Wikipedia, the Free Encyclopedia*. http://en.wikipedia.org/wiki/Encyclopedia (accessed February 8, 2010).

Wikimedia Foundation. 2009. "Wikipedia Statistics: Thursday, December 31, 2009." http://stats.wikimedia.org/EN (accessed February 4, 2010).

Denis Diderot (1713-1784) and Jean le Rond d'Alembert (1717-1783): Encyclopédie, 1751-1777

References

d'Alembert, Jean le Rond, and Denis Diderot. 1751. *Encyclopédie, ou dictionnaire raisonne des sciences, des arts et des metiers*. Paris: Briasson, David, Le Breton, and Durand.

Barzun, Jacques and Ralph H. Bowen, trans. 1956. *Diderot: Rameau's Nephew and Other Works*. Indianapolis: Hackett Publishing. [Quotation, p. 283]

Eisenstein, Elizabeth L. 1980. *The Printing Press as an Agent of Change*. Cambridge: Cambridge University Press.

Grimsley, Ronald. 1963. *Jean d'Alembert*. Oxford: Clarenden Press.

Rorty, Amélie. 2003. *The Many Faces of Philosophy: Reflections from Plato to Arendt*, 237. Oxford: Oxford University Press.

Scholarly Publishing Office of the University of Michigan Library. 2010. "The Encyclopedia of Diderot & d'Alembert: Collaborative Translation Project." http://quod.lib.umich.edu/d/did/intro.html; http://quod.lib.umich.edu/d/did (accessed February 4, 2010).

Scott, Michon. 2010. "Rocky Road: Diderot and d'Alembert." http://www.strangescience.net/didalem.htm (accessed February 4, 2010).

Wernick, Robert. 1997. "Declaring an Open Season on the Wisdom of the Ages." *Smithsonian Magazine*, May.

Wikimedia Foundation. 2010. "Encyclopédie. The 18th-Century French Encylopaedia." http://en.wikipedia.org/wiki/Encyclopédie (accessed February 4, 2010).

Image Credits

Diderot portrait: public domain image from *The Hundred Greatest Men*. 1885. New York: D. Appleton & Company.

d'Alembert portrait: public domain image, engraving by William Hopwood, 1784–1853.

Système Figuré: public domain image.

Contributors

Loet Leydesdorff and Peter A. Hook.

Jimmy Wales (b. 1966): Founder of Wikipedia, 2001

References

Lewine, Edward. 2007. "The Encylopedist's Lair." *New York Times Magazine*, November 17.

Sanger, Larry. 2001. "Let's Make a Wiki." Nupedia-L: Internet Archive, January 10. http://web.archive.org/web/20030414014355/; http:/www.nupedia.com/pipermail/nupedia-l/2001-January/000676.html (accessed January 1, 2008).

Wikimedia Foundation. 2010. "Jimmy Wales." *Wikipedia, the Free Encyclopedia*. http://en.wikipedia.org/wiki/Jimmy_Wales (accessed February 4, 2010). [Quotation, sourced material]

Wikimedia Foundation. 2010. Wikipedia, the Free Encyclopedia Home Page. http://wikipedia.org (accessed February 4, 2010).

Image Credits

Wales portrait: public domain image, licensed under the Creative Commons Attribution ShareAlike 2.5 License.

19 Knowledge Dissemination

Johann Gutenberg (1398–1468): Printing Press, 1455

References

Baldwin Project. "John Gutenberg and the Invention of Printing." From *Great Inventors and Their Inventions* by Frank P. Bachman. http://www.mainlesson.com/display.php?author=bachman&book=inventors&story=gutenberg (accessed October 30, 2007).

Rees, Fran. 2006. *Johannes Gutenberg: Inventor of the Printing Press*. The Signature Lives Series. Mankato, MN: Compass Point Books.

Image Credits

Public domain image: Meisner, Heinrich, and Johannes Luther. 1900. *Die Erfindung der Buchdruckerkunst*. Bielefeld and Leipzig, Germany: Verlag von Velhagen & Klafing.

Theodor Holm Nelson (b. 1937): Xanadu, 1960; Hypertext 1963

References

Nelson, Theodor Holm. 1982. *Literary Machines*. Sausalito, CA: Mindful Press.

Nelson, Theodor Holm. 1987. *Computer Lib/Dream Machines*. Redmond, WA: Tempus Books of Microsoft Press.

Nelson, Theodor Holm. 2006. "My Life and Work, Very Brief." http://hyperland.com/mlawLeast.html (accessed January 1, 2008). [Quotation, self]

Rayward, W. Boyd. 1994. "Visions of Xanadu: Paul Otlet (1868–1944) and Hypertext." *Journal of the American Society for Information Science* 45, 4: 235–250. http://people.lis.illinois.edu/~wrayward/otlet/xanadu.htm (accessed February 10, 2010).

Image Credits

Nelson portrait courtesy of Matt Locke, http://www.test.org.uk.

Vinton Gray Cerf (b. 1943): Internet, 1973

References

Abbate, Janet. 1999. *Inventing the Internet*. Cambridge, MA: MIT Press.

Cerf, Vinton G. 2005. "Letter to the Honorable Joe Barton and the Honorable John D. Dingell on Internet Neutrality." Posted to the Official Google Blog by Alan Davidson, November 8, 2005. http://googleblog.blogspot.com/2005/11/vint-cerf-speaks-out-on-net-neutrality.html (accessed January 1, 2008). [Quotation]

Cerf, Vinton G., and Bob E. Kahn. 1974. "A Protocol for Packet Network Interconnections." *IEEE Transactions on Communications COM* 22, 5: 637–648.

Hafner, Katie. 1998. *Where Wizards Stay Up Late: The Origins of the Internet*. Carmichael, CA: Touchstone Books.

Massachusetts Institute of Technology. 2009. "Inventor of the Week: Vinton Cerf—Internet Protocols (TCP/IP)." Lemelson MIT Program, MIT School of Engineering. http://web.mit.edu/invent/iow/cerf.html (accessed November 12, 2009).

Miniwatts Marketing Group. 2008. "World Internet Usage Statistics and Populations Stats." http://www.internetworldstats.com/stats.htm (accessed June 2, 2008).

Image Credits

Cerf portrait courtesy of Vinton G. Cerf.

Timothy Berners-Lee (b. 1955): World Wide Web, 1989

References

Berners-Lee, Tim. 1999. *Weaving the Web: The Original Design and Ultimate Destiny of the World Wide Web*. San Francisco: HarperCollins.

Berners-Lee, Tim. 2010. "The WorldWideWeb Browser." http://www.w3.org/People/Berners-Lee/WorldWideWeb (accessed February 15, 2010).

Berners-Lee, Tim, Dieter Fensel, James A. Hendler, Henry Lieberman, and Wolfgang Wahlster. 2005. *Spinning the Semantic Web: Bringing the World Wide Web to Its Full Potential*. Cambridge, MA: MIT Press.

Berners-Lee, Tim, Wendy Hall, and James A. Hendler. 2006. *A Framework for Web Science*. Boston: Now Publishers.

Berners-Lee, Tim, James A. Hendler, and O. Lassila. 2001. "The Semantic Web." *Scientific American*, May. http://www.sciam.com/2001/0501issue/0501berners-lee.html (accessed January 1, 2008). [Quotation]

Soylent Communications. 2008. "Tim Berners-Lee." NNDB. http://www.nndb.com/people/573/000023504 (accessed October 30, 2007).

Image Credits

Berners-Lee portrait courtesy of Tim Berners-Lee and the World Wide Web Consortium (W3C).

Contributors

Amy van der Hiel.

20 Knowledge Classification

Melvil Dewey (1851–1931): Dewey Decimal Classification System (DDC), 1876

References

BrainyMedia.com. 2008. "Melvil Dewey Quotes." Brainy Quote. http://www.brainyquote.com/quotes/authors/m/melvil_dewey.html (accessed June 2, 2008).

Dewey, Melvil. 1876. "The Profession." *American Library Journal* 1. [Quotation]

Foskett, Anthony Charles. 1973. *The Universal Decimal Classification*. London: Clive Bingley Ltd.

OCLC. 2008. "Melvil Dewey Biography." OCLC. http://www.oclc.org/dewey/resources/biography (accessed June 2, 2008).

Wiegand, Wayne A. 1996. *Irrepressible Reformer: A Biography of Melvil Dewey*. Chicago: American Library Association.

Wikimedia Foundation. 2008. "List of Dewey Decimal Classes." *Wikipedia, the Free Encyclopedia*. http://en.wikipedia.org/wiki/List_of_Dewey_Decimal_classes (accessed June 2, 2008).

Image Credits

Dewey portrait: public domain image.

Paul Otlet (1868-1944) and Henri-Marie Lafontaine (1854-1943)

References

Nobel Web AB. 2009. "Henri La Fontaine: The Nobel Peace Prize 1913." http://nobelprize.org/nobel_prizes/peace/laureates/1913/fontaine-bio.html (accessed August 12, 2009).

Noorederlicht (production company). 1998. *Alle Kennis van de Wereld* (Biography of Paul Otlet). http://www.archive.org/details/paulotlet (accessed October 30, 2007).

Image Credits

Otlet portrait: public domain image; search assisted by Michael Buckland of the University of California, Berkeley.

Lafontaine portrait: public domain image; search assisted by the Library of Congress.

Designing the Universal Decimal Index (UDC), 1905

References

Otlet, Paul. "Encyclopedia Mundaneum Universalis." Unfinished manuscript.

Otlet, Paul. 1935. *Monde; Essai d'Universalisme*. Brussels: Editions Mundaneum.

Wright, Alex. 2003. "Forgotten Forefather: Paul Otlet." *Boxes and Arrows*, November. http://www.boxesandarrows.com/view/forgotten_forefather_paul_otlet (accessed October 30, 2007).

Contributors

Charles van den Heuvel.

The Mundaneum, 1910

References

Encyclopædia Britannica. 2010. *Encyclopædia Britannica Online*, s.v. "Henri-Marie Lafontaine." http://www.britannica.com/EBchecked/topic/327724/Henri-Marie-Lafontaine (accessed March 30, 2010).

Otlet's "Traité de Documentation," 1934

References

Otlet, Paul. 1934. *Traité de Documentation: Le Livre sur le Livre—Théorie et Pratique*. Brussels: Editiones Mundaneum, Palais Mondial, Van Keerberghen & Fils. [Quotation]

Otlet, Paul. 1990. "The Science of Bibliography and Documentation." In *The International Organization and Dissemination of Knowledge: Selected Essays of Paul Otlet*, edited and translated by W. Boyd Rayward. Amsterdam: Elsevier.

Rayward, W. Boyd. 1994. "Visions of Xanadu: Paul Otlet (1868–1944) and Hypertext." *Journal of the American Society for Information Science* 45, 4: 235–250. http://people.lis.uiuc.edu/~wrayward/otlet/xanadu.htm (accessed October 30, 2007).

University of Illinois at Urbana-Champaign. 2004. "European Modernism and the Information Society: Informing the Present, Understanding the Past." Conference hosted by the Graduate School of Library and Information Science, May 6–8, 2005. http://www.lis.uiuc.edu/conferences/EuroMod.05 (accessed October 30, 2007).

van den Heuvel, Charles. 2007. "Building Society, Constructing Knowledge, Weaving the Web: Otlet's Visualizations of a Global Information Society and His Concept of a Universal Civilization. In *European Modernism and the Information Society*, edited by W. Boyd Rayward, 127–153. London: Ashgate Publishers.

Wright, Alex. 2003. "Forgotten Forefather: Paul Otlet." *Boxes and Arrows*, November. http://www.boxesandarrows.com/view/forgotten_forefather_paul_otlet (accessed October 30, 2007).

Image Credits

Two images, *see* Otlet 1934.

Contributors

Ron Day and Charles van den Heuvel.

21 Knowledge Interlinkage

References

Rozenberg, Michal, Miri Munk, and Anat Kainan. 2006. "A Talmud Page as a Metaphor of a Scientific Text." *International Journal of Qualitative Methods* 5, 4.

Weinberg, Bella Hass. 1997. "The Earliest Hebrew Citation Indexes." *Journal of the American Society for Information Science* 48, 4: 318–330.

Contributors
Blaise Cronin.

Frank Shepard–Legal Citation Index, 1873

References
Garfield, Eugene. 1954. "Association of Ideas Techniques in Documentation: Shepardizing the Literature of Science." http://www.garfield.library.upenn.edu/papers/assocofideasy1954.html (accessed June 10. 2008).

Shapiro, Fred R. 1992. "Origins of Bibliometrics, Citation Indexing, and Citation Analysis: The Neglected Legal Literature." *Journal of the American Society for Information Science and Technology* 43, 5: 337–339.

Shepard, Frank. 1875. *Illinois Citations: Legal Citators*. Chicago: Author.

Image Credits
Shepard portrait used with permission of LexisNexis; search assisted by Jane Morris.

Public domain images of Shepard's Citations arranged and digitized by Peter A. Hook, Indiana University.

Contributors
Peter A. Hook.

Eugene Garfield (b. 1925): "Citation Indexes for Science," 1955

References
Adair, W. C. 1955. "Citation Indexes for Scientific Literature." *American Documentation* 6: 31–32. http://www.garfield.library.upenn.edu/papers/adaircitationindexesforscientificliterature1955.html (accessed October 30, 2007).

Cronin, Blaise, and Helen Barsky Atkins, eds. 2000. *The Web of Knowledge: A Festschrift in Honor of Eugene Garfield*. ASIS Monograph Series. Medford, NJ: Information Today.

Garfield, Eugene. 1955. "Citation Indexes for Science: A New Dimension in Documentation through Association of Ideas." *Science* 122: 108–111.

Garfield, Eugene. 1955. "Needed—a National Science Intelligence and Documentation Center." Paper presented at the Symposium on Storage and Retrieval of Scientific Information of the annual meeting of the American Association for the Advancement of Science, Atlanta, Georgia, December 28, 1955. http://www.garfield.library.upenn.edu/papers/natlsciinteldoccenter.html (accessed October 30, 2007). [Quotation]

Garfield, Eugene. 1974. "ISI's Atlas of Science May Help Students in Choice of Career in Science." *Essays of an Information Scientist 1974–1976* 2, 38: 311–314. http://www.garfield.library.upenn.edu/essays/v2p311y1974-76.pdf (accessed October 30, 2007).

Garfield, Eugene. 1981. *ISI Atlas of Science: Biochemistry and Molecular Biology 1979–1980*, 112. Philadelphia, PA: ISI.

Garfield, Eugene. 1987. "Mapping the World of Nutrition: Citation Analysis Helps Digest the Menu of Current Research." *Essays of an Information Scientist* 10: 349. http://www.garfield.library.upenn.edu/essays/v10p349y1987.pdf (accessed October 30, 2007).

Garfield, Eugene. 1988. "Mapping the World of Epidemiology: Part 2. The Techniques of Tracking Down Disease. *Essays of an Information Scientist* 11: 290. http://www.garfield.library.upenn.edu/essays/v11p290y1988.pdf (accessed October 30, 2007).

Garfield, Eugene, and James E. Shea, eds. 1984. *ISI Atlas of Science: Biotechnology and Molecular Genetics 1981/82*. Philadelphia: ISI.

Marshakova, Irina V. 1973. "A System of Document Connections Based on References." *Scientific and Technical Information Serial of VINITI* 6: 3–8.

Small, Henry. 1973. "Co-Citation in Scientific Literature: A New Measure of the Relationship between Publications." *Journal of the American Society for Information Science* 24: 265–269.

Szigeti, Helen. 2001. "The ISI Web of Knowledge Platform: Current and Future Directions." Published online by ISI Inc, Philadelphia, PA.

Wells, Herbert George. 1938. *World Brain*. New York: Doubleday & Company, Inc.

Wouters, Paul. 1999. "The Citation Culture." PhD diss., University of Amsterdam. http://garfield. library.upenn.edu/wouters/wouters.pdf (accessed October 30, 2007).

Image Credits
Garfield portrait courtesy of Eugene Garfield and Thomson Reuters.

Larry Page (b. 1973) and Sergey Brin (b. 1973): Google 1998

References
Brin, Sergey, and Lawrence Page. 1998. "The Anatomy of a Large-Scale Hypertextual Web Search Engine." *Proceedings of the 7th International World Wide Web Conference*. http://infolab.stanford.edu/~backrub/google.html (accessed September 30, 2009).

Darnton, Robert. 2009. Google and the New Digital Future. *The New York Review of Books* 56: 20. http://www.nybooks.com/articles/23518 (accessed January 25, 2010).

Giles, C. Lee, Kurt D. Bollacker, and Steve Lawrence. 1998. "CiteSeer: An Automatic Citation Indexing System." *Proceedings of the 3rd ACM Conference on Digital Libraries (DL'98)*, edited by Ian Witten, Rob Akscyn, and Frank M. Shipman III, 89–98. New York: ACM Press. http://clgiles.ist.psu.edu/papers/DL-1998-citeseer.pdf (accessed October 30, 2007).

Page, Lawrence, Sergey Brin, Rajeev Motwani, and Terry Winograd. 1998. "The PageRank Citation Ranking: Bringing Order to the Web." Stanford Digital Library Technologies Project. http://infolab.stanford.edu/~backrub/pageranksub.ps (accessed June 2, 2008).

Toobin, Jeffrey. 2007. "Google's Moon Shot." *The New Yorker*, February 5. http://www.newyorker.com/reporting/2007/02/05/070205fa_fact_toobin (accessed October 30, 2007).

Image Credits
Page and Brin portraits: Google. "Corporate Information: Management." http://www.google.com/corporate/execs.html (accessed October 30, 2007). (c) Google Inc., 2007. Reprinted with Permission.

22 Knowledge Visualization

References
Alexander, Christopher, Sara Ishikawa, and Murray Silverstein. 1977. *A Pattern of Language*. New York: Oxford University Press.

Bendavid-Val, Leah, ed. 2003. *Through the Lens: National Geographic's Greatest Photographs*. New York: National Geographic.

Berghaus, Heinrich Karl Whilhelm. 1862. *Physikalischer Atlas oder Sammlung von Karten, auf denen die hauptsächlichsten Erscheinungen der anorganischen und organischen Natur nach ihrer geographischen Verbreitung und Vertheilung bildlich dargestellt sind.* Gotha, Germany: Verlag von Justus Perthes.

Holmes, Nigel. 1993. *The Best in Diagrammatic Graphics*. London: Quarto Publishing.

Kepes, Gyorgy. 1956. *The New Landscape in Art and Science*. Chicago: Paul Theobald and Co.

Klee, Paul. 1961. *The Thinking Eye*. New York: G. Wittenborn.

Otlet, Paul. 1990. "Transformations in the Bibliographical Apparatus of the Sciences." In *The International Organization and Dissemination of Knowledge: Selected Essays of Paul Otlet*, edited and translated by W. Boyd Rayward, 78. Amsterdam: Elsevier.

Williats, John. 1997. *Art and Representation: New Principles in the Analysis of Pictures*. Princeton, NJ: Princeton University Press.

Kees Boeke (1884–1966): "Cosmic View: The Universe in 40 Jumps," 1957

References
Billington, James H. 1997. *The Work of Charles and Ray Eames: A Legacy of Invention*, 15–26. New York: Harry N. Abrams, Inc.

Boeke, Kees. 1957. *Cosmic View: The Universe in 40 Jumps*. http://nedwww.ipac.caltech.edu/level5/Boeke/Boekem14.html (accessed January 1, 2008). [Quotation]

Boeke, Kees. 1957. *Cosmic View: The Universe in 40 Jumps*. New York: John Day Company. http://www.vendian.org/mncharity/cosmicview (accessed October 30, 2007).

Eames, Lucia (dba Eames Office). 2006. "Celebrate Powers of 10 Day." http://powersof10.com (accessed October 30, 2007).

Morrison, Philip, Phylis Morrison, and the Office of Charles and Ray Eames. 1983. *Powers of Ten: A Book about the Relative Size of Things in the Universe and the Effect of Adding Another Zero*. New York: W. H. Freeman & Company.

Wikimedia Foundation. 2007. "Kees Boeke." *Wikipedia, the Free Encyclopedia*. http://en.wikipedia.org/wiki/Kees_Boeke (accessed October 30, 2007).

Image Credits
Boeke portrait: public domain image.

Cosmic View, *see* Boeke 1957.

Derek John de Solla Price (1922–1983): "Networks of Scientific Papers," 1965

References
de Solla Price, Derek J. 1961. *Science Since Babylon*. New Haven: Yale University Press.

de Solla Price, Derek J. 1965. *Little Science, Big Science*. New York: Columbia University Press.

de Solla Price, Derek J. 1965. "Networks of Scientific Papers." *Science*, 149, 3683: 510–515.

Garfield, Eugene. 1984 "A Tribute to Derek John de Solla Price: A Bold, Iconoclastic Historian of Science." *Essays of an Information Scientist* 7: 213–217. http://garfield.library.upenn.edu/essays/v7p213y1984.pdf (accessed October 30, 2007). [Quotation. p. 241]

Garfield, Eugene. "Essays on Derek John de Solla Price." http://www.garfield.library.upenn.edu/price/derekprice.html (accessed October 30, 2007).

Tukey, J. W. 1962. "Keeping Research in Contact with the Literature: Citation Indices and Beyond." *Journal of Chemical Documentation* 2, 34-37.

Xhignesse, Louis V., and Charles E. Osgood. 1967. "Bibliographical Citation Characteristics of the Psychological Journal Network in 1950 and in 1960." *American Psychologist* 22, 9: 778-791.

Image Credits
Price portrait: public domain image, released by Mark de Solla Price, photo by Malcolm S. Kirk.

Edward Tufte (b. 1943): "The Visual Display of Quantitative Information," 1983

References
Smith, Fran. 2007. "Intelligent Designs: When Information Needs to be Communicated, Edward Tufte Demands both Truth and Beauty." *Stanford Magazine*, March/April.

Tufte, Edward R. 1983. *The Visual Display of Quantitative Information*. Cheshire, CT: Graphics Press.

Tufte, Edward R. 1990. *Envisioning Information*. Cheshire, CT: Graphics Press.

Tufte, Edward R. 1997. *Visual Explanations: Images and Quantities, Evidence and Narrative*. Cheshire, CT: Graphics Press.

Tufte, Edward R. 2006. *Beautiful Evidence*. Cheshire, CT: Graphics Press.

Zachry, Mark, and Charlotte Thralls. 2004. "An Interview with Edward R. Tufte." *Technical Communication Quarterly* 13, 4: 447-462. [Quotation, p. 452]

Image Credits
Tufte portrait used by permission.

23 Man-Machine Symbiosis

References
Preece, J., Y. Rogers, and H. Sharp. 2002. *Human-Computer Interaction: Concepts and Design*. Boston: Addison-Wesley.

Shneiderman, Ben, and Catherine Plaisant. 2009. *Designing the User Interface: Strategies for Effective Human-Computer Interaction*. 5th ed. Boston, MA: Addison-Wesley.

Ada Lovelace (1815–1852): First Computer Program, 1843

References
Ada Project. 2007. "Ada Byron King." Women at SCS, Carnegie Mellon University. http://www.cs.yale.edu/homes/tap/ada-lovelace.html (accessed October 30, 2007).

Baum, Joan. 1986. *The Calculating Passion of Ada Byron*. Hamden, CT: Archon Books.

Moore, Doris Langley. 1977. *Ada: Countess of Lovelace*. London: John Murray. [Quotation]

Toole, Betty A. 1992. *Ada, the Enchantress of Numbers*. Mill Valley, CA: Strawberry Press.

Wikimedia Foundation. 2007. "Ada Lovelace." *Wikipedia, the Free Encyclopedia*. http://en.wikipedia.org/wiki/Ada_Lovelace (accessed October 30, 2007).

Image Credits

Lovelace portrait: public domain image, published November 1, 1838.

Vannevar Bush (1890–1974): Memex, 1945

References

Buckland, Michael K. 1992. "Emanuel Goldberg, Electronic Document Retrieval, and Vannevar Bush's Memex." *Journal of the American Society for Information Science* 43, 4: 284–294.

Bush, Vannevar. 1945. "As We May Think." *The Atlantic Monthly* 176, 1: 101–108.

Bush, Vannevar. 1945. "As We May Think." *Life Magazine,* November 19.

Bush, Vannevar. 1945. *Science, the Endless Frontier: A Report to the President of the United States of America*. http://www.nsf.gov/od/lpa/nsf50/vbush1945.htm (accessed October 30, 2007). [Quotation]

Wikimedia Foundation. 2007. "Vannevar Bush." *Wikipedia, the Free Encyclopedia*. http://en.wikipedia.org/wiki/Vannevar_Bush (accessed October 30, 2007).

Zachary, G. Pascal. 1997. *Endless Frontier: Vannevar Bush, Engineer of the American Century*. New York: Free Press.

Image Credits

Bush portrait courtesy of the National Science Foundation.

Cyclops camera, mechanical supersecretary, and Memex images, *see* Bush 1945 (November 19).

Grace Murray Hopper (1906-1992): First Compiler, 1952

References

History of Computing Project. 2004. "Grace Murray Hopper." http://www.thocp.net/biographies/hopper_grace.html (accessed October 30, 2007). [Quotation]

Hopper, Grace. 1952. "The Education of a Computer." *Proceedings of the Association for Computing Machinery Conference*, Pittsburgh, May. Reprinted in *IEEE Annals of the History of Computing*, 1998, 9, 3–4: 271–281.

Whitelaw, Nancy. 1995. *Grace Hopper: Programming Pioneer*. New York: W. H. Freeman and Company.

Image Credits

Hopper portrait: public domain image from the Library of Congress.

J. C. R. Licklider (1915–1990): "Man-Computer Symbiosis," 1960

References

Campbell-Kelly, Martin. 2006. "From World Brain to the World Wide Web." Annual Gresham College BSHM Lecture, November 9. http://www.gresham.ac.uk/event.asp?PageId=39&EventId=486 (accessed October 30, 2007).

Hauben, Michael, and Ronda Hauben. 1995-1996. "Netizens: An Anthology," ch. 5-7. http://www.columbia.edu/~hauben/netbook (accessed February 5, 2010).

Licklider, Joseph Carl Robnett. 1960. "Man-Computer Symbiosis." *IRE Transactions on Human Factors in Electronics* HFE-1: 4–11. http://groups.csail.mit.edu/medg/people/psz/Licklider.html (accessed October 30, 2007). [Quotation, Summary]

Licklider, Joseph Carl Robnett. 1965. *Libraries of the Future*. Cambridge, MA: MIT Press.

Licklider, J.C.R., and R. W. Taylor. 1968. "The Computer as a Communication Device." *Science and Technology*, April: 21–31.

Rheingold, Howard. 2000. *Tools for Thought: The History and Future of Mind-Expanding Technology*. Cambridge, MA: MIT Press.

Wikimedia Foundation. 2010. "J. C. R. Licklider." *Wikipedia, the Free Encyclopedia*. http://en.wikipedia.org/wiki/J.C.R._Licklider (accessed February 8, 2010).

Image Credits

Licklider portrait: public domain image.

Douglas C. Engelbart (b. 1925): "Augmenting Human Intellect: A Conceptual Framework," 1962

References

Bootstrap Alliance. 2008. "Bootstrap Institute: History in Pictures." http://www.bootstrap.org/chronicle/pix/pix.html (accessed June 2, 2008).

Engelbart, Douglas C. 1962. "Augmenting Human Intellect: A Conceptual Framework." Summary Report AFOSR-3233, October. Stanford Research Institute, Menlo Park, CA. [Quotation, Introduction: A. General]

Sutherland, I. 1965. "The Ultimate Display." *Proceedings of IFIP* 65, 2: 506–508, 582–583.

Image Credits

Engelbart portrait, hardware and interface images are courtesy of Douglas Engelbart and Bootstrap Alliance, *see* Bootstrap Alliance 2008.

24 Global Brain

References

Bernal, John D. 1939. *The Social Function of Science*. London: Routledge & Kegan Ltd.

Bloom, H. 2000. *Global Brain: The Evolution of Mass Mind from the Big Bang to the 21st Century*. New York: John Wiley & Sons, Inc.

De Reuck, A., and J. Knight, eds. 1967. *Communication in Science: Documentation and Automation.* London: Churchill.

Garfield, Eugene, and Irving H. Sher. 1963. *Science Citation Index*. Philadelphia: Institute for Scientific Information.

Garvey, William D., and Belver C. Griffith. 1964. "Scientific Information Exchange in Psychology: The Immediate Dissemination of Research Findings Is Described for One Science." *Science* 146, 3652: 1655–1659.

Goffman, W., and K. S. Warren. 1969. "Dispersion of Papers Among Journals Based on a Mathematical Analysis of 2 Diverse Medical Literatures." *Nature* 221, 5187: 1205–1207.

Herring, C. 1968. "Distill or Drown—Need for Reviews." *Physics Today* 21, 9: 27.

McDougall, William. 1920. *The Group Mind*. New York: G. P. Putnam's Sons.

Merton, Robert K. 1969. "The Matthew Effect in Science. The Reward and Communication Systems of the Science are Considered" *Science* 159, 3810: 56–63.

Moravcsik, Michael J., and Simon Pasternack. 1966. "A Debate on Preprint Exchange." *Physics Today* 19: 60–73.

Watson, J. 1968. *The Double Helix.* London: Weidenfeld and Nicolson.

Ziman, J. M. 1968. *Public Knowledge*. London: Cambridge University Press.

Ziman, J. M. 1969. "Information, Communication, Knowledge." *Nature* 224: 318–324.

H. G. Wells (1866–1946): "World Brain," 1938

References

Abbot, Robert. 1999. *The World as Information: Overload and Personal Design*, 110. Exeter, UK: Intellect Books.

Wells, H. G. 1923. *The Outline of History, Being a Plain History of Life and Mankind*. 4th ed., vol. 4, 1305. New York: The Review of Reviews Company. [Quotation, Chapter 41]

Wells, Herbert George. 1897. *The Invisible Man*. London: C. Arthur Pearson.

Wells, Herbert George. 1901. *First Men in the Moon*. London: George Newnes.

Wells, Herbert George. 1938. *World Brain*. New York: Doubleday & Company, Inc.

Image Credits

Wells portrait: public domain image, assistance provided by The Lilly Library, Indiana University.

Buckminster Fuller (1895–1983): World Game, 1960s

References

Buckminster Fuller Institute. 2007. "Introduction to Buckminster Fuller's World Game." http://www.bfi.org/taxonomy/term/170/all (accessed October 30, 2007). [Quotation, make the world work]

Fuller, Buckminster. 1969. *Operating Manual for Spaceship Earth*. Carbondale, IL: Southern Illinois University Press.

Fuller, Buckminster. 1973. *Earth, Inc*. Garden City, NY: Anchor Press.

Fuller, Buckminster. 1981. *Critical Path*, xxv, 124. New York: St. Martin's Press. [Quotation, Introduction, War is obsolete.]

Fuller, Buckminster, and E. J. Applewhite. 1975. *Synergetics: Explorations in the Geometry of Thinking*. New York: Macmillan.

Fuller, Buckminster, and Anwar Dil. 1983. *Humans in Universe*. New York: Mouton.

Wikimedia Foundation. 2007. "Buckminster Fuller." *Wikipedia, the Free Encyclopedia*. http://en.wikipedia.org/wiki/Buckminster_Fuller (accessed October 30, 2007). [Quotation, benefiting all humanity]

Wikimedia Foundation. 2007. "World Game." *Wikipedia, the Free Encyclopedia*. http://en.wikipedia.org/wiki/World_Game (accessed October 30, 2007).

Zung, Thomas T. K., ed. 2001. *Buckminster Fuller: Anthology for the New Millennium*. New York: St. Martin's Press.

Image Credits

Fuller portrait courtesy of the estate of R. Buckminster Fuller.

James Burke (b. 1936): Connections 1979

References

Burke, James. 1978. *Connections*. Boston: Little, Brown and Company.

Burke, James. 2000. *Circles: Fifty Round Trips Through History Technology Science Culture*. New York: Simon & Schuster.

Burke, James. 2000. *The Knowledge Web: From Electronic Agents to Stonehenge and Back—and Other Journeys through Knowledge*. New York: Simon & Schuster.

Delio, Michelle. "Help Build the Web of Knowledge." *CondéNet, Inc: Wired*. http://www.wired.com/culture/lifestyle/news/2002/05/52594 (accessed October 30, 2007).

James Burke Institute. 2007. "James Burke Institute's Knowledge Web." http://www.k-web.org (accessed October 30, 2007).

James Burke Institute. 2007. "James Burke Institute's Knowledge Web—Vision." http://k-web.org/public_html/vision.htm (accessed October 30, 2007). [Quotation].

Wikimedia Foundation. 2007. "James Burke (science historian)." *Wikipedia, the Free Encyclopedia*. http://en.wikipedia.org/wiki/James_Burke_(science_historian) (accessed October 30, 2007).

Image Credits

Burke portrait courtesy of James Burke.

Contributors

Luis Rocha, Ronald E. Day, Blaise Cronin, Lokman I. Meho, Kevin W. Boyack, Peter A. Hook, Patrick McKercher, Stevan Harnad, and Ben Shneiderman.

26 Milestones in Mapping Science

All references in this section are organized in order of their appearance in the timeline. For each double page spread all algorithm, visualization, tool, and book references are listed together with image credits.

Image Credits

Milestones in Mapping Science Timeline: Elisha F. Hardy (design), Katy Börner (concept). Image credits for individual photos and graphics are given by section.

Introduction

References

Freeman, Linton C. 2000. "Visualizing Social Networks." *Journal of Social Structure* 1. http://www.cmu.edu/joss/content/articles/volume1/Freeman.html (accessed October 30, 2007).

Börner, Katy, Soma Sanyal, and Alessandro Vespignani. 2007. "Network Science." In *Annual Review of Information Science and Technology,* edited by Blaise Cronin, vol. 41: 537–607. Medford, NJ: Information Today/American Society for Information Science and Technology.

Börner, Katy, Chaomei Chen, and Kevin W. Boyack. 2003. "Visualizing Knowledge Domains." In *Annual Review of Information Science and Technology*, edited by Blaise Cronin, vol. 37: 179–255. Medford, NJ: Information Today/American Society for Information Science and Technology.

Hearst, Marti A. 1999. "User Interfaces and Visualization." In *Modern Information Retrieval*, edited by R. Baeza-Yates and B. Ribeiro-Neto, 257–323. New York: ACM Press and Addison-Wesley. http://people.ischool.berkeley.edu/~hearst/irbook/chapters/chap10.html (accessed January 1, 2008).

26 Timeline: 1930–1960

Algorithms

References

Thurstone, L. L. 1931. "Multiple Factor Analysis." *Psychological Review* 38: 406–427.

Tryon, R. C. 1939. *Cluster Analysis*. New York: McGraw-Hill.

Shannon, C. E. 1948. "A Mathematical Theory of Communication." *The Bell System Technical Journal* 27, 379–423, 623–656.

Visualizations

References

Moreno, Jacob L. 1933. "Emotions Mapped by New Geography." *New York Times*, April 3, 17.

Moreno, Jacob L. 1934. *Who Shall Survive?* Washington, DC: Nervous and Mental Disease Publishing Company.

Lundberg, G. A., and M. Steele. 1938. "Social Attraction-Patterns in a Village." *Sociometry* 1: 375–419.

Wells, Herbert George. 1938. *World Brain*. New York: Doubleday & Company, Inc.

Bernal, John D. 1939. *The Social Function of Science*. London: Routledge & Kegan Ltd.

Northway, M. L. 1940. "A Method for Depicting Social Relationships Obtained by Sociometric Testing." *Sociometry* 3: 144–150.

Ellingham, H.J.T. 1948. "Divisions of Natural Science and Technology." *Reports and Papers of the Royal Society Scientific Information Conference*. London: The Royal Society, Burlington House.

Bock, R. D., and S. Z. Husain. 1952. "Factors of the Tele: A Preliminary Report." *Sociometry* 15: 206–219.

Allen, Gordon. "Citation Network." *Citation Index to Genetics and General Science Literature*, edited by Eugene Garfield. Research proposal by the Institute for Scientific Information, submitted July 15, 1960 to the National Science Foundation.

Image Credits—Maps

"Sociometry" is a public domain image; research assistance by Jaime Panzanella at the American Sociological Association and the Moreno family.

"Lady Bountiful" is a public domain image; research assistance from the American Sociological Association.

"World Encyclopedia" reproduced with permission of Ayer Publishers.

"1939 Map of Science" reproduced from *The Social Function of Science*, by permission of Taylor and Francis Books.

"Target Sociogram" is a public domain image; research assistance by Jaime Panzanella at the American Sociological Assocation.

Ellingham's "Natural Science and Technology Chart," © The Royal Society, used by permission.

"Plot of Schoolchildren" is a public domain image; research assistance by Jaime Panzanella at the American Sociological Assocation.

Allen's "First Citation Network" reproduced from Eugene Garfield's "Citation Index to Genetics and General Science Literature," 1960, by permission of Thomson Reuters.

Image Credits—Portraits

Thurstone portrait of Thurstone is a public domain image.

Northway portrait courtesy of the Trent University Archives.

Allen portrait courtesy of College Archives and Special Collections, SUNY Brockport Drake Memorial Library.

Books

References

Otlet, Paul. 1934. *Traité de Documentation: Le Livre sur le Livre—Théorie et Pratique*. Brussels: Editiones Mundaneum, Palais Mondial, Van Keerberghen & Fils.

Moreno, Jacob L. 1934. *Who Shall Survive?* Washington, DC: Nervous and Mental Disease Publishing Company.

Wells, Herbert George. 1938. *World Brain*. New York: Doubleday & Company, Inc.

Bernal, John D. 1939. *The Social Function of Science*. London: Routledge & Kegan Ltd.

Bradford, Samuel C. 1948. *Documentation*. London: Crosby Lockwood.

Brillouin, Léon. 1956. *Science and Information Theory*. New York: Academic Press.

de Solla Price, Derek J. 1961. *Science Since Babylon*. New Haven: Yale University Press.

Dijksterhuis, E. J. 1961. *The Mechanisation of the World Picture*, C. Dikshoom, trans. Oxford, UK: Clarendon Press.

28 Timeline: 1960–1982

Algorithms

References

Kessler, Michael M. 1963. "Bibliographic Coupling Between Scientific Papers." *American Documentation* 14, 1: 10–25.

Tutte, W. T. 1963. "How to Draw a Graph." *Proceedings of the London Mathematical Society* 13, 3: 743–768.

Ward, Joe H. 1963. "Hierarchical Grouping to Optimize an Objective Function." *Journal of American Statistical Association* 58, 301: 236–244.

Kruskal, J. B. 1964. "Multidimensional Scaling: A Numerical Method." *Psychometrika* 29: 115–129.

Kruskal, J. B. 1964. "Multidimensional Scaling by Optimizing Goodness of Fit to a Nonmetric Hypothesis." *Psychometrika* 29: 1–27.

Kruskal, J. B., and M. Wish. 1978. *Multidimensional Scaling*. London: Sage.

Torgerson, W. S. 1952. "Multidimensional Scaling: 1. Theory and Method." *Psychometrika* 17: 401–419.

Shepard, R. N. 1962. "Analysis of Proximities: Multidimensional Scaling with an Unknown Distance Function I & II." *Psychometrika* 27: 125–139, 219–246.

Levenshtein, V. I. 1965. "Binary Codes Capable of Correcting Spurious Insertions and Deletions of Ones (original in Russian)." *Russian Problemy Peredachi Informatsii* 1: 12–25.

Salton, Gerard. 1971. "Automatic Indexing Using Bibliographic Citations." *Journal of Documentation* 27: 98–110.

Anthonisse, J. M. 1971. "The Rush in a Directed Graph." *Technical Report BN9/71*. Amsterdam: Stichting Mathematisch Centrum.

Freeman, L. C. 1977. "A Set of Measures of Centrality Based on Betweenness." *Sociometry* 40, 1: 35-41.

Small, Henry. 1973. "Co-Citation in Scientific Literature: A New Measure of the Relationship between Publications." *Journal of the American Society for Information Science* 24: 265–269.

Marshakova, Irina V. 1973. "A System of Document Connections Based on References." *Scientific and Technical Information Serial of VINITI* 6: 3–8.

Salton, G., C. Yang, and A. Wong. 1975. "A Vector Space Model for Automatic Indexing." *Communications of the ACM* 18, 1: 613–620.

Lee, R.C.T., J. R. Slagle, and H. Blum. 1977. "A Triangulation Method for the Sequential Mapping of Points from N-Space to Two-Space." *IEEE Transactions on Computers* 26, 3: 288–292.

White, H. D. 1981. "Cocited Author Retrieval Online: An Experiment with the Social-Indicators Literature." *Journal of the American Society for Information Science* 32 1: 16–21.

White, H. D., and B. C. Griffith. 1981. "Author Cocitation: A Literature Measure of Intellectual Structure." *Journal of the American Society for Information Science* 32 3: 163–171.

Visualizations

References

Garfield, Eugene, Irving H. Sher, and Richard J. Torpie. 1964. "The Use of Citation Data in Writing the History of Science." Report for Air Force Office of Scientific Research under contract AF49 (638)-1256. Philadelphia, PA: Institute for Scientific Information. http://www.garfield.library.upenn.edu/papers/useofcitdatawritinghistofsci.pdf (accessed September 30, 2009).

de Solla Price, Derek J. 1965. "Networks of Scientific Papers." *Science* 149, 3683: 510–515.

Narin, Francis. 1968. "TRACES—Technology in Retrospect and Critical Events in Science." National Science Foundation Contract NSF C-535, IIT Research Institute, vols. 1 and 2.

Small, Henry, and E. Greenlee. 1986. "Collagen Research in the 1970s." *Scientometrics* 10, 1–2: 95–117.

Small, Henry. 1973. "Co-Citation in Scientific Literature: A New Measure of the Relationship between Publications." *Journal of the American Society for Information Science* 24: 265–269.

White, Howard D., and Belver C. Griffith. 1981. "Author Cocitation: A Literature Measure of Intellectual Structure." *Journal of the American Society for Information Science* 32 3: 163–171.

Griffith, Belver C., ed. 1980. *Key Papers in Information Science*. White Plains, NY: Knowledge Industry Publications.

Marshakova, Irina V. 1981. "Citation Networks in Information Science." *Scientometrics* 3, 1: 13–26.

Marshakova, Irina V. 1973. "A System of Document Connections Based on References." *Scientific and Technical Information Serial of VINITI* 6: 3–8.

Garfield, Eugene. 1981. *ISI Atlas of Science: Biochemistry and Molecular Biology 1979–1980*, 112. Philadelphia, PA: ISI.

Image Credits—Maps

See Garfield et al. 1964, courtesy of Thomson Reuters.

See de Solla Price 1965, courtesy of Eugene Garfield.

See Narin 1968, public domain.

See Small 1986, with kind permission from Springer Science and Business Media.

See Small 1973, reprinted with permission of John Wiley & Sons, Inc.

White images, *see* White and Griffith 1981; reproduced from Howard D. White's files.

See Marshakova 1981, 1973, with kind permission from Springer Science and Business Media

See Garfield 1981, courtesy of Thomson Reuters.

Image Credits—Portraits

Garfield portrait courtesy of Eugene Garfield and Thomson Reuters.

de Solla Price portrait is a public domain image by Malcolm S. Kirk and released into the public domain by Mark de Solla Price.

Narin portrait courtesy of Francis Narin.

Small portrait © Balázs Schlemmer.

White portrait courtesy of Howard D. White.

Tools

References

ESRI. 2008. ESRI Home Page. http://www.esri.com (accessed June 2, 2008).

Image Credits

Screenshot of Geographic Information System, courtesy of ESRI.

Books

References

Kuhn, Thomas Samuel. 1962. *The Structure of Scientific Revolutions*. Chicago: The University of Chicago Press.

Rogers, Everett M. 1962. *Diffusion of Innovations*. New York: Free Press.

de Solla Price, Derek J. 1965. *Little Science, Big Science*. New York: Columbia University Press.

Licklider, Joseph Carl Robnett. 1965. *Libraries of the Future*. Cambridge, MA: MIT Press.

Cronin, Blaise, ed. *Annual Review of Information Science and Technology*. Medford, NJ: Information Today, Inc. (Published annually 1966–2011).

Hägerstrand, Torsten. 1967. *Innovation Diffusion as a Spatial Process*. Chicago: The University of Chicago Press.

Nalimov, Vasily V., and Z. M. Mulchenko. 1969. *Naukometriya: Izuchenie Razvitiya Kauki ka Informatsionnogo Protessa* [Scientometrics: The Study of the Development of Science as an Information Process]. Moscow: Nauka.

Dobrov, G. M., V. N. Klimenjuk, L. P. Smirnov, and A. A. Savelev. 1969. *Potensial Nauki* [Capacity of Science]. Kiev: Naukova Dumka.

Dobrov, G. M. 1970. *Aktuelle Probleme der Wissenschaftswissenschaft*. Berlin: Dietz-Verlag.

Menard, Henry W. 1971. *Science: Growth and Change*. Cambridge, MA: Harvard University Press.

Crane, Diana. 1972. *Invisible Colleges: Diffusion of Knowledge in Scientific Communities*. Chicago: The University of Chicago Press.

National Science Board. 1972. *Science Indicators 1972.* Washington, DC: Superintendent of Documents, U.S. Government Printing Office.

Merton, Robert. 1973. *The Sociology of Science.* Chicago: The University of Chicago Press.

Garfield, Eugene. 1979. *Citation Indexing: Its Theory and Application in Science, Technology, and Humanities.* New York: Wiley.

Laitko, H. 1979. *Wissenschaft als allgemeine Arbeit.* Berlin: Akademie Verlag.

Popper, Karl. 1979. *Truth, Rationality, and the Growth of Scientific Knowledge.* Frankfurt am Main: Vittorio Klostermann.

Institute for Scientific Information. 1981. *ISI Atlas of Science Biochemistry and Molecular Biology 1978/80.* Philadelphia: ISI.

Fuller, Buckminster. 1981. *Critical Path.* New York: St. Martin's Press.

30 Timeline: 1982–1998

Algorithms

References

Garfield, Eugene. 1980. "ABCs of Cluster Mapping. Part 1. Most Active Fields in the Life Sciences in 1978." *Essays of an Information Scientist* 4, 1979–1980: 634. http://www.garfield.library.upenn.edu/essays/v4p634y1979-80.pdf (accessed October 30, 2007).

McCain, K. W. 1991. "Mapping Economics Through the Journal Literature: An Experiment in Journal Cocitation Analysis." *Journal of the American Society for Information Science* 42, 4: 290–296.

Callon, M., J. P. Courtial, W. A. Turner, and S. Bauin. 1983. "From Translations to Problematic Networks: An Introduction to Co-Word Analysis." *Social Science Information* 22: 191–235.

Eades, P. 1984. "A Heuristic for Graph Drawing." *Congressus Numerantium* 42: 149–160.

Kohonen, Teuvo. 1995. *Self-Organizing Maps.* Heidelberg: Springer-Verlag.

Kamada, T. and S. Kawai. 1989. "An Algorithm for Drawing General Undirected Graphs." *Information Processing Letters* 31: 7–15.

Garfield, Eugene and Henry G. Small. 1989. "Identifying the Changing Frontiers of Science." *Conference Proceedings of Innovation: At the Crossroads Between Science and Technology*, edited by M. Kranzberg, Y. Elkana, and Z. Tadmor, 51–65. Haifa, Israel: The S. Neaman Press. http://www.garfield.library.upenn.edu/papers/362/362.html (accessed October 30, 2007).

Garfield, Eugene. 1982. "Computer-Aided Historiography—How ISI Uses Cluster Tracking to Monitor the 'Vital Signs' of Science." *Essays of an Information Scientist* 5: 473. http://www.garfield.library.upenn.edu/essays/v5p473y1981-82.pdf (accessed February 9, 2010).

Schvaneveldt, R., ed. 1990. *Pathfinder Associative Networks: Studies in Knowledge Organization.* Norwood, NJ: Ablex Publishing Corp.

Deerwester, S., S. T. Dumais, G. W. Furnas, T. K. Landauer, and R. Harshman. 1990. "Indexing by Latent Semantic Analysis." *Journal of the American Society for Information Science* 41, 6: 391–407.

McCain, K. W. 1991. "Mapping Economics Through the Journal Literature: An Experiment in Journal Cocitation Analysis." *Journal of the American Society for Information Science* 42, 4: 290–296.

Johnson, B., and Ben Shneiderman. 1991. "Treemaps: A Space-Filling Approach to the Visualization of Hierarchical Information Structures." *Proceedings of the 2nd International IEEE Visualization Conference*, 284–291, San Diego, CA.

Shneiderman, Ben. 1992. "Tree Visualization with Tree-Maps: 2-D Space-Filling Approach." *ACM Transactions on Graphics* 11 1: 92–99. *See also* http://www.cs.umd.edu/hcil/treemap-history (accessed February 10, 2010).

Fruchterman, T. M.J., and E. M. Reingold. 1991. "Graph Drawing by Force-Directed Placement." *Software: Practice and Experience* 21, 11: 1129–1164.

Landauer, T. K., P. W. Foltz, and D. Laham. 1998. "Introduction to Latent Semantic Analysis." *Discourse Processes* 25: 259–284.

Small, Henry. 1995. "Navigating the Citation Network." *Proceedings of the 58th Annual Meeting of the American Society for Information Science* 32: 118–126, Chicago, October 9–12.

Small, Henry. 1997. "Update on Science Mapping: Creating Large Document Spaces." *Scientometrics* 38 2: 275–293.

Swanson, D. R. and N. R. Smalheiser. 1997. "An Interactive System for Finding Complementary Literatures: A Stimulus to Scientific Discovery." *Artificial Intelligence* 91: 183–203.

Visualizations

References

White, Howard D., and Belver C. Griffith. 1981. "Author Co-Citation: A Literature Measure of Intellectual Structure." *Journal of the American Society for Information Science* 32, 3: 163–171. "A Map of Information Science" (p. 165), reprinted by permission of John Wiley & Sons, Inc.

Halasz, Frank G., Thomas P. Moran, and Randall H. Trigg. 1987. "NoteCards in a Nutshell." *Proceedings of the ACM CHI+GI'87 Human Factors in Computing Systems and Graphics Interface Conference*, edited by J. M. Carroll and P. P. Tanner, 45–52. New York: ACM.

Ennis, J. G. 1992. "The Social Organization of Sociological Knowledge: Modeling the Intersection of Specialties." *American Sociological Review* 57, 2: 259–265.

Kohonen, Teuvo. 1995. *Self-Organizing Maps.* Heidelberg: Springer-Verlag.

WEBSOM. 1999. "WEBSOM map: Million documents." http://websom.hut.fi/websom/milliondemo/html/root.html (accessed October 30, 2007).

Mackinlay, J. D., R. Rao, and S. K. Card. 1995. "An Organic User Interface for Searching Citation Links." *Proceedings of the ACM SIGCHI Conference on Human Factors in Computing Systems*, Denver, CO, May 7–11. http://www.acm.org/sigchi/chi95/Electronic/documnts/papers/jdm_bdy.htm (accessed October 30, 2007).

Chen, Hsinchun, Andrea L. Houston, Robin R. Sewell, and Bruce R. Schatz. 1998. "Internet Browsing and Searching: User Evaluations of Category Map and Concept Space Techniques." Special Issue (Artificial Intelligence Techniques for Emerging Information Systems Applications), *Journal of the American Society for Information Science* 49, 7: 582–603.

Novak, Joseph D. 1998. *Learning, Creating, and Using Knowledge: Concept Maps as Facilitative Tools in Schools and Corporations.* Mahwah, NJ: Lawrence Erlbaum Associates.

Johnson, B., and Ben Shneiderman. 1991. "Treemaps: A Space-Filling Approach to the Visualization of Hierarchical Information Structures." *Proceedings of the 2nd International IEEE Visualization Conference*, 284–291, San Diego, CA.

Shneiderman, Ben. 1992. "Tree Visualization with Tree-Maps: 2-D Space-Filling Approach." *ACM Transactions on Graphics* 11, 1: 92–99. *See also* http://www.cs.umd.edu/hcil/treemap-history (accessed February 10, 2010).

White, Howard D., and Katherine W. McCain. 1998. "Visualizing a Discipline: An Author Co-citation Analysis of Information Science, 1972–1995." *Journal of the American Society for Information Science* 49, 4: 327–355.

Image Credits—Maps

See White & Griffith 1981 (p. 165), reprinted by permission of John Wiley & Sons, Inc.

See Halasz et al. 1987, courtesy of PARC (Palo Alto Research Center, Inc.).

See Ennis 1992 (table 1, p. 261), reproduced by permission of James G. Ennis.

See Kohonen 1995, with kind permission from Springer Science and Business Media.

See Mackinlay et al. 1995, © 1995 by Association for Computing Machinery, used by permission.

See Chen et al. 1998, courtesy of the Artificial Intelligence Lab, the University of Arizona.

See Novak 1998, courtesy of Joseph D. Novak.

See Shneiderman 1992, courtesy of the University of Maryland Human–Computer Interaction Lab.

See White and McCain 1998 (fig. 4), reprinted by permission of John Wiley & Sons, Inc.

Image Credits—Portraits

White portrait courtesy of Howard D. White.

Moran portrait courtesy of Lydia Moran.

Trigg portrait courtesy of Randy Trigg.

Kohonen portrait courtesy of Teuvo Kohonen.

Mackinlay portrait courtesy of Jock Mackinlay.

Rao portrait courtesy of Ramana Rao.

Card portrait courtesy of Stuart Card.

Chen portrait courtesy of the Artificial Intelligence Lab, the University of Arizona.

Novak portrait courtesy of Joseph Novak.

Shneiderman portrait courtesy of John Consoli, University of Maryland.

Tools

References

Tobler, W. 1987. "An Experiment in Migration Mapping by Computer." *The American Cartographer* 14, 2: 155–163.

Wise, J. A., J. J. Thomas, K. Pennock, D. Lantrip, M. Pottier, A. Schur, and V. Crow. 1995. "Visualizing the Non-Visual: Spatial Analysis and Interaction with Information from Text Documents." *Proceedings of the IEEE Symposium on Information Visualization*, edited by Nahum Gershon and Stephen G. Eick, 51–58. Los Alamitos, CA: IEEE Computer Society.

Batagelj, Vladimir, and Andrej Mrvar. 1998. "Pajek—Program for Large Network Analysis." *Connections* 21, 2: 47–57.

Nooy, W. D., Andrej Mrvar, and Vladimir Batagelj. 2005. *Exploratory Social Network Analysis with Pajek.* Cambridge: Cambridge University Press.

Swanson, D. R., and N. R. Smalheiser. 1997. "An Interactive System for Finding Complementary Literatures: A Stimulus to Scientific Discovery." *Artificial Intelligence* 91: 183–203.

University of Illinois at Chicago. Arrowsmith Project Home Page. http://arrowsmith.psych.uic.edu/arrowsmith_uic (accessed February 10, 2010).

Image Credits

"The Sandbox for Analysis—Concepts and Methods" by W. Wright et al. ACM CHI 2006. Oculus nSpace Sandbox® is a registered trademark of Oculus Info, Inc.

Screenshot of "Science and Technology Dynamics Toolbox" courtesy of Loet Leydesdorff.

Screenshot of "In Flow" courtesy of Valdis Krebs.

Screenshot of "Flow Mapper" courtesy of Waldo Tobler.

Screenshot of "IN-SPIRE" courtesy of Pacific Northwest National Laboratory.

Pajek courtesy of Andrej Mrvar, associate professor of Social Science Informatics, University of Ljubljana, Faculty of Social Sciences, Kardeljeva pl. 5, 1000 Ljubljana, Slovenia, and Vladimir Batagelj, University of Ljubljana FMF–Department of Mathematics and IMFM–Institute of Mathematics, Physics and Mechanics Department for Theoretical Computer Science Jadranska 19, 1000 Ljubljana, Slovenia.

Screenshot of Swanson and Smalheiser 1997 taken by Kristin Reed, Indiana University.

Data Credits

Kepler, T., and W. Wright. 2004. "GeoTime Information Visualization." *IEEE InfoVis 2004.* GeoTime is a registered trademark of Oculus Info, Inc. Used by permission.

Software Credits

Leydesdorff, Loet. 2005. "Software and Data of Loet Leydesdorff." University of Amsterdam. http://users.fmg.uva.nl/lleydesdorff/software.htm (accessed October 30, 2007).

Krebs, Valdis. 2008. "InFlow—Software for Social Network Analysis & Organizational Network Analysis." Orgnet.com. http://www.orgnet.com/inflow3.html (accessed June 2, 2008).

Regents of University of California, Santa Barbara, Center for Spatially Integrated Social Science. 2008. "Tobler's Flow Mapper." http://csiss.ncgia.ucsb.edu/clearinghouse/FlowMapper (accessed February 5, 2010). Tutorial from http://csiss.ncgia.ucsb.edu/clearinghouse/FlowMapper/MovementMapping.pdf (accessed June 2, 2008).

Berry, Michael, Theresa Do, Gavin O'Brien, Vijay Krishna, and Sowmini Varadhan. 1993. *SVDPACKC (Version 1.0) User's Guide.* University of Tennessee Technical Reports. http://www.netlib.org/tennessee/ut-cs-93-194.ps (accessed March 11, 2010).

Books

References

Boorstin, Daniel J. 1983. *The Discoverers.* New York: Random House.

Irvine, John, and Ben R. Martin. 1984. *Foresight in Science: Picking the Winners.* London: Frances Pinter Pub Ltd.

Cronin, Blaise. 1984. *The Citation Process: The Role and Significance of Citations in Scientific Communication.* London: Taylor Graham.

Whitley, Richard. 1984. *The Intellectual and Social Organization of the Sciences*. New York: Oxford University Press.

Garfield, Eugene, and James E. Shea, eds. 1984. *ISI Atlas of Science: Biotechnology and Molecular Genetics 1981/82 and Bibliographic Update for 1983/84*. Philadelphia: ISI.

Bourdieu, Pierre. 1984. *Homo Academicus*. Paris: Editions de Minuit.

de Solla Price, Derek J. 1986. *Little Science, Big Science … and Beyond*. New York: Columbia University Press.

Latour, Bruno, and Steve Woolgar. 1986. *Laboratory Life: The Construction of Scientific Facts*. Princeton, NJ: Princeton University Press.

Callon, Michel, John Law, and Arie Rip, eds. 1986. *Mapping the Dynamics of Science and Technology: Sociology of Science in the Real World*. London: The Macmillan Press.

Yablonskij, Anatolij Ivanovich. 1986. *Matematicheskije modeli v issledovanii nauki* [Mathematical Models in the Exploration of Science]. Moscow: Nauka.

Garfield, Eugene. 1986. *Essays of an Information Scientist: Towards Scientography*, vol. 9. Philadelphia, PA: ISI Press.

Latour, Bruno. 1987. *Science in Action: How to Follow Scientists and Engineers Through Society*. Milton Keynes, UK: Open University Press.

Dumenton, G. G. 1987. *Seti nauchnykh kommunikatsii* [Networks of Scientific Communication]. Moscow: Nauka.

van Raan, Anthony F. J., ed. 1988. *Handbook of Quantitative Studies of Science and Technology*. Amsterdam: Elsevier Science Publishers.

Marshakova, Irina V. 1988. *Sistema Tsitirovaniya* [System of Citing]. Moscow: Nauka.

Kara-Mursa, S. G. 1988. *Problemy intensifikatsii nauki: tekhnologiya nauchnykh issledovanii* [Problems of Intensification of Science: Scientific Research Technology]. Moscow: Nauka.

Turnbull, David. 1989. *Maps Are Territories: Science Is an Atlas*. Geelong, Australia: Deakin University Press.

Martin, Ben R., and John Irvine. 1989. *Research Foresight: Priority-Setting in Science*. London: Pinter Publishers.

Egghe, Leo, and Ronald Rousseau. 1990. *Introduction to Informetrics: Quantitative Methods in Library, Documentation and Information Science*. Amsterdam: Elsevier Science Publishers.

Braam, Robert R., H. F. Moed, and A. F. J. van Raan. 1991. *Mapping of Science: Foci of Intellectual Interest in Scientific Literature*. Leiden, Netherlands: DSWO Press.

Borgman, C. L., ed. 1992. *Scholarly Communication and Bibliometrics*. Newbury Park, CA: Sage Publications.

Scott, John. 1992. *Network Analysis: A Handbook*. Newbury Park, CA: Sage Publications.

Cole, S. 1992. *Making Science: Between Nature and Society*. Cambridge, MA: Harvard University Press.

Wasserman, Stanley, and K. Faust. 1994. *Social Network Analysis: Methods and Applications*. Structural Analysis in the Social Sciences 8. New York: Cambridge University Press.

Gibbons, Michael, Camille Limoges, Helga Nowotny, Simon Schwartzman, Peter Scott, and Martin Trow. 1994. *The New Production of Knowledge: The Dynamics of Science and Research in Contemporary Societies*. New York: Sage Publications.

Valente, Thomas W. 1995. *Network Models of the Diffusion of Innovations*. Cresskill, NJ: Hampton Press.

Leydesdorff, Loet. 1995. *The Challenge of Scientometrics: The Development, Measurement, and Self-Organization of Scientific Communications*. Leiden: DSWO Press/Leiden University.

Nagurney, Anna, and Stavros Siokos. 1997. *Financial Network: Statics and Dynamics*. Berlin: Springer-Verlag.

Christensen, Clayton M. 1997. *The Innovator's Dilemma*. Boston: Harvard Business School Press.

32 Timeline: 1998–2000

Algorithms

References

Giles, C. Lee, Kurt D. Bollacker, and Steve Lawrence. 1998. "CiteSeer: An Automatic Citation Indexing System." *Proceedings of the 3rd ACM Conference on Digital Libraries (DL'98)*, edited by Ian Witten, Rob Akscyn, and Frank M. Shipman III, 89–98. New York: ACM Press. http://clgiles.ist.psu.edu/papers/DL-1998-citeseer.pdf (accessed October 30, 2007).

Brin, Sergey, and Lawrence Page. 1998. "The Anatomy of a Large-Scale Hypertextual Web Search Engine." *Proceedings of the 7th International World Wide Web Conference*. http://infolab.stanford.edu/~backrub/google.html (accessed September 30, 2009).

Page, Lawrence, Sergey Brin, Rajeev Motwani, and Terry Winograd. 1998. "The PageRank Citation Ranking: Bringing Order to the Web." Stanford Digital Library Technologies Project. http://infolab.stanford.edu/~backrub/pageranksub.ps (accessed June 2, 2008).

Kleinberg, Jon M. 1998. "Authoritative Sources in a Hyperlinked Environment." *Proceedings of the Ninth Annual ACM-SIAM Symposium on Discrete Algorithms*, 668–677.

Kleinberg, Jon M. 1999. "Authoritative Sources in a Hyperlinked Environment." *Journal of the ACM* 46, 5: 604–632.

Visualizations

References

SmartMoney. 2007. "Map of the Market." http://www.smartmoney.com/marketmap (accessed February 10, 2010).

Dodge, Martin. 2001. "Show Me the Money: The Map of the Market." *Mappa.Mundi Magazine*. http://mappa.mundi.net/maps/maps_023 (accessed October 30, 2007).

Chen, C., L. Thomas, J. Cole, and C. Chennawasin. 1999. "Representing the Semantics of Virtual Spaces." *IEEE Multimedia* 6, 2: 54–63.

Small, Henry. 1999. "Visualizing Science by Citation Mapping." *Journal of the American Society for Information Science* 50, 9: 799–813.

Hobbs, Robert, and Mark Lombardi. 2003. *Global Networks*, edited by Judith Richards. New York: Independent Curators International.

Pierogi Gallery. 2005. "Mark Lombardi." http://www.pierogi2000.com/flatfile/lombardi.html (accessed October 30, 2007).

MicroPatent. 2005. "Aureka: Turning Information into Actionable Intelligence." Thomson Reuters Corporation. http://www.micropatent.com/static/aureka.htm; http://www.infovis.net/imagenes/T11_N2_Newsmaps.jpg (accessed October 30, 2007).

Leuski, Anton, and James Allan. 2000. "Lighthouse: Showing the Way to Relevant Information." *Proceedings of the IEEE Symposium on Information Visualization 2000*, 125–130. Los Alamitos, CA: IEEE Computer Society.

Börner, Katy, and Chaomei Chen. 2002. "Visual Interfaces to Digital Libraries: Motivation, Utilization and Socio-Technical Challenges." In *Lecture Notes in Computer Science: Visual Interfaces to Digital Libraries*, edited by Katy Börner and Chaomei Chen, vol. 2539: 1–9. Berlin: Springer-Verlag.

Image Credits—Maps

Map of the Market courtesy of Martin Wattenberg, IBM, and SmartMoney.com.

See Chen et al. 1999, copyright 1999, IEEE.

See Small 1999, reprinted with permission of John Wiley & Sons, Inc.

Lombardi visualization, collection of Mr. and Mrs. Michael Scott. Photo by John Berens.

Aureka Software screenshot courtesy of Thomson Reuters.

Lighthouse System by Anton Leuski.

Screenshot of Visual Net, courtesy of Antarctica Systems, Inc.

Image Credits—Portraits

Wattenberg portrait courtesy of Jonathan Feinberg.

Chen portrait courtesy of Chaomei Chen.

Small portrait © Balázs Schlemmer.

Lombardi portrait courtesy of Pierogi Gallery, Brooklyn, NY.

References

Klavans, Richard, and Kevin W. Boyack. 2006. "Quantitative Evaluation of Large Maps of Science." *Scientometrics* 68, 3: 475–499.

Davidson, George S., B. Hendrickson, David K. Johnson, C. E. Meyers, and Brian N. Wylie. 1998. "Knowledge Mining with VxInsight: Discovery Through Interaction." *Journal of Intelligent Information Systems* 11, 3: 259–285.

Davidson, George S., Brian N. Wylie, and Kevin W. Boyack. 2001. "Cluster Stability and the Use of Noise in Interpretation of Clustering." *Proceedings of IEEE Information Visualization 2001*, 23–30. Los Alamitos, CA: IEEE.

Informatik 5. 2008. DocMINER Home Page. RWTH Aachen University. http://www-i5.informatik.rwth-aachen.de/i5new/projects/DocMINER/DocMINER.html (accessed June 2, 2008).

CERN—European Organization for Nuclear Research (Hoschek). 1999. Colt Project Home Page. http://www-itg.lbl.gov/~hoschek/colt (accessed October 30, 2007).

Search Technology, Inc. 2008. Vantage Point Home Page. http://www.thevantagepoint.com (accessed June 2, 2008).

Gahegan, M., M. Takatsucka, M. Wheeler, and F. Hardisty. 2002. "GeoVISTA Studio: An Integrated Suite of Visualization and Computation Methods for Exploration and Knowledge Construction in Geography." *Computer, Environment and Urban Systems* 26 4: 267–292.

Image Credits

DrL, courtesy of Sandia National Laboratory, *see also* Davidson et al. 2001.

VxInsight, courtesy of Sandia National Laboratory, *see also* Davidson et al. 1998.

DocMINER (Document Maps for Information Elicitation and Retrieval) is hosted at Informatik 5, Aachen University.

Vantage Point Tool, *see also* Porter, Alan L., and Scott W. Cunningham. 2005. *Tech Mining*. Hoboken, NJ: John Wiley & Sons, Inc.

GeoVISTA Studio created by Mark Gahegan, Masa Takatsuma, Frank Hardisty, Alan MacEachren, and many other researchers at the GeoVISTA Center, Penn State University.

Books

References

Wilson, E. O. 1998. *Consilience: The Unity of Knowledge*. New York: Knopf.

Porter, Michael E. 1998. *The Competitive Advantage of Nations*. New York: The Free Press.

Wouters, Paul. 1999. "The Citation Culture." PhD diss., University of Amsterdam. http://garfield.library.upenn.edu/wouters/wouters.pdf (accessed October 30, 2007).

Gieryn, Thomas F. 1999. *Cultural Boundaries of Science: Credibility on the Line*. Chicago: The University of Chicago Press.

Watts, D. J. 1999. *Small Worlds*. Princeton, NJ: Princeton University Press.

Boutellier, Roman, Oliver Gassman, and Maximilian von Zedtwitz. 2000. *Managing Global Innovation: Uncovering the Secrets of Future Competitiveness*. Heidelberg: Springer-Verlag.

Cronin, Blaise, and Helen Barsky Atkins, eds. 2000. *The Web of Knowledge: A Festschrift in Honor of Eugene Garfield*. ASIS Monograph Series. Medford, NJ: Information Today.

Bloom, H. 2000. *Global Brain: The Evolution of Mass Mind from the Big Bang to the 21st Century*. New York: John Wiley & Sons, Inc.

Headrick, Daniel R. 2000. *When Information Came of Age: Technologies of Knowledge in the Age of Reason and Revolution, 1700–1850*. New York: Oxford University Press.

Geisler, Eliezer. 2000. *The Metrics of Science and Technology*. Westport, CT: Quorum Books.

34 Timeline: 2000–2001

Algorithms

References

Brandes, Ulrik. 2001. "A Faster Algorithm for Betweenness Centrality." *Journal of Mathematical Sociology* 25, 2: 163–177.

Kleinberg, J. M. 2002. "Bursty and Hierarchical Structure in Streams." *Proceedings of the 8th ACM SIGKDD International Conference on Knowledge Discovery and Data Mining*, 91–101. New York: ACM Press.

Griffiths, Thomas L., and Mark Steyvers. 2002. "A Probabilistic Approach to Semantic Representation." *Proceedings of the 24th Annual Conference of the Cognitive Science Society*, 381–386. George Mason University, Fairfax, VA, August 7–10.

Griffiths, Thomas L., and Mark Steyvers. 2004. "Finding Scientific Topics." *PNAS* 101, Suppl. 1: 5228–5235.

Richardson, Matthew, and Pedro Domingos. 2001. "The Intelligent Surfer: Probabilistic Combination of Link and Content Information in PageRank."

Proceedings of Advances in Neural Information Processing Systems 14. Cambridge, MA: MIT Press.

Girvan, M., and M. Newman. 2002. "Community Structure in Social and Biological Networks." *PNAS* 99: 7821–7826.

Visualizations

References

Chen, Chaomei, and R. J. Paul. 2001. "Visualizing a Knowledge Domain's Intellectual Structure." *IEEE Computer* 34, 3: 65–71.

Newman, Mark E. J. 2004. "Who Is the Best Connected Scientist? A Study of Scientific Coauthorship Networks." In *Complex Networks, Lecture Notes in Physics* 650, edited by E. Ben-Naim, H. Frauenfelder, and Z. Toroczkai, 337–370. Berlin: Springer-Verlag.

Newman, Mark E. J. 2000. "Who Is the Best Connected Scientist? A Study of Scientific Coauthorship Networks." SFI Working Paper 00-12-64. Santa Fe Institute. http://www.santafe.edu/media/workingpapers/00-12-064.pdf (accessed February 10, 2010).

Old, L. John. 1999. "Spatial Representation and Analysis of Co-Citation Data on the 'Canonical 75': Re-viewing White and McCain." PhD Research Project L710, Research in Library and Information Science, School of Library and Information Science, Indiana University.

Wattenberg, Martin. 2002. "Idea Line." Whitney Artport: The Whitney Museum Portal to Net Art. http://artport.whitney.org/commissions/idealine.shtml (accessed March 30, 2010).

Andrews, Keith, Vedran Sabol, Wilfried Lackner, Christian Gütl, and Josef Moser. 2001. "Search Result Visualisation with xFIND." *Proceedings of the Second International Workshop on User Interfaces to Data Intensive Systems (UIDIS'01)*, edited by Epaminondas Kapetanios and Hans Hinterberger, 50–58. Los Alamitos, CA: IEEE Computer Society.

Zhu, Donghua, and Porter, Alan L. 2002. "Automated Extraction and Visualization of Information for Technological Intelligence and Forecasting." *Technological Forecasting & Social Change* 69, 5: 495–506.

Paley, W. Bradford. 2002. "Illuminated Diagrams: Using Light and Print to Comparative Advantage." http://www.textarc.org/appearances/InfoVis02/InfoVis02_IlluminatedDiagrams.pdf (accessed October 30, 2007).

Image Credits—Maps

Kartoo Visual Interface to Search Results courtesy of Hamza Habib.

See also Chen and Paul 2001, © 2001, IEEE.

See Newman 2004, with kind permission from Springer Science and Business Media.

See Old 1999.

See Wattenberg 2002.

See Andrews et al. 2001, © 2001, IEEE.

See Zhu and Porter 2002, reprinted with permission from Elsevier.

Courtesy of W. Bradford Paley.

Image Credits—Portraits

Chen portrait courtesy of Chaomei Chen.

Old portrait courtesy of John Old.

Wattenberg portrait courtesy of Jonathan Feinberg.

Zhu portrait courtesy of Donghua Zhu.

Porter portrait courtesy of Alan Porter.

Paley portrait courtesy of W. Bradford Paley.

Software Credits

Kartoo. 2007. "Kartoo Visual Meta Search Engine." Previously available at http://www.kartoo.com (accessed October 30, 2007).

Tools

References

Baur, Michael, Marc Benkert, Ulrik Brandes, Sabine Cornelsen, Marco Gaertler, Boris Köpf, Jürgen Lerner, and Dorothea Wagner. 2001. "Visone." *Graph Drawing 2001*, edited by G. Goos, J. Hartmanis, and J. van Leeuwen, 463–464. Berlin: Springer-Verlag.

Kapler, Thomas and William Wright. 2005. "GeoTime Information Visualization." *Information Visualization* 4, 2: 136–146.

Image Credits

UCINet courtesy of Steve P. Borgatti.

Visone Tool courtesy of Ulrik Brandes.

E-Mail Analysis courtesy of ADVIZOR Solutions, Inc.

Infoscope courtesy of Macrofocus GmbH.

Visual Thesaurus, © 1998–2007 Thinkmap, Inc. All rights reserved.

They Rule courtesy of Josh On.

TouchGraph courtesy of TouchGraph LLC, http://touchgraph.com.

GeoTime is a registered trademark of Oculus Info, Inc., *see* Kapler and Wright 2004.

Issue Ticker, Touchscreen version (2005) courtesy of Govcom.org Foundation, Amsterdam.

Software Credits

Borgatti, S. P., M. G. Everett, and L. C. Freeman. 2002. "Ucinet for Windows: Software for Social Network Analysis." Harvard, MA: Analytic Technologies. http://www.analytictech.com/ucinet (accessed February 10, 2010).

Brandes, Ulrik, and Dorothea Wagner. 2010. "Visone: Analysis and Visualization of Social Networks." http://visone.info (accessed March 30, 2010).

© Advizor Solutions, Inc. Used by permission.

Macrofocus GmbH. 2008. "Macrofocus." http://www.macrofocus.com (accessed June 2, 2008).

Rogers, Richard. 2002. "The Issue Crawler: Towards a Live Social Science on the Web." *EASST Review* 21, 3/4: 8-11.

Thinkmap, Inc. 2008. "Thinkmap Visual Thesaurus." http://www.visualthesaurus.com (accessed June 2, 2008).

On, Josh. 2008. "They Rule." http://www.theyrule.net (accessed October 30, 2007).

TouchGraph, LLC. 2007. "TouchGraph Google Browser." http://www.touchgraph.com/TGGoogleBrowser.html (accessed June 10, 2008).

Oculus Info, Inc. 2006. "Oculus Excel Visualizer." http://www.oculusinfo.com/papers/ExcelVizWhitepaper-final.pdf (accessed January 1, 2008).

Books

References

Leydesdorff, Loet. 2001. *A Sociological Theory of Communications: The Self-Organization of the Knowledge-Based Society*. Boca Raton, FL: Universal Publishers.

Tuomi, Ilkka. 2002. *Networks of Innovation: Change and Meaning in the Age of the Internet*. New York: Oxford University Press.

Jaffe, Adam B., and Manuel Trajtenberg. 2002. *Patents, Citations and Innovations: A Window on the Knowledge Economy*. Cambridge, MA: MIT Press.

Börner, Katy, and Chaomei Chen. 2002. "Visual Interfaces to Digital Libraries: Motivation, Utilization and Socio-Technical Challenges." In *Lecture Notes in Computer Science: Visual Interfaces to Digital Libraries*, edited by Katy Börner and Chaomei Chen, vol. 2539: 1–9. Berlin: Springer-Verlag.

Chen, Chaomei. 2002. *Mapping Scientific Frontiers*. London: Springer-Verlag.

Barabási, Albert-László. 2002. *Linked: How Everything Is Connected to Everything Else and What It Means*. Cambridge, MA: Perseus Books Group.

36 Timeline: 2002–2004

Algorithms

References

Xing, Wenpu, and Ali Ghorbani. 2004. "Weighted PageRank Algorithm." In *Second Annual Conference on Communication Networks and Services Research (CNSR'04)*, 305–314. Los Alamitos, CA: IEEE Computer Society.

Visualizations

References

Skupin, André. 2002. "A Cartographic Approach to Visualizing Conference Abstracts." *IEEE Computer Graphics and Applications* 22, 1: 50–58.

Visualization and Information Retrieval Research Group. 2004. "VisualLink: Associative Information Visualizer." Drexel University. http://project.cis.drexel.edu/infovis/ (accessed March 30, 2010).

Visualization and Information Retrieval Research Group. 2010. "Citation Mapping and Visualization." Drexel University. http://project.cis.drexel.edu/authorlink/ (accessed March 30, 2010).

Lin, X., H. D. White, and J. Buzydlowski. 2003. "Real-Time Author Co-Citation Mapping for Online Searching." *Information Processing and Management* 39, 5: 689–706.

White, H. D., X. Lin, J. Buzydlowski, and Chaomei Chen. 2004. "User Controlled Mapping of Significant Literatures." *PNAS* 101, Suppl. 1: 5297–5302.

White, H. D. 2001. "Author-Centered Bibliometrics through CAMEOs: Characterizations Automatically Made and Edited Online." *Scientometrics* 51, 3: 607-637.

Lenoir, Timothy. 2004. "Emerging from the Digital Dark Ages: Challenges and Opportunities for the History of Science and Technology in the Information Age." In *Technikforschung: zwischen Reflexion und Dokumentation*, edited by Roland Ris, 11–26. Bern: Swiss Academy of Humanities and Social Sciences.

Lenoir, Timothy. 2007. "Making Studies in New Media Critical." In *MediaArtHistories*, edited by Oliver Grau, 355–380. Cambridge, MA: MIT Press.

Stanford University Biochemistry Department Genealogy. "hpsCollaboratory." Stanford University. http://hpslab.stanford.edu/projects/StanfordBiochemistry/genealogy.html (accessed October 30, 2007).

Batty, M. 2003. "The Geography of Scientific Citation." *Environment and Planning A* 35, 5: 761–765.

Batty, Michael. 2005. *Cities and Complexity: Understanding Cities with Cellular Automata, Agent-Based Models, and Fractals*. Cambridge, MA: MIT Press.

Boyack, Kevin W., and Katy Börner. 2003. "Indicator-Assisted Evaluation and Funding of Research: Visualizing the Influence of Grants on the Number and Citation Counts of Research Papers." Special issue (Visualizing Scientific Paradigms), *Journal of the American Society of Information Science and Technology* 54, 5: 447–461.

Kutz, Daniel. 2004. "Examining the Evolution and Distribution of Patent Classifications." *Proceedings of the 8th International Conference on Information Visualisation (IV'04)*, edited by Ebad Banissi, 983–988. Los Alamitos, CA: IEEE Computer Society.

Chen, Chaomei, and J. Kuljis. 2003. "The Rising Landscape: A Visual Exploration of Superstring Revolutions in Physics." *Journal of the American Society for Information Science and Technology* 54, 5: 435–446.

Chen, Chaomei. 2006. "CiteSpace II: Detecting and Visualizing Emerging Trends and Transient Patterns in Scientific Literature." *Journal of the American Society for Information Science and Technology* 57, 3: 359–377.

Chen, Chaomei. 2004. "Searching for Intellectual Turning Points: Progressive Knowledge Domain Visualization." *PNAS* 101, Suppl. 1: 5303–5310.

Moody, James. 2004. "The Structure of a Social Science Collaboration Network." *American Sociological Review* 69: 213–238.

Moody, James, and Ryan Light. 2006. "A View from Above: The Evolving Sociological Landscape." *The American Sociologist* 37: 67–86.

Mane, Ketan, and Katy Börner. 2004. "Mapping Topics and Topic Bursts in *PNAS*." *PNAS* 101, Suppl. 1: 5287–5290.

Image Credits—Maps

See Skupin 2002, © 2002, IEEE.

AuthorLink and ConceptLink courtesy of Xia Lin.

Scholarly Genealogies courtesy of Timothy Lenoir and Casey Alt.

Geography of Science, reprinted by permission of Pion Limited, London.

Linking Papers and Funding, *see* Boyack and Börner 2003.

Evolution and Distribution of Patent Classifications courtesy of Daniel O. Kutz.

Mapping Scientific Frontiers courtesy of Chaomei Chen.

See Moody and Light 2006, with kind permission from Springer Science and Business Media.

See Mane and Börner 2004, © 2004, National Academy of Sciences, USA.

Image Credits—Portraits

Skupin portrait courtesy of Marinta Skupin.

Lin portrait courtesy of Xia Lin.

White portrait courtesy of Howard D. White.

Buzydlowski portrait courtesy of Jan W. Buzydlowski.

Lenoir portrait courtesy of Timothy Lenoir.

Batty portrait courtesy of Michael Batty.

Boyack portrait © Balázs Schlemmer.
Kutz portrait courtesy of Daniel O. Kutz.
Chen portrait courtesy of Chaomei Chen.
Moody portrait courtesy of James Moody.
Mane portrait courtesy of Ketan Mane.

Tools

References

Information Visualization Laboratory, School of Library and Information Science, Indiana University. 2005. "InfoVis Cyberinfrastructure." **http://iv.slis.indiana.edu** (accessed October 30, 2007).

Bender-deMoll, Skye, and Daniel A. McFarland. 2003. "SoNIA: Social Network Image Animator." **http://www.stanford.edu/group/sonia**; **http://sourceforge.net/projects/sonia** (accessed October 30, 2007).

Auber, David. 2003. "Tulip: A Huge Graph Visualisation Framework." In *Graph Drawing Softwares, Mathematics and Visualization*, edited by Petra Mutzel and Michael Jünger, 105–126. Berlin: Springer-Verlag.

Chen, Chaomei. 2006. "CiteSpace II: Detecting and Visualizing Emerging Trends and Transient Patterns in Scientific Literature." *Journal of the American Society for Information Science and Technology* 57, 3: 359–377.

Garfield, Eugene. 2004. "Historiographic Mapping of Knowledge Domains Literature." *Journal of Information Science* 30, 2: 119–145. **http://garfield.library.upenn.edu/papers/jis30%282%29p119-145y2004.pdf** (accessed October 30, 2007).

HistCite Software LLC. 2008. "HistCite: Bibliometric Analysis and Visualization Software." **http://www.histcite.com** (accessed June 10, 2008).

Fekete, Jean-Daniel. 2004. "The Infovis Toolkit." *Proceedings of the 10th IEEE Symposium on Information Visualization*, 167–174. IEEE Press.

Image Credits

StoCNET courtesy of Mark Huisman.
JUNG courtesy of Joshua O'Madadhain
Information Visualization Cyberinfrastructure courtesy of Katy Börner, Indiana University.
SoNIA (Social Network Image Animator) courtesy of ATA SPA.
Refviz Overview of Literature Search Results courtesy of Thomson Reuters.
Tulip Software courtesy of David Auber, LaBRI Université Bordeaux 1.
Dynet Dynamic Network Software Package courtesy of ATA SPA.
Citespace Trend Analysis courtesy of Chaomei Chen.
HistCite Historiography courtesy of Eugene Garfield.
The Infovis Toolkit courtesy of Jean Daniel Fekete and Catherine Plaisant.

Software Credits

Huisman, M., and M.A.J. van Duijn. 2003. "StOCNET: Software for the Statistical Analysis of Social Networks." *Connections* 25, 1: 7–26.

O'Madadhain, Joshua, Danyel Fisher, and Scott White. 2008. "JUNG: Java Universal Network/Graph Framework." University of California, Irvine. **http://jung.sourceforge.net** (accessed May 29, 2008).

Baumgartner, Jason, Katy Börner, N. J. Deckard, and N. Sheth. 2003. "An XML Toolkit for an Information Visualization Software Repository." *IEEE Information Visualization Conference Poster Compendium*, 72–73.

Thomson Reuters. 2008. "RefViz: Explore Research Literature . . . Visually!" **http://www.refviz.com** (accessed June 2, 2008).

Advanced Technology Assessment SpA. "DyNet LS Tool." **http://www.atalab.com/software/dynet** (accessed October 30, 2007).

Chen, Chaomei. 2008. "CiteSpace: Visualizing Patterns and Trends in Scientific Literature." **http://cluster.cis.drexel.edu/~cchen/citespace** (accessed June 10, 2008).

Books

References

Monge, Peter R., and Noshir S. Contractor. 2003. *Theories of Communication Networks*. New York: Oxford University Press.

Watts, Duncan. 2003. *Six Degrees: The Science of a Connected Age*. New York: W. W. Norton & Company.

Livingstone, David N. 2003. *Putting Science in Its Place: Geographies of Scientific Knowledge*. Chicago: The University of Chicago Press.

Shiffrin, Richard M., and Katy Börner. 2004."Mapping Knowledge Domains." *PNAS* 101, Suppl. 1: 5183–5185.

Moed, H. F., W. Glänzel, and U. Schmoch. 2004. *Handbook of Quantitative Science and Technology Research: The Use of Publication and Patent Statistics in Studies of S&T Systems*. Dordrecht: Kluwer Academic Publishers/Springer-Verlag.

Freeman, Linton C. 2004. *The Development of Social Network Analysis: A Study in the Sociology of Science*. Vancouver, Canada: Empirical Press.

Scotchmer, Suzanne. 2004. *Innovation and Incentives*. Cambridge, MA: MIT Press.

38 Timeline: 2004-2005

Algorithms

References

Kobourov, Stephen G., and Kevin Wampler. 2005. "Non-Euclidean Spring Embedders." *IEEE Transactions on Visualization and Computer Graphics*, 11, 6: 757–767. Piscataway, NJ: IEEE Educational Activities Department.

Phan, D., L. Xiao, R. Yeh, P. Hanrahan, and T. Winograd. 2005. "Flow Map Layout." *Proceedings of the IEEE Symposium on Information Visualization*, 29: 219–224. **http://graphics.stanford.edu/papers/flow_map_layout/flow_map_layout.pdf** (accessed March 16, 2010).

Councill, Isaac G., C. Lee Giles, Hui Han, and Eren Manavoglu. 2005. "Automatic Acknowledgment Indexing: Expanding the Semantics of Contribution in the CiteSeer Digital Library." *Proceedings of the 3rd International Conference on Knowledge Capture*, 19–26. New York: ACM Press.

Cronin, Blaise, Gail McKenzie, Lourdes Rubio, and Sherrill Weaver-Wozniak. 1993 "Accounting for Influence: Acknowledgments in Contemporary Sociology." *Journal of the American Society for Information Science* 44, 7: 406–412.

Visualizations

References

Skupin, André. 2002. "A Cartographic Approach to Visualizing Conference Abstracts." *IEEE Computer Graphics and Applications* 22, 1: 50–58.

Skupin, André. 2004. "The World of Geography: Visualizing a Knowledge Domain with Cartographic Means." *PNAS* 101, Suppl. 1: 5274–5278.

Nesbitt, Keith V. 2004. "Getting to More Abstract Places Using the Metro Map Metaphor." *Proceedings of the 8th International Conference on Information Visualisation*, 488–493. Washington, DC: IEEE Computer Society.

Ke, Weimao, Katy Börner, and Lalitha Viswanath. 2004. "Analysis and Visualization of the IV 2004 Contest Dataset." *IEEE Information Visualization Conference 2004 Poster Compendium*, 49–50. **http://conferences.computer.org/InfoVis/files/compendium2004.pdf** and **http://iv.slis.indiana.edu/ref/iv04contest/Ke-Borner-Viswanath.gif** (animated version) (accessed October 30, 2007).

Bender-DeMoll, Skye, Nocola Nottoli, and J. A. Rodriguez. 2005. "Critical Paths and Trajectories in Networks." Lucca, Italy: Advanced Technology Assessment. **http://www.atalab.com/technology** (accessed May 29, 2008).

Moody, James, Daniel A. McFarland, and Skye Bender-DeMoll. 2005. "Dynamic Network Visualization: Methods for Meaning with Longitudinal Network Movies." *American Journal of Sociology* 110: 1206–1241.

Bender-DeMoll, Skye, and Daniel A. McFarland. 2006. "The Art and Science of Dynamic Network Visualization." *Journal of Social Structure* 7, 2. **http://www.cmu.edu/joss/content/articles/volume7/deMollMcFarland** (accessed October 30, 2007).

Coombs, C., J. Dawes, and A. Tversky. 1970. Mathematical Psychology: An Elementary Introduction, 73–75. Englewood Cliffs, NJ: Prentice-Hall.

Tobler, Waldo. 1995. "Migration: Ravenstein, Thornthwaite, and Beyond." *Urban Geography* 16, 4: 327–343.

Smith, Marc. 2002. "Tools for Navigating Large Social Cyberspace." *Communications of the ACM* 45, 4: 51–55.

Morris, Steven A., and Kevin W. Boyack. 2005. "Visualizing 60 Years of Anthrax Research." *Proceedings of the 10th International Conference of the International Society for Scientometrics and Informetrics*, edited by P. Ingwerson and B. Larsen, 45–55. Stockholm: Karolinska University Press.

Boyack, Kevin W., Richard Klavans, and Katy Börner. 2005. "Mapping the Backbone of Science." *Scientometrics* 64, 3: 351–374.

Image Credits—Maps

GIS Map of Geography, © 2005, André Skupin.
Subway Domain Map: Nesbitt, Keith V. 2004. PhD diss. map. Newcastle, Australia. Courtesy of IEEE and Keith V. Nesbitt, Charles Sturt University, Australia; 2004, IEEE.
Critical Paths and Trajectories of Individuals courtesy of ATA SPA.
See Moody et al. 2005.
Journal Flow Map of Data by Coombs et al., courtesy of Waldo Tobler, CSISS.org
Treemap View of 2004 Usenet Returnees courtesy of Marc A. Smith, Microsoft Research.
Crossmap of Anthrax Research courtesy of Steven Morris.
Backbone of Science courtesy of Kevin W. Boyack, Richard Klavans, and Katy Börner.

Image Credits—Portraits

Skupin portrait courtesy of Marinta Skupin.
Conway portrait courtesy of Dan Conway.
Ke portrait courtesy of Weimao Ke.
Moody portrait courtesy of James Moody.
McFarland portrait courtesy of Daniel A. McFarland.
Bender-DeMoll portrait courtesy of Skye Bender-DeMoll.
Tobler portrait courtesy of Waldo Tobler.
Smith portrait courtesy of Marc A. Smith.
Fisher portrait courtesy of Danyel Fisher.
Morris portrait © Balázs Schlemmer.
Boyack portrait © Balázs Schlemmer.
Klavans portrait © Balázs Schlemmer.

Tools

References

Heer, Jeffrey, Stuart K. Card, and James A. Landay. 2005. "Prefuse: A Toolkit for Interactive Information Visualization." *Proceedings of the 2005 Conference on Human Factors in Computing Systems, CHI 2005*, edited by Gerrit C. van der Veer and Carolyn Gale, 421–430. New York: ACM Press.

Adar, Eytan. 2006. "GUESS: A Language and Interface for Graph Exploration." *Proceedings of the SIGCHI Conference on Human Factors in Computing Systems*, edited by Rebecca Grinter, Thomas Rodden, Paul Aoki, Ed Cutrell, Robin Jeffries, and Gary Olso, 791–800. New York: ACM Press.

Torvik, Vetle I., Marc Weeber, Don R. Swanson, and Nell R. Smalheiser. 2005. "A Probabilistic Similarity Metric for Medline Records: A Model for Author Name Disambiguation." *Journal of the American Society for Information Science and Technology* 56, 2: 140–158.

Douglas, Shawn M., Gaetano T. Montelione, and Mark Gerstein. 2005. "PubNet: A Flexible System for Visualizing Literature-Derived Networks." *Genome Biology* 6, 9: R80.

Ichise, Ryutaro, Hideaki Takeda, and Kosuke Ueyama. 2005. "Community Mining Tool using Bibliography Data." *Proceedings of the 9th International Conference on Information Visualization*, 953–958. Washington, DC: IEEE Computer Society.

OmniViz, Inc. 2008. "Unpublished visualization." **http://www.biowisdom.com/tag/omniviz** (accessed February 10, 2010).

Tools—Graphics

NetVis Dynamic Visualization of Social Networks courtesy of Jonathon N. Cummings.
R Statistical Computing Language, © R Foundation, from **http://www.r-project.org**.
Prefuse Visualization API as captured by Russell J. Duhon.
GUESS Graph Exploration System, Eytan Adar, GUESS.
Author-Name Disambiguation courtesy of Vetle I. Torvik.
See Douglas et al. 2005, courtesy of Shawn M. Douglas.
CiNii Researchers Link Viewer courtesy of Ryutaro Ichise.
BioWisdom courtesy of OmniViz, Inc.

Software Credits

Cummings, Jonathon N. 2007. "NetVis Module: Dynamic Visualization of Social Networks." **http://www.netvis.org** (accessed October 30, 2007).

The R Foundation. 2003. "R Project for Statistical Computing." http://www.r-project.org (accessed October 30, 2007).

Adar, Eytan. 2007. "GUESS: The Graph Exploration System." http://graphexploration.cond.org (accessed October 30, 2007).

Douglas, Shawn M., Gaetano T. Montelione, and Mark Gerstein. 2005. "PubNet: Publication Network Graph Utility." http://pubnet.papers.gersteinlab.org (accessed October 30, 2007).

CiNii. 2004. "Researchers Link Viewer." http://ri-www.nii.ac.jp/ComMining1/index.cgi (accessed January 1, 2008).

Books

References

Carrington, P. J., J. Scott, and S. Wasserman, eds. 2005. *Models and Methods in Social Network Analysis*. Cambridge: Cambridge University Press.

Cronin, Blaise. 2005. *The Hand of Science: Academic Writing and Its Rewards*. Lanham, MD: Scarecrow Press.

Godin, Benoit. 2005. *Measurement and Statistics on Science and Technology: 1920 to the Present*. Routledge Studies in the History of Science, Technology, and Medicine. New York: Routledge.

40 Timeline: 2006-I

Visualizations

References

Boyack, Kevin W., Katy Börner, and Richard Klavans. 2007. "Mapping the Structure and Evolution of Chemistry Research." *Proceedings of the 11th International Conference of the International Society for Scientometrics and Informetrics*, edited by Daniel Torres-Salinas and Henk F. Moed, 112–123. Madrid: CSIC.

Garfield, Eugene, Irving H. Sher, and Richard J. Torpie. 1964. "The Use of Citation Data in Writing the History of Science." Report for Air Force Office of Scientific Research under contract AF49 (638)-1256. Philadelphia, PA: Institute for Scientific Information. http://www.garfield.library.upenn.edu/papers/useofcitdatawritinghistofsci.pdf (accessed September 30, 2009).

HistCite Software LLC. 2008. "HistCite: Bibliometric Analysis and Visualization Software." http://www.histcite.com (accessed June 10, 2008).

Viégas, Fernanda, Martin Wattenberg, and D. Kushal. 2004. "Studying Cooperation and Conflict Between Authors with History Flow Visualizations." *Proceedings of SIGCHI* 6, 1: 575–582. Vienna, Austria, April 24–29. http://alumni.media.mit.edu/~fviegas/papers/history_flow.pdf (accessed June 2, 2008).

Paley, W. Bradford. 2002. "TextArc: Revealing Word Associations, Distribution and Frequency." TextArc. http://textarc.org/TextArcOverview.pdf (accessed October 30, 2007).

Börner, Katy, Elisha F. Hardy, Bruce W. Herr II, Todd Holloway, and W. Bradford Paley. 2007. "Taxonomy Visualization in Support of the Semi-Automatic Validation and Optimization of Organizational Schemas." *Journal of Informetrics* 1, 3: 214–225.

Klavans, Richard, and Kevin W. Boyack. 2007. "Is There a Convergent Structure to Science?" *Proceedings of the 11th International Conference of the International Society for Scientometrics and Informetrics*, edited by Daniel Torres-Salinas and Henk F. Moed, 437–448. Madrid: CSIC.

Günther, Ingo. 2006. "WorldProcessor." http://www.WorldProcessor.org (accessed October 30, 2007).

Günther, Ingo. 2005. *WorldProcessor: Chiayi Art Festival on the Tropic of Cancer*. Taibao City, Taiwan.

Institute for the Future for the Horizon Scanning Centre of the UK Government's Office of Science and Innovation. 2006. "Delta Scan: The Future of Science and Technology, 2005–2055." Stanford Humanities Lab/Metamedia. http://humanitieslab.stanford.edu/2/Home (accessed March 30, 2010).

Image Credits—Maps

2002 Structure Map: Boyack, Kevin W., and Richard Klavans. 2005. "The Structure of Science." Albuquerque, NM, and Berwyn, PA. Courtesy of Kevin W. Boyack and Richard Klavans, SciTech Strategies, Inc., http://www.mapofscience.com (accessed June 10, 2008).

HistCite Visualization of DNA Development by Eugene Garfield (HistCite), Elisha Hardy, Katy Börner (graphic design), Ludmila Pollock (images), and Jan Witkowski (text). Philadelphia, PA. Courtesy of Eugene Garfield, Thomson Reuters ISI, Indiana University, and Cold Spring Harbor Laboratory.

History Flow courtesy of Fernanda Viégas and Martin Wattenberg, IBM, Inc.

TextArc Visualization of "The History of Science" courtesy of W. Bradford Paley.

Taxonomy Visualization courtesy of Katy Börner.

Boyack, Kevin W., and Richard Klavans. 2006. "Map of Scientific Paradigms." Albuquerque, NM, and Berwyn, PA. Courtesy of Kevin W. Boyack, Sandia National Laboratories and Richard Klavans, SciTech Strategies, Inc. http://www.mapofscience.com (accessed June 10, 2008).

Günther, Ingo. 2006. "WorldProcessor: Zones of Invention—Patterns of Patents." New York, NY. Courtesy of Ingo Günther.

Pang, Alex Soojung-Kim, David Pescovitz, Marina Gorbis, and Jean Hagan. 2006. "Science & Technology Outlook: 2005–2055." Palo Alto, CA. Courtesy of Institute for the Future.

Image Credits—Portraits

Boyack portrait © Balázs Schlemmer.

Klavans portrait © Balázs Schlemmer.

Garfield portrait courtesy of Eugene Garfield and Thomson Reuters.

Witkowski portrait courtesy of Cold Spring Harbor Laboratory.

Pollock portrait courtesy of Ludmilla Pollock.

Hardy portrait courtesy of Elisha F. Hardy.

Börner portrait courtesy of Katy Börner.

Viégas portrait courtesy of Fernanda Viégas.

Feinberg portrait courtesy of Jonathan Feinberg.

Boyack portrait © Balázs Schlemmer.

Klavans portrait © Balázs Schlemmer.

Günther portrait courtesy of Ingo Günther.

Pang portrait © 2007, Institute for the Future.

Pescowitz portrait © 2007, Institute for the Future.

Software Credits

Paley, W. Bradford. "TextArc." http://textarc.org (accessed October 30, 2007).

Tools

References

Bergström, Peter, and E. James Whitehead Jr. "CircleView: Scalable Visualization and Navigation of Citation Networks." http://www.cs.ucsc.edu/~ejw/papers/bergstrom_ivica2006_final.pdf (accessed October 30, 2007).

Compendium Institute. 2008. Compendium Institute Home Page. http://compendium.open.ac.uk/institute (accessed March 30, 2010).

Conklin, Jeffrey. 2005. *Dialogue Mapping: Building Shared Understanding of Wicked Problems*. Hoboken, NJ: Wiley.

Azoulay, Pierre, Andrew Stellman, and Joshua Graff Zivin. 2006. "PublicationHarvester: An Open-Source Software Tool for Science Policy Research." NBER Working Paper No. 12039. National Bureau of Economic Research, CUNY, and Stanford, CA.

Tools—Graphics

Smart Network Analyzer courtesy of Valdis Krebs.

CircleView Document Visualization courtesy of Peter Bergotrom/VC Santa Cruz.

Compendium Dialog Mapping courtesy of Jeff Conklin.

PublicationHarvester courtesy of Stellman & Greene Consulting. http://www.stellman-greene.com/PublicationHarvester.

Software Credits

Krebs, Valdis. 2008. "InFlow—Software for Social Network Analysis & Organizational Network Analysis." Orgnet.com. http://www.orgnet.com/inflow3.html (accessed June 2, 2008).

Books

References

Leydesdorff, Loet. 2006. *The Knowledge-Based Economy: Modeled, Measured, Simulated*. Boca Raton, FL: Universal Publishers.

Braun, Tibor, ed. 2006. *Evaluations of Individual Scientists and Research Institutions: A Selection of Papers Reprinted from the Journal* Scientometrics. Budapest, Hungary: Akadémiai Kiadó.

42 Timeline: 2006-II

Visualizations

References

Paley, W. Bradford. 2002. "Illuminated Diagrams: Using Light and Print to Comparative Advantage." http://www.textarc.org/appearances/InfoVis02/InfoVis02_IlluminatedDiagrams.pdf (accessed October 30, 2007).

Boyack, Kevin W., Richard Klavans, W. Bradford Paley, and Katy Börner. 2007. "Mapping, Illuminating, and Interacting with Science." Paper presented at *SIGGRAPH 2007: The 34th International Conference and Exhibition on Computer Graphics and Interactive Techniques* in San Diego, CA, August 5–9.

Börner, Katy, Fileve Palmer, Julie M. Davis, Elisha F. Hardy, Stephen M. Uzzo, and Bryan J. Hook. 2009. "Teaching Children the Structure of Science." *Proceedings of SPIE: Conference on Visualization and Data Analysis*, 7243, 1: 1-14. Bellingham, WA: SPIE.

Günther, Ingo. 2006. WorldProcessor Project Home Page. http://www.WorldProcessor.org (accessed October 30, 2007).

Günther, Ingo. 2005. *WorldProcessor: Chiayi Art Festival on the Tropic of Cancer*. Taibao City, Taiwan.

Simms, Andrew, Dan Moran, and Peter Chowla. 2006. *The U.K. Interdependence Report: How the World Sustains the Nation's Lifestyles and the Price It Pays*. London: New Economics Foundation. http://www.neweconomics.org/gen/uploads/f2abwpumbr1wp055y2l10s5514042006174517.pdf (accessed October 30, 2007).

SCImago Research Group. 2010. "Atlas of Science Project." Universities of Granada, Extremadura and Carlos III (Madrid). http://www.atlasofscience.net (accessed March 16, 2010).

Moya-Anegón, Félix, Benjamín Vargas-Quesada, Víctor Herrero-Solana, Zaida Chinchilla-Rodríguez, Elena Corera-Alvarez, and Francisco Muñoz-Fernández. 2004. "A New Technique for Building Maps of Large Scientific Domains Based on the Cocitation of Classes and Categories." *Scientometrics* 61, 1: 129–145.

SCImago Research Group. 2006. "Atlas of Science Quick Tour." Universities of Granada, Extremadura and Carlos III (Madrid). http://www.atlasofscience.net/pdf/atlas-of-science-quick-guide-v3.pdf (accessed October 30, 2007).

Vargas-Quesada, Benjamín, and Félix de Moya-Anegón. 2007. *Visualizing the Structure of Science*. Berlin: Springer-Verlag.

Image Credits—Maps

Geographic Map: Where Science Gets Done, the Illuminated Diagram Map used in *Places & Spaces: Mapping Science*, with input from Kevin W. Boyack, Richard Klavans, John Burgoon, and W. Bradford Paley.

Topic Map: How Scientific Paradigms Relate, the Illuminated Diagram Map used in *Places & Spaces: Mapping Science*, based on "Map of Scientific Paradigms Research" and Node layout by Kevin W. Boyack and Richard Klavans, SciTech Strategies, Inc. http://www.mapofscience.com with graphics and typography by W. Bradford Paley, Information Esthetics, http://didi.com/brad, conceptualized and commissioned by Katy Börner.

Hands-On Science Maps for Kids, featured in *Places & Spaces: Mapping Science*, World Map featuring original artwork by Fileve Palmer, Indiana University.

Hands-On Science Maps for Kids, *Places & Spaces: Mapping Science*, Science Map by Fileve Palmer (original artwork and design) and Katy Börner (concept), Indiana University, incorporating the structural elements of the "Map of Scientific Paradigms" by Kevin W. Boyack and Richard Klavans, SciTech Strategies, Inc. http://www.mapofscience.com (accessed June 10, 2008).

WorldProcessor Globe of Patents, © Ingo Günther, WorldProcessor.com, 1988–2007.

UK's Global Ecological Footprint, Ecological Footprint Trade Flow Map Visualization. Visualization by Doantam Phan, Stanford University, data by Dan Moran, Global Footprint Network, published in *The UK Interdependence Report* by New Economics Foundation.

Atlas of Science courtesy of Félix de Moya Anegón, SCImago Research Group.

Image Credits—Portraits

Boyack portrait © Balázs Schlemmer.

Klavans portrait © Balázs Schlemmer.

Burgoon portrait courtesy of Katy Börner.

Kennard portrait courtesy of Katy Börner.

Paley portrait courtesy of W. Bradford Paley.

Palmer portrait courtesy of Fileve Palmer.

Davis portrait courtesy of Julie M. Davis.

Börner portrait courtesy of Katy Börner.

Hardy portrait courtesy of Elisha F. Hardy.

Günther portrait courtesy of Ingo Günther.

Moya-Anegón portrait courtesy of Félix de Moya-Anegón.

Tools

References

Herr II, Bruce W., Weixia (Bonnie) Huang, Shashikant Penumarthy, and Katy Börner. 2007. "Designing Highly Flexible and Usable Cyberinfrastructures for Convergence." *Progress in Convergence—Technologies for Human Wellbeing*, edited by William S. Bainbridge and Mihail C. Roco, 1093: 161–179. Boston: Annals of the New York Academy of Sciences.

Mons, Albert. 2007. WikiProfessional Home Page. **https://www.wikiprofessional.org/portal** (accessed October 30, 2007).

Mons, Barend. 2006. "Beyond Text Mining." *Knowledge in Service to Health: Leveraging Knowledge for Modern Science Management.* Office of Extramural Research, National Institutes of Health. US Department of Health and Human Services. **http://grants.nih.gov/grants/KM/OERRM/OER_KM_events** (accessed October 30, 2007).

Image Credits

NWB Team. 2006. "Network Workbench Tool." Indiana University, Northeastern University, and University of Michigan, **http://nwb.slis.indiana.edu** (accessed March 15, 2010).

BibExcel Bibliographic Data Analyzer courtesy of Olle Persson.

Knowledge Spaces for Collective Annotation courtesy of Knewco, Inc.

Software Credits

Persson, Olle. 2010. "BIBEXCEL." **http://www8.umu.se/inforsk/Bibexcel** (accessed February 10, 2010).

Knewco, Inc. 2008. "Knewco: Knowledge Redesigned." **http://knewco.com** (accessed June 10, 2008).

44 Timeline: 2006–2007

Visualizations

References

Carvalho, Rui, and Michael Batty. 2006. "The Geography of Scientific Productivity: Scaling in US Computer Science." *Journal of Statistical Mechanics: Theory and Experiment* P10012. **http://arxiv.org/PS_cache/physics/pdf/0603/0603242v3.pdf** (accessed October 30, 2007).

Holloway, Todd, Miran Božičević, and Katy Börner. 2007. "Analyzing and Visualizing the Semantic Coverage of Wikipedia and Its Authors." *Complexity* 12, 3: 30–40.

Herr II, Bruce W., Russell Jackson Duhon, Katy Börner, Elisha F. Hardy, and Shashikant Penumarthy. 2008. "113 Years of Physical Review: Using Flow Maps to Show Temporal and Topical Citation Patterns." *Proceedings of the 12th Information Visualization Conference (IV'08)*, 421–426. Los Alamitos, CA: IEEE Computer Society Conference Publishing Services.

Chen, Chaomei, Jian Zhang, Weizhong Zhu, and Michael S. Vogeley. 2007. "Delineating the Citation Impact of Scientific Discoveries." *IEEE/ACM Joint Conference on Digital Libraries*, 19–28. Vancouver, Canada, June 17–22.

Herr II, Bruce W., Todd Holloway, Katy Börner, Elisha F. Hardy, and Kevin Boyack. 2007. "Science Related Wikipedian Activity." Information Visualization Laboratory, School of Library and Information Science, Indiana University, Bloomington. **http://www.scimaps.org/maps/map/science_related_wiki_49/** (accessed March 30, 2010).

Boyack, Kevin W. 2007. "Using Detailed Maps of Science to Identify Potential Collaborations." *Scientometrics* 79, 1, 27–44.

SciTech Strategies, Inc. 2008. "Maps of Science." **http://mapofscience.com** (accessed June 10, 2008).

Zeller, Daniel. 2007. "Hypothetical Model of the Evolution of Science." **http://scimaps.org/dev/map_detail.php?map_id=163** (accessed May 29, 2008).

Gerstein, Mark, and Shawn M. Douglas. 2007. "RNAi Development." *PLoS Computational Biology* 3, 4: 0774–0775.

Image Credits—Maps

See Carvalho and Batty 2006.

See Holloway et al. 2007. Reprinted with permission of John Wiley & Sons, Inc.

See Herr II et al. 2008, © 2008 IEEE.

Chen, Chaomei, Jian Zhang, Lisa Kershner, Michael S. Vogeley, J. Richard Gott III, and Mario Juric. 2007. "Mapping the Universe: Space. Time. Discovery!" Philadelphia, PA, and Princeton, NJ. Courtesy of Drexel University and Princeton University.

Herr II, Bruce W. (data mining and visualization), Todd Holloway (data mining), Elisha F. Hardy (graphic design), Kevin W. Boyack (graph layout), Katy Börner (concept). 2007. "Science Related Wikipedian Activity." Bloomington, IN, and Albuquerque, NM. Courtesy of Indiana University.

2003 ISI/Scopus/UCSD Map © 2007, The Regents of the University of California, all rights reserved.

Hypothetical Model of the Evolution and Structure of Science, 2007, courtesy of Daniel Zeller, Pierogi Gallery, Brooklyn, and G-Module, Paris.

RNAi Development and Nobel Prize courtesy of Shawn M. Douglas.

Image Credits—Portraits

Carvalho portrait courtesy of Rui Carvalho.

Batty portrait courtesy of Michael Batty.

Holloway portrait courtesy of Todd Holloway.

Herr II portrait courtesy of Bruce W. Herr II.

Duhon portrait courtesy of Russell J. Duhon.

Hardy portrait courtesy of Elisha F. Hardy.

Penumarthy portrait courtesy of Shashikant Penumarthy.

Börner portrait courtesy of Katy Börner.

Chen portrait courtesy of Chaomei Chen.

Vogeley portrait courtesy of Micheal S. Vogeley.

Boyack portrait © Balázs Schlemmer.

Klavans portrait © Balázs Schlemmer.

Zeller portrait courtesy Pierogi Gallery, Brooklyn, and G-Module, Paris.

Gerstein portrait courtesy of Shawn M. Douglas.

Douglas portrait courtesy of Shawn M. Douglas.

Tools

References

Lin, Xia, Yen Bui, and Dongming Zhang. 2007. "Visualization of Knowledge Structures." *Proceedings of Information Visualization 2007*, edited by Ebad Banissi et al., 476–481. Los Alamitos, CA: IEEE Computer Society.

Smith, Marc A. 2004. "Netscan: A Social Accounting Search Engine." Microsoft Corporation: Community Technologies Groups. Previously available at **http://netscan.research.microsoft.com**.

Harzing, Anne-Wil. 2008. Harzing.com Home Page. **http://www.harzing.com** (accessed June 2, 2008).

SparkIP. 2008. Spark IP Home Page. Previously available at **http://sparkip.com** (accessed January 1, 2008).

Cokol, Murat, Raul Rodriquez-Esteban. 2008. "Visualizing Evolution and Impact of Biomedical Fields." *Journal of Biomedical Informatics* 41, 6: 1050-1052.

Tools—Graphics

Visual Concept Explorer courtesy of Xia Lin.

Netscan Usenet Visualization courtesy of Marc A. Smith and Danyel Fisher, Microsoft Research.

Electronic Scientific Portfolio Assistant courtesy of Discovery Logic, Inc.

Treparel Patent Visualization courtesy of Anton Heijs, Treparel.

Publish or Perish Academic Citation Analyzer courtesy of Harzing.com and Tarma Software Research Pty Ltd.

SparkCluster Map courtesy of SparkIP.

SciTrends courtesy of www.scitrends.net (Murat Cokol and Raul Rodriguez-Esteban).

Software Credits

Lin, Xia. 2010. "Visual Concept Explorer." **http://cluster.cis.drexel.edu/vce** (accessed March 15, 2010).

Heijs, Anton, and Treparel Information Solutions, B.V. 2008. "Treparel." **http://www.treparel.com** (accessed June 2, 2008). Used by permission.

Books

References

Vargas-Quesada, Benjamín, and Félix de Moya Anegón. 2007. *Visualizing the Structure of Science.* Berlin: Springer-Verlag.

Contributors

We thank Mike Pollard, Julie D. Wugalter, and Laurel Haak for 2007 Electronic Scientific Portfolio Assistant; Project Key Indicators image from DiscoveryLogic.

46 Timeline: 2007

Visualizations

References

Hook, Peter A. 2007. "Network Derived Domain Maps of the United States Supreme Court: 50 Years of Co-Voting Data and a Case Study on Abortion." *Presentation at the International Workshop and Conference on Network Science 2007.* Queens, NY, May 22. **http://ella.slis.indiana.edu/~pahook/product/2007-05-22_NetSci07.ppt** (accessed October 30, 2007).

Boyack, Kevin W., Katy Börner, and Richard Klavans. 2007. "Mapping the Structure and Evolution of Chemistry Research." *Proceedings of the 11th International Conference of the International Society for Scientometrics and Informetrics,* edited by Daniel Torres-Salinas and Henk F. Moed, 112–123. Madrid: CSIC.

Hidalgo, César A., B. Klinger, Albert-L. Barabási, and R. Hausmann. 2007. "The Product Space Conditions the Development of Nations." *Science* 317, 5837: 482–487.

Igami, Masatsura, and Ayaka Saka. 2007. "Capturing the Evolving Nature of Science: The Development of New Scientific Indicators and the Mapping of Science." *OECD Science, Technology and Industry Working Papers,* 2007/1. Paris: OECD Publishing. **http://www.oecd.org/dataoecd/11/40/38134903.pdf** (accessed October 30, 2007).

Rosvall, M., and C. T. Bergstrom. 2008. "Maps of Random Walks on Complex Networks Reveal Community Structure." *PNAS* 105, 4: 1118–1123.

Leydesdorff, Loet, and Ismael Rafols. 2009. "A Global Map of Science Based on the ISI Subject Categories." *Journal of the American Society for Information Science and Technology* 60, 2: 348–362.

Herr II, Bruce W., Gully Burns, and David Newman. 2006. "Society for Neuroscience 2006." *Places & Spaces: Mapping Science.* **http://scimaps.org/maps/neurovis** (accessed November 27, 2007).

Image Credits—Maps

"Citation Network of Supreme Court Abortion and Birth Control Cases with KeyCite Status Flags" courtesy of Peter A. Hook, Indiana University.

"Mapping the Evolution of Chemistry Research" courtesy of Kevin W. Boyack, Katy Börner, and Richard Klavans.

"The Product Space" courtesy of César A. Hidalgo.

"Mapping Multidisciplinary Science," public domain image by Masatsura Igami and Ayaka Saka, National Institute of Science and Techology Policy, Japan.

See Rosvall and Bergstrom 2008, © 2008, National Academy of Sciences, USA.

"A Global Map of Science based on the ISI Subject Categories," courtesy of Loet Leydesdorff.

"Interactive Google Map of 2006 Society for Neuroscience Abstracts," courtesy of Bruce W. Herr II, Gully Burns (USC), and David Newman (UCI). *See also* **http://www.nihmaps.org**.

Herr II, Bruce W., Edmund M. Talley, Gully A.P.C. Burns, David Newman, and Gavin LaRowe. 2009. "The NIH Visual Browser: An Interactive Visualization of Biomedical Research." *Proceedings of the 13th International Conference on Information Visualization*, 505-509. Barcelona, Spain: IEEE Computer Society.

Image Credits—Portraits

Hook portrait courtesy of Peter A. Hook.

Boyack portrait © Balázs Schlemmer.

Klavans portrait © Balázs Schlemmer.

Börner portrait courtesy of Katy Börner.

Igami portrait courtesy of Masatsura Igami.

Saka portrait courtesy of Ayaka Saka.

Rosvall portrait courtesy of Martin Rosvall.

Bergstrom portrait courtesy of Carl Bergstrom.

Leydesdorff portrait courtesy of Loet Leydesdorff.

Herr II portrait courtesy of Bruce W. Herr II.

Discussion

Contributors

This timeline benefited from material compiled by Peter A. Hook, Kevin W. Boyack, and Loet Leydesdorff. Bryan J. Hook and Mark A. Price kindly retrieved books and papers.

49 Part 3: Toward a Science of Science

References

Santayana, George. 1905. *The Life of Reason*, 74–88. New York: C. Scribner's Sons.

Image Credits

From Zeller, Daniel. 2007. *Hypothetical Model of the Evolution and Structure of Science.* New York, NY. Courtesy of Daniel Zeller.

50 Building Blocks

References

Monmonier, Mark, and H. J. de Blij. 1996. *How to Lie with Maps*. Chicago: The University of Chicago Press. [Quotation]

Overview

References

Nielsen, Jakob. 1998. *Designing Web Usability*. Indianapolis, IN: New Riders Publishing.

Rubin, Jeffrey. 1994. *Handbook of Usability Testing: How to Plan, Design, and Conduct Effective Tests*. New York: John Wiley & Sons.

Shneiderman, Ben, and Catherine Plaisant. 2004. *Designing the User Interface: Strategies for Effective Human-Computer Interaction.* 4th ed. Boston, MA: Addison-Wesley.

Needs Analysis

References

Mayhew, Deborah J. 1999. *Usability Engineering Lifecycle: A Practitioner's Handbook for User Interface Design.* San Diego, CA: Academic Press.

Nielsen, Jakob. 1993. *Usability Engineering*. San Diego, CA: Academic Press.

Shneiderman, Ben. 2002. *Leonardo's Laptop: Human Needs and the New Computing Technologies*. Cambridge, MA: MIT Press.

Conceptualizing Science

No references or credits.

Data Acquisition and Preprocessing

No references or credits.

Data Analysis, Modeling, and Layout

References

Willinger, Walter, Ramesh Govindan, Sugih Jamin, Vern Paxson, and Scott Shenker. 2002. "Scaling Phenomena in the Internet: Critically Examining Criticality." *PNAS* 99, 1: 2573–2580.

Data Communication–Visualization Layers

References

Shneiderman, Ben. 1996. "The Eyes Have It: A Task by Data Type Taxonomy for Information Visualizations." *Proceedings of the IEEE Symposium on Visual Languages*, 336–343. Washington, DC: IEEE Computer Society.

Validation and Interpretation

Image Credits

Flowchart: Elisha F. Hardy (design), Katy Börner (concept), Indiana University.

Scholarly Marketplaces

No references or credits.

52 Conceptualizing Science: Basic Anatomy of Science

References

Garfield, Eugene. 1959. "A Unified Index to Science." *Proceedings of the International Conference on Scientific Information*, vol. 1. Washington DC: National Academy of Sciences National Research Council. [Quotation, p. 468]

Toward a Science of Science

References

Barnes, Barry, and David Edge, eds. 1982. *Science in Context: Readings in the Sociology of Science.* Cambridge, MA: MIT Press.

Ben-David, Joseph, and Teresa A. Sullivan. 1975. "Sociology of Science." *Annual Review of Sociology* 1: 203–222.

Bernal, John D. 1939. *The Social Function of Science.* London: Routledge & Kegan Ltd.

Borgman, Christine L., and Jonathan Furner. 2002. "Scholarly Communication and Bibliometrics." In *Annual Review of Information Science and Technology*, edited by Blaise Cronin, 36: 3–72. Medford, NJ: Information Today.

de Solla Price, Derek J. 1961. *Science Since Babylon.* New Haven: Yale University Press.

de Solla Price, Derek J. 1965. *Little Science, Big Science.* New York: Columbia University Press.

de Solla Price, Derek J. 1965. "Networks of Scientific Papers." *Science* 149, 3683: 510–515.

de Solla Price, Derek J. 1986. *Little Science, Big Science … and Beyond.* New York: Columbia University Press.

Egghe, Leo, and Ronald Rousseau. 1990. *Introduction to Informetrics: Quantitative Methods in Library, Documentation and Information Science.* Amsterdam: Elsevier Science Publishers.

Garfield, Eugene, Irving H. Sher, and Richard J. Torpie. 1964. "The Use of Citation Data in Writing the History of Science." Report for Air Force Office of Scientific Research under contract AF49 (638)-1256. Philadelphia, PA: Institute for Scientific Information. **http://www.garfield.library.upenn.edu/papers/useofcitdatawritinghistofsci.pdf** (accessed September 30, 2009).

Leydesdorff, Loet. 2005. "The Triple Helix Model and the Study of Knowledge-Based Innovation Systems." *International Journal of Contemporary Sociology* 42, 1: 12–27.

Merton, Robert King. 1973. *The Sociology of Science: Theoretical and Empirical Investigations*. Chicago: The University of Chicago Press.

Narin, Francis. 1976. *Evaluative Bibliometrics: The Use of Publication and Citation Analysis in the Evaluation of Scientific Activity*. Cherry Hill, NJ: Computer Horizons, Inc.

Narin, Francis, and Joy K. Moll. 1977. "Bibliometrics." In *Annual Review of Information Science and Technology*, edited by M. E. Williams, 12: 35–58. White Plains, NY: Knowledge Industry Publications.

Small, Henry, and E. Sweeney. 1985. "Clustering the Science Citation Index Using Co-citations. Part I. A Comparison of Methods." *Scientometrics* 7, 3–6: 391–409.

Small, Henry, E. Sweeney, and E. Greenlee. 1985. "Clustering the Science Citation Index Using Co-citations. Part II. Mapping Science." *Scientometrics* 8, 5–6: 321–340.

Thelwall, M. 2004. *Link Analysis: An Information Science Approach*. San Diego, CA: Academic Press.

Thelwall, M., L. Vaughan, and L. Björneborn. 2005. "Webometrics." In *Annual Review of Information Science and Technology*, edited by Blaise Cronin, 39: 81–135. Medford, NJ: Information Today/American Society for Information Science and Technology.

van Raan, Anthony F. J., ed. 1988. *Handbook of Quantitative Studies of Science and Technology*. Amsterdam: Elsevier Science Publishers.

White, Howard D., and Katherine W. McCain. 1989. "Bibliometrics." In *Annual Review of Information Science and Technology*, edited by M. E. Williams, 24: 119–186. Amsterdam: Elsevier Science Publishers.

Wouters, Paul. 1999. "The Citation Culture." PhD diss., University of Amsterdam. **http://garfield.library.upenn.edu/wouters/wouters.pdf** (accessed October 30, 2007).

Features and Norms of Science

References

Merton, Robert. 1973. *The Sociology of Science*. Chicago: The University of Chicago Press.

Wikimedia Foundation. 2009. "Robert K. Merton." *Wikipedia, the Free Encyclopedia*. **http://en.wikipedia.org/wiki/Robert_K._Merton** (accessed November 18, 2009).

Wilson, E. O. 1998. *Consilience: The Unity of Knowledge*, 53. New York: Knopf.

The Power of Visual Conceptualizations

References

Bernal, John D. 1939. *The Social Function of Science.* London: Routledge & Kegan Ltd.

de Solla Price, Derek J. 1965. *Little Science, Big Science.* New York: Columbia University Press.

Garfield, Eugene. 1979. *Citation Indexing: Its Theory and Application in Science, Technology, and Humanities.* New York: Wiley.

Goffman, William, and V. A. Newill. 1964. "Generalization of Epidemic Theory: An Application to the Transmission of Ideas." *Nature* 204: 225–228.

Kuhn, Thomas Samuel. 1962. *The Structure of Scientific Revolutions*. Chicago: The University of Chicago Press.

Lem, Stanisław. 1997. *Filozofia Przypadku*. **http://english.lem.pl** (accessed February 17, 2010).

Otlet, Paul. 1934. *Traité de Documentation: Le Livre sur le Livre—Théorie et Pratique*. Brussels: Editiones Mundaneum, Palais Mondial, Van Keerberghen & Fils.

Popper, Karl. 1979. *Truth, Rationality, and the Growth of Scientific Knowledge*. Frankfurt am Main: Vittorio Klostermann.

Wells, Herbert George. 1938. *World Brain*, 19. New York: Doubleday & Company, Inc.

Yablonskij, Anatolij Ivanovich. 1986. *Matematicheskije modeli v issledovanii nauki* [Mathematical Models in the Exploration of Science]. Moscow: Nauka.

Image Credits

Growth chart of scientific production, *see* de Solla Price 1965.

Visual Conceptualization of Science

Image Credits

Science Dynamics: Elisha F. Hardy (design), Katy Börner (concept), Indiana University; incorporated clips of work by Kevin W. Boyack, Richard Klavans (UCSD Map of Science), and Eugene Garfield (HistCite visualization); both used by permission.

(II) Cumulative Structure

References

de Solla Price, Derek J. 1965. *Little Science, Big Science*. New York: Columbia University Press.

de Solla Price, Derek J. 1965. "Networks of Scientific Papers." *Science* 149, 3683: 510–515, fig. 6.

Merton, Robert King. 1965. *On the Shoulders of Giants: A Shandean Postscript*, 290. New York: Harcourt, Brace & World.

(III) Research Specialty Tubes

References

Kuhn, Thomas Samuel. 1962. *The Structure of Scientific Revolutions*. Chicago: The University of Chicago Press.

(IV) Research Frontier Epidermis

References

de Solla Price, Derek J. 1982. "The Parallel Structures of Science and Technology." In *Science in Context: Readings in the Sociology of Science*, edited by Barry Barnes and David Edge, 176. Cambridge, MA: MIT Press.

(VI) Linkages

References

Granovetter, Mark. 1973. "The Strength of Weak Ties." *American Journal of Sociology* 78, 1360–1380.

Wouters, Paul. 1999. "The Citation Culture." PhD diss., University of Amsterdam. **http://garfield.library.upenn.edu/wouters/wouters.pdf** (accessed October 30, 2007).

(VII) Clustering in Time, Geography, and Topic Space

References

Batty, M. 2003. "The Geography of Scientific Citation." *Environment and Planning A* 35, 5: 761–765.

Börner, Katy, Shashikant Penumarthy, Mark Meiss, and Weimao Ke. 2006. "Mapping the Diffusion of Scholarly Knowledge Among Major U.S. Research Institutions." *Scientometrics* 68, 3: 415–426.

de Solla Price, Derek J. 1965. *Little Science, Big Science.* New York: Columbia University Press.

Wellman, B., H. D. White, and N. Nazer. 2004. "Does Citation Reflect Social Structure? Longitudinal Evidence from the 'Globenet' Interdisciplinary Research Group." *Journal of the American Society for Information Science and Technology* 55: 111–126.

Science Dynamics

References

Menard, Henry W. 1971. *Science: Growth and Change,* 25. Cambridge, MA: Harvard University Press.

Conceptualizing Science: Basic Units, Aggregate Units, and Linkages

References

de Solla Price, Derek J. 1965. "Networks of Scientific Papers." *Science* 149, 3683: 510–515, fig. 6. [Quotation]

Basic Units

Authors

References

Menard, Henry W. 1971. *Science: Growth and Change,* 10, 85. Cambridge, MA: Harvard University Press.

Scientific Papers

References

Otlet, Paul. 1934. *Traité de Documentation: Le Livre sur le Livre—Théorie et Pratique.* Brussels: Editiones Mundaneum, Palais Mondial, Van Keerberghen & Fils.

Cronin, Blaise, and Sherrill Weaver-Wozniak. 1992. "An Online Acknowledgment Index: Rationale and Feasibility." *Proceedings of the 16th International Online Meeting,* 281–290. London: Oxford Learned Information.

van Raan, Anthony F. J. 2004. "Sleeping Beauties in science." *Scientometrics* 59, 3: 461-466.

Journals

References

Beaver, Donald D. 2001. "Reflections on Scientific Collaboration (and Its Study): Past, Present, and Future." *Scientometrics* 52, 3: 365–377.

Patents and Technical Reports

References

de Solla Price, Derek J. 1982. "The Parallel Structures of Science and Technology." In *Science in Context: Readings in the Sociology of Science,* edited by Barry Barnes and David Edge, 176. Cambridge, MA: MIT Press.

Hardware, Software, Datasets

References

Barabási, A. L., H. Jeong, Z. Néda, E. Ravasz, A. Schubert, and T. Vicsek. 2002. "Evolution of the Social Network of Scientific Collaborations." *Physica A: Statistical Mechanics and its Applications* 311, 3-4: 590–614.

Ben-David, Joseph. 1971. *The Scientist's Role in Society: A Comparative Study.* Englewood Cliffs, NJ: Prentice-Hall.

Funding

References

de Solla Price, Derek J. 1965. *Little Science, Big Science.* New York: Columbia University Press.

Irvine, J., and B.R. Martin. 1985. "Evaluating Big Science: CERN's Past Performance and Future Prospects." *Scientometrics* 7, 3–6: 281–308.

Newman, Mark E. J. 2001. "Scientific Collaboration Networks. I. Network Construction and Fundamental Results." *Physical Review E* 64, 1: 016131.

Aggregated Units

References

Beaver, Donald D., and R. Rosen. 1978. "Studies in Scientific Collaboration, Part I. The Professional Origins of Scientific Co-Authorship." *Scientometrics* 1: 65–84.

Friedkin, Noah E. 1998. *A Structural Theory of Social Influence.* Cambridge: Cambridge University Press.

Invisible Colleges

References

Crane, Diana. 1972. *Invisible Colleges: Diffusion of Knowledge in Scientific Communities.* Chicago: The University of Chicago Press.

Menard, Henry W. 1971. *Science: Growth and Change.* Cambridge, MA: Harvard University Press.

Newman, Mark E. J. 2001. "Scientific Collaboration Networks. II. Shortest Paths, Weighted Networks, and Centrality." *Physical Review E* 64: 016132.

Newman, Mark E. J. 2004. "Coauthorship Networks and Patterns of Scientific Collaboration." *PNAS* 101, 1: 5200–5205.

Research Specialties

References

Chen, Chaomei. 2003. *Mapping Scientific Frontiers: The Quest for Knowledge Visualization.* New York: Springer-Verlag.

Chen, Chaomei. 2004. *Information Visualization: Beyond the Horizon.* New York: Springer-Verlag.

Crane, Diana. 1972. *Invisible Colleges: Diffusion of Knowledge in Scientific Communities.* Chicago: The University of Chicago Press.

Kuhn, T. S. 1970. *The Structure of Scientific Revolutions.* 2nd ed. Chicago: The University of Chicago Press.

Thagard, P. 1992. *Conceptual Revolutions.* Princeton, NJ: Princeton University Press.

Contributors

Chaomei Chen and Jan Witkowski.

Research Fronts

References

Chen, Chaomei, and S. Morris. 2003. "Visualizing Evolving Networks: Minimum Spanning Trees versus Pathfinder Networks." *Proceedings of IEEE Symposium on Information Visualization* 67–74. Seattle, WA: IEEE.

de Solla Price, Derek J. 1965. "Networks of Scientific Papers." *Science* 149, 3683: 510–515.

Garfield, Eugene. 1994. "Scientography: Mapping the Tracks of Science." Reprinted from *Current Contents: Social and Behavioural Sciences* 7: 5–10.

Garfield, Eugene, Irving H. Sher, and Richard J. Torpie. 1964. "The Use of Citation Data in Writing the History of Science." Report for Air Force Office of Scientific Research under contract AF49 (638)-1256. Philadelphia, PA: Institute for Scientific Information. http://www.garfield.library.upenn.edu/papers/useofcitdatawritinghistofsci.pdf (accessed September 30, 2009).

Morris, Steven A., Gary G. Yen, Zheng Wu, and Benyam Asnake. 2003. "Time Line Visualization of Research Fronts." *Journal of the American Society for Information Science and Technology* 54, 5: 413–422.

Small, Henry G., and Belver C. Griffith. 1974. "The Structure of Scientific Literatures I: Identifying and Graphing Specialties." *Science Studies* 4, 1: 17–40.

Linkages and Derived Networks

Direct Linkages

References

Nicolaisen, Jeppe. 2007. "Citation Analysis." In *Annual Review of Information Science and Technology,* edited by Blaise Cronin, 41: 609–641. Medford, NJ: Information Today/American Society for Information Science and Technology.

Image Credits

Aggregate Units: Elisha F. Hardy (design), Katy Börner (concept), Indiana University.

Co-Occurrence Linkages

References

Börner, Katy, Shashikant Penumarthy, Mark Meiss, and Weimao Ke. 2006. "Mapping the Diffusion of Information Among Major U.S. Research Institutions." *Scientometrics* 68, 3: 415–426.

Kessler, Michael M. 1963. "Bibliographic Coupling Between Scientific Papers." *American Documentation* 14, 1: 10–25.

Wellman, B., H. D. White, and N. Nazer. 2004. "Does Citation Reflect Social Structure? Longitudinal Evidence from the 'Globenet' Interdisciplinary Research Group." *Journal of the American Society for Information Science and Technology* 55: 111–126.

Co-Citation Linkages

References

Marshakova, Irina V. 1973. "A System of Document Connections Based on References." *Scientific and Technical Information Serial of VINITI* 6: 3–8.

Small, Henry. 1973. "Co-Citation in Scientific Literature: A New Measure of the Relationship Between Publications." *Journal of the American Society for Information Science* 24: 265–269.

Small, Henry, and E. Greenlee. 1986. "Collagen Research in the 1970s." *Scientometrics* 10, 1–2: 95–117.

56 Conceptualizing Science: Basic Properties, Indexes, and Laws

References

de Solla Price, Derek J. 1982. "The Parallel Structures of Science and Technology." In *Science in Context: Readings in the Sociology of Science,* edited by Barry Barnes and David Edge, 176. Cambridge, MA: MIT Press. [Quotation]

Basic Properties

Balance of Papers and Citations

References

de Solla Price, Derek J. 1965. "Networks of Scientific Papers." *Science* 149, 3683: 510–515, fig. 3.

Image Credits

de Solla Price, Derek J. 1965. "Networks of Scientific Papers." *Science* 149, 3683: 510–515, fig. 3. Reprinted with permission from AAAS.

Author Trajectories

References

Bala, Venkatesh, and Sanjeev Goyal. 2000. "Self-Organization in Communication Networks." *Econometrica* 68: 1181-1230.

Braun, Tibor, Wolfgang Glänzel, and András Schubert. 2001. "Publication and Cooperation Patterns of the Authors of Neuroscience Journals." *Scientometrics* 51 3: 499-510.

de Solla Price, Derek J. 1986. *Little Science, Big Science ... and Beyond.* New York: Columbia University Press.

de Solla Price, D. 1975. "Studies in Scientometrics I: Transience and Continuance in Scientific Authorship." *Ciência da informação* 4, 1: 27.

Katz, J. Sylvan, and Diana Hicks. 2000. "How Much Is a Collaboration Worth? A Calibrated Bibliometric Model." *Scientometrics* 40 3: 541–554.

Melin, Göran. 2000. "Pragmatism and Self-Organization: Research Collaboration at the Individual Level." *Research Policy* 29 1: 31–40.

Meyrowitz, Joshua. 1985. *No Sense of Place: The Impact of Electronic Media on Social Behavior.* New York: Oxford University Press.

Image Credits

From *Little Science, Big Science* by Derek J. de Solla Price, © 1986, Columbia University Press. Reprinted with permission of the publisher.

Scientific Growth

References

Menard, Henry W. 1971. *Science: Growth and Change.* 19–20, 126. Cambridge, MA: Harvard University Press.

Indexes

Herfindahl-Hirschman Index, 1945/1950

References

Herfindahl, O. C. 1950. "Concentration in the U.S. Steel Industry." PhD diss., Columbia University.

Hirschman, A. O. 1945. *National Power and the Structure of Foreign Trade.* Berkeley: University of California Press.

Hirschman, A. O. 1964. "The Paternity of an Index." *American Economic Review* 54, 5: 761–762.

U.S. Department of Justice. "The Herfindahl-Hirschman Index." http://www.justice.gov/atr/public/testimony/hhi.htm (accessed March 10, 2010).

Price Index, 1970

References

de Solla Price, Derek J. 1970. *Citation Measures of Hard Science, Soft Science, Technology and Non-Science*, 3–22. Lexington, MA: D. C. Heath & Co.

Hirsch Index, 2005

References

Barendse, William. 2007. "The Strike Rate Index: A New Index for Journal Quality Based on Journal Size and the h-index of Citations." *Biomedical Digital Libraries* 4, 3. http://www.bio-diglib.com/content/4/1/3 (accessed March 1, 2008).

Hirsch, J. E. 2005. "An Index to Quantify an Individual's Scientific Research Output." *PNAS* 102, 46: 16569–16572.

Manafy, Michell. "Scopus Harnesses the h-Index to Increase the Quality and Reliability of Citation Tracking." *E-Content Magazine*. April 2007. http://www.econtentmag.com/Articles/ArticleReader.aspx?ArticleID=35680&CategoryID=17 (accessed March 1, 2008).

Roediger III, Henry L. 2006. "The *h* Index in Science: A New Measure of Scholarly Contribution." *APS Observer: The Academic Observer* 19, 4. http://www.psychologicalscience.org/observer/getArticle.cfm?id=1971 (accessed March 1, 2008).

Saad, Gad. 2006. "Exploring the h-Index at the Author and Journal Levels using Bibliometric Data of Productive Consumer Scholars and Business-Related Journals Respectively." *Scientometrics* 69, 1: 117–120.

Van Noorden, Richard. 2007. "Hirsch Index Ranks Top Chemists." *Chemistry World*, April. http://www.rsc.org/chemistryworld/News/2007/April/23040701.asp (accessed March 1, 2008).

General Laws

Power Laws

References

Amaral, L. A. N., A. Scala, M. Barthelemy, and H. E. Stanley. 2000. "Classes of Small-World Networks." *PNAS* 97: 11149–11152.

Barabási, Albert-László, and Réka Albert. 1999. "Emergence of Scaling in Random Networks." *Science* 286, 5349: 509–512.

Bonitz, M, E. Bruckner, and A. Scharnhorst. 1999. "The Matthew Index—Concentration Patterns and Matthew Core Journals." *Scientometrics* 44, 3: 361–378.

de Solla Price, Derek J. 1976. "A General Theory of Bibliometric and other Cumulative Advantage Processes." *Journal of the American Society for Information Science* 27, 5: 292–306.

de Solla Price, Derek J. 1986. *Little Science, Big Science … and Beyond.* New York: Columbia University Press.

D'Souza, Raissa M., Christian Borgs, Jennifer T. Chayes, Noam Berger, and Robert D. Kleinberg. 2007. "Emergence of Tempered Preferential Attachment from Optimization." *PNAS* 104, 15: 6112–6117.

Eggenberger, F., and G. Pólya. 1923. "Über die Statistik Verketteter Vorgänge." *Zeitschrift für Angewandte Mathematik und Mechanik* 3: 279–289.

Glänzel, W. 2003. "Bibliometrics as a Research Field: A Course on Theory and Application of Bibliometric Indicators." Course Handouts. http://www.norslis.net/2004/Bib_Module_KUL.pdf (accessed June 10, 2008).

King, Jean. 1987. "A Review of Bibliometric and other Science Indicators and their Role in Research Evaluation." *Journal of Information Science* 13, 5: 261–276.

Merton, Robert K. 1969. "The Matthew Effect in Science. The Reward and Communication Systems of the Science are Considered" *Science* 159, 3810: 56–63.

Merton, Robert King. 1988. "The Matthew Effect in Science, II: Cumulative Advantage and the Symbolism of Intellectual Property." *ISIS* 79: 606–623.

Merton, Robert King. 1995. "The Thomas Theorem and the Matthew Effect." *Social Forces* 74, 2: 379–424. http://www.garfield.library.upenn.edu/merton/matthewii.pdf (accessed March 1, 2008).

Mitzenmacher, Michael. 2004. "A Brief History of Generative Models for Power Law and Lognormal Distributions." *Internet Mathematics* 1, 2: 226–251.

Newman, Mark E. J. 2001. "The Structure of Scientific Collaboration Networks." *PNAS* 98, 2: 404–409.

Newman, Mark E. J. 2005. "Power Laws, Pareto Distributions and Zipf's Law." *Contemporary Physics*, 46, 5: 323–351.

New Testament. Matthew 25:29.

Simon, Herbert A. 1955. "On a Class of Skew Distribution Functions." *Biometrika* 42, 3–4: 425–440.

Yule, G. 1925. "A Mathematical Theory of Evolution Based on the Conclusions of Dr. J. C. Willis." *F.R.S. Philosophical Transactions of the Royal Society of London*, Series B, 213: 21–87.

Zipf, George K. 1949. *Human Behavior and the Principle of Least Effort*. Cambridge, MA: Addison-Wesley.

Image Credits

From *Little Science, Big Science* by Derek J. de Solla Price, © 1986, Columbia University Press. Reprinted with permission of the publisher.

Pareto Principle, 1897

References

The New School. 2010. "Vilfreda Pareto." *History of Economic Thought.* http://cepa.newschool.edu/het/profiles/pareto.htm (accessed March 30, 2010).

Pareto, Vilfredo. 1897. *Cours d'économie politique professé à l'université de Lausanne, Vol II:* 299–334.

Wikimedia Foundation. 2008. "Pareto Principle." *Wikipedia, the Free Encyclopedia.* http://en.wikipedia.org/wiki/Pareto_principle (accessed June 10, 2008).

Lotka's Law, 1926

References

de Solla Price, Derek J. 1965. *Little Science, Big Science.* New York: Columbia University Press.

Lotka, A. J. 1926. "The Frequency Distribution of Scientific Productivity." *Journal of the Washington Academy of Sciences* 14: 317–324.

Bradford's Law, 1934

References

Bradford, S. C. 1934. "Sources of Information on Specific Subjects." *Engineering: An Illustrated Weekly Journal* 137: 85–86.

Hjørland, Birger, and Jeppe Nicolaisen. 2005. "Bradford's Law of Scattering: Ambiguities in the Concept of 'Subject.'" *Lecture Notes in Computer Science*, 3507. Berlin: Springer-Verlag. http://www.springerlink.com/content/mp7l1w4jh0gnxd11 (accessed May 21, 2008).

Zipf's Law, 1949

References

Zipf, George K. 1949. *Human Behavior and the Principle of Least Effort.* Cambridge, MA: Addison-Wesley.

Moore's Law, 1965

References

Intel, Inc. 2009. "Moore's Law: Made Real by Intel Innovation." http://www.intel.com/technology/mooreslaw (accessed November 18, 2009).

Wikimedia Foundation. 2009. "Moore's Law." *Wikipedia, the Free Encyclopedia.* http://en.wikipedia.org/wiki/Moore's_law (accessed November 18, 2009).

Metcalfe's Law, circa 1980

References

Gilder, George. 1993. "Metcalfe's Law and Legacy." *Forbes ASAP*, September 13.

Wikimedia Foundation. 2009. "Metcalfe's Law." *Wikipedia, the Free Encyclopedia.* http://en.wikipedia.org/wiki/Metcalfe's_law (accessed November 18, 2009).

Contributors

Blaise Cronin and Loet Leydesdorff.

58 Conceptualizing Science: Science Dynamics

References

Börner, Katy, Chaomei Chen, and Kevin W. Boyack. 2003. "Visualizing Knowledge Domains." In *Annual Review of Information Science and Technology*, edited by Blaise Cronin, vol. 37: 179–255. Medford, NJ: Information Today/American Society for Information Science and Technology.

Chen, Chaomei. 2006. "CiteSpace II: Detecting and Visualizing Emerging Trends and Transient Patterns in Scientific Literature." *Journal of the American Society for Information Science and Technology* 57, 3: 359–377.

Cronin, Blaise, and Carol A. Hert. 1995. "Scholarly Foraging and Network Discovery Tools." *Journal of Documentation* 51, 4: 388–403. [Quotation]

Shiffrin, Richard M., and Katy Börner. 2004. "Mapping Knowledge Domains." *PNAS* 101, Suppl. 1: 5183–5185.

Ziman, John. 1978. "Dynamic Equilibrium of Science." *Nature* 276: 127–128.

Properties of Basic Units

Scholarly Information Foraging

References

Pirolli, P., and S. Card. 1999. "Information Foraging." *Psychological Review* 106, 4: 643–675.

Sandstrom, Pamela E. 1994. "An Optimal Foraging Approach to Information Seeking and Use." *Library Quarterly* 64: 414–449.

Sandstrom, Pamela E. 1998. "Information Foraging Among Anthropologists in the Invisible College of Human Behavioral Ecology: An Author Co-citation Analysis." PhD diss., Indiana University.

Hypes, Fads, and Fashions

References

Adar, Eytan, Li Zhang, Lada A. Adamic, and Rajan M. Lukose. 2004. "Implicit Structure and the Dynamics of Blogspace." *Workshop on the Weblogging Ecosystem, 13th International World Wide Web Conference.* Hewlett-Packard Development Company, L.P. http://www.blogpulse.com/papers/Adar_blogworkshop2_ppt.pdf (accessed May 21, 2008).

BBC News. 2007. "Fans Snap up Spice Girls Tickets." October 1. http://news.bbc.co.uk/2/hi/entertainment/7021976.stm (accessed February 14, 2008).

Börner, Katy, Jeegar Maru, and Robert L. Goldstone. 2004. "The Simultaneous Evolution of Author and Paper Networks." *PNAS* 101, Suppl. 1: 5266–5273.

Dezsö, Z., E. Almaas, A. Lukács, B. Rácz, I. Szakadát, and A.-L. Barabási. 2005. "Fifteen Minutes of Fame: The Dynamics of Information Access on the Web." http://arxiv.org/abs/physics/0505087 (accessed March 1, 2008).

Image Credits

Citation Probability, Russell J. Duhon (data preparation), Elisha F. Hardy (design), Katy Börner (concept), Indiana University.

Evolving Networks

The "Rich Get Richer" Effect

References

Albert, R., and A.-L. Barabási. 2002. "Statistical Mechanics of Complex Networks." *Reviews of Modern Physics* 74, 1: 47–97.

Barabási, Albert-László, and Réka Albert. 1999. "Emergence of Scaling in Random Networks." *Science* 286, 5349: 509–512.

Dorogovtsev, Serguei N., and J. F. F. Mendes. 2002. "Evolution of Networks." *Advanced Physics* 51: 1079–1187. http://arxiv.org/abs/cond-mat/0106144v2 (accessed March 1, 2008).

Dorogovtsev, Serguei N., and J. F. F. Mendes. 2003. *Evolution of Networks from Biological Nets to the Internet and WWW.* London: Oxford University Press.

Coevolving Author-Paper Networks

References

Börner, Katy, Jeegar Maru, and Robert L. Goldstone. 2004. "The Simultaneous Evolution of Author and Paper Networks." *PNAS* 101, Suppl. 1: 5266–5273.

Gilbert, Nigel. 1997. "A Simulation of the Structure of Academic Science." *Sociological Research Online* 2, 2. http://www.socresonline.org.uk/2/2/3.html (accessed October 2, 2009).

Dynamics of Innovation Networks

References

Gilbert, Nigel, Andreas Pyka, and Petra Ahrweiler. 2001. "Innovation Networks—A Simulation Approach." *Journal of Artificial Societies and Social Simulation* 4, 3. **http://jasss.soc.surrey.ac.uk/4/3/8.html** (accessed March 1, 2008).

Diffusion Patterns

References

Chen, Chaomei, and Diana Hicks. 2004. "Tracing Knowledge Diffusion." *Scientometrics* 59, 2: 199–211.

Gladwell, Malcolm. 2000. *The Tipping Point: How Little Things Can Make a Big Difference*. Boston: Little, Brown.

Mahajan, Vijay, and Robert A. Peterson. 1985. *Models for Innovation Diffusion*. Newbury Park, CA: Sage.

Tabah, Albert N. 1999. "Literature Dynamics: Studies on Growth, Diffusion, and Epidemics." In *Annual Review of Information Science and Technology*, edited by M. E. Williams, vol. 34: 249–286. Amsterdam: Elsevier Science Publishers.

Valente, Thomas W. 2005. "Models and Methods for Innovation Diffusion." In *Models and Methods in Social Network Analysis*, edited by Peter J. Carrington, John Scott, and Stanley Wasserman, 98–116. Cambridge: Cambridge University Press.

Diffusion of Innovations

References

Rogers, Everett M. 1962. *Diffusion of Innovations*. New York: Free Press.

Image Credits

Cumulative Frequency Distribution: Russell J. Duhon (data preparation), Elisha F. Hardy (design), Katy Börner (concept), Indiana University.

Idea Spreading as Epidemic Process

References

Bailey, Norman T. J. 1957. *The Mathematical Theory of Epidemics*. New York: Hafner Publishing Inc.

Bettencourt, Luís M. A., Carlos Castillo-Chávez, David Kaiser, and David E. Wojick. 2006. "Report for the Office of Scientific and Technical Information: Population Modeling of the Emergence and Development of Scientific Fields." Office of Scientific and Technical Information, U.S. Department of Energy. **http://www.osti.gov/innovation/research/diffusion/epicasediscussion_lb2.pdf** (accessed June 2, 2008).

Bettencourt, Luís M. A., Ariel Cintrón-Arias, David I. Kaiser, and Carlos Castillo-Chávez. 2006. "The Power of a Good Idea: Quantitative Modeling of the Spread of Ideas from Epidemiological Models." *Physica A: Statistical Mechanics and its Applications*, 364: 513–536.

Garfield, Eugene. 1980. "The Epidemiology of Knowledge and the Spread of Scientific Information." *Current Comments* 35: 5–10. Reprinted in *Essays of an Information Scientist* 4, 1979–80, 586–591. **http://www.garfield.library.upenn.edu/essays/v4p586y1979-80.pdf** (accessed March 1, 2008).

Goffman, William. 1966. "Mathematical Approach to the Spread of Scientific Ideas—the History of Mast Cell Research." *Nature* 212, 5061: 449–452.

Goffman, William, and V. A. Newill. 1964. "Generalization of Epidemic Theory: An Application to the Transmission of Ideas." *Nature* 204: 225–228.

U.S. Department of Energy, Office of Scientific and Technical Information. "Diffusion of Research." **http://www.osti.gov/innovation/research/diffusion/#casestudies; http://www.osti.gov/innovation/research/diffusion/birdflurun2.pdf** (accessed March 1, 2008).

Feedback Cycles

Educational Supply and Demand Cycles

References

Menard, Henry W. 1971. *Science: Growth and Change*. Cambridge, MA: Harvard University Press.

Symbolic Capital & the Cycle of Credibility

References

Bourdieu, Pierre. 1986. "The Forms of Capital." *Handbook of Theory and Research for the Sociology of Education*, edited by John G. Richardson, 243–248. Westport, CT: Greenwood Press.

Latour, Bruno, and Steve Woolgar. 1982. "The Cycle of Credibility." *Science in Context: Readings in the Sociology of Science*, edited by Barry Barnes and David Edge, 35–43. Cambridge, MA: MIT Press.

Latour, Bruno, and Steve Woolgar. 1986. *Laboratory Life: The Construction of Scientific Facts*. Princeton, NJ: Princeton University Press.

Whitley, Richard. 1984. *The Intellectual and Social Organization of the Sciences*. New York: Oxford University Press.

Image Credits

Image adapted from Latour and Woolgar 1982, "The Cycle of Credibility," fig. 1, p.35; Elisha F. Hardy (design), Katy Börner (concept), Indiana University.

60 Data Acquisition and Preprocessing

References

Gray, Jim, and Prashant Shenoy. 1999. "Rules of Thumb in Data Engineering." *Proceedings of 16th International Conference on Data Engineering*, 3–12. Washington, DC: IEEE Computer Society. **http://research.microsoft.com/en-us/um/people/gray; ftp://ftp.research.microsoft.com/pub/tr/tr-99-100.** pdf (accessed February 10, 2010).

Sanger, Lawrence M. 2006. "The Future of Free Information." *History and Memory: Present Reflections on the Past to Build Our Future*. Macau Ricci Institute Studies, Vol. 5. **http://www.riccimac.org/eng/mris/5.htm**. (accessed November 19, 2009). [Quotation]

Data Types, Sizes, and Formats

No references or credits.

Data Quality

References

Bar-Ilana, Judit, Mark Leveneb, and Ayelet Lina. 2007. "Some Measures for Comparing Citation Databases." *Journal of Informetrics* 1, 1: 26-34.

Buneman, Peter, Sanjeev Khanna, and Wang-Chiew Tan. 2000. "Data Provenance: Some Basic Issues." *Lecture Notes in Computer Science*, vol. 1974. Berlin: Springer-Verlag. **http://db.cis.upenn.edu/DL/fsttcs.pdf** (accessed March 1, 2008).

Hall, Bronwyn H., Adam B. Jaffe, and Manuel Trajtenberg. 2001. "The NBER Patent Citations Data File: Lessons, Insights and Methodological Tools." *CEPR Discussion Papers* 3094. **http://ideas.repec.org/p/cpr/ceprdp/3094.html** (accessed March 1, 2008).

Norris, Michael, and Charles Oppenheim. 2007. "Comparing Alternatives to the Web of Science for Coverage in the Social Sciences' Literature." *Journal of Informetrics* 1: 161-169.

Wouters, Paul. 1999. "The Citation Culture." PhD diss., University of Amsterdam, 169. **http://garfield.library.upenn.edu/wouters/wouters.pdf** (accessed October 30, 2007).

Data Coverage

References

Belew, Richard K. 2005. "Scientific Impact Quantity and Quality: Analysis of Two Sources of Bibliographic Data." **http://arxiv.org/abs/cs.IR/0504036** (accessed February 5, 2010).

Bosman, Jeroen, Ineke van Mourik, Menno Rasch, Eric Sieverts, and Huib Verhoeff. 2006. "Scopus Reviewed and Compared: The Coverage and Functionality of the Citation Database Scopus, Including Comparisons with Web of Science and Google Scholar." Utrecht University Library. **http://igitur-archive.library.uu.nl/DARLIN/2006-1220-200432/Scopus doorgelicht & vergeleken - translated.pdf**. (accessed March 1, 2008).

Boyack, Kevin W. 2004. "Mapping Knowledge Domains: Characterizing *PNAS*." *PNAS* 101, Suppl. 1: 5192-5199.

Boyack, Kevin W., and Katy Börner. 2003. "Indicator-Assisted Evaluation and Funding of Research: Visualizing the Influence of Grants on the Number and Citation Counts of Research Papers." Special Issue (Visualizing Scientific Paradigms), *Journal of the American Society of Information Science and Technology* 54, 5: 447–461.

de Moya-Anegón, Félix, Zaida Chinchilla-Rodríguez, Benjamín Vargas-Quesada, Elena Corera-Álvarez, Francisco José Muñoz-Fernández, Antonio González-Molina, and Victor Herrero-Solana. 2007. "Coverage Analysis of Scopus: A Journal Metric Approach." *Scientometrics* 73, 1: 53–78.

Elsevier B.V. 2008. Scopus Home Page. **http://www.scopus.com/scopus/home.url** (accessed June 10, 2008).

Fingerman, Susan. 2006. "Electronic Resources Reviews: Web of Science and Scopus: Current Features and Capabilities." *Issues in Science and Technology Librarianship*. **http://www.istl.org/06-fall/electronic2.html** (accessed March 1, 2008).

Google, Inc. 2008. Google Scholar Beta Home Page. **http://scholar.google.com** (accessed June 10, 2008).

Hicks, Diana. 1999. "The Difficulty of Achieving Full Coverage of International Social Science Literature and the Bibliometric Consequences." *Scientometrics* 44, 2: 193–215.

Meho, Lokman I., and Kiduk Yang. 2007. "Impact of Data Sources on Citation Counts and Rankings of LIS Faculty: Web of Science versus Scopus and Google Scholar." *Journal of the American Society for Information Science and Technology* 58, 13: 2105–2125.

Nisonger, T. E. 2004. "Citation Autobiography: An Investigation of ISI Database Coverage in Determining Author Citedness." *College & Research Libraries* 65, 2: 152–163.

Pauly, Daneil, and Konstantinos I. Stergiou. 2005 "Equivalence of Results from Two Citation Analyses: Thomson ISI's Citation Index and Google Scholar's Service." *Ethics in Science and Environmental Politics*, 33–35. **http://www.int-res.com/articles/esep/2005/E65.pdf** (accessed March 1, 2008).

ProQuest LLC. 2008. Ulrichsweb.com Home Page. **http://www.ulrichsweb.com/ulrichsweb** (accessed June 10, 2008).

Science Policy Support Group. 2010. "SPSG—The PARIS/ISSC Project." **http://www.stage-research.net/SPSG/paris-issc/BerlinHealeyRothman.html** (accessed March 16, 2010).

Thomson Reuters. 2008. "Web of Science." **http://scientific.thomsonreuters.com/products/wos** (accessed June 10, 2008).

Data Credits

Science Citation Index by Thomson Reuters.

Image Credits

Science map with topic coverage overlay: Russell J. Duhon (data preparation), Elisha F. Hardy (design), Katy Börner (concept), Indiana University, featuring data from Thomson Reuters, Web of Science, and the science paradigm map outline from Boyack and Klavans (2008), used by permission.

Data Acquisition

References

Domingos, Pedro, and Matt Richardson. 2002. "The Intelligent Surfer: Probabilistic Combination of Link and Content Information in PageRank." *Advances in Neural Information Processing Systems 14*, 1441–1448. Cambridge, MA: MIT Press.

Menczer, Filippo. 2002. "Growing and Navigating the Small World Web by Local Content." *PNAS* 99, 22: 14014–14019.

Page, Lawrence, Sergey Brin, Rajeev Motwani, and Terry Winograd. 1998. "The PageRank Citation Ranking: Bringing Order to the Web." Stanford Digital Library Technologies Project. **http://infolab.stanford.edu/~backrub/pageranksub.ps** (accessed June 2, 2008).

Data Preprocessing

References

Moed, Henk F. 2005. *Citation Analysis in Research Evaluation*, 194. Berlin: Springer-Verlag.

Newman, Mark E. J. 2005. "Power Laws, Pareto Distributions and Zipf's Law." *Contemporary Physics*, 46, 5: 323–351.

Porter, Martin. 2006. "The Porter Stemming Algorithm." **http://tartarus.org/~martin/PorterStemmer** (accessed August 16, 2009).

Stevens, S. S. 1946. "On the Theory of Scales of Measurement." *Science* 103: 677–680.

Torvik, Vetle I., Marc Weeber, Don R. Swanson, and Nell R. Smalheiser. 2005. "A Probabilistic Similarity Metric for Medline Records: A Model for Author Name Disambiguation." *Journal of the American*

Society for Information Science and Technology 56, 2: 140–158.

Wooding, S., K. Wilcox-Jay, G. Lewison, and J. Grant. 2006. "Co-Author Inclusion: A Novel Recursive Algorithmic Method for Dealing with Homonyms in Bibliometric Analysis." *Scientometrics* 66, 1: 11–21.

Data Augmentation

References

Garfield, Eugene. 2005. "The Agony and the Ecstasy—the History and Meaning of the Journal Impact Factor." *Presentation at the International Congress on Peer Review and Biomedical Publication.* Chicago, IL, September 16. http://garfield.library.upenn.edu/papers/jifchicago2005.pdf (accessed June 10, 2008).

Software Credits

MelissaData. 2008. "Geocoder (Lat/Long) Lookup." http://www.melissadata.com/lookups/geocoder.asp (accessed June 10, 2008).

Baseline Statistics

No references or credits.

Data Integration and Federation

References

Bollen, Johan, and Herbert Van de Sompel. 2006. "An Architecture for the Aggregation and Analysis of Scholarly Usage Data." *Proceedings of the 6th ACM/IEEE-CS Joint Conference on Digital Libraries 298-307.* New York: ACM Press.

Börner, Katy. 2006. "Semantic Association Networks: Using Semantic Web Technology to Improve Scholarly Knowledge and Expertise Management." In *Visualizing the Semantic Web,* 2nd ed., edited by Vladimir Geroimenko and Chaomei Chen, 183–198. London: Springer-Verlag.

Deep Web Technologies. 2008. "Explorit™ Research Accelerator." http://www.deepwebtech.com (accessed June 10, 2008).

Google, Inc. 2008. Google Scholar Beta Home Page. http://scholar.google.com (accessed June 10, 2008).

La Rowe, Sumeet Adinath Ambre, John W. Burgoon, Weimao Ke, and Katy Börner. 2007. "The Scholarly Database and Its Utility for Scientometrics Research." *Proceedings of the 11th Annual Information Visualization International Conference*, edited by D. Torres-Salinas and H. F. Moed, 459–464. Washington DC: IEEE Computer Society.

Los Alamos National Laboratory. 2008. "MESUR: Metrics from Scholarly Usage of Resources." http://www.mesur.org/MESUR.html (accessed June 10, 2008).

Office of Science and Technical Information, U.S. Department of Energy. 2010. "WorldWideScience: The Global Science Gateway." http://worldwidescience.org (accessed March 30, 2010).

Science.gov. 2010. "USA.gov for Science." http://www.science.gov (accessed March 16, 2010).

Contributors

Eleanor G. Frierson and Walter L. Warnick.

Evolving Interoperability Standards and Tools

References

Cyganiak, Richard. 2008. "The Linking Open Data Dataset Cloud." http://richard.cyganiak.de/2007/10/lod (accessed June 10, 2008).

Geonames.org. 2008. GeoNames Home Page. http://www.geonames.org (accessed June 10, 2008).

International DOI Foundation. 2008. The DOI System Home Page. http://www.doi.org (accessed June 10, 2008).

Library of Congress: Standards. 2007. "IFLA/CDNL Alliance for Bibliographic Standards: Library of Congress Action Area." http://www.loc.gov/standards/uri/info.html (accessed June 10, 2008).

National Science Digital Library. 2010. NSDL Home Page. http://nsdl.org (accessed March 16, 2010).

Openarchives.org. 2008. Open Archives Initiative Home Page. http://www.openarchives.org (accessed June 10, 2008).

Open Archives Initiative. 2008. "Object Reuse and Exchange: ORE Specification and User Guide—Table of Contents." http://www.openarchives.org/ore/0.9/toc (accessed June 10, 2008).

Openarchives.org. 2008. "Registered Data Providers." http://www.openarchives.org/Register/BrowseSites (accessed June 10, 2008).

Openarchives.org. 2008. "The Open Archives Initiative Protocol for Metadata Harvesting." http://www.openarchives.org/OAI/openarchivesprotocol.html (accessed June 10, 2008).

Rodriguez, Marko A., Johan Bollen, and Herbert Van de Sompel. 2007. "Mesur's OWL RDF/XML Ontology." http://tweety.lanl.gov/public/schemas/2007-01/mesur.owl (accessed February 5, 2010).

Talis Developer Network. 2008. "The Talis Community Licence" (draft). http://www.talis.com/tdn/tcl (accessed June 10, 2008).

The Trustees of Princeton University. 2010. "About WordNet." http://wordnet.princeton.edu (accessed March 30, 2010).

Wikimedia Foundation. 2010. *Wikipedia, the Free Encyclopedia*. http://wikipedia.org (accessed February 4, 2010).

Image Credits

Image of Linked Data sets, by Richard Cyganiak, Anja Jetntzsch, and Chris Bizer, using OmniGraffle software: http://www.omnigroup.com/applications/OmniGraffle. *See also* the evolving linked data sets graphic at http://esw.w3.org/topic/SweoIG/TaskForces/CommunityProjects/LinkingOpenData. *See* Cyganiak 2008.

CrossRef Web Services

References

Bizer, Chris, Richard Cyganiak, and Tom Heath. 2007. "How to Publish Linked Data on the Web." Berlin: Web-based Systems Group, Freie Universität. http://www4.wiwiss.fu-berlin.de/bizer/pub/LinkedDataTutorial (accessed June 10, 2008).

Pila Inc. 2002. "Crossref Metadata Services." http://www.crossref.org/metadata_services.html (accessed February 10, 2010).

Contributors

Mark Doyle, Duane Degler, and Herbert Van de Sompel.

The W3C SWEO Linking Open Data Community Project

References

Creative Commons. "Share, Remix, Reuse—Legally." http://creativecommons.org (accessed June 10, 2008).

ESW Wiki. 2010. "LinkingOpenData: W3C SWEO Community Project." http://esw.w3.org/topic/SweoIG/TaskForces/CommunityProjects/LinkingOpenData (accessed March 30, 2010).

Geonames.org. 2008. GeoNames Home Page. http://www.geonames.org (accessed June 10, 2008).

Ley, Michael. 2008. "The DBLP Computer Science Bibliography." Universität Trier. http://www.informatik.uni-trier.de/~ley/db (accessed June 10, 2008).

MetaBrainz Foundation. 2008. MusicBrainz Home Page. http://musicbrainz.org (accessed June 10, 2008).

Talis Developer Network. 2008. "The Talis Community Licence" (draft). http://www.talis.com/tdn/tcl (accessed June 10, 2008).

The Trustees of Princeton University. 2010. "About WordNet." http://wordnet.princeton.edu (accessed March 30, 2010).

W3C: Technology and Society Domain. 2004. "Semantic Web Activity: RDF." http://www.w3.org/RDF (accessed June 10, 2008).

Wikimedia Foundation. 2009. "OpenURL." *Wikipedia, the Free Encyclopedia*. http://en.wikipedia.org/wiki/OpenURL (accessed February 5, 2010).

Wikimedia Foundation. 2008. Wikibooks Home Page. http://en.wikibooks.org/wiki/Main_Page (accessed June 10, 2008).

Practical Ontology of Scholarly Data

References

Rodriguez, Marko A., Johan Bollen, and Herbert Van de Sompel. 2007. "Mesur's OWL RDF/XML Ontology Description." http://tweety.lanl.gov/public/schemas/2007-01/mesur.owl. (accessed February 5, 2010).

Rodriguez, Marko A., Johan Bollen, and Herbert Van de Sompel. 2007. "A Practical Ontology for the Large-Scale Modeling of Scholarly Artifacts and Their Usage." *Proceedings of the 7th ACM/IEEE-CS Joint Conference on Digital Libraries.* New York: ACM Press.

Digitometic Services and Tools

References

Brody, Tim, Simon Kampa, Stevan Harnad, Les Carr, and Steve Hitchcock. 2003. "Digitometric Services for Open Archives Environments." In *Research and Advanced Technology for Digital Libraries (ECDL'03)*, edited by T. Koch and I. Sølvberg, 207–220. Berlin: Springer-Verlag. http://eprints.ecs.soton.ac.uk/7503 (accessed October 30, 2007).

Data Preservation

Contributors

Stacy Kowalczyk, Ying Ding, Nianli Ma, and Herbert Van de Sompel.

62 Data Analysis, Modeling, and Layout

References

Börner, Katy, Chaomei Chen, and Kevin W. Boyack. 2003. "Visualizing Knowledge Domains." In *Annual Review of Information Science and Technology*, edited by Blaise Cronin, vol. 37: 179–255. Medford, NJ: Information Today/American Society for Information Science and Technology.

Börner, Katy, Soma Sanyal, and Alessandro Vespignani. 2007. "Network Science." In *Annual Review of Information Science and Technology*, edited by Blaise Cronin, vol. 41: 537–607. Medford, NJ: Information Today/American Society for Information Science and Technology.

R. W. Hamming. 1962. *Numerical Methods for Scientists and Engineers.* New York: McGraw-Hill. [Quotation]

Algorithm Functionality

References

Willinger, Walter, Ramesh Govindan, Sugih Jamin, Vern Paxson, and Scott Shenker. 2002. "Scaling Phenomena in the Internet: Critically Examining Criticality." *PNAS* 99, 1: 2573–2580.

Algorithm Scopes

No references or credits.

Algorithm Implementations

No references or credits.

Workflow Design

References

Börner, Katy, Chaomei Chen, and Kevin W. Boyack. 2003. "Visualizing Knowledge Domains." In *Annual Review of Information Science and Technology*, edited by Blaise Cronin, vol. 37: 179–255. Medford, NJ: Information Today/American Society for Information Science and Technology.

Callon, M., J. P. Courtial, W. A. Turner, and S. Bauin. 1983. "From Translations to Problematic Networks: An Introduction to Co-Word Analysis." *Social Science Information* 22: 191–235.

Cambia, Open Initiative Institute. 2008. Patentlens Home Page. http://www.patentlens.net/daisy/patentlens/patentlens.html (accessed June 10, 2008).

Di Battista, G., P. Eades, R. Tamassia, and I. G Tollis. 1999. *Graph Drawing: Algorithms for the Visualization of Graphs*. New York: Prentice-Hall.

Elsevier B.V. 2008. Scopus Home Page. http://www.scopus.com/scopus/home.url (accessed June 10, 2008).

Fruchterman, T. M.J., and E. M. Reingold. 1991. "Graph Drawing by Force-Directed Placement." *Software: Practice and Experience* 21, 11: 1129–1164.

Google, Inc. 2008. Google Scholar Beta Home Page. http://scholar.google.com (accessed June 10, 2008).

Griffiths, Thomas L., and Mark Steyvers. 2002. "A Probabilistic Approach to Semantic Representation." *Proceedings of the 24th Annual Conference of the Cognitive Science Society*, 381–386. George Mason University, Fairfax, VA, August 7–10.

Kamada, T. and S. Kawai. 1989. "An Algorithm for Drawing General Undirected Graphs." *Information Processing Letters* 31: 7–15.

Khosla, Nitin. 2004. "Dimensionality Reduction Using Factor Analysis." Master's thesis, Griffith University, Australia.

Kohonen, Teuvo. 1995. *Self-Organizing Maps*. Heidelberg: Springer-Verlag.

Kruskal, J. B., and M. Wish. 1978. *Multidimensional Scaling*. London: Sage.

Labelle, François. 2003. "Dimensionality Reduction Using Principal Components Analysis." http://www.cs.mcgill.ca/~sqrt/dimr/dimreduction.html (accessed June 10, 2008).

Landauer, T. K., P. W. Foltz, and D. Laham. 1998. "Introduction to Latent Semantic Analysis." *Discourse Processes* 25: 259–284.

Lee, R.C.T., J. R. Slagle, and H. Blum. 1977. "A Triangulation Method for the Sequential Mapping of Points from N-Space to Two-Space." *IEEE Transactions on Computers,* 26, 3: 288–292.

ProQuest LLC. 2010. COS (Community of Science) Home Page. **http://www.cos.com** (accessed March 16, 2010).

Salton, G., C. Yang, and A. Wong. 1975. "A Vector Space Model for Automatic Indexing." *Communications of the ACM* 18, 1: 613–620.

Schvaneveldt, R., ed. 1990. *Pathfinder Associative Networks: Studies in Knowledge Organization.* Norwood, NJ: Ablex Publishing Corp.

Thomson Reuters. 2008. "ISI Web of Knowledge." **http://isiwebofknowledge.com** (accessed August 14, 2009).

Tryon, R. C. 1939. *Cluster Analysis.* New York: McGraw-Hill.

White, H. D., and B. C. Griffith. 1981. "Author Cocitation: A Literature Measure of Intellectual Structure." *Journal of the American Society for Information Science* 32, 3: 163–171.

Will, Todd. 1999. "Introduction to the Singular Value Decomposition." University of Wisconsin, La Crosse. **http://www.uwlax.edu/faculty/will/svd** (accessed June 10, 2008).

Image Credits

Workflow for Mapping Science, *see* Börner et al. 2003.

Layout

References

Di Battista, G., P. Eades, R. Tamassia, and I. G Tollis. 1999. *Graph Drawing: Algorithms for the Visualization of Graphs.* New York: Prentice-Hall.

Fruchterman, T. M.J., and E. M. Reingold. 1991. "Graph Drawing by Force-Directed Placement." *Software: Practice and Experience* 21, 11: 1129–1164.

Kamada, T., and S. Kawai. 1989. "An Algorithm for Drawing General Undirected Graphs." *Information Processing Letters* 31: 7–15.

Lee, R.C.T., J. R. Slagle, and H. Blum. 1977. "A Triangulation Method for the Sequential Mapping of Points from N-Space to Two-Space." *IEEE Transactions on Computers* 26, 3: 288–292. .

Temporal Analysis

References

Kleinberg, J. M. 2002. "Bursty and Hierarchical Structure in Streams." *Proceedings of the 8th ACM SIGKDD International Conference on Knowledge Discovery and Data Mining,* 91–101. New York: ACM Press.

Kleinberg, J. M. 2002. "Sample Results from a Burst Detection Algorithm." **http://www.cs.cornell.edu/home/kleinber/kdd02.html** (accessed May 30, 2008).

Image Credits

Screen capture of data from The Nielsen Company, 2008. "BlogPulse: a service of Nielsen Buzzmetrics." **http://www.blogpulse.com**. Used by permission, all rights reserved.

Geographic Analysis

References

Kraak, Menno-Jan, and Ferjan Ormeling. 2003. *Cartography: Visualization of Geospatial Data.* Upper Saddle River, NJ: Prentice-Hall.

Skupin, A. 2000. "From Metaphor to Method: Cartographic Perspectives on Information Visualization." *Proceedings of InfoVis 2000,* 91–97. Salt Lake City, UT: IEEE Computer Society.

Tobler, W. A. 1973. "Continuous Transformation Useful for Districting." *Science* 219: 215–220.

Töpfer, F. 1974. *Kartographische Generalisierung.* Gotha, Germany: VEB Herrmann Haack/Geographisch-Kartographische Anstalt.

Töpfer, F., and W. Pillewizer. 1996. "The Principles of Selection." *Cartographic Journal* 3: 10–16.

Data Credits

Science Citation Index by Thomson Reuters.

Image Credits

Understanding Outside Collaborations of the Chinese Academy of Sciences: Russell J. Duhon (data preparation), Elisha F. Hardy (design), Katy Börner (concept), Indiana University.

Contributors

The National Science Library of Chinese Academy of Sciences compiled the Chinese coauthorship data for the *Places & Spaces* exhibition in Beijing, China.

Topical Analysis

References

Callon, M., J. P. Courtial, W. A. Turner, and S. Bauin. 1983. "From Translations to Problematic Networks: An Introduction to Co-word Analysis." *Social Science Information* 22: 191–235.

Callon, M., J. Law, and A. Rip, eds. 1986. *Mapping the Dynamics of Science and Technology.* London: Macmillan.

Deerwester, S., S. T. Dumais, G. W. Furnas, T. K. Landauer, and R. Harshman. 1990. "Indexing by Latent Semantic Analysis." *Journal of the American Society for Information Science* 41, 6: 391–407.

Griffiths, Thomas L., and Mark Steyvers. 2002. "A Probabilistic Approach to Semantic Representation." *Proceedings of the 24th Annual Conference of the Cognitive Science Society,* 381–386. George Mason University, Fairfax, VA, August 7–10.

Griffiths, Thomas L., and Mark Steyvers. 2004. "Finding Scientific Topics." *PNAS* 101, Suppl. 1: 5228–5235.

Kohonen, Teuvo. 1995. *Self-Organizing Maps.* Heidelberg: Springer-Verlag.

Kruskal, J. B., and M. Wish. 1978. *Multidimensional Scaling.* London: Sage.

Landauer, T. K., P. W. Foltz, and D. Laham. 1998. "Introduction to Latent Semantic Analysis." *Discourse Processes* 25, 259–284.

Landauer, T. K., and S. T. Dumais. 1997. "A Solution to Plato's Problem: The Latent Semantic Analysis Theory of the Acquisition, Induction, and Representation of Knowledge." *Psychological Review* 104: 211–240.

Porter, Martin. 2006. "The Porter Stemming Algorithm." **http://tartarus.org/~martin/PorterStemmer** (accessed August 16, 2009).

Salton, Gerard, and C.S. Yang. 1973. "On the Specification of Term Values in Automatic Indexing." *Journal of Documentation* 29: 351–372.

Skupin, André. 2002. "A Cartographic Approach to Visualizing Conference Abstracts." *IEEE Computer Graphics and Applications* 22, 1: 50–58.

Smith, Lindsay. 2002. "A Tutorial on Principal Components Analysis." **http://www.cs.otago.ac.nz/cosc453/student_tutorials/principal_components.pdf** (accessed June 10, 2008).

Image Credits

Visualization of 2,220 Conference Abstracts, *see* Skupin, 2002.

Skupin, André. "A Cartographic Approach to Visualizing Conference Abstracts." *IEEE Computer Graphics and Applications* 22, 1: 50–58, © 2002, IEEE.

Network Analysis

References

Barabási, A.-L. 2002. *Linked: The New Science of Networks.* Cambridge, MA: Perseus Publishing.

Borgman, Christine L., and Jonathan Furner. 2002. "Scholarly Communication and Bibliometrics." In *Annual Review of Information Science and Technology,* edited by Blaise Cronin, 36: 3–72. Medford, NJ: Information Today.

Börner, Katy, Luca Dall'Asta, Weimao Ke, and Alessandro Vespignani. 2005. "Studying the Emerging Global Brain: Analyzing and Visualizing the Impact of Co-Authorship Teams." *Complexity: Special Issue on Understanding Complex Systems* 10, 4: 58–67.

Börner, Katy, Soma Sanyal, and Alessandro Vespignani. 2007. "Network Science." In *Annual Review of Information Science and Technology,* edited by Blaise Cronin, vol. 41: 537–607. Medford, NJ: Information Today/American Society for Information Science and Technology.

Carrington, P. J., J. Scott, and S. Wasserman, eds. 2005. *Models and Methods in Social Network Analysis.* Cambridge: Cambridge University Press.

Lenoir, Timothy. 2002. "Quantitative Foundations for the Sociology of Science: On Linking Blockmodeling with Co-Citation Analysis." *Social Networks: Critical Concepts in Sociology,* edited by John Scott. New York: Routledge.

Merton, Robert. 1973. *The Sociology of Science.* Chicago: The University of Chicago Press.

Monge, Peter R., and Noshir S. Contractor. 2003. *Theories of Communication Networks.* New York: Oxford University Press.

Narin, Francis, and Joy K. Moll. 1977. "Bibliometrics." In *Annual Review of Information Science and Technology,* edited by M. E. Williams, 12: 35–58. White Plains, NY: Knowledge Industry Publications.

Nicolaisen, Jeppe. 2007. "Citation Analysis." In *Annual Review of Information Science and Technology,* edited by Blaise Cronin, 41: 609–641. Medford, NJ: Information Today/American Society for Information Science and Technology.

Scott, J. P. 2000. *Social Network Analysis: A Handbook.* London: Sage Publications.

Thelwall, M., L. Vaughan, and L. Björneborn. 2005. "Webometrics." In *Annual Review of Information Science and Technology,* edited by Blaise Cronin, 39: 81–135. Medford, NJ: Information Today/American Society for Information Science and Technology.

Wasserman, Stanley, and K. Faust. 1994. *Social Network Analysis: Methods and Applications.* Structural Analysis in the Social Sciences 8. New York: Cambridge University Press.

White, Howard D., and Katherine W. McCain. 1989. "Bibliometrics." In *Annual Review of Information Science and Technology,* edited by M. E. Williams, 24: 119–186. Amsterdam: Elsevier Science Publishers.

Wilson, C. S. 2001. "Informetrics." In *Annual Review of Information Science and Technology,* edited by M. E. Williams, vol. 34: 107–286. Medford, NJ: Information Today/American Society for Information Science and Technology.

Image Credits

Weighted coauthor network, *see* Börner et al. 2005.

64 Data Communication–Visualization Layers

References

Abrams, Janet, and Peter Hall. 2006. *Else/Where: Mapping New Cartographies of Networks and Territories.* Minneapolis: University of Minnesota Design Institute.

Card, Stuart, Jock Mackinlay, and Ben Shneiderman, eds. 1999. *Readings in Information Visualization: Using Vision to Think.* San Francisco: Morgan Kaufmann.

Chen, Chaomei. 2004. *Information Visualization: Beyond the Horizon.* New York: Springer-Verlag.

ESRI. 2008. ESRI Home Page. **http://www.esri.com** (accessed June 2, 2008).

Frankel, Felice. 2002. *Envisioning Science: The Design and Craft of the Science Image.* Cambridge, MA: MIT Press.

Harris, Robert L. 2000. *Information Graphics: A Comprehensive Illustrated Reference.* New York: Oxford University Press.

Kosslyn, Stephen M. 1993. *Elements of Graph Design,* 89. New York: W. H. Freeman & Company.

MacEachren, Alan M. 2004. *How Maps Work: Representation, Visualization, and Design.* New York: Guilford Press.

Spence, Robert. 2007. *Information Visualization: Design for Interaction.* Upper Saddle River, NJ: Prentice-Hall.

Thomas, James J., and Kristin A. Cook, eds. 2005. *Illuminating the Path: The Research and Development Agenda for Visual Analytics.* National Visualization and Analytics Center. **http://nvac.pnl.gov/agenda.stm** (accessed March 1, 2008).

Tobler, Waldo. 1970. "A Computer Model Simulating Urban Growth in the Detroit Region." *Economic Geography* 46, 2: 234–240. [Quotation]

Tollis, Ioannis G., Giuseppe Di Battista, Peter Eades, and Roberto Tamassia. 1998. *Graph Drawing: Algorithms for the Visualization of Graphs.* Upper Saddle River, NJ: Prentice-Hall.

Tufte, Edward R. 1990. *Envisioning Information.* Cheshire, CT: Graphics Press.

Tufte, Edward R. 1997. *Visual Explanations: Images and Quantities, Evidence and Narrative.* Cheshire, CT: Graphics Press.

Tufte, Edward R. 2001. *The Visual Display of Quantitative Information.* 2nd ed. Cheshire, CT: Graphics Press.

Tufte, Edward R. 2006. *Beautiful Evidence.* Cheshire, CT: Graphics Press.

Wurman, Richard Saul. 1997. *Information Architects*. New York: Watson-Guptill Publications.

Wurman, Richard Saul, and Robert Jacobson. 2000. *Information Design*. Cambridge, MA: MIT Press.

Visual Perception and Cognitive Processing

References

Arnheim, Rudolf. 1954. *Art and Visual Perception: A Psychology of the Creative Eye*. Berkeley, CA: University of California Press.

Arnheim, Rudolf. 1969. *Visual Thinking*. Berkeley, CA: University of California Press.

Csikszentmihalyi, Mihaly. 1991. *Flow: The Psychology of Optimal Experience*. New York: HarperCollins.

Goldstone, Robert L., and Larry Barsalou. 1998. "Reuniting Perception and Conception." *Cognition* 65: 231–262.

Healey, Christopher G. 2007. "Perception in Visualization." North Carolina State University, Department of Computer Science. http://www.csc.ncsu.edu/faculty/healey/PP (accessed March 1, 2008).

Koffka, K. 1935. *The Principles of Gestalt Psychology*, 720. London: Lund Humphries.

Maeda, John. 2006. *The Laws of Simplicity*. Cambridge, MA: MIT Press.

Palmer, S. E. 1999. *Vision Science: From Photons to Phenomenology*. Cambridge, MA: MIT Press.

Ware, Colin. 2000. *Information Visualization: Perception for Design*. San Francisco: Morgan Kaufmann.

Usable Interfaces and Flow Experiences

References

Bederson, Benjamin B. 2004. "Interfaces for Staying in the Flow." *Ubiquity* 5, 27, Sept. 1-7. http://www.acm.org/ubiquity/views/v5i27_bederson.html (accessed February 5, 2010).

Csikszentmihalyi, Mihaly. 1991. *Flow: The Psychology of Optimal Experience*. New York: HarperCollins.

Image Credits

Elisha F. Hardy (design), Katy Börner (concept), Indiana University, adapted from source material in Csikszentmihalyi 1991.

Sociable Interfaces

References

Börner, Katy, and Raquel Navarro-Prieto. 2005. "Guest Editors' Introduction: Special Issue: Collaborative Information Visualization Environments." *Presence: Teleoperators and Virtual Environments* 14, 1: 3. Cambridge, MA: MIT Press.

Jo Kim, Amy. 2000. *Community Building on the Web: Secret Strategies for Successful Online Communities*. Berkeley, CA: Peachpit Press.

Preece, Jenny. 1998. "Empathic Communities: Reaching Out Across the Web." *Interactions* 5, 2: 32–43.

(Visual) Scalability Issues

References

Eick, Stephen G., and Alan F. Karr. 2000. "Visual Scalability." Technical Report Number 106, National Institute of Statistical Sciences. *IEEE Transactions on Visualization and Computer Graphics*.

Shneiderman, Ben. 2002. *Leonardo's Laptop: Human Needs and the New Computing Technologies*. Cambridge, MA: MIT Press.

Modular Visualization Design

References

Adobe Systems, Inc. 2008. Adobe Photoshop Home Page. http://www.adobe.com/products/photoshop (accessed June 10, 2008).

Adobe Systems, Inc. 2008. Dreamweaver Home Page. http://www.adobe.com/products/dreamweaver/?sdid=CKIQP (accessed May 30, 2008).

Alexander, Christopher, S. Ishikawa, and M Silverstein. 1977. *A Pattern Language: Towns, Buildings, Construction*. Center for Environmental Structure Series 2. New York: Oxford University Press.

ESRI. 2008. ESRI Home Page. http://www.esri.com (accessed June 2, 2008).

Shneiderman, Ben, Gerhard Fischer, Mary Czerwinski, Mitch Resnick, Brad Myers, Linda Candy, Ernest Edmonds, Mike Eisenberg, Elisa Giaccardi, Tom Hewett, Pamela Jennings, and Bill Kules, et al. 2006. "Creativity Support Tools: Report From a US National Science Foundation Sponsored Workshop." *International Journal of Human-Computer Interaction* 20, 2: 61–77.

Communication Versus Exploration

References

MacEachren, Alan M. 2004. *How Maps Work: Representation, Visualization, and Design*. New York: Guilford Press.

Interpretation

References

Huff, Darell. 1954. *How to Lie with Statistics*. New York: Norton.

Jones, Gerald E. 1995. *How to Lie with Charts*. San Francisco: Sybex.

Monmonier, Mark, and H. J. de Blij. 1996. *How to Lie with Maps*. Chicago: The University of Chicago Press.

Visualization Layers

Reference System

References

Fabrikant, Sara Irina, Daniel R. Montello, and David M. Mark. 2006. "The Distance-Similarity Metaphor in Region-Display Spatializations." *Special Issue on Exploring Geovisualization: IEEE Computer Graphics & Application*, edited by T. M. Rhyne, A. MacEachren, and J. Dykes, vol. 26, 4: 34–44. Los Alamitos, CA: IEEE Computer Society Press.

Fabrikant, S. I., D. R. Montello, M. Ruocco, and R. S. Middleton. 2004. "The Distance-Similarity Metaphor in Network-Display Spatializations." *Cartography and Geographic Information Science* 31, 4: 237–252.

Tobler, Waldo. 1970. "A Computer Model Simulating Urban Growth in the Detroit Region." *Economic Geography* 46, 2: 234–240.

Projection / Distortion

References

Furnas, George W., and Benjamin B. Bederson. 1995. "Space-Scale Diagrams: Understanding Multiscale Interfaces." In *Human Factors in Computing Systems CHI '95 Conference Proceedings*, 234–241. New York: ACM Press.

Leung, Y. K., and M. D. Apperley. 1994. "A Review and Taxonomy of Distortion-Oriented Presentation Techniques." *ACM Transactions on Computer-Human Interaction* (TOCHI) 1, 2: 126–160.

Aggregation and Clustering

References

Schvaneveldt, R., ed. 1990. *Pathfinder Associative Networks: Studies in Knowledge Organization*. Norwood, NJ: Ablex Publishing Corp.

Combination

References

Tufte, Edward R. 1990. *Envisioning Information*. Cheshire, CT: Graphics Press.

Interaction

References

Shneiderman, Ben. 1996. "The Eyes Have It: A Task by Data Type Taxonomy for Information Visualizations." *Proceedings of the IEEE Symposium on Visual Languages*, 336–343. Washington, DC: IEEE Computer Society.

Deployment

Image Credits

Data Communication Visualization Layers: Elisha F. Hardy (design), Katy Börner (concept), Indiana University.

66 Exemplification

References

Börner, Katy, Jeegar Maru, and Robert L. Goldstone. 2004. "The Simultaneous Evolution of Author and Paper Networks." *PNAS* 101, Suppl 1: 5266–5273.

Börner, Katy, Shashikant Penumarthy, Mark Meiss, and Weimao Ke. 2006. "Mapping the Diffusion of Information among Major U.S. Research Institutions." *Scientometrics* 68, 3: 415–426.

Mane, Ketan, and Katy Börner. 2004. "Mapping Topics and Topic Bursts in *PNAS*." *PNAS* 101, Suppl. 1: 5287–5290.

Quotation popularized by Benjamin Franklin in the United States, attributed to a Chinese proverb.

Data Acquisition

Image Credits

PNAS Papers: Elisha F. Hardy (design), Katy Börner (concept), Indiana University.

Data Preparation

References

NCBI. 2008. MeSH Home Page. http://www.ncbi.nlm.nih.gov/sites/entrez?db=mesh (accessed June 10, 2008).

Data Analysis, Modeling, and Layout

References

Adobe Systems. 2008. Photoshop CS3 Home Page. http://www.adobe.com/products/photoshop (accessed June 10, 2008).

Callon, Michel, John Law, and Arie Rip, eds. 1986. *Mapping the Dynamics of Science and Technology: Sociology of Science in the Real World*. London: The Macmillan Press.

Fruchterman, T.M.J., and E. M. Reingold. 1991. "Graph Drawing by Force-Directed Placement." *Software: Practice and Experience* 21, 11: 1129–1164.

Kleinberg, J. M. 2002. "Bursty and Hierarchical Structure in Streams." *Proceedings of the 8th ACM SIGKDD International Conference on Knowledge Discovery and Data Mining*, 91–101. New York: ACM Press.

Schvaneveldt, R. W., ed. 1990. *Pathfinder Associative Networks: Studies in Knowledge Organization*. Norwood, NJ: Ablex.

Vlado, Andrej, and Vladimir Batagelj. 2008. "Networks/Pajek: Program for Large Network Analysis." http://vlado.fmf.uni-lj.si/pub/networks/pajek (accessed June 10, 2008).

Data Communication–Visualization Layers

References

Adobe Systems. 2008. Photoshop CS3 Home Page. http://www.adobe.com/products/photoshop (accessed June 10, 2008).

Barabási, A. L., H. Jeong, Z. Néda, E. Ravasz, A. Schubert, and T. Vicsek. 2002. "Evolution of the Social Network of Scientific Collaborations." *Physica A: Statistical Mechanics and its Applications* 311, 3-4: 590–614.

Börner, Katy, Jeegar Maru, and Robert L. Goldstone. 2004. "The Simultaneous Evolution of Author and Paper Networks." *PNAS* 101, Suppl. 1: 5266–5273.

Börner, Katy, Shashikant Penumarthy, Mark Meiss, and Weimao Ke. 2006. "Mapping the Diffusion of Information Among Major U.S. Research Institutions." *Scientometrics* 68, 3: 415–426.

Callon, Michel, John Law, and Arie Rip, eds. 1986. *Mapping the Dynamics of Science and Technology: Sociology of Science in the Real World*. London: The Macmillan Press.

Fruchterman, T.M.J., and E. M. Reingold. 1991. "Graph Drawing by Force-Directed Placement." *Software: Practice and Experience* 21, 11: 1129–1164.

Mane, Ketan, and Katy Börner. 2004. "Mapping Topics and Topic Bursts in *PNAS*." *PNAS* 101, Suppl. 1: 5287–5290.

Newman, Mark E. J. 2001. "Scientific Collaboration Networks. I. Network Construction and Fundamental Results." *Physical Review E* 64, 1: 016131.

Newman, Mark E. J. 2001. "The Structure of Scientific Collaboration Networks." *PNAS* 98, 2: 404–409.

Redner, S. 1998. "How Popular Is Your Paper? An Empirical Study of the Citation Distribution." *The European Physical Journal B* 4, 2: 131–134.

Vlado, Andrej, and Vladimir Batagelj. 2008. "Networks/Pajek: Program for Large Network Analysis." http://vlado.fmf.uni-lj.si/pub/networks/pajek (accessed June 10, 2008).

Data Credits

PNAS data set provided by the *PNAS* for the Arthur M. Sackler Colloquium, "Mapping Knowledge Domains," May 9–11, 2003.

Image Credits
I(a–c) and II(a–b): Russell J. Duhon (data preparation), Elisha F. Hardy (design), Katy Börner (concept), Indiana University; adapted from Börner et al. 2006.
II(a–b): Russell J. Duhon (data preparation), Elisha F. Hardy (design), Katy Börner (concept), Indiana University.
II(c): *See* Mane and Börner 2004. © 2004, National Academy of Sciences. U.S.A.
III(a–c): Elisha F. Hardy (design), Katy Börner (concept), Indiana University; adapted from Börner et al. 2004.

Interpretation

Study (I) Mapping Knowledge Diffusion and the Importance of Space

References
Batty, M. 2003. "The Geography of Scientific Citation." *Environment and Planning A* 35, 5: 761–765.
Börner, Katy, Shashikant Penumarthy, Mark Meiss, and Weimao Ke. 2006. "Mapping the Diffusion of Information among Major U.S. Research Institutions." *Scientometrics* 68, 3: 415–426.
Katz, J. S. 1994. "Geographical Proximity and Scientific Collaboration." *Scientometrics* 31: 31–43.
Thelwall, M. 2002. "Evidence for the Existence of Geographic Trends in University Web Site Interlinking." *Journal of Documentation* 58: 563–574.

Study (II) Identifying Research Topics and Trends

References
Mane, Ketan, and Katy Börner. 2004. "Mapping Topics and Topic Bursts in *PNAS*." *PNAS* 101, Suppl. 1: 5287–5290.

Study (III) Modeling the Co-Evolution of Author-Paper Networks

References
Börner, Katy, Jeegar Maru, and Robert L. Goldstone. 2004. "The Simultaneous Evolution of Author and Paper Networks." *PNAS* 101, Suppl 1: 5266–5273.

68 Scholarly Marketplaces

References
Harnad, Stevan, Tim Brody, Les Carr, Yves Gingras, Chawki Hajjem, and Alma Swan. 2007. "Incentivizing the Open Access Research Web." *CTWatch Quarterly* 3, 3. http://www.ctwatch.org/quarterly/articles/2007/08/incentivizing-the-open-access-research-web (accessed March 1, 2008).
Shulman, Polly. 2007. "The Player." *Smithsonian*, October. http://www.smithsonianmag.com/specialsections/innovators/von-ahn.html (accessed May 30, 2008). [Quotation from interview and profile of Luis von Ahn]

Designing Vibrant Marketplaces

Using a Million Minds

References
Facebook. 2010. Facebook Home Page. http://www.facebook.com (accessed April 14, 2010).
Wikimedia Foundation. 2010. Wikipedia Home Page. http://wikipedia.org (accessed April 14, 2010).
Yahoo!, Inc. 2010. Flickr Home Page. http://flickr.com (accessed April 14, 2010).
YouTube, LLC. 2010. Youtube Home Page. http://wwww.youtube.com (accessed April 14, 2010).
Moveon Political Action. 2010. "MoveOn.org: Democracy in Action." http://www.moveon.org (accessed April 14, 2010).
Digg, Inc. 2010. Digg Home Page. http://digg.com (accessed April 14, 2010).
Spikesource, Inc. 2010. Spikesource Home Page. http://www.spikesource.com (accessed April 14, 2010).
Zopa, Ltd. 2005. "Zopa: Loans from People, not Banks." http://uk.zopa.com/ZopaWeb (accessed April 14, 2010).
eBay, Inc. 2010. eBay Home Page. http://www.ebay.com (accessed April 14, 2010).
Online Data Services, Ltd. 2010. "IdeaConnection: Build on the Genius of Others." http://www.ideaconnection.com (accessed April 14, 2010).
Linden Research, Inc. 2010. "Second Life." http://secondlife.com (accessed April 14, 2010).
Williams, Anthony. 2010. "The New Alexandrians: Sharing for Science and The Science of Sharing." Wikinomics. http://www.socialtext.net/wikinomics/index.cgi?the_new_alexandrians (accessed April 14, 2010).
Amazon Web Services LLC. 2010. "Amazon Web Services." http://aws.amazon.com (accessed April 14, 2010).
Amazon Web Services LLC. 2010. "Amazon Elastic Compute Cloud (Amazon EC2)." http://aws.amazon.com/ec2 (accessed April 14, 2010).
IBM Corporation. 2010. "Many Eyes beta: for shared visualization and discovery." http://manyeyes.alphaworks.ibm.com/manyeyes (accessed April 14, 2010).
Swivel, LLC. 2010. "Swivel: See, understand & share numbers." http://www.swivel.com (accessed April 14, 2010).
Google, Inc. 2010. "Google Image Labeler." http://images.google.com/imagelabeler (accessed April 14, 2010).

Software Bazaars and Crowdsourcing

References
Howe, Jeff. 2006. "The Rise of Crowdsourcing." *Wired*, June 14. http://www.wired.com/wired/archive/14.06/crowds.html (accessed May 30, 2008).
Page, Scott E. 2007. *The Difference: How the Power of Diversity Creates Better Groups, Firms, Schools and Societies*. Princeton, NJ: Princeton University Press.
Raymond, Eric S. 2000. *The Cathedral and the Bazaar*. Sebastopol, CA: O'Reilly Media, Inc. Online at http://catb.org/~esr/writings/cathedral-bazaar/cathedral-bazaar (accessed November 3, 2009). [Quotation; *mages* was changed to *magi*]
Travis, John. 2008. "Science and Commerce: Science by the Masses." *Science* 319, 5871: 1750–1752.

Social Engineering

Understanding Human Needs

References
Csikszentmihalyi, Mihaly. 1991. *Flow: The Psychology of Optimal Experience*. New York: HarperCollins.
Maslow, Abraham. 1954. *Motivation and Personality*. New York: HarperCollins.
Maslow, Abraham. 1971. *The Farther Reaches of Human Nature*. New York: Viking Press.

Image Credits
"Needs" adapted from Maslow's hierarchy: Elisha F. Hardy (design), Katy Börner (concept), Indiana University.

Contributors

Ying Ding and Linda Stone.

Socio-Technical Cloud Design

No references or credits.

Internet

References
Geobytes, Inc. 2003. Geobytes Home Page. http://www.geobytes.com/IpLocator.htm. (accessed June 10, 2008).

Storage Clouds

References
Amazon Web Services LLC. 2010. "Amazon Simple Storage Service." http://aws.amazon.com/s3 (accessed March 11, 2010).
Gilheany, Steve. 2010. "Paper Sizes and Paper Weight: Metric and US Standards." Archives Builders, Inc. Manhattan Beach, CA. http://www-tcad.stanford.edu/~goojs/REFERENCES/paper_standard.html (accessed May 30, 2010).

Compute Clouds

References
Allen, Bruce. 2008. LIGO Scientific Collaboration. "Einstein@Home." http://einstein.phys.uwm.edu (accessed May 30, 2008).
Amazon Web Services LLC. 2010. "Amazon Elastic Compute Cloud." http://aws.amazon.com/ec2 (accessed March 10, 2010).
Astropulse. 2008. Astropulse Home Page. http://www.astropulse.com (accessed May 30, 2008).
Gray, Jim. 2003. *Distributed Computing Economics*. Technical Report MSR-TR-2003-24, Microsoft Research. http://research.microsoft.com/apps/pubs/default.aspx?id=70001 (accessed February 10, 2010).
Grid.org. 2008. Grid.org Home Page. http://grid.org (accessed May 30, 2008).
IBM Thomas J. Watson Research Center. 2006. "Deep Thunder: Precision Forecasting for Weather-Sensitive Business Operations." http://www.research.ibm.com/weather/DT.html (accessed May 30, 2008).
INRIA/IN2P3. 2008. XtremWeb Home Page. http://www.xtremweb.net (accessed May 30, 2008).
Pande, Vijay. 2008. Stanford University. "Folding@Home: Distributed Computing." http://folding.stanford.edu (accessed May 30, 2008).
University of California. 2010. "SETI@Home." http://setiathome.berkeley.edu (accessed March 19, 2010).

Cyberinfrastructures

References
Atkins, Daniel E., Kelvin K. Droegemeier, Stuart I. Feldman, Hector Garcia-Molina, Michael L. Klein, David G. Messerschmitt, Paul Messina, Jeremiah P. Ostriker, and Margaret H. Wright. 2003. "Revolutionizing Science and Engineering Through Cyberinfrastructure: Report of the National Science Foundation Blue-Ribbon Advisory Panel on Cyberinfrastructure." National Science Foundation. http://www.nsf.gov/od/oci/reports/atkins.pdf (accessed March 25, 2008).
Berman, Fran, Geoffrey Fox, and Anthony J. G. Hey, eds. 2003. *Grid Computing: Making the Global Infrastructure a Reality*. New York: Wiley.

Service Oriented Architectures

References
OSGi Alliance. 2008. OSGi Alliance Home Page. http://www.osgi.org (accessed June 10, 2008).
Herr II, Bruce W., Weixia (Bonnie) Huang, Shashikant Penumarthy, and Katy Börner. 2007. "Designing Highly Flexible and Usable Cyberinfrastructures for Convergence." *Progress in Convergence—Technologies for Human Wellbeing*, edited by William S. Bainbridge and Mihail C. Roco, 1093: 161–179. Boston: Annals of the New York Academy of Sciences.
ProgrammableWeb.com. 2010. ProgrammableWeb Home Page. http://www.programmableweb.com (accessed March 16, 2010).
ProgrammableWeb.com. 2010. "API Dashboard." http://www.programmableweb.com/apis (accessed March 16, 2010).

Web Services

References
IP Address Locator. 2010. What Is My IP Address? Home Page. http://www.ip-adress.com (accessed March 12, 2010).
ProgrammableWeb.com. 2010. ProgrammableWeb Home Page. http://www.programmableweb.com (accessed March 16, 2010).

Application Design

The Wiki Way

References
Giles, Jim. 2007. "Key Biology Databases Go Wiki." *Nature* 445: 691–691.
Google, Inc. 2010. Google Maps Home Page. http://maps.google.com (accessed March 11, 2010).
Koriakine, Alexandre, and Evgeniy Saveliev. Wikimapia Home Page. http://wikimapia.org (accessed March 1, 2008).
Mons, Albert. 2007. WikiProfessional Home Page. https://www.wikiprofessional.org/portal (accessed October 30, 2007).
Mons, Barend, Michael Ashburner, Christine Chicester, Erik van Mulligen, Marc Weeber, Johan den Dunnen, Gert-Jane van Ommen, Mark Musen, Matthew Cockerill, Henning Hrmjakob, Albert Monds, Abel Pakcer, Robert Pacheco, Suzanna Lewis, Alfred Berkeley, William Melton, Nikolas Barris, Jimmy Wales, Gerard Meijssen, Erik Moeller, Peter Jan Roes, Katy Börner, and Amos Bairoch. 2008. "Calling on a Million Minds for Community Annotation in WikiProteins." *Genome Biology* 9: R89.
OpenStreetMap. 2010. "The Free Wiki World Map." http://www.openstreetmap.org (accessed March 16, 2010).

Science News. 2007. "Mapmaking for the Masses: User-Generated Content Can Profoundly Impact Geographic Information Systems." *Science Daily*, December 31. http://www.sciencedaily.com/releases/2007/12/071203111251 (accessed May 30, 2008).

Tapscott, Don, and Anthony D. Williams. 2006. *Wikinomics: How Mass Collaboration Changes Everything*. New York: Portfolio/Penguin.

ConceptWiki. 2010. "WikiProteins." http://proteins.wikiprofessional.org/index.php/WikiProteins (accessed March 30, 2010).

Gapminder, Swivel and Many Eyes

References

Creative Commons. 2010. Gapminder Home Page. http://www.gapminder.org (accessed March 11, 2010).

Heer, Jeffrey, and Maneesh Agrawala. 2006. "Software Design Patterns for Information Visualization." *IEEE Transactions on Visualization and Computer Graphics* (TVCG) 12, 5: 853-860.

IBM, Inc. 2008. Many Eyes Beta Home Page. http://services.alphaworks.ibm.com/manyeyes/home (accessed October 30, 2007).

MarketWire Incorporated. 2010. "Swivel Launches Swivel Geography." http://www.marketwire.com/mw/rel.jsp?id=738241 (accessed March 15, 2010).

Swivel, LLC. 2010. Swivel Home Page. http://www.swivel.com (accessed March 16, 2010).

Plug-and-Play Data and Software

References

Herr II, Bruce W., Weixia (Bonnie) Huang, Shashikant Penumarthy, and Katy Börner. 2007. "Designing Highly Flexible and Usable Cyberinfrastructures for Convergence." *Progress in Convergence—Technologies for Human Wellbeing*, edited by William S. Bainbridge and Mihail C. Roco, 1093: 161–179. Boston: Annals of the New York Academy of Sciences.

NWB Team. 2006. "Network Workbench Tool." Indiana University, Northeastern University, and University of Michigan. http://nwb.slis.indiana.edu (accessed March 15, 2010).

NWB Team. 2008. "Network Workbench Community Wiki." https://nwb.slis.indiana.edu/community (accessed March 30, 2010).

Data Credits

UN. 2009. "Avianan Influenza and the Pandemic Threat." http://www.un-influenza.org (accessed March 30, 2010).

Image Credits

Cloud architecture: Elisha F. Hardy (design), Katy Börner (concept), Indiana University.

Screen capture of Network Workbench Tool, *see* NWB Team 2006.

Screen capture of Flu Season data, courtesy of Swivel and Swoodie (graph creator).

Contributors

Bruce W. Herr II and Weixia (Bonnie) Huang.

71 Part 4: Science Maps in Action

References

Moreno, Jacob L. 1933. "Emotions Mapped by New Geography." *New York Times*, April 3. [Quotation]

Data Credits

Science Citation Index (SCI), Social Sciences Citation Index (SSCI), and Arts & Humanities Index (A&HI) by Thomson Reuters, 2001–2004; Scopus Database, 2001–2005.

All world and science map overlays for each of the 30 maps: 2002 Base Map, *see* Boyack et al. 2009: Science location of map significance by Elisha F. Hardy (design), Katy Börner (concept).

World Map by Russell J. Duhon, overlay of geographical influence and significance by Elisha F. Hardy (design), Katy Börner (concept).

Image Credits

Extracted from: Skupin, André. 2005. *In Terms of Geography*. New Orleans, LA. In Katy Börner & Deborah MacPherson (eds.), First Iteration (2005): The Power of Maps, *Places & Spaces: Mapping Science.* http://scimaps.org (accessed May 4, 2009).

72 Places & Spaces: Mapping Science

Motivation & Goal

Image Credits

Börner portrait courtesy of Katy Börner.

McPherson portrait courtesy of Katy Börner.

Boyack portrait courtesy of Katy Börner.

10 Iterations in 10 Years

No references or credits.

Elements of the Exhibit

References

Boyack, Kevin W., Katy Börner, and Richard Klavans. 2009. "Mapping the Structure and Evolution of Chemistry Research." *Scientometrics* 79, 1: 45–60.

Boyack, Kevin W., Richard Klavans, W. Bradford Paley, and Katy Börner. 2007. "Mapping, Illuminating, and Interacting with Science." Paper presented at *SIGGRAPH 2007: The 34th International Conference and Exhibition on Computer Graphics and Interactive Techniques* in San Diego, CA, August 5–9.

Börner, Katy, Fileve Palmer, Julie M. Davis, Elisha F. Hardy, Stephen M. Uzzo, and Bryan J. Hook. 2009. "Teaching Children the Structure of Science." *Proceedings of SPIE: Conference on Visualization and Data Analysis*, 7243, 1: 1-14. Bellingham, WA: SPIE.

Data Credits

Indiana University School of Library and Information Science Information Visualization Laboratory. 2010. "Places & Spaces: Mapping Science." http://scimaps.org (accessed March 12, 2010).

Image Credits

Intro panel: Elisha F. Hardy (design), Katy Börner (concept).

Illuminated Diagrams: *see* Boyack 2007.

Hands-On Maps for Kids: *see* Börner 2009 and http://scimaps.org/http://scimaps.org/flat/exhibit_info (accessed March 23, 2010).

WorldProcessor Globes: © Ingo Günther, WorldProcessor.com, 1988–2007.

DVD: cover art by Ingo Günther and Stephen Oh, back cover by Elisha F. Hardy, concept by Katy Börner.

Web site: design and programming by Elisha F. Hardy and Russell J. Duhon, concept by Katy Börner, *see* http://scimaps.org.

74 Venues

References

Boyack, Kevin W., Katy Börner, and Richard Klavans. 2009. "Mapping the Structure and Evolution of Chemistry Research." *Scientometrics* 79, 1: 45–60.

Data Credits

Indiana University School of Library and Information Science Information Visualization Laboratory. 2010. "Places & Spaces: Mapping Science." http://scimaps.org (accessed March 12, 2010).

Image Credits

All photos of *Places & Spaces* venues taken by Katy Börner.

Geographic Map of Exhibit Venues: Russell J. Duhon (data preparation), Elisha F. Hardy (design), Katy Börner (concept), Indiana University.

Science and Geography Maps: Elisha F. Hardy (design), Katy Börner (concept), Indiana University; adapted from Boyack and Klavans 2002 Base Map, *see* Boyack et al. 2009.

Contributors

Bryan J. Hook compiled the data used in the geographic and science map of exhibit venues.

76 1st Iteration of Exhibit (2005): The Power of Maps

Image Credits

Overview of the first iteration: Elisha F. Hardy (design), Katy Börner (concept), Indiana University.

78 Cosmographia World Map

References

Ptolemaeus, Claudius. 1478. *Cosmographia.* Rome: Arnoldus Buckinck.

Skelton, R. A. 1963. "Bibliographical Note." Reproduction of Claudius Ptolemaus, *Cosmographia*, Ulm, 1482. Amsterdam: Theatrum Orbis Terrarum.

Thomson, J. O. 1948. *History of Ancient Geography*. Cambridge: Cambridge University Press.

Image Credits

Mare Indicvm, from the Ulm edition of *Cosmographia.* Courtesy of the Lilly Library, Indiana University, Bloomington, IN.

Ptolemy (woodcut) from the collection of the James Ford Bell Library, University of Minnesota, Minneapolis, MN.

Ptolemy portrait: Courtesy of the James Ford Bell Library, University of Minnesota, Minneapolis, MN.

Contributors

Deborah MacPherson and Bonnie DeVarco coauthored the original biography and description of this map.

80 Nova Anglia, Novvm Belgivm et Virginia

References

Hermon Dunlap Smith Center for the History of Cartography. 2010. "The John Smith Map of Virginia: Derivations and Derivatives." The Newberry Library. http://www.newberry.org/smith/slidesets/ss24.html (accessed July 15, 2007).

Smith, John. 1612. "Virginia … Discovered and Described by Captayn John Smith." In *A Map of Virginia with a Description of the Countrey, the Commodities, People, Government and Religion* by Joseph Barnes. Oxford: Joseph Barnes.

Image Credits

Johannes Janssonius portrait is original artwork by Karl Marti, Indiana University.

Smith's map of Virginia is a public domain image; assistance from Kristin Lehner at George Mason University.

Nova Anglia Novvm Belgivm et Virginia courtesy of the Library of Congress, Geography and Map Division.

Contributors

Deborah MacPherson and Bonnie DeVarco coauthored the original biography and description of this map.

82 A New Map of the Whole World with the Trade Winds According to the Latest and Most Exact Observations

References

Moll, Herman. 1736. "Atlas Minor. Or a New and Curious Set of Sixty-Two Maps …" London: Thos. Bowles and John Bowles.

Moll, Herman. 1736. "A New Map of the Whole World with the Trade Winds According to Ye Latest and Most Exact Observations." David Rumsey Map Collection, Cartography Associates. http://www.davidrumsey.com/maps4691.html (accessed July 15, 2007).

Image Credits

Moll portrait is original artwork by Karl Marti, Indiana University.

Moll Map detail courtesy of the Lilly Library, Indiana University, Bloomington, IN.

A New Map of the Whole World with the Trade Winds courtesy of the David Rumsey Map Collection, Cartography Associates, San Francisco, CA.

Contributors

Deborah MacPherson and Bonnie DeVarco coauthored the original biography and description of this map.

84 Napoleon's March to Moscow

References

Corbett, John. 2007. "Charles Joseph Minard: Mapping Napoleon's March, 1861." Center for Spatially Integrated Social Science. http://www.csiss.org/classics/content/58 (accessed April 2, 2007).

Robinson, Arthur H. 1967. "The Thematic Maps of Charles Joseph Minard." In *Imago Mundi: A Review of Early Cartography* 21, 95–108.

Tufte, Edward R. 1983. *The Visual Display of Quantitative Information*. Cheshire, CT: Graphics Press.

Tufte, Virginia, and Dawn Finley. 2002. "Minard's Sources." Edward Tufte.com: New ET Writings, Artworks & News. http://www.edwardtufte.com/tufte/minard (accessed April 2, 2007).

Image Credits

Minard signature is courtesy of the Library of Congress.

Map, *see* Tufte 2001.

Contributors

Deborah MacPherson, Bonnie DeVarco, and John Corbett coauthored the original biography and description of this map.

86 1996 Map of Science: A Network Representation of the 43 Fourth Level Clusters Based on Data from the 1996 Science Citation Index

References

Garfield, Eugene. 1998. "AAAS Talk—Mapping the World of Science." *Presentation at the 150th Anniversary Meeting of the AAAS*. Philadelphia, PA, February 14. http://www.garfield.library.upenn.edu/papers/mapsciworld.html. (accessed February 5, 2010.)

Lee, R.C.T., J. R. Slagle, and H. Blum. 1977. "A Triangulation Method for the Sequential Mapping of Points from N-Space to Two-Space." *IEEE Transactions on Computers* 26, 3: 288–292.

Marshakova, Irina V. 1973. "A System of Document Connections Based on References." *Scientific and Technical Information Serial of VINITI* 6: 3–8.

Small, Henry. 1973. "Co-citation in Scientific Literature: A New Measure of the Relationship between Publications." *Journal of the American Society for Information Science* 24: 265–269.

Small, Henry. 1997. "Update on Science Mapping: Creating Large Document Spaces." *Scientometrics* 38, 2: 275–293.

Small, Henry. 1999. "Visualizing Science by Citation Mapping." *Journal of the American Society for Information Science* 50 9: 799–813.

Small, Henry. 2000. "Charting Pathways Through Science: Exploring Garfield's Vision of a Unified Index to Science." In *Web of Knowledge: A Festschrift in Honor of Eugene Garfield*, edited by Helen Barsky Atkins and Blaise Cronin, 449–473. Medford, NJ: Information Today.

Small, Henry, and Eugene Garfield. 1985. "The Geography of Science: Disciplinary and Natural Mapping." *Journal of Information Science,* 11, 4: 147–159.

Small, Henry, and Phineas Upham. 2009. "Citation Structure of an Emerging Research Area on the Verge of Application." *Scientometrics* 79, 2: 365-375.

Thomson Reuters. 2008. "Organic Thin-Film Transistors. A Research Front Map Interview with Professor John A. Rogers." http://esi-topics.com/otft/interviews/rfm3_JohnARogers.html (accessed August 15, 2009).

Data Credits

Map by Henry G. Small, with data sourced from Thomson Reuters.

Science Citation Index (SCI) by Thomson Reuters, 36, 720 highly cited, multidisciplinary papers 1981–1995.

Image Credits

Small portrait © Balázs Schlemmer.

1996 Map of Science courtesy of Henry G. Small, Thomson Reuters.

Highly Cited Multi-Disciplinary Documents from Thomson Reuters, reproduced by permission

Exploring the 1996 Map of Science courtesy of Henry G. Small, Thomson Reuters ISI. *See also* Thomson Reuters 2008.

Courtesy of Henry G. Small, Thomson Reuters, *see also* Thomson Reuters 2008. "Highly Cited Multi-Disciplinary Documents" from Thomson Reuters, reproduced by permission.

Software Credits

Courtesy of Henry G. Small, Thomson Reuters.

Contributors

Henry G. Small provided feedback on the text, images, and references.

90 PhD Thesis Map

References

Garland, K. 1994. *Mr. Beck's Underground Map*, 20. Middlesex, UK: Capitol Transport Publishing.

Hong, Seok-Hee, Damian Merrick Hugo, and A. D. do Nascimento. 2004. "The Metro Map Layout Problem." *Proceedings of the 2004 Australasian Symposium on Information Visualisation* 35, 91–100. Christchurch, New Zealand.

Nesbitt, Keith V. 2003. "Multi-sensory Display of Abstract Data." PhD diss., University of Sydney.

Nesbitt, Keith V. 2004. "Getting to More Abstract Places Using the Metro Map Metaphor." *Proceedings of the 8th International Conference on Information Visualisation*, 488–493. Washington, DC: IEEE Computer Society.

Nesbitt, Keith V. 2004. Ph.D. Thesis Map. Newcastle, Australia. Courtesy of IEEE and Keith V. Nesbitt, Charles Sturt University, Australia; © 2004, IEEE.

Nesbitt, Keith V. 2006. "Designing Multi-Sensory Displays of Abstract Data—with Stock Market Trading Examples." *Presentation at Indiana University*. Bloomington, IN, October 9. http://vw.indiana.edu/talks-fall06/nesbitt.ppt (accessed October 5, 2009).

Stott, Jonathan M., and Peter Rodgers. 2004. "Metro Map Layout Using Multicriteria Optimization." *Proceedings of the 8th International Conference on Information Visualisation (IV"04)*, 355–362. London, July 14–16. Washington, DC: IEEE Computer Society.

Image Credits

Nesbitt portrait courtesy of Dan Conway.

Ph.D. Thesis Map, *see* Nesbitt 2003.

Subway and train maps, *see* Nesbitt 2004.

Software Credits

Microsoft PowerPoint

Contributors

Keith V. Nesbitt coauthored the biography as well as the description of this map.

94 Timeline of 60 Years of Anthrax Research Literature

References

Crane, Diana. 1972. *Invisible Colleges: Diffusion of Knowledge in Scientific Communities*. Chicago: The University of Chicago Press.

Kessler, Michael M. 1963. "Bibliographic Coupling Between Scientific Papers." *American Documentation* 14, 1: 10–25.

Koffka, K. 1935. *The Principles of Gestalt Psychology*, 720. London: Lund Humphries

Kuhn, T. S. 1970. *The Structure of Scientific Revolutions*. 2nd ed. Chicago: The University of Chicago Press.

Morris, Steven A. 2005. "Unified Mathematical Treatment of Complex Cascaded Bipartite Networks: The Case of Collections of Journal Papers." PhD diss., Oklahoma State University. http://eprints.rclis.org/4661 (accessed February 10, 2010).

Morris, Steven A., Benyam Asnake, and Gary G. Yen. 2003. "Dendrogram Seriation Using Simulated Annealing." *Information Visualization* 2, 95-104.

Morris, Steven A., Camille DeYong, Zheng Wu, Sinan Salman, and Dagmawi Yemenu. 2002. "DIVA: A Visualization System for Exploring Document Databases for Technology Forecasting." *Computers and Industrial Engineering* 43, 841–862.

Morris, Steven A., and Kevin W. Boyack. 2005. "Visualizing 60 Years of Anthrax Research." *Proceedings of the 10th International Conference of the International Society for Scientometrics and Informetrics*, edited by P. Ingwerson and B. Larsen, 45–55. Stockholm: Karolinska University Press.

Morris, Steven A., and Gary G. Yen. 2004. "Crossmaps: Visualization of Overlapping Relationships in Collections of Journal Papers." *PNAS* 101, Suppl. 1: 5291–5296.

Morris, Steven A., Gary G. Yen, Zheng Wu, and Benyam Asnake. 2003. "Time Line Visualization of Research Fronts." *Journal of the American Society for Information Science and Technology* 54, 5: 413–422.

Image Credits

Morris portrait courtesy of Steven A. Morris.

See Morris and Boyack 2005; courtesy of Steven A. Morris.

Technique adapted from Steven A. Morris, *see* Morris 2005.

Papers plotted as circles adapted from Morris and Boyack 2005; courtesy of Steven A. Morris.

Authors, References, Index Terms graphic courtesy of Steven A. Morris, *see also* Morris and Yen 2004.

Software Credits

The software is the latest revision of DIVA, which was described in Morris et al. 2002 and Morris and Boyack 2005.

Data Credits

Science Citation Index by Thomson Reuters.

98 Treemap View of 2004 Usenet Returnees

References

Bruls, M., K. Huising, and J. J. van Wijk. 2000. "Squarified Treemaps." *Proceedings of Joint Eurographics and IEEE TCVG Symposium on Visualization (TCVG 2000)*, 33–42. Washington, DC: IEEE Computer Society.

Fiore, A., and Marc A. Smith. 2001. "Tree Map Visualizations of Newsgroups." Technical Report MSR-TR-2001-94, October 4. Microsoft Corporation Research Group. http://research.microsoft.com/apps/pubs/default.aspx?id=69889 (accessed August 15, 2009).

Fisher, D., M. Smith, and H. Welser. 2006. "You Are Who You Talk To: Detecting Roles in Usenet Newsgroups." *Proceedings of the 39th Hawaii International Conference on System Sciences,* 59b. Koloai, HI, January 4-7.

Shneiderman, B. 1992. "Tree Visualization with Tree-Maps: 2-D Space-Filling Approach." In *ACM Transactions on Graphics* 11, 92–99. New York: ACM Press.

Shneiderman, B., and M. Wattenberg. 2001. "Ordered Treemap Layouts." *Proceedings IEEE Symposium on Information Visualization 2001*, 73–79. Washington, DC: IEEE Computer Society.

Smith, Marc A. 2004. "Netscan: A Social Accounting Search Engine." Microsoft Corporation: Community Technologies Groups. Previously available at http://netscan.research.microsoft.com.

Smith, Marc. 2002. "Tools for Navigating Large Social Cyberspace." *Communications of the ACM* 45, 4: 51–55.

Smith, Marc. 2008. "Mapping Usenet: Visualization." Previously available at http://research.microsoft.com/~masmith/all_map.jpg.

Turner, T. C., Marc A.Smith, Danyel Fisher, and Howard T. Welser. 2005. "Picturing Usenet: Mapping Computer-Mediated Collective Action." *Journal of Computer-Mediated Communication* 10, 4, article 7.

Welser, Howard T., Eric Gleave, Danyel Fisher, and Marc A. Smith. 2007. "Visualizing the Signatures of Social Roles in Online Discussion Groups." *Journal of Social Structure* 8, 2.

Image Credits

Smith portrait courtesy of Marc A. Smith.

Fisher portrait courtesy of Danyel Fisher.

See Smith, Marc A. 2004.

Treemap set up, visualization, and Netscan graphics courtesy of Marc A. Smith, Microsoft Research.

Data Credits

See Smith, Marc A. 2004.

Software Credits

See Smith, Marc A. 2004.

Contributors

Marc A. Smith and Danyel Fisher coauthored the biographies as well as the description of this map.

102 In Terms of Geography

References

Fabrikant, S., and André Skupin. 2005. "Cognitively Plausible Information Visualization." In *Exploring GeoVisualization*, edited by J. Dykes, A. M. MacEachren, and M. J. Kraak, sect. E: 35. Amsterdam: Elsevier.

Kohonen, Teuvo. 1995. *Self-Organizing Maps*. Heidelberg: Springer-Verlag.

National Academies Board on Mathematical Sciences and Applications. 2005. "Toward Improved Visualizations of Uncertain Information." http://sites.nationalacademies.org/DEPS/BMSA. (accessed February 10, 2010).

Skupin, André. 2009. André Skupin Home Page. http://geography.sdsu.edu/People/Pages/skupin (accessed August 15, 2009).

Skupin, André. 2004. "The World of Geography: Visualizing a Knowledge Domain with Cartographic Means." *PNAS* 101, Suppl. 1: 5274–5278.

Skupin, André, and Sarah Fabrikant. 2007. "Spatialization." In *The Handbook of Geographical Information Science*, edited by J. Wilson and S. Fotheringham, 61–79. Boston: Blackwell.

Skupin, A., and R. Hagelman. 2005. "Visualizing Demographic Trajectories in Attribute Space." *GeoInformatica* 9, 2: 159–179.

Image Credits

Skupin portrait courtesy of Marinta Skupin.

In Terms of Geography © 2005, André Skupin; data processing and coding by Shujing Shu.

Graphics of the geographical reference system, SOM processing, and close-up of the map details courtesy of André Skupin, all rights reserved.

Graphics of map nodes and applied labels courtesy of André Skupin, all rights reserved.

Photo, *see* Skupin 2004, © 2004, IEEE.

Data Credits

Association of American Geographers. http://www.aag.org (accessed October 30, 2007).

Software Credits

AbstractMap and other proprietary software by André Skupin. Final rendering executed in ESRI ArcGIS.

Kohonen Neural Networks Research Centre, Helsinki University of Technology. 2010. "A Self-Organizing Map Program." http://www.cis.hut.fi/research/som_pak (accessed March 15, 2010).

Contributors

André Skupin coauthored the description of this map.

106 The Structure of Science

References

Boyack, Kevin, and Katy Börner. 2001. "Mapping Aging Research Handout." Prepared for the *Mapping Aging Research Presentation at the NIA Planning Workshop: Assessments of Behavioral and Social Science Research Vitality*. Washington DC, November 9. http://ella.slis.indiana.edu/~katy/research (accessed June 2, 2008).

Boyack, Kevin W., Katy Börner, and Richard Klavans. 2007. "Mapping the Structure and Evolution of Chemistry Research." *Proceedings of the 11th International Conference of the International Society for Scientometrics and Informetrics*, edited by Daniel Torres-Salinas and Henk F. Moed, 112–123. Madrid: CSIC.

Davidson, George S., B. Hendrickson, David K. Johnson, C. E. Meyers, and Brian N. Wylie. 1998. "Knowledge Mining with VxInsight: Discovery Through Interaction." *Journal of Intelligent Information Systems* 11, 3: 259–285.

Davidson, George S., Brian N. Wylie, and Kevin W. Boyack. 2001. "Cluster Stability and the Use of Noise in Interpretation of Clustering." *Proceedings of IEEE Information Visualization 2001*, 23–30. Los Alamitos, CA: IEEE.

Kessler, Michael M. 1963. "Bibliographic Coupling Between Scientific Papers." *American Documentation* 14, 1: 10–25.

Klavans, Richard, and Kevin W. Boyack. 2006. "Quantitative Evaluation of Large Maps of Science." *Scientometrics* 68, 3: 475–499.

Image Credits

Portraits of Kevin W. Boyack and Richard Klavans © Balázs Schlemmer.

Boyack, Kevin W., and Richard Klavans. 2005. "The Structure of Science." Albuquerque, NM, and Berwyn, PA. Courtesy of Kevin W. Boyack and Richard Klavans, SciTech Strategies, Inc., http://www.mapofscience.com (accessed June 10, 2008).

Technique graphics, disciplinary model, community model, and data overlays of the 2002 Base Map courtesy of Kevin W. Boyack and Richard Klavans, SciTech Strategies, Inc., http://www.mapofscience.com (accessed June 10, 2008).

Algorithm visualization, VxOrd, VxInsight, and zoom views courtesy of Kevin W. Boyack and Richard Klavans, SciTech Strategies, Inc., http://www.mapofscience.com (accessed June 10, 2008).

Data Credits

Science Citation Index and Social Sciences Citation Index by Thomson Reuters.

Software Credits

VxOrd, *see* Davidson et al. 2001.

VxInsight, *see* Davidson 1998.

110 2nd Iteration of Exhibit (2006): The Power of Reference Systems

Image Credits

Overview of the second iteration: Elisha F. Hardy (design), and Katy Börner (concept), Indiana University.

112 U.S. Frequency Allocations Chart

References

National Telecommunications and Information Administration. 2003. "U.S. Frequency Allocation Chart." http://www.ntia.doc.gov/osmhome/Allochrt.html (accessed July 15, 2007).

Image Credits

Courtesy of National Telecommunications and Information Administration with redesigned elements by Shravan Rajagopal, Indiana University.

National Telecommunications and Information Administration. 2003. "U.S. Frequency Allocation Chart." Courtesy of the Office of Spectrum Management, Washington, DC.

Data Credits

National Telecommunications and Information Administration

Contributors

Deborah MacPherson and Bonnie DeVarco coauthored the original biography and description of this map.

114 The Visual Elements Periodic Table

References

Dürsteler, Juan C. 2007. "The Periodic Table of the Elements." *Inf@Vis! The Digital Magazine of Infovis.net*. http://www.infovis.net/printMag.php?num=188&lang=2 (accessed July 15, 2007).

Mendeleyev, Dmitry Ivanovich, and William B. Jensen. 2005. *Mendeleev on the Periodic Law: Selected Writings, 1869–1905*. Mineola, NY: Dover Publications.

Scerri, E. R. 2006. *The Periodic Table: Its Story and Its Significance*. Oxford: Oxford University Press.

Image Credits

Robertson portrait courtesy of Murray Robertson.

Emsley portrait courtesy of John Emsley.

Robertson, Murray, and John Emsley. 2005. Visual Elements Periodic Table. London, UK. Courtesy of the Royal Society of Chemistry Images, © Murray Robertson, 1999–2006.

Contributors

Deborah MacPherson and Bonnie DeVarco coauthored the original biography and description of this map.

116 Cartographica Extraordinaire: The Historical Map Transformed

References

Rumsey, David, and Edith M. Punt. 2004. *Cartographica Extraordinaire: The Historical Map Transformed*. Redlands, CA: ESRI Press.

Image Credits

Rumsey portrait courtesy of David Rumsey.

Punt portrait courtesy of Edith M. Punt.

Courtesy of ESRI Press, © 2004 David Rumsey, ESRI, DigitalGlobe, Inc., MassGIS. All rights reserved.

Contributors

Deborah MacPherson and Bonnie DeVarco coauthored the original biography and description of this map. David Rumsey provided valuable feedback.

118 Sky Chart of New York City in April 2006

References

Hirshfield, Alan, and Roger W. Sinnott, eds. 1999. *Sky Catalogue 2000.0. Vol. 2, Galaxies, Double and Variable Stars, and Star Clusters: Stars to Visual Magnitude 2000.0.* Cambridge, MA: Sky Publishing Corporation/Cambridge University Press.

Robinson, Leif J. 2010. "A Brief History of *Sky and Telescope*." http://www.skyandtelescope.com/about/generalinfo/3305301.html (accessed February 12, 2010)

Image Credits

Portrait of Roger W. Sinnott, the iFactory logo, and Sky & Telescope logo image, courtesy of *Sky & Telescope* magazine.

Sinnott, Roger W. 2006. "The Interactive Factory." Sky Chart of New York City in April 2006. Cambridge, MA. Courtesy of *Sky & Telescope* magazine.

Contributors

Deborah MacPherson and Bonnie DeVarco coauthored the original biography and description of this map.

120 HistCite Visualization of DNA Development

References

Asimov, Isaac. 1963. *The Genetic Code*. New York: Signet/Penguin.

Garfield, Eugene. 2004. "Historiographic Mapping of Knowledge Domains Literature." *Journal of Information Science* 30, 2: 119–145.

Garfield, Eugene, A. I. Pudovkin, and V. S. Istomin. 2003. "Mapping the Output of Topical Searches in the Web of Knowledge and the Case of Watson-Crick." *Information Technology and Libraries* 22, 4: 183–187.

Garfield, Eugene, A. I. Pudovkin, and V. S. Istomin. 2003. "Why Do We Need Algorithmic Historiography?" *Journal of the American Society for Information Science and Technology* 54, 5: 400–412.

Garfield, Eugene, Irving H. Sher, and Richard J. Torpie. 1964. "The Use of Citation Data in Writing the History of Science." Report for Air Force Office of Scientific Research under contract AF49 (638)-1256. Philadelphia, PA: Institute for Scientific Information. http://www.garfield.library.upenn.edu/papers/useofcitdatawritinghistofsci.pdf (accessed September 30, 2009).

Watson, J. D., and F.H.C. Crick. 1953. "A Structure for Deoxyribose Nucleic Acid." *Nature* 171: 737–738.

Wilkins, M.H.F., A. R. Stokes, and H. R. Wilson. 1953. "Molecular Structure of Deoxypentose Nucleic Acids." *Nature* 171, 738–740.

Image Credits

Garfield portrait courtesy of Eugene Garfield and Thomson Reuters.

Pollock portrait courtesy of Ludmilla Pollock.

Portrait of Jan Witkowski courtesy of Cold Spring Harbor Laboratory.

Garfield, Eugene. 2006. HistCite Visualization of DNA Development by Eugene Garfield (HistCite), Elisha Hardy and Katy Börner (graphic design), Ludmila Pollock (images), Jan Witkowski (text). Philadelphia, PA. Courtesy of Eugene Garfield, Thomson Reuters, Indiana University, and Cold Spring Harbor Laboratory.

Manually compiled historiography, *see* Garfield et al. 1964.

Automically generated historiograph and screen shots, Katy Börner, Indiana University, using HistCite Software.

Data Credits

Asimov data, *see* Asimov 1963.

Garfield, Eugene. 1963. Science Citation Index. *Science Citation Index 1961*, 1: v–xvi. http://garfield.library.upenn.edu/papers/80.pdf (accessed August 13, 2009).

Garfield, Eugene, and Irving H. Sher. 1963. *Genetics Citation Index*, 864. Philadelphia: Institute for Scientific Information.

Software Credits

HistCite Software LLC. 2008. "HistCite: Bibliometric Analysis and Visualization Software." http://www.histcite.com (accessed June 10, 2008).

Contributors

Diane Ippoldo prepared the network diagrams for the 1964 paper.

Soren Paris compiled the data from Wilkins et al. (1953) and Watson and Crick (1953) and the citing papers data set; Eugene Garfield coauthored this description. A major part of the HistCite description comes from http://www.histcite.com.

124 History Flow Visualization of the Wikipedia Entry on "Abortion"

References

Heckel, Paul. 1978. "A Technique for Isolating Differences between Files." *Communications of the ACM* 21, 4: 264–268.

IBM Collaborative User Experience Research Group. 2003. "History Flow: Gallery." http://www.research.ibm.com/visual/projects/history_flow/gallery.htm (accessed June 2, 2008).

IBM Collaborative User Experience Research Group. 2003. "History Flow: Visualizing the Editing History of Wikipedia Pages." http://www.research.ibm.com/visual/projects/history_flow (accessed March 31, 2007).

Viégas, Fernanda, Martin Wattenberg, and D. Kushal. 2004. "Studying Cooperation and Conflict Between Authors with History Flow Visualizations." *Proceedings of SIGCHI* 6, 1: 575–582. Vienna, Austria, April 24–29. http://alumni.media.mit.edu/~fviegas/papers/history_flow.pdf (accessed June 2, 2008).

Image Credits

Wattenberg portrait courtesy of Jonathan Feinberg.

Viégas portrait courtesy of Fernanda Viégas.

All HistFlow maps courtesy of Fernanda Viégas and Martin Wattenberg, IBM, *see* Viégas et al. 2004.

Data Credits

Wikimedia data downloaded from http://download.wikipedia.org in May 2003.

128 TextArc Visualization of "The History of Science"

References

Williams, Henry Smith. 1904. *A History of Science*. New York: HarperCollins.

Image Credits

Paley portrait courtesy of W. Bradford Paley.

Paley, W. Bradford. 2006. TextArc Visualization of "A History of Science." New York, NY. Courtesy of W. Bradford Paley.

All TextArc maps courtesy of W. Bradford Paley.

Data Credits

Alice's Adventures in Wonderland by Lewis Carroll (originally published in 1865), text downloaded from Project Gutenberg, http://www.gutenberg.org/etext/928 (accessed June 10, 2008).

A History of Science by Henry Smith Williams (originally published in 1904), text downloaded from Project Gutenberg online: http://www.gutenberg.org/etext/1705 (Volume 1: *The Beginnings of Science*); http://www.gutenberg.org/etext/1706 (Volume 2: *The Beginnings of Modern Science*); http://www.gutenberg.org/etext/1707 (Volume 3: *Modern Development of the Physical Sciences*); http://www.gutenberg.org/etext/1708 (Volume 4: *Modern Development of the Chemical and Biological Sciences*). (accessed June 10, 2008.)

Software Credits

Paley, W. Bradford. 2002. TextArc Home Page. http://textarc.org (accessed June 10, 2008).

Contributors

Some text was taken from http://textarc.org and http://textarc.org/TextArcOverview.pdf, with permission by W. Bradford Paley.

132 Taxonomy Visualization of Patent Data

References

Börner, Katy, Elisha F. Hardy, Bruce W. Herr II, Todd Holloway, and W. Bradford Paley. 2007. "Taxonomy Visualization in Support of the Semi-Automatic Validation and Optimization of Organizational Schemas." *Journal of Informetrics* 1, 3: 214–225.

Image Credits

Börner portrait courtesy of Katy Börner.

Holloway portrait courtesy of Todd Holloway.

Map and two sketches, *see* Börner et al. 2007.

Zoom into Bar Graphs, Zoom into References, and Zoom into Citations by Elisha F. Hardy (design), Katy Börner (concept), Indiana University.

Data Credits

United States Patent and Trademark Office, data sets downloadable from ftp://ftp.uspto.gov/pub/patdata.

The number of references a patent shares was downloaded from http://patft.uspto.gov/netahtml/PTO/search-bool.html.

Contributors

Josh Bonner and Alaa Elie Abi Haidar programmed initial TV prototypes; Shashikant Penumarthy gave expert advice; Eric Giannella helped with the selection of patent examples.

136 Map of Scientific Paradigms

References

Davidson, George S., B. Hendrickson, David K. Johnson, C. E. Meyers, and Brian N. Wylie. 1998. "Knowledge Mining with VxInsight: Discovery Through Interaction." *Journal of Intelligent Information Systems* 11, 3: 259–285.

Davidson, George S., Brian N. Wylie, and Kevin W. Boyack. 2001. "Cluster Stability and the Use of Noise in Interpretation of Clustering." *Proceedings of IEEE Information Visualization 2001*, 23–30. Los Alamitos, CA: IEEE.

Dickinson, Boonsri. 2007. "Map: Science's Family Tree." *Discover*, May 31. http://discovermagazine.com/2007/jun/map-science2019s-family-tree (accessed June 10, 2008).

Editorial Staff. 2007. "Scientific Method: Relationships among Scientific Paradigms." *Seed*, March 7. http://www.seedmagazine.com/news/2007/03/scientific_method_relationship.php (accessed June 10, 2008).

Klavans, Richard, and Kevin W. Boyack. 2006. "Quantitative Evaluation of Large Maps of Science." *Scientometrics* 68, 3: 475–499.

Klavans, Richard, and Kevin W. Boyack. 2007. "Is There a Convergent Structure to Science? A Comparison of Maps using the ISI and Scopus Databases." *Proceedings of the 11th International Conference of the International Society for Scientometrics and Informetrics*, edited by Daniel Torres-Salinas and Henk F. Moed, 437-448. Madrid, Spain, June 25-27.

Boyack, Kevin W., and Richard Klavans. 2008. "Measuring Science-Technology Interaction Using Rare Inventor-Author Names." *Journal of Informetrics* 2, 3: 173-183.

Marris, Emma. 2006. "Brilliant Display." *Nature* 444: 985.

National Science Foundation. 2006. *Investing in America's Future: Strategic Plan, FY 2006–2011*, 10. Washington, DC: US Government Printing Office.

Paley, W. Bradford. 2007. "Map of Science in the Journal *Nature*, *SEED* and *Discover* Magazines." http://wbpaley.com/brad/mapOfScience (accessed August 16, 2009).

Committee on Science, Engineering, and Public Policy (COSEPUP), and Policy and Global Affairs (PGA). 2007. *Rising Above the Gathering Storm: Energizing and Employing America for a Brighter Economic Future*. Washington, DC: National Academies Press. http://www.nap.edu/catalog.php?record_id=11463 (accessed June 2, 2008).

Slashdot. 2007. "How Scientific Paradigms Relate," March 20. http://science.slashdot.org/article.pl?sid=07/03/20/2347203&from=rss (accessed June 10, 2008).

Image Credits

Portraits of Kevin W. Boyack and Richard Klavans © Balázs Schlemmer.

Boyack, Kevin W., and Richard Klavans. 2006. "Map of Scientific Paradigms." Albuquerque, NM, and Berwyn, PA. Courtesy of Kevin W. Boyack, Sandia National Laboratories and Richard Klavans, SciTech Strategies, Inc. http://www.mapofscience.com (accessed June 10, 2008).

All VxOrd maps (Country Profiles, Institutional Strengths) courtesy of Richard Klavans and Kevin W. Boyack, SciTech Strategies, Inc., http://www.mapofscience.com (accessed June 10, 2008).

Patent yield, *see* Boyack and Klavans 2008.

Paley, W. Bradford. 2007. Topic Map: How Scientific Paradigms Relate, used by permission.

Data Credits

Science Citation Index and Social Sciences Citation Index by Thomson Reuters.

Software Credits

VxOrd, *see* Davidson et al. 2001.

VxInsight, *see* Davidson 1998.

140 WorldProcessor: Zones of Invention—Patterns of Patents

References

Birchall, Ian. 2003. "Michael Kidron (1930–2003)." *International Socialism: A Quarterly Journal of Socialist Theory*. http://pubs.socialistreviewindex.org.uk/isj99/birchall.htm (accessed June 2, 2008).

Günther, Ingo, User Science Institute, Kyushu University, Fukuoka City, Japan, and WorldProcessor Globes, Inc. 1998–2005. "Kodomo Project." http://kodomo-project.org/worldprocessor (accessed June 2, 2008).

Günther, Ingo. 2005. WorldProcessor. Chiayi Art Festival on the Tropic of Cancer. Ji-Tung Art Printing.

Smith, Dan. 2003. *The Penguin State of the World Atlas*. 7th ed. New York: Penguin.

Image Credits

Günther portrait courtesy of Ingo Günther.

WorldProcessor: Zones of Invention—Patterns of Patents (New York, NY, 2006) © Ingo Günther, WorldProcessor.com, 1988–2007.

Photos of WorldProcessor globes courtesy of Ingo Günther, *see also* http://WorldProcessor.org.

Data Credits

World Intellectual Property Organization Database (WIPO).

U.S. Patent and Trademark Office.

Contributors

Ingo Günther coauthored the description of the map and globes; the assistance of John Mahoney, Stephen Oh, Monika D. Zhu, and many others has been invaluable.

144 3rd Iteration of Exhibit (2007): The Power of Forecasts

Image Credits

Overview of the third iteration: Elisha F. Hardy (design), Katy Börner (concept), Indiana University.

146 Tectonic Movements and Earthquake Hazard Predictions

References

UNAVCO Facility. 2010. Jules Map Server Home Page. http://jules.unavco.org (accessed February 5, 2010).

UNAVCO Facility. 1994-2010. "Jules Verne Voyager Jr. Earth: World." http://jules.unavco.org/VoyagerJr/Earth (accessed February 5, 2010).

Image Credits

Hamburger portrait courtesy of Indiana University.

Meertens portrait courtesy of UNAVCO.

Michael W. Hamburger and Chuck Meertens (data and visualization), Elisha F. Hardy (graphic design). 2007. "Tectonic Movements and Earthquake Hazard Predictions." Bloomington, IN, and Boulder, CO. portrait of Indiana University and UNAVCO Consortium.

Software Credits

Jules Verne Voyager Map Tool courtesy of UNAVCO.

Contributors

Michael W. Hamburger coauthored this description.

148 The Oil Age: World Oil Production 1859–2050

References

Campbell, Colin J. 2003. *The Essence of Oil & Gas Depletion*. Essex, UK: Multi-Science Publishing Company.

Cutler, Cleveland. 2005. "Net Energy from the Extraction of Oil and Gas in the U.S." *Energy* 30, 5: 769–782.

Cleveland, Cutler J., Robert Costanza, Charles A. S. Hall, and Robert Kaufman. 1984. "Energy and US Economy: A Biophysical Perspective." *Science* 225, 4665: 890–897.

Duncan, Richard C. 2000. "The Peak of World Oil Production and the Road to the Olduvai Gorge." Pardee Keynote Symposia, Geological Society of America, Summit 2000. Reno, NV, November 13. http://dieoff.org/page224.htm (accessed June 2, 2008).

Hirsch, Robert, Roger Bezdek, and Robert Welding. 2005. *Peaking of World Oil Production: Impacts, Mitigation, and Risk Management*. United States: SAIC.

Ramage, Janet. 1997. *Energy: A Guidebook*. Oxford: Oxford University Press.

Shah, Sonia. 2004. *Crude: The Story of Oil*. New York: Seven Stories Press.

Yergin, Daniel. 1992. *The Prize: The Epic Quest for Oil, Money, & Power*. New York: Free Press.

Image Credits

Bracken portrait courtesy of Richard Katz, SF Informatics.

Menninger portrait courtesy of Richard Katz, SF Informatics.

Poremba portrait courtesy of Richard Katz, SF Informatics.

Katz portrait courtesy of Richard Katz, SF Informatics.

Rob Bracken (writer), Dave Menninger (graphic artist), Michael Poremba (statistician), Richard Katz (catalyst). 2006. "The Oil Age: World Oil Production 1859–2050." San Francisco, CA. Courtesy of San Francisco Informatics.

Data Credits

U.S. Energy Information Agency.

BP Statistical Review (June 2005).

Colin Campbell's Oil Depletion Model, *see* Campbell 2003.

Software Credits

Open Office spreadsheet

Adobe Illustrator

Contributors

Richard Katz authored the biographies and description and acknowledges Gregson Vaux and Jean Laherrere.

150 Impact of Air Travel on Global Spread of Infectious Diseases

References

Colizza, V., A. Barrat, M. Barthelemy, A. J. Valleron, and A. Vespignani. 2007. "Modeling the Worldwide Spread of Pandemic Influenza: Baseline Case and Containment Interventions." *PLoS Medicine* 4, 1.

Colizza, V., A. Barrat, M. Barthelemy, and A. Vespignani. 2006. "The Modeling of Global Epidemics: Stochastic Dynamics and Predictability." *Bulletin of Mathematical Biology* 68, 8: 1893–1921.

Colizza, V., A. Barrat, M. Barthelemy, and A. Vespignani. 2006. "The Role of the Airline Transportation Network in the Prediction and Predictability of Global Epidemics." *PNAS* 103: 2015.

Cx-Nets Collaborators. "Complex Networks Collaboratory." http://cxnets.googlepages.com (accessed July 20, 2007).

IATA. 2007. International Air Transport Association Home Page. http://www.iata.org (accessed July 20, 2007).

Image Credits

Portraits of Vittoria Colizza and Alessandro Vespignani courtesy of Alessandro Vespignani.

Vittoria Colizza (research and data), Alessandro Vespignani (research), Elisha F. Hardy (graphic design). 2007. Impact of Air Travel on Global Spread of Infectious Diseases. Bloomington, IN. Courtesy of Indiana University.

Data Credits

IATA data, *see* IATA, 2007.

U.S. Census database, http://www.census.gov, and the United Nations Statistics Database, http://unstats.un.org/unsd (accessed March 23, 2010).

Software Credits

ArcGIS software by ESRI (http://www.esri.com) was used for the visualization of simulated results. Simulation results were obtained with a simulation software written in C language developed by the authors, *see both* Colizza et al. 2006 references and Colizza et al. 2007.

Contributors

A. Barrat and M. Barthelemy (model development, data analysis, study design), A.-J. Valleron (study design), J. J. Ramasco and D. Balcan (data collection and analysis), K. Börner (discussions and comments).

152 [./logicaland] Participative Global Simulation

References

Brecke, Peter. 1993. "Integrated Global Models that Run on Personal Computers." *Simulation* 60, 2: 140–144.

Kile, Frederick, and Arnold Rabehl. 1977. "Evolution of an Integrated Modeling Approach." *IEEE Transactions on Systems, Man, and Cybernetics*, SMC-7, 12: 859–863.

Image Credits

Portraits of Michael Aschauer, Maia Gusberti, Nik Thoenen, and Sepp Deinhofer courtesy of Nik Thoenen, nt@re-p.org.

Michael Aschauer, Maia Gusberti, Nik Thoenen, Sepp Deinhofer (collaborator). 2002. "[./logicaland] Participative Global Simulation." Vienna, Austria. Courtesy of Michael Aschauer, Maia Gusberti, and Nik Thoenen, in collaboration with Sepp Deinhofer, http://re-p.org.

Data Credits

[./logicaland] http://logicaland.net/download.html (accessed June 10, 2008).

Software Credits

[./logicaland] is licensed under GNU GPL and source code is available at http://logicaland.net/download.html (accessed June 10, 2008).

Contributors

Michael Aschauer, Maia Gusberti, and Nik Thoenen, in collaboration with Sepp Deinhofer; http://re-p.org.

154 Science & Technology Outlook: 2005–2055

References

Foresight. 2007. "Horizon Scanning Centre." http://www.foresight.gov.uk/Horizon Scanning Centre (accessed March 23, 2010).

HM Treasury. 2010. "Opportunities and Challenges for the UK: Analysis for the 2007 Comprehensive Spending Review." http://webarchive.nationalarchives.gov.uk/+/http://www.hm-treasury.gov.uk/spending_review/spend_csr07/spend_csr07_index.cfm (accessed March 11, 2010).

Pang, Alex Soojung-Kim. 2006. "First Press on the Delta Scan." IIFTF's Future Now: Emerging Technologies and their Implications for the Future. http://future.iftf.org/2006/12/first_press_on_.html. (accessed March 16, 2010).

Phaal, Robert. 2006. "Public Domain Roadmaps." University of Cambridge and the Institute for Manufacturing. http://www.ifm.eng.cam.ac.uk/ctm/trm/documents/published_roadmaps.pdf (accessed March 16, 2010).

Phaal, Robert, Martin Eppler, and Alan Blackwell, organizers. 2007. "Workshop 4: Visualizing Strategy—Exploring Graphical Roadmap Forms." Visualization Summit, Eidgenössiche Technische Hochschule/Swiss Federal Institute of Technology, Zürich, Switzerland. http://www.ia.arch.ethz.ch/visualization_summit/workshop_04.htm (accessed June 2, 2008).

Stanford Humanities Laboratory. 2006. Delta Scan: The Future of Science and Technology, 2005–2055, Home Page. http://humanitieslab.stanford.edu/2/Home (archived forum accessed August 16, 2009).

Stanford Humanities Laboratory. 2006. "Delta Scan: The Future of Science and Technology, 2005–2055, Project Page." http://humanitieslab.stanford.edu/2/247. (accessed March 30, 2010).

Image Credits

Portraits of Alex Soojung-Kim Pang and David Pescovitz © 2007, Institute for the Future.

Alex Soojung-Kim Pang, David Pescovitz, Marina Gorbis, and Jean Hagan. 2006. "Science & Technology Outlook: 2005–2055." Palo Alto, CA. Courtesy of The Institute for the Future.

Photographs of IFTF workshops courtesy of Alex Soojung-Kim Pang.

Image of Delta Scan home page courtesy of Stanford Humanities Laboratory.

Data Credits

The database is hosted by the Stanford University Foresight Research group, housed in the University's Wallenberg Center.

IFTF database is hosted by the Stanford University Foresight Research group, housed in the University's Wallenberg Center, *see* Stanford Humanities Laboratory 2006.

Contributors

This description was coauthored by Alex Soojung-Kim Pang.

158 113 Years of Physical Review

References

American Physical Society. 2008. "Physics and Astronomy Classification Scheme." http://publish.aps.org/PACS (accessed June 2, 2008).

Börner, Katy, Shashikant Penumarthy, Mark Meiss, and Weimao Ke. 2006. "Mapping the Diffusion of Information among Major U.S. Research Institutions." *Scientometrics* 68, 3: 415–426.

Chicago Area Transportation Study. 1959. *Final Report: Survey Findings*, vol. 1: 46, fig. 23.

Coombs, C., J. Dawes, and A. Tversky. 1970. *Mathematical Psychology: An Elementary Introduction*, 73–75. Englewood Cliffs, NJ: Prentice-Hall.

Garfield, Eugene. 1992. "'Of Nobel Class': Part 1. An Overview of ISI Studies on Highly Cited Authors and Nobel Laureates." *Current Comments* 33: 3–13. http://www.garfield.library.upenn.edu/essays/v15p116y1992-93.pdf (accessed June 2, 2008).

Garfield, Eugene. 1992. "'Of Nobel Class': Part 2. Forecasting Nobel Prizes Using Citation Data and the Odds Against It." *Current Comments* 35: 3–12. http://www.garfield.library.upenn.edu/essays/v15p127y1992-93.pdf (accessed June 2, 2008).

Kessler, Michael M. 1963. "Bibliographic Coupling Between Scientific Papers." *American Documentation* 14, 1: 10–25.

Kohonen, Teuvo. 1995. *Self-Organizing Maps*. Heidelberg: Springer-Verlag.

Martello, Angela. 1990. "Prediction of the 2004 Nobel Prize." *The Scientist* 4, 17: 16 and 18: 16–17.

Pendlebury, David. 2008. "Choosing Thomson Citation Laureates: The Process and Results." Research Services, Thomson Scientific. http://scientific.thomson.com/nobel/essay (accessed June 10, 2008).

Phan, D., L. Xiao, R. Yeh, P. Hanrahan, and T. Winograd. 2005. "Flow Map Layout." *Proceedings of the IEEE Symposium on Information Visualization*, 29: 219–224. http://graphics.stanford.edu/papers/flow_map_layout/flow_map_layout.pdf (accessed March 16, 2010).

Redner, S. 2005. "Citation Statistics from 110 Years of Physical Review." *Physics Today* 58, 49–54.

Robinson, Arthur H. 1955. "The 1837 Maps of Henry Drury Harness." *The Geographical Journal* 121, 4: 440–450.

Tobler, W. 1981. "A Model of Geographical Movement." *Geographical Analysis* 13, 1: 1–20.

Tobler, W. 1987. "An Experiment in Migration Mapping by Computer." *The American Cartographer* 14, 2: 155–163.

Tobler, W. 2001. "Geographical Movement." http://www.geog.ucsb.edu/~tobler/presentations (accessed June 10, 2008).

Tobler, Waldo. 2004. "Tobler's Flow Mapper." Center for Spatially Integrated Social Science, Regents of the University of California, Santa Barbara. http://csiss.ncgia.ucsb.edu/clearinghouse/FlowMapper (accessed February 5, 2010).

Image Credits

Duhon portrait courtesy of Russell J. Duhon.

Penumarthy portrait courtesy of Shashikant Penumarthy.

Hardy portrait courtesy of Elisha F. Hardy.

Bruce W. Herr II and Russell Jackson Duhon (data mining and visualization), Elisha F. Hardy (graphic design), Shashikant Penumarthy (data preparation), Katy Börner (concept). 2007. "113 Years of Physical Review." Bloomington, IN. Courtesy of Indiana University.

Excerpted from "113 Years of Physical Review," courtesy of Indiana University. Screenshot courtesy of Russell J. Duhon, rendered by Elisha F. Hardy.

Zoom and page background excerpted from "113 Years of Physical Review," courtesy of Indiana University.

Image excerpted from "Europe Raw Cotton Imports in 1858, 1864, and 1865" by Charles Joseph Minard, courtesy of the Library of Congress, Geography and Map Division.

Continuous flow maps and discrete flow maps courtesy of Waldo Tobler @ CSISS.org.

Flow Map Layout images, *see* Phan et al. 2005; used by permission.

Data Credits

U.S. Patent and Trademark Office data sets downloadable from ftp://ftp.uspto.gov/pub/patdata (accessed March 30, 2010).

Physical Review data set was provided by the American Physical Society (APS), http://www.aps.org (accessed June 10, 2008).

Nobel Web AB. 2009. "All Nobel Laureates in Physics." http://nobelprize.org/nobel_prizes/physics/laureates (accessed August 16, 2009).

Thomson Reuters. 2010. "Web of Knowledge - Science." http://isiwebofknowledge.com (accessed March 30, 2010).

Thomson Reuters. 2009. "Thomson Scientific Predicts the Nobel Prizes in 2008." http://scientific.thomson.com/nobel (accessed August 16, 2009).

U.S. Census Bureau. 2000. "United States Census 2000." http://www.census.gov/main/www/cen2000.html (accessed June 10, 2008).

Software Credits

Center for Spatially Integrated Social Science, Regents of the University of California, Santa Barbara. 2008. "Tobler's Flow Mapper." http://csiss.ncgia.ucsb.edu/clearinghouse/FlowMapper (accessed February 5, 2010).

Phan, D., L. Xiao, R. Yeh, P. Hanrahan, and T. Winograd. 2005. "Flow Map Layout." *Proceedings of the IEEE Symposium on Information Visualization*, 29: 219–224. http://graphics.stanford.edu/papers/flow_map_layout/flow_map_layout.pdf (accessed March 16, 2010).

Contributors

W. Bradford Paley and Daniel Zeller helped conceptualize. Jan Witkowski (Cold Spring Harbor Laboratory) provided inspiration and feedback. Gavin LaRowe was involved in parsing the data. Mark Doyle was our primary contact at American Physical Society. Soma Sanyal retrieved ISI Prize Nobel data. Bruce and Russell coauthored the description. Doantam Phan rendered the two flow maps.

162 Mapping the Universe: Space, Time, and Discovery!

References

Anthonisse, J. M. 1971. "The Rush in a Directed Graph." *Technical Report* BN9/71. Amsterdam: Stichting Mahtematisch Centrum.

Brandes, Ulrik. 2001. "A Faster Algorithm for Betweenness Centrality." *Journal of Mathematical Sociology* 25, 2: 163–177.

Chen, Chaomei. 1999. "Visualising Semantic Spaces and Author Co-Citation Networks in Digital Libraries." *Information Processing and Management* 35, 2: 401–420.

Chen, Chaomei. 2004. "Searching for Intellectual Turning Points: Progressive Knowledge Domain Visualization." *PNAS* 101, Suppl. 1: 5303–5310.

Chen, Chaomei. 2005. "CiteSpace: Quick Guide." http://cluster.cis.drexel.edu/~cchen/citespace/doc/guide.ppt (accessed June 2, 2008).

Chen, Chaomei. 2006. "CiteSpace II: Detecting and Visualizing Emerging Trends and Transient Patterns in Scientific Literature." *Journal of the American Society for Information Science and Technology* 57, 3: 359–377.

Chen, Chaomei, F. Ibekwe-SanJuan, E. SanJuan, and C. Weaver. 2006. "Visual Analysis of Conflicting Opinions." *IEEE Symposium on Visual Analytics Science and Technology* (VAST 2006), 59–66. Baltimore, MD, October 31–November 2.

Chen, Chaomei, and S. Morris. 2003. "Visualizing Evolving Networks: Minimum Spanning Trees versus Pathfinder Networks." *Proceedings of IEEE Symposium on Information Visualization*, 67–74. Seattle, WA: IEEE.

Chen, Chaomei, and R. J. Paul. 2001. "Visualizing a Knowledge Domain's Intellectual Structure." *IEEE Computer* 34, 3: 65–71.

Chen, Chaomei, Jian Zhang, Weizhong Zhu, and Michael S. Vogeley. 2007. "Delineating the Citation Impact of Scientific Discoveries." *IEEE/ACM Joint Conference on Digital Libraries*, 19–28. Vancouver, Canada, June 17–22.

Freeman, L. C. 1977. "A Set of Measures of Centrality Based on Betweenness." *Sociometry* 40, 1: 35-41.

Girvan, M., and M. Newman. 2002. "Community Structure in Social and Biological Networks." *PNAS* 99: 7821–7826.

Gott III, J. Richard, Mario Juric, David Schlegel, Fiona Hoyle, Michael S. Vogeley, Max Tegmark, Neta Bahcall, and Jon Brinkmann. 2005. "A Map of the Universe." *The Astrophysical Journal* 624: 463–484.

Kleinberg, J. M. 2002. "Bursty and Hierarchical Structure in Streams." *Proceedings of the 8th ACM SIGKDD International Conference on Knowledge Discovery and Data Mining*, 91–101. New York: ACM Press.

NASA. 2004. "Hubble Digs Deeply, Toward Big Bang." http://www.nasa.gov/vision/universe/starsgalaxies/hubble_UDF.html (accessed June 2, 2008).

Schvaneveldt, R. W., ed. 1990. *Pathfinder Associative Networks: Studies in Knowledge Organization*. Norwood, NJ: Ablex.

Sloan Digital Sky Survey. 2008. "Mapping the Universe." http://www.sdss.org (accessed June 2, 2008).

Small, Henry G., and Belver C. Griffith. 1974. "The Structure of Scientific Literatures I: Identifying and Graphing Specialties." *Science Studies* 4, 1: 17–40.

Image Credits

Chen portrait courtesy of Chaomei Chen.

Vogeley portrait courtesy of Michael S. Vogeley.

Chaomei Chen, Jian Zhang, Lisa Kershner, Michael S. Vogeley, J. Richard Gott III, and Mario Juric. 2007. "Mapping the Universe: Space, Time, Discovery!" Philadelphia, PA, and Princeton, NJ. Courtesy of Drexel University and Princeton University.

All diagrams and CiteSpace screenshots courtesy of Chaomei Chen. *See also* Chen 2006.

Data Credits

Sloan Digital Sky Survey. 2003. "SDSS Data Release 1." http://www.sdss.org/dr1 (accessed June 10, 2008).

Thomson Reuters. 2008. "Web of Science." http://scientific.thomsonreuters.com/products/wos (accessed June 10, 2008).

Software Credits

Chen, Chaomei. 2008. "CiteSpace: Visualizing Patterns and Trends in Scientific Literature." http://cluster.cis.drexel.edu/~cchen/citespace (accessed June 10, 2008).

Contributors

This description was coauthored by Chaomei Chen.

166 Science Related Wikipedian Activity

References

Davidson, George S., Brian N. Wylie, and Kevin W. Boyack. 2001. "Cluster Stability and the Use of Noise in Interpretation of Clustering." *Proceedings of IEEE Information Visualization 2001*, 23–30. Los Alamitos, CA: IEEE.

Giles, Jim. 2007. "Power Struggle. Second Sight." *New Scientist* 2604: 55.

Holloway, Todd, Miran Božičević, and Katy Börner. 2007. "Analyzing and Visualizing the Semantic Coverage of Wikipedia and Its Authors." *Complexity* 12, 3: 30–40.

Kleinberg, J. M. 2002. "Bursty and Hierarchical Structure in Streams." *Proceedings of the 8th ACM SIGKDD International Conference on Knowledge Discovery and Data Mining*, 91–101. New York: ACM Press.

Korfiatis, N., and Marios Poulos. 2006. "Evaluating Authoritative Sources Using Social Networks: An Insight from Wikipedia." *Online Information Review* 30, 3: 252–262.

Leuf, B., and W. Cunningham. 2001. *The Wiki Way*. Boston: Addison-Wesley Professional.

Pullen, J. P. 2005. "Freedom of Information: Wikipedia's Jimmy Wales Emancipates the Encyclopedia." *Continental*, July. http://magazine.continental.com/archive/072005/content (accessed October 9, 2009).

Sarwar, Badrul, George Karypis, Joseph Konstan, and John Riedl. 2001. "Item-Based Collaborative Filtering Recommendation Algorithms." *Proceedings of the 10th International Conference on World Wide Web*, 285–295. Hong Kong: ACM Press.

Waldman, S. 2004. "Who Knows? (Criticism of Wikipedia)" *The Guardian*, October 26.

Wikimedia Foundation. 2010. "EasyTimeline." Wikimedia Meta-Wiki. http://en.wikipedia.org/wiki/Wikipedia:EasyTimeline (accessed February 10, 2010).

Wikimedia Foundation. 2009. "Wikipedia Statistics." *Wikipedia, the Free Encyclopedia*. http://en.wikipedia.org/wikistats/EN (accessed February 5, 2010).

Wikimedia Foundation. 2009. "Wikipedia: Multilingual Statistics." *Wikipedia, the Free Encyclopedia*. http://en.wikipedia.org/wiki/Wikipedia:Multilingual_statistics (accessed February 5, 2010).

Zlatić, V., Miran Božičević, H. Štefančić, and M. Domazet. 2006. "Wikipedias: Collaborative Web-Based Encyclopedias as Complex Networks." *Physical Review E*, 74, 1: 016115. http://arxiv.org/pdf/physics/0602149 (accessed October 6, 2009).

Image Credits

Herr II portrait courtesy of Bruce W. Herr II.

IVL Logo: Shravan Rajagopal (design), Katy Börner (concept), Indiana University.

CNS Logo: Elisha F. Hardy (design), Katy Börner (concept), Indiana University.

Bruce W. Herr II (data mining and visualization), Todd Holloway (data mining), Elisha F. Hardy (graphic design), Kevin W. Boyack (graph layout), and Katy Börner (concept). 2007. "Science-Related Wikipedian Activity." Bloomington, IN, and Albuquerque, NM. Courtesy of Indiana University.

VxOrd layouts courtesy of Kevin W. Boyack.

Screenshot from Science-Related Wikipedian Activity courtesy of Bruce W. Herr II.

Excerpted from "An Emergent Mosaic of Wikipedian Activity," by Bruce W. Herr II and Todd Holloway, courtesy of Indiana University.

Screenshot courtesy of Bruce W. Herr II.

Data Credits

Wikimedia Foundation. 2008. *Wikipedia, the Free Encyclopedia*. http://wikipedia.org (accessed June 10, 2008).

Software Credits

Wikimedia Foundation. 2008. "Index of /tools/: MWDumper.jar." http://download.wikimedia.org/tools (accessed June 10, 2008).

Kleinberg, Jon. "Sample Results from a Burst Detection Algorithm." http://www.cs.cornell.edu/home/kleinber/kdd02.html (accessed June 10, 2008).

Information Visualization Cyberinfrastructure, School of Library and Information Science, Indiana University, Bloomington. 2004. "Software>Burst Detection." http://iv.slis.indiana.edu/sw/burst.html (accessed June 10, 2008).

VxOrd, *see* Davidson et al. 2001.

Contributors

Wikimedia Foundation for making data dumps freely available for research. Miran Božičević introduced us to Wikipedia data.

170 Maps of Science: Forecasting Large Trends in Science

References

Boyack, Kevin, and Katy Börner. 2001. "Mapping Aging Research Handout." Prepared for the *Mapping Aging Research Presentation at the NIA Planning Workshop: Assessments of Behavioral and Social Science Research Vitality*. Washington DC, November 9. http://ella.slis.indiana.edu/~katy/research (accessed June 2, 2008).

Davidson, George S., B. Hendrickson, D. K. Johnson, C. E. Meyers, and Brian N. Wylie. 1998. "Knowledge Mining with VxInsight: Discovery Through Interaction." *Journal of Intelligent Information Systems* 11, 3: 259–285.

Davidson, George S., Brian N. Wylie, and Kevin W. Boyack. 2001. "Cluster Stability and the Use of Noise in Interpretation of Clustering." *Proceedings of IEEE Information Visualization 2001*, 23–30. Los Alamitos, CA: IEEE.

Klavans, Richard, and Kevin W. Boyack. 2006. "Quantitative Evaluation of Large Maps of Science." *Scientometrics* 68, 3: 475–499.

Kobourov, Stephen G., and Kevin Wampler. 2004. "Non-Euclidean Spring Embedders." *Proceedings of IEEE Symposium on Information Visualization*, 207–214.

Image Credits

Portraits of Kevin W. Boyack and Richard Klavans © Balázs Schlemmer.

"Maps of Science: Forecasting Large Trends in Science," © 2007, The Regents of the University of California, all rights reserved.

Three graph layouts, *see* Kobourov and Wampler 2004, fig. 1, p. 208. © 2004, IEEE used by permission.

UCSD Map of Science and Mercator projection, Visualization of the ISI and Scopus Databases, and legend courtesy of Richard Klavans and Kevin W. Boyack, SciTech Strategies, Inc., http://www.mapofscience.com (accessed June 10, 2008).

Screenshots of UCSD Map of Science with data overlays available at http://www.mapofscience.com, reproduced with permission of Richard Klavans and Kevin W. Boyack, SciTech Strategies, Inc., http://www.mapofscience.com (accessed June 10, 2008).

Data Credits

Science Citation Index (SCI), Social Sciences Citation Index (SSCI), and Arts & Humanities Index (A&HI) by Thomson Reuters, 2001–2005. Scopus Database by Elsevier, 2001–2005.

Software Credits

VxOrd, *see* Davidson et al. 2001.

VxInsight, *see* Davidson 1998.

SciTech Strategies, Inc. 2008. "Maps of Science." http://mapofscience.com (accessed June 10, 2008).

Contributors

Coauthored by Kevin W. Boyack.

174 Hypothetical Model of the Evolution and Structure of Science

Image Credits

Zeller portrait courtesy of Daniel Zeller.

Hypothetical Model of the Evolution and Structure of Science, 2007, courtesy of Daniel Zeller, Pierogi Gallery, Brooklyn, and G-Module, Paris.

Superficial Inquiry, 2005; Microbial Interaction, 2005; Permeable Unit; Two Studies for the Hypothetical Model, 2007 courtesy of Daniel Zeller, Pierogi Gallery, Brooklyn, and G-Module, Paris.

Photograph of Daniel Zeller at work courtesy of Daniel Zeller.

178 Additional Elements of the Exhibit

Image Credits

Photographs of the exhibit setup courtesy of Katy Börner.

180 Illuminated Diagrams

References

Boyack, Kevin W., Richard Klavans, W. Bradford Paley, and Katy Börner. 2007. "Mapping, Illuminating, and Interacting with Science." Paper presented at *SIGGRAPH 2007: The 34th International Conference and Exhibition on Computer Graphics and Interactive Techniques* in San Diego, CA, August 5–9.

Nature Editorial Staff, News. 2006. "2006 Gallery: Brilliant Display." *Nature* 444: 985.

Paley, W. Bradford. 2006. W. Bradford Paley Home Page. http://www.wbpaley.com/brad (accessed June 10, 2008).

Paley, W. Bradford. 2007. "Map of Science in the Journal *Nature*, *SEED* and *Discover* Magazines." http://wbpaley.com/brad/mapOfScience (accessed August 16, 2009). [Quotation]

Paley, W. Bradford. 2007. "*Nature, Seed* Map of Science Reprint." *Information Aesthetics*.

Paley, W. Bradford. 2007. "Scientific Method: Relationships Among Scientific Paradigms." *SEED*, March 7. http://www.seedmagazine.com/news/2007/03/scientific_method_relationship.php; accessed at http://www.flickr.com/photos/7446536@N03/430561725/; http://informationesthetics.org/documents/scienceMapPrintMockupEd2.jpg (accessed June 2, 2008).

Sandia National Laboratories. 2007. "A Map of Science." http://www.sandia.gov/news/features/mapping_science.html (accessed June 2, 2008).

SourceForge, Inc. 2008. Slashdot Home Page. http://Slashdot.org (accessed June 10, 2008).

Image Credits

Photographs of Peter Kennard, Richard Klavans, W. Bradford Paley (kneeling), Kevin Boyack (standing), and John Burgoon and of exhibit setup courtesy of Katy Börner.

Photographs courtesy of Katy Börner, with design overlay by Elisha F. Hardy.

Kevin W. Boyack (scientometrics and data shaping), Dick Klavans (scientometrics and node layout), W. Bradford Paley (typography, graphics, and interaction design), John Burgoon (geographic mapmaking), and Peter Kennard (system design and programming). 2006. Illuminated Diagram of Topic Map and Geographic Map.

A: Boyack, Kevin W., and Richard Klavans. 2006. "Map of Scientific Paradigms." Albuquerque, NM, and Berwyn, PA. Courtesy of Kevin W. Boyack, Sandia National Laboratories and Richard Klavans, SciTech Strategies, Inc. http://www.mapofscience.com (accessed June 10, 2008).

B: "Illuminated Diagram Topic Map," based on "Map of Scientific Paradigms Research" and Node layout by Kevin W. Boyack and Richard Klavans, SciTech Strategies, Inc., http://www.mapofscience.com. Graphics and typography by W. Bradford Paley, Information Esthetics, http://didi.com/brad. Conceptualized and commissioned by Katy Börner.

C: "Relationships Among Scientific Paradigms," as displayed in *Nature*, December 21, 2006, vol. 444, p. 985: 2006 Gallery.

D: "Relationships Among Scientific Paradigms," as displayed in *SEED* magazine, March 7, 2007.

Photograph of the Illuminated Diagrams setups, one with Eric Wernert, Indiana University, courtesy of Katy Börner.

Contributors

New ID software was implemented by Jagan Lakshmipathy. Eric Wernert, Mike Boyles, Chris Eller, and Pooja Gupta contributed to the new ID setup. Kevin W. Boyack and W. Bradford Paley coauthored the text.

186 Hands-On Science Maps for Kids

References

Palmer, Fileve, Julie Smith, Elisha F. Hardy, and Katy Börner. 2007. "Hands-On Science Maps for Kids." *Places & Spaces: Mapping Science*. http://scimaps.org/exhibit/kids (accessed June 10, 2008).

Image Credits

Palmer portrait courtesy of Fileve Palmer.

Davis portrait courtesy of Julie M. Davis.

Photograph courtesy of Katy Börner, with design overlay by Elisha F. Hardy.

World Map: Inventions and Science Map: Inventions by Fileve Palmer (original artwork and design) and Katy Börner (concept), Indiana University, using Illuminated Diagram of Topic Map and Geographic Map by Kevin W. Boyack, Dick Klavans, W. Bradford Paley, John Burgoon, and Peter Kennard, 2006.

Photographs courtesy of Katy Börner.

Timeline of Inventors and Inventions by Fileve Palmer (artwork) and Elisha F. Hardy (graphic design and layout).

Contributors

Base maps taken from Illuminated Diagram display by Kevin W. Boyack, Richard Klavans and W. Bradford Paley. Stephen Miles Uzzo (director of technology, New York Hall of Science) and Michael Lane (director of exhibit services, New York Hall of Science) provided expert advice and manufactured the physical maps.

192 WorldProcessor Globes

Image Credits

Photograph of John Burgoon, Ingo Günther, Stephen C. Oh, and Dongxia Monika Zhu courtesy of Katy Börner.

Photographs courtesy of Katy Börner, with design overlay by Elisha F. Hardy.

Globes © Ingo Günther, WorldProcessor.org, 1988–2007.

Contributors

Special thanks to John Burgoon, Monika Zhu, and Stephen C. Oh. © 2006 Ingo Günther.

194 Video of the Exhibition

References

Roberg, Nicole A. 2005. "My Science Story." *Places & Spaces: Mapping Science*. http://scimaps.org (accessed June 10, 2008).

Image Credits

Chad Redmon and Aaron Raskin (of Harbinger Media, Inc.) portraits courtesy of Chad Redmon and Aaron Raskin.

Herzig portrait © Yelena Yahontova, Photographer of Joy.

Photograph of the cover; screenshots of Katy Börner, Deborah MacPherson, John Ganly, Kevin Boyack, W. Bradford Paley, Richard Klavans, André Skupin, and Henry G. Small; and acknowledgments of the *Places & Spaces Mapping Science* DVD courtesy of Katy Börner and Harbinger Media, Inc., used by permission.

Contributors

Nicole (Nikki) Roberg designed the "Science Maps for Kids Coloring Book."

197 Part 5: The Future of Science Maps

References

U.S. Army. 2010. "CSI Reprint: After Action Report by Joshua L. Chamberlain." Command and General Staff College. http://www.cgsc.edu/carl/resources/csi/Chamberlain/CHAMBERLAIN.asp (accessed March 30, 2010). [Quotation]

Image Credits

Excerpted from: Palmer, Fileve, Julie M. Smith, Elisha F. Hardy, and Katy Börner. 2007. "Hands-On Science Maps for Kids." *Places & Spaces: Mapping Science*. http://scimaps.org/flat/exhibit_info/#Kids (accessed March 23, 2010).

198 Science Maps as Visual Interfaces to Scholarly Knowledge

References

Kahle, Brewster. 2007. "Universal Access to All Knowledge Is Within Our Grasp." *Presentation at Microsoft Faculty Summit 2007*. Redmond, WA, July 16. Previously available at http://research.microsoft.com/workshops/FS2007/presentations/Kahle_Brewster_Faculty_Summit_071607.ppt (accessed January 1, 2008). [Quotation]

Otlet, Paul. 1934. *Traité de Documentation: Le Livre sur le Livre—Théorie et Pratique*. Brussels: Editiones Mundaneum, Palais Mondial, Van Keerberghen & Fils.

Identification and Preservation of Valuable Knowledge

References

Campbell, Laura. 2010. "Q&A: Digital Library Director Says Innovation, Leadership Require More Than a Vision." *ComputerWorld*. http://www.computerworld.com/s/article/294195/Q_A_Digital_Library_Director_Says_Innovation_Leadership_Require_More_Than_a_Vision?intsrc=hm_list (accessed February 10, 2010).

Wesch, Michael. 2007. "Information R/Evolution." http://www.youtube.com/watch?v=-4CV05HyAbM (accessed January 1, 2008).

Adding Value–Ratings & Context

References

Alexa Internet, Inc. 2008. Alexa Home Page. http://www.alexa.com (accessed March 1, 2008).

CondeNet, Inc. 2008. Reddit Home Page. http://reddit.com (accessed March 1, 2008).

Digg, Inc. 2008. Digg Home Page. http://digg.com (accessed March 1, 2008).

Garfield, Eugene. 2005. "The Agony and the Ecstasy—the History and Meaning of the Journal Impact Factor." *Presentation at the International Congress on Peer Review and Biomedical Publication*. Chicago, IL, September 16. http://garfield.library.upenn.edu/papers/jifchicago2005.pdf (accessed June 10, 2008).

Icerocket. 2008. Icerocket Home Page. http://www.icerocket.com (accessed March 1, 2008).

Microsoft. 2010. Bing Home Page (*formerly* Microsoft Live Search). http://bing.com (accessed March 15, 2010).

Page, Lawrence, Sergey Brin, Rajeev Motwani, and Terry Winograd. 1998. "The PageRank Citation Ranking: Bringing Order to the Web." Stanford Digital Library Technologies Project. http://infolab.stanford.edu/~backrub/pageranksub.ps (accessed June 2, 2008).

Journal Ranking. 2006. http://www.journal-ranking.com/ranking/web (accessed March 1, 2008).

Social Meter. 2010. Social Meter Home Page. http://www.socialmeter.com (accessed March 16, 2010).

Spurl ehf. 2008. Spurl.net 1.0 Home Page. http://www.spurl.net (accessed February 8, 2010).

Technorati, Inc. 2008. Technorati Home Page. http://technorati.com (accessed March 1, 2008).

Tucows, Inc. 2008. BlogRolling Home Page. http://blogrolling.com (accessed March 1, 2008).

WhoLinksToMe.com. 2010. "WhoLinksToMe: Reputation Management Reimagined." http://wholinkstome.com (accessed March 30, 2010).

Yahoo! Inc. 2010. Yahoo! Home Page. http://www.yahoo.com (accessed March 30, 2010).

Yahoo!, Inc. 2008. Del.icio.us Home Page. http://del.icio.us (accessed March 1, 2008).

Collaborative Knowledge Production and Consumption

References

Connexions Project. 2008. Connexions Home Page. http://cnx.org (accessed March 1, 2008).

Industrial Memetics Institute. 2008. "MemeStreams: Science." http://www.memestreams.net/topics/science (accessed March 1, 2008).

Ion Channel Media Group, Ltd. 2007. Biolicious Home Page. http://www.biolicious.com (accessed March 1, 2008).

Knowledge and Data Engineering Group, University of Kassel. 2008. BibSonomy Home Page. http://www.bibsonomy.org (accessed March 1, 2008).

Nature Publishing Group. 2008. Connotea Home Page. http://www.connotea.org (accessed March 1, 2008).

Odlyzko, Andrew. 2001. "Content Is Not King." Technical report in *First Monday* 6. 2: February. http://www.firstmonday.org/issues/issue6_2/odlyzko (accessed January 1, 2008). [Quotation]

Oversity Limited. 2008. CiteULike Home Page. http://www.citeulike.org (accessed March 1, 2008).

Psychologie Information, ZPID Liebniz Institut und Saarländische Universitats und Landsbibliothek. 2008. PsychLinker Home Page. http://www.zpid.de/redact/category.php?cat=1 (accessed March 1, 2008).

Spalding, Tim. 2008. LibraryThing Beta Home Page. http://www.librarything.com (accessed March 1, 2008).

Warlick, David, and the Landmark Project. 2006. Landmark's Son of Citation Machine Home Page. http://www.citationmachine.net (accessed March 1, 2008).

Yahoo!, Inc. 2008. MyBlogLog Home Page. http://www.mybloglog.com (accessed March 1, 2008).

Contributors

Filippo Menczer and Ben Markines.

Supporting Social Navigation

References

Ackerman, M. S. 1994. "Augmenting the Organizational Memory: A Field Study of Answer Garden." *ACM Transactions on Information Systems* 16, 3: 203–224.

Ackerman, M. S., and D. W. McDonald. 1996. "Answer Garden 2: Merging Organizational Memory with Collaborative Help." *Proceedings of the 1996 ACM Conference on Computer Supported Cooperative Work:* 97–105. Boston: ACM Press.

Fesenmaier, J., and N. Contractor. 2001. "Inquiring Knowledge Networks on the Web (IKNOW): The Evolution of Knowledge Networks for Rural Development." *The Journal of the Community Development Society* 32, 1: 160–175.

Linden, G., B. Smith, and J. York. 2003. "Amazon.com Recommendations: Item-to-Item Collaborative Filtering." *Internet Computing IEEE* 7, 1: 76–80.

Rocha, L. M., T. Simas, A. Rechtsteiner, M. DiGiacomo, and R. Luce. 2005. "MyLibrary@LANL: Proximity and Semi-Metric Networks for a Collaborative and Recommender Web Service." *Proceedings of 2005 IEEE/WIC/ACM International Conference on Web Intelligence (WI'05)*: 565–571. IEEE Press

Combining Top-Down and Bottom-Up Knowledge Organization

References

Library of Congress. 2008. "Library of Congress Classification Outline." http://www.loc.gov/catdir/cpso/lcco (accessed March 1, 2008).

National Center for Biotechnology Information. 2008. MeSH Home Page. http://www.ncbi.nlm.nih.gov/sites/entrez?db=mesh (accessed March 1, 2008).

OCLC. 2008. Dewey Services: Dewey Decimal Classification Home Page. http://www.oclc.org/dewey (accessed March 1, 2008).

U.S. Patent and Trademark Office. 2008. "US Classes by Number with Title." http://www.uspto.gov/go/classification/selectnumwithtitle.htm (accessed March 1, 2008).

Visual Interfaces to Digital Libraries

References

Börner, Katy, and Chaomei Chen. 2002. "Visual Interfaces to Digital Libraries: Motivation, Utilization and Socio-Technical Challenges." In *Lecture Notes in Computer Science: Visual Interfaces to Digital Libraries*, edited by Katy Börner and Chaomei Chen, vol. 2539: 1–9. Berlin: Springer-Verlag.

CiNii. 2004. "Researchers Link Viewer." http://ri-www.nii.ac.jp/ComMining1/index.cgi (accessed January 1, 2008).

Ellingham, H.J.T. 1948. "Divisions of Natural Science and Technology." *Reports and Papers of the Royal Society Scientific Information Conference.* London: The Royal Society, Burlington House.

Hearst, Marti A. 1999. "User Interfaces and Visualization." *Modern Information Retrieval*, edited by R. Baeza-Yates and B. Ribeiro-Neto, 257–323. New York: ACM Press/Addison-Wesley. http://people.ischool.berkeley.edu/~hearst/irbook/chapters/chap10.html (accessed January 1, 2008).

Ichise, Ryutaro, Hideaki Takeda, and Kosuke Ueyama. 2005. "Community Mining Tool using Bibliography Data." *Proceedings of the 9th International Conference on Information Visualization*, 953–958. Washington, DC: IEEE Computer Society.

Image Credits

From Ellingham 1948, © The Royal Society.

MEDLINE browser and screenshot, Antarctica Systems, Inc.; used by permission.

CiNii Link Viewer, used by permission, *see also* CiNii 2004 and Ichise et al. 2005.

Memory Palaces & Mirror Gardens

References

Anders, Peter. 1998. *Envisioning Cyberspace: Designing 3-D Electronic Spaces,* 34. New York: McGraw-Hill Professional Publishing.

Spatial Memory Organization

References

Lakoff, G. 1987. *Women, Fire, and Dangerous Things: What Categories Reveal About the Mind.* Chicago: The University of Chicago Press.

Yates, Frances A. 1999. *The Art of Memory.* New York: Routledge.

Virtual Online Spaces

References

Activeworlds Corporation. 2008. ActiveWorlds Home Page. http://www.activeworlds.com (accessed January 1, 2008).

Colbert, Jack. 2008. "Jack Colbert: ZoomInfo Business People Information." http://www.zoominfo.com/people/Colbert_Jack_201967016.aspx (accessed March 30, 2010).

Puterbaugh, Mark. 2010. "Social Worlds, Libraries, the Future and Beyond!" http://vbiworld.blogspot.com/2007_01_01_archive.html (accessed March 30, 2010).

van den Besselaar, Peter, Isabel Melis, and Dennis Beckers. 2000. "Digital Cities: Organization, Content, and Use." In *Digital Cities: Technologies, Experiences, and Future Perspectives: Lecture Notes in Computer Science*, edited by Toru Ishida and Katherine Isbister, vol. 1765: 18–32. Berlin: Springer-Verlag.

Collaborative Memory Palaces

Image Credits

Eduverse and Mirrorgarden images courtesy of Katy Börner.

Evolving Mirror Gardens

References

Börner, Katy, and Shashikant Penumarthy. 2003. "Social Diffusion Patterns in Three-Dimensional Virtual Worlds." *Information Visualization* 2, 3: 182–198.

Gelernter D. H. 1992. *Mirror Worlds: Or the Day Software Puts the Universe in a Shoebox … How It Will Happen and What It Will Mean.* Oxford: Oxford University Press.

Penumarthy, Shashikant, and Katy Börner. 2005. "Analysis and Visualization of Social Diffusion Patterns in Three-Dimensional Virtual Worlds." In *Avatars at Work and Play: Collaboration and Interaction in Shared Virtual Environments (Computer Supported Cooperative Work)*, edited by Ralph Schroeder and Ann-Sofie Axelsson, 39–61. Amsterdam: Springer-Verlag.

Xiong, R., and J. Donath. 1999. "PeopleGarden: Creating Data portraits for Users." *Proceedings of the 12th Annual ACM Symposium on User Interface Software and Technology*, 37–44.

Implementation

References

Börner, Katy. 2002. "Twin Worlds: Augmenting, Evaluating, and Studying Three-Dimensional Digital Cities and Their Evolving Communities." In *Digital Cities II: Computational and Sociological Approaches. Lecture Notes in Computer Science*, edited by Makoto Tanabe, Peter van den Besselaar, and Toru Ishida, vol. 2362: 256–269. Berlin: Springer-Verlag.

200 Mapping Intellectual Landscapes for Economic Decision Making

References

Zachary, G. Pascal. 1997. *Endless Frontier: Vannevar Bush, Engineer of the American Century.* New York: Free Press. [Quotation, Vannevar Bush]

Monitoring Customer Activity

References

Burke, Raymond R. 2005. "The Third Wave of Marketing Intelligence." In *Retailing in the 21st Century: Current and Future Trends*, edited by Manfred Krafft and Murali Mantrala, 113–125. New York: Springer-Verlag.

Chi, Ed H., P. Pirolli, K. Chen, and J. Pitkow. 2001. "Using Information Scent to Model User Information Needs and Actions and the Web." *Proceedings of the CHI Conference on Human Factors in Computing Systems*, 490–497.

Eick, Stephen G. 2002. "Visual Analysis of Website Browsing Patterns." In *Visual Interfaces to Digital Libraries. Lecture Notes in Computer Science*, edited by Katy Börner and Chaomei Chen, vol. 2539: 65–80. Berlin: Springer-Verlag.

Goodchild, Michael F. 2007. "Citizens as Sensors: The World of Volunteered Geography." *GeoJournal* 69, 4: 211–221.

Google, Inc. 2010. Google Analytics Home Page. http://www.google.com/analytics (accessed March 11, 2010).

Leykin, Alex. 2007. *Visual Human Tracking and Group Activity Analysis: A Video Mining System for Retail Marketing.* PhD diss., Indiana University. http://www.cs.indiana.edu/cgi-pub/oleykin/website/download/thesis2007.pdf (accessed March 1, 2008).

Leykin, Alex, and Mihran Tuceryan. 2007. "Detecting Shopper Groups in Video Sequences." *Advanced Video and Signal Based Surveillance (AVSS).* http://www.cs.indiana.edu/cgi-pub/oleykin/website/download/avss2007.pdf (accessed January 1, 2008).

Leykin, Alex, Yang Ran, and Riad Hammoud. 2007. "Thermal-Visible Video Fusion for Moving Target Tracking and Pedestrian Classification." *Computer Vision and Pattern Recognition*, 1–8, June 17–22.

Exploiting Social Networks

References

Facebook, Inc. 2009. Facebook Home Page. http://facebook.com (accessed August 16, 2009).

Hobbs, Robert, and Mark Lombardi. 2003. *Global Networks*, edited by Judith Richards. New York: Independent Curators International.

LinkedIn Corporation. 2008. LinkedIn Home Page. **http://www.linkedin.com** (accessed January 1, 2008).

On, Josh. 2008. "They Rule." **http://www.theyrule.net** (accessed October 30, 2007).

SourceForge, Inc. 2008. GenIsis Value Networks Suite Home Page. **http://sourceforge.net/projects/genisis** (accessed January 1, 2008).

Visible Path Corporation. 2007. VisualPath+ Home Page. Previously available at **https://www.visiblepath.com/registration/vpHomePage.action** (accessed January 1, 2008).

Image Credits

They Rule courtesy of Josh On.

Identifying Experts and Innovations

References

Cohen, Wesley M., and Daniel A. Levinthal. 1990. "Absorptive Capacity: A New Perspective on Learning and Innovation." *Administrative Science Quarterly* 35, 1: 128–152.

Griliches, Zvi. 1979. "Issues in Assessing the Contribution of R&D to Productivity Growth." *Bell Journal of Economics* 10, 1: 92–116.

Jaffe, Adam B., and Manuel Trajtenberg. 1996. "Flows of Knowledge from Universities and Federal Labs: Modeling the Flow of Patent Citations Over Time and Across Institutional and Geographic Boundaries." *National Bureau of Economic Research Working Paper Series* 5712. **http://www.nber.org/papers/w5712.pdf** (accessed January 1, 2008).

Jaffe, Adam B., Manuel Trajtenberg, and Rebecca Henderson. 1993. "Geographic Localization of Knowledge Spillovers as Evidenced by Patent Citations." *Quarterly Journal of Economics* 108, 3: 577–598.

Pakes, Ariel, and Kenneth L. Sokoloff. 1996. "Science, Technology, and Economic Growth." *PNAS* 93, 23: 12655–12657.

Zucker, L. G., and M. R. Darby. 1996. "Star Scientists and Institutional Transformation: Patterns of Invention and Innovation in the Formation of the Biotechnology Industry." *PNAS* 93, 23: 12709–12716.

Zucker, L. G., M. R. Darby, and J. Armstrong. 1998. "Geographically Localized Knowledge: Spillovers or Markets?" *Economic Inquiry* 36, 1: 65–86.

Communicating Intelligently and Effectively

References

Conklin, Jeffrey. 1987. "Hypertext: An Introduction and Survey." *Computer* 20, 9: 17–41.

Conklin, Jeffrey. 2005. *Dialogue Mapping: Building Shared Understanding of Wicked Problems*. Hoboken, NJ: Wiley.

Image Credits

Courtesy of Iceberg: Build Custom Workflow applications, **http://geticeberg.com** (accessed March 30, 2010).

Mapping the Future

References

Phaal, Robert. 2006. "Public Domain Roadmaps." University of Cambridge and the Institute for Manufacturing. **http://www.ifm.eng.cam.ac.uk/ctm/trm/documents/published_roadmaps.pdf** (accessed March 16, 2010).

Mapping the Market

References

Hemscott, Inc. 2006. Hemscott Data Home Page. **http://www.hemscottdata.com** (accessed January 1, 2008).

Interactive Real-Time Data Services. 2008. "ComStock." **http://www.interactivedata-rts.com** (accessed October 30, 2007).

Shneiderman, Ben. 1992. "Tree Visualization with Tree-Maps: 2-D Space-Filling Approach." *ACM Transactions on Graphics* 11, 1: 92–99. *See also* **http://www.cs.umd.edu/hcil/treemap-history** (accessed February 10, 2010).

Shneiderman, Ben. 2006. "Treemaps for Space-Constrained Visualization of Hierarchies." **http://www.cs.umd.edu/hcil/treemap-history** (accessed March 26, 2008).

SmartMoney. 2007. "Map of the Market." **http://www.smartmoney.com/marketmap** (accessed February 10, 2010).

Thomson Reuters. 2008. "Thomson Reuters Financial. (Insider Trading Data)." **http://thomsonreuters.com/products_services/financial** (accessed February 10, 2010).

Zacks Investment Research. 2008. Zacks Investment Research Home Page (Earnings Estimates). **http://www.zacks.com** (accessed January 1, 2008).

Image Credits

Image obtained from **http://www.smartmoney.com/marketmap**; originally created by Martin Wattenberg, reseacher, IBM Watson Research Center.

Mapping the Economic Landscape

References

Feenstra, Robert C., Robert E. Lispsey, and Harry P. Bowen. 1997. "World Trade Flows, 1970–1992, with Production and Tariff Data." *National Bureau of Economic Research Working Paper Series* 5910. **http://www.nber.org/papers/w5910.pdf** (accessed January 1, 2008).

Feenstra, Robert C., Robert E. Lipsey, Haiyan Deng, Alyson C. Ma, and Hengyong Mo. 2005. "World Trade Flows: 1962–2000." *National Bureau of Economic Research Working Paper Series* 11040. **http://cid.econ.ucdavis.edu/data/undata/NBER-UN_Data_Documentation_w11040.pdf** (accessed January 1, 2008).

Hidalgo, Cesar A., B. Klinger, Albert-L. Barabási, and R. Hausmann. 2007. "The Product Space Conditions the Development of Nations." *Science* 317, 5837: 482–487.

Kutz, Daniel. 2004. "Examining the Evolution and Distribution of Patent Classifications." *Proceedings of the 8th International Conference on Information Visualisation (IV'04)*, edited by Ebad Banissi, 983–988. Los Alamitos, CA: IEEE Computer Society.

Simms, Andrew, Dan Moran, and Peter Chowla. 2006. *The U.K. Interdependence Report: How the World Sustains the Nation's Lifestyles and the Price It Pays*. London: New Economics Foundation. **http://www.neweconomics.org/gen/uploads/f2abwpumbr1wp055y2l10s5514042006174517.pdf** (accessed October 30, 2007).

Data Credits

NBER–United Nations Trade Data, 1962–2000. **http://cid.econ.ucdavis.edu/data/undata/undata.html** (accessed March 1, 2008).

Image Credits

The Product Space map courtesy of Cesar A. Hidalgo.

Malaysia export maps courtesy of Cesar A. Hidalgo.

202 Science of Science Policy (Maps) for Government Agencies

References

Kelly, Kate, ed. 2007. *Rising Above the Gathering Storm: Energizing and Employing America for a Brighter Economic Future*. Washington, DC: National Academy of Sciences.

Marburger III, John H. 2005. "Address to the AAAS Forum on Science and Technology Policy, May 2005." **http://www.aip.org/fyi/2007/055.html** (accessed January 1, 2008). [Quotation]

Toward a "Science of Science"

References

Bernal, John D. 1939. *The Social Function of Science*. London: Routledge & Kegan Ltd.

de Solla Price, Derek J. 1961. *Science Since Babylon*. New Haven: Yale University Press.

de Solla Price, Derek J. 1965. *Little Science, Big Science*. New York: Columbia University Press.

Garfield, Eugene. 1955. "Citation Indexes for Science: A New Dimension in Documentation Through Association of Ideas." *Science* 122: 108–111.

Garfield, Eugene. 2007. "From the Science of Science to Scientometrics. Visualizing the History of Science with HistCite Software." *Proceedings of ISSI 2007: 11th International Conference of the International Society for Scientometrics and Informetrics* I & II, edited by D. Torres-Salinas and H. F. Moed, 21–26. Madrid, Spain, June 25–27.

Glänzel, Wolfgang. 2006. "A Concise Introduction to Bibliometrics & Its History." *Steunpunt O&O Statistieken*. **http://www.steunpuntoos.be/bibliometrics.html** (accessed January 1, 2008).

Pritchard, A. 1969. "Statistical Bibliography or Bibliometrics?" *Journal of Documentation* 25, 4: 348–349.

Nalimov, Vasily V., and Z. M. Mulchenko. 1969. *Naukometriya: Izuchenie Razvitiya Kauki ka Informatsionnogo Protessa* [Scientometrics: The Study of the Development of Science as an Information Process]. Moscow: Nauka.

Wouters, Paul. 1999. "The Citation Culture." PhD diss., University of Amsterdam. **http://garfield. library.upenn.edu/wouters/wouters.pdf** (accessed October 30, 2007).

Funding Science

References

AAAS. 2007. *AAAS Report XXXII: Research and Development Fiscal Year 2008*. **http://www.aaas.org/spp/rd/rd08main.htm** (accessed January 6, 2008).

AAAS. 2008. "Trends in Federal Research by Discipline, FY 1970–2006." **http://www.aaas.org/spp/rd/discip06.pdf** (accessed March 1, 2008).

AAAS. 2008. *AAAS Report XXXIII: Research & Development Fiscal Year 2009*. **http://www.aaas.org/spp/rd/rd09main.htm** (accessed October 6, 2009).

Lewison, Grant, Jonathan Grant, and Robert M. May. 1997. "Government Funding of Research and Development." *Science* 278, 5339: 878–880. **http://www.sciencemag.org/cgi/content/full/278/5339/878** (accessed January 1, 2008).

OECD. 2007. "Main Science and Technology Indicators (MSTI): 2007/2 edition." **http://www.oecd.org/document/26/0,3343,en_2649_201185_1901082_1_1_1_1,00.html** (accessed March 1, 2008).

Image Credits

GERD as a Percentage of GDP source: OECD, Main Science and Technology Indicators, October 2007, © OECD 2007.

AAAS Research and Development data visualized by Russell J. Duhon (data preparation), Elisha F. Hardy (design), Katy Börner (concept), Indiana University.

Judging Research Quality

References

Royal Academy of Engineering. 2000. "Measuring Excellence in Engineering Research." **http://www.raeng.org.uk/news/publications/list/reports/Measuring_Excellence.pdf** (accessed January 1, 2008).

SCImago Research Group. 2007. SCImago Journal & Country Rank Home Page. **http://www.scimagojr.com** (accessed January 1, 2008).

Thomson ISI. 2008. "ISI Web of Knowledge - ISIHighlyCited.com." **http://isihighlycited.com** (accessed February 8, 2010).

Computing Funding Impact

References

Center for Economic and Policy Research. 2006. "Cost Effectiveness of 25 Most Cited Think Tanks." **http://www.cepr.net/documents/cost_effectiveness_think_tanks_2006.pdf** (accessed March 1, 2008).

Boyack, Kevin W., and Katy Börner. 2003. "Indicator-Assisted Evaluation and Funding of Research: Visualizing the Influence of Grants on the Number and Citation Counts of Research Papers." *Journal of the American Society of Information Science and Technology* 54, 5: 447–461.

Lewison, Grant, Jonathan Grant, and Robert M. May. 1997. "Government Funding of Research and Development." *Science* 278, 5339: 878–880. **http://www.sciencemag.org/cgi/content/full/278/5339/878** (accessed January 1, 2008).

Image Credits

Boyack and Börner 2003: *Journal of the American Society for Information Science and Technology: Special Topic Issue on Visualizing Scientific Paradigms* 54, 5: 447-461. Reprinted with permission of John Wiley & Sons, Inc.

Geography of Science

References

Leydesdorff, Loet. 2000. "Is the European Union Becoming a Single Publication System?" *Scientometrics* 47, 2: 265–280. [Quotation]

Small, Henry, and Eugene Garfield. 1985. "The Geography of Science: Disciplinary and National Mappings." *Journal of Information Science* 11, 4: 147–159.

Wagner, Caroline S., and Loet Leydesdorff. 2005. "Mapping the Network of Global Science: Comparing International Co-authorships from 1990 to 2000." *International Journal of Technology and Globalisation* 1 2: 185–208.

Wagner, Caroline S., and Loet Leydesdorff. 2005. "Network Structure, Self-Organization, and the Growth of International Collaboration in Science." *Research Policy* 34, 10: 1608–1618.

Wagner, Caroline S., and Loet Leydesdorff. 2008. "International Collaboration in Science and the Formation of a Core Group." *Journal of Informetrics* 2, 4: 317–325. http://users.fmg.uva.nl/lleydesdorff/cswagner07 (accessed April 7, 2008).

Image Credits

Batty, M. 2003. "The Geography of Scientific Citation." *Environment and Planning A* 35, 5: 761–765. Used by permission of Pion Limited, London.

Science Dynamics

References

Boyack, Kevin W., Katy Börner, and Richard Klavans. 2007. "Mapping the Structure and Evolution of Chemistry Research." *Proceedings of the 11th International Conference of the International Society for Scientometrics and Informetrics*, edited by Daniel Torres-Salinas and Henk F. Moed, 112–123. Madrid: CSIC.

Burke, James. 1978. *Connections*. New York: Simon & Schuster.

Chen, Chaomei, and Diana Hicks. 2004. "Tracing Knowledge Diffusion." *Scientometrics* 59, 2: 199–211.

Georghiou, L. 1998. "Global Cooperation in Research." *Research Policy* 27, 4: 611–626.

Image Credits

Images adapted from Boyack et al. 2007.

Scientific Wealth of Nations

References

Horta, Hugo, and Francisco M. Veloso. 2007. "Opening the Box: Comparing EU and US Scientific Output by Scientific Field." *Technological Forecasting and Social Change* 74, 8: 1334–1356.

Igami, Masatsura, and Ayaka Saka. 2007. "Capturing the Evolving Nature of Science: The Development of New Scientific Indicators and the Mapping of Science." *OECD Science, Technology and Industry Working Papers*, 2007/1. Paris: OECD Publishing. http://www.oecd.org/dataoecd/11/40/38134903.pdf (accessed October 30, 2007).

Ingwersen, P., and B. Larsen. 2001. "Mapping National Research Profiles in Social Science Disciplines." *Journal of Documentation* 57, 6: 715–740.

King, D. A. 2004. "The Scientific Impact of Nations." *Nature* 430: 311–316.

Kostoff, R. N. 2004. "The (Scientific) Wealth of Nations." *The Scientist* 18, 18: 10. http://www.the-scientist.com/article/display/14941 (accessed January 1, 2008).

May, Robert M. 1997. "The Scientific Wealth of Nations." *Science* 275, 5301: 793–795.

Paley, W. Bradford. 2007. "Map of Science in the Journal *Nature*, *SEED* and *Discover* Magazines." http://wbpaley.com/brad/mapOfScience (accessed August 16, 2009).

Paley, W. Bradford. 2007. "The Strength of Nations." http://wbpaley.com/brad/mapOfScience/nations15_50pct.jpg (accessed June 10, 2008).

Zhou, Ping, and Loet Leydesdorff. 2006. "The Emergence of China as a Leading Nation in Science." *Research Policy* 32, 1: 83–104.

Image Credits

Images courtesy of Kevin W. Boyack and Richard Klavans, SciTech Strategies, Inc., http://www.mapofscience.com, with source data from Thomson Reuters.

Data Credits

SciTech Strategies, Inc., based on data sourced from Thomson Reuters.

Contributors

Loet Leydesdorff, Caroline Wagner, Richard Klavans, Kevin W. Boyack, Ann Carlson, Maria Zemankova, Janice Hicks, Kaye Husbands, Arthur Ellis, Mike Pollard, Matt Probus, Israel Lederhendler, Timothy Lenoir, Bill Valdez, Karl Schroder, Kelly Streepy, Joan Shigekawa, John T. Bruer, and Yuko Harayama.

204 Professional Knowledge Management Tools for Scholars

References

Cronin, Blaise, and Carol A. Hert. 1995. "Scholarly Foraging and Network Discovery Tools." *Journal of Documentation* 51, 4: 388. [Quotation]

Scholars as Researchers/Authors

References

Bar-Ilan, Judit. 2008. "Informetrics at the Beginning of the 21st Century—A Review." *Journal of Informetrics* 2, 1: 1–52.

Harzing, Anne-Wil. 2008. "Research in International and Cross-cultural Management." Harzing.com—Web Resources Page. http://www.harzing.com/resources.htm#/pop.htm (accessed March 1, 2008).

Leydesdorff, Loet, and Thomas Schank. 2008. "Dynamic Animations of Journal Maps: Indicators of Structural Change and Interdisciplinary Developments." *Journal of the American Society for Information Science and Technology* 59, 11: 1810–1818.

SCImago Research Group. 2008. "SJR—Country Ranking." http://www.scimagojr.com/countryrank.php (accessed March 1, 2008).

Scholars as Editors

References

Goldstone, Robert L. 2005. "Returning to a New Home." *Cognitive Science* 29: 1–4.

Scholars as Reviewers

No references or credits.

Scholars as Teachers

References

Chronicle of Higher Education. 2008. "Facts and Figures." http://chronicle.com/section/Facts-Figures/58 (accessed February 10, 2010).

CINDOC-CSIC. 2008. "Webometrics Ranking of World Universities: Jan 08." http://www.webometrics.info (accessed March 1, 2008).

QS Quacquarelli Symonds Limited. 2008. "QS World University Rankings." http://www.topuniversities.com/university-rankings (accessed February 10, 2010).

U.S. News and World Report L.P. 2008. "America's Best Colleges 2008." http://grad-schools.usnews.rankingsandreviews.com/best-graduate-schools (accessed February 10, 2010).

Scholars as Inventors

No references or credits.

Scholars as Investigators

References

ProQuest LLC. 2010. COS (Community of Science) Home Page. http://www.cos.com (accessed March 16, 2010).

Scholars as Team Leaders and Science Administrators

No references or credits.

Effective Knowledge Management Tools for Scholars

References

Neirynck, Thomas, and Katy Börner. 2007. "Representing, Analyzing, and Visualizing Scholarly Data in Support of Research Management." In *Conference Proceedings of 11th Annual Information Visualization International Conference (IV'07)*, 124–129. Los Alamitos, CA: IEEE Computer Society Conference Publishing Services.

Image Credits

Coauthor Network, Investigator-Project Network, Travels and Project Timeline: Russell J. Duhon (data preparation), Elisha F. Hardy (design), Katy Börner (concept), Indiana University.

Contributors

Sumeet Ambre, Gavin LaRowe, and Russell J. Duhon set up the IVL PHP code base and provided expert advice; Elisha F. Hardy designed the Web interface; Russell J. Duhon and Elisha F. Hardy produced visualizations.

Storing and Accessing Information

References

Börner, Katy. 2006. "Semantic Association Networks: Using Semantic Web Technology to Improve Scholarly Knowledge and Expertise Management." In *Visualizing the Semantic Web*, 2nd ed., edited by Vladimir Geroimenko and Chaomei Chen, 183–198. London: Springer-Verlag.

Mapping Geographic Knowledge Diffusion

References

Batty, M. 2003. "The Geography of Scientific Citation." *Environment and Planning A* 35, 5: 761–765.

Börner, Katy, Shashikant Penumarthy, Mark Meiss, and Weimao Ke. 2006. "Mapping the Diffusion of Scholarly Knowledge Among Major U.S. Research Institutions." *Scientometrics* 68, 3: 415–426.

206 Science Maps for Kids

References

Boecke, Kees. 1957. *Cosmic View: The Universe in 40 Jumps*. http://nedwww.ipac.caltech.edu/level5/Boeke/Boekem14.html (accessed January 1, 2008). [Quotation]

de Rosnay, Joël. 1975. *Le macroscope. Vers une vision globale*. Editions du Seuil. Translated by Robert Edwards in 1979 as *The Macroscope: A New World Scientific System*. New York: Harper & Row Publishers, Inc. http://pespmc1.vub.ac.be/macroscope (accessed January 1, 2008).

Science from Above

References

Austhink Software Pty Ltd. 2008. Austhink Home Page. http://www.austhink.com (accessed January 1, 2008).

Compendium Institute. 2008. Compendium Institute Home Page. http://compendium.open.ac.uk/institute (accessed March 30, 2010).

Conklin, Jeffrey. 2005. *Dialogue Mapping: Building Shared Understanding of Wicked Problems*. Hoboken, NJ: Wiley.

Hook, Peter A., and Katy Börner. 2005. "Educational Knowledge Domain Visualizations: Tools to Navigate, Understand, and Internalize the Structure of Scholarly Knowledge and Expertise." *New Directions in Cognitive Information Retrieval*, edited by Amanda Spink and Charles Cole, 187–208. Amsterdam: Springer-Verlag.

Inspiration Software, Inc. 2008. Inspiration Software Home Page. http://www.inspiration.com (accessed January 1, 2008).

L'Università della Svizzera italiana. 2008. Let's Focus Home Page. http://www.lets-focus.com (accessed January 1, 2008).

McKim, Robert H. 1980. *Thinking Visually: A Strategy Manual for Problem Solving*. Lifetime Learning Publications.

Novak, Joseph D. 1994. "A Science Education Research Program that Led to the Development of the Concept Mapping Tool and a New Model for Education." *Proceedings of the First International Conference on Concept Mapping*. Pamplona, Spain, September 14–17. http://cmc.ihmc.us/papers/cmc2004-286.pdf (accessed March 25, 2008).

Novak, Joseph D. 1998. *Learning, Creating, and Using Knowledge: Concept Maps as Facilitative Tools in Schools and Corporations*. Mahwah, NJ: Lawrence Erlbaum Associates.

Novak, Joseph D., and Alberto J. Cañas. "The Theory Underlying Concept Maps and How to Construct and Use Them." Institute for Human and Machine Cognition. http://cmap.ihmc.us/Publications/ResearchPapers/TheoryCmaps/TheoryUnderlyingConceptMaps.htm (accessed March 25, 2008).

West, C. K., J. A. Farmer, and P. M. Wolff. 1991. *Instructional Design: Implications from Cognitive Science* 58. Englewood Cliffs, NJ: Prentice-Hall. [Quotation]

Mapping Inventions and Inventors

Image Credits

"Nature—Science Is All Around Us." Original artwork by JoHanna M. Sanders.

Visual Interfaces to Educational Resources

References

Börner, Katy, and Chaomei Chen. 2002. "Visual Interfaces to Digital Libraries: Motivation, Utilization and Socio-Technical Challenges." In *Lecture Notes in Computer Science: Visual Interfaces to Digital Libraries*, edited by Katy Börner and Chaomei Chen, vol. 2539: 1–9. Berlin: Springer-Verlag.

Institute for the Study of Knowledge Management in Education (ISKME). 2007. "Open Educational Resources (OER) Commons." **http://www.oercommons.org** (accessed January 1, 2008).

National Science Digital Library. 2010. NSDL Home Page. **http://nsdl.org** (accessed March 16, 2010).

Science Maps in Teaching

References

Ke, Weimao, Katy Börner, and Lalitha Viswanath. 2004. "Analysis and Visualization of the IV 2004 Contest Dataset." IEEE Information Visualization Conference 2004 Poster Compendium, 49–50. **http://conferences.computer.org/InfoVis/files/compendium2004.pdf** and **http://iv.slis.indiana.edu/ref/iv04contest/Ke-Borner-Viswanath.gif** (animated version) (accessed October 30, 2007).

Roberg, Nicole A. 2006. "Science Maps for Kids." *Presentation at the 10th International Information Visualisation Conference*. London, UK, July 5–7.

Designing Macroscopes

Child Scientists

References

Benko, Hrvoje, et al. "Collaborative Visualization of an Archaelogical Excavation." Columbia University. **http://graphics.cs.columbia.edu/projects/ArcheoVis** (accessed February 10, 2010).

Cornell Lab of Ornithology and Audubon. 2008. "Great Backyard Bird Count." **http://www.birdsource.org/gbbc** (accessed January 1, 2008).

Dalton School. 2008. "Archaeotype." Digital Dalton. **http://intranet.dalton.org/DigitalDalton/projects/archaeotype.html** (accessed March 25, 2008).

Groundspeak, Inc. 2008. Geocaching Home Page. **http://www.geocaching.com** (accessed March 25, 2008).

Indiana University Geological Sciences. 2010. "Princeton Earth Physics Project: About PEPP." **http://www.indiana.edu/~pepp/about.html** (accessed March 12, 2010).

Schneider, Adam. 2008. "GPS Visualizer." **http://www.gpsvisualizer.com/examples** (accessed March 25, 2008).

UCAR and Colorado State University. 2008. The Globe Program Home Page. **http://viz.globe.gov** (accessed March 25, 2008).

Wikimedia Foundation. 2010. "Encyclopedia." *Wikipedia, the Free Encyclopedia*. **http://en.wikipedia.org/wiki/Encyclopedia** (accessed February 8, 2010).

Exploring the Web of Knowledge

References

Angliss, Sarah and Maggie Hewson. 2001. *Matter and Materials: Hands-On Science*. Boston: Kingfisher.

Angliss, Sarah and Maggie Hewson. 2002. *Hands-On Science*. Boston: Kingfisher.

Burke, James. 1999. *The Knowledge Web: From Electronic Agents to Stonehenge and Back—and Other Journeys through Knowledge*, 21. New York: Simon & Schuster. [Quotation]

Burke, James. 2008. "The Knowledge Web: A Project of the James Burke Institute." **http://www.k-web.org** (accessed April 7, 2008).

Chapman, Gillian, Pam Robson. 1995. *Exploring Time*. Minneapolis: Milbrook Press.

CBS Television. "Numbers." **http://www.cbs.com/primetime/numb3rs** (accessed January 1, 2008).

Gombrich, E. H. 2005. *A Little History of the World*. New Haven, CT: Yale University Press.

Hakim, Joy. 2004. *The Story of Science, Book One: Aristotle Leads the Way*. Washington, DC: Smithsonian Books.

Hakim, Joy. 2005. *The Story of Science, Book Two: Newton at the Center*. Washington, DC: Smithsonian Books.

Hakim, Joy. 2007. *The Story of Science, Book Three: Einstein Adds a New Dimension*. Washington, DC: Smithsonian Books.

Shasha, Dennis. 1997. *The Puzzling Adventures of Dr. Ecco*. Mineola, NY: Dover Publications.

Shasha, Dennis. 2004. *Dr. Ecco: Mathematical Detective (Codes, Puzzles, and Conspiracy)*. Mineola, NY: Dover Publications.

Wikimedia Foundation. 2010. "Connections (TV Series)." *Wikipedia, the Free Encyclopedia*. **http://en.wikipedia.org/wiki/Connections_(TV_series)** (accessed February 8, 2010).

Image Credits

Three images from James Burke, The Knowledge Web, used by permission.

Screenshot of the Knowledge Web interface courtesy of the Knowledge Web.

Computer Games

References

Activeworlds.com, Inc. 2008. Active Worlds: Educational Universe Home Page. **http://edu.activeworlds.com** (accessed January 1, 2008).

Anderson, Jacqueline M., Jeet Atwal, Patrick Wiegand, and Alberta Auringer Wood. 2005. *Children Map the World: Selections for the Barbara Petchenik Children's World Map Competition*. Redlands, CA: ESRI Press.

Barab, Sasha, Tyler Dodge, Hakan Tuzun, Kirk Job-Sluder, Craig Jackson, Anna Arici, Laura Job-Sluder, Roberts Carteaux Jr., Jo Gilbertson, and Conan Heiselt. 2007. "The Quest Atlantis Project: A Socially-Responsive Play Space for Learning." In *The Educational Design and Use of Simulation Computer Games*, edited by Brett E. Shelton and David Wiley. Rotterdam, The Netherlands: Sense Publishers. **http://inst.usu.edu/~bshelton/simcompgames** (accessed January 1, 2008).

Electronic Arts. 2001. SimCity Home Page. **http://simcity3000unlimited.ea.com/us/guide** (accessed January 1, 2008).

Electronic Arts. 2008. Spore Home Page. **http://www.spore.com** (accessed January 1, 2008).

Indiana University Learning Sciences. 2008. Quest Atlantis Home Page. Indiana University School of Education. **http://atlantis.crlt.indiana.edu** and **http://atlantis.crlt.indiana.edu/public/welcome.pl** (accessed January 1, 2008).

Take-Two Interactive Software, Inc. 2006. Civilization Home Page. **http://civilization.com** (accessed January 1, 2008).

Tom Snyder Productions. 2008. Tom Snyder Productions Home Page. **http://www.tomsnyder.com** (accessed March 25, 2008).

TryScience/New York Hall of Science. 2008. PowerUp Home Page. **http://www.powerupthegame.org** (accessed March 25, 2008).

Contributors

Patrick McKercher, Bonnie DeVarco, Stephen Uzzo, and Sasha Barab.

Daily Science Forecasts

References

Burke, James. 2007. *Connections*. New York: Simon & Schuster. [Quotation]

Everyday S&T for Everybody

No references or credits.

MarketSite Map

References

NASDAQ Stock Market. 2010. NASDAQ Home Page. **http://www.nasdaq.com** (accessed March 16, 2010).

Oculus Info, Inc. 2006. "Oculus Excel Visualizer." **http://www.oculusinfo.com/papers/ExcelVizWhitepaper-final.pdf** (accessed January 1, 2008).

Image Credits

Image © 2000, The Nasdaq Stock Market, Inc. Reprinted with the permission of the Nasdaq Stock Market, Inc. Photo credit: Peter Aaron/Esto. Visualization software and implementation by Oculus Info, Inc.

Industry Pull and Science Push

References

National Science Foundation. 2006. "Sustaining America's Competitive Edge: Enhancing Innovation and Competitiveness Through Investments in Fundamental Research." NSF Workshop, Arlington, VA, December 3–5. **http://enhancinginnovation.wustl.edu/Home.htm** (accessed March 1, 2008).

SparkIP. 2008. Spark IP Home Page. Previously available at **http://sparkip.com** (accessed January 1, 2008).

S&T Forecast News Online and in TV

References

Apple, Inc. 2008. iTunes Home Page. **http://www.iTunes.com** (accessed February 21, 2008).

Bloomberg L. P. 2008. Bloomberg Television Home Page. **http://www.bloomberg.com/media/tv** (accessed January 1, 2008).

Chastang, Julien. 2007. WeatherMole Home Page. **http://weathermole.com/WeatherMole/index.html** (accessed March 30, 2010).

Monmonier, Mark. 2000. *Air Apparent: How Meteorologists Learned to Map, Predict, and Dramatize Weather*. Chicago: The University of Chicago Press.

NASA. 2008. Science@NASA Home Page. **http://science.nasa.gov** (accessed January 1, 2008).

New York Academy of Sciences. 2008. "Science & the City: Webzine of the New York Academy of Sciences." **http://www.nyas.org/snc** (accessed January 1, 2008).

The Scientist. 2008. The Scientist Home Page. **http://www.the-scientist.com** (accessed January 1, 2008).

Thomson Scientific. 2008. "Thomson Science Products & Services." **http://www.scientific.thomson.com/products** (accessed January 1, 2008).

WGBH Educational Foundation. 2008. "NOVA Science Now." **http://www.pbs.org/wgbh/nova/sciencenow** (accessed January 1, 2008).

YouTube, LLC. 2008. YouTube Home Page. **http://www.youtube.com** (June 2, 2008).

Image Credits

Forecast image from istockphoto.com.

Public domain image from the National Oceanic and Atmospheric Administration.

Anatomy of S&T Forecasts

Learning from Weather Forecasts

References

Crowder, Bob, William J. Burroughs, Ted Robertson, Elinor Vallier-Talbot, and Richard Whitaker. 1996. *Weather*. Fairfax, VA: Time Life Education.

Richardson, Lewis F. 1922. *Weather Prediction by Numerical Process*. Cambridge: Cambridge University Press.

World Meteorological Organization. 2010. World Meteorological Organization (WMO) Home Page. **http://www.wmo.ch/pages/index_en.html** (accessed March 30, 2010).

Patchwork of Multiple Forecasts

Image Credits

Excerpted from: Michael W. Hamburger, Chuck Meertens (data and visualization), Elisha F. Hardy (graphic design). 2007. "Tectonic Movements and Earthquake Hazard Predictions." Bloomington, IN, and Boulder, CO.

Excerpted from: Michael Aschauer, Maia Gusberti, Sepp Deinhofer, and Nik Thoenen. 2006. [./logicaland] Home Page. **http://www.logicaland.net** (accessed October 30, 2007).

Designing a Scalable Infrastructure

References

Atkins, Daniel E., Kelvin K. Droegemeier, Stuart I. Feldman, Hector Garcia-Molina, Michael L. Klein, David G. Messerschmitt, Paul Messina, Jeremiah P. Ostriker, and Margaret H. Wright. 2003. "Revolutionizing Science and Engineering Through Cyberinfrastructure: Report of the National Science Foundation Blue-Ribbon Advisory Panel on Cyberinfrastructure." National Science Foundation. **http://www.nsf.gov/od/oci/reports/atkins.pdf** (accessed March 25, 2008).

Barroso, Luiz André, Jeffrey Dean, and Urs Hölzle. 2003. "Web Search for a Planet: The Google Cluster Architecture." *IEEE Micro* 23, 2: 22–28. http://www.search3w.com/Siteresources/data/MediaArchive/files/Google%2015000%20servers%20secrest.pdf (accessed January 1, 2008).

Brin, Sergey, and Lawrence Page. 1998. "The Anatomy of a Large-Scale Hypertextual Web Search Engine." *Proceedings of the 7th International World Wide Web Conference*. http://infolab.stanford.edu/~backrub/google.html (accessed September 30, 2009).

Cornell University Library. ArXiv.org Home Page. http://arxiv.org (accessed October 30, 2007).

Dimov, Dmitry, and Brian Mulloy. 2008. "Swivelpreview: About Us." http://www.swivel.com/about (accessed June 2, 2008).

IBM, Inc. 2008. Many Eyes Beta Home Page. http://services.alphaworks.ibm.com/manyeyes/home (accessed October 30, 2007).

Knowledge and Data Engineering Group, University of Kassel. 2008. BibSonomy Home Page. http://www.bibsonomy.org. (accessed March 1, 2008).

Nature Publishing Group. 2008. Connotea Home Page. http://www.connotea.org (accessed March 1, 2008).

Ning, Inc. 2010. Ning Home Page. http://www.ning.com (accessed March 16, 2010).

Oversity Limited. 2008. CiteULike Home Page. http://www.citeulike.org (accessed March 1, 2008).

Pierce, Marlon E., Geoffrey C. Fox, Joshua Rosen, Siddharth Maini, and Jong Y. Choi. 2008. "Social Networking for Scientists Using Tagging and Shared Bookmarks: A Web 2.0 Application." *2008 International Symposium on Collaborative Technologies and Systems (CTS 2008)*. http://grids.ucs.indiana.edu/ptliupages/publications/MSINetworkPortalFinal.pdf (accessed January 1, 2008).

Purdue University Network for Computational Nanotechnology. 2008. NanoHUB Home Page. http://www.nanohub.org (accessed March 1, 2008).

SciVee. 2008. SciVee Home Page. http://www.scivee.tv (accessed March 1, 2008).

SourceForge, Inc. 2008. SourceForge.net Home Page. http://sourceforge.net (accessed October 30, 2007).

SparkIP. 2008. Spark IP Home Page. Previously available at http://sparkip.com (accessed January 1, 2008).

Universities of Manchester and Southampton. 2008. My Experiment Beta Home Page. http://www.myexperiment.org (accessed March 1, 2008).

U.S. National Library of Medicine. 2008. MedlinePlus Home Page. http://medlineplus.gov (accessed March 1, 2008).

WebUpon Blog. 2006. Comment on "Google and Google Cluster Architecture." Comment posted August 15, 2006. http://webupon.com/search-engines/google-and-google-cluster-architecture (accessed February 10, 2010).

Wikimedia Foundation. 2006-2009. "Wikimedia Servers/Hardware Orders—Meta." http://meta.wikimedia.org/wiki/Image:Wikimedia-servers-2006-05-09.svg; http://meta.wikimedia.org/wiki/Wikimedia_servers (accessed January 1, 2008).

Wikimedia Foundation. 2007. "Ada Lovelace." *Wikipedia, the Free Encyclopedia*. http://en.wikipedia.org/wiki/Ada_Lovelace (accessed October 30, 2007).

Yahoo!, Inc. 2008. Flickr Home Page. http://www.flickr.com (accessed June 2, 2008).

YouTube, LLC. 2008. YouTube Home Page. http://www.youtube.com (June 2, 2008).

Valuing Collective Interpretation and Prediction

References

Aschauer, Michael, Maia Gusberti, Sepp Deinhofer, and Nik Thoenen. 2006. [./logicaland] Home Page. http://www.logicaland.net (accessed October 30, 2007).

Bettencourt, Luís M. A., Ariel Cintrón-Arias, David I. Kaiser, and Carlos Castillo-Chávez. 2006. "The Power of a Good Idea: Quantitative Modeling of the Spread of Ideas from Epidemiological Models." *Physica A: Statistical Mechanics and its Applications*, 364: 513–536.

Electronic Arts. 2001. SimCity Home Page. http://simcity3000unlimited.ea.com/us/guide (accessed January 1, 2008).

Science and Technology Forecast Advantage

References

Bettencourt, Luís M. A., Carlos Castillo-Chávez, David Kaiser, and David E. Wojick. 2006. "Report for the Office of Scientific and Technical Information: Population Modeling of the Emergence and Development of Scientific Fields." Office of Scientific and Technical Information, U.S. Department of Energy. http://www.osti.gov/innovation/research/diffusion/epicasediscussion_lb2.pdf (accessed June 2, 2008).

Ishizaki, Suguru. 2003. *Improvisational Design: Continuous, Responsive Digital Communication*. Cambridge, MA: MIT Press.

Contributors

Stephen Uzzo, Ryutaro Ichise, Shravan Rajagopal, Ingo Günther, Elizabeth Kerr, and Peter A. Hook.

210 Growing a "Global Brain and Heart"

References

ThinkExist. 2010. "Dalai Lama Quotes." http://en.thinkexist.com/quotes/dalai_lama (accessed February 16, 2010). [Quotation]

Facing Mankind's Global Challenges

References

Diamond, Jared. 2004. *Collapse: How Societies Choose to Fail or Succeed*. New York: Viking.

Glenn, Jerome C., and Theodore J. Gordon. 2007. *2007 State of the Future*. "Millennium Project." http://www.millennium-project.org/millennium/issues.html (accessed January 1, 2008).

International Union for Conservation of Nature and Natural Resources. 2010. "The IUCN List of Threatened Species." http://www.iucnredlist.org (accessed March 30, 2010).

Lomborg, Bjorn. 2004. *Global Crisis Global Solutions*. Cambridge: Cambridge University Press and Environmental Assessment Institute.

Sachs, Jeffrey D. 2005. *The End of Poverty: Economic Possibilities for Our Time*. New York: The Penguin Press.

United Nations. 2008. "The UN Millennium Development Goals." http://www.un.org/millenniumgoals (accessed March 1, 2008).

U.S. Central Intelligence Agency. 2010. "The World Factbook: United States." https://www.cia.gov/library/publications/the-world-factbook/geos/us.html (accessed March 30, 2010).

World Bank. 2005. *MiniAtlas of Millennium Development Goals: Building a Better World*. Washington, DC: The World Bank Group. http://publications.worldbank.org/ecommerce/catalog/product?item_id=4724647 (accessed June 2, 2008). [Quotation]

World Resources Institute. 2007. "EarthTrends Database." http://earthtrends.wri.org. (accessed June 2, 2008).

Aligning Science and Technology with Global Challenges

References

United Nations. 2008. "The UN Millennium Development Goals." http://www.un.org/millenniumgoals (accessed March 1, 2008).

Union of International Associations. 2008. "Union of International Associations: UIA Online Databases." http://www.diversitas.org/db/x.php (accessed January 1, 2008).

World Bank Group. 2009. "Online Atlas of the Millennium Development Goals." http://devdata.worldbank.org/atlas-mdg (accessed March 16, 2010).

Contributors

Kevin W. Boyack and Patrick M. McKercher.

Bridging Disciplinary Boundaries

References

Burke, James. 1978. *Connections*. Boston: Little, Brown and Company. [Quotation]

Pink, Daniel. 2005. *A Whole New Mind*. New York: Riverhead Hardcover. [Quotation]

Image Credits

Global Brain Image from istockphoto.com, incorporating features from the Map of Scientific Paradigms (Boyack, Klavans, and Paley), with additional design by Elisha F. Hardy.

Growing a Global Brain

References

Bloom, H. 2000. *Global Brain: The Evolution of Mass Mind from the Big Bang to the 21st Century*. New York: John Wiley & Sons, Inc.

Börner, Katy, Luca Dall'Asta, Weimao Ke, and Alessandro Vespignani. 2005. "Studying the Emerging Global Brain: Analyzing and Visualizing the Impact of Co-Authorship Teams." *Complexity: Special Issue on Understanding Complex Systems* 10, 4: 58–67.

Guimera, Roger, Brian Uzzi, Jarrett Spiro, and Luís A. Nunes Amaral. 2005. "Team Assembly Mechanisms Determine Collaboration Network Structure and Team Performance." *Science* 308, 5722: 697–702.

Growing a Global Heart

References

BrainyMedia.com. 2008. "Rabindranath Tagore Quotes." Brainy Quote. http://www.brainyquote.com/quotes/authors/r/rabindranath_tagore.html (accessed June 2, 2008). [Quotation, Rabindranth Tagore]

The World Is Flat

References

Airports Council International. 2007. "Airports Welcome Record 4.4 Billion Passengers in 2006." http://www.airports.org/cda/aci_common/display/main/aci_content07_c.jsp?zn=aci&cp=1-5-54_666_2__ (accessed June 2, 2008).

Friedman, Thomas L. 2005. *The World Is Flat: A Brief History of the Twenty-First Century*. New York: Farrar, Straus and Giroux.

Glenn, Jerome C., and Theodore J. Gordon. 2007. *2007 State of the Future*. "Millennium Project." http://www.millennium-project.org/millennium/issues.html (accessed January 1, 2008).

Collective Survival of the Fittest

No references or credits.

Competition and Cooperation

References

Axelrod, Robert. 1981. "The Evolution of Cooperation." *Science* 211, 4489: 1390–1396.

Axelrod, Robert. 1984. *The Evolution of Cooperation*. New York: Basic Books.

Axelrod, Robert. 2006. *The Evolution of Cooperation*. Rev. ed. Cambridge, MA: Perseus Books Group.

Menard, Henry W. 1971. *Science: Growth and Change*, 195. Cambridge, MA: Harvard University Press.

Nowak, Martin A. 2006. "Five Rules for the Evolution of Cooperation." *Science* 314, 5805: 1560–1563.

Sober, Elliott, and David Sloan Wilson. 1988. *Unto Others: The Evolution and Psychology of Unselfish Behavior*. Cambridge, MA: Harvard University Press.

Earth as Common-Pool Resource

References

Anodea, Judith. 2006. *Waking the Global Heart*. New York: Midpoint Trade Books. [Quotation]

Hardin, Garrett. 1968. "The Tragedy of the Commons." *Science* 162: 1243–1248.

Lloyd, William Forster. 1833. *Two Lectures on the Checks to Population*. Oxford: Oxford University Press.

Ostrom, Elinor. 1990. *Governing the Commons: The Evolution of Institutions for Collective Action*. Cambridge: Cambridge University Press.

Ostrom, Elinor, Roy Gardner, and James Walker. 1994. *Rules, Games, and Common Pool Resources*. Ann Arbor, MI: University of Michigan Press.

Mind-Heart Symbiosis

References

Hofstadter, Douglas. 2007. *I Am a Strange Loop*. New York: Basic Books. [Quotation]

Image Credits

Global Heart & Brain images from istockphoto.com, with additional design by Elisha F. Hardy.

Index

A

B

C

H

I

K

L

M

N

O

P

R

X